D1271550

Particulate Matter Science for Policy Makers

Particulate Matter Science for Policy Makers: A NARSTO Assessment was commissioned by NARSTO, a cooperative public-private sector organization of Canada, Mexico and the United States. It is a concise and comprehensive discussion of the current understanding by atmospheric scientists of airborne particulate matter (PM). Its goal is to provide policy makers who implement air-quality standards with this relevant and needed scientific information.

The assessment is organized using the following considerations for dealing with air-quality issues:

- Perspective for Managing PM
- Health Effects Context
- Atmospheric Aerosol Processes
- Emission Characterization
- Particle and Gas Measurements
- Spatial and Temporal Characterization of PM
- Receptor Methods
- Chemical Transport Models
- Visibility and Radiative Balance Effects
- Conceptual models of PM for nine geographic regions in North America

The primary audience for this volume will be regulators, scientists, and members of industry, all of whom have a stake in effective PM management. It will also inform exposure and health scientists, who investigate causal hypotheses of health impacts, characterize exposure, and conduct epidemiological and toxicological studies.

Peter H. McMurry is the Kenneth T. Whitby Professor of Mechanical Engineering at the University of Minnesota, where he has served as Department Head since 1977. He is a past President of the American Association for Aerosol Research. His research focuses on phenomena including nucleation, growth, water uptake, and light scattering. With his research colleagues he has developed a condensation nucleus counter to detect particles as small as 3 nm, aerodynamic lenses to produce tightly collimate particle beans, and instruments to measure particle properties that include density, refractive index, and water content. He has written recent review articles on atmospheric aerosol measurement and nucleation.

Marjorie F. Shepherd is Senior Science Advisor at the Atmospheric and Climate Science Directorate, Meteorological Service of Canada. Her work involves investigating sampling and analysis methodologies for ambient volatile organic compounds. She has coordinated several projects within the Canadian multi-stakeholder science assessment for ground-level ozone. As science advisor for the Meteorological Service of Canada, she co-lead, with Health Canada, the development of science assessments for HF, CO, ozone and particulate matter – all in support of developing Canadian air quality objectives and standards.

James S. Vickery is one of the founding members of NARSTO and a member of the Executive Steering Committee. He is a Co-Leader of the US Committee on Environment and Natural Resources - Particulate Matter Research Work Group, which coordinates all U.S. Federally sponsored research concerning Particulate Matter. He is Special Assistant to the Director of EPA's National Exposure Research Laboratory, where he has also held the positions of Division Director and Assistant Laboratory Director. He has also managed several different program and policy offices in EPA's Washington, DC headquarters.

Particulate Matter Science for Policy Makers

A NARSTO Assessment

Edited by

Peter H. McMurry
University of Minnesota

Marjorie F. Shepherd
Environment Canada

James S. Vickery
U.S. Environmental Protection Agency

PUBLISHED BY THE PRESS SYNDICATE OF THE UNIVERSITY OF CAMBRIDGE
The Pitt Building, Trumpington Street, Cambridge, United Kingdom

CAMBRIDGE UNIVERSITY PRESS
The Edinburgh Building, Cambridge CB2 2RU, UK
40 West 20th Street, New York, NY 10011-4211, USA
477 Williamstown Road, Port Melbourne, VIC 3207, Australia
Ruiz de Alarcón 13, 28014 Madrid, Spain
Dock House, The Waterfront, Cape Town 8001, South Africa

http://www.cambridge.org

First published 2004

Printed in the United States of America

Typeface Times 11/13 pt. *System* LaTeX 2ε [AU]

A catalog record for this book is available from the British Library.

Library of Congress Cataloging in Publication Data

ISBN 0 521 84287 5 hardback

Publication of this NARSTO Assessment was supported by the U.S. National Science Foundation under Research Grant ATM-0337151.

NARSTO wishes to express its special appreciation to Dr. George Hidy of Envair, for his substantial contributions to this Assessment. In his role as Co-Chair of NARSTO's Analysis and Assessment Team, Dr. Hidy has played a major role in setting the vision for this Assessment and leading its implementation, as well as contributing as a principal author.

TABLE OF CONTENTS

LIST OF FIGURES xvii

LIST OF TABLES xxiii

PREFACE xxvii

Commissioning of the Assessment xxvii
Verifying the Needs of Policy Makers xxviii
Preparing the Assessment xxix
Assessment Co-chairs xxix
Lead Authors xxix
Contributing Authors xxix
NARSTO Staff xxx
Acknowledgements xxx
MEMORIAL TO PROFESSOR GLEN CASS (1947-2001) xxxi

EXECUTIVE SUMMARY 1

$PM_{2.5}$ MASS AND COMPOSITION RESPONSES TO CHANGING EMISSIONS 1
LOCAL, REGIONAL, AND CONTINENTAL MANAGEMENT OF $PM_{2.5}$ 2
PREDICTIVE CAPABILITIES 2
COPOLLUTANT INTERACTIONS 3
NEW INSIGHTS FROM IMPROVED MEASUREMENTS AND MONITORING 3
LINKAGES BETWEEN THE HEALTH AND ATMOSPHERIC SCIENCE COMMUNITIES 3
TRACKING PROGRESS AND SUCCESS OF AIR-QUALITY MANAGEMENT 4
SCIENTIFIC UNCERTAINTIES AND LOOKING FORWARD 4

PARTICULATE MATTER SCIENCE FOR POLICY MAKERS: SYNTHESIS 7

THE INFORMATION FRAMEWORK 7
AIR-QUALITY MANAGEMENT: A POLICY OVERVIEW 9
THE NATURE OF PM: A BRIEF OVERVIEW 9
PM ISSUES AND POLICY QUESTIONS 12

CONTENTS

Policy Question #1 - Is there a significant PM problem and how confident are we? 12
Key science findings 14
Spatial and temporal characterization 14
Measurements 15
Health effects 17
Chemical deposition 18
Visibility and Climate 18
New science to improve implementation approaches 19
Policy Question #2 - Where there is a PM problem, what is its composition and what factors contribute to elevated concentrations? 19
Key science findings 19
Composition and the factor of seasonality 19
Emissions and atmospheric processes as contributing factors 21
Regional contributions 21
PM_{10} 24
New science to improve implementation approaches 24
Policy Question #3 - What broad, pollutant-based, approaches might be taken to fix the problem? 24
Key science findings 24
Coincident reduction of precursors 24
Reduction of carbonaceous PM 25
Reduction of nitrates 25
Differing regional strategies 25
The $SO_4^{=}$- NO_3^{-}- NH_4^{+}equilibrium 26
Area-specific insights 27
Regional reductions 28
PM_{10} 29
New science to improve implementation approaches 29
Policy Question #4 - What source-specific options are there for fixing the problem given the broad control approaches above? 29
Key science findings 30
Source-attribution approaches 30
Source-based modeling: Chemical-Transport Models 30
Receptor modeling 32
Emission inventories and source insights 33
Sources of PM 34
New science to improve implementation approaches 35
Policy Question #5 - What is the relationship between PM, its components, and other air-pollution problems on which the atmospheric-science community is working? 36
Key science findings 36
Visibility (regional haze) 36
Ozone 38
Chemical deposition (from PQ #1) 39
Climate change (from PQ#1) 39
New science to improve implementation approaches 39

Policy Question #6 - How can progress be measured? How can we determine the effectiveness of our actions in bringing about emission reductions and air-quality improvements, with their corresponding exposure reductions and health improvements? 39
Key science findings 40
Tracking emission changes 40
Tracking trends in ambient air-quality 40
Tracking changes in visibility 43
Tracking changes in exposure and health 44
New science to improve implementation approaches 44
Policy Question #7 - When and how should implementation programs be reassessed and updated to adjust for any weaknesses, and to take advantage of advances in science and technology? 44
Key science findings 44
New science to improve implementation approaches 46
Policy Question #8 - What further atmospheric-sciences information will be needed in the periodic reviews of national standards? 46
Key science findings 46
PM measurement considerations 47
Exposure-science considerations 47
Health-science considerations 48
Air-quality and exposure-modeling consi-derations 48
New science to improve implementation approaches 49
BENEFITS TO THE POLICY COMMUNITY OF NEW SCIENCE 50
REFERENCES 52

CHAPTER 1 PERSPECTIVE FOR MANAGING PM 53

1.1 THE NATURE OF AMBIENT PARTICLES 54
1.2 SCALE, SOURCES, AND MAN-AGEMENT OF THE PROBLEM 56
1.3 HEALTH IMPACTS 59
1.4 VISIBILITY IMPACTS 61
1.5 FEDERAL AMBIENT PM STANDARDS AND POLICY CONTEXTS 62
1.5.1 Canada 62
1.5.2 United States 63
1.5.3 Mexico 64
1.5.4 Implementation of Scientific Information into Decision Making 65
1.6 THE STRUCTURE OF THE ASSESSMENT 66
1.7 REFERENCES 68

CHAPTER 2 HEALTH CONTEXT FOR MANAGEMENT OF PARTICULATE MATTER 69

2.1 OVERVIEW 69
2.2 HISTORICAL PERSPECTIVE 69
2.3 EXPOSURE ASSESSMENT 72
2.4 SOURCES OF INFORMATION ON HEALTH RESPONSES TO AIR POLLUTION 76
2.4.1 Epidemiological Approaches 77
2.4.2 Controlled-Exposure Studies with Human Subjects 81
2.4.3 Laboratory Animal Studies 82
2.4.4 Tissue and Cell Studies 82
2.5 EPIDEMIOLOGICAL FINDINGS 83
2.5.1 Acute Exposure 83
2.5.2 Chronic Exposures 86
2.6 INTERVENTION STUDIES 90
2.7 TOXICOLOGICAL EVIDENCE 92
2.8 POLICY-RELEVANT FINDINGS AND FUTURE OPPORTUNITIES 94
2.9 REFERENCES 95

CHAPTER 3 ATMOSPHERIC AEROSOL PROCESSES 103

3.1 THE LIFE OF AN ATMOSPHERIC PARTICLE 103
3.2 A PARTICLE IS BORN: NUCLEATION 107
3.3 HOW LONG DOES IT TAKE FOR A PARTICLE TO COLLIDE WITH ANOTHER? 108
3.4 PARTICLES AND WATER 108
3.5 SECONDARY PM FORMATION 109
3.5.1 Sulfate 110
3.5.2 Nitrate 111
3.5.3 Secondary Organic Aerosol (SOA) Formation 111
3.5.4 Interactions of Primary and Secondary PM Components 112
3.6 FROM PRECURSOR EMISSIONS TO AEROSOL COMPONENT CONCENTRATION 113
3.6.1 Linearity 114
3.6.2 An Application to Southern California 115
3.6.3 Effectiveness of SO_2 Emission Reductions 115
3.6.4 VOC Emission Reductions and SOA 117
3.6.5 Limiting Reactants 117
3.7 REMOVAL AND LONG-RANGE TRANSPORT OF PM 118
3.8 PM AND OTHER POLLUTANTS 120
3.9 POLICY IMPLICATIONS 120
3.10 REFERENCES 123

CHAPTER 4 EMISSION CHARACTERIZATION 127

4.1 TYPES OF EMISSION INVENTORIES AND THEIR USES 129
4.2 THE NATURE AND CHARACTERISTICS OF EMISSION INVENTORIES 130
4.2.1 Characterization by Source Category 132
4.2.2 Geographical Distribution of Emissions 137
4.2.3 Temporal Variations, Trends and Forecasts in Emissions 138
4.2.3.1 Temporal Variations 138
4.2.3.2 Trends and Projections 140
4.3 ESTIMATING UNCERTAINTY IN EMISSION INVENTORIES 143
4.4 IMPROVING ESTIMATION METHODS 146
4.4.1 Methodological Improvements 148
4.4.2 The Source/Ambient-Air Interface 149
4.5 HOW WELL DO THE EMISSION INVENTORIES ADDRESS APPLICATION NEEDS? 151
4.6 SUMMARY 153
4.7 POLICY IMPLICATIONS 155
4.8 REFERENCES 156

CHAPTER 5 PARTICLE AND GAS MEASUREMENTS 159

5.1 CURRENTLY AVAILABLE TECHNOLOGY AND INSTRUMENT CAPABILITIES 161
5.1.1 Size-Selective Inlets 161
5.1.2 Integrated Denuder and Gravimetric Filter-Based Systems (substrate- and absorbent-based measurements) for Mass and Composition Sampling 163
5.1.3 Continuous and Semi-continuous Real-time Measurements 166
5.1.4 Personal Exposure Monitors 169
5.1.5 Single-Particle Measurement Capabilities 170
5.1.6 Optical Properties of Aerosols and Long-Path Optical Measurements 170
5.1.7 Chemical Analysis of Cloud and Fog Chemical Composition 173
5.1.8 Gas-Phase PM Precursors, Ozone, Ozone Precursors, and Oxidants 173
5.1.9 Meteorological Measurements 173
5.2 MEASUREMENT UNCERTAINTY AND VALIDATION 174
5.2.1 Estimation of Particle Measurement Uncertainty 174
5.2.2 Mass and Size Distribution 175
5.2.3 Aerosol Chemical Composition 176
5.2.4 Uncertainties in Routine Gas-phase Measurements Used for Network Monitoring 177
5.3 MEASUREMENT STRATEGIES AND NETWORK ISSUES 178
5.3.1 Deployment of Measurement Technology 178
5.3.2 Future Requirements for Measurement Strategies 182
5.4 SUMMARY 183

5.5 POLICY IMPLICATIONS 186
5.6 REFERENCES 187

CHAPTER 6 SPATIAL AND TEMPORAL CHARACTERIZATION OF PARTICULATE MATTER 191

6.1 INTRODUCTION 191
6.1.1 General Features Affecting Particulate Levels in North America 191
6.1.2 Spatial and Time Scales of Interest 192
6.1.3 Monitoring Capabilities 193
6.2 CONTINENTAL AND REGIONAL VARIATIONS OF PM CONCENTRATIONS 193
6.2.1 Spatial Variations of PM_{10} Mass: Where and When are PM_{10} Concentrations Highest? 195
6.2.2 Spatial Patterns of $PM_{2.5}$ Mass 198
6.2.3 Seasonal Variations of $PM_{2.5}$ Mass 201
6.2.4 The Composition of $PM_{2.5}$ and Its Geographical Variation 205
6.3 REGIONAL AND URBAN CONTRIBUTIONS TO PM 209
6.3.1 Comparisons Between Rural and Urban Sites 209
6.3.2 Evidence for Local PM Sources: Temporal Variations 209
6.3.3 PM Mass Concentrations at Remote Locations 210
6.3.4 Regional Transport 214
6.4 THE INFLUENCE OF INTERCONTINENTAL AEROSOL TRANSPORT ON PM MASS CONCENTRATIONS IN NORTH AMERICA 215
6.5 TRENDS AND THEIR IMPLICATIONS 217
6.6 COVARIATION OF PM WITH OZONE 223
6.7 SUMMARY 224
6.8 POLICY IMPLICATIONS 229
6.9 REFERENCES 231

CHAPTER 7 RECEPTOR METHODS 235

7.1 INTRODUCTION AND OVERVIEW 235
7.2 RECEPTOR MODEL TYPES 239
7.2.1 Chemical Mass Balance 241
7.2.2 Enrichment Factors 245
7.2.3 Multiple Linear Regression on Marker Species 245
7.2.4 Temporal and Spatial Correlation Eigenvectors 246
7.2.5 Time Series 246
7.2.6 Neural Networks 247
7.2.7 Aerosol Evolution and Equilibrium 247
7.3 RECEPTOR-MODEL INPUT MEASUREMENTS 247
7.3.1 Particle Size 248

7.3.2 Chemical Composition 248
7.3.2.1 Soil, Dust, and Industrial Markers 249
7.3.2.2 Combustion Markers 249
7.3.2.3 Secondary Sulfate and Nitrate 249
7.3.2.4 Carbonaceous Particles 252
7.3.2.5 Secondary Organic Aerosol 255
7.3.2.6 Other Chemical Markers 255
7.3.3 Temporal and Spatial Variability 256
7.3.4 Combining Size, Composition, Space, and Time 256
7.4 RECEPTOR MODELS AND DECISION-MAKING 257
7.4.1 Sulfur Reductions in Canadian Gasoline 257
7.4.2 $PM_{2.5}$ and Urban Haze in Denver, CO 260
7.4.3 Haze in the Grand Canyon 263
7.4.4 Understanding the Sources of PM_{10} and $PM_{2.5}$ in Mexico City 265
7.5 DEVELOPING PM MANAGEMENT STRATEGIES 266
7.5.1 Manageable and Unmanageable Source Contributions 266
7.5.2 Main Contributors to Manageable PM 267
7.6 SUMMARY 269
7.7 POLICY IMPLICATIONS 270
7.8 REFERENCES 271

CHAPTER 8 CHEMICAL-TRANSPORT MODELS 283

8.1 INTRODUCTION 283
8.2 CURRENT STATUS OF PM CHEMICAL-TRANSPORT MODELS 287
8.2.1 Emissions 287
8.2.2 Meteorology 288
8.2.3 Transport and Diffusion Processes 290
8.2.4 Chemical Transformations 290
8.2.5 Representation of PM 291
8.2.6 Deposition Processes 292
8.2.7 Computational Aspects 293
8.3 APPLICATIONS OF CHEMICAL-TRANSPORT MODELS TO THE SIMULATION OF EPISODIC AND LONG-TERM PM CONCENTRATIONS 294
8.3.1 Episodic Simulations 294
8.3.2 Long-Term Simulations 294
8.4 WHAT QUESTIONS CAN CHEMICAL-TRANSPORT MODELS ADDRESS AND HOW WELL? 295
8.4.1 Can the Contributions of Various Precursors and Source Types to PM Be Quantified? 295
8.4.2 Can the Relative Contributions of Long-Range Transport and Local Emissions Be Quantified? 296
8.4.3 Can the Relative Magnitude of Seasonal Contributions to PM Concentrations

Be Represented? 297
8.4.4 Can the Response of PM Levels to Changes in Emissions and Upwind Concentrations Be Predicted? 297
8.4.5 Can the Relationships Between PM and Other Air-Pollution Problems Be Quantified? 298
8.4.6 Can Other PM Properties that are Potentially Relevant to Health Effects Be Calculated? 298
8.4.7 Can PM Episodes Be Forecast in Real Time? 299
8.5 EVALUATION PROCESS FOR CHEMICAL-TRANSPORT MODELS 299
8.5.1 Model Simulations versus Ambient Measurements 299
8.5.2 Overview of the Performance-Evaluation Process 300
8.5.3 Data Needs for CTM Performance Evaluation 301
8.5.4 Corroboration of CTM Results with Indicator-Species Methods 301
8.6 CURRENT STATUS OF CTM PERFORMANCE AND INTERCOMPARISONS 302
8.7 USE OF CTMS TO COMPLEMENT MONITORING NETWORKS 306
8.8 USE OF CTMS TO SUPPORT ESTIMATIONS OF EXPOSURE 306
8.9 POLICY-RELEVANT RESULTS FROM CTM APPLICATIONS 308
8.9.1 PM CTMs 308
8.9.2 Acid-Deposition CTMs 310
8.9.3 Photochemical CTMs 313
8.10 CRITICAL UNCERTAINTIES 313
8.11 SUMMARY 316
8.12 POLICY IMPLICATIONS 318
8.13 REFERENCES 319

CHAPTER 9 VISIBILITY AND RADIATIVE BALANCE EFFECTS 325

9.1 HOW IS VISIBILITY LINKED TO PM? 325
9.1.1 How Is Visibility Distributed and How Has It Varied over the Years? 327
9.1.2 Factors Affecting the Relationship between PM and Visibility 329
9.1.3 Empirical Relationships between PM and Visibility 330
9.1.4 What are Some Special Issues with Visibility? 333
9.2 ROLES AND USES OF PM AND OPTICAL MEASUREMENTS IN VISIBILITY ASSESSMENT AND MANAGEMENT 334
9.2.1 Long-Term Monitoring Programs 334
9.2.2 Short-Term Measurement Programs 336
9.2.3 An Example of a Scenic Visibility Setting – The Colorado Plateau 336
9.2.4 Can One Use PM Studies for Visibility? 338
9.3 HOW ARE MODELS USED IN VISIBILITY MANAGEMENT? 338
9.3.1 What Specific Features are Required when Modeling Visibility? 339
9.3.2 Are Current Models Able to Simulate Visibility Conditions? 339
9.3.3 What Would Improve the Capacity to Model Visibility? 341
9.4 ATMOSPHERIC PARTICLES AFFECT THE GLOBAL RADIATION

BALANCE ... 342
9.5 VISIBILITY MANAGEMENT ISSUES AND APPROACHES ... 345
9.5.1 What Is Being Done to Manage Visibility? ... 345
9.5.2 Alignment of Visibility and PM Control Programs ... 346
9.5.3 Regional Planning Organizations ... 347
9.5.4 Point-Source Control Programs ... 347
9.5.5 International Programs ... 347
9.6 SUMMARY AND CONCLUSIONS ... 348
9.7 POLICY IMPLICATIONS ... 350
9.8 REFERENCES ... 351

CHAPTER 10 CONCEPTUAL MODELS OF PM FOR NORTH AMERICAN REGIONS ... 355

10.1 OVERVIEW ... 355
10.2 SUMMARY ... 357
10.3 CONCEPTUAL MODEL OF PM OVER THE SAN JOAQUIN VALLEY OF CALIFORNIA ... 359
10.3.1 Annual and Seasonal Levels of $PM_{2.5}$ and PM_{10} in Relation to Mass-Based Standards ... 359
10.3.2 Compositional Analysis of PM ... 364
10.3.3 Meteorological Influences ... 364
10.3.4 Atmospheric Processes Contributing to PM ... 365
10.3.5 Sources and Source Regions Contributing Principal Chemicals of Concern ... 366
10.3.6 Implications for Policy Makers ... 367
10.4 CONCEPTUAL MODEL OF PM OVER LOS ANGELES, CALIFORNIA ... 367
10.4.1 Annual and Seasonal Levels of $PM_{2.5}$ and PM_{10} in Relation to Mass-Based Standards ... 367
10.4.2 Compositional Analysis of PM ... 368
10.4.3 Meteorological Influences ... 370
10.4.4 Atmospheric Processes Contributing to PM ... 370
10.4.5 Sources and Source Regions Contributing Principal Chemicals of Concern ... 371
10.4.6 Implications for Policy Makers ... 373
10.5 CONCEPTUAL MODEL OF PM OVER MEXICO CITY ... 373
10.5.1 Annual and Seasonal Levels of $PM_{2.5}$ and PM_{10} in Relation to Mass-Based Standards ... 373
10.5.2 Compositional Analysis of PM ... 374
10.5.3 Meteorological Influences on PM ... 376
10.5.4 Atmospheric Processes Contributing to PM ... 377
10.5.5 Sources and Source Regions Contributing Principal Chemicals of Concern ... 377
10.5.6 Implications for Policy Makers ... 378
10.6 CONCEPTUAL MODEL OF PM OVER THE SOUTHEASTERN UNITED STATES ... 379
10.6.1 Annual and Seasonal Levels of $PM_{2.5}$ and PM_{10} in Relation to Mass-Based Standards ... 379
10.6.2 Compositional Analysis of PM ... 380
10.6.3 Meteorological Influences on PM ... 382

10.6.4 Atmospheric Processes Contributing to PM 383
10.6.5 Sources and Source Regions 383
10.6.6 Implications for Policy Makers 383
10.7 CONCEPTUAL MODEL OF PM OVER THE NORTHEASTERN UNITED STATES 385
10.7.1 Annual and Seasonal Levels of $PM_{2.5}$ and PM_{10} in Relation to Mass-based Standards 385
10.7.1.1 Annual Mean Concentrations of $PM_{2.5}$ 385
10.7.1.2 24-hr-Mean Concentration of $PM_{2.5}$ 385
10.7.1.3 Annual and Daily PM_{10} 386
10.7.1.4 Seasonal-Mean Concentrations of $PM_{2.5}$ 386
10.7.2 Compositional Analysis of PM 386
10.7.2.1 Seasonal Mean $PM_{2.5}$ Composition 388
10.7.3 Meteorological Influences on PM 389
10.7.4 Atmospheric Processes Contributing to PM 389
10.7.5 Sources and Source Regions Contributing Principal Chemicals of Concern 390
10.7.6 Implications for Policy Makers 390
10.8 CONCEPTUAL MODEL OF PM OVER THE WINDSOR-QUEBEC CITY CORRIDOR 391
10.8.1 Annual and Seasonal Levels of $PM_{2.5}$ and PM_{10} in Relation to Mass-Based Standards 391
10.8.2 Compositional Analysis of PM 392
10.8.3 Meteorological Influences on PM 394
10.8.4 Atmospheric Processes Contributing to PM 394
10.8.5 Sources and Source Regions Contributing Principal Chemicals of Concern 395
10.8.6 Implications for Policy Makers 395
10.9 CONCEPTUAL MODEL OF PM OVER THE U.S. UPPER MIDWEST – GREAT LAKES AREA 396
10.9.1 Annual and Seasonal Levels of $PM_{2.5}$ and PM_{10} in Relation to Mass-Based Standards 396
10.9.2 Compositional Analysis of PM 396
10.9.3 Meteorological Influences 396
10.9.4 Atmospheric Processes Contributing to PM 396
10.9.5 Sources and Source Regions Contributing Principal Chemicals of Concern 397
10.9.6 Implications for Policy Makers 398
10.10 CONCEPTUAL MODEL OF PM OVER THE CANADIAN PRAIRIE AND U.S. CENTRAL PLAINS 398
10.10.1 Annual and Seasonal Levels of $PM_{2.5}$ and PM_{10} in Relation to Mass-Based Standards 398
10.10.2 Compositional Analysis of PM 399
10.10.3 Meteorological Influences on PM 399
10.10.4 Atmospheric Processes Contributing to PM 401
10.10.5 Sources and Source Regions Contributing Principal Chemicals of Concern 401
10.10.6 Implications for Policy Makers 401
10.11 CONCEPTUAL DESCRIPTION OF PM OVER THE LOWER FRASER VALLEY AIRSHED 402
10.11.1 Annual and Seasonal Levels of $PM_{2.5}$ and PM_{10} in Relation to Mass-Based Standards 402
10.11.2 Compositional Analysis of PM 404
10.11.3 Meteorological Influences on PM 404

10.11.4 Atmospheric Processes Contributing to PM ... 407
10.11.5 Sources and Source Regions Contributing Principal Chemicals of Concern ... 407
10.11.6 Implications for Policy Makers ... 408
10.12 REFERENCES ... 409

CHAPTER 11 RECOMMENDED RESEARCH TO INFORM PUBLIC POLICY ... 415

11.1 RECOMMENDATIONS ... 416
11.2 FUTURE NARSTO PM ASSESSMENTS ... 431
11.3 REFERENCES ... 432

GLOSSARY ... 433

ACRONYMS AND ABBREVIATIONS ... 433
DEFINITIONS ... 435

APPENDIX A. EMISSION CALCULATIONS AND INVENTORY LISTINGS ... 439

A.1 HOW ARE EMISSIONS CALCULATED? ... 439
A1.1 Emission and Emission Reduction Factors ... 439
A1.2 Activity Patterns ... 440
A1.3 Spatial Allocation ... 442
A1.4 Processing for Model Applications. ... 443
A1.5 Limitations and Uncertainties ... 443
A.2 EMISSION INVENTORIES BY DETAILED SOURCE CATEGORY ... 445
A.3 REFERENCES ... 458

APPENDIX B. MEASUREMENTS ... 459

B.1 APPLICATIONS OF DATA FROM AIR-QUALITY MEASUREMENTS ... 459
B.2 CURRENTLY AVAILABLE TECHNOLOGY AND INSTRUMENT CAPABILITIES ... 459
B.2.1 Inlets ... 459
B.2.2 Integrated Denuder and Filter Systems (substrate- and absorbent-based measurements) for Mass and Composition Sampling ... 459
Denuders ... 463

Filters 465
Impactors 465
Chemical Analysis Methods for PM Collected on Filters 465
B.2.3 Continuous and Semi-continuous Real-time Measurements 467
Mass and Mass Equivalent 467
Inertial Methods 467
Pressure-Drop Method 471
Electron-Attenuation Method 471
Size Distribution and Mobility 471
Bulk Chemical Composition Methods 472
Black Carbon (BC) and Organic Carbon (OC) 472
Ionic Component of Aerosol Particles 473
Particulate Metals 473
B.2.4 Single-Particle Measurements 474
B.2.5 Optical Properties of Aerosols and Long-Path Optical Measurements 475
In-situ Measurements of Light Scattering and Light Absorption 475
Long-Path Measurement Techniques: Remote Sensing and Visibility 475
Satellite Measurements 477
B.2.6 Gas-Phase Aerosol Precursors, Ozone, Ozone Precursors and Oxidants 478
B.2.7 Meteorological Measurements 485
B.3 MEASUREMENT UNCERTAINTY AND VALIDATION 485
B.4 REFERENCES 487

APPENDIX C. MONITORING DATA: AVAILABILITY, LIMITATIONS, AND NETWORK ISSUES 493

C.1 MONITORING PROGRAMS AND OBJECTIVES 493
C.2 NETWORK DESIGN 494
C.3 NETWORK NEEDS 497
C.4 REFERENCES 498

APPENDIX D. GLOBAL AEROSOL TRANSPORT 501

APPENDIX E. PREPARATION OF THIS ASSESSMENT 509

LIST OF FIGURES

SYNTHESIS

Figure S.1 (1.1). Framework for informing PM management. 8
Figure S.2 (1.6). Development of a conceptual model. 9
Figure S.3 (3.2). Typical number and volume distributions of atmospheric particles. 11
Figure S.4 (1.3). Illustrative corroborative approach of providing atmospheric science input to the policy questions. 13
Figure S.5 (6.7). Average $PM_{2.5}$ concentrations.......... 15
Figure S.6 (6.3). Average annual PM_{10} mass concentrations. 16
Figure S.7 (6.12). Composition of $PM_{2.5}$ mass at representative urban and rural locations. 20
Figure S.8 (3.16). Chemical links between the ozone and PM formation processes. 22
Figure S.9 (6.15). Comparisons of average $PM_{2.5}$ mass concentrations and species concentrations. 23
Figure S.10 (6.21a). Trends in concentrations of SO_2 and $SO_4^=$ at CAPMoN sites, 1980 - 2000. 27
Figure S.11 (10.28). Reconstructed eastern U.S. fine mass partitioned into summer and winter components. 28
Figure S.12 (7.4). Three-day back-trajectories of Ontario Air-Quality Index site at Simcoe. 29
Figure S.13 (7.7). CMB source-contribution estimates for $PM_{2.5}$ in two regions of Canada. 33
Figure S.14 (4.3) Distribution of SO_2 sources in Canada and the United States. 34
Figure S.15. (6.21b). Trends in concentrations of SO_2 and $SO_4^=$ at CASTNet sites, 1980 - 2000. 42
Figure S.16 (1.5). Iterative communication for managing air-quality. 45
Figure S.17 (1.5). NRC framework for evaluating risks. 47

CHAPTER 1

Figure 1.1. Framework for informing PM management. 54
Figure 1.2. Representative composition of PM. 56
Figure 1.3. Illustrative transport scales for PM and other atmospheric pollutants. 59
Figure 1.4. NRC framework for evaluating risks extending from pollutant sources to responses. 61
Figure 1.5. Iterative communication for managing air quality. 65
Figure 1.6. Development of conceptual models of PM behavior. 67

CHAPTER 2

Figure 2.1. Schematic rendering of relationship among outdoor and indoor sources. 74
Figure 2.2. Regression analysis of daytime personal exposure to PM_{10}. 75
Figure 2.3. Particle deposition in the major regions of the human respiratory tract. 76
Figure 2.4. Schematic rendering of basic design for epidemiological studies. 77
Figure 2.5. Stylized summary of acute exposure studies. 83
Figure 2.6. Stylized summary of chronic exposure studies. 89

CHAPTER 3

Figure 3.1. Schematic of the life cycle of atmospheric particles and their interactions with the gas and aqueous phases. 104
Figure 3.2. Typical number and volume distributions of atmospheric particles with the different modes. 105
Figure 3.3. Electron micrographs of selected particles. 106
Figure 3.4. Evolution of a typical urban aerosol size distribution subject to coagulation. 108
Figure 3.5. Growth and evaporation of a $NaCl/Na_2SO_4$ particle with changing relative humidity. 109
Figure 3.6. Predicted particulate nitrate concentration as a function of relative humidity. 110
Figure 3.7. Schematic of the three pathways for the formation of $SO_4^=$ in the atmosphere. 110
Figure 3.8. Schematic of the formation of HNO_3 and particulate NO_3^- in the atmosphere. 111
Figure 3.9. Schematic of the formation of secondary organic aerosol in the atmosphere. 112
Figure 3.10. Speciation results for organic aerosol in Southern California. 113
Figure 3.11. Response of the fine PM mass to changes in $SO_4^=$ concentration. 116
Figure 3.12. Estimated response of fine PM to changes in $SO_4^=$ during winter. 117
Figure 3.13. Isopleths of predicted particulate NO_3^- concentration. 118
Figure 3.14. Schematic diagram of the processes of in-cloud scavenging. 119
Figure 3.15. Typical lifetime of atmospheric particles in the lower troposphere. 120
Figure 3.16. Chemical links between ozone and PM formation processes. 121
Figure 3.17. Percentage changes of maximum 1-hour average concentrations of ozone and $PM_{2.5}$ 121

CHAPTER 4

Figure 4.1. Distribution of sources of primary PM_{10} in southern Canada and the United States. 139
Figure 4.2. Distribution of sources of primary $PM_{2.5}$ in southern Canada and the United States. 140
Figure 4.3. Distribution of SO_2 sources in Canada and the United States. 141
Figure 4.4. Distribution of NH_3 sources in the United States and Canada 142
Figure 4.5. Flow diagram for development of the U.S. National Emission Inventory. 145

CHAPTER 5

Figure 5.1. Relationships of data-measurement techniques, air-quality management tools, and user applications. 160
Figure 5.2. Inlets used for particle size separation. 162
Figure 5.3. Representative denuder and filter sampler for collecting $PM_{2.5}$. 164

CHAPTER 6

Figure 6.1. Average air mass source regions and transport during July. 192
Figure 6.2. Locations of study regions. 194
Figure 6.3. Average annual PM_{10} mass concentrations. 196
Figure 6.4. Seasonal variation of PM_{10} mass concentrations. 197
Figure 6.5. Variations of average PM_{10} mass concentrations 198
Figure 6.6. Variations of average PM_{10} mass concentrations , Mexico City. 199
Figure 6.7. Average $PM_{2.5}$ concentrations. 200
Figure 6.8. Example urban-scale variations of average $PM_{2.5}$ mass concentrations in southern California. 201
Figure 6.9. Numbers of $PM_{2.5}$ monitoring sites exceeding three benchmark levels. 202

Figure 6.10. Statistical distribution of 24-hr $PM_{2.5}$ concentrations at 20 sites in CA. 203
Figure 6.11. Seasonal variations of $PM_{2.5}$ mass concentrations at selected IMPROVE monitoring locations. 204
Figure 6.12. Composition of $PM_{2.5}$ at representative urban and rural locations. 206
Figure 6.13. Mean annual $SO_4^=$ concentration in eastern North America during 1997-98. 207
Figure 6.14. Mean NO_3^- and NH_4^+ concentrations during 1997-98. 208
Figure 6.15. Comparisons of average $PM_{2.5}$ mass and species concentrations at paired urban and rural locations. 210
Figure 6.16. Comparisons of average $PM_{2.5}$ mass and species concentrations at urban and rural locations. 211
Figure 6.17. Example day-of-week variations of $PM_{2.5}$ mass and coarse mass concentrations. 212
Figure 6.18. Example diurnal variations of PM_{10} mass concentration at ten sites in Mexico City. 213
Figure 6.19. Trends in $PM_{2.5}$ mass at IMPROVE sites, 1988-98. 218
Figure 6.20. Trends in annual particulate $SO_4^=$ and NO_3^- concentrations. 220
Figure 6.21. Trends in concentrations of SO_2 and $SO_4^=$ at CAPMoN and CASTNet sites. 221
Figure 6.22. Seasonal variation of ozone, $PM_{2.5}$ mass, and PM_{10} mass at four locations. 224
Figure 6.23. Daily variations of ozone, $PM_{2.5}$ mass, and PM_{10} mass in Los Angeles. 225
Figure 6.24. Hourly variations of ozone and $PM_{2.5}$ mass in Ontario and Atlanta.. 226

CHAPTER 7

Figure 7.1. Five-minute averages of BC measured with an aethalometer in Mexico City. 238
Figure 7.2. Average PM_{10} at urban and non-urban locations in the Las Vegas, NV. 239
Figure 7.3. Material balance for PM_{10} and $PM_{2.5}$. .... .240
Figure 7.4. Three-day back-trajectories at Simcoe, ON, during May-September of 1998 and 1999. 241
Figure 7.5. Examples of $PM_{2.5}$ source profiles from emitters typical of Denver, CO. 250
Figure 7.5. (continued) Examples of $PM_{2.5}$ source profiles. 251
Figure 7.6. Examples of $PM_{2.5}$ source profiles. 254
Figure 7.7. CMB $PM_{2.5}$ source-contribution estimates for two regions of Canada.. 258
Figure 7.8. Conceptual model describing links between urban particles and tailpipe emissions. 259
Figure 7.9. Comparison of fractional source contributions to $PM_{2.5}$ during 1996 in Denver, CO. 261
Figure 7.10. Effects of changes in (a) NH_3, (b) HNO_3, and (c) $SO_4^=$ levels. 262
Figure 7.11. Comparison of fractions of $SO_4^=$ at Meadview, AZ. 266

CHAPTER 8

Figure 8.1a. Schematic of 3-D Eulerian framework for a chemical transport model. 285
Figure 8.1b. Schematic of 1-D Lagrangian framework for a chemical transport model on an Eulerian reference grid system. 285
Figure 8.2. Schematic description of the components of a PM modeling system. 287
Figure 8.3. Approaches to calculating long-term PM concentrations. 295
Figure 8.4. Comparison of measured and simulated 24-hr average $SO_4^=$ values concentrations in Big Bend National Park, Texas. 303
Figure 8.5. Comparison of measured and simulated chemical compositions of 24-hr average $PM_{2.5}$ in Big Bend National Park, Texas on 12 October 1999. 304

Figure 8.6. Comparison of measured and simulated 24-hr average $PM_{2.5}$ $SO_4^=$ concentrations over eastern North America for an EMEFS episode, 1 Aug. 1988. 305
Figure 8.7a. Application of two distinct CTMs to northeastern North America: AURAMS. 309
Figure 8.7b. Application of two distinct CTMs to northeastern North America: Models-3/CMAC. 310
Figure 8.8a. Annual predicted percentage contribution of the Ohio/WestVirginia/ Pennsylvania border SO_2 to total sulfur deposition in eastern North America.. 311
Figure 8.8b. Annual predicted percentage contribution of the Canadian SO_2 sources to total dry sulfur deposition in eastern North America. 312

CHAPTER 9

Figure 9.1. Illustration of the three processes by which particles and gases in the atmosphere affect visibility. 326
Figure 9.2. Distribution of average light extinction in national parks and wilderness areas of the United States. 327
Figure 9.3. Light extinction statistics for Canadian urban regions. 328
Figure 9.4. Calculated light scattering efficiency vs. particle diameter for several chemical species. 330
Figure 9.5. Effect of relative humidity on light scattering. 331
Figure 9.6. Relative contributions of PM constituents to particle-caused light extinction for several regions of the U.S. 333
Figure 9.7. Global, annual mean radiative forcings (W/m^2) for the period from pre-industrial (1750) to the present. 343
Figure 9.8. Measured change in annual surface air temperature 1961-1980. 345

CHAPTER 10

Figure 10.1 Development of a conceptual model by applying atmospheric science analyses. 355
Figure 10.2. The use of conceptual model to identify needed ambient concentration information and CTM applications. 356
Figure 10.3. North American areas covered by conceptual models. 357
Figure 10.4. Simplified conceptual model for the San Joaquin Valley. 360
Figure 10.5. Simplified conceptual model for the Los Angeles Air Basin. 360
Figure 10.6. Simplified conceptual model for Mexico City. 361
Figure 10.7. Simplified conceptual model for the Southeastern United States. 361
Figure 10.8. Simplified conceptual model for the Northeastern United States 362
Figure 10.9. Simplified conceptual model for the Windsor - Quebec City Corridor 362
Figure 10.10. Simplified conceptual model for the U.S. Upper Midwest— Great Lakes. 363
Figure 10.11 Simplified conceptual model for the Canadian Southern Prairie and the U.S. Northern Plains. 363
Figure 10.12. Simplified conceptual model for the Lower Fraser Valley. 364
Figure 10.13. Temporal variations of PM during IMS95. 365
Figure 10.14. Temporal variations of monthly-average $PM_{2.5}$ at sites in the South Coast Air Basin, 1993. 368
Figure 10.15. Temporal variations of monthly-average $PM_{2.5}$ at sites in the South Coast Air Basin, 1998-1999. 369

Figure 10.16. Temporal variations of monthly-average PM_{10} at sites in the South Coast Air Basin, 1998-1999. 369
Figure 10.17. Annual $PM_{2.5}$ and PM_{10} composition averaged across all monitoring sites in the South Coast Air Basin. 370
Figure 10.18. Individual source contributions to the 24-hr averaged size distribution of airborne PM. 372
Figure 10.19. Average annual concentrations of PM_{10} at five stations in the Mexico City Metropolitan Area, 2000. 374
Figure 10.20. Monthly average concentrations of PM_{10} at five stations in the Mexico City Metropolitan Area, 1995-2000. 375
Figure 10.21. Hourly average concentrations of PM_{10} at five stations in the Mexico City Metropolitan Area in 2000. 375
Figure 10.22. Concentrations of $PM_{2.5}$ at the Six Core Sampling Sites. 376
Figure 10.23. Seasonal $PM_{2.5}$ at the Fire Station monitor in Atlanta. 380
Figure 10.24. Annual-average PM fine composition at three sites in Atlanta. 381
Figure 10.25. $PM_{2.5}$ composition at Great Smoky Mountains, 1999. 382
Figure 10.26. Preliminary source-area impact analysis of $SO_4^=$ PM at Great Smoky Mountain National Park. 384
Figure 10.27. Seasonal behavior of the Reconstructed Fine Mass across the northeastern United States. 387
Figure 10.28. Reconstructed Fine Mass partitioned into the individual components. 388
Figure 10.29. Three-year mean, $PM_{2.5}$ concentrations for NAPS. 392
Figure 10.30. Bar charts of seasonal variations in $PM_{2.5}$ chemical composition. 393
Figure 10.31. The relative composition of $PM_{2.5}$ mass at Egbert, Ontario. 393
Figure 10.32. $PM_{2.5}$ measurements at Boundary Waters Canoe Area and at Shenandoah National Park. 397
Figure 10.33. Chemical speciation of $PM_{2.5}$ in national parks and wilderness areas of the Great Plains. 400
Figure 10.34. Annual variations in 24-hr average $PM_{2.5}$ measurements from the Chilliwack TEOM site. 403
Figure 10.35. Seasonal variations in 24-hr average $PM_{2.5}$ measurements from the Chilliwack TEOM. 403
Figure 10.36. Annual variations in 24-hr average PM_{10} measurements from the Chilliwack TEOM site. 405
Figure 10.37. Seasonal variations in 24-hr average PM_{10} measurements from the Chilliwack TEOM site. 405
Figure 10.38. The percentage distribution of the main $PM_{2.5}$ chemical constituents, urban Vancouver. 406
Figure 10.39. Percent contribution to reconstructed fine mass from the five dominant modes of fine particle composition for Clearbrook 406
Figure 10.40. Percent contribution to reconstructed fine mass from the five dominant modes of fine particle composition for Chilliwack. 406
Figure 10.41. Year 2000 Emission Inventory for PM_{10} and $PM_{2.5}$ for the Lower Fraser Valley. 408

APPENDIX D.

Figure D.1. Global distribution of aerosols and their likely source regions during June and December. ... 502
Figure D.2. Satellite data from July 1998 illustrating the extent of the Sahara dust plume. ... 503
Figure D.3. Seasonal pattern of fine particulate soil (dust) concentration at IMPROVE monitoring sites in the southeastern United States. ... 504
Figure D.4. Aerosol optical thickness and illustration of the positions of a dust plume during April 1998. ... 505
Figure D.5. $PM_{2.5}$ dust concentrations at IMPROVE monitoring sites on three dates. ... 505
Figure D.6. Satellite imagery indicating the locations of major fires during April and July 1998. ... 506
Figure D.7. Hourly PM_{10} concentrations at six locations during May 1998. ... 507

LIST OF TABLES

PM SCIENCE FOR POLICY MAKERS: SYNTHESIS

Table S.1 (1.3). Existing national PM standards and their implementation timetables. 10
Table S.2 (5.1 - 5.3). Measurement uncertainty in PM physical and chemical characteristics. 17
Table S.3 (8.2). Levels of confidence in aspects of chemical-transport model simulations. 31
Table S.4 (3.2). Typical pollutant/atmospheric issue relationships. 37
Table S.5 (9.2). Responses of regional haze and climate to reductions in the emissions of secondary PM precursors and primary PM. 38
Table S.6 (4.8). Estimated confidence level of emission estimates. 41
Table S.7. Typical periods for acquisition of information on progress. 46
Table S.8 (adapted from Textbox 2.5). Availability of ambient measurement methods for hypothesized causal elements of PM-induced health effects. 49
Table S.9. Policy benefits of the specific research directions. 52

CHAPTER 1

Table 1.1. Comparison of ambient particle fractions. 55
Table 1.2. General descriptions of PM emissions and source types. 60
Table 1.3. Existing PM standards and their implementation timetables. 63

CHAPTER 2

Table 2.1. Comparison of Mortality Risk Ratios for Smoking and Air Pollution from the Six Cities and ACS Prospective Cohort Studies. 87

CHAPTER 3

Table 3.1. Predicted changes in aerosol component concentrations for a reduction in precursor emissions for Southern California. 115
Table 3.2. Typical pollutant/atmospheric issue relationships. 122

CHAPTER 4

Table 4.1. Global sources of airborne particles roughly less than 10 µm in diameter 128
Table 4.2. Illustrative linkages between current emission information used for estimating ambient concentrations relevant to hypothesized causal elements for PM health effects. 131
Table 4.3. Summary of nationwide 1995 Canadian and 1999 U.S. emissions by similar categories. 133
Table 4.4. Demographics of example North American cities. 134
Table 4.5. Comparison of PM_{10} and $PM_{2.5}$ emissions between Atlanta, Toronto, Mexico City, and Los Angeles. 135
Table 4.6. Comparison of precursor gas emissions for Atlanta, Los Angeles, Mexico City, and Toronto. 136
Table 4.7. Summary of historical and projected national emissions for Canada and the United States. 143
Table 4.8. Estimated confidence level of emission estimates. 147

CHAPTER 5

Table 5.1. Estimated uncertainty in measurements of the physical properties of PM. 175
Table 5.2. Uncertainty in measurements of the chemical composition of PM: acids and inorganics. 176
Table 5.3. Uncertainty in measurements of the chemical composition of PM: carbon and organics. 177
Table 5.4. Uncertainty in routine measurements of gas-phase compounds. 179

CHAPTER 7

Table 7.1. Summary of receptor model source-apportionment models. .... 242 et seq.
Table 7.2. Receptor and source methods used to attribute $SO_4^=$ in the Grand Canyon to Mohave. .... 264 et seq.

CHAPTER 8

Table 8.1. Performance evaluations of PM grid models for $PM_{2.5}$ and components with the SCAQS data base in the Los Angeles basin. 303
Table 8.2. Present qualitative levels of confidence in various aspect of PM CTM simulations. .... 314 et seq.

CHAPTER 9

Table 9.1. Recent short-term visibility measurement programs. 336
Table 9.2. Responses of regional haze and climate to reductions in the emissions of secondary PM precursors and primary PM from present-day levels. 344

CHAPTER 10

Table 10.1. Average chemical composition of $PM_{2.5}$ at four sites, from March 2 to March 19, 1997. 376
Table 10.2. 1998 Emission Inventory for the MCMA 378
Table 10.3. Difference in annual average (1999) $PM_{2.5}$ levels in the urban-rural pairs of the SEARCH network. 380
Table 10.4. Long-term average values reported by national and state or provincial agencies for $PM_{2.5}$ and PM_{10} concentrations. 399
Table 10.5. Comparison of urban and rural $PM_{2.5}$ speciation measurements. 400
Table 10.6. Period-of-record mean levels measured in Alberta, Canada for the types of sites as indicated. 401

CHAPTER 11

Table 11.1. Summary of Recommendations .... 416 et seq.

APPENDIX A.

Table A.1. 1999 U.S national emissions for PM and related pollutants (thousand short tons). 446
Table A.2. 1999 Canadian emission inventory for PM and related pollutants (Ktonnes/yr). 454

APPENDIX B.

Table B.1. Quantifiable properties for particle and particle-related measurements. 460
Table B.2. Major components of selected integrated particulate samplers. 464
Table B.3. Summary of real-time particle monitoring techniques. 469
Table B.4. Summary of real-time single particle measurement techniques. 475
Table B.5.a. In-situ measurements of aerosol optical properties. 477
Table B.5.b. Long path measurements of aerosol optical properties. 477
Table B.6. Summary of real-time gas-phase precursor measurements. a. Sulfur compounds. 479
Table B.6. b. Ammonia. 480
Table B.6. c. Ozone. 480
Table B.6. d. Carbon monoxide. 481
Table B.6. e. Speciated volatile organic compounds. 482
Table B.6. f. Nitric oxide, nitrogen dioxide , and total reactive nitrogen. 483
Table B.6. g. Other nitrogen oxides. 484
Table B.6. h. Peroxides. 485
Table B.6. i. Odd hydrogen species. 485
Table B.7 Instruments used to measure meteorological parameters over a
long-path and/or above the surface. 486

APPENDIX C.

Table C.1 General specifications for PM observation and monitoring networks. 495

PREFACE

Regulatory agencies in Canada, the United States, and Mexico are entering a new phase of implementing ambient air-quality standards for fine and coarse particulate matter (PM). Efficiency and cost-effectiveness of this implementation process depend strongly on the prudent application of scientific knowledge by the PM management community - a feature made especially challenging by currently rapid scientific developments in the PM field. As a consequence there is a strong need for a comprehensive description of PM science, which presents current knowledge in a form that is convenient for use by PM managers and policy analysts.

This NARSTO[1] PM assessment responds to this need, with the main objective of interpreting complex and new atmospheric science in a manner that is useful for PM management applications. A secondary objective is to provide atmospheric-science and PM information useful to exposure and health scientists in their efforts to investigate causal hypotheses of health impacts, characterize exposure, and conduct epidemiological studies.

Commissioning of the Assessment

Members of the NARSTO Executive Assembly commissioned the preparation of a PM science assessment at their March 2000 meeting. The specific charge was to assess the state of scientific understanding of atmospheric PM as it relates to policy questions and to the implementation of new PM standards and goals. The science to be addressed included the description of air quality based on information from physical, chemical, meteorological, emission, and deposition measurements and projections. It also included the mathematical simulation of atmospheric processes as an integrated means for designing strategies that will reduce health impacts and improve visibility. The NARSTO PM Assessment was to evaluate critically the reliability and applicability of the technical and scientific tools currently available to support policy makers. The Assessment was also to address the requirements perceived necessary for substantially improving these tools in the next five to ten years.

In support of the charge for this PM Assessment, six objectives were identified:

1. Gain an understanding from policy makers of information needs and constraints, including economic, policy, and implementation boundaries.

2. Provide a comprehensive conceptual model of PM formation and distribution for science-policy analysts and air-quality policy makers. The model is to accommodate changing knowledge about atmospheric processes, emission sources, emission-control technology, exposure, and human-health and environmental impacts. It is to address existing limits in information and forecast the implications of expected results from ongoing and future research.

3. Provide a plain-language conceptual description of PM air quality for the public that describes the relevance of the atmospheric-science research with its recent progress and findings.

4. Recommend atmospheric-science and related emission research, with priorities tied to the policy-making process, to research managers developing a coordinated research strategy for PM.

5. Provide a framework for atmospheric scientists that relates their work to standards, implementation, and air-quality management, and to health, exposure, and environmental impact research for standard setting.

[1] NARSTO, originally the North American Research Strategy for Tropospheric Ozone, a multi-stakeholder entity, was organized in 1994 to sponsor cooperative, public/private, policy-relevant research on tropospheric ozone. In 1999, NARSTO was rechartered with its name unchanged though no longer an acronym and given an expanded charge to include airborne particulate matter. Currently its research programs focus on both ozone and PM, including their combined atmospheric chemistry and physics. As a major part of its charter, periodic policy-relevant science assessments are commissioned. Its first assessment was the 2000 Ozone Assessment.

6. Provide a context for researchers in related fields to link their work to that of the atmospheric-science community, supplying important information on the current state of knowledge of PM formation and distribution and offering opportunities for future research coordination.

Nine assumptions were also part of the charge, including the definition of the problem being addressed, linkages to other pollutants, and the balance to be achieved in the Assessment:

1. The definition of the problem is reducing levels of PM. Regulatory agencies are expected to recommend both $PM_{2.5}$ (fine PM) and $PM_{10\text{-}2.5}$ (coarse PM) standards over the next two years. This Assessment will encompass a review of PM characteristics relevant to both size fractions.

2. The PM Assessment will contain contextual information on exposure, health, and environmental impacts. This information will come from the extensive and in-depth science reviews, without update, prepared by Canada, Mexico, and the United States during their air-quality goals and standards setting process.

3. A summary of related environmental issues, including deposition, climate change, and air toxins, will be included to provide context for a discussion of the science of atmospheric aerosol occurrence and exposure.

4. Linkages between PM, ozone, and other copollutants are an important feature of the PM issue. The Assessment will address these linkages.

5. It is appropriate for the PM Assessment to address contextual aspects of emission-control technologies mainly by reference to existing publications and information.

6. It is necessary to achieve a balanced assessment picture for the PM issues facing the NARSTO member countries.

7. The PM Assessment process should be fully open to public participation.

8. The PM Assessment should explore accountability approaches that directly relate observations to determining the effectiveness of air-quality management plans.

9. Given the continuing advancement of information on PM, this Assessment will be the first in a series of periodic NARSTO PM assessments.

Verifying the Needs of Policy Makers

Between September 1999 and October 2000, the NARSTO Analysis and Assessment Team co-chairs prepared and confirmed policy questions for which scientific information supplied a critical need.

Box P.1. Policy questions confirmed with the policy community as central to implementing programs to achieve air-quality standards:

PQ1. Is there a significant PM problem and how confident are we?

PQ2. Where there is a PM problem, what is its composition and what factors contribute to elevated concentrations?

PQ3. What broad, pollutant-based, approaches might be taken to fix the problem?

PQ4. What source-specific options are there for fixing the problem given the broad control approaches above?

PQ5. What is the relationship between PM, its components, and other air-pollution problems on which the atmospheric science community is working?

PQ6. How can we measure our progress? How can we determine the effectiveness of our actions in bringing about emission reductions and air-quality improvements, with their corresponding exposure reductions and health improvements?

PQ7. When and how should we reassess and update our implementation programs to adjust for any weaknesses in our plan, and take advantage of advances in science and technology?

PQ8. What further atmospheric sciences information will be needed in the periodic reviews of our national standards?

To reach confirmation of these questions, hour-long interviews were held with 67 senior policy makers at 49 federal, state, and provincial environment departments and with private industries in Canada, the United States, and Mexico. The interview questions were based on guidance from the U.S. National Research Council's peer review of the Ozone Assessment (NARSTO 2000)[2]. The questions sought to identify senior policy-maker goals in managing air quality, to discover the type of information required to meet those goals, and to gauge policy makers' impressions of uncertainties in the atmospheric science related to managing PM.

Preparing the Assessment

The Assessment was written by a team of authors and directed by three co-chairs, as identified below:

Assessment Co-chairs

Peter McMurry, University of Minnesota, Dept. of Mechanical Engineering

Marjorie Shepherd, Environment Canada, Meteorological Service of Canada

James Vickery, U.S. Environmental Protection Agency, Office of Research and Development

Lead Authors

Chapter 1: Perspective for Managing PM: Marjorie Shepherd, Environment Canada, Meteorological Service of Canada

Chapter 2: Health Effects Context: Roger McClellan, Chemical Industries Institute of Toxicology (Emeritus); Barry Jessiman, Health Canada, Health Protection Branch

Chapter 3: Atmospheric Aerosol Processes: Spyros Pandis, Carnegie Mellon University

Chapter 4: Emission Characterization: George Hidy, Envair / Aerochem; David Niemi, Environment Canada, Environmental Protection Service; Thompson Pace, U.S. Environmental Protection Agency, Office of Air Quality Planning and Standards

Chapter 5: Particle and Gas Measurements: Fred Fehsenfeld, National Oceanic and Atmospheric Administration; Don Hastie, York University, Centre for Atmospheric Chemistry; Paul Solomon, U.S. Environmental Protection Agency, Office of Research and Development; Judith Chow, Desert Research Institute

Chapter 6: Spatial and Temporal Characterization of PM: Charles Blanchard, Envair

Chapter 7: Receptor Methods: Jeff Brook, Environment Canada, Meteorological Service of Canada; Elizabeth Vega, Instituto Mexicano del Petroleo; John Watson, Desert Research Institute

Chapter 8: Chemical Transport Models: Christian Seigneur, AER Inc.; Michael Moran, Environment Canada, Meteorological Service of Canada

Chapter 9: Visibility and Radiative-Balance Effects: Ivar Tombach, Consultant; Karen McDonald, Concordia University College of Alberta

Chapter 10: Conceptual Models of PM for North American Areas: James Vickery, U.S. Environmental Protection Agency, Office of Research and Development

Chapter 11: Recommended Research to Inform Public Policy: Peter McMurry, University of Minnesota, Dept. of Mechanical Engineering

Contributing Authors

Chapter 1: Hilda Martinez Salgado, INE Mexico; Terry Keating, U.S. Environmental Protection Agency, Office of Air and Radiation

[2] NARSTO, 2000. An assessment of tropospheric ozone pollution – a North American perspective. NARSTO Management Office (Envair), Pasco, Washington. http://www.cgenv.com/Narsto/

Chapter 3: Leonard Barrie, Pacific Northwest National Laboratory, U.S. Department of Energy

Chapter 4: Jason West, U.S. Environmental Protection Agency, Office of Air and Radiation; Michael Kleeman,University of California – Davis.

Chapter 6: Rudolph Husar, Washington University; Robert Vet, Environment Canada, Meteorological Service of Canada; Thomas Dann, Environment Canada, Environmental Protection Service; Graciela Raga, Universidad Nacional Autonoma de Mexico; Paul Solomon, U.S. Environmental Protection Agency; Elizabeth Vega, Instituto Mexicana del Petroleo

Chapter 8: Praveen Amar, NESCAUM; Jason West, U.S. Environmental Protection Agency, Office of Air and Radiation; Rafael Villasenor, IMP Mexico

Chapter 10 area descriptions as follows:

San Joaquin Valley - Betty Pun and Christian Seigneur, AER Inc.

Windsor-Quebec City Corridor - Michael Moran and Jeff Brook, Environment Canada, Meteorological Service of Canada

Mexico City - Sylvia Edgerton, Pacific Northwest National Laboratory, U.S. Department of Energy; Jason West, U.S. Environmental Protection Agency, Office of Air and Radiation; Hilda Martinez Salgado, INE Mexico; and Elizabeth Vega, Instituto Mexicano del Petroleo

Los Angeles - Michael Kleeman and Michael Hannigan, University of California – Davis, Department of Environmental Engineering

Lower Fraser Valley - Bruce Thomson and Bill Taylor, Environment Canada, Pacific and Yukon Region

Upper Mid-West - Betty Pun, Christian Seigneur, and Mark Leidner, AER Inc.

Canadian Southern Prairies and U.S. Northern Plains - Karen McDonald, Concordia University College of Alberta

Northeastern United States - Robin Dennis, U.S. Environmental Protection Agency, Office of Research and Development

Southeastern United States - Ted Russell, Georgia Institute of Technology, Department of Civil and Environmental Engineering

NARSTO Staff

Jeremy Hales and Diane Fleshman, Envair

Jeffrey West, U.S. EPA / NOAA

Technical Editor: Elizabeth Owczarski, Dom Entropji

Like the NARSTO 2000 Ozone Assessment, this PM Assessment has been reviewed extensively within the NARSTO community and externally through the U.S. National Academy of Sciences, the Royal Society of Canada, and the Red de Investigación y Desarrollo sobre Calidad del Aire en Grandes Ciudades of FUMEC (Fundación México-Estados Unidos para la Ciencia). The authors and the NARSTO Executive Steering Committee are grateful to the many reviewers for their constructive comments and suggestions during the preparation of this Assessment.

A valuable and comprehensive amount of information was assembled to produce this Assessment. The key findings and recommendations are presented in the Executive Summary and Synthesis immediately following. Detailed information appears in Chapters 1 through 11, as well as in the Appendices.

ACKNOWLEDGMENTS

A number of sponsors made the preparation of this Assessment possible. Their support is gratefully acknowledged.

Direct funding for the scientific effort and evaluation was provided by the American Petroleum Institute, the California Air Resources Board, the U.S. Department of Energy-Office of Science, the Meteorological Service of Canada, the Environmental Protection Service of Environment Canada, the U.S. Environmental Protection Agency

– Office of Research and Development and Office of Air and Radiation, EPRI, the U.S. National Science Foundation, the New York State Energy R and D Authority, the U.S. National Oceanic and Atmospheric Administration-Aeronomy Laboratory, the North American Commission on Environmental Cooperation, the Western Energy Supply and Transmission (WEST) Associates, the Utility Air Regulatory Group (UARG), the Southern Company, and the Midwest Ozone Group. Funding sources through the NARSTO Infrastructure Pool and in-kind support included most of the above institutions, and the Dunn-Edwards Corporation, Health Canada, the Ford Motor Company, the General Motors Corporation, Instituto Mexicano del Petroleo, Instituto Nacional de Ecologia, the Mid-Atlantic Regional Air Management Association (MARAMA), the New Jersey Department of Environmental Protection and Energy, and PPL.

MEMORIAL TO PROFESSOR GLEN CASS (1947-2001)

We dedicate *Particulate Matter for Policy Makers: A NARSTO Assessment* to the memory of Dr. Glen R. Cass, Chair and Professor of the School of Earth and Atmospheric Sciences at the Georgia Institute of Technology, formerly a member of the California Institute of Technology faculty, and Vice President-Elect of the American Association for Aerosol Research.

Glen represents the spirit of this assessment. He had an unusual ability to integrate the best of science with exquisite engineering judgment to arrive at sound approaches to air-quality management. Glen never lost sight of his goal to seek well-informed, practical approaches to improving ambient air quality. His innovative work to establish methods for quantifying PM source-receptor relationships used a sophisticated combination of emission inventories, laboratory and field experiments, and source-based and receptor modeling. His work exemplified the contemporary methods of analysis that we have reviewed and advocate in this assessment.

In addition to his prolific and insightful contributions as an outstanding scholar and educator, Glen was a close friend to many in NARSTO. He is sorely missed.

EXECUTIVE SUMMARY

Particulate Matter Science for Policy Makers: A NARSTO Assessment was commissioned by NARSTO, a cooperative public-private sector organization of Canada, Mexico and the United States. It is a concise and comprehensive discussion of the current understanding of airborne particulate matter (PM) among atmospheric scientists.[1] Its goal is to provide policy makers who implement air-quality standards with this relevant and needed scientific information. Policy development and revision is an ongoing process, and scientific understanding is continually improving. This assessment describes current science in a manner that reflects the needs of policy makers in addressing current and anticipated standards.

This assessment is organized using the following considerations for dealing with air-quality issues:

- Perspective for Managing PM
- Health Effects Context
- Atmospheric Aerosol Processes
- Emission Characterization
- Particle and Gas Measurements
- Spatial and Temporal Characterization of PM
- Receptor Methods
- Chemical Transport Models
- Visibility and Radiative Balance Effects.

Conceptual models of PM for nine geographic regions in North America illustrate the application of these considerations for developing management strategies.

This **Executive Summary** condenses the current understanding of the PM problem, focusing on the following key topics:

- $PM_{2.5}$ responses to changes in emissions
- Local, regional, and continental management
- Predictive capabilities
- Copollutant interactions
- Improved measurement tools
- Links with health-effect and climate-change studies
- Tracking success of air-quality management.

The **Synthesis** following this Executive Summary answers specific policy questions, developed through interviews with policy makers, with summary information from this assessment.

$PM_{2.5}$ MASS AND COMPOSITION RESPONSES TO CHANGING EMISSIONS

Ambient $PM_{2.5}$ results from direct particle emissions (carbon and soil dust, for example) and secondary particles (generated by atmospheric reactions of precursor gas emissions). Ambient PM is also influenced by meteorology. The major precursor gases are sulfur dioxide (SO_2), nitrogen oxides (NO_x), ammonia (NH_3), and volatile organic compounds (VOCs). Ambient PM mass is thus a mixture composed mostly of sulfate ($SO_4^=$), nitrate (NO_3^-), ammonium (NH_4^+), organic carbon (OC), black carbon (BC), and soil dust.

The relative mass fractions of secondary PM can change in a nonlinear manner with changing precursors owing to the complex chemistry of secondary PM formation. As SO_2 emissions decline, so will particulate $SO_4^=$ concentrations. The presence of NH_3 allows the formation of ammonium nitrate (NH_4NO_3) and ammonium sulfate [$(NH_4)_2SO_4$]. As $SO_4^=$ is removed, more NH_4NO_3 can be formed, provided sufficient NH_3 and nitric acid (HNO_3) are present. Consequently, the response of particulate NO_3^- to changing NO_x or VOCs is less clear, as it depends upon the amount of NH_4^+ and $SO_4^=$ present in particles in addition to VOC- or NO_x-limiting processes. In most locations, insufficient information exists on both emissions and ambient levels of NH_3 to predict how particle mass and composition would change in response to changing NH_3 emissions.

[1] The major emphasis of this assessment is on fine particles ($PM_{2.5}$), although larger particle sizes (PM_{10} and $PM_{10-2.5}$) are discussed.

The BC fraction will decrease in direct response to reductions in source emissions from, for example, diesel combustion. The OC fraction stems from both direct emissions of organic PM and the oxidation and polymerization of VOC emissions. Consequently, the extent to which the organic fraction may be managed depends upon both direct OC emissions and the attendant occurrence of oxidants and acids. As the particle-forming VOCs, acids, and oxidants decrease, so would the secondary OC fraction. Considerable uncertainty exists about the chemical composition of organic carbon particles, both at the source and in the atmosphere. Individual organic-carbon species are not measured routinely. Thus, explaining the response of the OC fraction of PM to changing precursor emissions is in its very early stages.

What we know and expect…

Based upon current understanding of secondary particle formation, it is anticipated that the existing management strategies in North America focused on the reduction of SO_2 will reduce $PM_{2.5}$ mass concentration, as will reductions in direct particle emissions, notably of BC and OC. The benefits of reducing nitrogen oxides or VOCs are uncertain.

LOCAL, REGIONAL, AND CONTINENTAL MANAGEMENT OF $PM_{2.5}$

Particles typically remain in the atmosphere for days to a few weeks, depending on their size and the rates at which they are removed from the atmosphere, for example, by precipitation. Particles in any given area may originate locally or from sources hundreds to thousands of kilometers away, or be formed during atmospheric transport from precursor gases originating from sources locally or far away.

Prevailing meteorology, topography and seasonal influences also alter the bulk $PM_{2.5}$ mass concentration and composition of particles.

Intercontinental transport of dust from both Asia and Africa occurs. It does not contribute significantly to annual-average concentrations in North America, but may occasionally contribute significantly to 24-hr concentrations. Forest fires and biomass burning both contribute locally and regionally during certain seasons. Satellite imagery and PM-composition measurements are used to identify the influence of these large-scale events.

Both local and regional emissions contribute to local concentrations in many urban areas. Regional contributions from sources distant to eastern North American urban areas can account for 50% to 75% of the total observed $PM_{2.5}$ mass concentration within a specific urban area.

What we know and expect …

$PM_{2.5}$ management strategies need to include the impact of both local and distant sources, within the context of prevailing meteorology and seasonal variability. Consequently, a management approach for one region or airshed may not be applicable in another. Each needs to be assessed individually to develop the most effective management approach.

PREDICTIVE CAPABILITIES

Several strategy-development tools are available utilizing analysis (e.g., receptor modeling) and simulation (e.g., chemical-transport modeling). Receptor models and chemical-transport models can be used in a complementary fashion to develop advice for policy makers, as part of a corroborative approach to providing guidance based on the best scientific understanding available. Receptor models are useful in selecting scenarios and identifying contributing sources and/or source types.

Current chemical-transport models are one useful tool for guiding policy as part of the collective scientific analysis, being most informative regarding the inorganic fraction ($SO_4^{=}$, NO_3^-, and NH_4^+) on regional and episodic (days to weeks) scales.

What we know and expect…

Source-specific options to reduce PM concentrations are best approached through corroborative analyses using emission inventories, ambient concentration measurements, and air-quality modeling. Capabilities of chemical-transport models, as well as being bound by current understanding of atmospheric processes, are limited by the type and quality of the ambient, meteorological, and emission information used to evaluate and run them. Evaluated, policy-ready models that can be applied routinely will become available gradually as these data needs are met; such models can be applied with growing confidence over the next five to ten years.

COPOLLUTANT INTERACTIONS

$PM_{2.5}$ and ground-level ozone, with other air contaminants, are closely related through common precursors, sources, physicochemical pathways, and meteorological processes. Thus, changes in the emissions of one pollutant can lead to changes in the concentrations of other pollutants. All fine particles, including those composed of carbonaceous materials, $SO_4^{=}$, and NO_3^{-}, scatter incoming and, lesser so, outgoing radiation. Black carbon and other dark particles absorb radiative energy. Coarse particles and cloud droplets formed by the condensation of water vapor on particles also have radiative effects. These radiative effects of PM can have local and global impacts on photochemistry and climate change. The scattering and absorption of visible radiation (light) can also obscure visibility and result in regional haze.

What we know and expect…

The current understanding of atmospheric processes shows that $PM_{2.5}$ problems are related to ground-level ozone, acid rain, and climate issues and share many of the same sources. This recognition provides the impetus for integrated and optimized management strategies that accommodate different atmospheric responses for each pollutant.

NEW INSIGHTS FROM IMPROVED MEASUREMENTS AND MONITORING

Considerable advances have been achieved in the real-time measurement of particle size and composition. Many of the new measurements are still research activities, but they demonstrate how more detailed measurements can provide insights to the fundamental atmospheric chemical processes and source contributions, and be used in chemical-transport model development and evaluation. For example, size-resolved composition measurements can help differentiate between local and regional source contributions, primary vs. secondary particle components, and natural vs. anthropogenic contributions.

Insufficient ambient data are available to examine fully all of the hypothesized causal elements for health effects, either due to lack of data or lack of suitable measurement technology. As the measurement techniques evolve toward more robust methods, suitable for mid- to long-term monitoring, it will become possible to include many more aspects of PM composition and characteristics in studies of population exposure and health outcomes.

What we know and expect…

Health-impact indicators based on epidemiology can be derived only for PM characteristics for which ambient measurements are available, and effects can be assigned only to those specific source types represented in the ambient concentration measurements. Linking of health effects to sources is being explored by way of source apportionment and toxicological studies. The rapid evolution in ambient and source monitoring methods will improve the capability to assign exposure and health links to sources.

LINKAGES BETWEEN THE HEALTH AND ATMOSPHERIC SCIENCE COMMUNITIES

A considerable and growing body of evidence shows an association between adverse health outcomes, especially of the cardio-respiratory system, and short- and long-term exposures to ambient PM, especially $PM_{2.5}$. Evidence, from both opportunistic studies of significant reductions in ambient PM over short time periods as well as long-term studies, also suggests that health improves when the ambient concentration of PM is reduced.

Hypotheses that explain how the various chemical and physical characteristics of PM cause specific adverse health impacts have been proposed and partially tested. Continuing to test these hypotheses will require ongoing close collaboration between the atmospheric and health-science communities.

What we know and expect...

Stronger linkages between the health, exposure, and atmospheric-science communities will strengthen our understanding. Examples include joint planning of field studies and projects, development of coordinated long-term research strategies that take into account the major areas of focus in each community, presentations/lectures at one another's conferences, and inclusion of contextual information on human health, exposure and atmospheric science within all major assessments/reports.

TRACKING PROGRESS AND SUCCESS OF AIR-QUALITY MANAGEMENT

Long-term data sets of a basic suite of measurements (PM mass concentration and composition) are needed to track the impact of changing emissions and management actions. The information from such long-term monitoring, combined with model predictions, will provide the iterative guidance needed to revise and focus management strategies.

At present, considerable PM data for North America exist in many different data sets (both short term and long term), obtained from different networks or studies, using several measurement techniques. Investigations of regional and continental PM transport and formation have been limited, however, by discontinuities in these data sets as well as by inconsistent methodologies in their preparation.

What we know and expect...

Sulfate particle loadings have responded to SO_2 emission reductions in both eastern North America and California. It is anticipated that more responses to currently planned changes in emission rates will appear in the next five to ten years. This information will be needed to revise and optimize PM management approaches.

SCIENTIFIC UNCERTAINTIES AND LOOKING FORWARD

The physics and chemistry of aerosols have been studied since before the 20th century. The in-depth study of aerosols in the context of air pollution is much more recent. The science community has a fundamental understanding of PM formation and transport, yet there remain several significant areas of uncertainty.

Qualitative uncertainties have been identified in emission-inventory information, monitoring, and chemical-transport models. Pollutant emissions from point sources which have been of management interest the longest are best characterized, while those from dispersed area sources for chemically complex pollutants or pollutants of recent interest, such as VOCs and NH_3, are much more uncertain. Confidence in the measurements of bulk PM mass concentration varies between measurement methods. Confidence in characterizing the carbon fraction is currently low because of unresolved sampling artifacts as well as ambiguities in determining the BC and OC fractions.

The current chemical-transport models for PM provide conceptually acceptable, though unevaluated, simulations for inorganic particulate species over episodic periods and regional scales.

What we know and expect...

Policy makers are currently benefiting from research initiated five to ten years ago, or longer. This research provides a basic understanding of PM formation, transport, and its major contributing sources. It characterizes the areas of North America where PM concentrations, visibility reduction, and potential population exposure are the greatest. Despite considerable scientific uncertainties, sufficient scientific confidence exists to devise management actions likely to improve air quality.

Policy makers are faced with the need to make decisions based on imperfect understanding of the atmosphere and human-health impacts. Therefore, this Assessment concludes with research recommendations that will substantially improve the ability of the atmospheric-science community to provide guidance for improving plans to implement air-quality standards and goals. Furthermore, continuing research will reduce the risk of erroneous decisions. Each recommendation addresses a consideration that is critical for improved understanding of the atmospheric environment. Because the individual recommendations cover areas that are respectively supportive, they should progress simultaneously. The result will be better science tools and stronger atmospheric-science analysis. The recommendations focus on:

- Improving the understanding of the carbonaceous fraction
- Performing long-term monitoring of PM, gaseous precursors, and copollutants
- Performing further evaluation and development of chemical-transport models
- Developing improved emission estimates (including chemical speciation)
- Making a commitment to the analysis of ambient data and fostering interactions between atmospheric, climate, and health-science communities
- Developing more systematic approaches for integrating diverse types of knowledge to guide development of PM management practices and tracking progress toward protecting health.

The atmospheric-science community has a sufficient understanding of particle formation and transport to support current directions for achieving air-quality standards and goals. The new science that is on the horizon will help refine the management actions for more effective approaches to address PM air-quality problems.

PARTICULATE MATTER SCIENCE FOR POLICY MAKERS: SYNTHESIS

Particulate Matter Science for Policy Makers: A NARSTO Assessment presents an eleven-chapter survey of current knowledge regarding the atmospheric behavior of suspended particulate matter (PM). This Assessment was commissioned by NARSTO[1] in early 2000, with the primary objectives of compiling and summarizing currently available information on PM, and presenting it in a manner that is useful for scientists and administrators charged with implementing associated air-quality standards.[2]

This Synthesis presents a condensed summary of the Assessment, focusing on eight key information needs - phrased as policy questions - identified through previous interviews with policy makers in Canada, the United States, and Mexico. In addition this Synthesis illustrates elements of the implementation decision process, provides overviews of the regulatory context, and indicates linkages across the science/policy interface. Reflecting the Assessment's general structure, it concludes with a summary of research recommendations for strengthening the scientific knowledge available to the policy community. The reader will find extensive linkages to the Assessment chapters throughout this discussion.

THE INFORMATION FRAMEWORK

The Assessment's structure reflects the information-flow framework shown in Figure S.1.[3] Illustrating the evolving interrelationship between air-quality standards and atmospheric science, this framework contains the following key components of the environmental-management decision process:

- The Atmospheric Environment: Understanding the relationships among pollutant emissions, their interactions with meteorology and other atmospheric processes, and the resultant atmospheric pollution loadings.
- Exposure and Impacts: Understanding of cause-effect relationships among atmospheric pollutants, exposures and deposition, and impacts on human health, visibility, and ecological systems. This includes the important risk-characterization step directly supporting standard and environmental goal setting.
- Analysis and Public Policy: Analysis and decision making by the policy community, considering atmospheric science and societal factors in relation to environmental goals, in formulating emission-reduction programs.

Reflecting its emphasis on atmospheric processes and the structure of Figure S.1, this Assessment compiles and summarizes current PM-related knowledge concerning:

- **Gaseous and particulate pollution emissions,** from both natural and anthropogenic sources, including the temporal and spatial nature of these emissions. (Chapter 4)
- **Atmospheric processes,** the chemical and physical reactions and interactions that take place in the atmosphere, affecting the composition and distribution of gaseous and particulate pollutants. These processes are influenced by meteorology and topography, and determine the short- and long-term chemical composition of the atmosphere. (Chapter 3)

---

[1] NARSTO, originally the North American Research Strategy for Tropospheric Ozone, a multi-stakeholder entity, was organized in 1994 to sponsor cooperative, public/private, policy-relevant research on tropospheric ozone. In 1999, NARSTO was rechartered with its name unchanged though no longer an acronym and given an expanded charge to include airborne particulate matter. Currently its research programs focus on both ozone and PM, including their combined atmospheric chemistry and physics. As a major part of its charter, periodic policy-relevant science assessments are commissioned. Its first assessment was the 2000 Ozone Assessment.

[2] The purpose, charge, goals, and objectives of this Assessment are discussed in the Preface

[3] This Synthesis is drawn from material presented in the complete PM Assessment. If an illustration is found in the full discussion, it is labeled here with its original figure or table number in parentheses, to assist the reader in locating additional information from the Assessment.

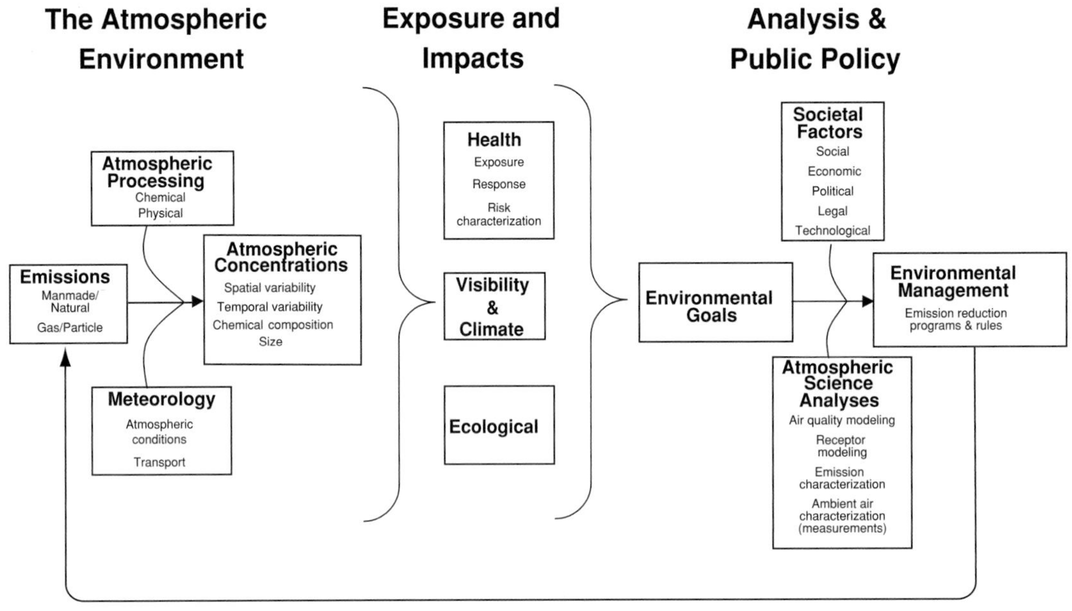

Figure S.1 (1.1). Framework for informing PM management.

- **Meteorology,** the fundamental atmospheric conditions (temperature, solar irradiation, humidity, precipitation, wind speed and direction) that determine the dispersion and transport of natural and man-made pollutant species throughout the global atmosphere. (Chapter 3)

- **Atmospheric concentrations,** the spatial and temporal characterization, primarily at the surface, of particle composition and size, and the magnitude of particle and precursor or co-pollutant gas concentrations. (Chapter 6)

- **Human health response to PM**, highly summarized knowledge from the health and exposure research communities setting a context for this atmospheric science assessment, touching on exposures and their relation ambient PM, susceptible populations, observed health responses tied to exposures, and current risk characterization that has guided development of ambient air-quality standards. (Chapter 2)

- **Visibility and Climate** impacts that occur as a result of PM absorbing and/or scattering visible radiation, obscuring one's ability to view objects or scenes and perturbing the Earth's radiation balance. Such impacts result directly from the particles themselves, as well as indirectly because of PM effects on cloud-formation processes. (Chapter 9)

- **Atmospheric science analyses,** the application of current understanding by tools, such as emission characterization, ambient characterization, receptor modeling, and chemical-transport modeling, to elucidate the relative contributions of different sources, both local and distant, to a region's PM problem. (Chapters 4, 5, 6, 7 and 8)

Consolidated scientific understanding for specific North American regions is presented in Chapter 10 in terms of **Conceptual Models**; i.e., synopses of the best understanding of the influence of emissions, meteorology, and atmospheric processes on ambient PM concentrations. Figure S.2 illustrates how the application of this consolidated information can provide a state-of-knowledge understanding for policy makers.

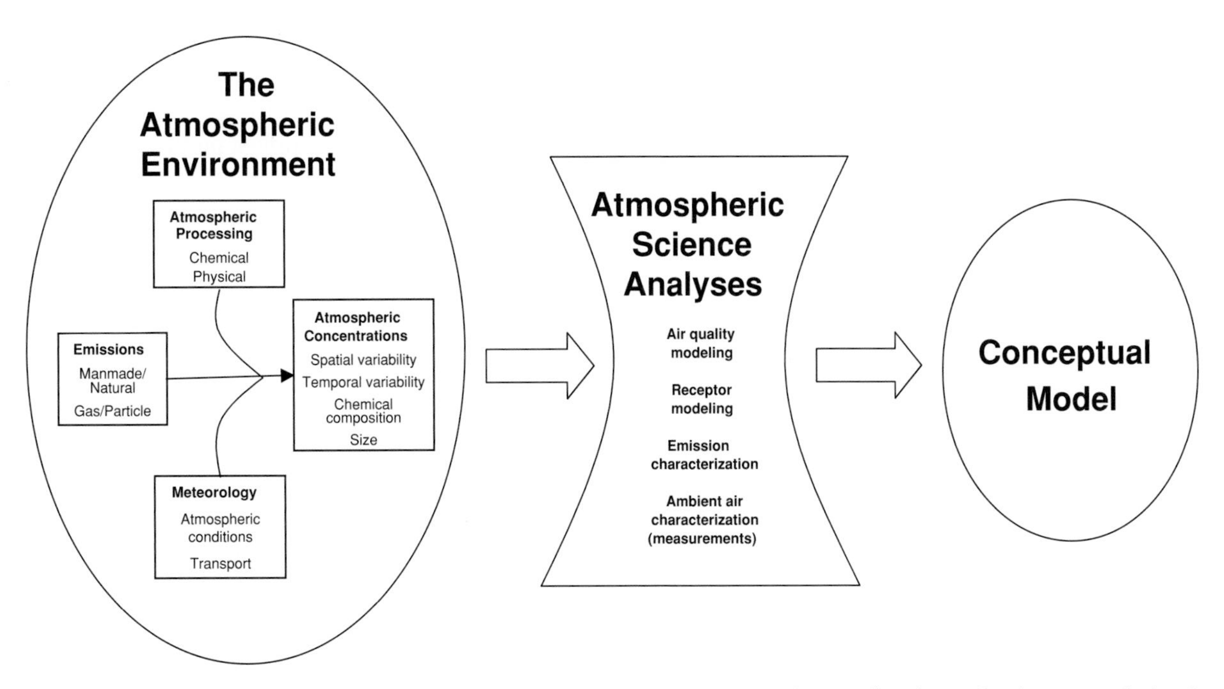

Figure S.2 (1.6). Development of a conceptual model based upon the application of science tools in the context of the best understanding of the atmospheric environment.

AIR-QUALITY MANAGEMENT: A POLICY OVERVIEW

Table S.1 summarizes current PM standards for Canada, the United States, and Mexico. Although these standards are under continuing review, this Assessment accepts them in their present form as appropriate goals: it makes no comment regarding the significance of health or environmental impacts associated with PM concentrations below the level of the standards, or possible problems associated with PM in areas where standards have not been set.

The atmospheric-science community continues to work closely with the health and environmental-effects communities as they examine the causal effects of PM, associated biochemical mechanisms, susceptibility, and other questions that are part of ongoing reviews of jurisdictional standards. Although the information presented in this assessment aids this standards development work, its primary focus is on atmospheric phenomena and the attainment of existing standards.

THE NATURE OF PM: A BRIEF OVERVIEW

The answers to the Policy Questions, which summarize this Assessment's key findings, require the definition of key terminology as well as a brief discussion of salient physical and chemical aspects of tropospheric PM. Strictly speaking, the term "aerosol" refers to PM and the gas in which it is suspended. Suspended particles can be solid, liquid, and/or multiphase in nature. Frequently "aerosol" is used imprecisely in the literature to denote only the particles; however the more accurate definition, which includes the suspending gas, is applied throughout this Assessment.

As indicated by Figure S.3, tropospheric particles span a wide size range from molecular clusters smaller than 1 nanometer to "giants" of 100 micrometers (μm) or more in diameter (human hairs are about 70 μm). Particle number concentrations are dominated by the "ultrafine particles," which are nominally smaller than 0.1 μm in diameter. Aerosol mass distributions are dominated by particles larger than 0.1 µm, which include "fine particles" up to 2.5 µm diameter and "coarse particles" larger than 2.5 µm. $PM_{2.5}$ designates the mass concentration of particles smaller than 2.5 µm (the fine particle mass), and $PM_{10-2.5}$ designates the total mass concentration of coarse particles between 2.5 and 10 µm.

Current regulatory concern focuses on the fine and coarse particles. In the simplest picture, particles in

Table S.1 (1.3). Existing national PM standards and their implementation timetables. [a]

Country	Current Standards	Implementation Timing
Canada	$PM_{2.5}$ of 30 µg/m^3 24-hr averaging time to be achieved by 2010, achievement to be based upon the 98th percentile ambient measurement annually, measured over three consecutive years.[b]	In June 2000, Ministers agreed to a set of initial actions to reduce pollutants that cause particulate matter and ozone, which will be undertaken by provincial/territorial and federal governments. The delivery date for completed reduction measures under the initial actions is 2005.
United States	For $PM_{2.5}$, the 3-yr average of the 98th percentile of 24-hr average concentrations at each population-oriented monitor must not exceed 65 µg/m^3, and the 3-yr average of the annual arithmetic mean concentration from single or multiple community-oriented monitors must not exceed 15 µg/m^3.[c] For PM_{10}, the fourth highest 24-hr concentration over 3 years must not exceed 150 µg/m^3, and the 3-yr average annual mean concentration must not exceed 50 µg/m^3.[c] Improve visibility on the haziest days and ensure no degradation on the clearest days, with the ultimate goal of reaching natural background conditions in 60 years.	For PM no later than the end of 2005, the U.S. EPA plans to designate and from the date of designation, states will have 3 years to develop and submit implementation plans for non-attainment areas, attainment deadlines to be set as early as 5 years and as late as 12 years after designation. To achieve visibility goals, states are required to develop an initial 10-15 year plan, in the same time frame as their PM state implementation plans, that is to be revised every 10 years starting in 2018.
Mexico	PM_{10} maximum allowable concentration of 150 µg/m^3 24-hr mean and an annual arithmetic mean of 50 µg/m^3.[d] Total Suspended Particles (TSP) maximum allowable concentration of 260 µg/m^3 24-hr mean and an annual arithmetic average of 75 µg/m^3.[d]	The Secretariat of Health has set air-quality Normas Oficiales Mexicanas (NOMs) that now must be met by all jurisdictions of the country.

[a] Some states and provinces have standards more stringent than national standards, for example California's PM standards.
[b] By the end of 2005 complete analysis to reduce information gaps and uncertainties and revise or supplement the PM and Ozone CWSs as appropriate for the year 2015; and report to Ministers in 2003 on the findings of the PM and Ozone environmental and health science, including a recommendation on a $PM_{10-2.5}$ Canada-Wide-Standard.
[c] The Clean Air Act requires that the NAAQS be reviewed every 5 years and a review process for the PM NAAQS is underway. As part of this review process, an updated Criteria Document, which summarizes the relevant scientific information about the sources, transformations, concentrations, and health and environmental impacts of PM, has been developed. Based on the information in the updated Criteria Document and supplementary analyses, a decision whether to retain or revise the NAAQS is now expected in late 2003 to early 2004.
[d] The Secretariat of Health is in the process of reviewing existing NOMs. The new NOM for particulate matter is anticipated to include three size fractions: TSP, PM_{10} and $PM_{2.5}$.

different size ranges originate from different sources. The ultrafines originate from combustion and nucleation. The fine particles derive principally from combustion and chemical transformations of gases to produce secondary products including sulfates ($SO_4^=$), nitrates (NO_3^-), and organics. The most important precursor gases for secondary aerosols are sulfur dioxide (SO_2), nitrogen oxides (NO_x), and certain volatile organic compounds (VOC). Ammonia (NH_3) reacts with $SO_4^=$ and NO_3^- to form particulate salts. The coarse mode originates mainly from mechanical resuspension of dust or plant-derived materials, or from sea spray.

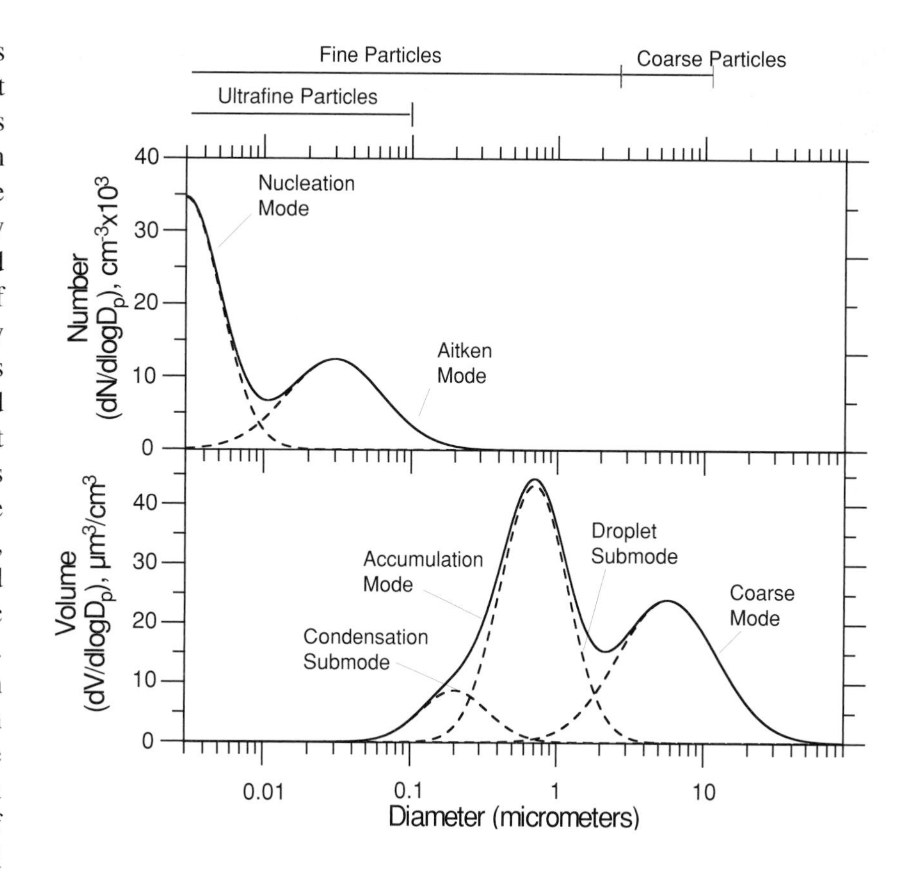

Figure S.3 (3.2). Typical number and volume distributions of atmospheric particles.

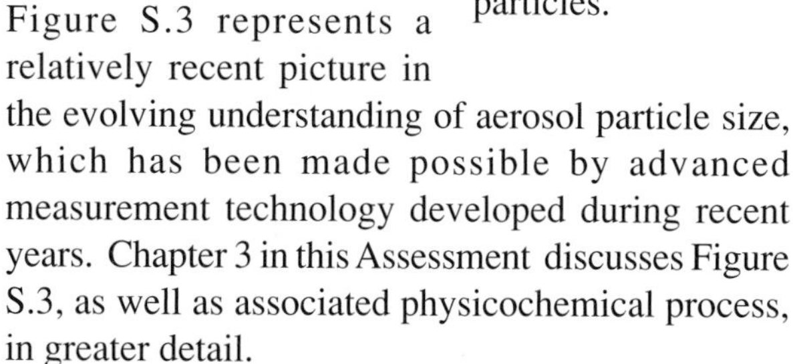

Figure S.3 represents a relatively recent picture in the evolving understanding of aerosol particle size, which has been made possible by advanced measurement technology developed during recent years. Chapter 3 in this Assessment discusses Figure S.3, as well as associated physicochemical process, in greater detail.

Aerosol particle size and composition are strong determinants of numerous physical phenomena and physiological impacts, as well as useful indicators of a pollutant's origin and atmospheric history. Important phenomena influenced by particle size include respiratory-tract deposition, visibility reduction, cloud formation, solar forcing, atmospheric photochemistry, source-receptor transport distances, and wet- and dry-deposition.

These features constitute key elements in the information-exchange pathways between air-pollution scientists and workers in associated fields, particularly the health-sciences, visibility, and climate-change communities. From the health-sciences standpoint, particle size is an extremely important determinant of respiratory impacts, and - although evidence is less complete in this area - particle composition is suspected to be important as well. This situation results in a need for an active and continuous, two-way information-exchange process between the air-pollution and health-research communities. Health scientists have a strong need for more directed and comprehensive measurements by their air-pollution counterparts, both in the form of broader chemical speciation and to help link ambient concentrations to personal exposure. Air-pollution scientists, on the other hand, need access to developing information on compositional importance from the health-sciences community, in order to focus their measurement-development efforts in the most beneficial channels.

From a visibility and climate-change standpoint, particle size and PM loading directly affect radiation transport, leading to alterations of the Earth's radiation balance as well as to visibility deterioration.

Particle composition plays a role in direct radiation impacts as well. For example, $SO_4^=$ and organic-carbon (OC) particles largely scatter light, leading to atmospheric cooling, whereas black carbon (BC) particles absorb light, with a corresponding warming tendency. So-called "indirect effects" depend on both particle size and composition as well. Acting as cloud-condensation nuclei, greater abundances of aerosol particles generally create clouds with greater droplet populations, resulting in "brighter" clouds that reflect more solar radiation, thus leading to a global cooling tendency (Type 1 indirect forcing). In addition, higher cloud-condensation nuclei populations - especially those composed of hygroscopic substances - are generally believed to result in more persistent, stable clouds thus leading again to a cooling effect (Type 2 indirect forcing). Currently, uncertainties associated with these forcing effects are extremely high. Within this context, it is clear that further progress in these areas will be aided substantially by active information-exchange linkages between air-pollution scientists and their counterparts in the climate-change and visibility fields. The importance of these and other linkages are further amplified in the following summary of key findings, as well as in the text of this Assessment.

PM ISSUES AND POLICY QUESTIONS

The eight Policy Questions used to focus this Assessment are summarized in Box S.1. The best current answers to these questions, summarized below, are based on a thorough review of the state of PM atmospheric science by the chapter authors. Given the large investment in research today, the quality of answers to these questions is expected to improve significantly over the next 3 to 5 years. The insights presented here provide only partial answers to several of these questions, as many involve considerations beyond the science scope of this Assessment.

The process of preparing scientific answers for the policy questions demonstrated that a corroborative approach is necessary for managing the PM problem. For example, Policy Question #1 regarding the significance of the PM problem (see Figure S.4a) relies on the science inputs from two disciplines, measurements and characterization. However for most others, for example Policy Question #3 regarding broad approaches to fixing the PM problem (see Figure S.4b), science inputs are needed from many disciplines.

Policy Question #1 - Is there a significant PM problem and how confident are we?

The existence of a significant PM problem is evident. PM levels that persistently exceed existing standards have been observed in urban areas throughout North America.

Box S.1. Policy questions confirmed with the policy community as central to implementing programs to achieve air-quality standards:

PQ1. Is there a significant PM problem and how confident are we?

PQ2. Where there is a PM problem, what is its composition and what factors contribute to elevated concentrations?

PQ3. What broad, pollutant-based, approaches might be taken to fix the problem?

PQ4. What source-specific options are there for fixing the problem given the broad control approaches above?

PQ5. What is the relationship between PM, its components, and other air-pollution problems on which the atmospheric science community is working?

PQ6. How can progress be measured? How can we determine the effectiveness of our actions in bringing about emission reductions and air-quality improvements, with their corresponding exposure reductions and health improvements?

PQ7. When and how should implementation programs be reassessed and updated to adjust for any weaknesses, and to take advantage of advances in science and technology?

PQ8. What further atmospheric sciences information will be needed in the periodic reviews of national standards?

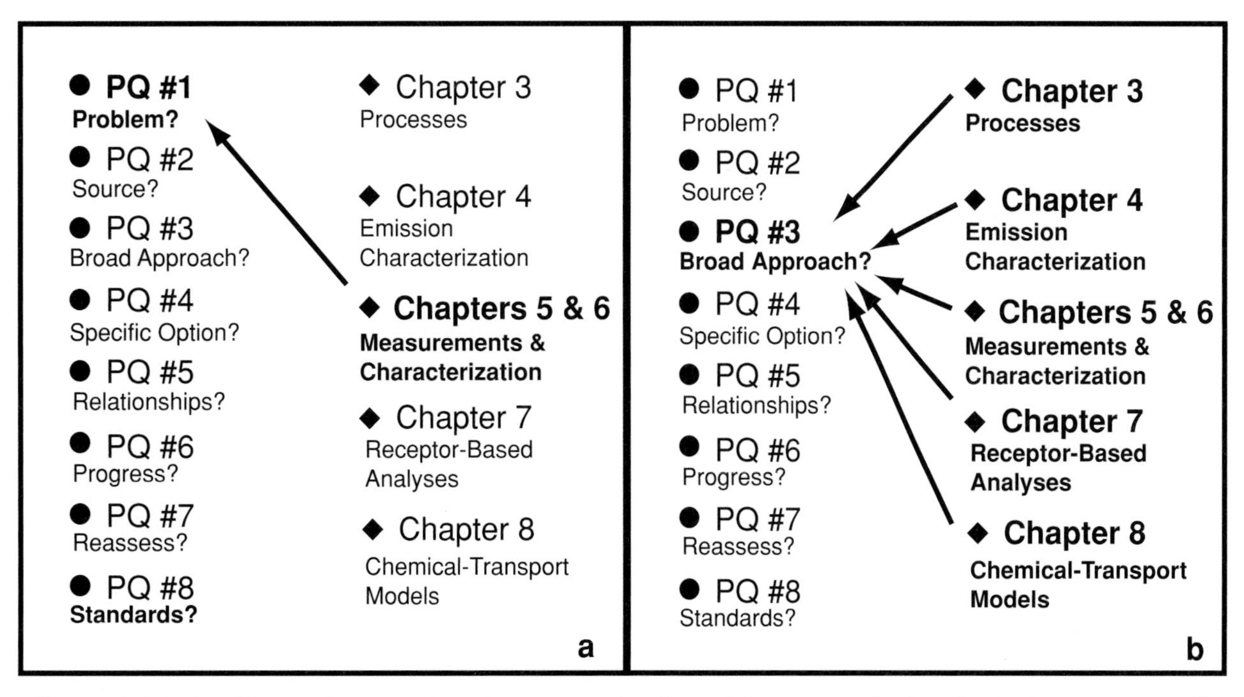

Figure S.4 (1.3). Illustrative corroborative approach of providing atmospheric science input to the policy questions. Input not shown also includes consideration of health and visiblity impacts.

- Annual average $PM_{2.5}$ mass concentrations at about half the urban sites in California, the southeastern United States, the northeastern United States, and the Ohio Valley – Great Lakes states, exceeded 15 µg/m^3 in 1999 and 2000. The 24-hr $PM_{2.5}$ mass concentrations at the majority of urban sites in Ontario exceeded 30 µg/m^3 at the 98th percentile during this period.
- The 24-hr PM_{10} concentrations in Mexico City and parts of California can exceed 150 µg/m^3 as often as 30 percent of the time.

Large portions of the population in North America are exposed to PM concentrations above the current standards.

- There are about 65 million people in the United States that live with annual average $PM_{2.5}$ levels above 15 µg/m^3. There are about 13 million Canadians that live in regions which can experience concentrations above the 24-hr standard of 30 µg/m^3. Eighteen million people in Mexico City live with average PM_{10} levels greater than 50 µg/m^3.

A considerable and growing body of evidence shows an association between adverse health effects, especially of the cardiorespiratory system, and exposure to ambient levels of PM.

- Personal exposure of individuals is related to both ambient (outdoor) and indoor environments.
- Epidemiological studies of large populations have shown statistically significant increases in indices of adverse health outcomes over the background incidence with increased levels of ambient PM.
- Increases in adverse health outcomes have been observed for both short- and long-term exposures to PM.
- Thresholds in ambient PM concentration-health response relationships have generally not been observed in epidemiology studies.
- Some evidence associates improved health with reductions in PM exposure.
- Hypotheses explaining how chemical and physical properties of PM may interact with the body to cause health outcomes have neither been proven nor eliminated from consideration.

In addition to health effects, ambient particles are associated with chemical deposition, deterioration of ecosystems, visibility impairment, and climate change. Other periodic national and international assessments address deposition and climate change. This assessment focuses on achieving the mass-based ambient air standards designed to protect human health and to a lesser degree, visibility impairment associated with PM.

- Some gaseous precursors (SO_2 and NO_x) and atmospheric processes contributing to particulate $SO_4^=$ and NO_3^- are the same as those leading to chemical deposition .
- Visibility impairment is sensitive to the chemical composition of $PM_{2.5}$, and also depends strongly on ambient relative humidity. Secondary particles tend to be in the size range that is most effective at scattering visible light.
- Particles also affect climate through mechanisms of light scattering and absorption similar to those that affect visibility.

Key science findings

Spatial and temporal characterization

The highest observed annual-mean $PM_{2.5}$ concentrations in North America occur at sites in California and many urban sites of the eastern, and especially the southeastern United States (Figure S.5). Shorter-term monitoring data from Mexico City suggest that local annual-mean $PM_{2.5}$ concentrations could be among the highest in North America.

- $PM_{2.5}$ mass measurements typically exhibit strongly skewed concentration frequency distributions, dominated by a large number of low values and a smaller number of high concentrations. Annual-average $PM_{2.5}$ can vary by up to a factor of two across distances of 50 to 100 km in some large metropolitan regions.
- In California, the southeastern United States, the northeastern United States, and the Ohio River Valley - Great Lakes states, annual-mean $PM_{2.5}$ mass concentrations at about half the urban sites exceeded the U.S. 3-year average annual-mean $PM_{2.5}$ mass standard of 15 μg/m³ in 1999 and 2000.

The 24-hr average $PM_{2.5}$ observations above standards are generally infrequent but not rare events.

- In Canada, 24-hr average concentrations greater than 30 μg/m³ occur over most of southern Ontario and Quebec more than 2 percent of the year.
- Similarly, many areas in California experience 24-hr concentrations above 65 μg/m³ more than two percent of the year. Occasionally, sites in the southeastern United States experience 24-hr concentrations greater than 65 μg/m³.
- Mexico has no $PM_{2.5}$ standard and limited monitoring, but levels seen in a spring intensive study (24-hr levels frequently greater than 65 μg/m³, maximum 185, average 35) are greater than North American levels recorded elsewhere.

Annual-average concentrations of PM_{10} across North America illustrate isolated locations of high annual PM_{10} levels that can exceed the level of the U.S. and Mexican annual standards (see Figure S.6).

- Annual PM_{10} concentrations in Los Angeles are 45-60 μg/m³ at most locations. In the San Joaquin Valley area concentrations are greater than 50 μg/m³.
- In Mexico City and parts of California, 24-hr PM_{10} concentrations above the standards may occur on a few (less than 5) to many (more than 30) days each year.

Peak 24-hr PM_{10} concentrations in Los Angeles have been reported at 187 μg/m³, and Corcoran in the San Joaquin Valley reported 175 μg/m³. In urban areas having multiple monitoring sites, 24-hr average PM_{10} mass can vary by up to roughly a factor of two over distances as small as 10 to 20 km.

The typically smaller spatial variations of $PM_{2.5}$ than PM_{10} are consistent with the well-known long residence time of fine particles, which permits transport over distances of 10 to 1000 km and leads to more spatially uniform mass concentrations.

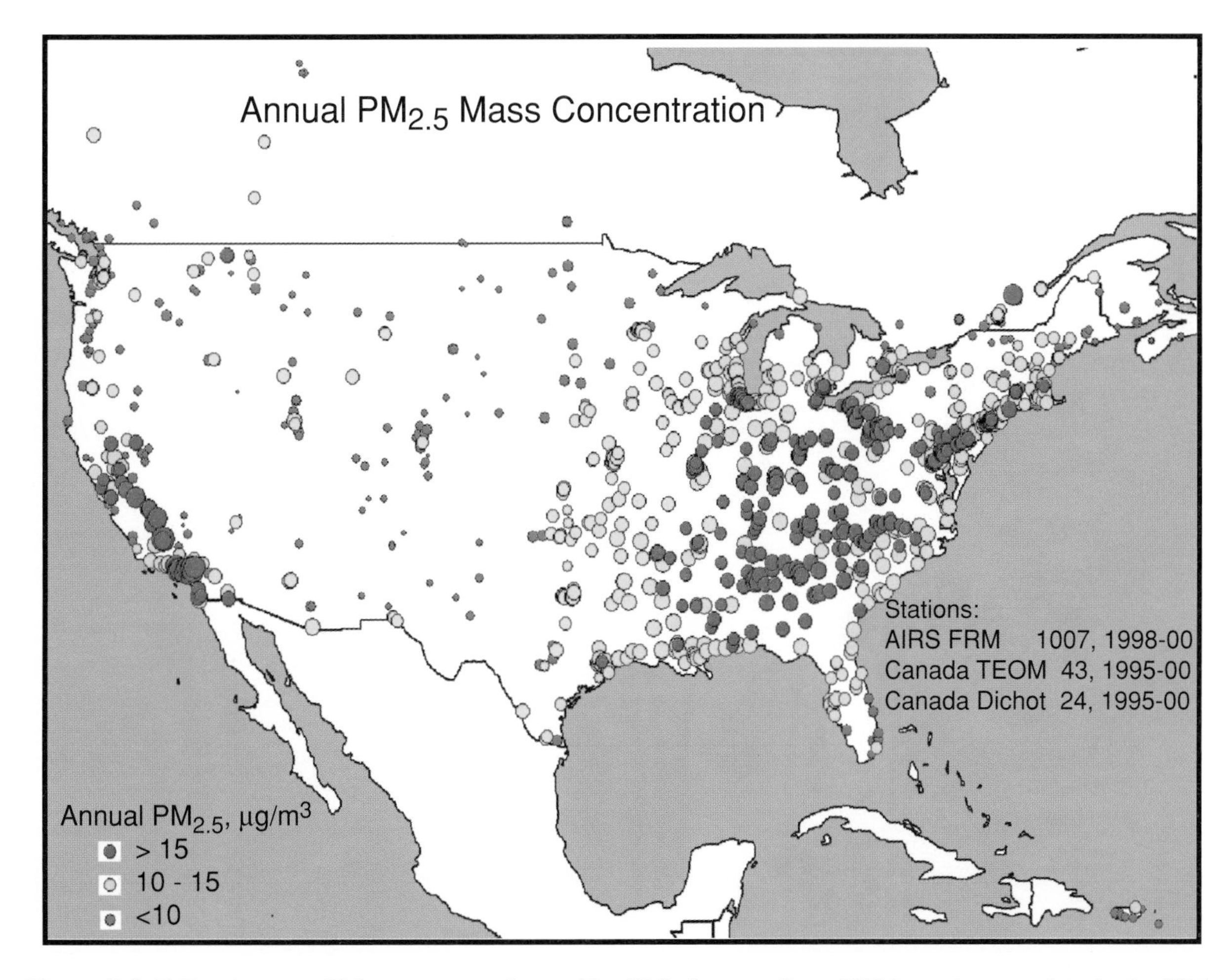

Figure S.5 (6.7). Average $PM_{2.5}$ concentrations. The U.S. data are from FRM monitors at sites in the EPA AIRS database for July 1998 through July 2000. Canadian data are from TEOM and dichotomous samplers operating from 1995 through 2000. The currently available data from sites in Mexico represented less than one year of sampling and were excluded from the computation of annual averages. Spot diameter varies in proportion to concentration.

Monitoring data also indicate that significant short-term (24-hr) variations of $PM_{2.5}$ mass may occur within urban areas. Monitoring data show typically larger inter-site differences in trace-element concentrations than in $PM_{2.5}$ mass concentrations. For many locations in North America, on average, PM_{10} is composed of 40 to 60 percent $PM_{2.5}$.

Measurements

The large number of properties needed to fully define PM (e.g., size, mass, number, and composition) makes the determination of the uncertainty inherent in any of the particle measurements more challenging than for gas-phase measurements. Appropriate reference materials or standards for many particle-phase components are limited. For this reason, measurement uncertainty (or perhaps more appropriately, consistency) is often evaluated by comparisons among several methods measuring similar PM attributes. Hence, for PM at present, convergence of methods is relied on rather than calibration to a known reference standard as a guide to measurement uncertainty. Ranges of uncertainty are noted in Table S.2.

The measurement of PM mass concentration, the basis of current PM standards and the source of the information presented here is accurate to within ± 5 percent, with precision of ± 10 percent, and a range of comparability between methodologies of 20 and 30 percent for $PM_{2.5}$ and PM_{10}. Direct measurements of the PM_{coarse} fraction, or $PM_{10\text{-}2.5}$, are

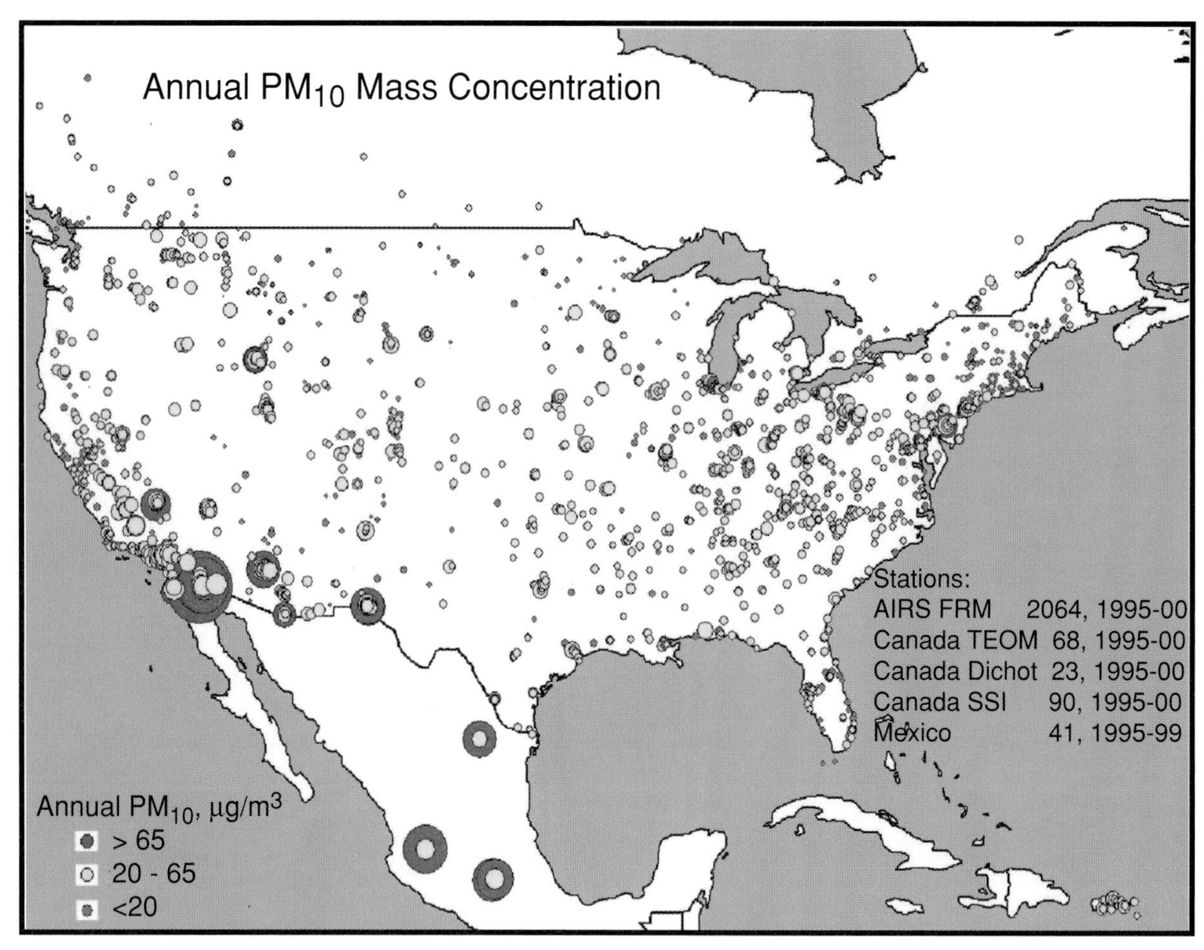

Figure S.6 (6.3). Average annual PM_{10} mass concentrations. The U.S. data are from sites in the EPA AIRS database. Canadian data are from TEOM and dichotomous samplers provided by Environment Canada. PM_{10} data were available for five cities in Mexico. Spot diameter varies in proportion to concentration.

generally not made, but this fraction is inferred by taking the difference between observed PM_{10} and $PM_{2.5}$ concentrations, if both have been obtained using the same methodologies.

PM composition measurements have widely divergent levels of accuracy, precision and comparability. Those for inorganic species, including $SO_4^=$, NO_3^-, NH_4^+ and metals, are similar to PM mass and those for speciated carbonaceous PM are essentially unknown. The measurement of BC and chemically speciated OC is especially problematic. Most variability in BC measurements results from differences in analytical technique, whereas most OC variability is attributed to differences in sampling methods. Measurements of organic compounds in particles are prone to errors due to volatility, interference from gaseous organic species, limitations of analytical methods and lack of calibration standards. It is likely that both positive and negative interferences occur during sampling, and there is no consensus as to the net effect of these interferences. The relative magnitudes of these effects also vary with sampling location and time. Current organic speciation only explains 10 to 20 percent of total organic composition.

Confidence in observations is also affected by sampling frequency. Where monitors are in place, sampling frequencies of 1 in 6 days or better provide annual mean estimates of ±10 percent, but they do not yield a representative range of 24-hr average data. Daily and 1-in-3-day monitoring data now becoming available at sites across the United States and parts of Canada are addressing this need.

Table S.2 (5.1 - 5.3). Measurement uncertainty in PM physical and chemical characteristics.

Compared Property	Analytical Accuracy [a]	Precision [b]	Range of Comparability [c]
Physical Properties of PM			
Mass ($PM_{2.5}$)	± 5%	± 10%	≤ 20%
Mass (coarse)	± 5%	± 10%	30%
Chemical Composition of PM			
Sulfate ($SO_4^=$)	± 5%	± 10%	25%
Nitrate (NO_3^-)	± 5%	± 10%	30%
Ammonium (NH_4^+)	± 5%	± 10%	15%
Total organic carbon (OC)	-- [d]	+-20% [d]	50% [d]
Black carbon (BC)	-- [d]	± 20% [d]	≥ 50% [d]
Total carbon	± 10%	± 10%	15%
Organic (speciated)	unknown	unknown	unknown
Trace elements	± 5%	± 10%	10 – 50%
Transition metals	± 5%	± 10%	10 – 50%
Biological aerosols	Unknown	unknown	unknown

[a] *Accuracy* is the ability of the laboratory methods to correctly measure the samples of a standard reference material.
[b] *Precision* is the standard deviation of repeated measurements of the same observable with similar collocated instrument.
[c] The range of observed levels of *Comparability* across a number of field comparisons designed to test the combined sampling *and* analytical comparability of several methods.
[d] Present practice operationally defines the BC and OC fractions mainly based on thermal differentiation followed by different detection methods for the thermal products. Without an appropriate standard accuracy and precision of the reported determinations are problematic.

Health effects

There is a considerable and growing body of evidence showing an association between adverse health effects, especially of the cardiorespiratory system, and exposure to ambient levels of PM.

The total personal PM exposure of individuals, which includes both their ambient (outdoor) environment and indoor environments, is related to the ambient PM concentration. The ambient air concentrations of PM and other pollutants have been extensively studied as potentially controllable variables that influence human health.

Epidemiological studies of large populations have shown statistically significant increases in indices of adverse health outcomes over the background incidence and increased levels of ambient PM. The adverse health outcomes observed in these studies are commonly (though not exclusively) cardiac and respiratory in nature. Such diseases are also relatively common in the general population and the increased incidences seen in these studies translate into a significant number of additional cases.

Increases in adverse heath outcomes have been observed for both short- and long-term exposures to PM. The increases in adverse health outcomes have been observed for a range of ambient particulate matter indicators including TSP, PM_{10}, $PM_{10-2.5}$ and $PM_{2.5}$. A higher potency has typically been observed for $PM_{2.5}$ compared to other PM indicators, consistent with the concept that smaller particles penetrate further down the respiratory tract. Nonetheless, the epidemiological literature does demonstrate some level of association with the larger PM material, which deposits more readily in the upper respiratory tract.

Certain population subgroups appear to have heightend susceptibility to PM, such as those with preexisting cardiac and respiratory disease (seen frequently in smokers), asthmatics, and the elderly. Increases in adverse health effects appear to occur without a threshold in the ambient concentration-health response relationship and appear to increase in a near-linear fashion from the baseline incidence of health effects with increasing PM concentrations. Some evidence suggests that there may be regional differences in the potency of PM indicators, possibly a reflection of regional differences in the composition of PM. Additionally, some evidence exists linking increases in adverse health outcome and specific sources of PM.

There is also some evidence that improved health is associated with reductions in PM exposure. Such studies have been both opportunistic in nature, taking advantage of significant reductions in PM pollution over short time periods to examine mortality and other adverse health endpoints, or have been designed to follow individuals or groups as their exposure to air-pollution has changed. These studies, while requiring careful interpretation, offer the opportunity to examine the benefits of specific air-quality interventions, as well as the opportunity to compare the benefits of specific changes in the air-pollution mix. Pursuit of these situations in a strategic and proactive manner offers the potential for gaining insights into the benefits of control activities.

A number of hypotheses have been advanced to explain how chemical and physical parameters of PM may interact with the body to provide mechanistic explanations for the health outcomes. These hypotheses are being evaluated in toxicological studies using laboratory animals and controlled exposure of human subjects. However, to date none of these hypotheses has been proven or eliminated from consideration. Continuing tests of these hypotheses will require collaboration between atmospheric scientists and health scientists in order to identify and characterize the hypothesized PM constituent or parameter, the PM indicators usually monitored, and other pollutants. Such collaboration will enhance efforts in the fields of toxicology, clinical effects, and in both short-term and long-term epidemiological investigations.

Further progress in understanding health effects of ambient PM and various constituents depends on achieving a high level of collaboration between health scientists and atmospheric scientists to obtain detailed characterization of PM in multiple communities over decades to match with health data.

Chemical deposition

The relationships of the chemical and meteorological processes of $PM_{2.5}$ $SO_4^=$ and NO_3^- formation and fate, the deposition of metallic and other cation components of PM, and the processes contributing to chemical deposition and ecosystem effects are noted in this Assessment. However, they are not reviewed here as they are the subject of in-depth periodic science assessments by other science assessment bodies.

The reader is referred to an extensive discussion of the effects of acidic deposition presented in the U.S. National Acid Precipitation Assessment Program (NAPAP) Biennial Report to Congress: An Integrated Assessment (National Science and Technology Council, 1998) and the 1997 Canadian Acid Rain Assessment (Environment Canada, 1997-1998). In addition, the effects of particulate pesticides, metal compounds, and chlorinated organic compounds are discussed in Deposition of Air Pollutants to the Great Waters, Third Report to Congress (U.S. Environmental Protection Agency, 2000).

Visibility and Climate

Perhaps the most common symptom of air-pollution recognizable to the public is obscured visibility, usually referred to as haze. Optically, PM interferes with visibility by absorbing or scattering; e.g., BC absorbs light, and $SO_4^=$, OC, and NO_3^- are highly reflective. Light scattering is roughly proportional to the mass concentration of fine particles, while light absorption is roughly proportional to the mass concentration of light-absorbing species such as BC. Haze is of concern for urban areas such as Mexico City and Los Angeles, and over much larger geographic regions in parts of North America, especially over the eastern half. In the east, an accumulation of particulate air pollution frequently results in haze extending over thousands of square kilometers. In the western United States, considerable controversy has persisted over the deterioration of visibility associated with pollution from isolated sources and urban areas. This problem is a chronic one, especially in the Desert Southwest of the United States and Rocky Mountain areas of Canada and the United States, where air quality was once considered pristine limited only by the "physical limit" of the light-scattering properties of gas molecules in air. See the more complete discussion of the relationship between PM and visibility in the answer to Policy Question #5.

On a global scale, the scattering and absorption of radiation by airborne particles has been identified as a potentially important factor in Earth's radiative energy balance and climate. As well, the major sources of PM are related to fossil-fuel combustion, which are also linked to climate and radiation-balance impacts. Direct impacts on the radiative balance are caused by $SO_4^=$ and OC (local cooling influences) and BC (warming influences). There are also indirect forcings due to the impact of elevated particle concentrations on cloud-formation processes, currently understood to be negative (cooling) but not well quantified. The radiative forcing due to particles depends upon the size, shape, chemical composition and spatial distribution (both vertical and horizontal). These effects vary in time and space, and are superimposed on the climate impacts of absorptive gases like methane and carbon dioxide in the atmosphere. This relationship is thoroughly discussed in the Intergovernmental Panel on Climate Change's Third Assessment Report (IPCC, 2002).

New science to improve implementation approaches

A commitment to long-term, continuous monitoring of PM mass, composition (notably the carbon fraction), gas-phase precursors, and copollutants, and the subsequent commitment to analyze these observations, will provide a more complete picture of the extent of the PM problem and the processes that contribute to it. Further discussion of the benefits of new science is provided at the end of the Synthesis.

Policy Question #2 - Where there is a PM problem, what is its composition and what factors contribute to elevated concentrations?

Locally observed PM is composed of multiple chemicals, largely OC, $SO_4^=$, and NO_3^- in combinations that differ by geographic region. PM composition is influenced by sources and seasonal meteorology, and has substantial regional contributions. The following typical differences have been observed:

- Sulfate is a major fraction of $PM_{2.5}$ in eastern North America, the carbonaceous fraction is significant everywhere, and NH_4NO_3 is a major contributor in California. Eastern and western coastal regions of the United States and Canada show marked seasonality in concentration and composition while central interior regions do not. Winter NO_3^- concentrations are greater than summer, and urban greater than rural in eastern North America. Maximum regional $PM_{2.5}$ concentrations occur during the summer over most of the East and during the fall and winter in the West.

- Local and regional emissions from upwind urban areas and rural sources can account for 50 to 75 percent of total observed local PM mass concentrations.

- Generally, PM_{10} consists of 40 to 60 percent $PM_{2.5}$, and the remainder is primarily locally generated, crustal/geological and biological material.

Key science findings

Composition and the factor of seasonality

In most of North America, $PM_{2.5}$ is composed of six major fractions. Sulfate and organic compounds can each account for 20 to 50 percent of $PM_{2.5}$ mass (Figure S.7). Non-coastal rural areas are dominated

[4]Currently reported OC and BC measurements are method-dependent and relatively imprecise (See Box 2, Chapter 1). This should be borne in mind when interpreting measurements such as those reported in Figure S.7.

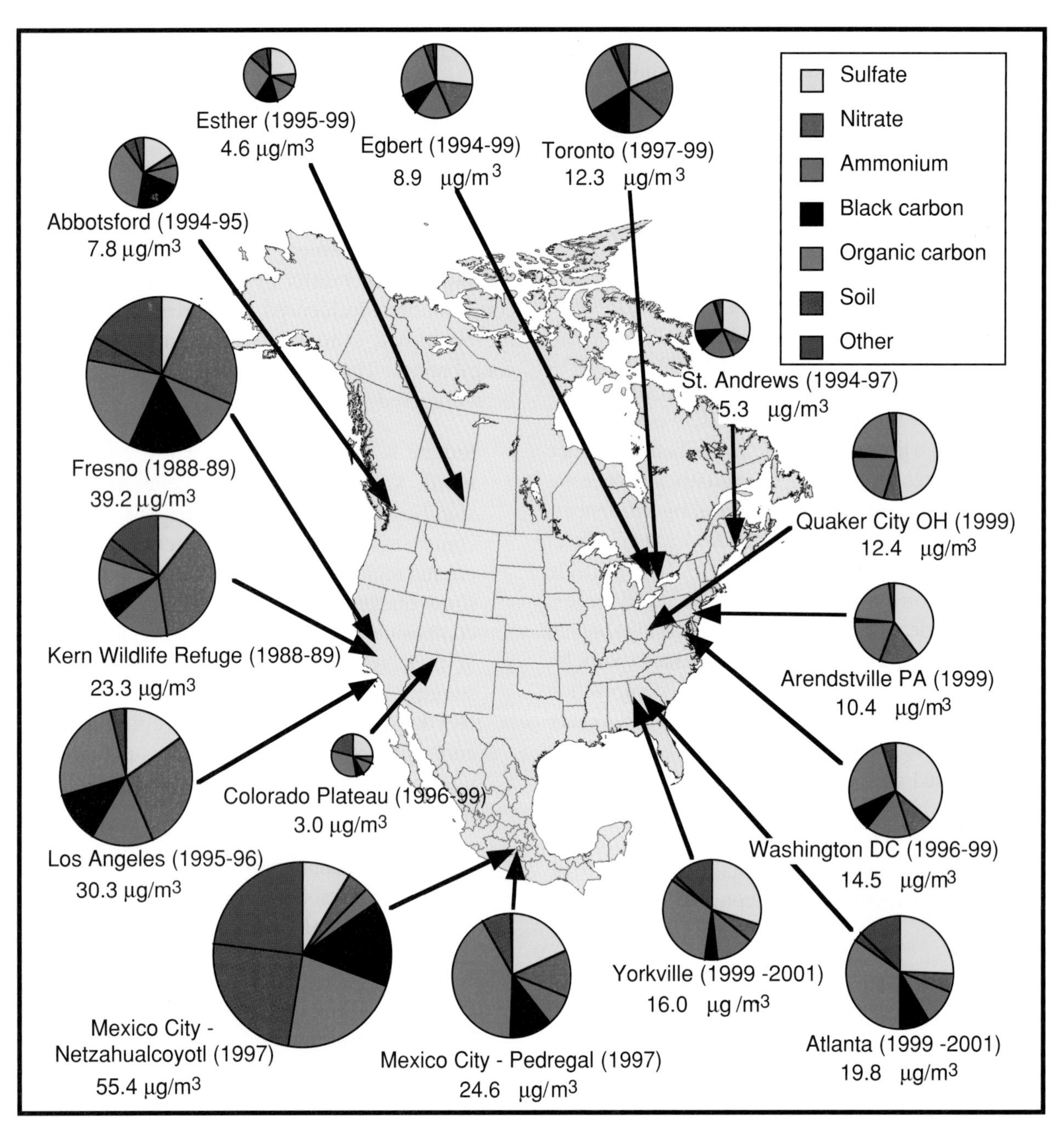

Figure S.7 (6.12). Composition of $PM_{2.5}$ mass at representative urban and rural locations, based on annual averages except for Mexico City.

by $SO_4^{=}$, OC, and BC.[4] Nitrate-containing particles appear to be important in parts of the West. They account for more than one quarter of the $PM_{2.5}$ in populated areas of southern and central California. Nitrate concentrations are important in the urban areas of eastern United States and southeastern Canada during cold weather months. Black carbon and soil dust are usually lesser contributors, except during short-term episodes.

Differences in $PM_{2.5}$ composition by region and season are apparent from observations supporting conceptual models prepared for representative North American areas.

- Los Angeles experiences its most intense and frequent episodes during the late summer and early fall. In California's San Joaquin Valley, episodes are most severe in intensity and frequency during the fall and winter. Both areas have large NH_4NO_3 composition and lesser but substantial contributions from OC. Crustal/ geological material can be a significant fraction of Los Angeles' $PM_{2.5}$.

- The Lower Fraser Valley of the Pacific Northwest experiences its maximum $PM_{2.5}$ in the late summer or fall. Mobile source emissions, agricultural emissions (NH_3) and road dusts account for about 70 percent of $PM_{2.5}$ mass.

- Mexico City's $PM_{2.5}$ is composed of primary OC and BC, with $(NH_4)_2SO_4$ and to a lesser extent NH_4NO_3 also being significant components.

- $PM_{2.5}$ of the Canadian Southern Prairie – U.S. Northern Plains and U.S. Upper Midwest-Great Lakes rural regions has little seasonal variation, is dominated by $SO_4^=$ and OC, and has important local and regional transport contributions.

- Summer regional $PM_{2.5}$ in the northeastern United States and Windsor-Quebec City Corridor (WQC) on average are twice winter concentrations. However, in large urban areas such as Philadelphia, New York City, and those of the eastern WQC, peak $PM_{2.5}$ occurs in winter when mass concentrations on average are slightly higher than summer. Summer regional $SO_4^=$ in the northeastern United States is more than twice the next nearest component, OC, and more than four times the NO_3^- and BC combined. Winter urban OC and $SO_4^=$ in Philadelphia and New York City each account for about a third of $PM_{2.5}$ mass. Nitrate is a significant component of northeastern United States and WQC urban $PM_{2.5}$ during the winter. In the WQC, summer $SO_4^=$ is also an important fraction though OC and NO_3^- are also significant fractions throughout the year. Both local and regional $SO_4^=$ and OC are important in both regions.

- The southeastern U.S. summer levels of PM are 1 to 3 times higher than winter levels. As with the northeastern United States, summer $SO_4^=$ and OC concentrations dominate, but in contrast to the Northeast these two are nearly balanced throughout the Southeast. Both local and regional contributions are important.

Emissions and atmospheric processes as contributing factors

Fine $SO_4^=$ and NO_3^- particles result principally from the oxidation of gaseous SO_2 and NO_x emissions, forming sulfuric and nitric acids, H_2SO_4 and HNO_3, with subsequent neutralization by NH_3. (Figure S.8). Organic particles are both emitted directly from sources and are formed in the atmosphere from VOC emissions. Black carbon and soil dust are emitted as primary particulate matter.

- Almost all particle $SO_4^=$ originates from SO_2 oxidation and is associated with NH_4^+. Ninety-five percent of SO_2 sources are anthropogenic, from fossil fuel combustion, and the majority of NH_3 sources are related to agricultural activities and, to a lesser extent, to transportation in some areas.

- Essentially all particle NO_3^- is derived from atmospheric oxidation of NO_x. The major anthropogenic source of NO_x is fossil-fuel combustion. Particle NO_3^- formation is affected by the availability of $SO_4^=$ and NH_3 if the process is limited by NH_3, and also by either VOC or NO_x if HNO_3 is the limiting factor in NH_4NO_3 formation.

- Organic carbon is primary and/or secondary, of biogenic (vegetative material, biogenic gases, spontaneous forest fires) and anthropogenic (fossil-fuel combustion, prescribed fires, cooking) origin. At present, it is not possible to generalize whether or not the majority of OC is of primary or secondary origin. Active research on OC sources will shed light on this question in the next few years.

- Black carbon originates as ultrafine or fine particles from primary sources during incomplete combustion of carbon-based fuels, for example, diesel engines, wood burning, and poorly maintained industrial and residential heating.

Regional contributions

Particles typically remain in the atmosphere for days to a few weeks, depending on particle size and the rate at which they are removed by precipitation and

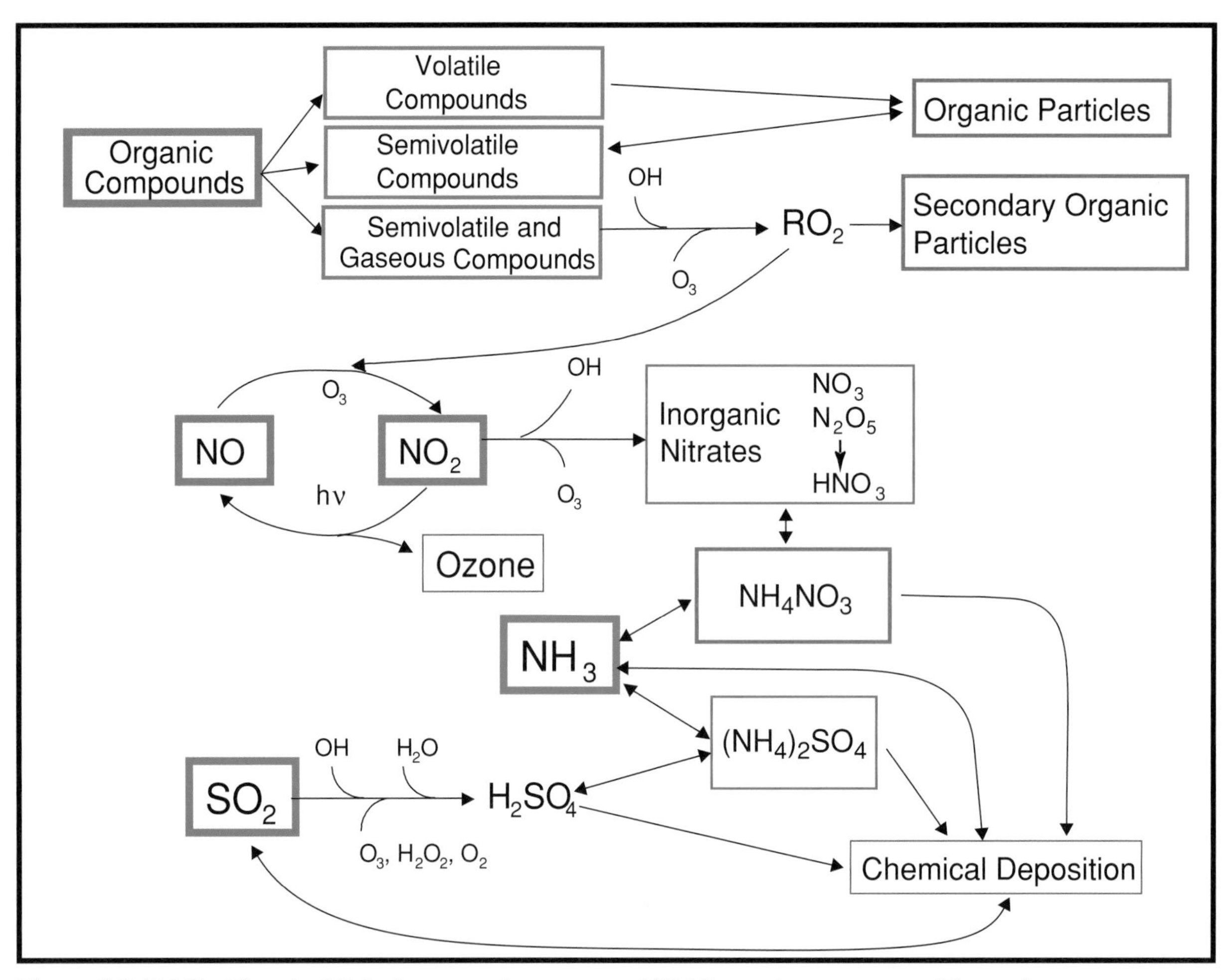

Figure S.8 (3.16). Chemical links between the ozone and PM formation processes. The major precursors are shown in green squares. The organic compounds can be gaseous (always in the gas phase), non-volatile (always in the condensed phase), and semivolatile (partitioned between the gas and condensed phases).

dry deposition. It follows that particles detected at a site may have been formed or emitted locally or may have originated at a site located hundreds or even thousands of kilometers away. Particles in the 0.1 to 2.5 μm diameter range have longer lifetimes than particles that are smaller or larger than this range. Very small (<0.1 μm) particles are more likely to be influenced by local emissions and very large (>2.5 μm) particles removed by deposition processes than particles in the intermediate size range.

$PM_{2.5}$ comes from both local and regional sources. Local sources cause highly variable distribution of mass concentration and composition between urban and surrounding regional areas (Figure S.9). Regional contributions to mass concentrations include interurban or long-range transport as well as non-anthropogenic background contributions. Urban areas show mean $PM_{2.5}$ levels exceeding those at nearby rural sites. In eastern North America, the differences imply that local urban contributions account for roughly 25 percent of the annual mean urban concentrations, with regional aerosol contributing the remaining, and larger, portion.

On average, summertime $SO_4^=$ and OC are strongly regional in eastern North America, with 75 to 95 percent of the urban $SO_4^=$ concentrations and 60 to 75 percent of the urban OC concentrations being the result of cumulative region-wide contributions.

The regional contribution to $PM_{2.5}$ varies among different geographical areas. On a continental scale, the background or baseline level of $PM_{2.5}$ ranges from an annual average of 3 to 7 μg/m³. Data from the U.S. IMPROVE network indicate that mean annual

$PM_{2.5}$ levels exceed 10 µg/m^3 at rural locations in much of the southeastern United States. Rural locations in other portions of the eastern United States, California, Ontario, and Quebec exhibit mean annual $PM_{2.5}$ levels of 5 to 10 µg/m^3.

Global-scale long-range transport can also affect PM mass concentrations. From a review of a few global transport events several statements can be made. For example, significant springtime intercontinental transport of fine dust from Asia to western North America occurs a few times per decade and may

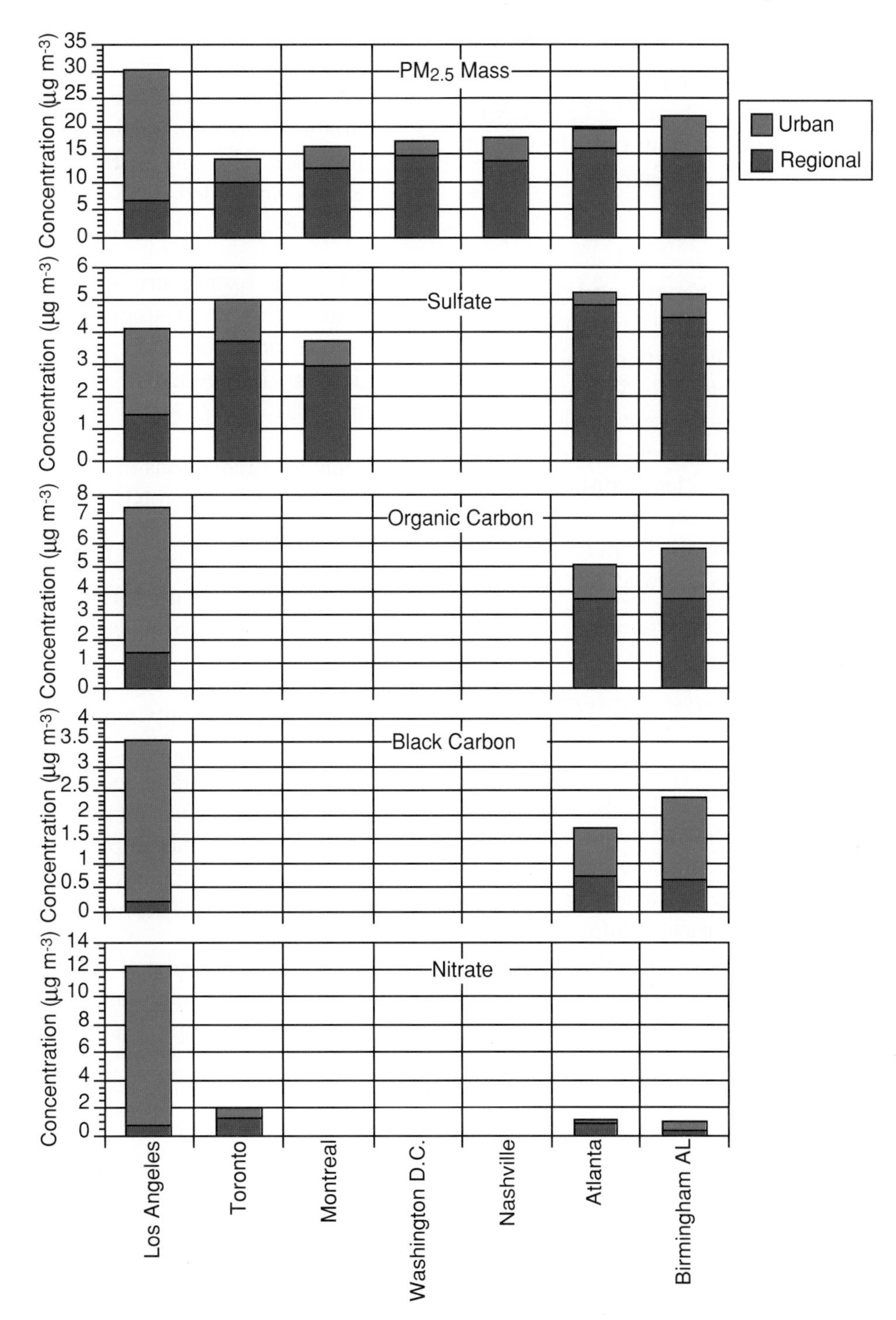

Figure S.9 (6.15). Comparisons of average $PM_{2.5}$ mass concentrations and species concentrations at paired urban and rural locations.

result in concentrations in excess of 24-hr standards. In the East, transported African dust is a summertime occurrence that can on occasion contribute elevated 24-hr averages in the southeastern United States. These events can be identified by the coincident use of satellite and surface measurements of optical depth or turbidity, particle mass concentration and composition. Such events appear to contribute <1 $\mu g/m^3$ to annual $PM_{2.5}$ averages. Large-scale forest fires have similarly been tracked in North America and on occasion can contribute significantly to the daily average $PM_{2.5}$ concentration.

PM_{10}

California's San Joaquin Valley and Los Angeles basin have annual-average and 24-hr concentrations among the highest in the United States. PM_{10} composition there is typically 40 to 55 percent crustal/geological material, about 20 percent OC and BC, 15 to 20 percent NO_3^-, and about 5 percent $SO_4^=$. Mexico City's PM_{10} also has a large crustal/geological fraction, at around 50 percent, with the remainder being OC and BC, around 32 percent, and a combination of $SO_4^=$, NO_3^-, and NH_4^+, around 17 percent.

New science to improve implementation approaches

The commitment to long-term, continuous monitoring of PM mass concentration, composition, gas-phase precursors, copollutants, and surface and aloft meteorology with the subsequent commitment to analyze these observations is as important here as it is to add certainty to the answer for Policy Question #1. The notable difference is the high importance placed here on better measurement and understanding of the carbon fraction of PM. In combination these investments will provide a much more complete picture of the PM problem and the processes that contribute to it, especially the poorly described OC and BC piece of the picture. Further discussion of the benefits of new science is provided at the end of this Synthesis.

Policy Question #3 - What broad, pollutant-based, approaches might be taken to fix the problem?

The current understanding of PM formation and composition offers both general and area-specific insights:

- $PM_{2.5}$ differs in its composition and its seasonal variation across the continent such that regional strategies targeting different precursors and seasons are needed. One uniform approach will not be optimum for all locations.

- $PM_{2.5}$, like ozone, has both regional and local contributions. Strategies that address both regional and local $PM_{2.5}$ and its precursors are likely to be needed in most areas of North America.

- Coincident reductions of $PM_{2.5}$ precursors (i.e., SO_2, NO_x, VOC, and NH_3) should be beneficial in most parts of North America in achieving desired PM mass concentrations, but some of those reductions may lead to temporary and/or localized counterproductive impacts in some areas.

- The current air-quality management approaches focusing on SO_2, NO_x, and VOC emission reductions are anticipated to be effective first steps toward reducing $PM_{2.5}$ across North America, noting that in parts of California and some eastern urban areas VOC emissions could be important to NO_3^- formation.

- The local suppression of mineral material such as soil dust and road dust continues to be the most beneficial approach to reducing $PM_{10.}$

Key science findings

Coincident reduction of precursors

Reduction of $SO_4^=$-containing particles by controlling SO_2 precursor emissions will be effective in reducing $PM_{2.5}$ mass in the majority of locations. Where $SO_4^=$ reacts with most available NH_3, reductions in $SO_4^=$ may be partially offset by increases in NH_4NO_3 that form as NH_3 is freed up. (See the discussion of the linearity concept in box S.2.)

Nitrates are another major fraction in some locations, though management of the NO_3^- fraction via NO_x emission reductions needs to be addressed in combination with VOCs, NH_3 and SO_2. This must be done in consideration of the relationship between NO_3^- production and local or regional ozone formation.

Reduction of carbonaceous PM

Carbonaceous $PM_{2.5}$, which current measurement methodologies somewhat arbitrarily separate into BC and OC fractions, is another major target. The OC fraction is usually much larger than the BC fraction and can consist of both primary and secondary material (formed in the troposphere from oxidation/condensation of primary gaseous emissions). The relative abundances of primary and secondary OC have been studied only in a few locations, mostly in California, and are not generally well understood. However, there are airsheds such as Mexico City and Los Angeles, where the secondary OC may be important at least during periods of high photochemical processing.

Secondary OC is formed from both anthropogenic and biogenic gaseous precursors. The most significant anthropogenic precursors are aromatics emitted by transportation and industrial sources (e.g., toluene, xylenes, trimethyl-benzenes). While reducing emissions of these compounds will reduce secondary OC, the precise benefits of such control are not presently known. Biogenic precursors including terpenes (e.g., α- and β-pinene, limonene) and the sesquiterpenes are expected to be major contributors to secondary OC in areas with significant vegetative cover. The significance of secondary OC produced from biogenic precursors is also not well understood

Because primary carbonaceous particles often contribute significantly to PM mass, reducing primary emissions of carbonaceous particles will reduce PM mass concentrations.

Reduction of nitrates

Reductions of NO_x, VOCs, and associated NH_3 throughout the year in and around large urban areas may be important in the East as well as the West to bring down the NO_3^- contributions to the 24-hr and/or annual PM averages. While summer NO_3^- concentrations in eastern North America are low in comparison with other $PM_{2.5}$ components, higher winter NO_3^- concentrations occur in northern urban areas. On an annual basis, particulate NO_3^- levels in eastern North America show a geographical pattern that closely parallels the particulate NH_4^+ pattern. Some analyses suggest that higher NO_3^- levels in and around large northern urban areas during winter months result from HNO_3 being readily available and more NH_3 being available when $SO_4^=$ levels are lower. Application of chemical-transport models for both warm and cold seasons as well as comprehensive ambient air monitoring will provide useful insights to guide effective particulate NO_3^- management.

Differing regional strategies

Conceptual models for nine North American regions show that a single, uniform approach to reducing $PM_{2.5}$ levels will not be effective for all areas of North America. For example, the high levels of $PM_{2.5}$ occurring in California's Los Angeles basin and San Joaquin Valley are dominated by winter NH_4NO_3, so that balanced reductions in NO_x and VOCs appear appropriate. In contrast, large regions of the eastern United States and southeastern Canada have high $PM_{2.5}$ concentrations driven by $SO_4^=$ and OC

Box S.2 Source-Receptor Linearity

Source-receptor linearity implies that a given percent reduction in emissions of a precursor gas in a region will result in the same percent change in ambient concentrations of its secondary aerosol products, taking into account the presence of a baseline or background level which is assumed to be irreducible. Although linearity is often assumed in order to estimate the benefits of emission controls, actual changes often deviate from linearity. Chemical-transport models can be used to obtain more quantitative insights into relationships between emissions and ambient concentrations.

concentrations in summer, pointing to the need for reductions in SO_x and OC.

The $SO_4^=$- NO_3^-- NH_4^+equilibrium

One major result of investigation of the complex chemistry of secondary particle formation has been the greatly improved knowledge of the linkages between the oxidation of SO_2, NH_3 and the equilibrium relationship between HNO_3 (from the oxidation of NO_x) and NH_4NO_3 as shown in Figure S.8. Results of modeling applications indicate that, depending on the magnitude of ambient precursor concentrations, temperature, and humidity, the relative mass fractions of NO_3^- and $SO_4^=$ can change in a nonlinear way with changing precursor (SO_2, NO_x and NH_3) concentrations. That is, as $SO_4^=$ concentrations decrease, NO_3^- reacts with available NH_3 to increase the NH_4NO_3 fraction relative to the $SO_4^=$ fraction of the particle mass. Reductions in SO_2 emissions will result in lower $(NH_4)_2SO_4$ concentrations. In some areas, the $SO_4^=$ reductions may be accompanied by localized or temporary increases in NH_4NO_3 concentrations especially when the atmospheric temperature is low (winter, fall, spring, nighttime).

Ammonia reacts preferentially with H_2SO_4, and, if sufficient NH_3 is available, it also combines with HNO_3 to form particulate NO_3^-. Declining $SO_4^=$ levels in eastern North America therefore have the potential to cause increasing NH_4NO_3 concentrations until particle NO_3^- formation is limited by the availability of NH_3. Existing observations have illustrated this phenomenon but are insufficient to explain how broadly it occurs over all seasons or in a variety of geographical regions. At rural (CAPMoN) monitoring locations in Canada, particulate $SO_4^=$ and NH_4^+ concentrations decreased from the early to late 1990s. During this same time period, particulate NO_3^- concentrations increased. Either, or both, increasing NO_x emissions or increasing rates of particulate NO_3^- formation (caused by increasing availability of NH_3 as $SO_4^=$ levels declined) may have contributed; both causes are considered probable. Seasonal data from the CAPMoN sites shows that particulate NO_3^- concentrations were highest when particulate $SO_4^=$ concentrations were lowest and the levels of NH_4^+ were highest. In the cases where the NO_3^- concentration increases significantly, additional controls of NO_x and/or VOC may be required with the SO_2 controls. Available chemical-transport models can be used for estimating relative changes in concentrations from these interactions.

Several additional cases illustrate the potential impacts of these nonlinearities:

- In Los Angeles, a 70 percent reduction of SO_2 emissions between 1977 and 1995 led to a halving of ambient $SO_4^=$ concentrations, and a time-lag of about 5 years occurred between the onset of emission reductions in 1978 and the first observable declines in ambient $SO_4^=$ concentrations. This historical example illustrates the potentially non-proportional response between ambient $SO_4^=$ concentrations and SO_2 emissions.

- A pronounced decrease of particulate $SO_4^=$ concentrations occurred in the eastern United States during the 1990s, and its timing indicates that it is a reflection of the U.S. SO_2 emission controls that were implemented as of the end of 1994. From 1989 to 1998, SO_2 emissions in the states east of and including Minnesota to Louisiana declined by about 25 percent. Average SO_2 and $SO_4^=$ concentrations at CASTNet monitoring sites in the same region declined in a similar way by about 40 percent, and at the same time exhibited correlation with the SO_2 emission trend. At the prevailing levels of SO_2 loading, the magnitudes of the emissions and concentration changes were not statistically different, supporting the utility of regional reductions of SO_2 emissions for affecting near-proportional reductions of particulate $SO_4^=$ in the eastern United States.

- In Canada, $PM_{2.5}$ concentrations in 6 eastern cities showed about a 40 percent decline from 1992 through 1996, and a 14 percent increase from 1996 through 1999. Although the temporal pattern was consistent with the timing of the U.S. SO_2 emission reductions, the particulate $SO_4^=$ concentrations did not exhibit a corresponding decline during the 1992-96 time period. However, the rural CAPMoN sites did exhibit declining concentrations of SO_2 and particulate $SO_4^=$ between 1990 and 1999 (see Figure S.10), suggesting that the effects of the U.S. control program may have been masked by local urban influences.

Improved understanding of the $SO_4^=$- NO_3^-- NH_4^+- equilibrium in $PM_{2.5}$ has resulted in major improvements in chemical-transport models in the last decade. The ability to estimate the response of PM to SO_2 emission reductions is considered satisfactory given evaluations based on ambient concentration observations. Predicted responses to NO_x reductions are less reliable, based on observational comparisons. This comparison with observations has some limitations due to NO_3^- measurement uncertainties and the unavailability of suitable information on NH_3 and HNO_3 concentrations.

Area-specific insights

- During the winter periods of peak $PM_{2.5}$ concentrations in Los Angeles and the San Joaquin Valley, NH_4NO_3 is the dominant component and is HNO_3 limited. Nitric acid can be reduced via VOC and NO_x emission reductions. The possibility of seasonal strategies that emphasize different sets of VOC and NO_x controls for $PM_{2.5}$ mass in winter and ozone in summer require optimization with the assistance of chemical-transport and receptor models.
- Mexico City's PM problem is amenable to control of diesel vehicles to reduce primary OC and BC emissions. The potential benefit of SO_2 and NO_x controls may be more effective than NH_3

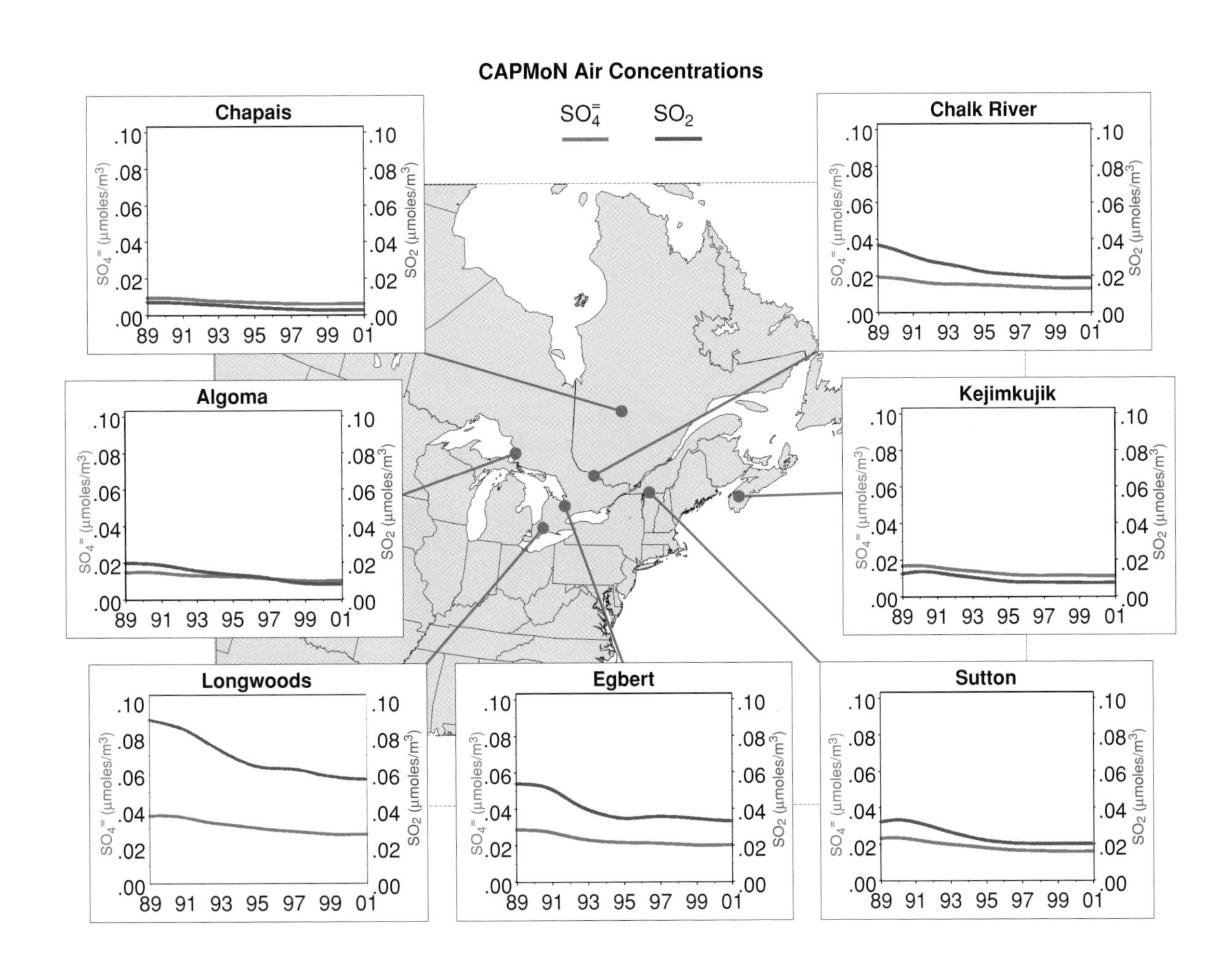

Figure S.10 (6.21a). Trends in concentrations of SO_2 and $SO_4^=$ at CAPMoN sites, 1980 - 2000.

controls in reducing secondary PM based on experience to date.

- Areas in the Canadian Southern Prairie, U.S. Northern Plains and Upper Midwest-Great Lakes region with $PM_{2.5}$ concentrations near applicable standards can limit further air-quality deterioration using a combination of local and regional controls for SO_2 and OC emissions.

- For the urban areas of the WQC and the northeastern and southeastern United States, regional control of SO_2 to reduce summer $PM_{2.5}$ concentrations and local control of SO_2 to reduce winter concentrations along with local control of OC emissions to reduce year-round concentrations is believed to be an effective approach (see Figure S.11). For areas in these regions also concerned about wintertime PM levels, e.g., cities in southeastern Canada, strategies that involve NO_x reductions may be effective.

Regional reductions

In some cases, addressing regional contributions to $PM_{2.5}$ is as important as addressing local contributions for reducing $PM_{2.5}$ concentrations. For instance, in the northeastern United States, average regional $PM_{2.5}$ concentrations can contribute 30 to 60 percent of the total levels seen in the large urban centers, particularly along the coast. In the southeastern United States, the regional $PM_{2.5}$ contribution is 10 to 40 percent. Even in the Canadian Southern Prairie and U.S. Northern Plains, levels of $PM_{2.5}$ found in upwind rural areas can contribute up to a third of peak levels in urban centers.

Achievement of the U.S. annual $PM_{2.5}$ standard in urban areas where the regional component is at present close to the level of the ambient standard is likely to require regional $PM_{2.5}$ management. As seen previously in Figure S.9, regional contribution to PM varies among different geographical areas. On a continental scale, the background or baseline level of $PM_{2.5}$ ranges from an annual average of 3 to 7 μg/m^3. Data from the U.S. IMPROVE network indicate that mean annual $PM_{2.5}$ levels exceed 10 μg/m^3 at rural locations in much of the southeastern United States. Rural locations in other portions of the eastern United States, California, Ontario, and Quebec exhibit mean annual $PM_{2.5}$ levels of 5 to 10 μg/m^3.

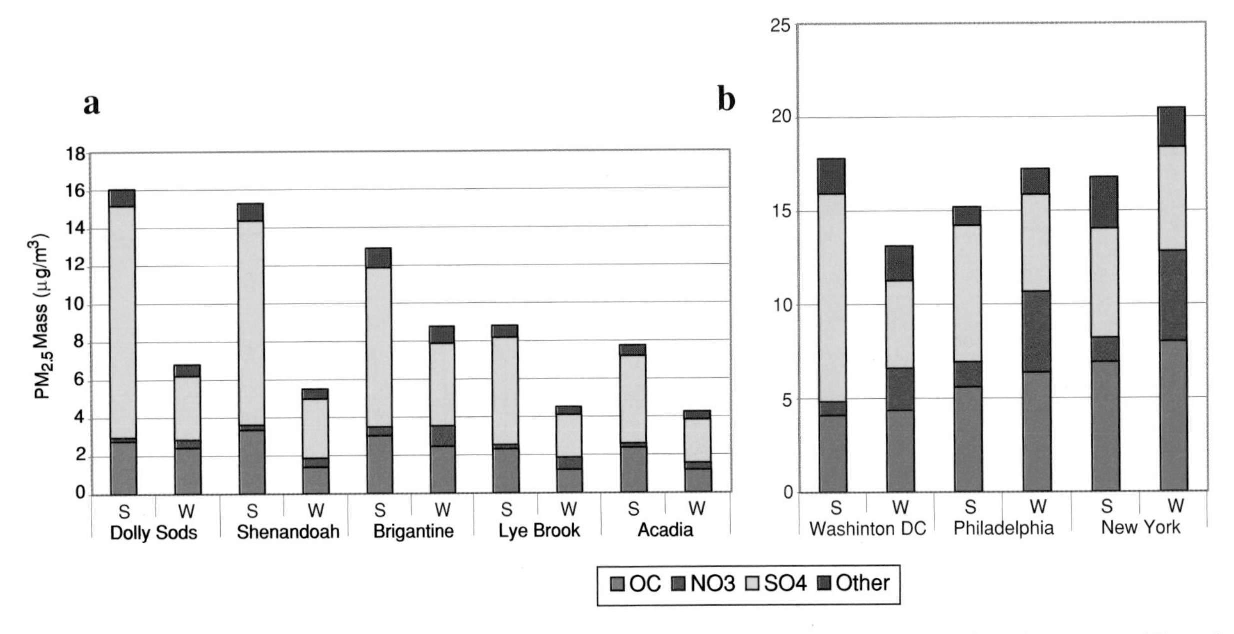

Figure S.11 (10.28). Reconstructed eastern U.S. fine mass partitioned into the individual summer (S) and winter (W) components at (a) rural sites - left and (b) urban sites – right

Information about source regions and pollutant transport in air masses can be obtained using air-mass trajectories and classifying meteorological conditions associated with ranges of PM mass concentration and composition. Pathways or trajectories of air masses alone do not establish the actual occurrence of pollution transport, since they do not indicate contaminant deposition or local vs. distant source contributions. However, this type of analysis (see Figure S.12) can test whether higher concentrations are associated with longer and more frequent transport pathways over specific source regions.

PM_{10}

Historically PM_{10} problems were the result of a wide range of unmanaged sources. Present-day high PM_{10} concentrations are more often associated with meteorological conditions conducive to the local suspension of mineral material such as soil dust and road dust.

New science to improve implementation approaches

Improved understanding of PM sources and processes (notably for carbonaceous PM), including the meteorology and chemistry that contributes to regional and local concentrations and nonlinearities, will help fine-tune PM management strategies that balance regional/urban and chemical precursor reductions to meet area needs. Further discussion of the benefits of new science is provided at the end of this Synthesis.

Policy Question #4 - What source-specific options are there for fixing the problem given the broad control approaches above?

Receptor and emission-based analyses to date point out that on average, greater than two-thirds of $PM_{2.5}$ is traceable back to anthropogenic sources. Several major source categories are important

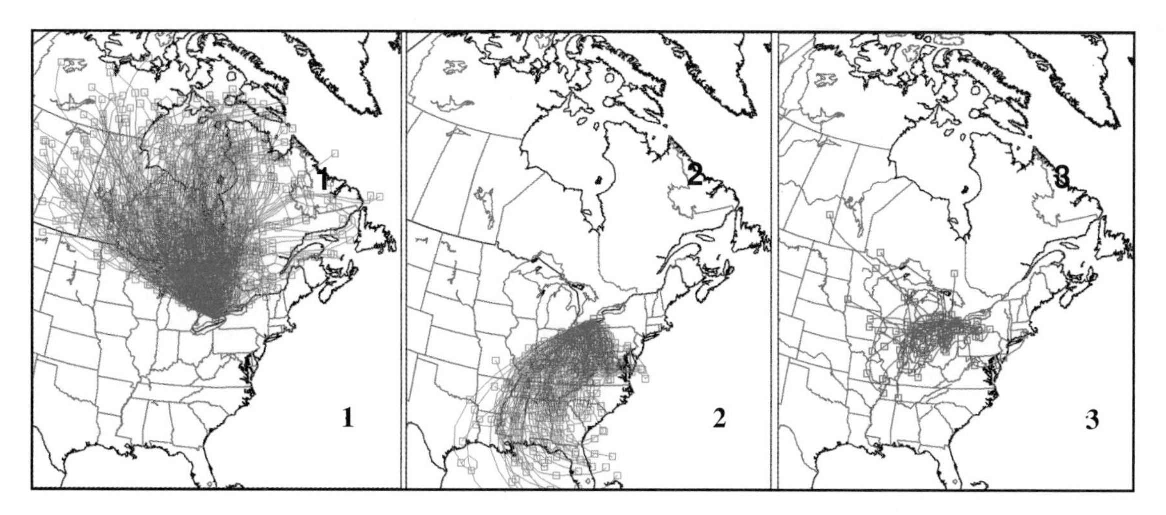

Figure S.12 (7.4). Three-day back-trajectories arriving at the Ontario Air-Quality Index site of Simcoe during May-September of 1998 and 1999 were sorted by transport sector. Back-trajectories represent the most probable path that the air mass followed en route to Simcoe. The sectors shown represent: 1) northerly flow over predominantly Canadian source regions, and 2) southerly flow over U.S. source regions. Six-hr average $PM_{2.5}$ from a TEOM were 6.7 ±6.1 $\mu g/m^3$ (±1SD) for sector 1 and 22.4 ±11.7 $\mu g/m^3$ for sector 2. Sector 3 groups the trajectories corresponding to $PM_{2.5}$ >30 $\mu g/m^3$, the Canadian standard for 24-hr average $PM_{2.5}$. These high-concentration cases were associated with very short transport distances, indicating stagnant conditions in the Midwest and Great Lakes region. There were 5 six-hour measurements >30 $\mu g/m^3$ in sector (1) and 51 in sector (2). This information is being used to estimate what portion of the $PM_{2.5}$ in the Toronto area is due to local emissions vs. transport from other parts of Canada and the United States.

contributors to PM and its precursors and should be the focus of further regional and local policy analysis, in particular:

- Fossil-fuel combustion sources, including electric utilities and internal-combustion engines
- Residential wood burning, wildfires, and other biomass burning
- Ammonia from intensive agricultural operations and, to a lesser extent, transportation.

Source-specific options to reduce PM concentrations are best approached through corroborative analyses using emission inventories, ambient concentration measurements, and air-quality modeling. Given the strengths and limitations in any one of these science tools, it is recommended that they be used in an integrated manner to provide science guidance to develop specific options for local and regional PM management.

Key science findings

Source-attribution approaches

Two principal methods of PM source attribution are 1) receptor modeling, which uses direct information on chemical composition of ambient PM combined with information on the chemical composition of source emissions and atmospheric-transport characteristics to identify and estimate the contributions of existing source categories to PM concentrations, and 2) source-based modeling, which involves the estimation of ambient $PM_{2.5}$ concentration distributions using chemical-transport models (CTMs), which estimate ambient concentration from emission inventories and numerical simulations of atmospheric processes. Receptor models are useful in understanding current conditions, notably relating primary particles to their source. Only CTMs are able to predict the response of ambient concentrations to changes in emission rates, and to directly estimate the contributions of sources to secondary particles. The use of these tools requires a combination of information about $PM_{2.5}$ and its precursors, meteorological data, emission inventories and chemical-source fingerprints, and numerical models.

Uncertainties associated with the results of current CTM simulations arise from the scientific formulation and numerical solution of the CTMs, as well as from the input data (i.e., meteorology, emissions, upwind or boundary air-pollutant concentrations). Other complementary techniques (receptor modeling or tracer studies) should be used to corroborate CTM results, particularly for those PM components known to be highly uncertain in CTM simulations.

Because the uncertainties involved in any particular analysis method are usually large or ill-defined, it is preferable to develop PM management strategies with inputs from multiple analyses using several approaches. The integrated outcome of these analyses can be combined into a "conceptual model" for PM. These analyses require user experience and atmospheric-science expertise, and make use of a combination of ambient concentration observations with emission inventories and meteorological data, chemical-transport modeling, and observationally based models.

Source-based modeling: Chemical-Transport Models

CTMs simulate PM composition and concentration using natural and anthropogenic emissions of PM and PM precursors as inputs. These models can be used to estimate the fraction of PM that is transported into an area from upwind sources and the fraction that is emitted or generated locally. CTMs also can be applied to predict the future impact of changes in emissions on ambient concentrations, to identify major contributing sources, and to quantify the relationships between PM and other air-pollution problems. There are a number of models presently in use or soon to be ready for policy applications, which can be applied to estimating 24-hr episodic PM conditions as well as annual averages. These include Canadian models, such as AURAMS, and U.S. models, such as CMAQ and CAMx.

For CTMs to be most useful to the policy community, their performance must be evaluated using comprehensive ambient data sets, including speciated size-resolved PM measurements and measurements of associated gas-phase species for varying seasons and regions. CTM evaluations for PM have been very limited to date outside of the Los Angeles area due to a lack of suitable evaluation data sets. Intensive field studies in Los Angeles, Texas, and to a lesser extent in eastern North America and the Lower Fraser Valley, have begun to provide some of the needed evaluation data sets. The recently conducted California Regional $PM_{10}/PM_{2.5}$ Air-Quality Study is the largest such field study to date. Application of these models should be reviewed thoroughly by experts to ensure again that the results are well understood prior to their use in policy development. A general sense of the confidence in current CTM simulations is presented in adapted Table S.3.

Advanced CTMs for PM can currently predict the changes in concentration resulting from formation of $SO_4^=$ and HNO_3 with moderate confidence (to within ±50 percent). The accuracy of particulate NO_3^- predictions is likely to be limited by the reliability of NH_3 emission information. CTMs can also predict concentrations of primary PM (e.g., BC, crustal material) with moderate confidence provided the emissions are well characterized; however, CTMs cannot predict with certainty the concentrations of primary PM from sources whose emissions are currently highly uncertain (e.g., fugitive dust, biomass burning). The largest uncertainties are associated with OC predictions because of uncertainty in 1) the emissions of primary OC and condensable organic gases, 2) the emissions of VOCs that form secondary particulate OC (particularly high molecular weight VOC $>C_6$), 3) the chemical reactions leading to secondary organic vapor-phase compounds, and 4) the partitioning of those compounds between the gas phase and the liquid/solid phases. As a consequence, CTMs are currently most useful addressing the $SO_4^=/NO_3^-/NH_4^+$ component of PM. CTM predictions of the organic fraction of PM (e.g., contributions of primary vs. secondary and anthropogenic vs. biogenic) will be much less certain, but can yield preliminary qualitative information for policy making.

Table S.3 (8.2). Levels of confidence in aspects of chemical-transport model simulations.

CTM Aspect	Confidence Level [a]	CTM Aspect	Confidence Level [a]
PM Mass Components		**Gases**	
PM ultrafine	VL	SO_2	H
PM fine	M	NO_x	H
PM coarse	M	NH_3	M
PM Composition		VOC	M
$SO_4^=$	M - H	HNO_3	M
NO_3^-	M	O_3	M
NH_4^+	M	**Spatial Scale**	
OC primary	L	Continental	L
OC secondary	VL	Regional	M
BC	L	Urban	L - M
Crustal	L	**Temporal Scale**	
Water	L	Annual	L
Metals, biologicals, peroxides	VL	Seasonal	L
		Episodic	M

[a] H: high, M: medium, L: low, VL: very low

Modelers have the most confidence in CTM predictions for $PM_{2.5}$ for episodic time scales (days) on a regional basis. However, the CTMs typically provide better results for long-term periods (e.g., one year) than for short-term periods (e.g., 24 hours or less) but this is thought to be the result of "averaging out" the influence of short-term, small-scale meteorological fluctuations and compensating errors within the computer simulation or in the understanding of the atmospheric processes. Thorough model-performance evaluations are being conducted to identify any such compensating errors. The ability of CTMs to reproduce seasonal variations needs to be tested to complement annual averaging. Longer-term CTM simulations also require additional approximations (e.g., simplified parameterizations or episode aggregation) to be made at present, due to computer-resource limitations and the impracticality of creating multi-year meteorological input data.

CTMs have been used effectively to identify the major source types contributing to local PM. However, the impact of specific point sources at distances of 100 km or more cannot be assigned as reliably. Instead, source attribution in regional-scale studies focuses on source regions and source categories rather than on individual sources. Uncertainty in the assignment of specific source-contributions results in part from our inability to predict wind flow and precipitation accurately over long distances and periods. It results also from compounding uncertainties in the model formulation, which increase as the simulation proceeds (i.e., increasing with distance). Uncertainties in meteorological aspects can be reduced to some extent by using data assimilation (of actual meteorological observations) in meteorological simulations. Advances in the current formulation of CTMs (e.g., by carefully evaluating specific components such as cloud processing and dry deposition against ambient data) will help reduce many sources of uncertainty.

Receptor modeling

Analysis of ambient and source data using receptor models is an important means of quantitatively identifying sources of primary particles. In some circumstances these approaches can provide at least qualitative information on sources of secondary particles. Such information is expected to improve in the future as new methods are developed and tested. The reliability of receptor methods depends on having appropriate speciated ambient PM data as well as on independent knowledge of speciated source emissions, often called source profiles.

Receptor models are ambient and emission chemical-composition intensive. A few receptor-model based studies have been performed with existing data, or from investment in modest, short-term measurement campaigns. Others that involve elaborate tracer releases, spatial and temporal coverage with a number of measurement sites, and aircraft overflights, cost several million dollars. Studies of this kind have been conducted in urban areas such as Los Angeles and Phoenix, and around large isolated electric power plants in the western United States. They have been used successfully to provide insight for state implementation planning and rule-making in these locations, as well as to reconcile apparent important biases in emission inventories. Receptor modeling has certain limitations. Perhaps most important of these stem from the fact that most current models can deal only with primary emissions, and require source profiles for application.

The Chemical Mass Balance (CMB) model is one of the historically applied receptor-modeling tools attributing PM components to specific source categories. Positive Matrix Factorization (PMF) is another receptor-model category also in common use. Application of these tools works best for primary, non-reactive, slowly depositing particles, which are directly related to the chemical composition of source categories. These receptor models cannot identify specific source categories for secondary PM components. Information on secondary particles from observational data is typically combined with the results of receptor-model analysis to complete a particle-composition mass balance. Figure S.13 illustrates the results of CMB modeling for two major urban centers in Canada. They both show large contributions from transportation sources and secondary inorganic components, pointing to a supplemental analysis of these subjects for further corroboration.

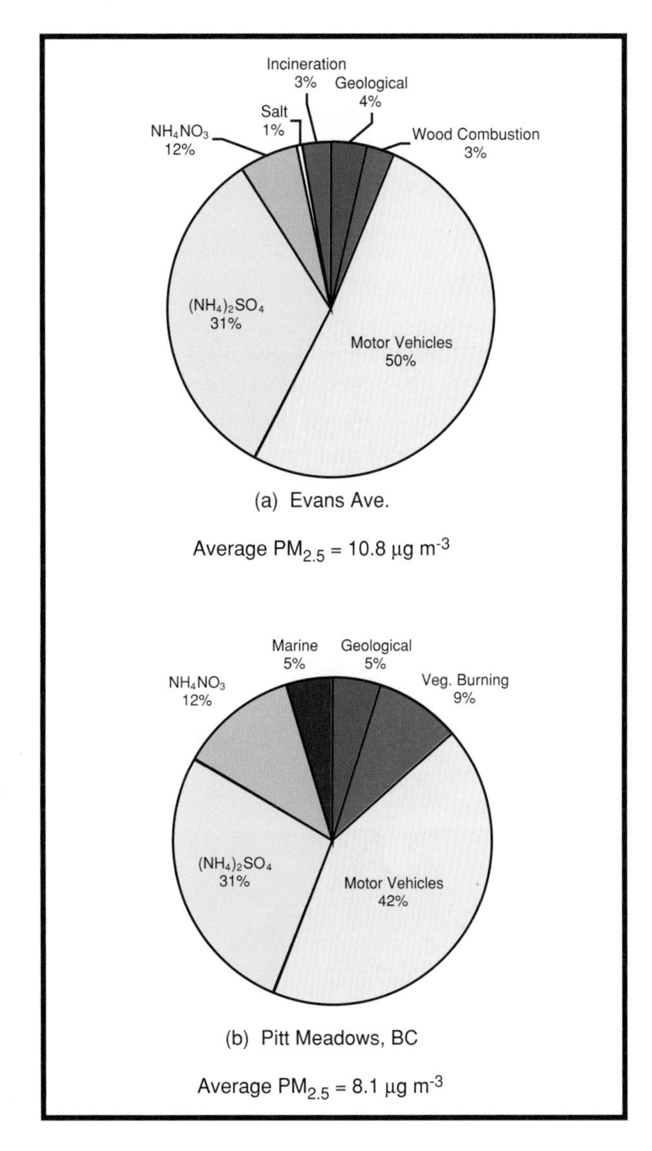

Figure S.13 (7.7). CMB source-contribution estimates for $PM_{2.5}$ in two regions of Canada. The analyses demonstrate that, given the range of sources assumed to be important, motor vehicles contribute a relatively large fraction. This information was used to support policy development focusing on sulfur in gasoline. a) Average motor-vehicle exhaust CMB contributions were 50 percent at Evans Ave., Toronto (average of twelve 24-hr observations from September 1995), which is within the range estimated for U.S. urban areas, but above the average. The Evans Ave. monitor is near high traffic volumes. b) Average vehicle exhaust $PM_{2.5}$ contributions in the Lower Fraser Valley of BC, at a site more distant from dense traffic, constituted ~42 percent of $PM_{2.5}$ (average for twenty-six 24-hr observations from July-August 1993).

Emission inventories and source insights

Currently, national emission inventories identify sources and emission rates of PM and precursor gases, including SO_2, NO_x, VOC and NH_3. In Canada and the United States, the inventories are updated and revised in detail over periods of two to five years, with annual review of certain species or source types. Mexico has completed an inventory for Mexico City, and has limited inventories for six other cities and certain large industrial sources. Mexico has initiated efforts to prepare a national inventory for the near future.

Emission inventories are most accurate and comprehensive for emission estimates of sources lumped together by category, with their individual differences averaged out over annual periods and segregated by geopolitical jurisdictions. This aggregate form serves well for most applications, but additional calculations are required to translate inventory data into specific spatial and temporal detail for CTM applications, and eventually for the estimation of trends. This post-inventory processing is generally done using emission-processing systems specific to intended CTM applications.

Comparisons of national emission inventories and their spatial distributions indicate similarities and differences between Canada and the United States, which follow lines of demographics, industrial activity, transportation, and open or fugitive sources associated with land-use practices. Spatial distributions of $PM_{2.5}$ and the gaseous precursors NO_x, SO_2, and VOC largely follow population centers and are the result of intensive energy use, industrial activity, or concentrations of transportation sources (see for example Figure S.14). PM_{10} and NH_3 patterns differ somewhat because of the contribution of airborne dust sources, and emissions from agricultural operations.

Comparison of major urban emissions in Atlanta, Los Angeles, Mexico City, and Toronto suggest generally similar patterns from common source categories except for Los Angeles, whose metropolitan emissions of NH_3, NO_x and VOC are large compared with other cities. The four major contributing source categories based on magnitude of PM and precursor emissions are industrial processes, electric power generation, transportation, and open sources.

The record of national emission estimates for precursor gases is long enough, from the mid 1980s to present, in Canada and the United States to make some qualitative statements about recent changes. SO_2 emissions are estimated to have declined by a third to a half, largely over the eastern half of North America. NO_x emissions have risen in Canada and have fallen slightly in the United States. National forecasts estimate that in Canada SO_2 and NO_x emissions will rise slightly or remain stable in the next decade, while VOC emissions will increase by about 10 percent. U.S. national forecasts estimate that SO_2 emissions will drop by 5 to10 percent, VOC emissions will drop by 15 to 20 percent, and NO_x emissions will drop by nearly 30 percent over the next 15 years.

Sources of PM

Receptor-based analysis indicates that greater than two-thirds of observed average $PM_{2.5}$ mass concentrations can be traced back to anthropogenic sources of primary PM and precursor gases.

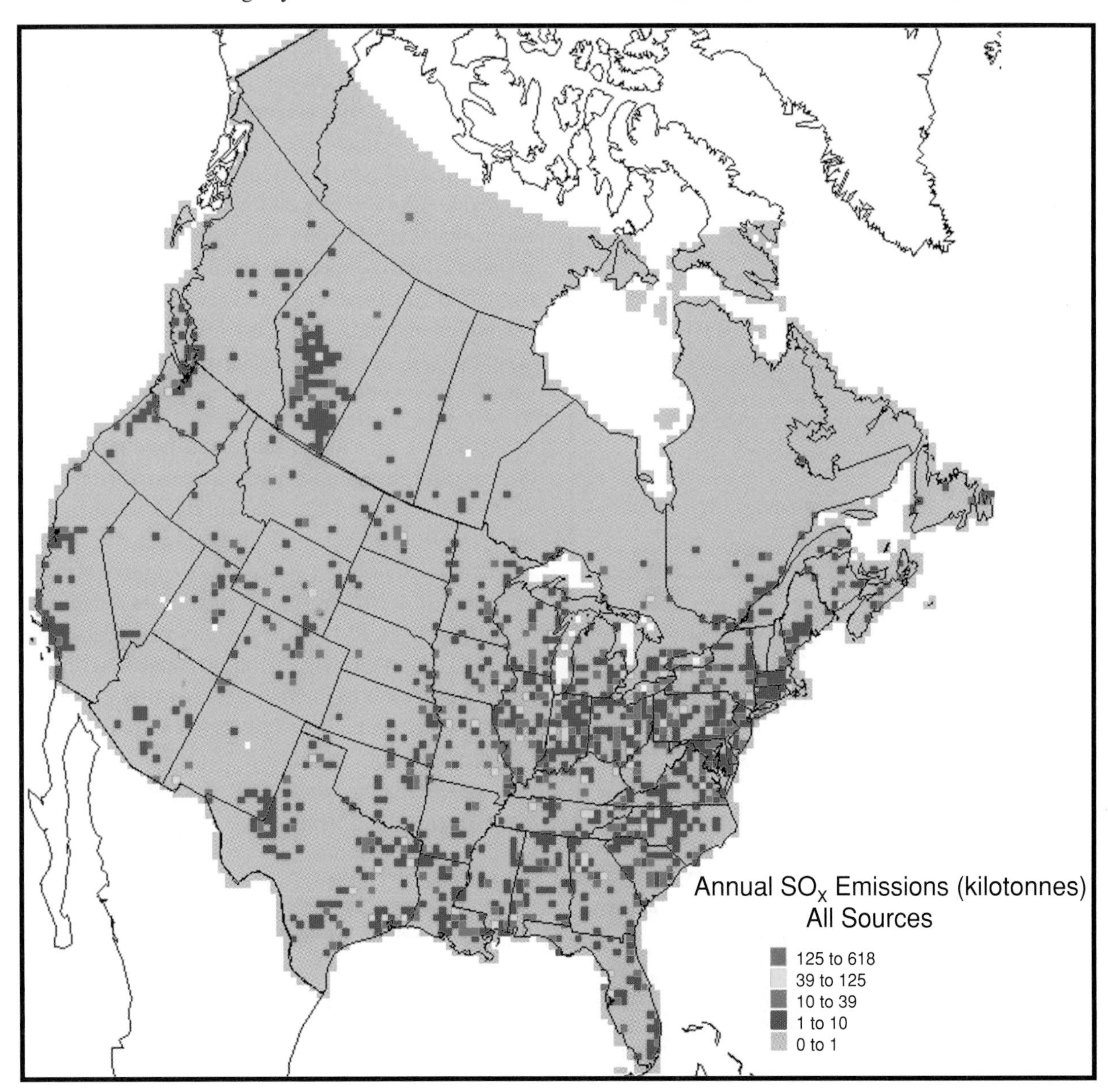

Figure S.14 (4.3) Distribution resolved to 1600 km^2 areas of SO_2 sources in Canada and the United States (Tonnes/yr). (This distribution is similar to NO_x and VOC emission distributions.)

Almost all particle $SO_4^=$ originates from SO_2 oxidation and is associated with NH_4^+. Ninety-five percent of SO_2 sources are anthropogenic, from fossil-fuel combustion, and the majority of NH_3 sources are believed to be related to agricultural activities.

PM_{10} particles in the atmosphere are strongly influenced by open or fugitive emissions (e.g., agricultural operations, road- and soil-dust suspension, sea salt, and vegetation detritus) and supplemented by $SO_4^=$ from SO_2 and NO_3^- from NO_x as part of the 40 to 60 percent fraction typically found to be $PM_{2.5}$.

Conceptual models based on currently available information have been prepared for nine North American areas. They yield the following source-specific insights:

- The Lower Fraser Valley of the Pacific Northwest will likely need future controls on mobile sources, agricultural NH_3, and road-dust emissions to offset future growth where levels are currently below standards.
- For the San Joaquin Valley of California, reduction of secondary particles via VOC and/or NO_x source controls appears important during peak periods. Uncertainties remain regarding the relative importance of VOC or NO_x reductions. Motor vehicles are key contributors and biomass burning may also be a significant. Both urban and regional source reductions are needed.
- For the Los Angeles basin, reduction of secondary particles via controls on VOC and/or NO_x appears important with transportation and agriculture being the key sources to be addressed. Primary organic compounds emitted from transportation, wood burning, and food cooking may contribute significantly to annual-average $PM_{2.5}$ and PM_{10} concentrations. Sulfate particles associated with regional transport are also a significant source to be considered.
- Mexico City's PM problems could benefit from control of diesel vehicles to reduce primary OC and BC emissions. SO_2 and NO_x controls may be more important than NH_3 controls for reducing secondary PM, and thus also should be examined.
- For the southeastern United States, high regional levels combined with local urban sources point to within-region source reductions needed from coal-powered utilities, gasoline and diesel vehicles, and residential wood burning. Rural areas can have both important local and distant source contributions, and some $SO_4^=$ reduction will be offset by NO_3^-, with likely increasing NH_3 emissions.
- Median $SO_4^=$ in the northeastern United States continues to drop from 1990 levels, likely due to SO_2 precursor reductions; but peaks remain, and regional transport in the summer from the Ohio River Valley (when $PM_{2.5}$ is at its peak) suggests further reduction in regional and local SO_2 would be beneficial. Control of local $SO_4^=$, OC, and NO_3^- in coastal urban areas will be important for winter $PM_{2.5}$ mass concentration reductions.
- Reducing $PM_{2.5}$ in the southwest Windsor-Quebec Corridor will require both local measures and cooperation with the United States, likely aimed at both SO_2 and NO_x controls. Similar source controls in Ontario will be needed to reduce Quebec's $PM_{2.5}$. Further consideration is warranted for OC source reductions in cities and wood combustion on local scales.
- For the Upper Midwest-Great Lakes region, reducing local and long-range contributions of OC and $SO_4^=$ should be a consideration for urban areas.
- For the Canadian Southern Prairie and U.S. Northern Plains, a potential increase in urban winter NO_3^- and $SO_4^=$ with population growth should encourage energy efficiency. Reduction of NH_3 from fertilizer applications may be supported, and smoke management is important to regional haze in these clean-air areas.

New science to improve implementation approaches

Improved chemical-transport models that incorporate a better understanding of PM processes are required. These should evolve through a thorough analysis of long-term continuous monitoring, utilize improved emission inventories, and be evaluated using a systematic approach for integrating ambient and modeled data. Such models will be instrumental in setting and evaluating progress against source-specific local and regional PM reduction programs. Further discussion of the benefits of new science is provided at the end of this Synthesis.

Policy Question #5 - What is the relationship between PM, its components, and other air-pollution problems on which the atmospheric-science community is working?

A quantitative understanding exists of the relationship between PM, its components, and visibility (regional haze), and a qualitative understanding exists of the relationship between PM and ground-level ozone and PM and climate change.

- Various species of PM, especially those that comprise $PM_{2.5}$, interfere with the transmission of light and impair visibility through light scattering, and to a lesser extent, absorption. The quantitative relationships of PM, relative humidity and visibility impairment are well understood and reasonably well characterized in mathematical form.
- Regional $PM_{2.5}$ and ground-level ozone share common precursors and similar influences of transport and meteorology in eastern North America, with the highest concentrations of both occurring in the summer months. In California, ozone-$PM_{2.5}$ chemical linkages also exist, but peak levels do not often occur during the same periods indicating differences in their source influences and dominant atmospheric-chemistry processes. The Ozone Assessment (NARSTO, 2000) conclusion that "...a more complete understanding of the $PM_{2.5}/O_3$ system will require focused, process-oriented field studies," still holds true.
- Wet and dry deposition of $SO_4^=$ and NO_3^- contributes to the acidification of ecosystems. Decreasing NH_3 emissions, where $SO_4^=$ concentrations are high, can reduce $PM_{2.5}$ mass concentration but may also increase particle and precipitation acidity.
- The relationship between $PM_{2.5}$ and atmospheric processes affecting the radiative balance and climate change are not well understood quantitatively, especially for "indirect effects' associated with cloud reflectivity and cloud-formation processes.

Key science findings

Tables S.4 and S.5 present our understanding of the complex interrelationships of PM, its components, and other air-pollution problems discussed in detail below. These tables are based on current understanding of atmospheric chemistry and physical processes, and show that potential trade-offs in particle interaction will vary with season and location.

Visibility (regional haze)

Particles in the air reduce visibility, the most readily apparent indication of air-pollution to the public. Visibility impairment is especially sensitive to the species that comprise $PM_{2.5}$, and also depends strongly on the ambient relative humidity. Theoretical or empirical relationships can be used to estimate the visibility impairment attributable to given $PM_{2.5}$ and $PM_{10-2.5}$ fractions of $SO_4^=$, NO_3^-, OC, BC, and soil at a specific relative humidity. The optical effects of particles depend on size, shape and chemical composition. Secondary particles tend to be of the size range (particle diameters of 0.3 to 1.0 μm) that is most effective at scattering visible light. Relative humidity is a particularly important atmospheric factor, since light scattering at high relative humidity is many times that of the dry aerosol for hygroscopic species, such as $SO_4^=$, NO_3^-, and some organic compounds.

Evidence indicates that non-urban visibility impairment in eastern North America is predominantly due to $SO_4^=$ particles, with organic particles generally second in importance. In the West, the contributions of $SO_4^=$ and organics are comparable, and NO_3^- plays a significant role in the populated areas of parts of California. Black carbon is a relevant contributor in some urban areas. Soil particles can be significant contributors to visibility impairment in areas susceptible to windblown dust.

PM-visibility relationships suggest that efforts to mitigate visibility impairment will require control of emissions of different chemical species in different geographic areas. Also, because visibility in the clearest areas is sensitive to even minute increases in PM concentrations, strategies to preserve visibility on the clearest days may require stringent limitations on emission growth, and even the dilute impacts from distant sources can be important.

Modeling visibility impacts is based on predicting aerosol composition using a CTM and then calculating the associated light extinction. Thus the quality of the visibility prediction depends greatly on the performance skill of the PM model. Models that predict secondary PM on a regional basis have limitations for some species as previously discussed, and the reliability of the latest PM models has not yet been evaluated extensively. In the absence of extensive evaluation with observations, the quality of modeled representations of light extinction based on these PM calculations is currently somewhat uncertain.

Despite limitations in the ability to model visibility, the U.S. National Academy of Sciences concluded nearly a decade ago (NRC, 1993) that existing knowledge and tools appeared to be adequate to develop strategies for managing visibility. On this basis and other information, the United States has initiated a 65-year program for mitigating regional haze in U.S. national parks and wilderness areas. The initial emission-management actions that may be implemented under this program are not likely to be sensitive to the uncertainties in current models.

Table S.4 (3.2). Typical pollutant / atmospheric issue relationships.[a]

Reduction in pollutant emissions	Change in associated pollutant or atmospheric issue					
		PM Composition				
	Ozone	Sulfate	Nitrate	Organic compounds	$PM_{2.5}$	Acid Deposition
SO_2		↓	↑ [f]		↓	↓
NO_x	↓↑ [b]	↑↓ [d]	↓ [g]	↓↑ [i]	↓↑	↓↑
VOC	↓	↑↓	↓↑ [h]	↓ [j]	↓↑	↓↑
NH_3		↓[e]	↓		↓	↑ [l]
Black Carbon	↑ [c]			↓ [k]	↓	
Primary Organic Compounds	↑ [c]			↓	↓	
Other primary PM (crustal, metals, etc.)	↑ [c]				↓	↑ [l]

[a] Arrow direction denotes increase (↑) or decrease (↓); arrow color denotes undesirable (red) or desirable (blue) response; arrow size signifies magnitude of change. Small arrows signify possible small increase or decrease. Blank entry indicates negligible response.
[b] In and downwind of some urban areas that are VOC limited.
[c] Effect on daytime ozone due to increase in solar flux and decrease in radical scavenging; effect on nighttime ozone unknown.
[d] Due to effect of NO_x on oxidant levels (OH, H_2O_2 and ozone); e.g., see SAMI modeling results.
[e] Due to effect of NH_3 on cloud/fog pH.
[f] Decrease in sulfate may make more NH_3 available for reaction with HNO_3 to form NH_4NO_3, more important when NH_4NO_3 is NH_3 limited.
[g] Decrease except special cases (e.g., SJV); decrease in NO_x may lead to increase in ozone with associated increase in HNO_3 formation.
[h] Increase due to less organic nitrate formation and more OH available for reaction with NO_2; decrease due to decrease in oxidant levels.
[i] Related to effect of NO_x on oxidant levels (OH, ozone and NO_3).
[j] Decrease of secondary component; magnitude depends on OC fraction that is secondary anthropogenic.
[k] Reduction of OC adsorbed or emitted with black carbon.
[l] Refers to net acidity atmospheric deposition, not to acidification potential to ecosystem.

Table S.5 (9.2). Responses of regional haze and climate to reductions in the emissions of secondary PM precursors and primary PM.

Pollutant Emitted	Change In Associated Issue	
	Regional Haze[a]	Climate Impact[c]
SO_2	↓	↑
NO_x	↑↓[b]	↑↓
VOC	↑↓[b]	↑↓
NH_3	↓	↑[d]
Black Carbon	↓	↓
Primary Organic Compounds	↓	↑
Other primary PM (crustal, metals, etc.)	↓	↑

[a] Direction of arrow indicates increase (↑) or decrease (↓) and color signifies undesirable (red) or desirable (blue) impact; size of arrow signifies magnitude of change. Small arrows signify possible or small change.
[b] No change if little NH_3 available in atmosphere.
[c] Direct effects only; indirect effects through clouds and precipitation are highly uncertain. Note that the extent and possibly the scale of climate impacts for listed pollutants is quite different from CO_2 and CH_4. Direction of arrow indicates warming ↑ or cooling ↓.
[d] More accurately, decreased aerosol-induced cooling.

Programs comparable to those in the United States that address visibility impairment have not been proposed by Canada and Mexico, although efforts at limiting and reducing PM concentrations there for health reasons can be expected to provide benefits to improved visibility. Measurement networks to characterize visibility and its trends nationwide do not exist in either country, so visibility conditions there are not extensively documented.

Ozone

Reductions in NO_x and VOC emissions for ozone management in areas containing substantial SO_2 and NH_3 emissions may result in changes in $PM_{2.5}$ mass concentrations as a result of the complex interactions between oxidants, $SO_4^=$ and NO_3^-. The $PM_{2.5}$ concentration changes depend on the mix of pollutant gases in a specific geographical area and season. Reductions of primary $PM_{2.5}$, SO_2, and NH_3 emissions for $PM_{2.5}$ management are generally expected to have a small effect on ozone concentrations. Combined NO_x- and VOC-management strategies for both PM and ozone can result in optimal strategies that differ from the ones that would be adopted if these problems were examined separately. An optimal strategy would require balancing VOC and NO_x controls to obtain the desired reductions in ozone and $PM_{2.5}$, while minimizing the potential disbenefits.

For example, in portions of the California San Joaquin Valley where peak fall and winter PM concentrations occur, particulate-NO_3^- concentrations are potentially more influenced by the VOC reaction cycle of the oxidant chemistry process than by its NO_x counterpart. In comparison, the summer oxidant problem in the valley can be NO_x sensitive, pointing to need to examine the interplay of seasonal control strategies for these two air-pollutant classes.

The characterization of VOC interactions with PM is one of the poorly resolved issues facing atmospheric science. VOCs in the troposphere are intimately involved in the production of ozone and other oxidants. Certain VOCs, including anthropogenic aromatic emissions such as toluene, xylenes, trimethyl-benzenes and biogenic emissions such as terpenes (α- and β-pinene, limonene, carene, etc.) and the sesquiterpenes react in the atmosphere to produce secondary organic particles. In the troposphere, VOC with carbon numbers between 2 and 7 are most prevalent and of concern in primarily ozone formation. The high carbon-number species (>C_7) when oxidized tend to produce increasing amounts of condensed material with increasing VOC molecular weight.

The NARSTO (2000) Ozone Assessment and its associated Critical Review Paper (Hidy et al., 2000) present an in-depth discussion of the multiple possible responses of particulate $SO_4^=$ and NO_3^- to decreases in NO_x and VOC emissions. The summary finding of the Ozone Assessment is that while there are positive and negative benefits resulting from NO_x and VOC reductions, their magnitudes are uncertain. This Assessment further concludes that the development of a more complete understanding of the $PM_{2.5}$ / ozone system requires focused, process-oriented field studies. Some such studies have occurred and early indications are that $SO_4^=$ and NO_3^- controls for PM reduction will have a net beneficial effect on both PM and ozone, even though some localized and/or temporary counterproductive impacts may occur.

Chemical deposition (from PQ #1)

The relationships between the chemical and meteorological processes of $PM_{2.5}$ $SO_4^=$ and NO_3^- formation and fate, the deposition of metallic and other cation components of PM, and the processes contributing to acidic deposition and ecosystem effects are noted in this Assessment. However, they are not reviewed here as they are the subject of in-depth periodic science assessments by other science assessment bodies.

The reader is referred to an extensive discussion of the effects of acidic deposition presented in the U.S. National Acid Precipitation Assessment Program (NAPAP) Biennial Report to Congress: An Integrated Assessment (National Science and Technology Council, 1998), and in the Canadian 1997 Acid Rain Assessment (Environment Canada, 1997-1998). In addition, the effects of particulate pesticides, metal compounds, and chlorinated organic compounds are discussed in Deposition of Air Pollutants to the Great Waters, Third Report to Congress (U.S. Environmental Protection Agency, 2000).

Climate change (from PQ#1)

On a global scale, the scattering and absorption of radiation by airborne particles has been identified as a potentially important factor in Earth's radiative energy balance and climate. As well, the major sources of PM are related to fossil-fuel combustion, which are also linked to climate and radiation balance impacts. Direct impacts on the radiative balance are caused by $SO_4^=$ and OC (local cooling influences) and BC (warming influences). There are also indirect forcings due to the impact of particles on cloud formation processes, currently understood to be positive (warming) but not well quantified. The radiative forcing due to particles depends upon the size, shape, chemical composition and spatial distribution (both vertical and horizontal). These effects vary in time and space and are superimposed on the climate impacts of absorptive gases like methane and CO_2 in the atmosphere. This relationship is thoroughly discussed in the Intergovernmental Panel on Climate Change's Third Assessment Report (IPCC, 2002).

New science to improve implementation approaches

Improved understanding of the pollutant interrelationships, co-benefits and potential trade-offs involving PM and its precursors, and environmental issues such as tropospheric ozone and climate change can come from long-term, continuous monitoring of PM, combined with analyses using CTM simulations. These can be strengthened by a systematic approach for integrating diverse types of knowledge on sources, properties and effects of PM. Further discussion of the benefits of new science is provided at the conclusion of this Synthesis.

Policy Question #6 - How can progress be measured? How can we determine the effectiveness of our actions in bringing about emission reductions and air-quality improvements, with their corresponding exposure reductions and health improvements?

Direct progress in meeting public-health and environmental goals is difficult to assess. The exposure and health-effects research communities are working on new indicators and measurements to track progress in lowering human exposure and reducing health impacts from PM. Cooperative efforts continue among these communities and the atmospheric-science community to further relate changes in exposure and health directly to changes in ambient air-quality.

Generally, improvement is measured against intermediate objectives of achieving emission-reduction targets and providing cleaner air.

- Emission changes can be measured directly for large point sources through continuous emission monitoring. However, changes in emissions from other types of sources including transportation and open sources are difficult to document consistently.
- Ambient air-quality monitoring networks generally provide regulatory-progress information and do not provide the spatial and temporal resolution required to reliably determine response to emission changes on other than broad regional scales.
- Regional-haze improvements resulting from PM changes can be measured directly or tracked reasonably well using estimates of indirect light extinction from $PM_{2.5}$ component concentrations.

Key science findings

Tracking emission changes

Present emission-inventory programs in Canada, Mexico, and the United States provide snapshots of provincial-, state-, and national-level emissions to meet implementation-planning needs. However, they are minimally effective for quantifying changes in emissions over time because of historical changes in emission-estimation methods. Routinely updated and detailed revision of annual, local, and regional emission inventories using self-consistent methods continue to be needed at intervals more frequent than the multiyear periods adopted for current emission inventories.

Estimation of emissions involves calculations based on emission factors (average emission rates per unit process input over a range of operating conditions) and activity factors (from estimates of consumption or use rates). Both emission factors and activity estimates vary within source categories and across time and geographical regions. Since it is impractical to measure every source to derive these factors, source-category averages are derived based on small numbers of samples taken from specific source types. The accuracy and representativeness of these factors are often limited due to the small sampling on which they are based. This leads to differing levels of uncertainty depending on the nature of the source and its activity patterns. Tracking emission changes over time is particularly difficult since both emission factors and activity patterns change with growth in areas, emission controls, and updated technology.

Uncertainties in emissions are difficult to estimate, but generally the highest uncertainties are associated with NH_3 and carbon sources, including VOC, motor-vehicle emissions (because of unknown in-use vehicle conditions), and open or fugitive sources. The most certain estimates are associated with point sources, such as continuous industrial processes and electrical power generation. Table S.6 describes, in qualitative terms, the current understanding of emission-information uncertainty.

Reconciliation of emissions from inventory estimates using CTMs and ambient data with receptor modeling, or in-use activity is a potentially useful means of verifying inventory estimates, but this is seldom done. Verification by independent methods is important to ensure the reliability of the inventories and subsequently the confidence in the CTM simulations.

Tracking trends in ambient air-quality

Ambient air-quality monitoring networks generally do not provide the spatial and temporal resolution required to reliably determine response to emission changes. PM monitoring programs in all three countries are being improved by adding high temporal-resolution and composition measurements, as well as new sites.

Table S.6 (4.8). Estimated confidence level of emission estimates.

Pollutant	Source	Method of Estimation	Estimated Confidence Of Category in Overall Inventory		
			Canada	USA	Mexico City
SO_2	Electric Utility	CEM/AP-42	H	H	H
	Ind. / Comm. Fuel Comb.	AP-42	M	M	M/L
	Other Fuel Comb.	AP-42	M	M	L
	Transportation	NR [c], Mobile, Mass Balance	M	M	L
	Industrial Processes	AP-42	M	M	N/A
	Other Man-made (Non Comb.)	AP-42	L	L	L
	Natural	Literature	L	L	L
NO_x	Electric Utility	CEM; AP-42	M-H	H	M
	Ind. / Comm. Fuel Comb.	AP-42	M	M	M
	Other Fuel Comb.	AP-42	M	M	M
	Transportation	MOBILE, NR [c]	H	H	M
	Industrial Processes	AP-42	M	M	L
	Other Man-made (Non Comb.)	AP-42	L	L	L
	Natural	BEIS	M	M	L
VOC [a]	Electric Utility	AP-42	M-H	M-H	M
	Ind. / Comm. Fuel Comb.	AP-42	M	M	M
	Other Fuel Comb.	AP-42	L	L	L
	Transportation	MOBILE, NR [c], Literature	M	H	L
	Industrial Processes	AP-42	M	M	L
	Other Man-made (Non Comb.)	AP-42	L	L	L
	Natural	BEIS	M	M	L
NH_3	Electric Utility	AP-42; literature	M	M	L
	Ind. / Comm. Fuel Comb.	AP-42; literature	L	L	L
	Other Fuel Comb.	AP-42;literature	L	L	L
	Transportation	Literature	M	M	L
	Industrial Processes	AP-42; literature	L	L	L
	Other Man-made (Non Comb.)	AP-42; literature	L-M	L	L
	Natural	Literature	L	L	NA
PM_{10} [b]	Electric Utility	AP-42	M	M	M
	Ind. / Comm. Fuel Comb.	AP-42	M	M	M
	Other Fuel Comb.	AP-42	L	L	L
	Transportation	NR [c], Mobile, Literature	M	M	L
	Industrial Processes	AP-42	M	M	L
	Other Man-made (Non Comb.)	AP-42	L	L	L
	Natural	AP-42/Literature	L	L	L
$PM_{2.5}$ [b]	Electric Utility	AP-42	M/L	M/L	NA
	Ind. / Comm. Fuel Comb	AP-42	M/L	M/L	NA
	Other Fuel Comb.	AP-42	L	L	NA
	Transportation	NR [c], Mobile, Literature	L	M	NA
	Industrial Processes	AP-42	L	L	NA
	Other Man-made (Non Comb.)	AP-42	L	L	NA
	Natural	AP-42/literature	L	L	NA

[a]For total VOC; speciation estimates rated low confidence level.
[b]For total PM; composition profiles rated low to medium confidence level
[c]EPA's Non-Road Model

Though trends have been observed over the past decade, they are inconsistent between sites and networks, and cannot be attributed to management action other than for the case of regional reductions of SO_2 (see for example Figure S.15). Work continues on optimizing national monitoring networks to serve the multiple purposes of trend detection, regulatory-progress and compliance determination, and scientific-information gathering.

A relationship exists between the spatial scale over which particulate pollutants are distributed and the spatial and temporal scales of measurements needed to assess their impacts. Relatively coarse spatial and temporal measurement resolution is typically adequate for regional pollutants. For locally emitted or reactive pollutants, measurements may be required with higher spatial and temporal resolution.

Long-term measurements, typically at least 5 to 10 years, are required to assess trends. Information from long-term routine measurements of gases and PM properties at representative sites is fundamental to successful study of population exposures, effects, model performance, and the efficacy of emission-control measures.

A long-term, basic suite of chemical measurements at spatially representative sites is needed to provide the full picture of PM formation, transport, trends and impacts. For example a minimum suite should measure: PM mass (fine and coarse), speciation of the six major components ($SO_4^=$, NO_3^-, NH_4^+, OC, BC, crustal material), ground-level ozone, the precursor gases (NO_x, SO_2, VOCs, and NH_3), HNO_3, and CO, along with meteorological measurements. Information on organic speciation, including gas/particle distributions of semivolatile organic compounds, is also desirable. High-resolution measurements of size distributions would also provide valuable complementary information on aerosol sources and formation processes.

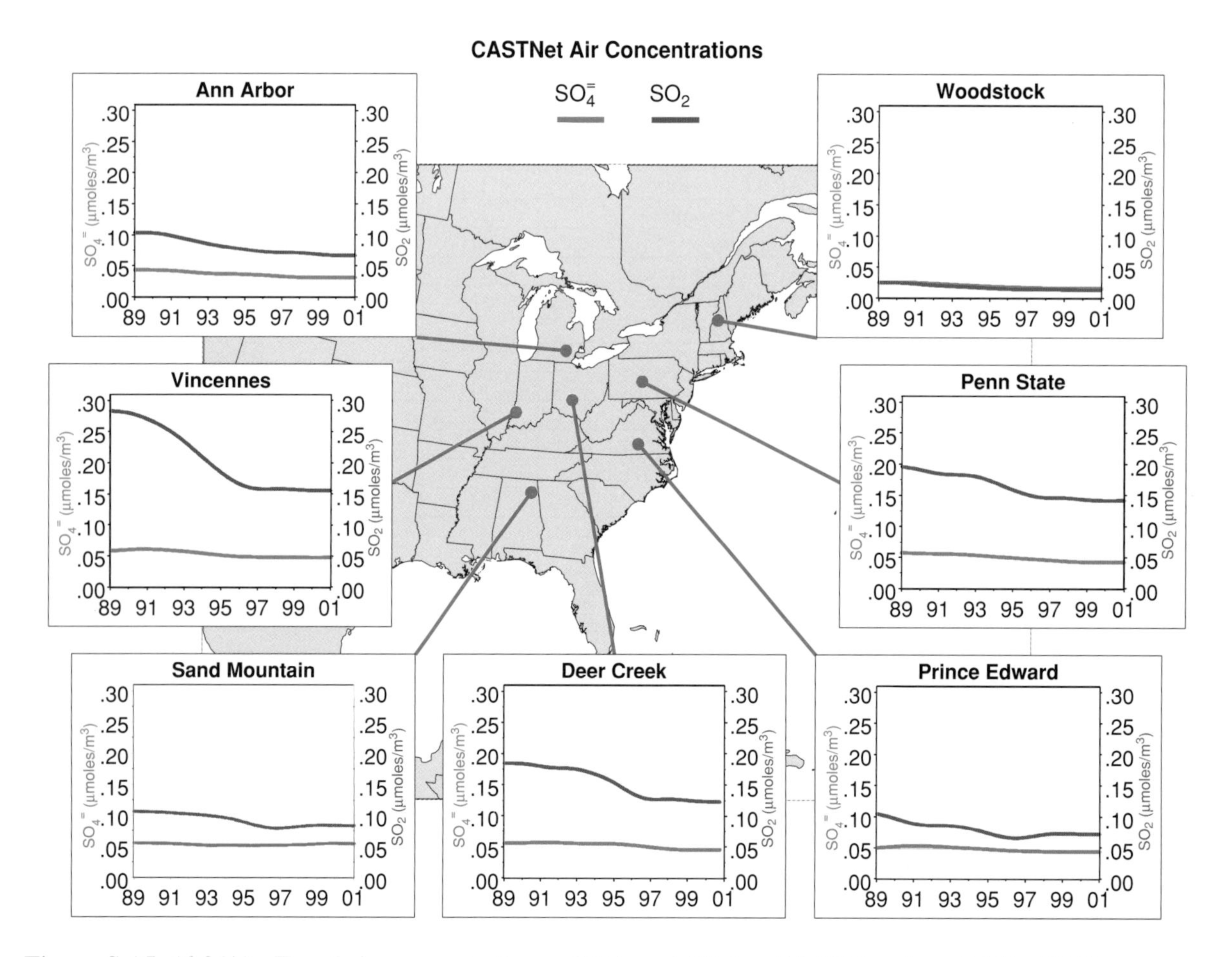

Figure S.15. (6.21b). Trends in concentrations of SO_2 and $SO_4^=$ at CASTNet sites, 1980 - 2000.

Gas-phase measurements of NH_3, HNO_3 or speciated VOC (for example semivolatile organic compounds - SVOC) are usually not included within most PM networks, yet they are necessary for fully characterizing the behavior of the organic component of aerosols. The NH_3 and HNO_3 observations are important for estimating the response of PM mass concentrations to changes in SO_2 and NO_x. Ammonia and HNO_3 observations in parallel with SO_2, water-soluble NH_4^+, $SO_4^=$, and NO_3^-, temperature and humidity are recommended for the sites at which chemical speciation measurements are being made to complete the picture of secondary particle formation.

Accurate measurements of meteorological parameters are essential for understanding aerosol sources, chemistry, transport and deposition. New insight is being gained regarding the complexity of the meteorological processes that influence the formation of secondary air pollutants and the transport of air pollutants, as well as the seasonal and annual meteorological variability that influences long-term trends in air quality. Measurements of conditions aloft, both meteorological and chemical, are more complicated than at ground level. Recent progress in remote sensing is providing significant opportunities for advancing knowledge of meteorological conditions aloft on space and time scales needed to interpret air-quality information.

Significant progress has been made in the past several years on developing instrumentation for real-time, continuous measurements of particle mass and composition. Comparability studies indicate that gravimetric filter-based methods and continuous measurements both have appropriate and useful roles to play in managing and tracking PM. However, the continuous real-time methods can supply additional information needed to improve understanding of the formation, redistribution, and loss of particles in the atmosphere, the sources of those particles, and the relation between PM and human health. Over the next few years the availability and capability of routine real-time, continuous instrumentation will evolve for regulatory applications. Efforts should be made to augment and/or replace the gravimetric filter-based methods with these continuous methods.

Tracking changes in visibility

Measuring visibility trends can be done reasonably well provided that light extinction is measured or estimated at some representative locations over a long time period. Tracking visibility, or regional haze, indirectly using light extinction estimated from $PM_{2.5}$ and $PM_{10-2.5}$ concentrations is appropriate as long as large pollutant-concentration changes do not occur over the evaluation period.

The comprehensive characterization of Class I (scenic) area visibility and PM throughout the United States began with the IMPROVE (Interagency Monitoring of Protected Visual Environments) program, which initiated operation in 1988. IMPROVE was substantially expanded in 2000 to cover a representative sampling of the entire United States. One product of the IMPROVE measurements is a database that is sufficiently long to infer trends.

Measurements of aerosol properties and light extinction were made at a few locations in Canada, near the border with the United States, under a program called GAViM (Guelph Aerosol and Visibility Monitoring) using part of the IMPROVE sampling system, and thus can be compared with some U.S. measurements.

The earliest routine measurements of visibility were those by human observers at airports. A data base of more than five decades is now available for the United States and Canada, and has been used to determine the spatial distribution and trends of visibility. Current airport visibility data have even been used to characterize visibility over the entire globe. Automatic observations of airport visibility by the Automated Surface Observing System (ASOS) and the Automated Weather Observation System (AWOS) are replacing human observers in many locations, and will eventually bring an end to the human-observer data base in most jurisdictions. Some ASOS optical data are being archived at the full instrument resolution. ASOS and AWOS light extinction measurements from all locations with time resolutions of one-hour average or shorter are especially useful for purposes of tracking air-pollution-related visibility to the extent they are made available.

Tracking changes in exposure and health

Ongoing examination of the relation between ambient concentration and total human exposure will likely lead to development of new exposure indicators and measurements to help assess progress in providing cleaner air. Similarly, the ongoing examination of health effects associated with PM, its components, and associated copollutants through the disciplines of epidemiology, toxicology, and molecular biology are leading to new insights of cause-effect and dose-response relationships, which will likely produce new indicators and measurements to assess progress in reducing human-health risks.

New science to improve implementation approaches

Our ability to measure progress against environmental and health goals would be greatly aided by the development of an accountability framework that incorporates a systematic approach to integrate diverse types of information on aerosol origins and properties. Further discussion of the benefits of new science is provided at the conclusion of this Synthesis.

Policy Question #7 - When and how should implementation programs be reassessed and updated to adjust for any weaknesses, and to take advantage of advances in science and technology?

A 5 or 7- year cycle should be anticipated for major assessments of new science and the application of improved science tools for PM management. An exception to this guideline would occur if and when the health-research community identifies specific causal agent(s) of harmful effects associated with PM. Such an event should prompt an immediate new set of analyses or new atmospheric science assessments using available tools.

Tracking improvements in air quality against changes in emissions is integral to the policy-making process. Optimal air-quality management programs need to incorporate the results of regularly scheduled science reviews including: reviews of emission trends, characterization of current ambient air-quality and trends, and analysis with receptor models and state-of-the art CTMs for an updated analysis of source-receptor relationships. The results of these analyses should feed into the reevaluation of management strategies to allow mid-course modifications where improvements can be made. An ongoing exchange of information between the science and policy community should be promoted to take full advantage of these reviews and the information they produce.

Key science findings

The timing for evaluation of progress towards achieving PM standards depends upon the rate of implementation of emission-reduction actions and the time it takes for its impact to be seen within the natural variability of ambient data. Large changes in emissions depend on implementation of regulations and voluntary actions and technological response, typically requiring a minimum of several years. Emission inventories reflect changes that occur over 5- to 10-year periods. The commensurate air-quality data record needed to reflect those changes can take up to ten years to obtain. Thus trends will ordinarily require at least 5 years and perhaps a decade of sustained record to evaluate an emission-ambient air response.

The same conclusions were reached in the Ozone Assessment (NARSTO, 2000) when reviewing atmospheric science in the context of implementing ozone standards. That assessment concluded that the establishment of an iterative progress-driven environmental management process is necessary for credible pollution mitigation.

The NARSTO Ozone Assessment identified three major steps in a context of accountability for air-quality management. These three apply equally well to PM:

- Verification of the effectiveness of controls
- Verification that air quality is responding to emission changes
- Verification that abatement measures have results in health and environmental benefits.

Optimal air-quality management programs need to incorporate the results of regularly scheduled science reviews including: reviews of emission trends, characterization of current ambient air quality and trends, and analysis with receptor models and state-of-the art CTMs for updated analyses of source-receptor relationships. The results of these analyses should feed into the reevaluation of management strategies to allow mid-course modifications where improvements can be made.

The exchange between the science and policy community should be iterative, with the aim of reliably establishing the cause and effect between lowered emissions and improved air quality. Verification of expected progress should be used to establish expected changes in exposure levels and health outcomes at different locations (Figure S.16). Developing robust methods to translate air-quality changes into changes in health metrics remains a major challenge facing the science community.

In the past, from the first acknowledgement that information is required and a comprehensive assessment undertaken, the process of evaluating current knowledge with the application of current scientific tools typically has taken 6 to 8 years (NAPAP, 1991; Environment Canada, 1997-1998). To assess new information, apply the science tools, and communicate the results in a usable fashion to the policy community more frequently than every 5 to 7 years is not likely to be effective or productive (Table S.7).

History has shown that the results of major field studies and special projects continue to feed the development of science tools and air-quality policy 10 to 15 years after the study is completed. In fact, the greatest contribution from these efforts usually occurs 5 to 10 years later, as the knowledge is assimilated across the community. Because air-quality management decisions today are based upon the science of 5 to 10 or more years ago, it is important to take a long-term view to supporting innovative

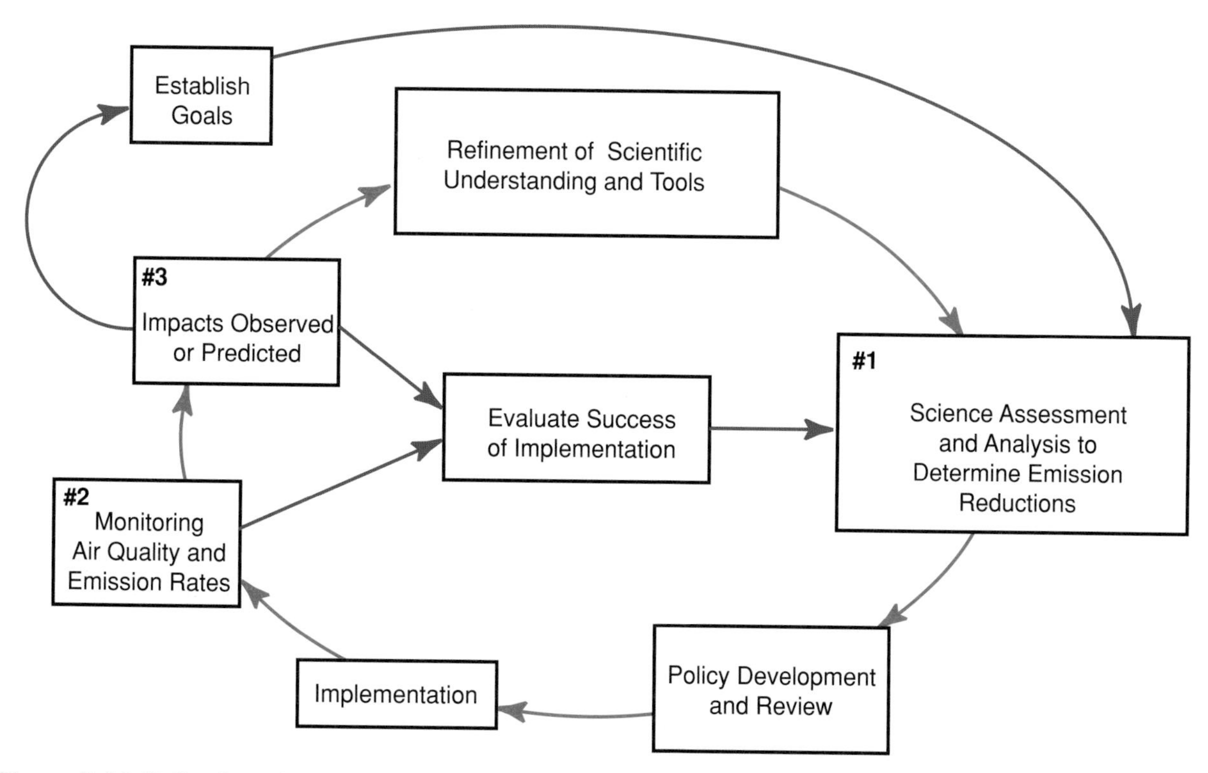

Figure S.16 (1.5). Iterative communication for managing air-quality to reduce health and environmental impacts in the context of current scientific understanding and accountability. The process involves moving through stages of gathering new scientific information, performing assessment, developing policy, tracking outcomes, improving the knowledge base, and revising the policies and implementation plans.

research and exploring new approaches to improving scientific understanding. The process of knowledge being assimilated by the science community needs to be factored into policy analysis. Research information and benefits coming out of recent or current scientific work to fill science gaps may not be immediately recognized, and require review, testing, and vetting before being accepted in a subsequent round of policy analysis.

New science to improve implementation approaches

A commitment to developing the accountability framework recommended in Policy Question #6 together with commitment to provide the needed resources to evaluate and synthesize the information obtained can lead directly to improved answers to how and when to reassess and update implementation programs. Further discussion of the benefits of new science is provided at the conclusion of this Synthesis.

Policy Question #8 - What further atmospheric-sciences information will be needed in the periodic reviews of national standards?

The exposure and health-sciences communities have three fundamental needs for knowledge from the atmospheric-science community:

1. Characterization of PM air quality and meteorology through development and application of advanced measurements and analyses.
2. Identification of the sources of observed PM concentration and composition through an understanding of emission sources of PM and its gaseous precursors and the mechanisms of chemical transformation and atmospheric transport.
3. Reliable ambient air-quality models, i.e., CTMs, with defined uncertainty that can be linked to human exposure and health-effects models.

Substantial progress for all three of these needs is being made by ongoing and evolving development efforts in atmospheric science and intensive laboratory and field studies.

Key science findings

Figure S.17 presents a "pollutant source to receptor response" paradigm suggested by the National Research Council for guiding the elucidation of scientific understanding of PM health effects and integrating information for establishing ambient air-quality standards. This paradigm provides a useful framework for identifying information needs of the atmospheric-science community. The first two boxes with connecting arrows represent the focal points of atmospheric-science contributions. Here the sources and emissions of PM and its precursors, its atmospheric transport and transformation, and its ambient concentration and composition are understood and identified. The means by which this is accomplished is through measurements, including

Table S.7. Typical periods for acquisition of information on progress.

Progress Metric and Acquisition of Information	Time Required
Track changes in emission rates	1 – 3 yr
Track changes in ambient concentrations	3 – 5 to 5 – 10 yr
Plan, execute and report on major field studies and research projects	5 – 10 yr
Production of science assessments	2 – 3 yr
Full state-of-knowledge assessments applying current science tools	6 – 8 yr
Next NARSTO PM Assessment recommended for…	End of 2008

the development of needed methods where they do not exist or are inadequate, laboratory and field studies of chemical and physical processes, and the development of source- and receptor-oriented models. Further progress in reducing the uncertainties in the linkages within the paradigm will require continued close collaboration between atmospheric and health scientists.

PM measurement considerations

Exposure to ambient air pollution depends on individual activity patters over multiple environments that require knowledge of geographical and temporal variability of PM concentrations and composition. Historically this kind of knowledge has been lacking because of limited community-scale monitoring.

In urban areas having multiple monitoring sites, average PM_{10} mass may vary by up to roughly a factor of two over distances as small as approximately 10 to 20 km, though the observations are spatially correlated. In some cities, average $PM_{2.5}$ mass concentrations may show factor-of-two variations, though such variations are more typical over distances of 50 to 100 km or more. Monitoring data indicate that significant variations of daily $PM_{2.5}$ mass concentrations may occur over scales of a few kilometers in urban areas. Monitoring data show typically larger intersite differences in composition than in fine-particle mass concentrations. This non-uniformity of PM and its potential undersampling is a limitation for health and exposure scientists.

In many cities, average $PM_{2.5}$ concentrations are more uniform than PM_{10} concentrations. This evidently results from homogenization influenced by a relatively large regional component on which local source contributions are superimposed. This phenomenon is also enhanced by the relatively long residence times of fine particles. The tendency towards more uniform concentrations of $PM_{2.5}$ in the absence of strong local source influences aids in rationalizing the use of a limited number of monitoring sites to characterize concentration and composition over a few kilometers extent.

Exposure-science considerations

Exposure evaluations generally require measurements that allow direct comparisons across three elements: ambient air quality, indoor air quality, and personal exposure to ambient and indoor environments. Measurements in the personal-exposure category are typically performed using small, specially designed sampling instruments, which are worn by test subjects over multi-day periods. Usually these personal-exposure samplers are miniaturized adaptations of conventional air-quality measurement techniques, and have focused

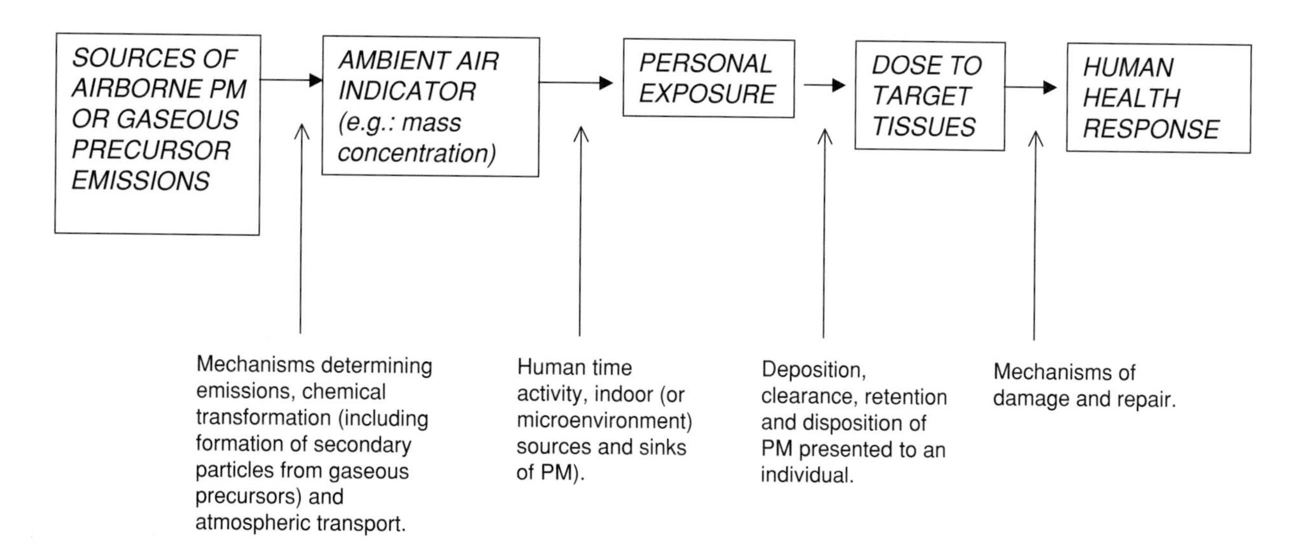

Figure S.17 (1.5). NRC framework for evaluating risks extending from sources of toxicants to responses. (NRC, 1998)

primarily on PM mass and other environmental constituents having significant human-health effects. Compared to their more conventional counterparts these small devices are less consistent in performance, and are typified by higher measurement uncertainty.

Table S.8 presents the list of hypothesized causal agents that are the primary focus of current health and exposure studies, and indicates current availability of corresponding measurement instrumentation. From this one can note that methods are available to perform nearly all of the required measurements. Many of these methods, however, are available on only limited bases and require specially trained personnel.

Study design is another area of consideration in the context of methods, when it comes to ambient and exposure measurement and data comparability. As exposure scientists study people of differing subpopulations, who live in different geographic regions, in different urban or rural settings, and spend their days in and out of different microenvironments, it is important they work closely with atmospheric scientists to come to a clear understanding of the surrounding ambient environment. Translation of personal activity patterns and exposure measurements into an exposure model that applies widely for estimating exposure in urban settings is now under development and testing. This effort will involve both exposure constructs and CTM simulations, requiring joint atmospheric-science and exposure-science effort.

Health-science considerations

The health-research community (including epidemiological, toxicological and clinical researchers) is restricted to investigations of the impacts of those particle characteristics that can be measured. Of the many hypothesized causal elements listed, for example in Table S.8, there are only sufficient ambient measurements to test half of these hypotheses, because population-based health studies typically require relatively long-term data sets in order to achieve statistically significant results.

Investigation and assessment of health impacts typically have been based upon metrics (dry bulk size fractionated mass by gravimetry) that are associated with the concentrations of ambient particles. Data from the gravimetric filter-based methods have been used to assess health impacts over the longest time period and are the basis of most health-impact studies used to set current ambient PM standards. Any new PM mass and composition measurements that are related to health impacts need to be linked to this historical context.

Statistically robust estimates of excess risk to ambient PM exposure typically require studies documenting daily disease events in association with monitoring days. PM associated risks are typically a small percentage increase over baseline mortality or morbidity rates for North America; thus PM observations taken over a large number of events, perhaps exceeding 10,000 events (e.g., respiratory deaths per day times the number of monitored days) are desirable for these studies. Atmospheric scientists providing ambient data for these studies need to keep this data-sufficiency requirement in mind.

A number of hypotheses have been advanced to explain how chemical and physical parameters of PM may interact with the body to provide mechanistic explanations for health outcomes. These hypotheses are being evaluated in toxicological studies using laboratory animals and controlled exposures of human subjects. However, to date none of these hypotheses has been proven or excluded. The ultimate test of any of these hypotheses may lead to long-term studies and if so will require close collaboration between atmospheric scientists and health scientists. Such studies will require ambient monitoring of the hypothesized PM constituent or parameter, the PM indicators usually monitored, and other pollutants to test for associations with health indices in large populations.

Air-quality and exposure-modeling considerations

CTM technology has advanced substantially in recent years. However, the capabilities and uncertainties of these models remain bound by the current understanding of atmospheric processes, as well as the ambient and meteorological data and emission information used to run and evaluate them. Current

Table S.8 (adapted from Textbox 2.5). Availability of ambient measurement methods for hypothesized causal elements of PM-induced health effects.

Hypothesized Element(s) - Rationale	**Ambient Air Measurement Capability**
1. **Particle Mass Concentration -** Non-chemically specific mass cardio-pulmonary loading response associated with a complex chemical mixture of wide range of particle size.	Routine
2. **Particle Size/Surface Area -** Response to fine particles with major surface area for adsorption of chemical species and subsequent desorption in lower lungs.	Research
3. **Ultrafine PM -** Animal experiments suggest particles less than 0.1 µm diameter may have a strong physiological effect on the respiratory system.	Research
4. **Metals or Metal Compounds -** Certain metals like V, Cu, Fe, Zn, and Ni have cytotoxic or inflammatory properties. These may catalyze an adverse respiratory response.	Research
5. **Acids -** Acidic particles have been shown to have toxic properties in some animal studies based on hydrogen ion delivered to respiratory surfaces.	Research
6. **Organic Compounds -** There are a large number of organic compounds found in PM, some of which are known to be carcinogenic.	Research
7. **Biogenic Particles -** There are a variety of particles that are found from biogenic sources, including spores, fungi, bacteria and viruses.	Research
8. **Sulfate and Nitrate Salts -** These compounds are believed to be mainly ammonium salts in PM.	Routine
9. **Peroxides -** The presence of peroxides in particles and their toxic properties provide a hypothetical pathway to health effects.	Unavailable
10. **Soot -** Soot particles (or black carbon) potentially can stimulate a toxic response in themselves or carry adsorbed material that can initiate a response.	Research
11. **Copollutant Interactions -** Some epidemiological and/or laboratory exposure studies have suggested that a synergistic response may take place when PM and gases such as SO_2, NO_2, O_3 or CO are present.	Routine

Routine = Instrumentation and protocols are commonly available and can be maintained by typical technical personnel. Research = instrumentation and protocols are available, though either on a limited basis or under development for some constituents, and often require specially trained personnel.

spatial and temporal resolution of CTMs does not yet match the requirements for estimating human exposure. Ambient air-quality modeling typically has been performed with spatial resolution of about four kilometers. However, exposure-modeling inputs require finer spatial resolution.

CTMs will need to be adapted to provide relevant predictions as the health community focuses on specific particle characteristics. CTM modeling capability is currently available for only a few of the proposed causal agents noted in Table S.8. Coordinated work by the two communities will need to continue to advance the development of sophisticated exposure assessment.

New science to improve implementation approaches

All the new science previously recommended to strengthen answers to Policy Questions #1-7 will directly benefit the exposure and health-science communities and the information they supply to policy makers periodically reviewing national standards. Further discussion of the benefits of new science is provided in the conclusion to this Synthesis.

BENEFITS TO THE POLICY COMMUNITY OF NEW SCIENCE

This Assessment concludes with a discussion of major research recommendations (Chapter 11) that will substantially improve the ability of the atmospheric-science community to provide guidance on developing effective implementation plans. They are presented here without priority, as simultaneous progress in each area is needed. Policy makers are faced with the need to make decisions based on imperfect understanding of the atmosphere and human-health impacts. Acting on these recommendations will incrementally reduce the uncertainties in decision making over time. The benefits to the air-quality policy community of achieving the science understanding recommended here are broadly applicable to many aspects of managing PM.

1. Improved understanding of carbonaceous aerosols (emissions, measurements of properties and chemical speciation, and atmospheric transformations):

    - Will improve the ability to tie specific sources and source types back to ambient levels and effects via receptor modeling and CTMs.

    - Will improve understanding, through OC speciation, of secondary organic-aerosol formation processes, leading to higher confidence in PM mass and speciation outputs from CTMs, and further improve CTM predictive capacity for related co-pollutants, especially ground-level ozone.

    - Will provide higher confidence, through more and better OC and BC mass measurements, in the total PM mass-concentration data and improve the quantity and type of information available to the health community to investigate health impacts due to PM.

    - Will enable progress in managing particulate carbon and VOC emissions to be tracked through both emission rates and ambient concentrations.

2. Long-term (multi-decade) monitoring of PM mass, composition, and gas/particle distributions, and gas-phase precursors and copollutants in parallel with health impacts studies:

    - Will provide robust, consistent and credible trend information for direct measurement of progress against ambient standards, in response to PM and precursor-gas emission changes.

    - Will provide critical inputs (continuous PM mass, composition, copollutants) for acute and chronic epidemiological health-impact studies, and establish associations between PM properties and health effects.

    - Will improve understanding of the limiting atmospheric processes that provide insight to targeting precursor-emission reductions to reduce $PM_{2.5}$ mass.

3. Evaluating and further developing the performance of chemical-transport models:

    - Will provide greater confidence in predictions of PM mass and composition in response to multi-pollutant (PM and precursor-gas) emission reductions for all relevant time scales (few days to seasonal to annual) and space scales (from neighborhood to regional to continental scales). Specifically:

        > Determine multisource influences on PM mass and composition for a given location to provide insight on relative source contributions, currently and in the future.

        > Define a benchmark to assess progress against, by predicting how far current implementation plans will go in achieving standards.

    - Will provide predictive inputs to exposure models to estimate population health impacts under future emission-reduction scenarios.

    - Will identify limitations in current emission-inventory information and provide guidance to improving emission information.

- Will provide an essential component of the corroborative science analyses (in conjunction with receptor-modeling techniques and evaluation of ambient data) to provide guidance for implementation planning.

4. Improved emission inventories and emission models:

   - Will improve CTM predictions of PM and composition, and receptor-modeling analysis of current relative source contributions, by supplying more accurate inputs.
   - Will provide better tracking of emission changes as implementation plans take effect.
   - Will provide necessary inputs to link health and visibility impacts back to specific sources and source-types via CTM applications.

5. Commitment to the analysis, synthesis and archiving of ambient data and fostering interactions between atmospheric, climate and health-science communities:

   - Will ensure that the full value of the investment in field studies, monitoring networks, and research programs are realized.
   - Will ensure that the data become available for widespread use by all stakeholders, and members of the atmospheric-, health-, climate-, and ecological-science communities, fostering interdisciplinary collaborations.
   - Will ensure that measurement programs are optimally designed to address multiple information objectives effectively and cost-efficiently.
   - Will focus the different science communities on interdisciplinary projects, working to achieve greater understanding of source-receptor relationships and gas-particle processes to integrate research activities across atmospheric issues.
   - Will be necessary to meet the objectives of long-term monitoring effectively (i.e., Recommendation 2 above).

6. More systematic approaches for integrating diverse types of knowledge on sources, properties, and effects of PM to assist with the development of management practices and tracking their progress towards protecting health:

   - Will enable the air-quality policy community to take full advantage of the investment of financial and intellectual resources, and integrate the knowledge in the context of solving the PM air-quality problem.
   - Will improve the ability of the implementation community to make mid-course corrections as emission management actions unfold over time.
   - Will provide to the public and stakeholders the evidence that the costs and efforts in managing air pollution are leading to better protection of public health and the environment.

These recommendations cannot be acted upon in isolation and progress in each area needs to proceed simultaneously. Each recommendation addresses a critical component in understanding the atmospheric environment and is mutually supportive of contributions to developing better science tools and atmospheric-science analysis. The interconnectedness of the recommendations is illustrated in Table S.9, which also links the anticipated new knowledge to be gained by following the recommendations with an improved ability to respond to Policy Questions 1 through 8.

Table S.9. Policy benefits of the specific research directions: ◍ major benefits, ● modest benefits.

Recommendation	Policy Question							
	1	2	3	4	5	6	7	8
1	●	◍	◍	●	◍			◍
2	◍	◍			◍	◍		◍
3			◍	◍	◍	◍		●
4			◍	◍	●			◍
5	◍	◍	●	●	●	◍	◍	●
6	●	●	◍	●		◍	◍	◍

REFERENCES

Environment Canada, 1997-1998. 1997 Canadian Acid Rain Assessment, Vol. 1-5, Environment Canada, Ottawa.

Hidy, G.M., et al., 2000. Fine particles and oxidant pollution: developing an agenda for cooperative research, Journal of the Air and Waste Management Association 50, 613-632.

IPCC, 2002. Climate Change 2001, Synthesis Report. Intergovernmental Panel on Climate Change, Geneva, Switzerland. http://www.ipcc.ch/pub/reports.htm

NARSTO, 2000. An Assessment of Tropospheric Ozone Pollution– A North American Perspective, NARSTO Management Office (Envair), Pasco, Washington. http://www.cgenv.com/Narsto/

National Research Council (NRC), 1993. Protecting Visibility in National Parks and Wilderness Areas. National Academy Press, Washington, D.C.

National Research Council (NRC), 1998. Research Priorities for Airborne Particulate Matter 1. Immediate Priorities and a Long-Range Research Portfolio, National Academy Press, Washington, D.C., 1998.

National Science and Technology Council, 1998. U.S. National Acid Precipitation Assessment Program (NAPAP) Biennial Report to Congress: An Integrated Assessment, National Science and Technology Council, Washington, D.C.

U.S. EPA, 2000. Deposition of Air Pollutants to the Great Waters, Third Report to Congress, U.S. Environmental Protection Agengy, Washington, D.C.

U.S. National Acid Precipitation Assessment Program (NAPAP), 1991. The U.S. National Acid Precipitation Assessment Program, Vol. 1, U.S. Government Printing Office, Washington DC.

CHAPTER 1

Perspective for Managing PM

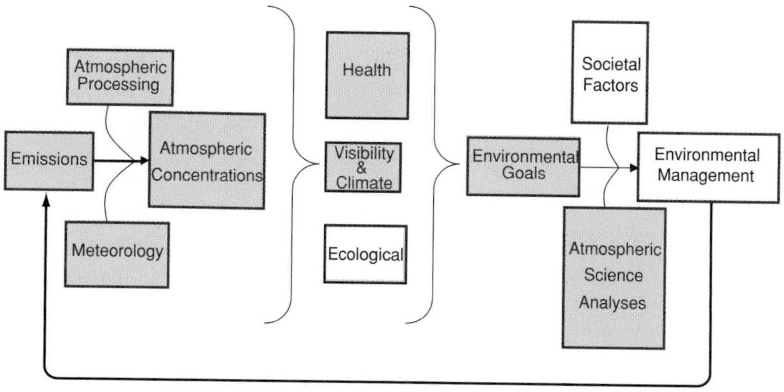

Principal Author: Marjorie Shepherd

Particulate Matter Science for Policy Makers: A NARSTO Assessment is a concise and comprehensive discussion of the current understanding of airborne particulate matter (PM) among atmospheric scientists. Its goal[1] is to provide policy makers who implement current air-quality standards[2] with relevant scientific information.

This chapter introduces the framework used to organize the Assessment, as well as the scientific scope and the regulatory setting. Background discussions of the PM problem address the nature of ambient particles; the scale, sources, and management of the problem; health and visibility impacts; and federal ambient PM standards and policy contexts for Canada, the United States, and Mexico.

The information contained in this Assessment addresses three key applications:

- Implementing fine and coarse PM standards.
- Designing implementation plans for PM management in concert with those for ozone and other ambient pollutants (e.g., NO_x, SO_2, VOCs, and CO), leading to effective multi-pollutant air-quality management approaches.
- Coordinating linkages between the atmospheric- and health-science communities to investigate the causal hypotheses of health impacts, characterize exposure, and contribute to long-term epidemiological studies.

Several other PM-related aspects, although important, are not addressed directly in this Assessment. Climate-change and ecosystem impacts, particularly noteworthy in this regard, are discussed extensively in other reports[3], and thus not covered here.

The interface between health and atmospheric sciences is essential both because of the importance of health effects in establishing air-quality standards and because of the critical need for information exchange between the two scientific communities. Chapter 2 of this Assessment provides a brief overview of PM impacts on human health, to set a context for the remainder of the document.

The framework developed for organizing this Assessment is illustrated in Figure 1.1. Emphasizing the delivery of information for use by the policy community, the three principal components of the framework are:

The Atmospheric Environment: Understanding the relationships among pollutant emissions, their interactions with meteorology and other

---

[1] Specific guidance pertaining to this Assessment's objectives and target audience given to the PM Assessment Team by NARSTO's Executive Assembly and Executive Steering Committee appears in Appendix E.

[2] Canadian, U.S., and Mexican air-quality standards undergo continuous review as a part of ongoing evaluation procedures. While future modifications of these standards are possible, this Assessment accepts them in their current forms to help focus the presentation. Science review documents pertinent to current PM standards are CEPA (2000), Diario Oficial de la Federación (1994), NOM (1993), U.S. EPA (1996), and WGAQOG (1999).

[3] For comprehensive reviews of climate-change and ecosystem impacts in the context of PM refer to IPCC (2002), Environment Canada (1998), and NAPAP (1998).

atmospheric processes, and the resultant atmospheric-pollution loadings.

Exposure and Impacts: Understanding of cause-effect relationships among atmospheric pollutants, exposures and deposition, and impacts on human health, visibility, and ecological systems. This includes the important risk-characterization step directly supporting standard and environmental goal setting.

Analysis and Public Policy: Analysis and decision making by the policy community, considering atmospheric science and societal factors in relation to environmental goals, in formulating emission-reduction programs.

Because of its focus on atmospheric phenomena, this Assessment deals primarily with components on the left of Figure 1.1, although health and visibility effects are discussed.

1.1 THE NATURE OF AMBIENT PARTICLES

Individual particles are characterized by their sizes, shapes, and chemical composition. They can be solid or liquid, spherical or irregularly shaped, and can contain internal mixtures of species and phases. Particle surface composition may differ from bulk composition, and it is common for PM falling within any particular size range to include particles having a variety of shapes and compositions. "Semivolatile" species such as water, NH_4NO_3, and many organic compounds continually undergo exchanges between the gas and condensed phases, often resulting in pronounced compositional changes of PM with time.

PM is typically composed of a complex mixture of chemicals, a mixture strongly dependent on source characteristics. The terminology used to describe PM incorporates source characteristics, formation

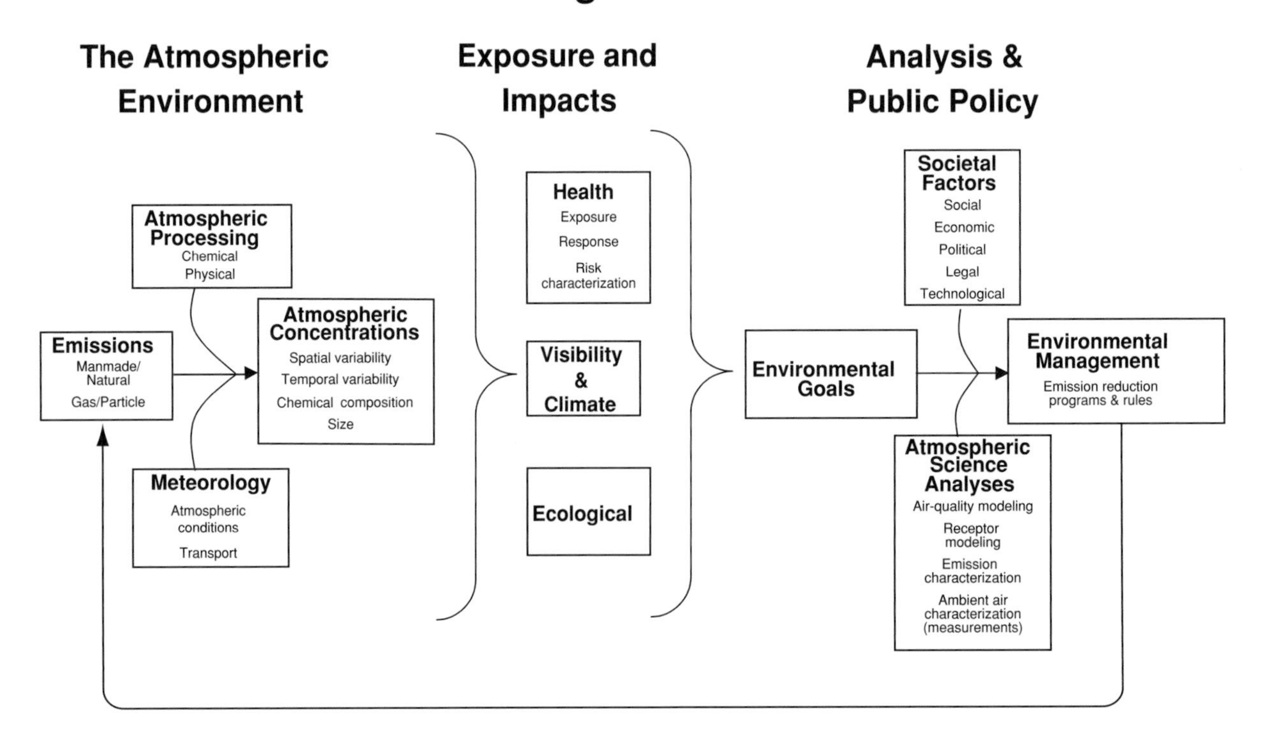

Figure 1.1. Framework for informing PM management.

mechanisms, particle sizes, and chemical composition. Other distinguishing features of PM, such as particle size, origin (primary or secondary), mode, and carbon content, also are used to describe PM. The terminology used in this Assessment is defined in Textbox 1.1 and in the glossary.

As indicated in Table 1.1 and Figure 1.2, PM composition varies between the fine and coarse fractions. The fine fraction is characterized by constituents such as $SO_4^=$, NO_3^-, NH_4^+, metals, and hundreds of different organic carbon compounds. The coarse fraction is characterized by materials typical of the earth's crust (primarily suspended dust and construction debris) and grinding processes (metals).

Because of practical necessity it is common to characterize PM in terms of simplified, aggregate properties. Common characterizations include total mass concentration and total number concentration. Often such characterizations pertain to specified size ranges: PM_x, for example, is used to signify the mass concentration of particles falling within size-range x.

Table 1.1. Comparison of ambient particle fractions.

	Fine (≤2.5 μm)		Coarse (2.5 – 10 μm)
	Ultrafine (<0.1 μm)	**Accumulation (0.1 – 2.5 μm)**	
Formed from:	Combustion, high-temperature processes, and atmospheric reactions		Break-up of large solids/droplets
Formed by:	Nucleation Condensation Coagulation	Condensation Coagulation Evaporation of fog and cloud droplets in which gases have dissolved and reacted	Mechanical disruption (crushing, grinding, and abrasion of surfaces) Evaporation of sprays Suspension of dusts Reactions of gases in or on particles
Composed of:	Sulfate, $SO_4^=$ Black carbon Metal compounds Low-volatility organic compounds	Sulfate, $SO_4^=$ Nitrate, NO_3^- Ammonium, NH_4^+ Hydrogen ion, H^+ Black carbon Large variety of organic compounds Metals: compounds of Pb, Cd, V, Ni, Cu, Zn, Mn, Fe, etc. Particle-bound water	Suspended soil or street dust Fly ash from uncontrolled combustion of coal, oil, and wood Nitrates and chlorides from HNO_3 and HCl Oxides of crustal elements (Si, Al, Ti, and Fe) $CaCO_3$, NaCl, and sea salt Pollen, mold, and fungal spores Plant and animal fragments Tire, brake-pad, and road-wear debris
Typical Atmospheric half-life:	Minutes to hours	Days to weeks	Minutes to hours
Important Removal processes:	Growth into accumulation mode Wet and dry deposition	Wet and dry deposition	Wet and dry deposition
Typical Travel distance:	<1 to 10s of km	100s to 1000s of km	<1 to 10s of km (100s to 1000s in dust storms)

Source: Adapted from Wilson and Suh (1997).

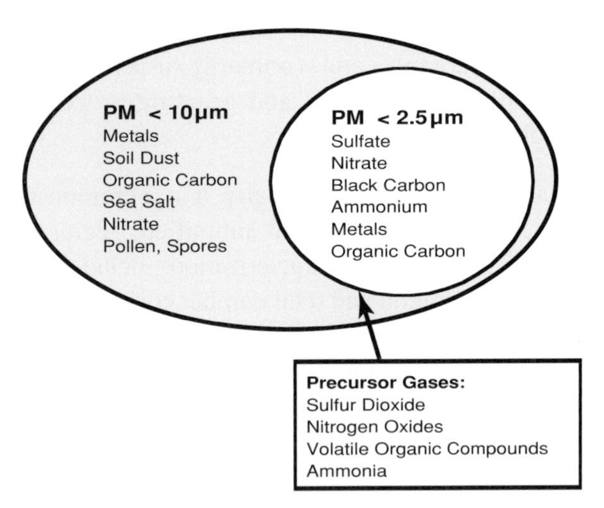

Figure 1.2. Representative composition of PM.

Particles directly emitted to the atmosphere are termed "primary" particles. Such particles can be coarse (e.g., dust, plant debris, pollen, sea spray) or fine (e.g., combustion products, a small fraction of sea spray). "Secondary" PM results from the condensation/deposition of gaseous precursors to the particulate phase. Although direct nucleation from the gas phase definitely is a contributing factor, most secondary material accumulates on pre-existing particles in the 0.1 to 1.0 µm range and typically accounts for significant fractions of the $PM_{2.5}$ mass. Ultrafine particles originate from combustion sources and from homogeneous nucleation of low vapor-pressure compounds. Although ultrafine particles contribute only a small fraction of PM mass, they often account for most of the particle number concentration. It is not unusual, for example, for the particle number concentration to increase by factors to 2 to 10 during nucleation events.

Measurements show that PM can be classified into four modes, which reflect particle origins. These are the coarse mode (particle diameter larger than ~2 µm), the accumulation mode (0.1 to 2 µm), the Aitken mode (0.01 to 0.1 µm), and the nucleation mode (<0.01 µm). Coarse-mode particles are lost rapidly by sedimentation, and nucleation-mode particles grow and coagulate, typically migrating into the Aitken mode as a result. Accumulation-mode particles have longer lifetimes than larger or smaller particles because they settle slowly and have low diffusivities. Because of this, accumulation-mode particles can be transported extended distances – on the order of 1000 km or more – before they are removed from the atmosphere.

The carbon fraction, which on average may represent from 1/5 to 1/2 of $PM_{2.5}$ mass, is currently not well characterized by routine measurements. It typically consists of elemental carbon plus large numbers of different organic species and arises from direct emissions and as a result of transformation and condensation of gaseous organic compounds. Combined with technical difficulties in measuring the carbon fraction, this compositional complexity has resulted in the PM carbon component being poorly understood at present.

Terms such as "black carbon (BC)," "elemental carbon (EC)," and "soot" are often used interchangeably by air-quality, atmospheric, health, and industrial researchers, reflecting the presently high degree of ambiguity in this particular area. Textbox 1.2 describes the terminology used in this Assessment. Scientific considerations associated with this issue are discussed at length in later chapters.

1.2 SCALE, SOURCES, AND MANAGEMENT OF THE PROBLEM

PM mass concentrations vary significantly on both temporal and spatial scales. The highest PM loadings, excluding dust storms and fires, usually occur in major urban centers and small industrial areas where local sources strongly influence air quality. Long-range transport of PM and precursor gases has been documented for ground-level ozone and its precursors, as well as nitrate and sulfate compounds on regional, continental, and transoceanic scales. Figure 1.3 illustrates the potential for PM to be an issue at all spatial scales depending on the relative contributions of precursor gases and primary particles.

Varying source contributions and meteorological influences drive a large dynamic range of seasonal and diurnal variations in PM mass concentration and composition. Consequently PM composition and its relationships to copollutants, especially ground-level ozone, also vary (see Chapter 6). Generally, the long-term behavior of daily-averaged PM mass

Box 1.1. Terminology

The aerosol literature has experienced a proliferation of diffusely defined terminology (e.g., fine particles, total suspended particulate matter, aerosols, superfine particles, ultrafine particles, hyperfine particles, nanoparticles, . . .), a situation which has served to obscure scientific communication in the field. In an attempt to limit this tendency, this Assessment restricts itself to a limited number of terms, defined below. This terminology is consistent with that used by the Intergovernmental Panel on Climate Change.

Suspended particulate matter (PM): Any non-gaseous material (liquid or solid) which, owing to its small gravitational settling rate, remains suspended in the atmosphere for appreciable time periods.

Aerosol: A mixture of suspended PM and its gaseous suspending medium.

The terminology denoting suspended PM subclasses is selected primarily, but not totally, on the basis of physicochemical processes involved in formation and growth of the particles (see Chapter 3, Figure 3.2, which describe four "modes": coarse, accumulation, Aitken, and nucleation).

Ultrafine particles: Particles operationally defined (mainly within the health-sciences community) as those having diameters less than 0.1 μm

Fine particles: Particles operationally defined as those smaller than 2.5 μm aerodynamic diameter. Fine-particle measurements include the accumulation mode (nominally 0.1 to 2.5 μm), where most of the submicron mass is found and, depending on measurement technique, may include ultrafine particles, where most of the particle number concentration is found. Because filter-based $PM_{2.5}$ sampling techniques collect all particles smaller than 2.5 μm, such "fine-particle" samples implicitly include ultrafine particles. However, because the properties and effects of ultrafine particles are different from those of larger particles, it is often useful to separately identify "fine" and "ultrafine" particles as distinct fractions of $PM_{2.5}$.

Coarse particles: Particles extending through the high end of the aerosol size distribution. This Assessment adopts the proposed regulatory definition, which includes those particles between 2.5 μm and 10 μm aerodynamic diameter ($PM_{10\text{-}2.5}$). Many "coarse-particle" sampling techniques also collect particles in the finer ranges. Thus reported data designated as "coarse-particle" data may or may not include contributions from finer modes, and these contributions can be significant.

PM_{10}: The mass concentration of particles smaller than 10 μm. In practice, PM_{10} samplers do not provide perfectly sharp cuts at 10 μm. Instead, size-dependent collection efficiencies typically decrease from 100 percent at ~ 1.5 μm to 0 percent at ~15 μm, and are equal to 50 percent at 10 μm.

Primary PM: PM that is emitted directly to the atmosphere in solid or liquid form.

Secondary PM: PM formed in the atmosphere through condensation/deposition of gaseous precursors.

concentrations exhibits a strongly skewed distribution, dominated by a large number of low values and a small number of high values. At urban sites, with multiple years of data, mean PM mass concentrations can range between 4 and 8 times the estimated natural background levels, implying that anthropogenic activities make substantial contributions to ambient PM loadings in both urban and non-urban locations. The relative source contributions to ambient PM vary both geographically and seasonally. Table 1.2 summarizes the major source contributions to primary and secondary PM.

Effective PM management usually requires consideration of both anthropogenic and natural sources. Within this context, this Assessment identifies two broad source categories: manageable and unmanageable. At one extreme, manageable emissions can be identified readily with industrial sources, commercial operations, power plants, residential dwellings, and transportation. At the other extreme, unmanageable emissions result from volcanic eruptions, windblown sea spray, dust storms from remote arid areas, and forest or brush fires initiated by lightning strikes or spontaneous combustion. Other unmanageable emissions include

Box 1.2. The Carbon Fraction

The "**carbon fraction**" may refer to black carbon and primary organic and/or secondary organic carbon.

Black carbon (BC) is the light-absorbing carbonaceous material in atmospheric particles; most BC is primary. Terms that are sometimes used interchangeably include elemental carbon (EC), soot, and graphitic carbon. BC is chemically complex: it can include light-absorbing solids or liquids, and its composition varies with the source. Conventional analytical methods[a] for BC do not measure its composition, but rather measure parameters that serve as indicators of BC concentrations. For historical reasons, specific measurement methods often use a particular term (e.g., BC, EC, soot, etc.) to identify the measured quantity; for example, BC and EC are often used to indicate optical and thermal measurement methods, respectively. In this Assessment, we use BC generically and apply other terms only when required by context

Organic carbon (OC) includes both primary emissions and secondary organic PM. Secondary organics are produced in the atmosphere by chemical transformations of volatile organic compounds (VOCs) or by the uptake of organic gases by particles.

Organic compounds play an important role in the PM problem:

(a) *Small organic molecules* (C_1 to C_6 compounds) occur in the atmosphere mainly as vapors. These reactive compounds contribute significantly to photochemical reactions leading to ozone generation, as well as the oxidation of SO_2, NO_x and VOCs. SO_2 and NO_x oxidation lead to $SO_4^=$ and HNO_3 formation, and contribute significantly to secondary inorganic PM. The oxidation of certain high molecular-weight VOCs can lead to the formation of products with low volatilities that contribute to secondary organic PM. The most significant of these precursors include aromatics (e.g., toluene, xylenes, trimethyl-benzenes) emitted by transportation and industrial sources and terpenes (e.g., α-and β-pinene, limonene) and the sesquiterpenes emitted by biogenic sources.

(b) *Very high molecular-weight organics* (C_{25} and greater) exist primarily in the condensed phase at ambient temperatures, and thus are emitted mainly as primary organic particles. These particles can be fine or ultrafine, depending on the emission source and ambient conditions.

(c) *Intermediate molecular-weight organic compounds* exhibit a range of volatilities, which depend both on the molecular weight and the molecular structure. Some of these compounds (for example four-ring polycyclic aromatic hydrocarbons) exist in both the gas and particle phases and are referred to as semivolatile compounds.

(d) The nucleation and condensation of high molecular-weight organic vapors can occur as hot combustion gases mix with cool ambient air. The effect of such processes on size distributions of organic particles in the atmosphere depends on the composition of the exhaust gases and on ambient conditions. These processes are described with emission models.

A major challenge for the scientific community is to resolve the carbon estimates predicted by chemical transport models or estimated by receptor models in conjunction with ambient observations. Currently, BC and OC measurements are highly method dependent and only 10 to 20 percent of the organic species can be identified. Typically OC measurements are scaled by an empirical multiplicative factor to account for the molecular form of the organic compounds. This factor may range from 2.0 in an aged air mass to 1.4 near fresh emission sources. Often a typical value of 1.4 is applied leading to potential underestimation of OC particle mass[b]. Such ambiguities could be resolved if there were analytical techniques available to measure organic particle speciation. Measurements of speciation would also provide insights into origins of ambient organic PM and would facilitate evaluations of chemical-transport models.

[a] There are multiple methodologies in use for BC, many of which also estimate total carbon (TC) and OC. By definition TC = OC + BC. TC methodologies agree well, but the OC + BC split may differ by ~30% depending on methodology.

[b] OM (organic matter) is the emerging terminology to represent reported concentrations for OC in recognition that they are manipulated, or adjusted, to account for the presence of hydrogen and oxygen and other non-carbon atoms. This Assessment uses OC synonymously with OM.

sulfur gases from terrestrial and marine sources, NO_x from soil respiration and lightning strikes, and organic vapors from vegetation. Residing between these extremes are sources/events that relate directly or indirectly to human activities, including prescribed burning, vegetation clearing for agriculture, and windblown dust from changes in land use (surface mining, agriculture, and construction). Some "fugitive" particle sources are included in emission inventories, while others are not. Sources that relate indirectly to human activities are considered manageable in the context of this Assessment.

Cases where unmanageable emissions cause exceedances of current PM air-quality standards are uncommon and usually identifiable as special events (e.g., visible smoke or dust plumes). These events can be recorded for explicit consideration in determining compliance with standards. Unmanageable natural emissions are usually a minor (<20 percent) portion of the $PM_{2.5}$ mass during average situations. However, at the present time, confidence in determining the overall importance of natural OC contributions is low.

From a practical standpoint, manageable sources can be classified as being either managed or unmanaged. Managed emitters are those arising from human activity that are subject to permits and/or to emission reduction strategies within a given political jurisdiction. Conversely, unmanaged emitters also arise from human activity, but are not easily subject to such measures. This may be due to the nature of the emissions (e.g., difficult to control) and possibly because they are assumed to be relatively small sources. In this context, measures can be undertaken to reduce unmanaged emissions, albeit at a cost that may not be considered acceptable.

1.3 HEALTH IMPACTS

A considerable and growing body of evidence shows an association between adverse health effects and exposure to ambient levels of PM. Epidemiological studies of large populations have frequently shown a statistical association between elevated levels of PM mass (PM_{10}, $PM_{2.5}$ and $PM_{10-2.5}$). These statistically stable estimates of excess risk most readily emerge in studies of large populations over long time periods. The adverse health outcomes observed in these studies are commonly (though not exclusively) cardiac and respiratory in nature. Such diseases are also relatively common in the general population, and the increased incidences seen in these studies correspond to statistically significant numbers of additional cases in observed populations. While the PM-associated risks represent a small percentage increase over baseline cardiorespiratory disease occurrence in North America, in aggregate, the absolute number of PM-associated health effects is large.

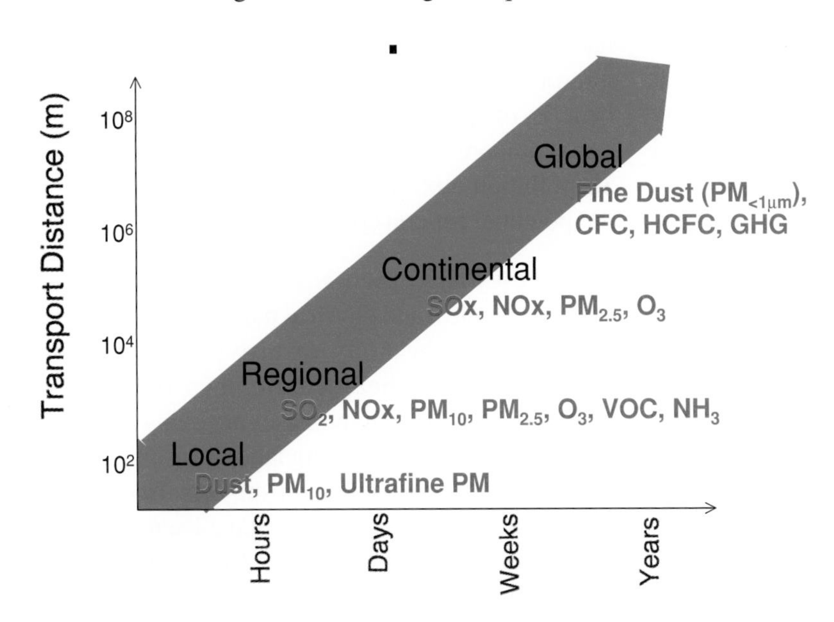

Figure 1.3. Illustrative transport scales for PM and other atmospheric pollutants.

The total personal PM exposure of individuals, which includes both their ambient (outdoor) and indoor environments, is related to the ambient-air content of PM. The ambient-air concentration of PM and other pollutants has been extensively studied as a potentially controllable variable that influences human health.

Increases in adverse heath outcomes have been observed for both short- and long-term

Table 1.2. General descriptions of PM emissions and source types.

	Emissions	General Source Types
Primary	Crustal / Soil Dust / Road Dust	Paved / unpaved roads, vehicle tire and brake wear, construction, agricultural and forestry operations, and high wind events.
Primary	Salt (NaCl)	Oceans, road salt and salt pans / dry lake beds.
Primary	Biogenic material	Pollen, spores and plant waxes.
Primary	Metals	Industrial processes and transportation
Primary	Black carbon	Fossil-fuel combustion (especially diesel engines).
Primary	Semivolatile organic compounds (direct condensation of organic vapors at ambient conditions) and non-volatile organic compounds	Fossil fuel combustion, surface coatings and solvents, cooking, and industrial processes. Forest fires and biomass burning.
Secondary	Semivolatile and volatile organic compounds (forming secondary organic aerosols)	
Secondary	Sulfur dioxide (forming sulfate particles)	Electrical utilities, transportation, mining and smelting, and industrial processes.
Secondary	Ammonia (contributing to formation of ammonium sulfate and ammonium nitrate)	Agriculture and animal husbandry, with minimal contributions from transportation and industrial processes.
Secondary	Nitrogen oxides (forming ammonium nitrate with ammonia)	All types of fossil-fuel combustion, and to a minor degree microbial processes in soils.

exposures to PM, and for a range of ambient PM indicators including total suspended particles (TSP), PM_{10}, $PM_{10-2.5}$ and $PM_{2.5}$. A higher potency has typically been observed for $PM_{2.5}$ compared to other PM indicators, consistent with the concept that smaller particles penetrate further down the respiratory tract. Nonetheless, the epidemiological literature does demonstrate some level of association with the larger PM metrics.

Thresholds in ambient PM concentration-health response relationships generally have not been observed in epidemiology studies. Some evidence suggests that there may be regional differences in the potency of PM indicators, possibly a reflection of regional differences in PM composition. Additionally, some evidence exists linking increases in adverse health outcomes to specific sources of PM.

Evidence of improved health is also associated with reductions in PM exposure. Studies have been both opportunistic in nature, taking advantage of significant reductions in PM pollution over short time periods to examine mortality and other health endpoints, or have been designed to follow individuals or groups as their exposure to air pollution has changed. Such studies, while requiring careful interpretation, offer the opportunity to examine the benefits of specific air-quality interventions, as well as to compare the benefits of specific changes in the air-pollution mix. Pursuit of these situations offers the potential for considerable insight.

Hypotheses have been advanced to explain how various chemical and physical parameters of PM may interact with the body to provide mechanistic explanations for health outcomes (Chapter 2, Textbox 2.6). These hypotheses are being evaluated in toxicological studies using laboratory animals and controlled exposure of human subjects. However, to date none of these hypotheses have been proven or eliminated from consideration. Continuing tests of these hypotheses will require collaboration between aerosol scientists and health scientists in order to identify and characterize the hypothesized PM constituent or parameter, the PM indicators usually monitored, and other pollutants. Such collaboration will enhance efforts in the fields of toxicology, clinical studies, and in both short-term and long-term epidemiological investigations.

The scientific understanding of PM health effects has been greatly facilitated using a "pollutant source-to-receptor response" paradigm as shown in Figure 1.4. This paradigm is useful for identifying information needs through targeted research efforts, for integrating information to establish ambient air-quality standards, and the concomitant control strategies for achieving such standards. Further progress in reducing the uncertainties in the linkages within the paradigm will require continued and closer collaboration between atmospheric and health scientists.

1.4 VISIBILITY IMPACTS

Perhaps the most common symptom of air pollution recognizable to the public is obscured visibility, usually referred to as haze. Optically, PM interferes with visibility by absorbing or scattering (i.e., reflecting) visible light. For example, BC absorbs light and $SO_4^=$ is highly reflective. Light scattering is roughly proportional to the mass concentration of fine particles, while light absorption is roughly proportional to the mass concentration of the light-absorbing species such as BC. This degradation in visibility is manifested as a reduction in the distance to which one can see and a decrease in the apparent contrast and color of distant objects, causing a washed-out or hazy appearance. The relationship between visibility and PM is discussed further in Chapter 9.

As a result of past efforts to reduce primary particle emissions from sources, observation of plumes from individual sources has declined significantly since the 1970s. However, haze is a concern for urban areas such as Mexico City and Los Angeles, and over much larger geographic regions in parts of North America, especially over the eastern half. In the East, an accumulation of ambient fine particles frequently results in haze extending over thousands of square kilometers. The haze problem, especially in the Desert Southwest of the United States and in the Rocky Mountain areas of Canada and the United States where visibility was once considered pristine - limited only by the "physical limit" of the light-scattering properties of air - is considered a regional issue, the result of additive contributions of multiple sources.

Visibility improvement, although secondary in importance to human health as a driver for air pollution regulation across North America, nevertheless will be a continuing consideration in the evolution of air-quality goals and standards, and is a continuing aesthetic concern of the public.

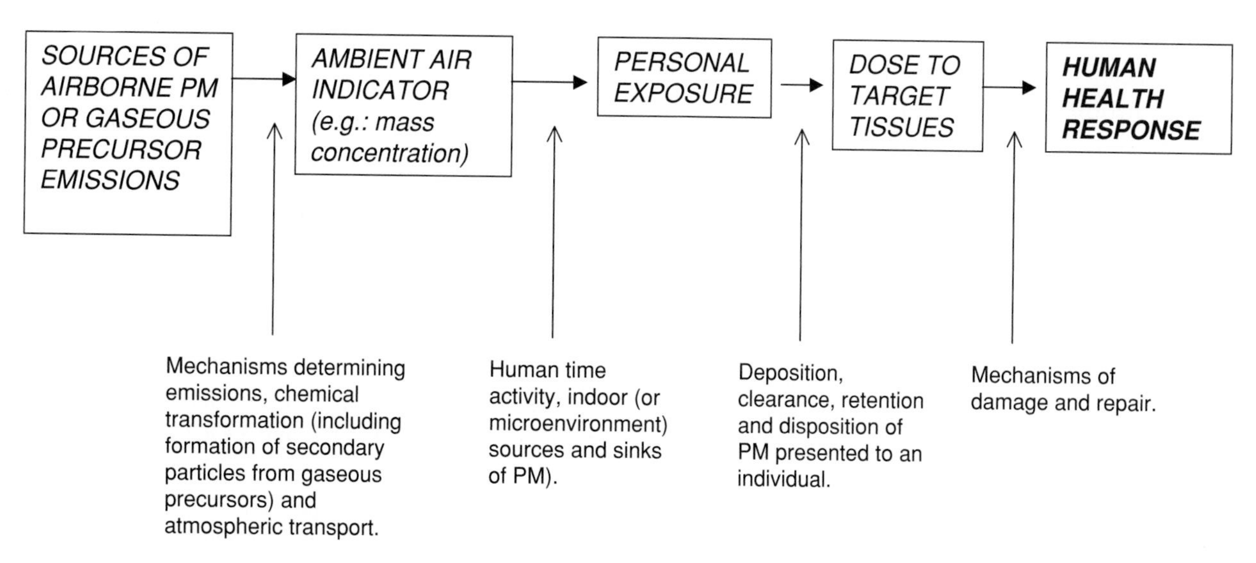

Figure 1.4. NRC framework for evaluating risks extending from pollutant sources to responses (NRC, 1998).

1.5 FEDERAL AMBIENT PM STANDARDS AND POLICY CONTEXTS

Canada, the United States, and Mexico have all adopted formal national ambient air-quality standards for particles[4]. In each case, these standards represent ambient concentrations and target achievement dates. National ambient air-quality standards are summarized in Table 1.3 and discussed more fully in the following sections.

1.5.1 Canada

In Canada, a federal/provincial Working Group on Air Quality Objectives and Guidelines prepares the science assessments outlining the current state of knowledge on effects, dose-response relationships, and risk information resulting in science-based recommendations for ambient objectives or standards. For particles, Canada Wide Standards (CWS) for PM were developed within a multi-stakeholder framework under the Canadian Council of Ministers for the Environment[5]. These standards represent a balance of factors (science, economics, technological feasibility, and societal interests) designed to select an achievable standard that will reduce human-health and environmental impacts. These standards may not be fully protective of public health and will be reviewed in 2005. Within the CWS is an obligation for Keeping Clean Areas Clean. Jurisdictions recognize that polluting "up to a limit" is not acceptable and that the best strategy to avoid future problems is keeping clean areas clean. Jurisdictions are working with stakeholders and the public to establish implementation programs that apply pollution prevention and best management practices.

In June 2000, Ministers agreed to a set of initial actions which will be undertaken by provincial/territorial and federal governments, to reduce pollutants that cause PM and ozone. The industrial sectors included in these actions have been selected because they are significant emitters of the relevant precursors are common to most jurisdictions and affect many communities across Canada, and because effective action requires a multi-jurisdictional approach. The industrial sectors covered include: transportation, residential wood burning, pulp and paper, lumber and allied wood products, electric power, iron and steel, base metals smelting, and concrete batch and asphalt mix plants. The delivery date for completed reduction measures under the initial actions is 2005.

The CWS for PM will be reviewed as follows:

- By the end of 2005 complete additional scientific, technological and economic analysis: 1) to reduce information gaps and uncertainties and revise or supplement the PM and ozone CWSs as appropriate for the year 2015, and 2) report to Ministers in 2003 on the findings of the PM and ozone environmental and health science, including a recommendation on a $PM_{10-2.5}$ CWS.
- By the end of year 2010 assess the need and, if appropriate, revise the CWSs for PM and ozone for target years beyond 2015.

Under Part 1, Section 64 (c) of the Canadian Environmental Protection Act (CEPA, 1999), PM_{10} and particularly $PM_{2.5}$, has been declared a toxic substance. This declaration states that "a substance is toxic if it is entering or may enter the environment in a quantity or concentration or under conditions that constitute or may constitute a danger in Canada to human life or health." The declaration is based principally on the sufficient weight of evidence of mortality and morbidity in the general population exposed to ambient concentrations of PM_{10} and $PM_{2.5}$ examined in recent extensive epidemiological analyses in Canada and in other countries (at ambient concentrations currently occurring in Canada), as well as on supporting data in experimental and controlled human exposure studies. To ensure that the Canadian federal government is fully able to manage PM in the context of the CEPA, the gaseous precursors (SO_2, NO_x, NH_3 and VOCs) to secondary

[4] In some countries, ambient air standards have also been adopted by other jurisdictions (e.g., at the municipal, state or provincial level). This assessment is also intended to speak to implementation at that level, though only national regulatory backgrounds are discussed.
[5] For additional information on Canada Wide Standards for Particulate Matter and Ozone refer to : http://www.ccme.ca/initiatives/standards.html?category_id=5.

Table 1.3. Existing PM standards and their implementation timetables.[a]

Country	Current Standards	Implementation Timing
Canada	$PM_{2.5}$ of 30 µg/m^3 24-hr averaging time to be achieved by 2010, achievement to be based upon the 98th percentile ambient measurement annually, measured over three consecutive years.[b]	In June 2000, Ministers agreed to a set of initial actions to reduce pollutants that cause PM and ozone, which will be undertaken by provincial/territorial and federal governments. The delivery date for completed reduction measures under the initial actions is 2005.
United States	For $PM_{2.5}$, the 3-yr average of the 98thpercentile of 24-hr average concentrations at each population-oriented monitor must not exceed 65 µg/m^3, and the 3-yr average of the annual arithmetic mean concentration from single or multiple community-oriented monitors must not exceed 15 µg/m^3. [c] For PM_{10}, the fourth highest 24-hr concentration over 3 years must not exceed 150 µg/m^3 , and the 3-yr average annual mean concentration must not exceed 50 µg/m^3.[c] Improve visibility on the haziest days and ensure no degradation on the clearest days, with the ultimate goal of reaching natural background conditions in 60 years.	For PM no later than the end of 2005, the U.S. EPA plans to designate non-attainment areas. From the date of designation, states will have 3 years to develop and submit implementation plans for attainment with deadlines to be set as early as 5 years and as late as 12 years after designation. To achieve visibility goals, states are required to develop an initial 10-15 year plan, in the same time frame as their PM state implementation plans, that is to be revised every 10 years starting in 2018.
Mexico	PM_{10} maximum allowable concentration of 150 µg/m ^{3} 24-hr mean and an annual arithmetic mean of 50 µg/m^3.[d] Total Suspended Particles (TSP) maximum allowable concentration of 260 µg/m^3 24-hr mean and an annual arithmetic average of 75 µg/m ^{3}.[d]	The Secretariat of Health has set air-quality Normas Oficiales Mexicanas (NOMs) that now must be met by all jurisdictions of the country.

[a] Some states and provinces have standards more stringent than national standards, for example California's PM standards.
[b] By the end of 2005 complete analysis to reduce information gaps and uncertainties and revise or supplement the PM and ozone CWSs as appropriate for the year 2015; and report to Ministers in 2003 on the findings of the PM and Ozone environmental and health science, including a recommendation on a $PM_{10\text{-}2.5}$ Canada-Wide-Standard.
[c] The Clean Air Act requires that the NAAQS be reviewed every 5 years and a review process for the PM NAAQS is underway. As part of this review process, an updated Criteria Document, which summarizes the relevant scientific information about the sources, transformations, concentrations, and health and environmental impacts of PM, has been developed. Based on the information in the updated Criteria Document and supplementary analyses, a decision whether to retain or revise the NAAQS is now expected in late 2003 to early 2004.
[d] The Secretariat of Health is in the process of reviewing existing NOMs. The new NOM for particulate matter is anticipated to include three size fractions: TSP, PM_{10} and $PM_{2.5}$.

PM formation are also being reviewed in the context of the CEPA toxic definition. Canada is in the process of developing risk-management instruments in response to the CEPA declaration.

1.5.2 United States

Under the U.S. Clean Air Act, the U.S. EPA has set National Ambient Air Quality Standards (NAAQS) for airborne particles since 1971.[6] Over time, the standards have evolved with an increasing focus on particle size, increasing stringency for smaller particles, and a move toward more robust statistical forms.

With advice from the Clean Air Scientific Advisory Committee, primary standards have been set for $PM_{2.5}$ and PM_{10} based upon scientific criteria indicating the kind and extent of all identifiable effects on public health that may be expected, and allowing an adequate margin of safety requisite to protect public health. The PM NAAQS reflect consideration of the fine and coarse fraction as separate pollutants, both of which are related to adverse health impacts. The standards reflect information from community studies showing statistical associations between health effects and both short- and long-term exposures to PM.

The Clean Air Act requires that the NAAQS be reviewed every 5 years and a review process for the PM NAAQS is underway. As part of this review process, an updated Criteria Document, which summarizes the relevant scientific information about the sources, transformations, concentrations, and health and environmental impacts of PM, has been developed.[7] Based on the information in the updated Criteria Document and supplementary analyses, a decision whether to retain or revise the NAAQS is now expected in late 2003 to early 2004.

After the U.S. EPA establishes a PM NAAQS and a 3-year period of monitoring leads to a determination of attainment status, states found in non-attainment are responsible for developing State Implementation Plans (SIPs) that demonstrate how they will attain and maintain air quality in compliance with the NAAQS. Following the 1997 PM NAAQS promulgation, a new national monitoring network using a reference monitor for $PM_{2.5}$ had to be installed across the country. The 3-year monitoring period began in 1998 and 1999. Attainment designations are expected to coincide with the completion of the 5-year review cycle of the PM NAAQS so that no later than the end of 2005, the U.S. EPA plans to designate areas as "attainment," "non-attainment," or "unclassifiable" with respect to the PM NAAQS. From the date of designation, states will have three years to develop and submit SIPs for non-attainment areas. Attainment dates for the PM NAAQS have not been set; however, the Clean Air Act allows attainment deadlines to be set as early as 5 years and as late as 12 years after designation.

In addition to the PM NAAQS, the U.S. EPA has also adopted regulations to decrease regional haze and improve visibility in national parks and wilderness areas.[8] These regulations require states to develop progress goals for each protected area that will improve visibility on the haziest days and ensure no degradation on the clearest days, with the ultimate goal of reaching natural background conditions in 60 years. To achieve these goals, states are required to develop an initial 10- to 15-year plan that is to be revised every 10 years starting in 2018.

The timelines for the development of PM SIPs and regional-haze plans are linked. Although the exact schedule for implementation is not finalized, the regional-haze plans could be due as early as 2004 and PM SIPs as early as 2006.

The Prevention of Significant Deterioration (PSD) program, aimed at keeping clean areas clean though new source review and approval procedures, is likely to change from the basis now used for PM analysis about the time PM SIPs are due. Current guidance directs that analysis be done using PM_{10} as a surrogate, given the absence of sufficient $PM_{2.5}$ monitoring, emissions, and modeling information. It is expected this information will become available over the next several years.

1.5.3 Mexico

All the environmental legislation in Mexico is covered through the General Law of Ecological Equilibrium for the Protection of the Environment by different regulations and official norms. The regulations pertaining to the Protection and Control

[6] For additional information on U.S. NAAQS for PM and ozone refer to: http://www.epa.gov/ttn/oarpg/naaqsfin/.

[7] Air Quality Criteria for Particulate Matter. Research Triangle Park, NC, USA: Office of Research and Development, U.S. Environmental Protection Agency.

[8] For additional information on the U.S. regional haze regulations, refer to: http://www.epa.gov/oar/vis/.

of Atmospheric Pollution indicate the goal that air quality should be satisfactory in all human-inhabited areas of the country illustrating the Secretariat of Health's concern for human health, and responsibility for coordinating and enforcing needed legislation.

All of the criteria pollutants have an official or legal norm (Normas Oficiales Mexicanas, NOMs)[9]. These explain the possible effects on the environment and public health while at the same time setting a maximum allowable limit, usually expressed as an ambient concentration limit.

1.5.4 Implementation of Scientific Information into Decision Making

In developing environmental policy, air-quality decision makers follow a decision path:

- Identify the problem
- Set a target
- Design a management plan
- Implement the plan
- Track success
- Revise the target
- Revise the management plan.

The path is applicable across regulatory systems and requires scientific information. To meet the decision-maker's information needs, an iterative flow of information between the science and policy communities should exist. Such an iterative process is illustrated in Figure 1.5. The NARSTO Ozone Assessment (NARSTO, 2000) identified three major steps in the context of accountability in air-quality management:

1. Verification of the effectiveness of controls
2. Verification that air quality is responding to emission changes

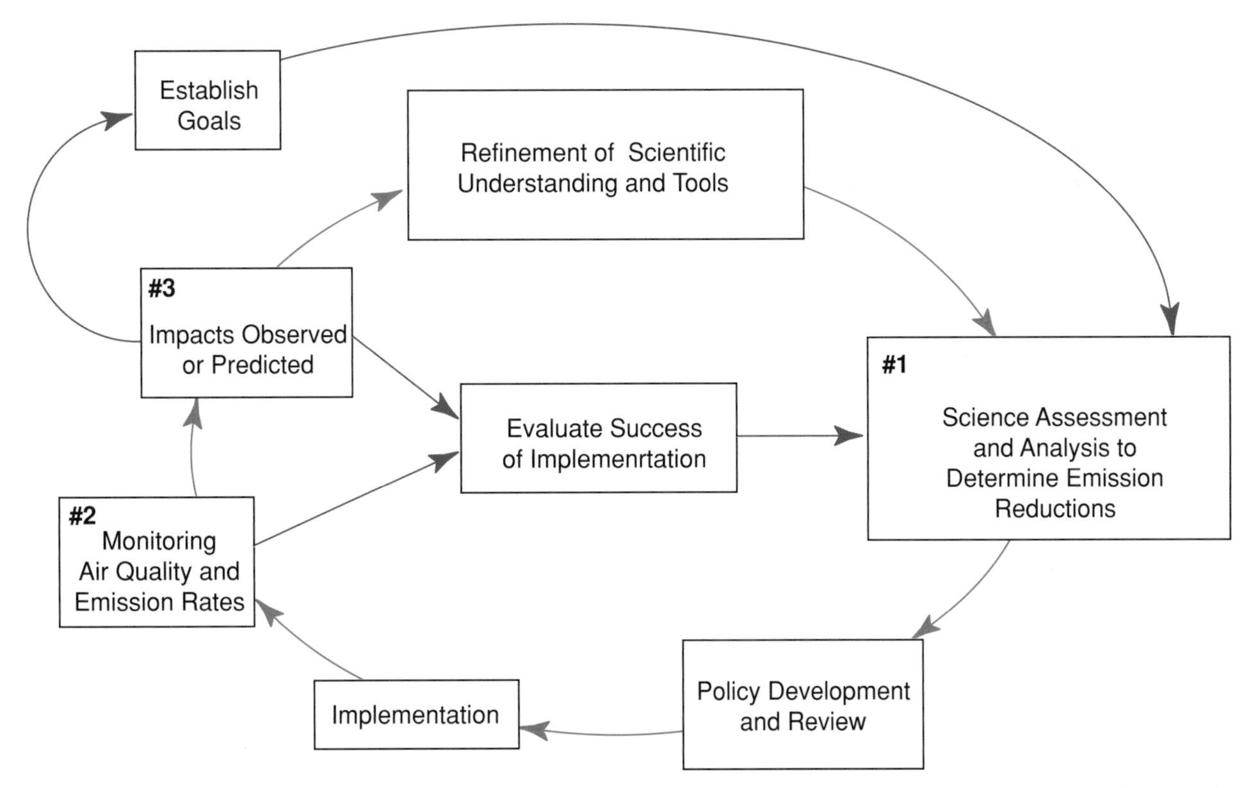

Figure 1.5. Iterative communication for managing air quality to reduce health and environmental impacts in the context of current scientific understanding and accountability.

[9] Information on Mexican air quality NOMs can be found at http://www.ssa.gob.mx/dirgsa/NOM.HTML.

3. Verification that abatement measures have results in health and environmental benefits.

These accountability steps also need to be iterative, and should include evaluation of the overall impact of the implementation program on ambient PM, provide ongoing application of science tools to refine implementation programs, and inform the public about progress of specific source types in reducing emissions ensuring a public accountability for the improvement of air quality. The communication regarding this progress in reducing emission rates can be complemented by discussion of improvements in public health and the environment according to the paradigm presented in Figure 1.4. An accountability framework that will enable measurement of progress towards the goal of protecting human health is needed.

The iterative process moves through stages of gathering new scientific information, performing assessments, developing policy, tracking outcomes, improving the knowledge base, and revising the policies and implementation plans and is consistent with the framework presented in Figure 1.1. Based upon North American experience in managing the acid rain and ground-level ozone problems, this cycle typically takes 8 to 10 years. The PM observations begun in the mid-1980s in Canada and the United States have improved upon previous observations. Ambient air-quality standards have been evolving with the most recent ones set in 1997 (United States) and 2000 (Canada) and implementation plans are being designed now. Thus, the PM management process has moved about two-thirds of the way through the first iteration.

History has shown that the results of major field studies and special projects continue to feed the development of science tools and air-quality policy 10 to 15 years after the specific studies are completed as the associated knowledge is assimilated across the community. Thus it is important to take a long-term view to supporting innovative research and exploring new approaches to improving scientific understanding. In the scientific discussions in Chapters 2 through 9 in this Assessment, such suggested improvements are highlighted in italicized text; a full discussion of of recommendations, along with cross-links to the Assessment's content is provided in Chapter 11.

The optimal mechanism for ensuring that policy-makers have the most recent scientific results is to improve the dialogue between the scientific research community and the policy community, making it as continuous as possible. This communication has been facilitated by public- and private-sector partnerships, which have been effective in bringing forward a variety of scientific views and collective understanding of the science at early stages before emissions management decisions are put in place. Continuation of the public-sector/private-sector partnerships for effective and timely communication should be continued and fostered. Over the past decade there have been major changes in the science/policy dialogue on air quality, which like air-quality management, is following a philosophy of "continual improvement."

1.6 THE STRUCTURE OF THE ASSESSMENT

This Assessment is organized along disciplinary lines, as illustrated in the framework of Figure 1.1. Topics addressed in the Assessment are:

- **Gaseous and particulate pollution emissions**, from both natural and anthropogenic sources, including the temporal and spatial nature of these emissions. (Chapter 4)

- **Atmospheric processes,** the chemical and physical reactions and interactions that take place in the atmosphere, affecting the composition and distribution of gaseous and particulate pollutants. These processes are influenced by meteorology and topography and determine the short- and long-term chemical composition of the atmosphere. (Chapter 3)

- **Meteorology**, the fundamental atmospheric conditions (temperature, solar irradiation, humidity, precipitation, wind speed and direction) that determine the dispersion and transport of natural and man-made pollutant species throughout the global atmosphere. (Chapter 3)

- **Atmospheric concentrations**, the spatial and temporal characterization, primarily at the surface, of particle composition and size, and the magnitude of particle and precursor or copollutant gas concentrations. (Chapter 6)
- **Visibility and Climate** impacts that occur as a result of PM absorbing and/or scattering visible radiation, obscuring one's ability to view objects or scenes and perturbing the Earth's radiation balance. Such impacts result directly from the particles themselves, as well as indirectly as the result of PM effects on cloud-formation processes. (Chapter 9)
- **Atmospheric science analyses**, the application of current understanding by tools, such as emission characterization, ambient characterization, receptor modeling, and chemical-transport modeling, to elucidate the relative contributions of different sources, both local and distant, to a region's PM problem. (Chapters 4, 5, 6, 7 and 8)

In addition to these scientific components, this Assessment presents region-specific analyses (Chapter 10), which examine nine North American regions and provide individual PM problem overviews for these areas. These overviews are framed as "conceptual models" of PM source-receptor behavior as shown schematically in Figure 1.6. Intended to compile and summarize the most advanced and pertinent knowledge of source-receptor behavior, these conceptual models are the outcome of the atmospheric analyses and combine information from a variety of sources, including emission

Box 1.3. The Value of a Conceptual Model

Conceptual models identify the most effective approaches for managing PM to achieve air-quality standards, potentially significant disbenefits (temporary or otherwise), copollutant responses, and opportunities for copollutant reductions, by facilitating an understanding of the balance between limiting chemical and physical processes in each source environment or airshed.

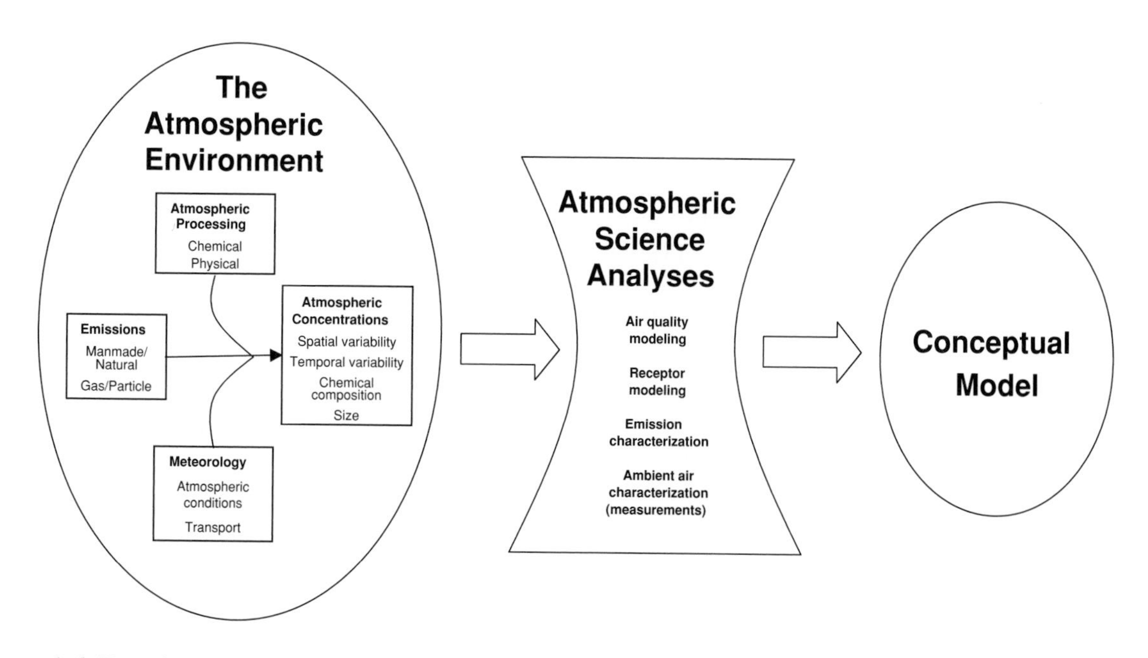

Figure 1.6. Development of conceptual models of PM behavior, based upon the application of science tools in the context of the best understanding of the atmospheric environment.

characterizations, air-quality measurements, and meteorological observations as well as analyses from chemical-transport and receptor model investigations.

1.7 REFERENCES

Canadian Environmental Protection Act (CEPA), 2000. Priority Substances List Assessment Report: Respirable Particulate Matter Less Than or Equal to 10 Microns, Environment Canada and Health Canada, 73 pp., May 2000.

Diario Oficial de la Federación, Mexico. 1994. CDXCV 16, 23 dic. p. 3b and 4ª

NOM 024 - SSA1-1993, NOM 025 - SSA1-1993

Environment Canada,1998. 1997 Canadian Acid Rain Assessment, Vol.1-5, Environment Canada, Ottawa.

IPCC, 2002. Climate change 2001. The scientific basis - Summary for policymakers and technical summary of the working group I contribution to the IPCC third assessment report. Intergovernmental Panel on Climate Change Secretariat, World Meteorological Organization, Geneva, Switzerland.

National Science and Technology Council, 1998. U.S. National Acid Precipitation Assessment Program (NAPAP) Biennial Report to Congress: An Integrated Assessment. National Science and Technology Council, Washington, DC.

NARSTO, 2000. An assessment of tropospheric ozone pollution – a North American perspective. NARSTO Management Coordinator's Office (Envair), Pasco, Washington. http://www.cgenv.com/Narsto

National Research Council (NRC), 1998. Research Priorities for Airborne Particulate Matter 1. Immediate Priorities and a Long-Range Research Portfolio. National Academy Press, Washington, D.C.

U.S. EPA, 1996. Air Quality Criteria for Particulate Matter. Office of Research and Development, EPA/600/P-95/001aF, Volumes I – III.

Wilson, W.E., Suh, H.H., 1997. Fine particles and coarse particles: concentration relationships relevant to epidemiologic studies. Journal of the Air and Waste Management Association 47 1238-1249.

Federal-Provincial Working Group on Air Quality Objectives and Guidelines (WGAQOG), 1999. National Ambient Air Quality Objectives for Particulate Matter: Science Assessment, Environment Canada and Health Canada, ISBN 0-662-26715-X, Catalogue No. H46-2/98-220-1E, December 1998.

CHAPTER 2

Health Context for Management of Particulate Matter

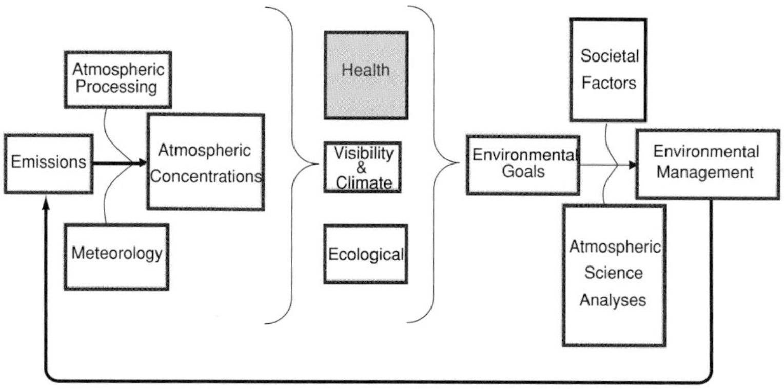

Principal Authors: Roger McClellan and Barry Jessiman

2.1 OVERVIEW

This chapter's objective is to provide a health context for this Assessment which, as noted in Chapter 1, focuses on atmospheric-science considerations related to the management of airborne PM. PM, unspecified as to chemical composition, is of major concern for producing health effects and, thus, subject to government regulation in many countries including Canada, Mexico, and the United States. There is a considerable and growing body of evidence showing an association between increases in adverse health effects, especially of the cardiac and respiratory systems, and elevated levels of PM in air. This evidence has served as a driver for promulgation of new standards and guidance for PM and the development of related control strategies.

The intent of this chapter is to provide the reader with an awareness of the critical interfaces between the atmospheric sciences and health sciences, and to identify opportunities for improved interactions and flow of information between the communities in both directions (opportunities appear in italic text). This awareness should facilitate the development of new information to be considered in future reviews of the health effects of PM and in the development of effective and efficient control strategies to minimize human health impacts of PM pollution.

The chapter builds on the overall framework (Figure 1.1) used for this Assessment by utilizing an expanded conceptual framework that links sources of air pollution to human-health responses (Figure 2.1). This paradigm is similar to that proposed for research on PM by the NRC (1998), shown in Figure 1.4.

The subsequent material in this chapter provides 1) a historical perspective on health and air pollution, 2) a review of exposure assessment and dosimetry, 3) sources of information for understanding the health effects of PM, emphasizing epidemiological approaches, 4) a review of key epidemiological findings, 5) consideration of supporting toxicological evidence, and 6) a summary of policy-relevant findings and opportunities for advancing the science in this field.

Specific studies are cited to provide key information and illustrate important concepts. However, this chapter is not intended to be a comprehensive review of the substantial literature available on the health effects of PM and its co-pollutants. The reader interested in more detailed and comprehensive coverage of PM health effects will find several recent publications useful (California EPA, 2002; CEPA/FPAC, 1999, 2002; Gehr and Heyder, 2000; Holgate et al., 1999; McClellan, 1999, 2002; Molina and Molina, 2002; NRC, 1998, 1999, 2001, 2002 and 2004; Phalen, 2002; U.S. EPA, 1996a, 1996b; 2003a, 2003b; Wilson and Spengler, 1996).

2.2 HISTORICAL PERSPECTIVE

The history of airborne materials impacting on health has been reviewed (Brimblecombe, 1999). Air-pollution problems were recognized very early in many industrial areas, but probably nowhere was the problem as apparent as in London. The high levels of pollution combined with the city's notorious fog created a serious problem and the term "smog"– from contraction of smoke and fog.

In the late 1800s and early 1900s, a series of incidents drew attention to air pollution as a serious health issue. The first observations were on acute air-pollution episodes, periods in which there was a short-term increase in ambient pollution greater than would normally be expected as part of day-to-day variation (Anderson, 1999). In December 1930, stable atmospheric conditions in the Meuse Valley in Belgium resulted in pollution causing 60 deaths and illnesses of hundreds of people. In October 1948, particularly calm and stable meteorology in Donora, Pennsylvania, resulted in severe air pollution and increased morbidity and mortality from respiratory effects. A serious pollution event with serious health impacts occurred in Pozo Rico, Mexico in 1950. In December 1952, smog in London resulted in about 4,000 deaths, principally among the infirm, the old, or those with respiratory disease. Another major smog episode in London in December 1962 resulted in 340 deaths.

The post-World War II era brought changes that impacted air quality. The primary fuel for railroad locomotives shifted from coal to diesel oil. Many businesses shifted from the use of coal to oil. The use of gasoline-powered passenger vehicles began to grow exponentially. Trucks became a major mode of long-distance transport of goods and there was a shift from gasoline to diesel engines for heavy-duty applications. Additionally at this time, large generating stations, using coal as a fuel, were built to meet the demand for electrical power. Tall stacks soon came in vogue as a means of releasing emissions above the inversion layer and enhancing dispersion.

In the 1960s, studies were reported indicating that ambient air pollution had effects beyond those associated with acute pollution episodes. Martin (1964) reported that overall annual mortality, as opposed to acute episodic mortality, in the Greater London area was significantly related to smoke levels. Holland and Reid (1965) conducted a cross-sectional study of lung function of postal workers in London compared to that of others in country towns with lower pollution levels. The socioeconomic level of the workers was the same and smoking intensity could be characterized. There was a clear decrement in function for the London workers compared to provincial workers.

It is informative to consider briefly the historical evolution of the metrics used in studies of air pollution and, especially, the role of PM. In the earliest episodes in the late 1800s, a typical metric was the frequency of "fog" or "stinking-fog" days. The latter were days of especially high pollution with the associated odor of sulfurous compounds. For episodes occurring in the mid-1900s, the air-pollution metrics were "Black Smoke" (measured by the reflectance method) and SO_2. The reported values were extraordinarily high as illustrated by the 1952 London episode. For it, mean maximum levels of Black Smoke were 1600 $\mu g/m^3$ with a single maximum value of 4460 $\mu g/m^3$ and for SO_2 a mean maximum level of 700 $\mu g/m^3$ and a single maximum of 1340 $\mu g/m^3$. Other studies in the mid-1900s, for example of the 1953 New York City episode, related at least in part to forest fires, reported smoke levels measured as coefficients of haze (Greenberg et al., 1962a,b).

A report of the 1975 Pittsburgh episode used Total Suspended Particulates (TSP) as an air-pollution metric (with a maximum value of 700 $\mu g/m^3$) along with SO_2 (130 ppb) (Stebbings et al., 1976). The first report from Steubenville for studies of air-pollution episodes conducted in 1978 and 1979 reported TSP values of 422 and 271 $\mu g/m^3$ and SO_2 of 281 and 455 $\mu g/m^3$ (Dockery et al., 1982).

The first National Ambient Air-Quality Standard (NAAQS) for PM in the United States was set in 1971 (Federal Register, 1971), based on a Department of Health Education and Welfare criteria document (U.S. HEW, 1969). It used TSP as an indicator and relied heavily on the acute pollution episode data from England with extrapolation from Black Smoke measurements.

In the 1950s and 1960s, concern for the health effects of airborne radioactive material and the availability of radioactive tracer technology served as a stimulus to the development of a large body of literature on the relationship between particle size and respiratory tract deposition (ICRP, 1994; NCRP, 1997). These data served to focus attention on particles of a size that were most readily inhaled. It was during this period the concept of aerodynamic size emerged and new instruments were developed for size-selective monitoring including monitors for PM_{15}, PM_{10}, $PM_{2.5}$ and PM_1.

The evolution of PM metrics continued as new studies were conducted on the health effects of both acute and chronic exposure. One of the first acute-exposure epidemiological studies using PM_{10} as an air-pollutant metric was conducted in 1985-86 (Dockery et al., 1992). Sulfate was reported as an air-pollutant metric in a study by Bates and Sizto (1987). Johnson et al. (1990) reported a study of lung function in children that used $PM_{2.5}$ as an exposure metric in addition to TSP and PM_{10}.

A review of the chronic exposure studies also reveals a similar shift in the PM metric. Lave and Seskin (1970), in their classic population-based study of mortality that considered U.S. Standard Metropolitan Statistical Areas, reported an association of mortality with TSP and $SO_4^{=}$. Miller et al. (1979), in a landmark paper, discussed the rationale for moving to size-selective mass-based standards. This included the merits of using PM_{15}, PM_{10}, $PM_{2.5}$ and PM_1 metrics. The use of PM_{10} as an air-pollutant metric in chronic exposure studies began to appear in papers in the late 1980s. For example, Özkaynak and Thurston (1987) reported on a population-based mortality study that used PM_{10}. Dockery et al. (1993) reported on mortality in the Six Cities cohort-based study, and Dockery et al. (1989) reported on symptoms and disease in children in the Six Cities Study. Interestingly, these latter studies used both PM_{15} or PM_{10} and $PM_{2.5}$ as metrics.

In 1987, the indicator for the U.S. NAAQS was changed from TSP to PM_{10}. The PM_{10} metric was considered to be more health protective than TSP because it was directed at particles that had a higher probability of being inhaled and deposited in the thorax. The 1987 PM_{10} standard was based largely on the acute pollution-episode data (Federal Register, 1987). Numerous epidemiological studies using PM_{10} as an indicator began appearing in the late 1980s. Use of the PM_{15} and PM_1 metrics was discontinued and only limited data were collected on the $PM_{2.5}$ metric after PM_{10} was established as the regulated indicator.

In 1997, the United States announced a new NAAQS for PM that retained the PM_{10} indicator and, most significantly, introduced a new $PM_{2.5}$ indicator (Federal Register, 1997). A national $PM_{2.5}$ monitoring plan was initiated in 1998 to meet a statutory requirement for three years of ambient data prior to determining attainment status and the need for implementation plans. With hundreds of sites going in across the nation, a wealth of new $PM_{2.5}$ information has begun to flow. In the late 1990s, an increasing number of papers began to appear using the $PM_{2.5}$ metric, a reflection of the addition of $PM_{2.5}$ as an indicator for the NAAQS and the availability of monitoring data. Canada established a Canada Wide Standard for $PM_{2.5}$ in 2000.

Following promulgation of the revised PM NAAQS in 1997, numerous petitions for legal review concerning a number of issues were filed. Consideration of the details of the legal actions by the U.S. Court of Appeals and the U.S. Supreme Court are beyond the scope of this chapter. However, several outcomes are worthy of note. First, the Court reaffirmed prior rulings holding that in setting the NAAQS, U.S. EPA is not permitted to consider the cost of implementing the standards; EPA can consider costs in establishing the schedule for implementing the standards. Second, the Court concluded that PM_{10} is a poorly matched indicator for coarse particulate pollution. The Court vacated the 1997 revised PM_{10} standard, resulting in the 1987 PM_{10} standards remaining in effect.

The U.S. EPA has interpreted the Court's ruling as precluding the promulgation of further revisions to the PM_{10} NAAQS. Alternatively, in the most recent Criteria Document (CD) and Staff Paper (SP) for particulate matter, EPA has laid the groundwork for proposing a coarse-particle $PM_{10-2.5}$ NAAQS to complement the fine-particle $PM_{2.5}$ NAAQS. The $PM_{10-2.5}$ indicator refers to particles with a mean aerodynamic diameter greater than 2.5 μm but less than or equal to 10 μm. At the time this Chapter was prepared, a Federal Reference Method for $PM_{10-2.5}$ had not yet been developed. The majority of measurements of $PM_{10-2.5}$ and related studies of health effects have been based on differences between concurrent measurements of PM_{10} and $PM_{2.5}$.

From the foregoing, it is apparent that the inputs of both atmospheric scientists, making measurements of air quality, and health scientists, evaluating health responses, have been required for the conduct of epidemiological studies. It is also apparent that epidemiological studies of associations between increased levels of adverse health responses and air pollution can be conducted only using the air-pollutant metrics that have been measured. Hence there has been a progression, from monitoring data and then epidemiological evaluations to using "stinking fog" days to Black Smoke and coefficients of haze to TSP to PM_{10} to $PM_{2.5}$. The latter metrics in the series are reflective of the metrics used in federal standards or guidance and, hence, in regulatory-compliance monitoring. The epidemiological studies have also considered pollutants other than PM for which measurements were available. This typically has included the other regulated criteria pollutants such as ozone, SO_2, NO_x, and CO. As will be discussed later, the opportunity is now at hand to select and evaluate in both toxicological and epidemiological studies PM metrics other than those that are reflective of current regulations. These studies will be designed to test whether these new metrics are more closely linked to adverse health effects than the existing PM indicators and, thus, warrant consideration in developing new regulations and control strategies for PM.

2.3 EXPOSURE ASSESSMENT

Major advances have been made in exposure assessment in recent years. This topic has been reviewed by Özkaynak (1999), the U.S. National Research Council (1991), and the U.S. EPA (1992).

A number of options exist for assessing exposure for use in the conduct of epidemiological studies (Textbox 2.1). These various indices can be placed into perspective by considering the schematic rendering of the relationship between ambient (outdoor) and indoor concentrations of pollutants and biologically effective dose as shown in Figure 2.1. In considering this figure, it is important to recognize that pollutants from outdoor sources can penetrate indoors and, thus, are a major source of indoor concentrations of pollutants including PM, especially in buildings that use outside air directly for heating or cooling. The vast majority of epidemiological studies of air pollution, including those of PM, have been conducted using outdoor fixed-location monitors to provide pollutant composition and concentration profiles for use as exposure metrics (see Figure 2.1).

The Total Exposure Assessment Methodology (TEAM) and National Human Exposure Assessment Study (NHEXAS) were human-exposure assessment field studies that provided a foundation for building models of human exposure to CO, VOCs, and PM (Özkaynak et al., 1996; Wallace, 2000). The relationship between personal exposure and ambient concentrations has been found to vary with the pollutant, climate, building practices, and personal activities. The validity of using ambient PM_{10} measurements as a surrogate for PM_{10} personal exposures was demonstrated for one location and a short period of observation by Clayton et al. (1993). This work is summarized in Figure 2.2. In this figure data are presented relating total personal exposure, personal exposure to ambient PM and personal exposure to non-ambient (i.e., indoor origin) PM to ambient concentrations. As may be noted (panel c) the personal exposure to non-ambient PM was not correlated with ambient PM as would be expected. On the other hand, personal exposure to ambient PM was very closely correlated with ambient PM concentrations (panel b). This is a reflection of both the high penetration of ambient PM_{10} in dwellings resulting in exposure indoors and the individual's exposure to ambient PM_{10} outdoors. Total personal exposure to total PM (panel a) was correlated with ambient PM concentrations. However, as expected the correlation was not as good as seen in panel b because of the influence of PM of non-ambient (indoor) origin. It should be noted that these data were collected during September in Riverside, CA, a locale with a very mild climate and unique PM composition. In Riverside, many homes open directly to the outdoors in September, creating significant opportunities for air exchange, compared with many other North American locations. Thus, caution should be exercised in extrapolating these findings to other areas.

Textbox 2.1. Indirect Methods of Assessing Exposure

Source of Information	Type of Information
Source strength	Emission rate (mass per time), traffic density
Geographical information	Distance of the place of residence from the source
Dispersion models	Spatiotemporal concentration distributions from modeling of emission rates, meteorology, air chemistry, geography
Ambient (Outdoor) monitoring	Ambient concentration for defined period
Outdoor-indoor penetration	Modeling from outdoor concentration, building and ventilation characteristics
Indoor monitoring	Indoor concentration for defined period
Questionnaires and interviews	Source strength, distance from the source, time-activity
Personal exposure	Concentration over time modeled from concentrations of pollutants in microenvironments and time-activity patterns
Personal monitoring	Continuous or cumulative concentrations measured over time
Human samples	Concentration of biomarkers of exposure in human tissues and hair
Toxicological models	Concentration and dose of pollutants in target organs modeling from concentration, breathing rate, and metabolism

[Modified from Samet and Jaakkola (1999)]

The NRC Committee on Research Priorities for Airborne Particulate Matter identified personal exposure studies of PM as deserving of high priority. In their 2001 report, the Committee (NRC, 2001) identified 15 studies as being underway on the outdoor-personal exposure topic. Special attention was being given to studying potential susceptible populations including the elderly, children, asthmatic children and adults, chronic obstructive pulmonary disease (COPD) patients, patients with myocardial infarcts and their spouses, and non-smoking healthy adults in different cities across the United States. Examples of such studies are the work of Rojas-Bracho et al. (2000) with individuals with COPD and of Sarnat et al. (2000) of senior citizens. *Further progress in understanding total personal exposure including the relationship between PM and various constituents, such as carbonaceous species in the ambient and indoor environments, will require continued collaboration between atmospheric scientists, health scientists and specialists in the field of exposure assessment.* This kind of research is just beginning to involve detailed characterization of PM constituents, an area that will continue to benefit from the input of atmospheric scientists.

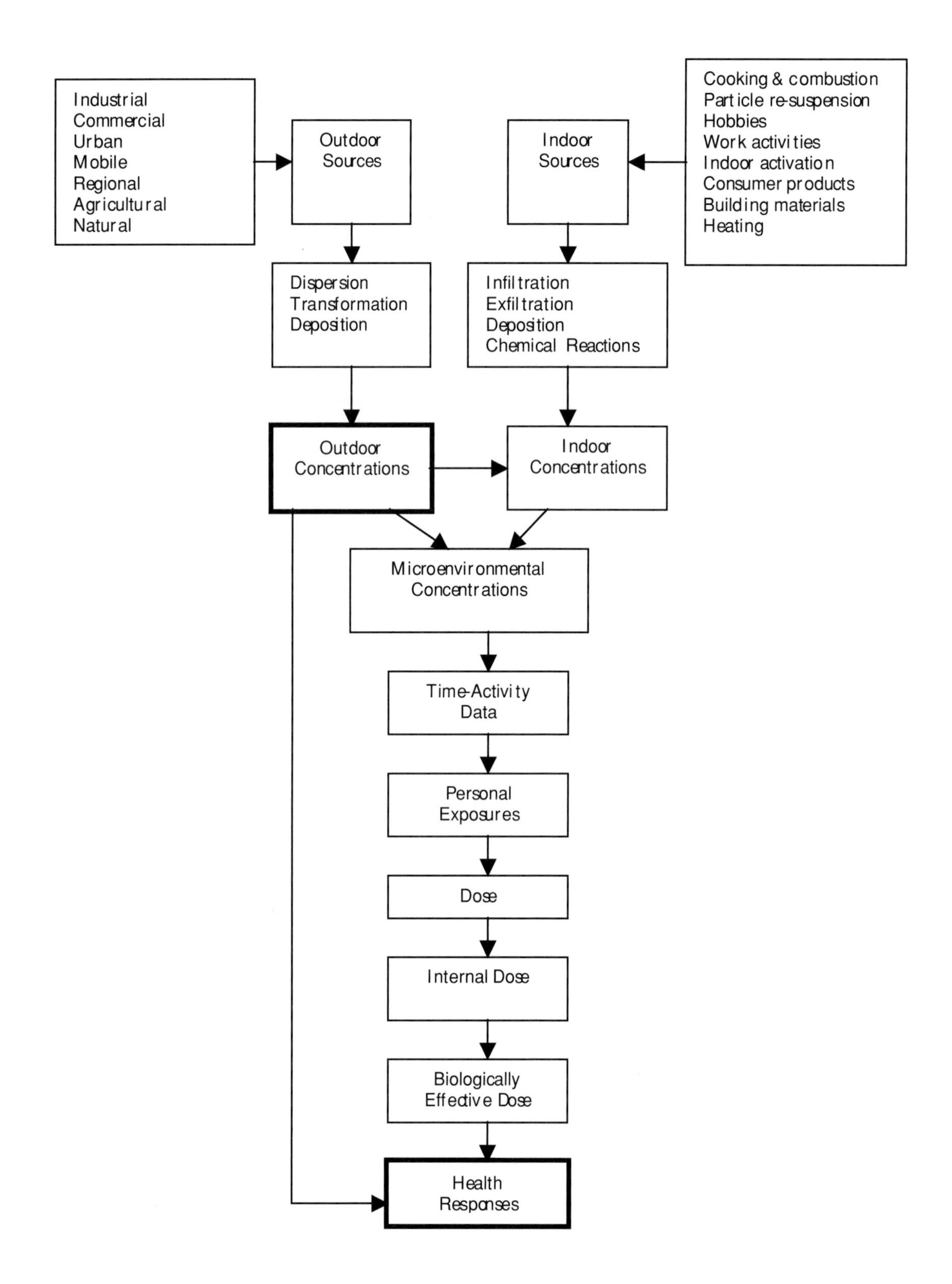

Figure 2.1. Schematic rendering of relationship among outdoor and indoor sources, personal exposure, biologically effective dose, and health responses (Adapted from Özkaynak, 1999).

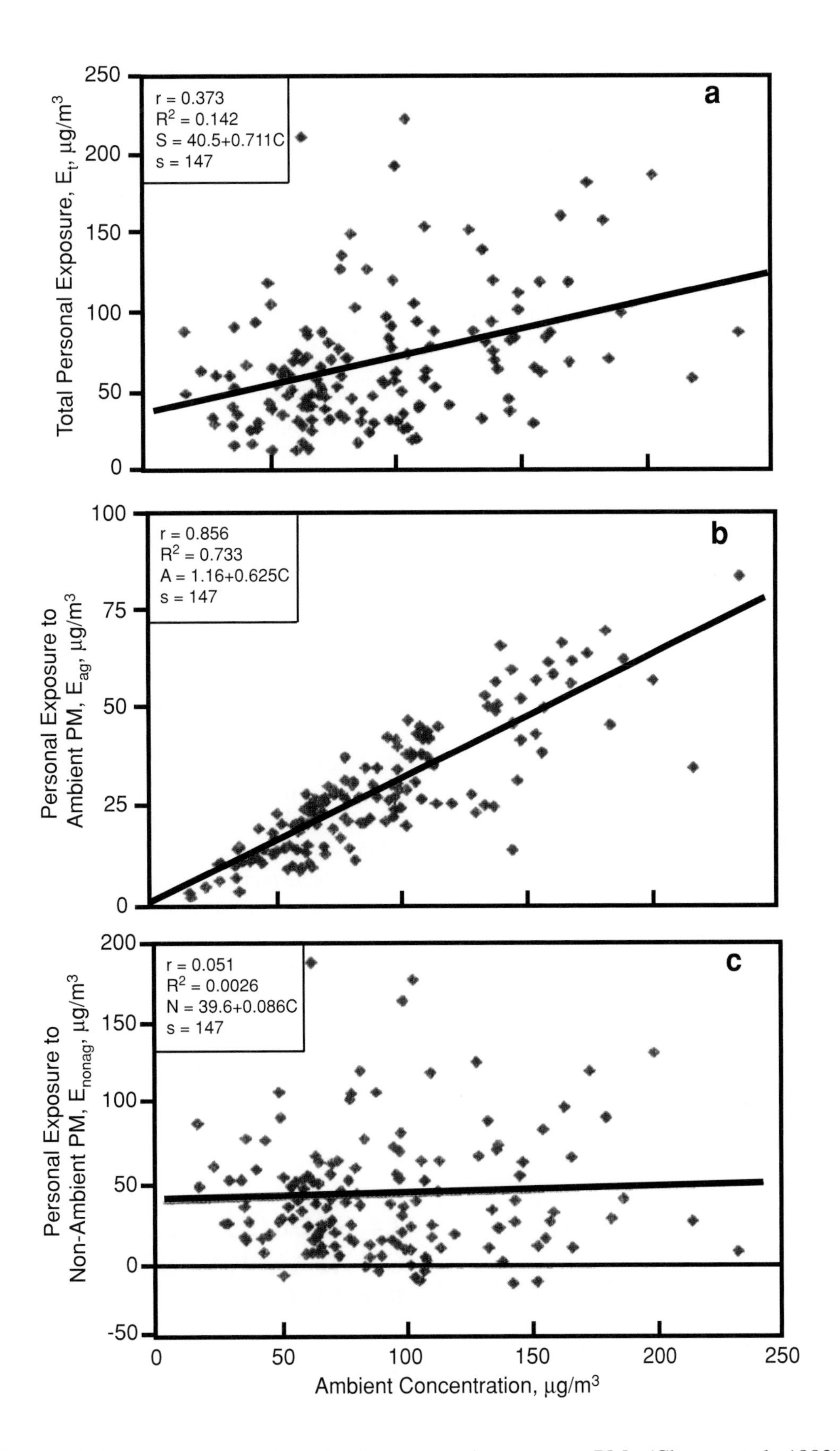

Figure 2.2. Regression analysis of daytime personal exposure to PM_{10} (Clayton et al., 1993).

In considering the relationship between various measures of exposure and internal dose from PM, it is important to consider the very important influence particle size has on the deposition of inhaled PM (Figure 2.3). Excellent summaries of these data are available (ICRP, 1994; NCRP, 1997; Miller, 1999; Schlesinger, 1995). The particle-deposition efficiencies shown in Figure 2.3 have been corrected to account for inhalability. Inhalability is the sampling efficiency of the nose and mouth for particles (Phalen et al., 1986). For 10 µm particles, inhalability is about 77 percent and for $PM_{2.5}$, inhalability is greater than 90 percent (Vincent, 1999). Particle size is plotted as aerodynamic diameter in the figure. Different, size-dependent ranges of aerodynamic behavior exist. The deposition of larger particles is influenced mainly by inertial forces and gravitational settling, while deposition of particles less than about 0.5 µm diameter is influenced by Brownian diffusion. Particle size influences both the fractional total deposition in the respiratory tract as well as the distribution of PM in subunits of the respiratory tract. Consideration of deposition data of this kind as well as the size distribution of ambient PM was a major factor leading to the establishment of PM_{10} and $PM_{2.5}$ as indicator metrics for PM standards.

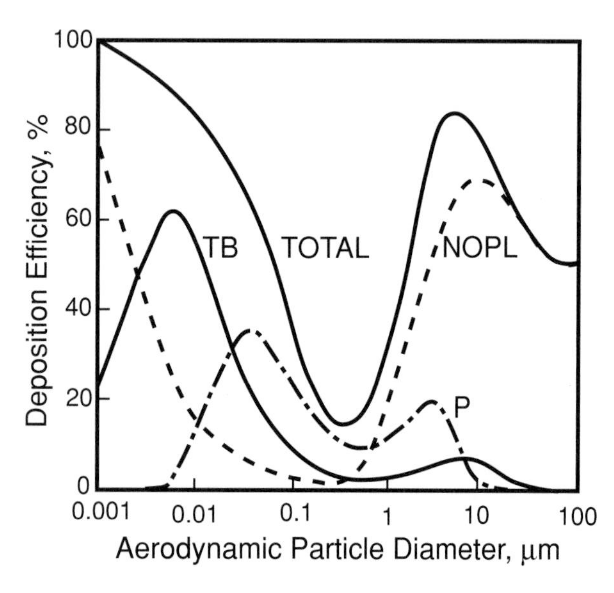

Figure 2.3. Particle deposition in the major regions of the human respiratory tract during normal respiration corrected for size-dependent inhalability. (NOPL, naso-oro-pharyngo-laryngeal) region; TB, tracheobronchial region; and P, pulmonary region). Developed from the National Council on Radiation Protection Model (NCRP, 1997) by Phalen (2002).

For any given particle size, some portion of the particles of that size will deposit in each of the three primary regions of the respiratory tract: naso-oro-pharyngo-laryngeal, tracheobronchial, and pulmonary. This is also the case when deposition is integrated for all particle sizes within a $PM_{2.5}$ or PM_{10} sample. Any given sample only represents a measure of what is in the air and is not necessarily reflective of any individual's exposure. For example, even if the ambient air were characterized with $PM_{2.5}$ and $PM_{10-2.5}$ samples, there may be substantial material present greater than 10 µm in size that has not been sampled and has the potential for being inhaled and deposited in the naso-oro-pharyngo-laryngeal region and, to a lesser extent, in the tracheobronchial region and influencing the development of disease.

Once particles are deposited, the real size, surface area, and chemical composition become major factors determining the clearance of PM from the respiratory tract, and the converse - retention. The clearance of relatively insoluble PM in pure forms (such as aluminosilicate, titanium dioxide, and black carbon) from the different regions of the respiratory tract is generally quite well understood (Wolff, 1996; Snipes, 1995; Miller, 1999). Unfortunately, only limited data are available on the clearance and retention patterns of specific chemical constituents of ambient PM. A lack of such data has been a factor impeding progress in understanding the various mechanisms by which PM and specific PM constituents may cause health effects. *Progress in understanding the disposition in the human body, and in laboratory animal species, of various key constituents of ambient PM can undoubtedly be aided by closer collaboration between atmospheric scientists and health scientists.*

2.4 SOURCES OF INFORMATION ON HEALTH RESPONSES TO AIR POLLUTION

Information on health responses to air pollution, including PM, are obtained from multiple sources: epidemiological studies, controlled exposure studies of human subjects, investigations using laboratory animals and research conducted in vitro with cells and tissues.

2.4.1 Epidemiological Approaches

Epidemiology comprises the scientific methods used to study the occurrence of disease in human populations, including description of the occurrence of disease and identification of the causes of disease such as air pollution. Detailed coverage of epidemiological approaches to studying air pollution is provided in recent reviews by Samet and Jaakkola (1999) and Pope and Dockery (1999). Epidemiological studies are of special value because they involve the study of people in their natural setting and exposures that occur in the course of everyday life. Their results can document increased occurrence of adverse effects of air pollution, describe the relationship between exposure and response, and characterize effects on susceptible groups within the population, e.g., persons with cardiorespiratory disease.

In general, epidemiological studies are carried out to 1) to determine if air pollution or a source of air pollution poses a hazard to human health, 2) to characterize the relationship between the level of exposure and the response, and 3) to examine responses of potentially sensitive groups to pollutant exposures. These objectives relate directly to the information needs of policy makers as they address key questions: 1) Does a particular pollutant pose a hazard to human health?, 2) What are the levels of risk at specific exposure levels? and 3) Which groups need special consideration because of their susceptibility? Epidemiological studies may also provide information to assist the policy makers in deciding on control strategies (What is the anticipated health impact of reductions in a particular pollutant or reduction of emissions from a given source?) and whether a particular control strategy has had impact (Are there reductions in health impacts related to control measures?).

The fundamental focus of epidemiological studies of air pollution and health as schematically depicted in Figure 2.4 is to evaluate the association between various health metrics (upper portion of figure) and various air-quality metrics (lower portion of figure). A glossary of some of the key terms used in the field of epidemiology is provided in Textbox 2.2. A listing of some of the indirect methods of assessing exposure was provided earlier in Textbox 2.1 and some of the common health metrics are provided in Textbox 2.3. The terms acute (i.e., of short duration) and chronic (i.e., of long duration) may be used in describing both exposure and health responses. Acute or short exposures usually give rise to acute effects. For example, an immediate symptom of cardiac or respiratory effects is death. However, these acute effects may be observed in individuals with chronic disease arising from a range of factors. Chronic, or long-duration exposure, perhaps over a lifetime, may

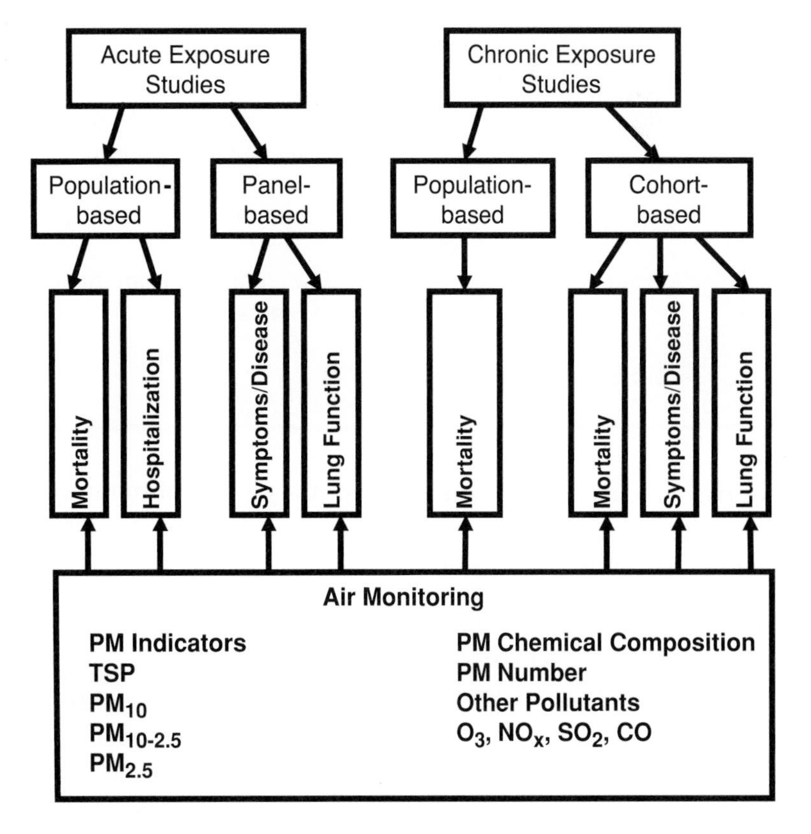

Figure 2.4. Schematic rendering of basic design for epidemiological studies (Adapted from Pope and Dockery, 1999).

Textbox 2.2. Glossary of Epidemiological Terms

Association	Non-random occurrence of disease in relation to exposure
Bias	Error in the measurements of an exposure's effect
Case-Control Study	An analytical design involving selection of diseased cases and non-diseased controls followed by assessment of prior exposures
Trial	An analytical design involving random assignment of exposure to two or more subject groups
Panel or Cohort Study	An analytical design involving selection of exposed and non-exposed subjects with subsequent follow-up for development of disease
Confounding	Bias resulting from the contamination of an exposure's effect by that of another risk factor
Cross-Sectional Study	Subjects are identified and exposure and disease status determined at one point in time
Incidence Rate	Ratio of number of new cases to population at risk during a specified time period
Misclassification	Bias from error in determining exposure or disease status
Mortality Rate	Ratio of number of deaths to population at risk during a specified time period
Prevalence	Proportion of population with disease at a particular time
Selection Bias	Bias resulting from the technique used to select a study's subjects

[From Samet and Jaakkola (1999)]

give rise to bouts of acute symptoms of disease or more typically diseases that develop and are manifested over long time periods.

The listing (Textbox 2.3) of health metrics focuses on the respiratory system because it is both the portal of entry for air pollutants and is frequently impacted by the inhaled materials. In considering these outcomes, it is important to recognize that they are not specific or unique in their relationship to either air pollution in general or to specific pollutant indicators such as PM. Indeed, cardiac and respiratory diseases are caused or aggravated by many factors and are among the most common ailments as will be detailed below. In epidemiological investigations, the models used are typically additive or relative-risk models with the results expressed as increased risk relative to the baseline morbidity or mortality associated with a given increment of exposure to a particular pollutant. As will be discussed later, the ambient air-pollution effect detected may represent a small percentage increase over the background level of disease. Thus, the effects of ambient PM are reflected as a statistical increase in morbidity and mortality in populations rather than as effects that can be identified in specific individuals.

It has been known for a long time that the health effects of inhaled pollutants are not restricted to the respiratory tract, but that inhaled materials can impact non-respiratory organs. Examples are 1) inhaled lead affecting both the hematopoietic and nervous systems, 2) carbon monoxide affecting the central nervous system, the oxygen-carrying capacity of blood and the heart, and 3) inhaled chemicals like benzene causing leukemia. Only very recently has an association been demonstrated between exposure to PM and heart disease including death (Utell et al., 2002).

Textbox 2.3. Health Outcome Measures in Studies of Air Pollution

General
- Overall mortality
- Morbidity index

Cardiac and Respiratory
- Acute and chronic symptoms
- Acute infections
- Chronic cardiac respiratory diseases
- Degree of non-specific airways responsiveness
- Reduced level of lung function
- Increased rate of lung function decline
- Decreased rate of lung function growth
- Altered cardiac functio
- Exacerbation of a chronic cardiac or respiratory disease
- Hospitalization for a chronic cardiac or respiratory disease
- Lung cancer
- Death secondary to a cardiac or respiratory disease

Neuropsychological
- Reduced performance on neurobehavioral testing
- Neuropsychological syndrome
- Neuropsychological disease

[Modified from Samet and Jaakkola (1999)]

The total burden of respiratory and cardiac disease in most countries including Canada, Mexico, and the United States is substantial. This can be illustrated using data from the United States. In 1997, there were 3,475,000 hospital discharges for respiratory disease (Lawrence and Hall, 1999). Of these, 38 percent were for pneumonia, 14 percent for asthma,13 percent for chronic bronchitis, 8 percent for acute bronchitis, and others not specified. There were 195,943 deaths recorded as caused by respiratory disease (Hoyert et al., 1999). Of these, 44 percent resulted from acute infections, 10 percent were for emphysema and bronchitis, 2.8 percent for asthma, and 42 percent from unspecified COPD. In 1997, there were 4,188,000 hospital discharges with heart diseases listed as the first diagnosis (Lawrence and Hall, 1999). Of these, 50 percent were for ischemic heart disease, 18 percent for myocardial infarction or heart attack, 23 percent for congestive heart failure, and 15 percent for cardiac dysrhythmias. In 1997, there were 726,974 deaths from heart disease (Hoyert et al., 1999). From consideration of these values, it is apparent that there is a large background of respiratory and cardiac disease arising from multiple risk factors, which makes it difficult to detect small signals of increased disease attributed to PM. Because these diseases are very common it is also apparent that even a very small percentage increase attributed to PM will represent a substantial number of cases.

The study designs of epidemiological investigations of air pollution can be grouped in several ways. In Figure 2.4, the focus is first on acute versus chronic exposure and second on the unit of observation: populations or individuals. Studies based on groups are referred to as ecological studies. Population studies may compare 1) indicators of adverse health effects across geographic areas with different pollutant levels, and 2) temporal associations between pollutant levels and measures of health outcome for a single or multiple communities. The National Morbidity, Mortality, and Air Pollution Study (NMMAPS; Samet, et al., 2000a, 2000b; HEI, 2000a, 2000b) is an excellent example of a retrospective time-series study of 90 communities in the United States.

Study designs that have individuals as a unit of observation are of three types: 1) cross-sectional studies, 2) cohort studies, and 3) case-control studies (see Textbox 2.2 for details). Of the designs having individuals as the unit of observation and analysis, the cross-sectional study is generally the most economical and feasible approach. Although the effect of ambient pollution may be studied on individuals, data are ultimately aggregated and statistical comparisons made on a group or subpopulation basis. It has not been possible to attribute the health status of any given individual to PM or other pollutants except in a statistical sense. Cross-sectional studies are often used to compare the health status of residents of more versus less polluted communities. The effects of exposure may be biased, for example, by the tendency of more susceptible persons to reduce their exposure by leaving the polluted community.

In a cohort study, exposures of individuals are assessed directly or indirectly and they are followed for the development of the outcomes of interest, for example, the occurrence of symptoms of disease or changes in lung function. Cohort studies may be prospective if the disease of interest will occur in the future, or retrospective if they have already taken place when the study is initiated. The Six Cities Study (Ferris et al., 1979) is an excellent example of a prospective cohort study that included from the beginning substantial involvement from atmospheric scientists. More than 8000 subjects were enrolled in 1974 through 1976 in six cities selected for a range of air-quality conditions and then followed with periodic measurements of lung function, respiratory symptoms, and other health indices, and simultaneous monitoring of air pollution. The cohort design has the advantage of direct estimation of disease rates and the opportunity to prospectively accumulate comprehensive exposure data including changes in exposure over time.

Panel studies represent a special type of short-term cohort study. The subjects, often individuals considered to be of greater susceptibility such as patients with COPD or heart disease, are enrolled and both health outcomes and exposure monitored intensively. Hence usually only a modest number of individuals can be studied, especially if an attempt is made to characterize (directly or indirectly) individual exposures.

The case-control study approach compares exposures of persons having the outcome of interest with those of controls, providing a measure of association between exposure and disease. This design has been widely used for studying lung cancer. It represents an optimum approach for studying uncommon diseases. It has been used infrequently for studying non-malignant respiratory diseases which are common, as was discussed earlier, and air pollution.

The successful conduct of epidemiological investigations is dependent upon the quality of both health-outcome data and exposure data. Shortcomings in health-outcome data cannot be made up for by giving increased emphasis to the exposure data or vice versa. All studies of environmental air pollution inevitably involve multiple pollutants. The atmosphere is a complex mixture of gases and PM that varies both spatially and temporally in chemical composition and in the size distribution of the particulate phase. While the focus in any given study may be on some specific PM indicator such as $PM_{2.5}$, it is necessary to consider the effects of other air pollutants, including other PM indicators, that may also adversely impact health. *Long-term monitoring data on PM mass concentrations, PM composition, gas-phase precursors of secondary particulate species, and gas-phase copollutants in multiple communities will provide the best opportunity for evaluation of associations with health data.*

It is also critical that other variables that can influence health be considered. This includes factors such as age, climate and temperature, socio-economic status, educational level, occupation, and lifestyle factors such as smoking. And finally, it is important to consider the total body of epidemiological and related evidence. This is especially the case when attempting to assess whether the statistical associations observed are causal. This issue has been addressed by Hill (1965) and Rothman (1986) in proposing criteria to move beyond statistical associations to consider whether a causal association has been demonstrated (Textbox 2.4).

Textbox 2.4. Criteria for Assessing Causality of Associations

Strength of association	Strong associations considered to be more likely causal than weak associations
Consistency	Repeated observation of the association in different studies strengthens the likelihood of causality
Specificity	A cause is associated with a single effect
Temporality	Exposure precedes effect
Biological gradient	An exposure-response relationship is present
Plausibility	The association should be consistent with relevant biologic data

[From Hill (1965) and Rothman (1986).]

Chemical composition is not considered in current North American PM standards. This poses a difficult problem in deciding what is an appropriate material for use in clinical studies. Specific PM constituents such as H_2SO_4 droplets and carbon black have been studied. In addition, in recent years technology has been developed to expose both laboratory animals and human subjects to concentrated ambient PM (Sioutas et al., 1995). Such studies have an advantage in that investigators are studying real-world PM; however, they must accept whatever the composition of PM is in the ambient air at the time of the study.

2.4.2 Controlled-Exposure Studies with Human Subjects

Studies with human subjects exposed to well-characterized test atmospheres provide an opportunity to obtain detailed information on the disposition of inhaled PM and clinical responses. The basic approaches to the conduct of such studies have been described by Frampton and Utell (1999), Frampton et al. (2000), and Utell et al. (2002). All such studies must be carried out under defined protocols that have been reviewed and approved by an Institutional Review Board to ensure conformance with all the applicable standards for human experimentation. The subjects in such studies are defined as to their health status and may be normal individuals or potentially sensitive individuals such as those with asthma or COPD. Ethical considerations dictate that the subjects not be the most ill or feeble individuals in a particular diagnostic category. Most controlled-exposure studies of air pollutants have been conducted with well-defined atmospheres of single pollutants such as radio-labeled particles, ozone, H_2SO_4, NO_2, or carbon black. The studies with inhaled radio-labeled particles and subsequent measurements of deposition and retention have provided a sound understanding of the influence of particle size on the disposition of particles (ICRP, 1994; NCRP, 1997).

A broad array of measurements can be made to assess changes in the clinical condition of the subjects related to the test exposure. This has typically included measurements of pulmonary function and in some recent studies an assessment of various parameters related to cardiac function (Frampton and Utell, 1999; Utell et al., 2002). Advances have been occurring at a phenomenal rate in the understanding of both normal biology and disease at the molecular level. The human genome is on the verge of being fully characterized and increased attention is being given to characterizing related proteins (the proteome) and how normal metabolism occurs as well as the handling of introduced materials (metaboleome). Many of these advances are now being used to study the effects of inhaled particles and other pollutants in controlled-exposure studies with human subjects. Some of these approaches ultimately may be adapted for use in large-scale population studies. This may include the development of new biomarkers of exposure, susceptibility, and health effects. A key consideration in such work will be the validation of a growing list of candidate biomarkers such that they can be used with confidence to define the point at which responses should be equated to adverse health effects that warrant preventive measures (ATS, 2000).

An obvious strength of clinical studies is that they involve human subjects, and thus the data generated are clearly of interest in setting standards to protect human health. Key limitations are the small number of subjects that can be studied, short periods of exposure and observation, the need to extrapolate from simple aerosols to complex ambient PM and from high concentrations of simple aerosols or concentrated ambient PM to the lower PM concentrations typical of ambient air. The conduct of controlled studies with human subjects requires the use of specialized equipment and facilities, capabilities that exist in only a few institutions around the world. *Atmospheric scientists, as active collaborators in controlled-exposure studies with human subjects and laboratory animals, can assist in producing and characterizing test atmospheres that will assist in testing various mechanistic hypotheses.*

2.4.3 Laboratory Animal Studies

Laboratory animal studies with controlled exposures typically play a central role in characterizing the hazardous properties of various substances including ambient PM. Approaches to the conduct of such studies have been reviewed (Valentine and Kennedy, 2001; McClellan and Henderson, 1995). In the absence of adequate human data, laboratory animal data may be used not only to describe hazard but may also play a crucial role in characterizing the exposure-response relationship for the substance. In the case of PM, laboratory animal studies share many of the limitations described earlier for controlled-exposure studies with human subjects. In addition, there is the issue of extrapolating from another species to humans. In the case of inhaled materials this extrapolation issue is complicated by differences in the dimensions of the respiratory tract of laboratory animals and humans and the impact of these differences on the deposition of inhaled PM. Fortunately, there is a large body of literature on the deposition and clearance of inhaled particles in the principal laboratory animal species for comparison to data on humans (Schlesinger, 1995; Miller, 1999; Snipes, 1995; Wolff, 1996). As with human studies, a serious limitation relates to the need for specialized equipment and facilities to conduct inhalation studies, capabilities that exist in only a few institutions around the world. If alternative approaches such as intratracheal instillation are used to introduce material into the respiratory tract, there is an additional challenge of extrapolating from non-physiological approaches of delivering PM to the inhalation route of interest.

Laboratory animal studies have been conducted using simple atmospheres containing well-defined PM such as carbon black or titanium dioxide. In addition, some studies have been done with complex atmospheres such as diluted vehicle exhaust. Other studies have been done using concentrated ambient particulate matter using the new particle concentrator technology (Clarke, et al., 2000; Godleski et al., 2000).

A strength of laboratory animal studies is the extent to which different species and strains can be used to take advantage of particular characteristics matched to an experimental need. In recent years this has included animals that have been genetically manipulated to either enhance or suppress certain biological characteristics. Recognizing the extent to which many of the uncertainties related to PM effects relate to responses of susceptible individuals, researchers have made on-going efforts to develop and use animal models that mimic some of the characteristics of humans considered to be especially susceptible to PM. Several different animal models have been developed for pulmonary and cardiac diseases including consideration of asthma and aging (Cantor, 1989; Bice et al., 2000; Mauderly, 2000; and Muggenburg et al., 2000). An obvious advantage of laboratory animal studies is the ability to carry out experimental manipulations that might not be feasible with human subjects.

2.4.4 Tissue and Cell Studies

A wide range of in vitro experiments can be conducted with tissues and cells obtained from both laboratory animals and human subjects. In some cases, cultured cells from established cell lines can be used allowing experiments to be performed under carefully replicated conditions to test the influence of multiple experimental variables. *Atmospheric scientists can aid in the design and use of novel systems to deliver PM with specific characteristics to cell populations in ways that will enhance extrapolation from the in vitro situation to the in vivo.*

As will be discussed later, controlled-exposure studies with human subjects, laboratory animals and cells provide valuable tools for evaluating and prioritizing the multiple hypotheses that have been advanced to explain the mechanisms of action of PM. *Atmospheric scientists can play a valuable collaborative role in helping health scientists understand the dynamic nature of ambient PM size and chemical composition and identify parameters for investigation in biological studies.*

2.5 EPIDEMIOLOGICAL FINDINGS

2.5.1 Acute Exposure

The designs for acute exposure studies and chronic exposure studies shown in Figure 2.4 will be used as a basis for discussion of the current status of our knowledge of health effects associated with PM exposure.

Numerous acute exposure, population-based mortality studies have been conducted as reviewed by Pope and Dockery (1999), CEPA/FPAC (1999), Anderson (1999), California EPA (2002) and the U.S. EPA (1996a,b, 2001; 2002). In recent years the population-based studies have been extended to evaluation of hospitalization and related health-care endpoints. These studies have been conducted most frequently using administrative morbidity and mortality data bases collected for other purposes and air-quality monitoring data collected for regulatory compliance purposes. The earliest and most methodologically simple studies focused on acute pollution episodes. They were simple because of the substantial effect compared to background mortality. A summary of the findings from acute exposure studies using PM_{10} as a metric developed by Pope and Dockery (1999) from a review of the literature is shown in Figure 2.5. All of the changes shown are expressed as percent change relative to the background rate of the morbidity or mortality indicator. These authors have cautioned that these are not precise estimates because of the difficulty of comparing across studies using different measures of pollution, differently defined health endpoints, and different models. However, the data are illustrative of the relative magnitude of epidemiological associations.

In the 1970s, investigators began to use formal time-series analyses to explore potential associations between daily mortality and air pollution at relatively low levels. Time-series studies have the advantage of using mortality and air-pollution data over much

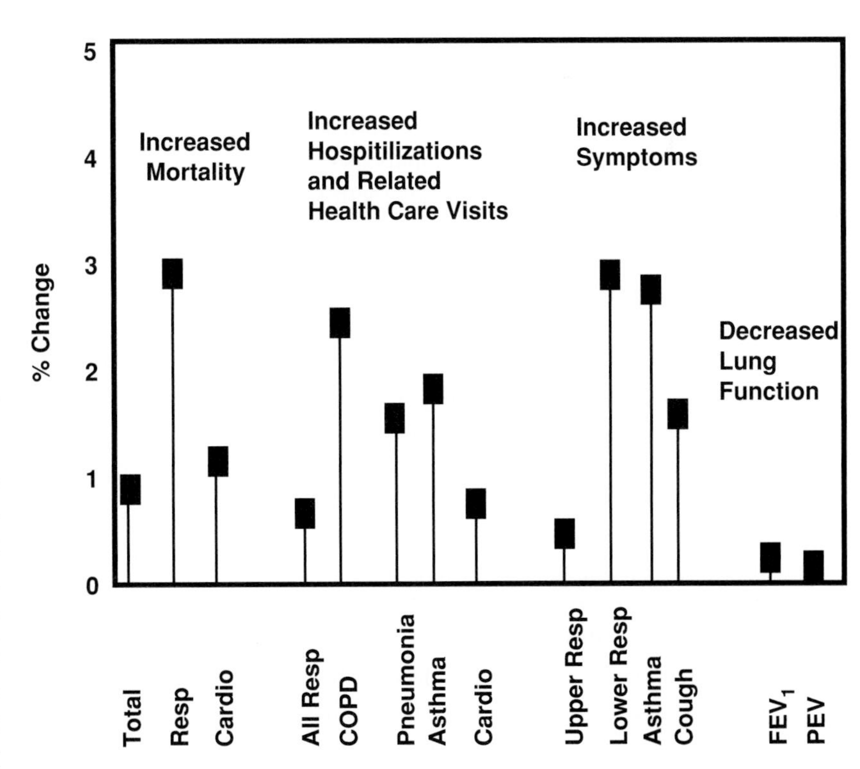

Figure 2.5. Stylized summary of acute exposure studies, percent change in health endpoint per 10 μg/m³ increase in PM_{10} (adapted from Pope and Dockery, 1999)

longer time periods than the few days of a single, acute pollution episode. This approach increased the statistical power of the studies by increasing the number of adverse outcomes (deaths usually identified by cause) and the number of monitored days. In time-series studies, each population essentially serves as its own control, i.e., it is assumed that various risk factors such as smoking, employment, and socio-economic status are not changing day to day.

In the early 1990s, investigators began to conduct time-series analyses using Poisson regression modeling. Many of the studies used the Generalized Additive Models (GAM) described by Hastie and Tibshirani (1990). These studies took advantage of computerized records of daily mortality, air pollution (typically the criteria pollutants), season, barometric pressure, temperature, and other weather metrics. Inclusion of other pollutants in the model generally reduced the estimated PM effect. Population-based socio-economic data were also available. Data on factors other than air pollutants are important because the relative risk associated with changes in these various confounders may be significant as well and need to be addressed in the course of estimating air-pollution risks. New statistical software became available in the 1990s for Poisson regression analysis of count data permitting statistical analysis of daily counts of death even for relatively small metropolitan areas.

Some of the time-series studies provided for classification of mortality by cause. PM had the largest effects on deaths related to respiratory and cardiovascular disease and, because the number of cardiovascular deaths were greater than the number of respiratory deaths, the majority of the effect was attributable to the former. Typically, the relative risk of mortality increased with increasing PM concentrations in a near-linear fashion. A lag between air pollution and mortality was usually observed, suggesting that mortality lagged 1 to 5 days following an increase in air pollution, though a few studies have suggested periods of slightly elevated risk for a period of up to several weeks. Because PM measurements were not always made using the same metrics, conversions were made by assuming that PM_{10} = 0.55 x TSP; PM_{10} = Coefficient of Haze/0.55; and PM_{10} = Black Smoke.

The most substantial time-series study conducted to date, NMMAPS, was carried out by investigators at Johns Hopkins University (Samet et al., 2000a, 2000b; HEI 2000a, 2000b). The study examined the association between mortality and air pollution for 90 U.S. cities over the period 1987 to 1994 using the GAM feature of S-PLUS, a statistical software package. The investigators originally reported a pooled estimate of a 0.41 percent increase (1.0041 times the baseline mortality) per 10 $\mu g/m^3$ increase in PM_{10} with a posterior standard error of 0.05 when the standard GAM convergence criteria were used. As an aside, the statistical association was based on PM_{10} rather than $PM_{2.5}$ because $PM_{2.5}$ data were not available. *In the future, as information becomes available on $PM_{2.5}$ and other PM indicators such as the carbonaceous fraction and inorganic elements from the speciation studies, it will be appropriate to extend the NMMAPS kind of analyses.*

The NMMAPS' investigators reported in 2002 that the standard convergence criteria in S-PLUS were not sufficient; the result was an overestimation of the effect of PM pollution and an underestimation of the variance for a given city (Dominici et al., 2002). Using substantially more restrictive convergence criteria, the investigators found the pooled estimate of the effect on mortality decreased to 0.27 percent (1.0027 times the baseline mortality) per 10 $\mu g/m^3$ increase in PM_{10}, again with a standard error of 0.05. The investigators also observed that using the more restrictive convergence criteria, the within-city variance increased compared to when the standard convergence criteria were used. The underestimation of the within-city variance may be balanced by the overestimation of between-city variance without impacting the total variance (Daniels et al., 2002). The existence of between-city differences may be reflective of pollution differences between the cities including PM composition. *The opportunity exists for atmospheric and health scientists to collaborate on the search for variations in health impacts that may relate to differences in PM characteristics.*

Ramsey et al. (2003) have also examined the analytical methods in detail and indicated that the standard errors of the measures of association were systematically underestimated, with the potential to increase the apparent level of statistical significance. This occurred in part because of a large correlation between nonlinear functions included in the GAM analyses. The reader interested in obtaining a better understanding of issues involved in software reliability in performing these types of analyses is referred to the papers of McCullough (1998, 1999).

Samet et al. (2003) have also evaluated the use of Generalized Linear Models (GLM) with natural cubic splines for confounder adjustment (Daniels et al., 2002). They found a pooled estimate of 0.21 percent (1.0021 times the baseline mortality) per 10 $\mu g/m^3$ increase in PM_{10}. Samet and colleagues have cautioned against selecting any particular model as correct, and urged researchers to explore the sensitivity of their findings to model selection.

The recent findings of issues related to model selection and statistical criteria are clearly not restricted to the NMMAP study. The findings have triggered re-evaluation of a number of key time-series studies conducted during the last decade. Because of the importance of the time-series studies in the review of the NAAQS for PM, arrangements were made for the Health Effects Institute to convene a special panel to peer review the re-evaluated time-series analyses. The report of the panel (HEI, 2003a) includes sections on each of the studies re-evaluated, such as the NMMAP study (HEI, 2003b), so this information could be considered in the Criteria Document (U.S. EPA, 2003a) and Staff Paper (U.S. EPA, 2003b) currently being prepared. In addition to considering the individual studies the special HEI panel provided a commentary on the strengths and weaknesses of the time-series methodology. An important conclusion was that difficulties remain in the handling of confounding by weather, which have important implications for the resulting estimated concentration-response coefficients for PM indicators and other air pollutants.

Pope and Dockery (1999), based on their review of the literature, provided a schematic summary of effects estimates of acute exposure to PM_{10} (Figure 2.5). They suggested that each 10 $\mu g/m^3$ increase in PM was associated with a 0.8 percent (1.008 times the baseline mortality) increase in daily mortality. This estimate, as well as those for other endpoints, may need to be revised depending on the outcome of the re-analysis of key time-series studies. Pope and Dockery (1999) noted that the associations for respiratory mortality were substantially larger, and those for cardiovascular mortality were larger than those for total mortality. Increases in PM_{10} exposure were also associated with increased hospitalizations and increased health-care visits for respiratory and, to a lesser extent, cardiovascular disease. Associations were also observed for lower respiratory symptoms, exacerbation of asthma and coughing and, to a lesser degree, upper respiratory symptoms and small declines in lung function.

While the results in Figure 2.5 are for PM_{10}, generally similar effects have been observed for $PM_{2.5}$. However, the percent change is usually larger per μg of $PM_{2.5}$ than per μg of PM_{10} reflecting the greater apparent potency of the $PM_{2.5}$ fraction. Limited data are available on the $PM_{10-2.5}$ metric. There are health effects attributed to this fraction although the findings are generally not as consistent as for $PM_{2.5}$. These inconsistencies may in part be a reflection of the sampling methods and calculation of $PM_{10-2.5}$ by subtracting $PM_{2.5}$ values from PM_{10} values. That $PM_{10-2.5}$ causes health effects is not surprising recognizing that a substantial portion of any material in this size fraction deposits in the naso-oro-pharyngeo-laryngeal region and, to a lesser extent, in the tracheobronchial and pulmonary regions. *As improved techniques are developed for quantifying the $PM_{10-2.5}$ fraction, it will be important to conduct additional epidemiological studies with the aim of better characterizing the concentration – response relationships for both the $PM_{10-2.5}$ and $PM_{2.5}$ fractions as well as for specific chemical constituents in the two fractions.*

Some attempts have been made to evaluate association between particular sources of pollution and health outcomes. Such studies offer a direct bridge between pollution sources, ambient air quality and apparent population health effects. One controversial attempt of this kind used air-quality modeling (Chapter 8) to estimate ambient PM concentrations from certain sources, then estimated potential morbidity and mortality in exposed populations (Levy et al., 2000). In another study, Laden et al. (2000) took advantage of the extensive air-monitoring data collected in the Six Cities Study to use factor-analysis methods, as noted in Chapter 7, to identify through chemical tracers different classes of sources. They identified combustion PM from mobile sources (Pb tracer) and coal-fired power plants (Se tracer) and crustal PM (Si tracer) and related these to mortality. An increased level of crustal $PM_{2.5}$ was not associated with increased mortality. However, they observed for a 10 $\mu g/m^3$ increase in $PM_{2.5}$ from mobile sources a 3.4 percent (Confidence Interval [C.I.] 1.7-5.2 percent) increase in mortality (1.034 times the baseline mortality). For an equivalent increase in $PM_{2.5}$ from coal combustion they observed a 1.1 percent (C.I. 0.3-2.0 percent) increase in daily mortality (1.011 times the baseline mortality).

A key to this study was an extensive data base on the size and chemical composition of PM and the ability to establish linkages to major sources of PM. The Laden et al. (2000) study aggregated the results of the six cities; however, it did not sufficiently examine the nature of sources of PM city by city. In particular, there are ambiguities associated with interpretation of data from one city, Boston, relating to a residual oil combustion source near the monitoring site, and the presence of Se in both coal and residual oil (Grahame and Hidy, 2003). While these studies point to the important future role of integration of source and ambient air characterization with health effects, they also indicate that each of the components needs to be analyzed with the most current detailed information available. Additional studies of this type involving close collaboration between atmospheric scientists and health scientists will help advance the understanding of PM and the role of particular kinds of sources of PM and influencing health. *The improvement of national emission inventories will greatly facilitate the conduct and interpretation of studies attempting to link sources, ambient concentrations, and health effects.*

2.5.2 Chronic Exposures

Acute exposure studies linked to indices of acute morbidity or mortality provide little information about how air pollution affects longer-term morbidity or mortality rates, how much life is shortened, or the potential role of air pollution in the process of inducing chronic disease. Chronic exposure studies as shown schematically in Figure 2.4 evaluate the effects of ambient exposure that persists for long periods of time as well as the cumulative effects of repeated acute exposure episodes. Chronic exposure has been reviewed by Pope and Dockery (1999), CEPA/FPAC (1999), California EPA (2002) and the U.S. EPA (1996a,b; 2001; 2002).

The two most comprehensive cohort-based studies of mortality are the Six Cities Study and the American Cancer Society Cohort. The Six Cities Study involved a 14-16 year follow-up of 8000 adults living in six communities selected to provide a gradient in several indices of air pollution (Ferris et al., 1979; Dockery et al., 1993). Extensive effort was made to characterize ambient air, including the use of TSP, PM_{10}, $PM_{2.5}$ and $SO_4^=$ as PM indicators. The original American Cancer Society (ACS) study (Pope et al., 1995) included 500,000 persons followed from 1982-1989 living in 151 metropolitan areas across the United States with $SO_4^=$ measurements (measured in 1986) and 240,000 individuals living in 50 metropolitan areas with $PM_{2.5}$ measurements (for 1979-1983). Both studies were based on prospective-cohort health and individual risk-factor data and could control for individual differences in age, gender, race, cigarette smoking and other risk factors. Strengths of the Six Cities Study relate to its balanced study design (approximately the same number of subjects in each community) and the planned prospective collection of air-pollution data. A primary limitation was the limited number of subjects (8000) in a limited area (six communities in mid- to northeastern United States). A strength of the ACS study was the larger number of subjects (up to 500,000) from communities across the United States. A limitation was that individuals self-enrolled, and there was a lack of provision for prospective collection of air-pollution data and, thus, the need to rely on air-pollution data collected primarily for regulatory compliance and other purposes.

Key results from the two studies are shown in Table 2.1. In the Six Cities Study, excess mortality associated with $PM_{2.5}$ was 26 percent greater for the most polluted city compared to the least polluted city. In the ACS study, the excess mortality was 15 percent and 17 percent greater for the most polluted compared to the least polluted area based on $SO_4^=$ and $PM_{2.5}$, respectively, as indicators. It is informative to consider the data for cigarette smokers in the two studies. In both studies, smokers, not unexpectedly, showed a substantial increase in all-cause mortality

Table 2.1. Comparison of Mortality Risk Ratios (and 95 percent C.I.) for Smoking and Air Pollution from the Six Cities and ACS Prospective Cohort Studies. (from Pope and Dockery, 1999)

Cause of Death	Current Smoker [a]		Particulate Air Pollution (Most vs least polluted city)		
	Six Cities	ACS	Six Cities ($PM_{2.5}$)	ACS ($PM_{2.5}$)	ACS (SO_4)
All	2.00 (1.51-2.65)	2.07 (1.75-2.43)	1.26 (1.08-1.47)	1.17 (1.09-1.26)	1.15 (1.09-1.22)
Cardiopulmonary	2.30 (1.56-3.41)	2.28 (1.79-2.91)	1.37 (1.11-1.68)	1.31 (1.17-1.46)	1.26 (1.16-1.37)
Lung Cancer	8.00 (2.97-21.6)	9.73 (5.96-15.9)	1.37 (0.81-2.31)	1.03 (0.80-1.33)	1.36 (1.11-1.66)
All Others	1.46 (0.89-2.39)	1.54 (1.19-1.99)	1.01 (0.79-1.30)	1.07 (0.92-1.24)	1.01 (0.92-1.11)

[a] Risk ratios for current cigarette smokers with approximately 25 pack-years (about average at enrollment for both studies) compared with never smokers.

(about 2.0 times that for never smokers) and cardiopulmonary mortality (about 2.30 times that for never smokers) and a very substantial excess risk for lung cancer (more than 8.0 times that of never smokers). The magnitude of the cigarette-smoking effects emphasizes the importance of accurate ascertainment of smoking history and status in all cohort studies so this risk factor can be adequately controlled for in the overall analysis. The Six Cities Study design also had provision for evaluation of respiratory disease and lung function. Other studies have also evaluated these parameters for an association with PM exposure.

Pope and Dockery (1999), based on their review of the literature, have provided a schematic summary of the effect of chronic exposure to $PM_{2.5}$ (Figure 2.6). They have emphasized that these are representative values and are not intended to be definitive point estimates. They elected to use $PM_{2.5}$ as the metric because it was the primary PM measurement in the most recent and most rigorous chronic studies. They noted that in these studies, and especially the mortality studies, $PM_{2.5}$ was more closely associated with the health outcomes than PM_{10} or TSP.

Several points should be kept in mind when comparing the results of the acute (Figure 2.5) versus chronic (Figure 2.6) exposure studies. As already noted, the acute exposure results are expressed per 10 $\mu g/m^3$ of PM_{10} while the chronic exposure results are per 5 $\mu g/m^3$. $PM_{2.5}$ frequently represents about 50 percent of the PM_{10} fraction, an observation that can be kept in mind in comparing the magnitude of the effects in the two figures. In addition, it should be recognized that the chronic effects are not merely a summing of the effects of a series of acute exposure episodes. It should also be kept in mind that the graphs are plots of relative risk, i.e., the percentage change in baseline occurrence associated with increases in ambient PM. In the absence of specific knowledge of the background or baseline rates it is not possible to calculate the absolute level of added risk and make comparisons on that basis.

Both the Six Cities and ACS data sets have been subjected to a new analysis (Krewski et al., 2000). This effort generally confirmed the original analyses and conclusions despite identifying some issues of data quality and completeness in both studies. The Krewski et al. (2000) analysis did find that when the

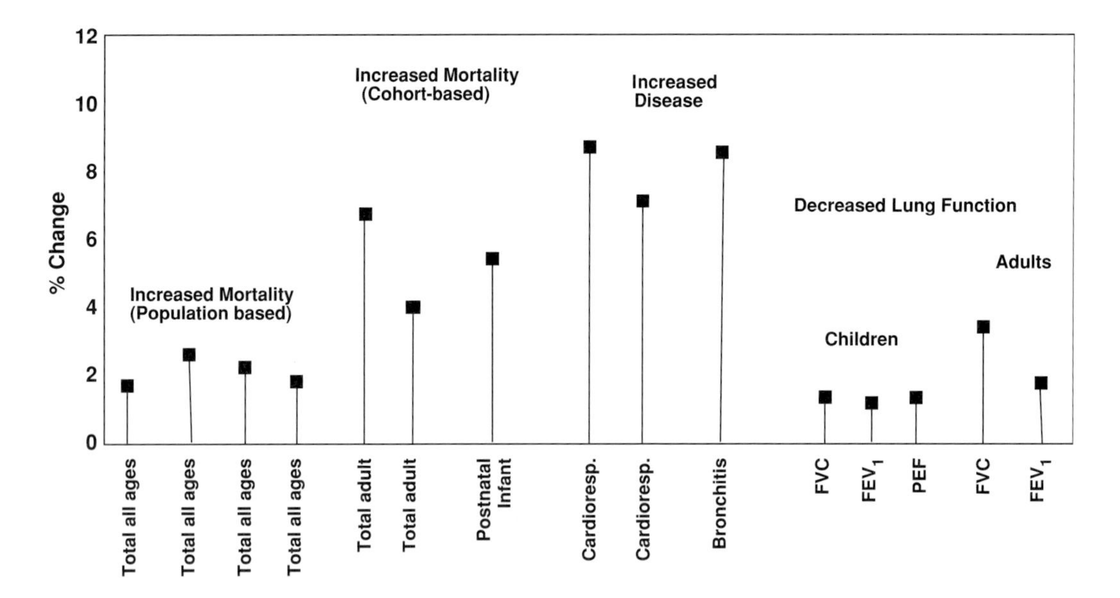

Figure 2.6. Stylized summary of chronic exposure studies, percent change in health endpoint per 5 µg/m^3 change in $PM_{2.5}$. Duplicative values are based on the results from different studies. (Adapted from Pope and Dockery, 1999).

general decline in $PM_{2.5}$ over the observation period was included as a time-dependent variable, the association between $PM_{2.5}$ and all-cause mortality reduced the relative risk to 1.22 (C.I. 1.03-1.45). The importance of independent confirmation of the basic validity of the original findings should not be underestimated in view of the central role of the results of these studies in the establishment of PM standards.

The ACS study has recently been updated using data on vital status and cause of death through December 31, 1998 (Pope et al., 2002). The new analysis also included additional air-monitoring data, most notably $PM_{2.5}$ data for 116 metropolitan areas for 1999-2000 (mean of 14.0 µg/m^3 compared to 21.1 µg/m^3 for 61 areas in 1979-1983), PM_{10} data for 102 areas for 1982-1998 (mean 28.8 µg/m^3) and TSP data for 150 areas for 1982-1998 (mean 56.7 µg/m^3). Each 10 µg/m^3 increase in $PM_{2.5}$ was associated with approximately a 4 percent, 6 percent, and 8 percent increased risk of all-cause, cardiopulmonary and lung-cancer mortality, respectively (1.04, 1.06, and 1.08 times the baseline respectively). Measures of the coarse-particle fraction and TSP were not consistently associated with mortality. As expected, the estimated relative risks for an average current smoker were equal to 2.58, 2.89 and 14.80 times that of a never smoker for all-cause, cardiopulmonary, and lung cancer mortality, respectively, values that were higher than that of the earlier analysis for the aging cohort (Table 2.1).

A critical issue in interpreting the results of both the Six Cities and the ACS data sets is the period of exposure that yielded the health results observed. Air quality has improved substantially over the lifetime of the individuals who have died during the studies. The specific levels of ambient PM and other air pollutants are not known for most of the earlier time the individuals lived. The air-quality data available are only present for relatively recent time periods and these are the data used to develop the denominators in the reported exposure concentration-response coefficients, as for example in Table 2.1. If early life exposures were higher and contributed to the observed PM-associated mortality, then the coefficients based on recent air-quality measurements are biased on the high side by some unknown amount.

A consistent finding from both the acute- and chronic-exposure epidemiological findings is that the PM effects are most substantial in susceptible subpopulations. These include individuals with pre-existing diseases of the respiratory and cardiac systems and children and the elderly.

A key issue in evaluating any hazardous material contained in PM is the nature of the ambient concentration – personal exposure – response relationship. Of particular interest is the threshold (if a threshold indeed exists) and shape of the concentration-response function. A detailed coverage of these topics is beyond the scope of this chapter. In most studies that have identified these issues, clear evidence of a threshold has not been evident and the added level of adverse health outcomes appeared to increase monotonically in a near-linear manner with increasing PM concentration (Pope, 2000; Daniels et al., 2000). It is not known if the evidence on absence of thresholds and linearity drawn from the time-series studies will change when the re-analyses are completed. These are issues being evaluated now.

It has been noted by others critiquing this topic (NRC, 2002) that the statistical power to assess the shape of ambient concentration-response functions is weakest at the lower and upper ends of the observed concentration ranges. Further, since the number of communities studied are limited, the ability to formally test for the absence or existence of a no-effects threshold is limited. In addition, even if thresholds exist, they may not be the same for all endpoints and all regions of the continent. It should be noted that in some large-scale studies such as NMMAP (Samet et al., 2000; Samet et al., 2003), there was variability in the PM_{10} response among the cities with some cities not exhibiting a statistically significant increase in the relative risk for PM_{10} exposure. This may be interpreted as representing differences in PM potency or concentrations or as evidence of a variation in PM_{10} composition or evidence of a practical threshold.

It is appropriate to consider both the strengths and weaknesses of epidemiological studies, as they supply the most critical data for setting health-based standards and developing controlled strategies. A major strength of the epidemiological studies is that they involve people carrying out their normal everyday activities and breathing ambient air both outdoors and within dwellings, work places, and vehicles. The epidemiological investigations are largely based on measurements of ambient air. Such measurements may have a variable degree of coherence with personal exposure, though work to date indicates that ambient measures do suffice in characterizing personal exposure to ambient PM. Public policy and pollution-control strategies typically, and necessarily, focus on ambient air.

Pope and Dockery (1999) have identified four limitations of epidemiological studies. These are worthy of consideration because in some cases coordinated activities between atmospheric scientists and health scientists can serve to minimize the limitations. *First,* they note that epidemiology provides little information about biological mechanisms. While this is true in a broad sense, epidemiological studies using panels of individuals numbering in the tens to perhaps hundreds of people can provide valuable insights into the mechanisms of action of air pollutants. This is especially the case if both detailed measurements of air quality and various health indicators are made.

Second, interpretation of epidemiological studies is frequently handicapped by a lack of understanding of the coherence between ambient measurements and personal exposure. However, most work to date indicates that ambient measures do suffice in characterizing personal exposure to PM. Again, coordinated efforts between atmospheric scientists, exposure-assessment specialists, and health scientists are already improving the level of understanding in this area.

Third, a basic limitation of epidemiological studies is the difficulty of disentangling independent effects or potential interactions between highly correlated risk factors. Many of these issues are beyond the field of atmospheric science, e.g., ascertainment of smoking status, socio-economic status, and educational level. However, many of the factors such as weather and barometric pressure are of concern to the atmospheric-science community. One key issue that does intersect the atmospheric-science community relates to the need for information on co-pollutants such as particulates and gaseous precursors, (e.g., ozone, SO_2, NO_x, CO) and other measures yet to be identified. Atmospheric scientists are knowledgeable of the dynamic nature of the atmosphere and its constituents. *The insights of atmospheric scientists will be helpful in planning and conducting the sophisticated analyses of air monitoring and health data necessary to identify small signals of the biological effects of pollution amongst the many other causal factors for the same diseases. Comprehensive plans for both data collection and analysis are needed to maximize the return on substantial investments.*

Fourth, Pope and Dockery (1999) noted that a basic limitation of epidemiological studies is the inability to fully explore the relative health impacts of various constituents of particulate air pollution. This is an area in which atmospheric scientists can have substantial impact by working with health scientists. In a coordinated multi-disciplinary effort they can identify key atmospheric constituents, for example, 1) specific chemical constituents such as the carbonaceous content or specific classes of chemicals, i.e., organics or metals by particle size fraction, or 2) physical parameters such as surface area or particle number that warrant detailed study. This information would provide guidance for adopting sampling strategies that could yield measurements of air constituents on both spatial and temporal scales matched to health data. As is readily apparent from consideration of the relative risks attributed to the various size-based PM indicators, there is a need in cross-sectional cohort investigations to study simultaneously many communities and populations measured in the tens to hundreds of thousands of individuals over extended periods of time.

The air-quality measurement needs in epidemiological studies are quite different than those of a typical "atmospheric science" field campaign conducted for short periods of time. However, future epidemiological studies can build on the experience gained from short-term intense campaigns with emphasis given in the larger- and long-term epidemiological studies to obtaining air-monitoring data at multiple sites in a highly efficient and effective manner. The nature of future studies can be anticipated by considering two past landmark studies: the ACS study and the NMMAPS.

One landmark study is the ACS study (Pope et al., 2002). In the most recent report, analyses were reported for up to 500,000 individuals for which mortality could be ascertained for a 16-year period (1982-1998) and fine-particle data were available using $PM_{2.5}$ for 1979-1983 and 1999-2000 and $SO_4^=$ data were available for 1980-1981 and 1990. The study would have been strengthened if fine-particle measurements had been available for all 16 years and if additional speciation data had been available.

A second landmark study is the NMMAPS time-series study (Samet et al., 2000; Samet et al., 2003) of 90 cities for the period 1987 to 1994 using PM_{10} measurements. The EPA Staff Paper (U.S. EPA, 2003b) analyzed the findings reported for the NMMAPS and found an association between the study size and stability of the risk coefficients. It appeared that in excess of 10,000 events (deaths per day times the number of monitored days) were necessary to obtain stable estimates of risk.

While any given epidemiological study may be addressing hypotheses related to PM, it is important to recognize that people are breathing "one atmosphere," which includes all particulate and gaseous constituents. Thus, it is of major importance to have reliable data on the gaseous pollutants as well as on PM. *Most of all, it is important to recognize that advancing scientific knowledge of the health effects of air pollution, and especially the role of PM and specific constituents, will require moving beyond dependence on "regulatory-compliance" based air-monitoring data with its emphasis on currently regulated pollutant indicators.*

2.6 INTERVENTION STUDIES

An accountability framework that will enable measurement of progress toward the goal of protecting public health is needed. Unfortunately, risk-management actions are often difficult to assess for their utility in actually reducing the health impact of interest. For most environmental actions, reduction in the pollutant of interest is often as far as one can go in discerning the benefit of control actions when direct measurement of changes in the adverse effects in the population is not possible. In the absence of direct evidence the benefits must be extrapolated from knowledge of changes in the concentration of PM and other pollutants and knowledge of the potency of the pollutant indicators gained from other studies. In the United States, a Committee of the National Research Council has recently provided a useful critique of methods used in estimating the public health benefits of air-pollution regulations (NRC, 2002).

However, for some PM air-pollution situations it has been possible to directly measure the health benefits of specific risk-management actions. Administrative databases (such as mortality and hospital admission records) have been used to examine the influence of day-to-day (and longer) increases in air pollution. This raises the possibility of using similar techniques to examine declines of PM indicators and any concomitant changes in health status for several endpoints.

Precipitous declines in air pollution, usually elicited by regulatory initiatives, offer an opportunity to observe changes in underlying risk at rates much higher than one would normally observe in PM epidemiological studies. Several such regulatory mechanisms have been instituted with sufficient subsequent time elapsed to examine the influence of resultant air-pollution changes on mortality and other health effects. Two of the most notable cases are the banning of coal use in Dublin, Ireland in 1990, and the switch to low-sulfur fuels in Hong Kong in the same year.

Clancy et al. (2002) examined the periods leading up to and following the ban on coal sales in Dublin, and compared non-trauma, cardiovascular, and respiratory death rates for the two periods. From an air-quality impact perspective, this intervention resulted in a decline of Black Smoke by approximately 50 percent, while SO_2 declined by about 30 percent. For all three measures of mortality, statistically significant declines in the expected death rate occurred.

Several researchers have examined aspects of health indicator response to a reduction of sulfur content of fuels (gasoline and fuel oil) in 1990 in Hong Kong. Sulfur dioxide and PM measures all declined significantly in the period following the application of this regulation. Wong et al. (1998, 1999) examined respiratory symptoms in women and children in high- and low-pollution neighborhoods, finding statistically significant declines in the endpoints measured. Similarly, Hedley et al. (2002) used this air-quality scenario to examine the impact of the sulfur reduction on mortality rates for the following five-year period. The authors stratified deaths by age and specific cause of death, finding statistically significant declines in most categories.

One of the earliest studies of the influence of an "intervention" on air-quality and health responses was the study of Pope (1989) on the impact of a closure of a steel mill. Pope demonstrated that closure of the steel mill directly impacted air quality and that there were associated changes in health indices. Laboratory studies using particulate matter collected on filters demonstrated a coherence between laboratory findings of pulmonary toxicity and epidemiological observations (Dye et al., 2001). The laboratory studies pointed to a role for metal compounds mediating the toxic effects.

Another model of utility to investigate potential health benefits is exemplified in the work of Avol et al. (2001) in the twelve-community Children's Health Study. As part of this larger study, the investigators examined the impact of moving to communities of higher or lower air pollution. As a group, subjects who had moved to areas of lower PM_{10} showed increased growth in lung function and subjects who moved to communities with a higher PM_{10} showed decreased growth in lung function.

While all the above studies may have limitations of one kind or another, the design and results indicate that observing the impact of air-quality interventions is both possible and useful. Whether the result of regulatory intervention or a physical change in location of individuals, these studies were able to observe health benefits using available techniques. It should be possible to apply such techniques to other situations of this type, as well as to air-quality regimes where decline is significant but more gradual. Expanding these types of studies to a wider range of interventions and resultant air-quality impacts would serve to broaden our understanding of the population response to air pollution as well as provide a retrospective evaluation of proposed regulatory and other actions.

However, the difficulty of conducting such studies should not be underestimated as is apparent from consideration of the results of NMMAPS. For the 88 cities in the contiguous United States, statistically significant effects of PM_{10} were observed in some individual cities but not in other cities. In about a third of the cities the apparent added risk from ambient PM_{10} was either negative or zero. This and other evidence of inter-city and inter-regional differences needs to be carefully investigated. The magnitude of the pooled estimate of 0.27 percent increase (1.0027 times the baseline mortality) per 10 μg/m^3 in PM_{10} illustrates the statistical challenge of demonstrating the effects of changes in PM.

2.7 TOXICOLOGICAL EVIDENCE

Extensive research has been initiated, especially during the last decade, to provide information on the biological mechanisms of action of PM and specific constituents. This research has been directed toward two key questions: 1) What are the potential mechanisms by which PM causes health effects? and 2) What specific component or components in ambient PM cause health effects? Information on the first question is important in providing support (or lack of support) for biological plausibility to explain the statistical associations observed in epidemiological studies. Information on the second question also addresses the causality issue (Textbox 2.4) and, in addition, may potentially provide valuable guidance for the development of pollution-control strategies if specific PM constituents or their precursors and their sources can be targeted.

To help guide research on the mechanisms linking ambient PM exposure and adverse health outcomes, Mauderly et al. (1998) compiled a list of hypotheses that have been advanced by multiple investigators (Textbox 2.5). All of these hypotheses have been addressed to varying degrees by recent research, with the greatest use made of animal studies with controlled exposure to various agents. Of necessity, many of these studies have used short-duration exposures and relatively high concentrations of the particular agent in an attempt to identify effects and mechanisms of action. In addition to the animal studies, a limited number of studies have been initiated with small panels of human subjects exposed under controlled conditions to defined test materials such as carbon black. In some cases, subjects have been exposed to test aerosols of concentrated ambient PM.

The research performed to date is best viewed as being of an exploratory hypothesis-generating nature, yielding valuable background information for future systematic investigations. Some of the key findings have been reviewed by MacNee and Donaldson (1999); Frampton et al. (2000); Utell et al. (2002); NRC (2003); Lippmann et al. (2003); Green and Armstrong (2003); and Walberg (2003). The results to date of the laboratory-animal and human studies have not been able to unequivocally determine the particle characteristics or toxicological mechanisms by which ambient PM affects the respiratory and cardiac systems. Thus, all of the suggested hypotheses in Textbox 2.5 remain viable and are deserving of additional research to test their validity. The studies conducted to date have been very useful in identifying a series of inter-related pathways by which inhaled PM can impact the occurrence of cardiac and respiratory effects. It is quite likely that some of the pathways may be affected by particles of a particular size and/or chemical composition while other pathways may not be affected by those particles but be affected by particles of other sizes or with other chemical constituents. With mechanistic pathways identified, it will be possible to plan and conduct studies using aerosols with particles that vary in size and composition to obtain direct comparisons of the relative effects of the different PM components. *The conduct of comparative studies will require collaboration between aerosol scientists and health scientists to develop appropriate techniques for the production, characterization and delivery of the specialized test aerosols with known particle-size distributions and size-resolved chemical composition.*

Laboratory experimentation with either animals or human subjects exposed under controlled conditions with PM concentrations typically higher than found in the ambient environment, can aid in establishing the plausibility of the various hypotheses. However, ultimately it will be desirable to obtain human data from epidemiological investigations that monitor the putative agents and compare health effects with the various PM indicators over ranges of concentrations typically found in the environment. For specific PM constituents, the test of the hypotheses will be similar to the situation for PM_{10} versus $PM_{2.5}$ – i.e., testing whether the putative toxicant has a greater potency (effect per unit mass) than $PM_{2.5}$ or PM_{10}.

Textbox 2.5. Hypothetical Interaction Between PM and Human Physiological or Toxic Responses

Hypothesis of PM Characteristic and Human Health Response	Rationale for Hypothesis
Particle Mass Concentration – Non-chemically specific mass cardiopulmonary loading response associated with a complex chemical mixture of wide range of particle size.	Based on evidence derived from a variety of epidemiological studies in North America and elsewhere. PM mass is contributed largely by particles over 0.1 μm.
Particle Size/Surface Area —Response to fine particles with major surface area for adsorption of chemical species and subsequent desorption in lower lungs.	The surface area per unit particle size has a multi-modal distribution analogous to the mass distribution. Since particles are non-spherical, methods need to be adopted that can measure at least total surface area directly noting that the actual surface for adsorption may vary with condensed material. Surface area is contributed to largely by particles of 0.1 to 0.3 μm.
Ultrafine PM – Animal experiments suggest that some particles less than 0.1 μm diameter may have a strong physiological effect on the respiratory system.	Ultrafine particles in the air are generally believed to be highly transient in character. They may derive directly from source emissions or from atmospheric chemical reactions. Estimating human exposure is problematic. Ultrafine particles (of the order of 0.01 μm) dominate particle number count.
Metals or Metal Compounds– Certain metals like V, Cu, Fe, Zn, and Ni have cytotoxic or inflammatory properties. These may catalyze biochemical reactions that result in an adverse respiratory response.	The human reaction to combined metal and acidic components such as H_2SO_4 has been hypothesized to be potentially important for PM since the 1960s, but remains unconfirmed.
Acids— Acidic particles have been shown to have toxic properties in some animal studies based on hydrogen ion delivered to respiratory surfaces.	Linkages to acid anions such as $SO_4^=$ and NO_3^- have not been unequivocally demonstrated in epidemiological or laboratory studies on humans. Organic acids may be a factor that requires sustained investigation.
Organic Compounds – There are a large number of organic compounds found in PM, some of which are known to be mutagenic or carcinogenic.	The carbonaceous fraction is recognized to be the least well characterized of the components of PM. The organic material found in the air ranges from paraffinic or olefinic compounds of varying molecular weight to highly oxygenated or nitrated material. Only 10 to 20 percent of OC is attributed to specific organic compounds with existing methods.
Biogenic Particles – A variety of particles originate from biogenic sources, including spores, fungi, bacteria and viruses.	The phrase "biogenic particles" refers to those formed biogenically, as opposed to those arising from reactions of organic vapors of biogenic origin. The physiological or toxic interactions of individual biogenic particles and pollutant material are unknown.
Sulfate and Nitrate Salts– These compounds are believed to be mainly ammonium salts in PM.	The direct linkage between sulfate and nitrate salts and adverse human health effects remains uncertain. Sulfate concentrations and hospital admissions are clearly linked in S. Ontario.
Peroxides– The presence of peroxides in particles and their toxic properties provide a hypothetical pathway to health effects.	The presence of highly oxidative species related to photochemical processes is well known in atmospheric chemistry. Their significance in producing human-health responses has not been extensively investigated.
Soot – Soot particles (or elemental carbon) potentially can stimulate a toxic response in themselves or carry adsorbed material that can initiate a response.	Soot in the air can be modified significantly by inorganic gas, water or organic vapor adsorption. Different extents of their interactions may have different effects on both.
Co-pollutant Interactions– Some epidemiological and/or laboratory exposure studies have suggested that a synergistic response may take place when PM and gases such as SO_2, NO_2, O_3 or CO are present.	In most places air contains multiple pollutants that vary in concentration over time and space, and vary in their potency for producing adverse health effects. Epidemiological studies show a distribution of risk among gaseous and particle aspects of the air pollution mix. The lack of clarify and consistency may be due to the variability of copollutant interactions.

(After Mauderly et al., 1998)

Collaboration is needed between health scientists and atmospheric scientists in designing and conducting epidemiological studies and associated long-term monitoring programs for key atmospheric constituents that are hypothesized to be associated with adverse health effects. The size of the populations studied and the duration of the studies can be informed by considering the findings in recent epidemiological studies as was discussed earlier. Past experience has clearly demonstrated the value of doing studies in multiple cities and making observations on very large populations over extended periods of time.

2.8 POLICY-RELEVANT FINDINGS AND FUTURE OPPORTUNITIES

There is a considerable and growing body of evidence showing an association between adverse health effects, especially of the cardiac and respiratory systems, and exposure to elevated ambient levels of PM.

The total personal PM exposure of individuals, which includes their ambient (outdoor) and indoor environments, is related to the PM content of ambient air. The ambient-air concentrations of PM, and other air pollutants, have been extensively studied as potentially controllable variables that influence total personal exposure and, thus, human health.

Epidemiological studies of large populations have shown statistically significant increases in various indices of adverse health outcomes over the background incidence and increased levels of ambient PM. The adverse health outcomes observed in these studies are commonly (though not exclusively) cardiac and respiratory in nature. Such diseases are also very common in the general population and even a small percentage increase in background incidence translates into a significant number of additional cases.

Increases in adverse health outcomes have been observed for both short- and long-term exposures to PM. The increases in adverse health outcomes have been observed for a range of ambient particulate matter indicators including TSP, PM_{10}, $PM_{10-2.5}$ and $PM_{2.5}$. A higher potency has typically been observed for $PM_{2.5}$ compared to other PM indicators consistent with the concept that smaller particles penetrate further down the respiratory tract. Nonetheless, the epidemiological literature does demonstrate some level of association with the larger particles, which deposit more readily in the upper respiratory tract.

Certain population subgroups appear to have heightened susceptibility to PM, such as those with pre-existing cardiac and respiratory disease (seen frequently in smokers), asthmatics, and the elderly. Increases in adverse health effects appear to occur without a threshold in the ambient concentration-health response relationships and appear to increase in a near-linear fashion from the baseline incidence of health effects with increasing PM concentrations. Some evidence suggests that there may be regional differences in the potency of various PM indicators, possibly a reflection of regional differences in the composition of PM. Additionally, some evidence exists linking increases in adverse health outcomes and specific sources of PM.

There is also some evidence of improved health associated with reductions in PM exposure. Such studies have been both opportunistic in nature, taking advantage of significant reductions in PM pollution over short time periods to examine mortality and other adverse health endpoints, or have been designed to follow groups as their exposure to air pollution has changed. These studies, while requiring careful interpretation, offer the opportunity to examine the benefits of specific air-quality interventions, as well as the opportunity to compare the benefits of specific changes in the air-pollution mix. Pursuit of these situations in a strategic and proactive manner offers the potential for gaining insights into the benefits of control activities.

A number of hypotheses have been advanced to explain how various chemical and physical parameters of PM may interact with the body to provide mechanistic explanations for the various health outcomes. These hypotheses are being evaluated in toxicological studies using laboratory animals and controlled exposure of human subjects. However, to date none of these hypotheses have been proven or eliminated from consideration. Continuing tests of these hypotheses will require collaboration between atmospheric scientists and health scientists in order to identify and characterize the hypothesized PM constituent or parameters, the PM indicators usually monitored, and other pollutants, and to relate these to health effects in multiple communities over decades. Such collaboration will enhance efforts in the fields of toxicology, clinical effects, and in both short-term and long-term epidemiological investigations. *Further progress in understanding health effects of current levels of ambient PM and various constituents depends on achieving a high level of collaboration between health scientists and atmospheric scientists, to obtain detailed characterization of PM and copollutants in multiple communities over decades to match with health data.*

2.9 REFERENCES

Anderson, H.R., 1999. Health effects of air pollution episodes. In: Air Pollution and Health. Holgate, S.T., Samet, J.M., Koren, H.S., Maynard, R.L., eds., Academic Press, London and San Diego, pp. 461-482.

ATS (American Thoracic Society), 2000. What constitutes an adverse health effect of air pollution? American Journal of Respiratory and Critical Care Medicine 161, 665-673.

Avol, E.L., Gauderman, W.J., Tan, S.M., London, S.J., Peters, J.M., 2001. Respiratory effects of relocating to areas of differing air pollution levels. American Journal of Respiratory and Critical Care Medicine 164, 2067-2072.

Bates, D.V., Sizto, R., 1987. Air pollution and hospital admissions in southern Ontario: The acid summer haze effect. Environmental Research 43, 317-331.

Bice, D.E., Seagrave, J.D., Green, F.H.Y., 2000. Animal models of asthma: Potential usefulness for studying health effects of inhaled particles. Inhalation Toxicology 12, 829-862.

Brimblecombe, P., 1999. Air Pollution and Health History. In: Air Pollution and Health, Holgate, S.T., Samet, J.M., Koren, H.S., Maynard, R.L., eds., Academic Press, New York, NY, pp. 5-18.

California Environmental Protection Agency, Air Resources Board, 2002. Staff Report: Public Hearing to Consider Amendment to the Ambient Air Quality Standards for Particulate Matter and Sulfates. May 3, 2002. El Monte, CA.

Cantor, J.O., ed., 1989. CRC Handbook of Animal Models of Pulmonary Disease, CRC Press, Boca Raton, FL.

CEPA/FPAC (Canadian Environmental Protection Act/Federal/Provincial Air Quality), 1999. National Ambient Air Quality Objectives for Particulate Matter. Part 1. Science Assessment Document, Ottawa, Ontario, Canada.

Clancy, L., Goodman, P., Sinclair, H., Dockery, D.W., 2002. Effect of Air-pollution Control on death rates in Dublin, Ireland: An intervention study. Lancet. 360,1210.

Clayton, C.A., Perritt, R.L., Pellizzari, E.D., Thomas, K.W., Whitmore, R.W., Wallace, L.A., Özkaynak, H., Spengler, J.D., 1993. Particle Total Exposure Assessment Methodology (PTEAM) study: Distributions of aerosol and elemental concentrations in personal, indoor, and outdoor air samples in a southern California community. Journal of Exposure Analysis and Environmental Epidemiology 3, 227-250.

Clarke, R.W., Catalano, P., Coull, B., Kontrakis, P., Krishna Murthy, G.C., Rice, T., Godleski, J.J., 2000. Age-related responses in rats to concentrated urban air particles (CAPs). Inhalation Toxicology 12 (Suppl. 1): 73-84.

Daniels, M.J., Dominici, F., Samet, J.M., Zeger, S.L., 2000. Estimating particulate matters-mortality dose-response curves and threshold levels: An analysis of daily time-series for the 20 largest cities. American Journal of Epidimiology 152, 397-406.

Daniels, M., Dominici, F., Zeger S.L., Samet J., 2002. Underestimation of Standard Errors in Multi-site Time-series Studies. Technical Report, Johns Hopkins University, http://biosun01.biostat.jhsph.edu/~fdominic/research.html.

Dockery, D.W., Ware, J.H., Ferris Jr., B.J., eds., 1982. Changes in pulmonary function associated with air pollution episodes. Journal of the Air Pollution Control Association 32, 937-942.

Dockery, D.W., Speizer, F.E., Stram, D.O., et al., 1989. Effects of inhalable particles in the respiratory health of children. American Review of Respiratory Disease 139, 587-594.

Dockery, D.W., Schwartz, J., Spengler, J.D., 1992. Air pollution and daily mortality: Associations with particulates and acid aerosols. Environmental Research 59, 362-372.

Dockery, D.W., Pope III, C.A., Xu, X., Spengler, J.D., Ware, J.H., Fay, M.E., Ferris Jr., B.G., Speizer F.E., 1993. An association between air pollution and mortality in six U.S. cities. New England Journal of Medicine 329, 1753-1759.

Dockery, D.W. and Pope, C.A. 1994. Acute Respiratory Effects of Particulate Air Pollution. Ann Rev Public Health 15, 107-132.

Dominici, F., Daniels, M., Zeger, S.L. and Samet, J.M. 2002. Air Pollution and Mortality: Estimating Regional and National Dose-response Relationships. J. Am. Stat. Assoc. 97,100-111.

Dye, J.A., Lehman, J.R., McGee, J.K., Winsett, D.W., Ledbetter, A.D., Everitt, J.I., Ghio, A.J., Costa, D.L., 2001. Acute pulmonary toxicity of particulate matter filter extracts in rats: Coherence with epidemiological studies in Utah Valley residents. Environ. Health Perspectives 109 (Suppl. 3): 395-403

Federal Register. 1971. National Primary and Secondary Ambient Air Quality Standards. Fed. Reg. (April 30) 36, 8186-8201.

Federal Register. 1987. Revisions to the National Ambient Air Quality Standards for Particulate Matter. Fed. Reg. (July 1) 52, 24,634-24,669.

Federal Register. 1997. National Ambient Air Quality Standards for Particulate Matter. Final Rule. Fed. Reg. (July 18) 62, 38,652-38,752.

Ferris, B.G. Jr., Speizer, F.E., Spengler, J.D., *et al.* 1979. Effects of Sulfur Oxides and Respirable Particles on Human Health: Methodology and Demography of Populations in Study. Am. Rev. Respir. Dis. 120, 767-779.

Frampton, M.W. and Utell, M.J. 1999. Clinical Studies of Airborne Pollutants. In: Toxicology of the Lung, Gardner, D.E., Crapo, J.D. and McClellan, R.O., editors, Taylor & Francis, Philadelphia, PA, pp 455-481.

Frampton, M.W., Utell, M.J. and Samet, J.M. 2000. Cardiopulmonary Consequences of Particulate Inhalation. In: Particle-Lung Interactions, Gehr, P. and Heyder, J., editors, Marcel Dekker, New York, NY, Pp 653-670.

Gehr, P. and Heyder, J. 2000. Particle-Lung Interactions. Marcel Dekker, New York.

Godleski, J.J., Verrier, R.L., Koutrakis, P., Catalano, P., Coull, B.A. and Reinisch, U. 2000. Mechanisms of Morbidity and Mortality from Exposure to Ambient Air Particulates. Research Report No. 91, Health Effects Institute, Cambridge, MA.

Grahame, T. and Hidy, G. W. 2003. Using Factor Analysis to Attribute Health Impacts to Particulate Pollution Sources. Submitted to Environ. Health Persp.

Green, L.C. and Armstrong, S.R., 2003. Particulate Matter in Ambient Air and Mortality: Toxicologic Perspectives. Regulatory Toxicology and Pharmacology (in press).

Greenberg, L., Field, F., Reed, J.I. and Erhardt, C.L. 1962a. Air Pollution and Morbidity in New York City. J. Am Med Assoc 182,159-162.

Greenberg, L., James, M.B., Droletti, B.M. *et al.* 1962b. Report of Air Pollution Incident in New York City, 1953. Public Health Reports 77, 7-16.

Hastie, T.J. and Tibshirani, R.J. 1990. Generalized Additive Models. Chapman and Hall, London.

Health Effects Institute, 2003a. Revised analyses of the National Morbidity, Mortality, and Air Pollution Study (NMMAPS), part II. In: Revised analyses of time-series studies of air pollution and health. Special report, Boston, MA: Health Effects Institute, pp. 9-72. Available: http://www.healtheffects.org/news.htm [16 May 2003].

Health Effects Institute, 2003b. Commentary on revised analyses of selectied studies. In: Revised analyses of time-series studies of air pollution and health. Special report, Boston, MA: Health Effects Institute, pp. 255-290. Available: http://www.healtheffects.org/news.htm [16 May 2003].

Hedley, A.J., Wong, C., Thach, T.Q., Ma, S., Lam, T. and Anderson, H.R. 2002. Cardiorespiratory and All-cause Mortality after Restrictions on Sulfur Content of Fuel in Hong Kong: An Intervention Study. Lancet 360, 1646-52.

Hill, A.B. 1965. The Environment and Disease: Association or Causation? Proc. Royal Soc. Med. 58, 295-300.

Holgate, S.T., Samet, J.M., Koren, H.S. and Maynard, R.L., editors. 1999. Air Pollution and Health. Academic Press, London.

Holland, W.W. and Reid, D.D. 1965. The Urban Factor in Chronic Bronchitis. Lancet 323, 445-448.

Hoyert, D.L., Kochanek, K.D. and Murphy, S.L. 1999. Deaths: Final Data for 1997. National Vital Stat. Rep. 47,1-104.

ICRP (International Commission on Radiological Protection). 1994. Human Respiratory Tract Model for Radiological Protection. ICRP Publication 66: Annals ICRP; 24, 1-482.

Johnson, K.G., Gideon, R.A. and Luftsgaarden, D.O. 1990. Montana Air Pollution Study: Children's Health Effects. J. Off. Stat. 5, 391-408.

Krewski, D., Burnett, R.T., Goldberg, M.S., Hoover, K., Siemiatycki, J., Jerrett, J., Abrahamowicz, M. and White, W.H. 2000. Reanalysis of the Harvard Six Cities Study and the American Cancer Society Study of Particulate Air Pollution and Mortality. A Special Report of the Institute's Particle Epidemiology Reanalysis Project, Health Effects Institute, Cambridge, MA.

Laden, F., Neas, L.M., Dockery, D.W., Schwartz, J., 2000. Association of fine particulate matter from different sources with daily mortality in six U.S. cities. Environmental Health Perspectives 108, 841-947.

Lave, L.D., Seskin, E.P., 1970. Air pollution and human health. The quantitative effect with an estimate of the dollar benefit of pollution abatement, is considered. Science 169, 723-733.

Lawrence, L., Hall, M.J., 1999. 1997 summary national hospital discharge survey. Advance Data 308,1-16.

Levy, J.I., Hammitt, J.K., Yanagisawa, Y., Spengler, J., 1999. Development of a new damage function for power plants: Methods and applications. Environmental Science and Technology 33, 4364-4372.

Lippmann, M., Frampton, M., Schwartz, J., Dockery, D., Schlesinger, R., Koutrakis, P., Fronies, J., 2003. The EPA's particulate matter (PM) health effects research program: A mid-course (2-1/2 year). Report of the Status Progress and Plans. Environmental Health Perspectives (in press).

MacNee, W., Donaldson, K., 1999. Particulate air pollution: Injurious and protective measurements in the lungs. In: Air Pollution and Health. Holgate, S.T., Samet, J.M., Koren, H.S., Maynard, R.L., eds., Academic Press, London, pp. 653-672.

Martin, A.E., 1964. Mortality and morbidity statistics and air pollution. Proceedings of the Royal Society of Medicine 57, 969-975.

Mauderly, J.L., Ness, L., Schlesinger, R., 1998. PM monitoring needs related to health effects. In: Atmospheric Observations. Helping Build the Scientific Basis for Decisions Related to Airborne Particulate Matter. Albritton, D.L., Greenbaum, D.S., eds., Health Effects Institute, Cambridge, MA, pp. 9-14.

Mauderly, J.L., 2000. Animal models for the effect of age on susceptibility to inhaled particulate matter. Inhalation Toxicology 12, 863-900.

McClellan, R.O., Henderson, R.F., 1995. Concepts in Inhalation Toxicology. 2nd Edition, Taylor and Francis, Washington, DC.

McClellan, R.O., 1999. Ambient air pollution matter: Toxicology and standards. In: Toxicology of the Lung. Gardner, D.E., Crapo, J.D., McClellan, R.O., eds., Francis and Taylor, Philadelphia, PA, pp. 289-342.

McClellan, R.O., 2002. Setting ambient air quality standards for particulate matter. Toxicology 181-182: 329-347.

McCullough, B., 1998. Assessing the reliability of statistical software: Part I. American Statistics 52, 358-366.

McCullough, B., 1999. Assessing the reliability of statistical software: Part II. American Statistics. 53, 149-159.

Miller, F.J., Gardner, D.E., Graham, J.A., Lee, R.E., Wilson, W.E., Bachman, J.D., 1979. Size consideration for establishing a standard on inhalable particles. Journal of the Air Pollution Control Association 29, 610-615.

Miller, F.J., 1999. Dosimetry of particles in laboratory animals and humans. In: Toxicology of the Lung. Gardner, D.E., Crapo, J.D., McClellan, R.O., eds., Taylor and Francis, Philadelphia, PA, pp. 513-555.

Molina, LT., Molina, M.J., eds., 2002. Air Quality in the Mexico Megacity: An Integrated Assessment. Kluwer Academic Publishers, Dordrecht.

Muggenburg, B.A., Tilley, L., Green, F.H.Y., 2000. Animal models of cardiac disease: Potential usefulness for studying effects of inhaled particles. Inhalation Toxicology 12, 901-925.

NCRP (National Council on Radiation Protection), 1997. National Council on Radiation Protection and Measurements, Deposition, Retention and Dosimetry of Inhaled Radioactive Substances. NCRP Report SC-72, Bethesda, MD.

NRC (National Research Council), 1991. Committee on Advances in Assessing Human Exposure to Airborne Pollutants. Human Exposure Assessment for Airborne Pollutants: Advances and Opportunities. National Academy Press, Washington, DC.

NRC (National Research Council), 1998. Research Priorities for Airborne Particulate Matter: I. Immediate Priorities and a Long-Range Research Portfolio. National Academy Press, Washington, DC.

NRC (National Research Council), 1999. Research Priorities for Airborne Particulate Matter: II. Evaluating Research Progress and Updating the Portfolio. National Academy Press, Washington, DC.

NRC (National Research Council), 2001. Research Priorities for Airborne Particulate Matter: III. Early Research Progress. National Academy Press, Washington, DC.

NRC (National Research Council), 2002. Estimating the Public Health Benefits of Proposed Air Pollution Regulations. The National Academy Press, Washington, DC [www.nap.edu].

NRC (National Research Council), 2003. Research Priorities for Airborne Particulate Matter. IV. Summary Report. The National Academy Press, Washington, DC (in press).

Özkaynak, H., Spengler, J.D., 1987. Associations between 1980 U.S. mortality rates and alternative measures of airborne particulate concentrations. Risk Analysis 7, 449-461.

Özkaynak, H., Xue, J., Spengler, J., Wallace, L., Pellizzari, E., Jenkins, P., 1996. Personal exposures to airborne particles and metals: Results from the particle TEAM study in Riverside, California. Journal of Exposure Analysis and Environmental Epidemiology 6, 57-78.

Özkaynak, H., 1999. Exposure assessment. In: Air Pollution and Health. Holgate, S.T., Samet, J.M., Koren, H.S., Maynard, R.L., eds., Academic Press, London and San Diego, pp. 149-162.

Phalen, R.F., Hinds, W.C., John, W., Lioy, P.J., Lippmann, M., McCawley, M.A., Raabe, O.G., Soderholm, S.C., Stuart, B.O., 1986. Rationale and recommendations for particle size-selective sampling in the workplace. Applied Industrial Hygiene 1, 3-14.

Phalen, R.F., 2002. The Particulate Matter Controversy: A Case Study and Lessons Learned. Kluwer Academic Publishers, Boston.

Pope III, C.A., 1989. Respiratory disease associated with community air pollution and a steel mill, Utah valley. American Journal of Public Health 79, 623-628.

Pope III, C.A., Thunm, M.H.J., Namboodiri, M.M., Dockery, D.W., Evans, J.S., Speizer, F.E., Heath Jr., C.W., 1995. Particulate air pollution as a predictor of mortality in a prospective study of U.S. adults. American Journal of Respiratory and Critical Care Medicine 151, 669-674.

Pope III, C.A., Dockery, D.W., 1999. Epidemiology of particle effects. In: Air Pollution and Health. Holgate, S.T., Samet, J.M., Koren, H.S., Maynard, R.L., eds., Academic Press, London and San Diego, pp. 673-705.

Pope III, C.A., 2000. Epidemiology of fine particulate air pollution and human health: Biologic mechanisms and who's at risk? Environmental Health Perspectives 108 (Suppl. 4), 713-723.

Pope III, C.A., Burnett, R.T., Thun, M.J., Calle, E.E., Krewski, D., Ito, K., Thurston, G.D., 2002. Lung cancer, cardiopulmonary mortality, and long-term exposure to fine particulate air pollution. Journal of the American Medical Association 287, 1132-1141.

Ramsey, T., Burnett, R., Krewski, D., 2003. The effect of concurvity in generalized additive models linking mortality to air pollution. Epidemiology (in press).

Rojas-Bracho, L., Suh, H.H., Koutrakis, P., 2000. Relationship among personal, indoor and outdoor fine and coarse particle concentrations for individuals with COPD. Journal of Exposure Analysis and Environmental Epidemiology 10, 294-306.

Rothman, K.J., 1986. Modern Epidemiology. 1st Edition, Little, Brown and Company, Boston, MA.

Samet, J.M., Jaakkola, J.J.K., 1999. The epidemiologic approach to investigating outdoor air pollution. In: Air Pollution and Health. Holgate, S.T., Samet, J.M., Koren, H.S., Maynard, R.L., eds., Academic Press, London and San Diego, pp. 431-460.

Samet, J.M., Zeger, S.L., Schwartz, J., Dockery, D.W., 2000a. The national morbidity, mortality, and air pollution study. Part I: Methods and methodologic issues. Research Reports of the Health Effects Institute.

Samet J.M., Zeger, S.L., Dominici, F., Curriero, F., Coursac, I., Dockery, D.W., Schwartz J., Zanobetti, A., 2000b. The national morbidity, mortality, and air pollution study. Part II: Morbidity and mortality from air pollution in the United States. Research Reports of the Health Effects Institute.

Samet, J.M., Dominici, F., McDermott, A., Zeger, S.L., 2003. New problems for an old design: Time series analyses of air pollution and health epidemiology. Epidemiology (in press).

Sarnat, J.A., Kontrakis, P., Suh, H.H., 2000. Assessing the relationship between personal particulate and gaseous exposures of senior citizens living in Baltimore, MD. Journal of the Air and Waste Management Assocation 50, 184-1198.

Schlesinger, R.B., 1995. Deposition and clearance of inhaled particles. In: Concepts in Inhalation Toxicology. McClellan, R.O., Henderson, R.F., eds., Francis & Taylor, Washington, DC, pp. 191-224.

Sioutas, C., Koutrakis, P., Burton, R.M., 1995. A technique to expose animals to concentrated fine ambient aerosols. Environmental Health Perspectives 103, 172-177.

Snipes, M.B., 1995. Pulmonary retention of particles and fibers: Biokinetics and effects of exposure concentrations. In: Concepts in Inhalation Toxicology, McClellan, R.O., Henderson, R.F., eds., Taylor and Francis, Philadelphia, PA.

Stebbings, J.H., Fogleman, D.G., McClain, K.E., Townsend, M.C., 1976. Effect of the Pittsburgh air pollution episode upon pulmonary function in schoolchildren. Journal of the Air Pollution Control Association 26, 547-553.

U.S. EPA (U.S. Environmental Protection Agency), 1992. Guidelines for Exposure Assessment. Federal Register 57, 22,888-22,938.

U.S. EPA (U.S. Environmental Protection Agency), 1996a. Air Quality Criteria for Particulate Matter, Vol. 1-3. EPA/600/p-95/001A-CF. Research Triangle Park, NC. Office of Research and Development, National Center for Environmental Assessment, U.S. Environmental Protection Agency.

U.S. EPA (U.S. Environmental Protection Agency), 1996b. United States Environmental Protection Agency Review of the National Ambient Air Quality Standards for Particulate Matter: Policy Assessment of Scientific and Technical Information, OAQPS Staff Paper EPA/452/R-96/013. Research Triangle Park, NC, Office of Air Quality Planning and Standards, U.S. Environmental Protection Agency.

U.S. EPA (U.S. Environmental Protection Agency), 2001. Review of the National Ambient Air Quality Standards for Particulate Matter: Policy Assessment of Scientific and Technical Information, Office of Air Quality Planning and Standards, U.S. Environmental Protection Agency.

U.S. EPA (U.S. Environmental Protection Agency), 2002. Third External Review Draft of Air Quality Criteria for Particulate Matter (April 2002). EPA/600/P-99/002ac. Office of Research and Development, National Center for Environmental Assessment, Research Triangle Park, NC. [Online]. Available: http://www.epa.gov/ordntrnt/ORD/archives/2002/june/htm/article2.htm. [September 6, 2002].

U.S. HEW (U.S. Department of Health, Education and Welfare), 1969. Air Quality Criteria for Particulate Matter. U.S. Government Printing Office, Washington, DC, AP-49.

Utell, M.J., Frampton, M.W., Zareba, W., Devlin, R.B., Cascio, W.E., 2002. Cardiovascular effects associated with air pollution: Potential mechanisms and methods of testing. Toxicology 14,101-117.

Valberg, P.A., 2003. Is PM more toxic than the sum of its parts? Risk-assessment toxicity factors versus PM-mortality "effect functions." Inhalation toxicology (in press).

Valentine, R., Kennedy, G.L., 2001. Inhalation toxicology. In: Principles and methods in toxicology, Hayes, A.W., ed., Taylor and Francis, Philadelphia, PA, pp. 1085-1144.

Vincent, J.H., 1999. Sampling criteria for the inhalable fraction. In: Particle size-selective sampling for particulate air contaminants, Vincent, J.H., ed., American Conference of Governmental Industrial Hygienists, Cincinnati, OH, pp. 51-72.

Wallace, L., 2000. Correlations of personal exposure to particles with outdoor air measurements: A review of recent studies. Aerosol Science and Technology 32,15-25.

Wilson, R., Spengler, J., 1996. Particles in our air: Concentrations and health effects. Harvard University Press, Cambridge, MA.

Wolff, R.K., 1996. Experimental investigation of deposition and fate of particles: Animal models and interspecies differences. In: Aerosol Inhalation: Recent Resarch Frontiers, Marijnissen, J.C.M., and Gradon, L., eds., Kluwer Academic Publishers, Harwell, MA, pp. 247-263.

Wong, C.M., Lam, T.H., Peters, J., Hedley, A.J., Ong, S.G , Tam, A.Y., Liu, J., Spiegelhalter, D.J., 1998. Comparison between two districts of the effects of an air pollution intervention on bronchial responsiveness in primary school children in Hong Kong. Journal of Epidemiology and Community Health. 52, 571-8.

Wong, C.M., Hu, Z G., Lam, T.H., Hedley, A.J., Peters, J., 1999. Effects of ambient air pollution and environmental tobacco smoke on respiratory health of non-smoking women in Hong Kong. International Journal of Epidemiology 28, 859-64.

CHAPTER 3

Atmospheric Aerosol Processes

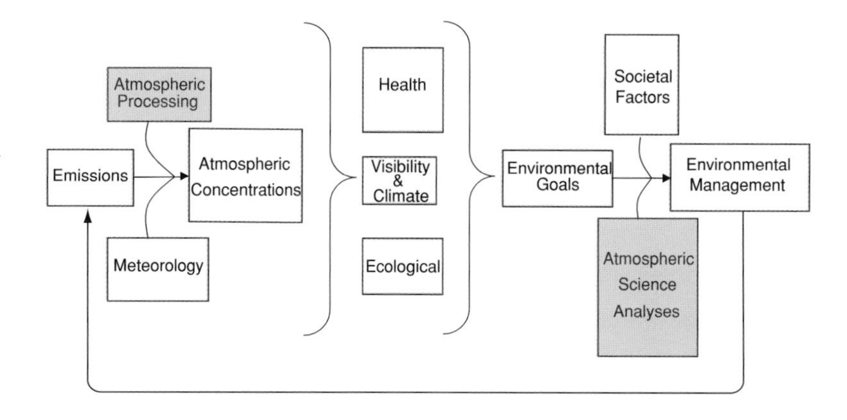

Principal Author: Spyros Pandis

3.1 THE LIFE OF AN ATMOSPHERIC PARTICLE

Atmospheric particles originate either as primary particles - by direct emission from a source or as secondary particles - through in-situ formation from the gas phase (nucleation). Particles vary in size from a few nanometers to tens of micrometers, with their composition reflecting their source. Secondary particles can be created in different parts of the atmosphere, sometimes high near a cloud or even the top of the troposphere and sometimes near the surface of the earth. After entering the lower atmosphere, new particles can exist for several days depending on removal processes. During their lifetime, they are changed by processes such as dilution, dispersion, coagulation, and chemical reaction.

Upon their emission to (or formation in) the atmosphere, particles move under the influence of local air currents, simultaneously diffusing and, possibly, colliding through turbulent and Brownian processes. These processes dilute the particles and mix them with other particles and gaseous compounds (Figure 3.1). Collisions between two or more particles typically result in coagulation, wherein the original particles adhere to form larger particles having the sum of the original masses. Coagulation effectively increases the mass of particles while depleting smaller particles, and often is an important mechanism for shifting the aerosol-size spectrum toward larger particle sizes.

If the particles avoid coagulation, which is relatively rapid near their source, they travel beyond the source region, interacting with vapors such as H_2SO_4, organics, HNO_3, and NH_3. These semivolatile or reactive vapors, when their concentration exceeds specific thresholds, condense upon available surfaces, including the surfaces of existing particles. Some condensed vapors react with other vapors and attract them to the condensed phase as well. H_2SO_4 reacts with NH_3, for example, and condensed organic compounds can dissolve other organic vapors. Particles form also as the consequence of gas-phase reactions such as the reaction of NH_3 with HNO_3 to form NH_4NO_3, thus transferring gaseous material to the particulate phase. Consequently the particles grow in size and contain material derived both from their origin and from the places where they have been. Some of this deposited material may return to the gas phase if the conditions are right. For instance, NH_4NO_3 can volatilize to produce NH_3 and HNO_3, and organic particles can volatilize to emit organic vapors. Because semivolatile particle components exchange continuously between the gas and condensed phases, it is difficult to measure PM concentrations in the atmosphere and to completely determine aerosol behavior and impact.

During their atmospheric lifetimes particles frequently encounter humidity environments exceeding 70 percent. Under such conditions and depending on their composition, they can absorb water vapor, consequently forming concentrated aqueous solutions. The amounts of water condensed by this process can be quite high, with particle-mass increases amounting to factors of three or four between low (<40 percent) and high (>80 percent) relative humidity. As relative humidity decreases, the water revolatilizes, resulting in particle drying.

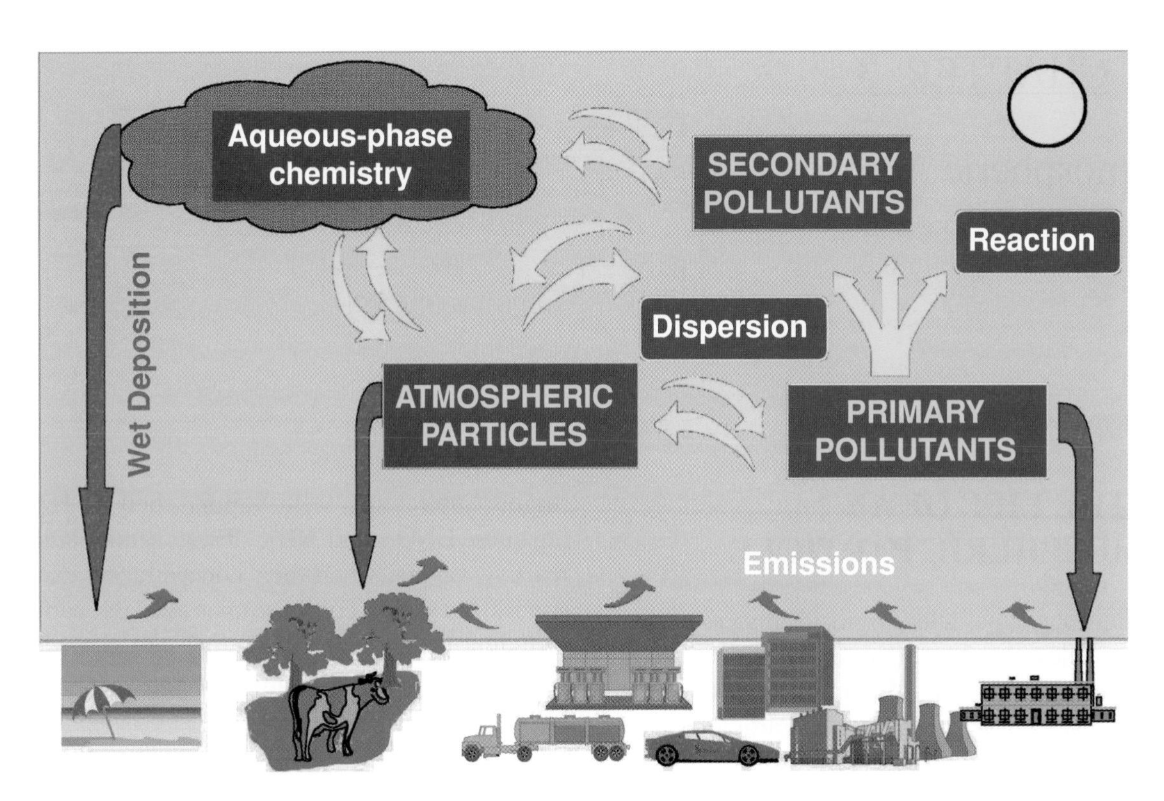

Figure 3.1. Schematic of the life cycle of atmospheric particles and their interactions with the gas and aqueous phases.

Typically, particles are transported extensive distances by the wind as the above-described physicochemical processes occur. At an average wind speed of a few meters per second, particles can travel a few hundred kilometers horizontally, and several kilometers vertically, in a period of one day.

When atmospheric particles experience relative humidities of around 100 percent or higher, they can absorb even larger amounts of water and form cloud droplets. This newly acquired water provides an environment suitable for a new suite of reactions, with the dissolution of SO_2 and its conversion to $SO_4^=$ being a prime example. In a majority of cases the cloud water subsequently re-evaporates, resulting in the re-formation of relatively dry particles. Such condensation-evaporation cycles usually leave their mark, however, with the progressive accumulation of condensed material, such as $SO_4^=$. Atmospheric particles eventually are removed by wet deposition in rain, snow, or fog water, or by dry deposition at the Earth's surface. Particles transported vertically to extended elevations generally have longer residence times and travel farther than particles at lower elevations.

As a result of particle emission, in-situ formation, and subsequent processes, the atmospheric particle distribution is characterized by a number of modes. The volume or mass distribution is dominated in most areas by two modes (Figure 3.2, lower panel): the accumulation mode (from around 0.1 to around 2 μm) and the coarse mode (from around 2 to around 50 μm). Accumulation-mode particles result from primary emissions, condensation of secondary sulfates, nitrates, and organics from the gas phase, and coagulation of smaller particles. In a number of cases the accumulation mode consists of two overlapping sub-modes, the condensation and droplet mode (Figure 3.2, lower panel) (John et al., 1990).

The condensation sub-mode results from primary particle emissions and growth of smaller particles by coagulation and vapor condensation. The droplet sub-mode is created during the cloud processing of some of the accumulation-mode particles. Particles in the coarse mode are usually produced by mechanical processes, such as wind or erosion (dust, sea salt, pollens, etc.). Most of the material in the coarse mode is primary, along with some secondary $SO_4^=$ and NO_3^-.

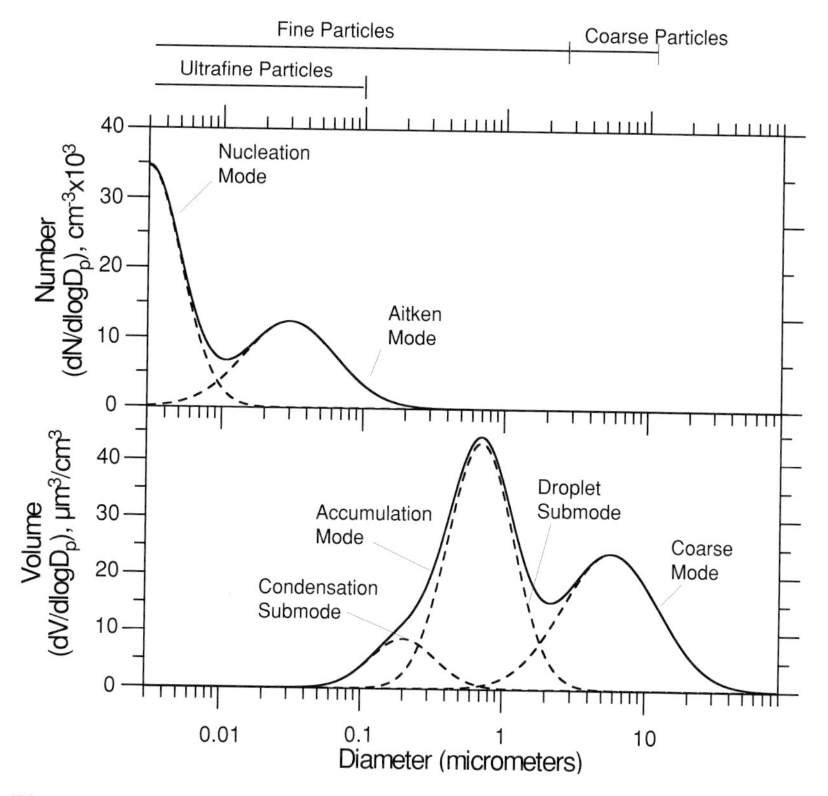

Figure 3.2. Typical number and volume distributions of atmospheric particles with the different modes.

The composition of individual particles changes continuously during their atmospheric lifetimes. Near a source, the composition of primary particles is determined by chemical and physical processes that occur within the source. For example, primary particulate emissions from diesel engines include mostly lubricating oil and solid black carbon, while primary particulate emissions from coal-fired power plants include fly ash which consists mostly of oxides and minerals. In the absence of processing, atmospheric particles of a given size would consist of an external mixture of chemically distinct particles emitted by each contributing source type. Atmospheric processing, however, leads to coagulation and the addition of secondary species to preexisting particles, with the result that particles of a given size become internally mixed and more similar chemically. Atmospheric measurements have shown that particles of a given size often include several chemically distinct types, and particles may include a mixture of compounds. Photographs of particles that were separated according to their tendency to absorb water are shown in Figure 3.3. The non-hygroscopic chain-agglomerate carbon particle shown in Figure 3.3a is similar in composition and morphology to particles emitted by diesel engines, while the hygroscopic particle shown in Figure 3.3b is nearly spherical and includes a mixture of sodium, sulfur, potassium and oxygen. Other work has shown that organics and sulfur are often found together in particles.

A different picture of the ambient PM distribution is obtained if one focuses on the number of particles instead of their mass (Figure 3.2, upper panel). The particles with diameter larger than 0.1 µm that contribute practically all the PM mass are negligible in number compared to the particles smaller than 0.1 µm. Two modes usually dominate the PM number distribution in urban and rural areas: the nucleation mode (particles smaller than 0.01 µm or so) and the Aitken nucleus mode (particles with diameters between 0.01 µm and 0.1 µm or so). The nucleation-mode particles are usually fresh, created in-situ from the gas phase by nucleation. A nucleation mode may or may not be present in a particular aerosol, depending on the atmospheric conditions. Most of the Aitken nuclei start their atmospheric life as primary particles, and secondary material condenses on them as they are transported through the atmosphere. Nucleation-mode particles have negligible mass (for example 100,000 particles per cubic centimeter with a diameter equal to 0.01 µm have a mass concentration of less than 0.05 µg/m^3) while the larger Aitken nuclei form the accumulation mode in the mass distribution.

As noted in Text Box 1.1, particles with diameters larger than 2.5 µm are operationally identified as coarse particles while those with diameters less than 2.5 µm are called fine particles. The fine-particle category typically includes most of the total number of particles and a large fraction of the mass. Fine particles with diameters smaller than 0.1 µm are often called ultrafine particles.

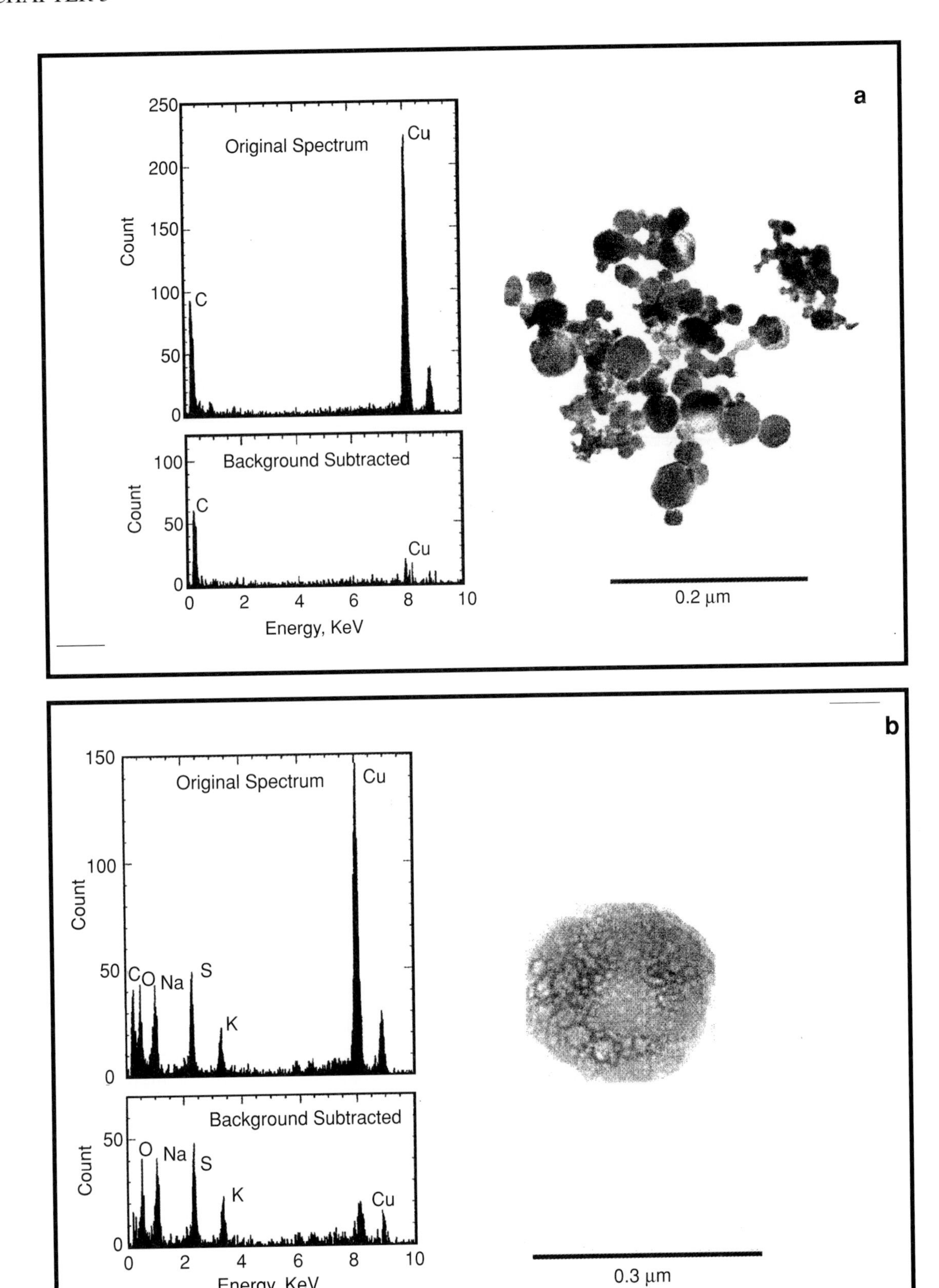

Figure 3.3. Electron micrographs of selected particles: a) carbon agglomerate and b) inorganic composite (McMurry et al., 1996).

3.2 A PARTICLE IS BORN: NUCLEATION

As indicated above, new atmospheric particles can form from gas-phase reactions in a number of different ways, with $SO_4^=$ particle formation being the best-known example. In this process SO_2 reacts with hydroxyl radical forming H_2SO_4 vapor as a product. The H_2SO_4 molecules thus formed either condense on pre-existing particles or combine with each other and form new particles (homogeneous nucleation). The condensation path is by far the easiest if sufficient numbers of pre-existing particles are present (Seinfeld and Pandis, 1998). However, if the quantity of such pre-existing particles is insufficient to accommodate the continuously formed H_2SO_4, its concentration increases, with the resulting formation of new particles. Water molecules participate in these reactions and it is speculated that NH_3 molecules play a role as well. A quantitative understanding of the role of organics in atmospheric nucleation does not exist even though it is known that they are prominent components of newly formed particles in forested regions. Under the right conditions (high H_2SO_4 production rate, high relative humidity, low temperature, low pre-existing PM concentration), tens of thousands of new particles per cubic centimeter can be formed in a matter of minutes. These particles have diameters of only a few nanometers and their mass is negligible compared to the rest of the particle distribution; however, their numbers can be huge and may dominate the total number concentration. Consequently homogeneous nucleation is not an important process if one is interested in PM_{10} or $PM_{2.5}$ or even PM_1 mass concentration; but it can be a major contributor to PM number concentration and formation of ultrafine particle mass. The sulfuric acid/water and sulfuric acid/water/ammonia systems are not the only ones of interest in this context. Newly formed organic vapors can also nucleate and form organic particles. These processes have been observed in the laboratory during the oxidation of biogenic hydrocarbons (Griffin et al., 1999) and there is evidence that they also occur near forests (Kavouras et al., 1998). Nucleation has been shown to be prevalent in the upper troposphere where the conditions (low temperatures, low particle loadings) are favorable. Recent studies have documented nucleation events in major cities like Atlanta, rural polluted areas, and forests (McMurry et al., 2000). Despite this progress, the importance of nucleation as a source of ultrafine particles in urban and rural areas is still not well understood.

Substantial formation of new particles can occur as fresh combustion emissions are entrained into the ambient air, for example near the tailpipes of diesel vehicles (Kittelson et al., 1999). As hot vapors emitted by such sources are cooled, the emitted vapors become supersaturated and new particles form. These ultrafine particles consist largely of organic compounds similar in composition to lubricating oils and fuels (with some sulfates). The numbers of these particles typically increase with increasing fuel sulfur content and decreasing ambient temperature. Concentrations of these ultrafine particles typically follow rush-hour patterns and are quite high near roadways in populated regions. Describing the emissions of these ultrafine particles will be a significant challenge for the next generation of particle emission inventories.

New-particle formation rates can increase if one removes existing particles without removing the corresponding vapors. Technologies that reduce particle emissions without reducing gas-phase precursor emissions have the potential to increase particle number concentrations while decreasing particle diameter and mass. Nucleation is one of the least understood atmospheric aerosol processes. Recent measurements have shown that, during the course of a day, nucleated particles can grow to sizes where they can serve as cloud-condensation nuclei (~0.05 µm). Therefore, it is likely that nucleation is significant to climate. Furthermore, if ultrafine particles are found to contribute significantly to health-effect impacts, the understanding of these processes will become critical. *Therefore, the ability to describe the formation of new particles by homogeneous nucleation should be improved.*

3.3 HOW LONG DOES IT TAKE FOR A PARTICLE TO COLLIDE WITH ANOTHER?

Coagulation occurs most efficiently between small particles (a few nanometers) and large ones (a few micrometers). The small particles experience rapid Brownian motion, while large ones provide big collision targets. Therefore, coagulation can be mainly viewed as a process for removing smaller atmospheric particles from the atmosphere. Coagulation has little or no effect on larger particles, because of the addition of a negligible mass to an already big particle. As a result, coagulation is an important process if one is interested in the small particles (or the total particle number) but it has little effect on PM_x mass concentrations with the exception of ultrafine particles. As a process, it is fairly well understood. The evolution of a typical urban particle-size distribution subject to coagulation is shown in Figure 3.4. Most particles with diameters smaller than 0.02 μm disappear after a few hours as they coagulate into larger particles. After a few days most of the particles with diameters smaller than 0.1 μm also disappear. During these few days, coagulation has for all practical purposes no effect on the size distribution of particles with diameters above 0.3 μm, which are responsible for most of the $PM_{2.5}$ mass concentration. If these particles are in a cleaner environment, at lower overall concentrations, their coagulation becomes less efficient with corresponding increases in particle lifetime (Schutz et al., 1990).

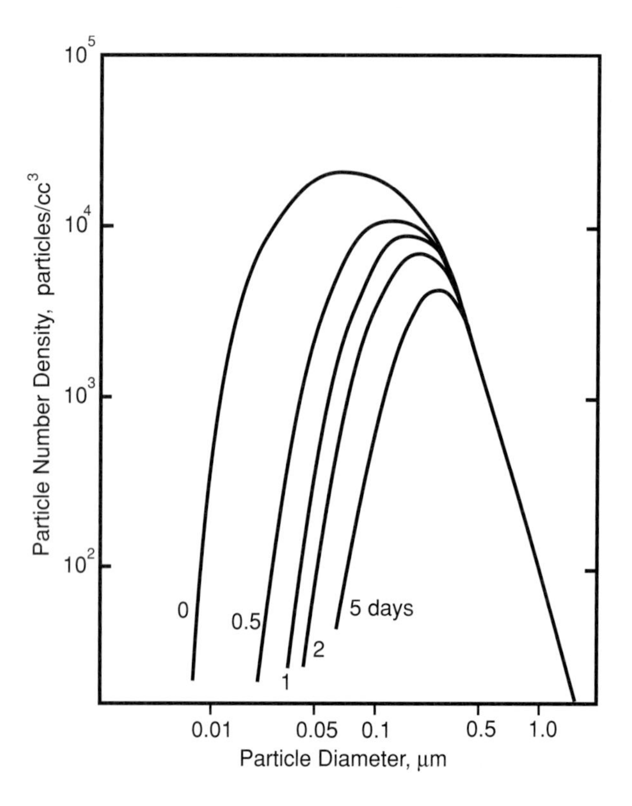

Figure 3.4. Evolution of a typical urban aerosol size distribution subject to coagulation.

3.4 PARTICLES AND WATER

At very low relative humidity, most inorganic atmospheric particles are solid with the exception of H_2SO_4. As the ambient relative humidity increases, the particles remain solid until the relative humidity reaches a critical threshold characteristic of the aerosol composition (illustrated as an example for salt, NaCl, in Figure 3.5). At this relative humidity (known as the deliquescence relative humidity), the solid salt particles spontaneously absorb water, producing a saturated aqueous solution. Further increase of the ambient relative humidity leads to additional condensation of water onto the salt solution. On the other hand, as the relative humidity over the wet particle is decreased, evaporation occurs. However, the solution generally does not crystallize at the deliquescence relative humidity, but remains supersaturated until a much lower relative humidity at which crystallization occurs. This hysteresis phenomenon is illustrated in Figure 3.5 for a mixed inorganic ($NaCl/Na_2SO_4$) particle, which deliquesces at 72 percent relative humidity and crystallizes at around 52 percent. For multi-component particles, the behavior becomes more complex but follows the same general principles. The ability of inorganic particles to absorb water is well understood (Figure 3.5), but questions still remain about the crystallization step. Some organic aerosol components are also hygroscopic, and there is growing evidence that organics contribute significantly to water uptake especially at low relative humidities (Saxena et al., 1995). Black carbon and OC are expected to react with OH and other oxidants, becoming hydrophilic; hence particles can absorb

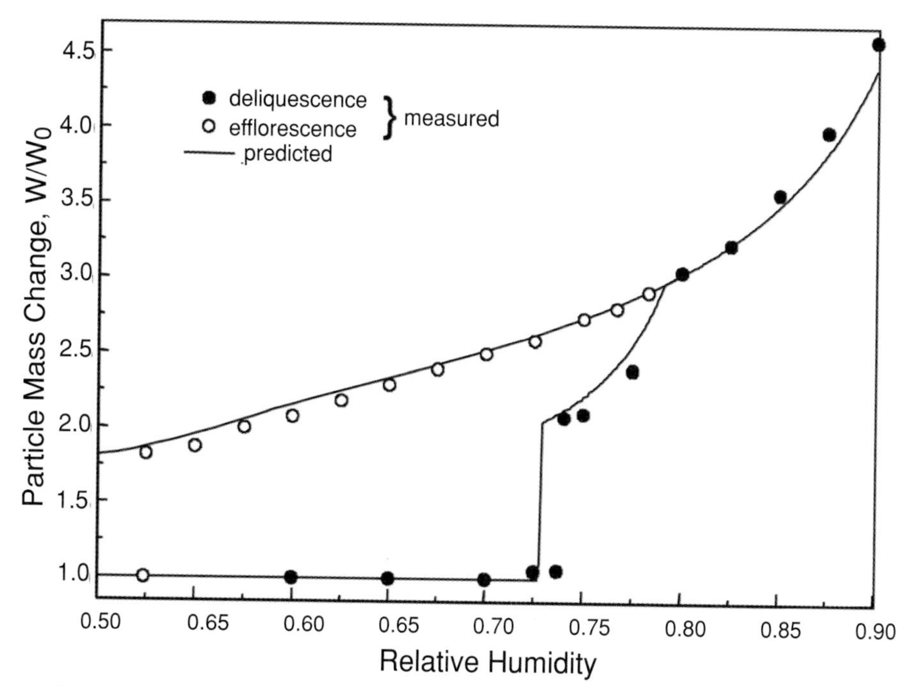

Figure 3.5. Growth (deliquescence) and evaporation (efflorescence) of a $NaCl/Na_2SO_4$ particle with changing relative humidity. The points are the laboratory measurements of Tang (1997) and the line the theoretical predictions of one of the aerosol thermodynamics models (Ansari and Pandis, 1999). The measurements indicate that the particle crystallized at 52 percent relative humidity.

water more efficiently as they age (Bertram et al., 2001). Despite these recent efforts, the effect of the OC fraction on the absorption of water by ambient particles is still not well understood. *More effort is needed to understand the properties of organic particles including factors that govern the hygroscopicity of organic compounds and their gas-particle partitioning.*

As the relative humidity and the water content of the particles change, so does the partitioning of unreactive semivolatile aerosol species. With everything else kept constant (atmospheric composition and temperature), a change in relative humidity changes the water content of atmospheric particles and therefore the preferences of aerosol components like nitrates or organics for the gas versus the condensed phases. Figure 3.6 shows that the NO_3^- concentration in a typical particle population can change from around 2 $\mu g/m^3$ to 7 $\mu g/m^3$ as the relative humidity of the environment increases from 30 to 80 percent.

The fact that atmospheric particles can absorb significant amounts of water at higher relative humidity and can retain some water at lower relative humidity creates a number of complications. As the "dry" PM mass (at 20 to 45 percent relative humidity) is the regulated quantity, the existence of water makes the measurement of this "dry" mass challenging. One needs to remove the water from the particles before the measurements. This change in particle composition can change the partitioning of the semivolatile aerosol components (usually reducing the PM mass because of losses of NO_3^- and organics). At the same time, it is possible that some amount of water may be left in the PM sample during the mass measurement. The discrepancies observed in the northeastern United States between the total PM mass and the sum of the masses of its components could be due to remaining condensed water. This unknown mass is usually reported as "other" in PM measurements. As the airways in the human body are environments of high relative humidity, similar changes happen to the particles during respiration.

3.5 SECONDARY PM FORMATION

Most of the observed ambient $PM_{2.5}$ mass usually originates as precursor gases (SO_2, NH_3, NO_x, VOC) and, through the physicochemical processes noted above, is transferred to the condensed phase. This fraction of the PM mass is called secondary.

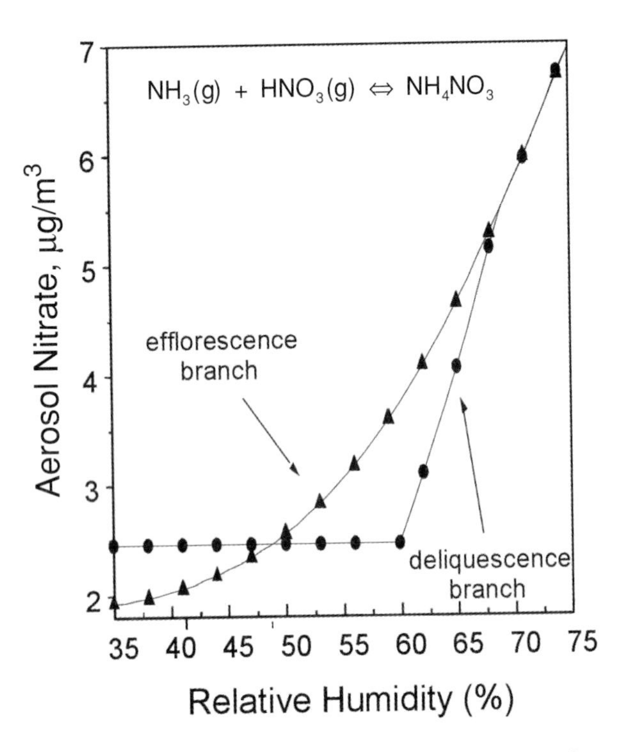

Figure 3.6. Predicted particulate nitrate concentration as a function of relative humidity for a typical environment. The actual value in the atmosphere will depend on the history of the aerosol particles.

3.5.1 Sulfate

The formation of $SO_4^=$ from the oxidation of SO_2 is an important process for most areas in North America. There are three different pathways for this transformation:

1. The oxidation of SO_2 in the gas phase by the hydroxyl radical, OH. This occurs at an average rate of 0.1 percent to 1 percent of SO_2 per hour (with peak rates up to 5 percent per hour) during the daytime. Nighttime conversion via this pathway is essentially negligible.

2. The dissolution of SO_2 in cloud, fog, or rain water and subsequent aqueous-phase oxidization (Figure 3.7). After the cloud evaporates, the $SO_4^=$ remains in the particles and is eventually deposited to the surface. Aqueous-phase production can be very fast: in some cases all the available SO_2 can be oxidized in less than an hour. Near SO_2 sources the process is usually limited by the availability of oxidants such as hydrogen peroxide, H_2O_2. An oxidation mechanism that generally takes place more slowly than oxidant-driven transformation of dissolved SO_2 is catalysis by transition heavy metals such as manganese and iron.

3. The oxidation of SO_2 in reactions in the water of the aerosol particles themselves. This process takes place continuously, but only produces appreciable $SO_4^=$ in alkaline (dust, sea salt) coarse particles (Sievering et al., 1992). Oxidation of SO_2 has been also observed on the surfaces of black carbon and metal oxide particles.

During the last twenty years, much progress has been made in understanding the first two major pathways, but some important questions still remain about the smaller third pathway. Models indicate that more than half of the $SO_4^=$ in the eastern United States and the overall atmosphere is produced in clouds (McHenry and Dennis, 1994; Langner and Rodhe, 1991). Processing of SO_2-rich air masses by clouds or fogs during stagnation periods can lead to elevated concentrations of $SO_4^=$ during specific days both in

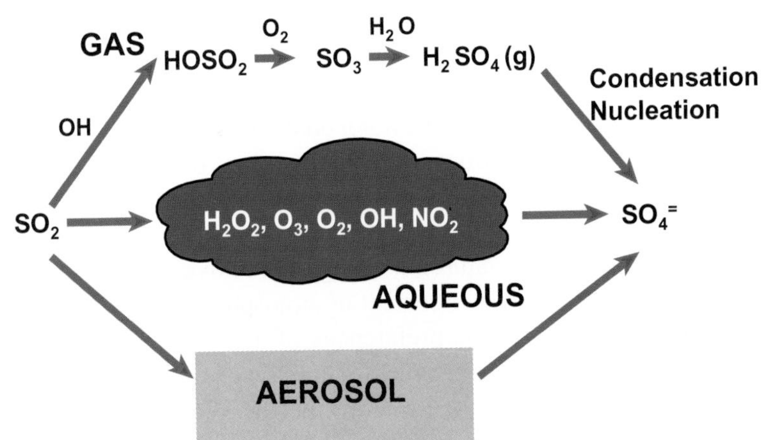

Figure 3.7. Schematic of the three pathways (reaction in the gas, cloud, and condensed phases) for the formation of $SO_4^=$ in the atmosphere. Some of the reactions in the aqueous and condensed phases (e.g., oxidation by oxygen) are catalyzed by trace amounts of metals.

the western (Pandis et al., 1992) and the eastern United States (Stein and Lamb, 2000).

The H_2SO_4 formed from the above pathways reacts readily with NH_3 to form ammonium sulfate, $(NH_4)_2SO_4$. If there is insufficient NH_3 present to fully neutralize the available H_2SO_4 (one molecule of H_2SO_4 requires two molecules of NH_3), part of the PM exists as ammonium bisulfate, NH_4HSO_4 (one molecule of H_2SO_4 and one molecule of NH_3) and the associated particles are acidic. In extreme cases, sulfates can exist in particles as H_2SO_4.

3.5.2 Nitrate

Nitrates are formed from the oxidation of NO and NO_2 (NO_x) either during the daytime (reaction with OH) or during the night (reactions with ozone and water) (Wayne et al., 1991). Nitric acid is continuously transferred between the gas and the condensed phases (condensation and evaporation) in the atmosphere (Figure 3.8). It naturally prefers the gas phase (when left alone), but reactions with gas-phase NH_3, sea salt, and dust result in its transfer to the condensed phase (Seinfeld and Pandis, 1998). The formation of aerosol NH_4NO_3 is favored by availability of NH_3, low temperatures, and high relative humidity. The resulting NH_4NO_3 is usually in the sub-micrometer particle range. Reactions with sea salt and dust lead to the formation of NO_3^- in the coarse particles. The availability of significant HNO_3 vapor in an area (even if the particulate NO_3^- concentrations are low or zero) is an indication of the potential for the future formation of NO_3^--containing particles. Heterogeneous reactions and perhaps photolysis lead to the reduction of particulate NO_3^- and return nitrogen oxides to the gas phase. These reactions are poorly understood but could be important (Honrath et al., 2000).

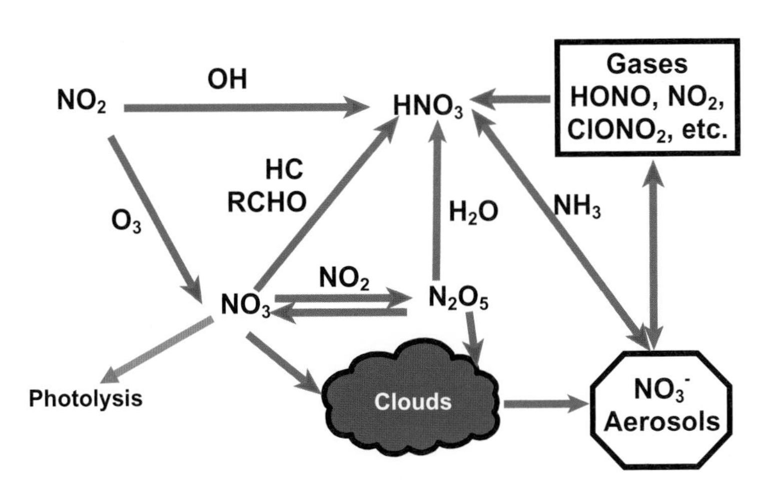

Figure 3.8. Schematic of the formation of HNO_3 and particulate NO_3^- in the atmosphere. Formation of particulate NO_3^- from HNO_3 requires either reaction with NH_3, sea salt or alkaline dust.

3.5.3 Secondary Organic Aerosol (SOA) Formation

The organic component of ambient particles is a complex mixture of hundreds or even thousands of organic compounds. These organic compounds are either emitted directly from sources (primary organic aerosol) or can be formed in-situ by condensation of low-volatility hydrocarbon-oxidation products (secondary organic aerosol). As organic gases are oxidized by species such as OH, ozone, and NO_3, their oxidation products accumulate. Some of these products have low volatilities and condense on available particles (Figure 3.9).

The ability of a given volatile organic compound (VOC) to produce SOA during its atmospheric oxidation depends on four factors: its atmospheric abundance, its chemical reactivity, the availability of oxidants, and the volatility of its products.

Many VOCs do not form PM under atmospheric conditions, owing to the high vapor pressure of their products. These include all alkanes with up to six carbon atoms (from methane to hexane isomers), all alkenes with up to six carbon atoms (from ethane to hexene isomers), benzene and many low molecular-weight carbonyls, chlorinated compounds, and oxygenated solvents (Figure 3.9). This Assessment refers to VOCs that produce SOAs as SOA precursors.

Aromatics are by far the most significant anthropogenic SOA precursors (Grosjean and Seinfeld, 1989). Compounds like toluene, xylenes, trimethyl-benzenes, emitted

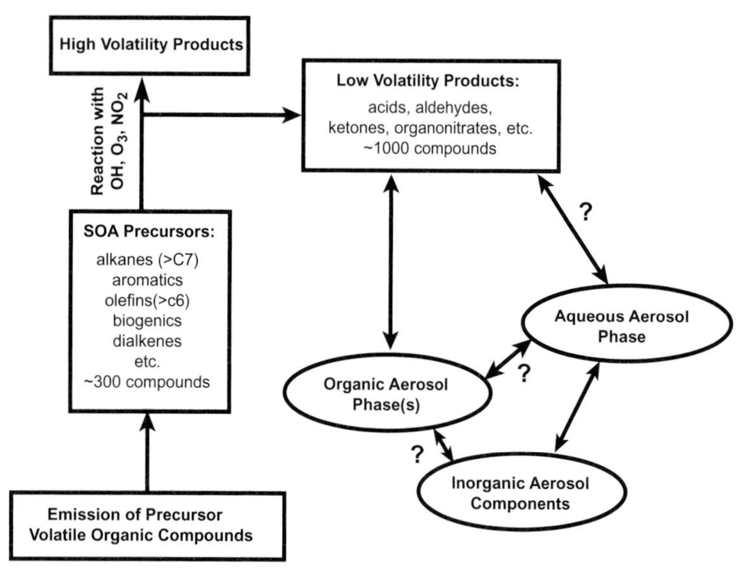

Figure 3.9. Schematic of the formation of secondary organic aerosol in the atmosphere.

by transportation and industrial sources, have been estimated to be responsible for 50 to 70 percent of the SOA. The experimental work of Odum et al. (1997) demonstrates that the SOA-formation potential of gasoline can be accounted for almost totally in terms of its aromatic fraction. Biogenic hydrocarbons emitted by trees are expected to be also an important source of secondary organic PM. Isoprene, an important biogenic VOC, does not form organic aerosol under ambient conditions; but terpenes (α- and β-pinene, limonene, carene, etc.) and the sesquiterpenes are expected to be major contributors to SOA in areas with significant vegetation cover. The rest of the anthropogenic hydrocarbons (higher alkanes, paraffins, etc.) have been estimated to contribute 5 to 20 percent to SOA concentrations, depending on the area.

The contribution of primary and secondary organic aerosol components to measured organic aerosol concentrations remains a controversial issue. Most of the relevant work in this area has been done in Southern and more recently Central California, and relatively little is known about the rest of North America. Early studies suggested that the majority of observed organic PM was secondary in nature. Later investigators focusing on the emissions of primary organic material proposed that 80 percent or so of the organic PM in Southern California on a monthly basis was primary (Hildemann et al., 1993). More recent studies suggest that the primary and secondary contributions are highly variable. Studies of pollution episodes (Turpin et al., 2000) indicated that the contribution of SOA to the organic PM varied from 20 percent to 80 percent during the same day. The contribution of vegetation to organic PM loadings is expected to depend on spatial vegetation coverage. This issue remains unresolved, and most of the existing estimates are highly uncertain. *The ability to model processes that involve carbonaceous PM should be improved.*

A multiplicative factor of 1.4 is commonly used to estimate organic-particle mass concentrations from raw OC measurements. The value of this factor has been the topic of considerable debate (Turpin et al., 2000). Turpin and Lim (2001) suggested that while 1.4 is a reasonable estimate of the average organic molecular weight per carbon for an urban aerosol, a ratio of 1.9-2.3 is more accurate for an aged aerosol, and 2.2-2.6 better represents an aerosol heavily impacted by wood smoke.

Despite the significant progress that has been made in understanding the origins and properties of organic PM, it remains the least understood component of $PM_{2.5}$. Its chemical complexity and the difficulty in determining its chemical composition, even by state-of-the-art methods, are limiting progress. The best efforts to unravel the chemical composition of organic PM have quantified the concentrations of hundreds of organic compounds, yet represent only 10 to 20 percent of the total organic aerosol mass (Figure 3.10). *Our understanding of the contributions of primary, secondary, biogenic, and anthropogenic organic PM components should be improved.*

3.5.4 Interactions of Primary and Secondary PM Components

Primary aerosol particles (e.g., BC, organics, dust, sea salt, fly ash) have both direct and indirect roles in the formation of secondary PM. Primary particles can serve as reaction sites for the formation of new

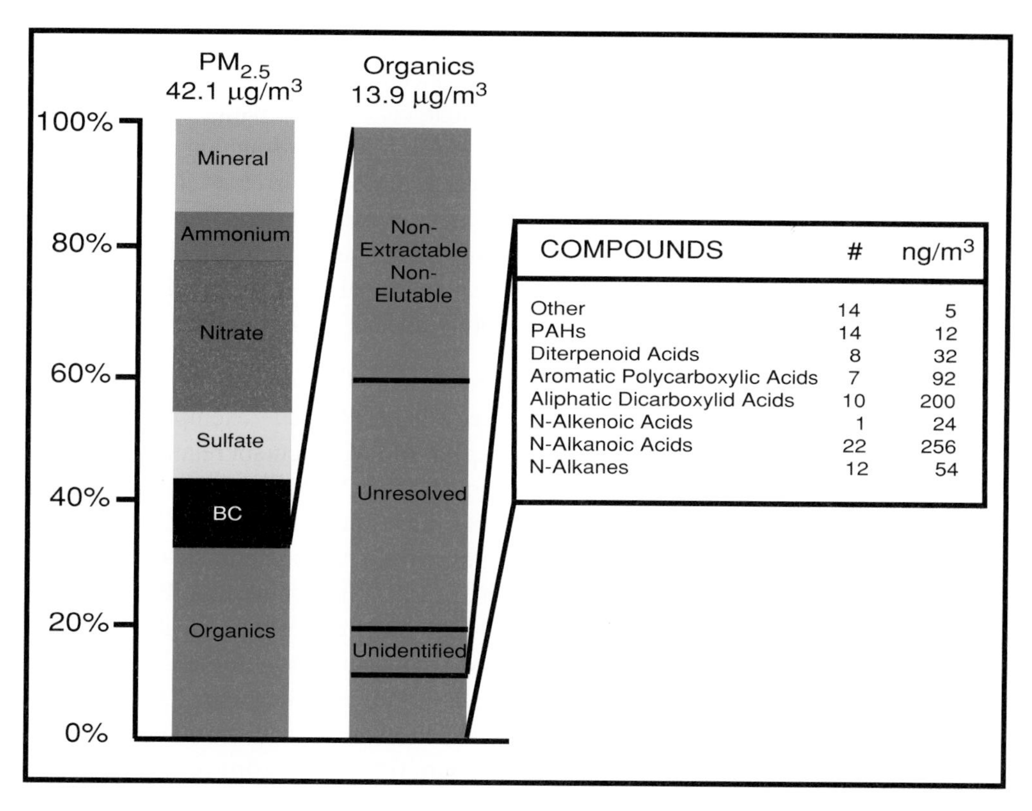

COMPOUNDS	#	ng/m3
Other	14	5
PAHs	14	12
Diterpenoid Acids	8	32
Aromatic Polycarboxylic Acids	7	92
Aliphatic Dicarboxylid Acids	10	200
N-Alkenoic Acids	1	24
N-Alkanoic Acids	22	256
N-Alkanes	12	54

Figure 3.10. Speciation results for organic aerosol in Southern California (Rogge et al., 1993). Even if a hundred or so individual organic compounds are identified and quantified, they represent only 15 percent or so of the total organic mass.

particulate material. The formation of NO_3^- by the reaction of sea salt or alkaline dust with HNO_3 vapor is an example of such reactions (Seinfeld and Pandis, 1998). Other important reactions include the formation of $SO_4^=$ on alkaline particles (discussed previously in Section 3.5.1) and the oxidation of primary organic aerosol compounds to more hydrophilic ones (Zhang et al., 1993). A number of studies have suggested that BC particles react with SO_2, ozone, and NO_x influencing both the gas- and particle-phase composition (Seinfeld and Pandis, 1998). The strong dependence of the often-conflicting results of these laboratory investigations on the nature of the BC surfaces used has prevented the extrapolation of their results to the atmosphere. As a result, the role of BC particles as sites for the production of $SO_4^=$ or NO_3^- remains not well understood.

3.6 FROM PRECURSOR EMISSIONS TO AEROSOL COMPONENT CONCENTRATION

As noted above, major precursors of secondary PM are SO_2, NO_x, selected VOCs, and NH_3. Other precursors like HCl, or dimethyl sulfide (DMS) are of secondary importance for most areas of interest in North America and will not be discussed further. Establishment of the relationship between the emissions of the above precursors and the PM concentrations in a given area is a necessary first step for the design of a $PM_{2.5}$ control strategy. Calculating the sensitivity of the concentration of $PM_{2.5}$ in a specific area to a reduction of SO_2 emissions is one example. Precursor emissions exhibit spatial and temporal variability, but for the sake of simplicity, the following discussion focuses on spatially and temporally uniform changes in precursor emissions. The relative contributions of long-range transport and local emissions are discussed in Chapter 8.

3.6.1 Linearity

The change in the emissions of a precursor and the change in the concentrations of its PM product(s), e.g., SO_2 and $SO_4^=$, can be related by:

$$\%\text{ change in PM concentration of component i} = (\text{Transfer Coefficient}) \times \left(\%\text{ change in emission rate of component i}\right)$$

This simplified relationship assumes that all of the sources of the precursor are reduced uniformly, and that primary emissions or transport of PM component i into the airshed are negligible. The transfer coefficient accounts for the effects of all the atmospheric processes (dispersion, advection, chemical reactions, removal, etc.) affecting the PM component between the source of the precursor and receptor. The value of this coefficient depends not only on the precursor-PM component combination (e.g., SO_2 and $SO_4^=$, VOC and SOA, NO_x and NO_3^-, etc.) but also on the location, season, magnitude of the emissions, averaging time, etc. The above equation can also be written as:

$$\frac{\text{Post-control concentration of component i}}{\text{Pre-control concentration of component i}} = \left(\frac{\text{Post-control precursor emission rate}}{\text{Pre-control precursor emission rate}}\right)^{\text{Transfer Coefficient}}$$

When the transfer coefficient is close to one, then atmospheric concentrations of the secondary PM vary in direct proportion to the emission rates of the precursor gases. In this case the relationship between secondary particles and gas-phase precursors is said to be linear. For nonlinear processes, the transfer coefficient is significantly different from one and is a strong function of the emissions. In such cases percent change in concentrations of secondary particulate matter may exceed or be less than the percent change in precursor-gas emissions.

The above approach is applicable to the direct relationship between precursor and aerosol components (e.g., SO_2 and $SO_4^=$ or NO_x and NO_3^-). However, each precursor may indirectly affect the other aerosol components. For example, changes in NO_x affect gas-phase chemistry and therefore change the concentrations of OH radical, H_2O_2, ozone, NO_3 radical, etc. These changes result in corresponding changes in $SO_4^=$ production rates (both in clean air because of OH and in clouds because of H_2O_2). These indirect effects can be quite significant: they result from complicated interactions and the major issue is not linearity of the corresponding processes (they are rarely linear), but rather their overall significance.

Previous applications of chemical-transport models (see Chapter 8) have exemplified the non-intuitive aspects of PM/precursor relationships. For example, reductions in VOC emissions have been predicted to lead sometimes to increases (Meng et al., 1997; Pai et al., 2000) and sometimes to decreases (Lurmann et al., 1997) in summertime $PM_{2.5}$ concentrations in the Los Angeles Basin, and to decreases in wintertime $PM_{2.5}$ concentrations in the California San Joaquin Valley (Pun and Seigneur, 1999). In addition, one must be aware that an emission-control strategy that is beneficial to ozone concentrations may be counterproductive for PM concentrations. Moreover, an emission-control strategy that benefits one PM component may adversely affect another. A summary of the possible responses of ozone and PM concentrations to changes in emission levels is presented in Section 3.8. Most of these complicated effects can be predicted by state-of-the-art chemical-transport models (see Chapter 8).

One approach that can be used to combine these direct and indirect effects is to examine the overall response of $PM_{2.5}$ concentrations (and not just the effect on one of the $PM_{2.5}$ components) to changes in the emissions of a precursor using the overall transfer coefficient (Seigneur et al., 2000):

[Overall Transfer Coefficient] = [Percent change of total $PM_{2.5}$] / [Percent change in source emission strength]

An overall transfer coefficient of unity represents a proportional relationship (e.g., a 50 percent decrease in precursor emissions leads to a 50 percent decrease in $PM_{2.5}$), an overall transfer coefficient of zero represents no change in $PM_{2.5}$, and a negative transfer coefficient represents an inverse relationship (i.e., a decrease in precursor emissions leads to an increase in $PM_{2.5}$ concentrations). The overall response of $PM_{2.5}$ to emission controls for a given source (overall transfer coefficient) can be calculated as the sum of the responses (individual transfer coefficients) of the various $PM_{2.5}$ components weighted by their fractional contribution to the total $PM_{2.5}$.

If the direct relationship of the precursor to the aerosol product is linear (transfer coefficient equal to one) and the precursor has no effect on the other aerosol components (transfer coefficient equal to zero) then the overall transfer coefficient (response of the total aerosol concentration to the emission change) is equal to the fraction of the component of interest in the ambient $PM_{2.5}$.

3.6.2 An Application to Southern California

Most of the available information about these relationships comes from model applications to Southern California. As an example, the transfer coefficients for a summer smog episode in Southern California are given in Table 3.1 (Lurmann et al., 1997). These were calculated for a uniform 50 percent reduction in the corresponding emissions throughout the basin and correspond to the average changes in concentrations in 8 locations. These results are theoretical estimates for a specific area and a given episode, and while by no means general are still instructive. None of the reported transfer coefficients (values in blue in Table 3.1) is equal to one. The reductions in ambient $SO_4^=$ concentrations are less than one because of background $SO_4^=$ levels and primary $SO_4^=$ emissions. The deviation of the transfer coefficients from unity during NO_x and NH_3 emission reductions results from gas- and aerosol-chemistry nonlinearity. Finally, the small sensitivity of organic PM to VOC reductions results from significant primary organic PM sources during that episode. The same study suggests that based on current understanding, a number of the indirect effects are of secondary importance, at least for this area. For example, the effect of NO_x on $PM_{2.5}$ $SO_4^=$ was calculated as –0.04 (a 50 percent reduction in NO_x resulted in a 2 percent increase in $SO_4^=$) (see Dennis et al., 2001 for a discussion of this effect for the eastern United States). However, these simulations indicate a number of additional benefits of some controls: VOC reductions in some cases also reduce NH_4NO_3 (sometimes they have the opposite effect). Some of the most counter-intuitive features of the system are also illustrated by this example (values in red in Table 3.1). Emission reductions of some of the precursors can indirectly increase other $PM_{2.5}$ components. In this case SO_2 controls lead to increased NO_3^- levels, and NO_x controls lead to increased OC concentrations. The change in organics is relatively small and results from the change in oxidation rates of the precursors at different VOC/NO_x ratios (Lurmann et al., 1997).

3.6.3 Effectiveness of SO_2 Emission Reductions

The increase in NO_3^- corresponding to SO_2 emission decreases can be a significant issue for the design of emission-control strategies. When the SO_2 concentrations decrease, then the major product, $SO_4^=$, also decreases. However, this $SO_4^=$ does not exist alone but is usually in the form of $(NH_4)_2SO_4$. In this case, when the $SO_4^=$ is reduced NH_4^+ volatilizes, leading to more NH_3 available in the gas phase. The additional gas-phase NH_3 can react with any available HNO_3 to produce NH_4NO_3-containing PM. Consequently the system can respond to reductions in $SO_4^=$ by substituting it with the available HNO_3 as NO_3^-. This replacement of the "lost $SO_4^=$" by NO_3^- is expected to happen practically always,

Table 3.1. Predicted changes (percent Concentration Reduction/percent Emission Reduction) in aerosol component concentrations for a uniform 50 percent reduction in precursor emissions for Southern California (June 1987) (Lurmann et al., 1997).

Emission Change	$SO_4^=$	NO_3^-	$NH_4^=$	OC	$PM_{2.5}$
SO_2	0.64	-0.14	0.10	< 0.05	0.06
NO_x	< 0.05	0.40	0.22	-0.08	0.12
NH_3	< 0.05	0.82	0.68	< 0.05	0.42
VOCs	< 0.05	0.16	0.10	0.16	0.08

sometimes with a negligible effect and sometimes to a significant extent. In the extreme case, if all the NH_3 returns to the condensed phase, removal of one $(NH_4)_2SO_4$ molecule (molecular weight of 132 g) can result in the formation of two NH_4NO_3 molecules (molecular weight of 80 g). So 132 mass units of PM are replaced by 2x80=160 mass units. In this extreme case, the reduction of SO_2 emissions by 64 g (1 mole) leads to a reduction of $(NH_4)_2SO_4$ by 132 g (1 mole) and an increase of NH_4NO_3 by 160 g (2 moles). The transfer coefficient of NH_4NO_3 for an SO_2 emission change therefore can be as low as –2.5 (an increase of NO_3^- 2.5 times the reduction of SO_2), and for the total $PM_{2.5}$ as low as –0.44. Obviously, in the other extreme reduction of SO_2 emissions by 1 mole leads to the removal of 1 mole of $(NH_4)_2SO_4$ and the transfer coefficient is as much as +2.06. These estimates illustrate the importance of NH_3 and its emissions in determining the effectiveness of SO_2 emission reductions.

Ansari and Pandis (1998) proposed a method for the quantification of this response for a given area based on measurements of the $SO_4^=$, total nitrate (gas and condensed phases) and total ammonium (gas and condensed phases). The system response, defined as the change in $PM_{2.5}$ for the change in $SO_4^=$, is shown in Figure 3.11. The green areas correspond to increases in $PM_{2.5}$ for reductions of $SO_4^=$, the blue areas correspond to nonlinear behavior of the system (reduction of 0-1 μg of $PM_{2.5}$ for each reduced μg of $SO_4^=$), and the white areas correspond to the near-linear behavior (reduction of 1-1.34 μg of $PM_{2.5}$ for each μg of reduced $SO_4^=$). The areas are defined as a function of the gas ratio:

$$GR = [NH_3]^F / [HNO_3]^T$$

where $[NH_3]^F$ is the "free" NH_3, that is NH_3 not associated with $SO_4^=$, and $[HNO_3]^T$ is the total nitrate (gas and aerosol). The concentration of free NH_3 is calculated by subtracting from the total NH_3 twice the moles of $SO_4^=$. Figure 3.11 indicates that the region of nonlinearity increases at low temperatures (wintertime and during the night) and at high relative humidity. This approach was applied to the eastern United States using data from the EMEFS study (1988-1990) (West et al., 1999). This work indicated that a significant area centered in the Ohio River valley, and extending from western Tennessee to Michigan to New Jersey was in the nonlinear or negative regime during the winter (Figure 3.12). In contrast, no site displayed nonlinearity in the summer, when higher temperatures inhibit the formation of NH_4NO_3, and the PM is commonly acidic. The spring and autumn results showed nonlinear responses in the upper Midwest (Ohio, Indiana and Illinois). While the method is general, the data used are dated. Application of such techniques to other areas or periods is currently limited by the lack of appropriate data. It is important to monitor concentrations of gas-phase precursors including NH_3 and HNO_3 as well as PM composition in order to understand the

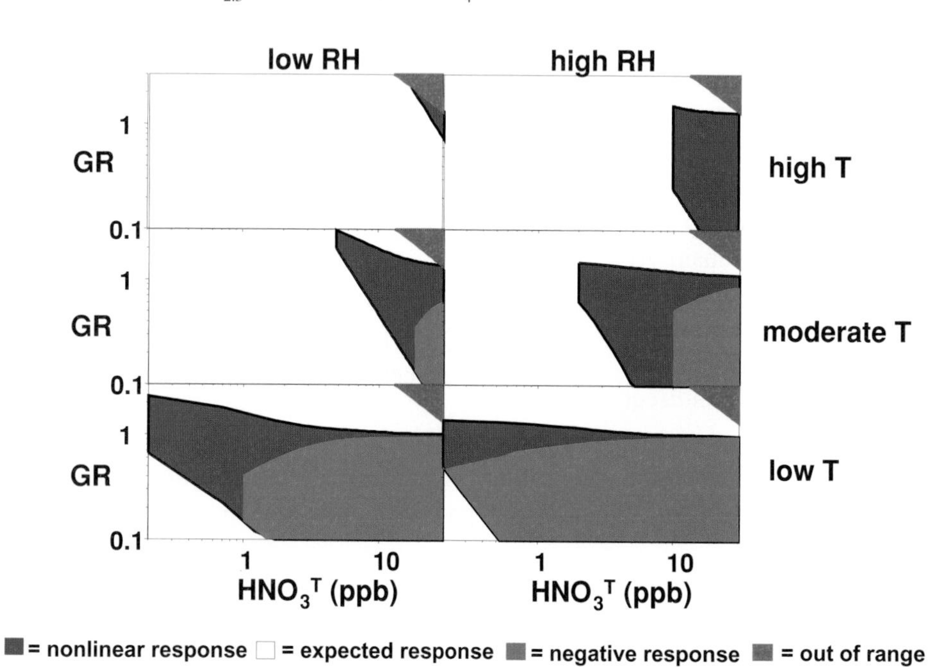

Figure 3.11. Response of the fine PM mass to changes in $SO_4^=$ concentration as a function of the available HNO_3, the gas ratio GR, temperature, and relative humidity. The white area (expected response) corresponds to linear behavior.

sensitivity of NH_4NO_3 concentrations to SO_2 emission rates.

SO_2 and $SO_4^=$ reductions correspond to movements vertically in the diagrams of Figure 3.11. Consequently areas currently in the linear regime may be approaching the nonlinear range, while others may be exiting the nonlinear response area and returning to the linear regime.

3.6.4 VOC Emission Reductions and SOA

The response of the organic component of atmospheric PM to emission changes is complicated by its dual origin. The primary component (direct emission of OC from combustion sources, etc.) is expected to respond practically linearly to emission changes. However, the secondary component is linked to VOC and NO_x emissions through strongly nonlinear chemical and thermodynamic processes. Quantification of the primary and secondary components of observed OC is the first task in the design of any OC control strategy. Unfortunately such analyses have been performed primarily for Southern and Central California, and little is known about OC origins in the rest of North America. Secondary organic PM is expected to respond only weakly to emission changes of the lower molecular-weight VOCs (below than six carbon atoms). As noted previously, these smaller compounds are not important organic PM precursors and therefore influence the organic PM only indirectly by affecting the levels of radicals and ozone. The formation of organic solutions by these compounds suggests that there is a super-linear relationship (the transfer coefficient is more than one) between OC precursors and SOA. Reduction of the emissions of a precursor will reduce not only the concentrations of the products of this precursor in the PM, but also transfer a fraction of the other organic compounds to the gas phase. Even if these processes are qualitatively understood, uncertainties about the details of the chemistry and thermodynamics of SOA do not currently allow the accurate quantification of these relationships. *The ability to model the chemical properties of secondary organic aerosol should be improved.*

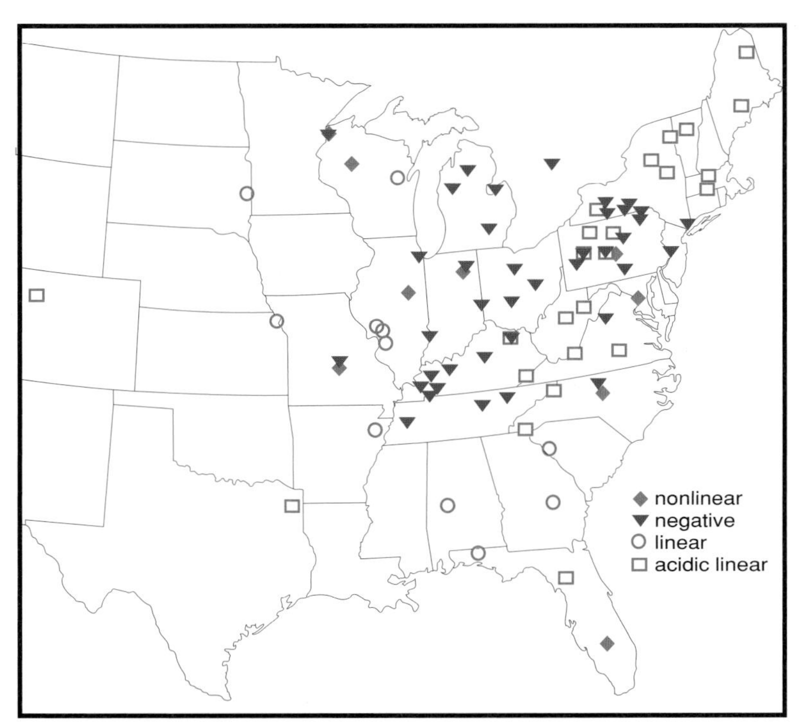

Figure 3.12. Estimated response of fine PM to changes in $SO_4^=$ during the winters of 1988-90 (West et al., 1999). Blue points correspond to increases of the $PM_{2.5}$ if $SO_4^=$ is reduced, and red points to significant replacement of $SO_4^=$ by NO_3 with a net reduction of $PM_{2.5}$. The open symbols reflect linear changes (an overall transfer coefficient of 1.0

3.6.5 Limiting Reactants

A useful concept for the design of control strategies for secondary PM is that of a limiting reactant. For cases where such a reactant exists, the secondary PM component responds readily to concentration changes of this reactant while at the same time it does not respond to changes in the concentrations of other precursors. This is similar in concept to the NO_x- and VOC-limited regimes in ozone formation. Available thermodynamic models can be used to determine if NH_3, H_2SO_4, or HNO_3 are limiting the formation of particulate NO_3^-, based on measurements of the availability of the gas and condensed-phase concentrations of these precursors (Blanchard et al., 2000; Watson et al., 1994). Figure 3.13 shows isopleths of particulate NO_3^- concentrations as a function of total HNO_3 and total NH_3 for given concentrations of the remaining aerosol components, temperature, and relative humidity (Blanchard et al., 2000). For a range of temperatures and relative humidities, these isopleths are approximately L-shaped, with a rather sharp transition between the horizontal and vertical segments. Where the isolines are horizontal, PM formation is not limited by NH_3 availability, as all nitrate exists in the condensed phase. The availability of HNO_3 is limiting the formation of particulate NO_3. Where the isolines are vertical, particulate NO_3^- formation is NH_3-limited but not HNO_3-limited.

Application of all these measurement-based techniques (which do not require emission inventories) is based on measurements of both gas and PM concentrations (NH_3, NH_4^+, HNO_3, NO_3^-). These are not routinely collected. Simpler computational approaches and rules of thumb have been proposed if one wants to avoid the use of an aerosol thermodynamic model (Blanchard et al., 2000). In order to improve the understanding of limiting reagents in NO_3^- formation, sampling networks should include collocated measurements of gas-phase NH_3 and HNO_3, as well as of particulate composition.

The above illustrative examples are based on numerical models representing the state-of-science in this area. These tools are by no means perfect (see discussion in Chapter 8) and neglect a number of potentially important processes (interactions between organic and inorganic aerosol components, different compositions of particles of the same size, etc.). The models have been able to reproduce daily-average $PM_{2.5}$ concentrations with errors of 30 to 50 percent, which are probably indicative of predictive ability. Finally, given the complexity of the system, direct extrapolation of these results for specific receptors to other areas in North America should be avoided. Application of the above modeling tools and chemical-transport models in other areas (see Chapter 8) can provide the corresponding insights about the responses of PM to changes in emissions of SO_2, NO_x, VOCs, and NH_3.

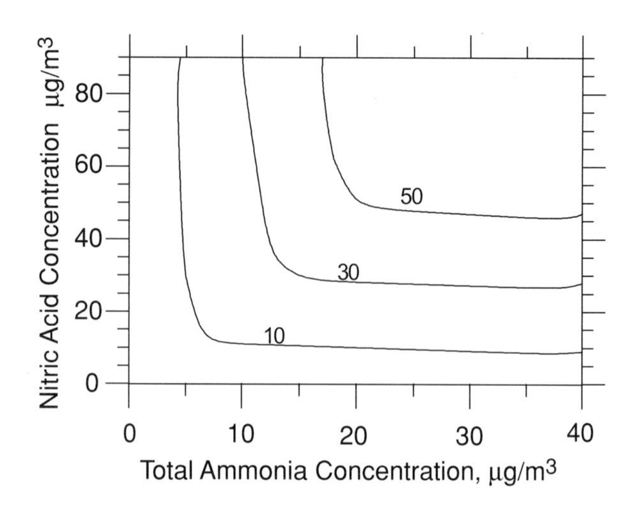

Figure 3.13. Isopleths of predicted particulate NO_3^- concentration ($\mu g/m^3$) as a function of total (gaseous + particulate) NH_3 and HNO_3 at 293 K and 80 percent relative humidity. The concentrations of other species were 25 $\mu g/m^3$ of $SO_4^=$, and 2 $\mu g/m^3$ of total chloride (Blanchard et al., 2000).

3.7 REMOVAL AND LONG-RANGE TRANSPORT OF PM

Wet and dry deposition are the ultimate paths by which particles are removed from the atmosphere. Dry deposition is, broadly speaking, the transport of gaseous and particulate species from the atmosphere onto surfaces in the absence of precipitation (including removal by fog deposition). Factors governing the dry deposition of atmospheric particles include the level of atmospheric turbulence, the properties of the particle (size, density, shape, composition), and the nature of the surface itself. The typical lifetimes of the sub-micrometer particles subject only to dry removal are of the order of a few weeks or even months. Particles with diameters in the 0.1 to 1 µm range live the longest because of their slow transfer through the quasi-laminar layer next to the ground. For larger particles, the lifetime, based solely on considerations of dry deposition, is of the order of a few days. While the above values are typical, actual particle lifetimes can be significantly longer or shorter depending on where the particles are emitted (e.g., close to the ground), and on atmospheric conditions.

Figure 3.14 presents a schematic depiction of PM wet-removal processes. It involves first the transformation of the particle to a new hydrometeor by condensation of water (cloud, rain, fog droplets or ice crystals), or its collision and capture by a pre-existing one, and then its delivery to the earth's surface. Both processes take place inside a cloud, while raindrops collect and remove particles below a raining cloud. The transformation of inorganic particles to cloud droplets (called cloud-droplet nucleation) is relatively well understood, while questions still remain about the ability of organic particles to serve as cloud condensation nuclei. The fundamentals of the removal of particles by the falling rain under a cloud are also relatively well understood (U.S. NAPAP, 1991). Large particles (larger than a few micrometers) are once more collected and removed efficiently by droplets both inside and below the cloud. For particles in the 0.5 to 2.5 μm range, nucleation to cloud droplets during the cloud-formation stage is the most efficient pathway for their incorporation to cloud and then rain droplets. Their composition (inorganic versus organic) is a critical factor for cloud-droplet nucleation and the droplet's subsequent lifetime.

The typical lifetime of a particle in the lower troposphere (Jaenicke, 1993) is shown in Figure 3.15. For particles responsible for most of the $PM_{2.5}$ mass concentration, the lifetime is estimated to be around a week on average. For individual particles the actual lifetime can vary from seconds to several weeks or months. Wet removal is usually the process that determines these lifetimes. During periods of frequent rain events, the lifetime of these particles is expected to be shorter than the value indicated in Figure 3.15, while during dryer periods $PM_{2.5}$ can spend even longer periods in the atmosphere. Particles of diameter less than 0.1 μm have shorter lifetimes because of their higher deposition velocities and also because of their coagulation with larger particles. These coagulation lifetimes are shorter in urban environments (see also Figure 3.4). Particles larger than a few micrometers have shorter residence times in the atmosphere as a result of their larger settling velocities and inertial properties.

Rough estimates of the mean transport distance of atmospheric particles can be obtained by multiplying the transport velocity by the mean atmospheric residence time. The climatologically averaged transport velocity within the mixed layer in North America is a few hundred kilometers per day. Based on an atmospheric lifetime of $PM_{2.5}$ of the order of a week, some of these particles could easily travel a few thousand kilometers from source to removal.

There is significant evidence that atmospheric particles can be transported over large distances, even intercontinentally under the right conditions. Well-documented examples include the transport of Saharan dust to Europe and North America (Schutz et al., 1990), dust transport from Asia to North America, and the transport of continental pollution to create the Arctic Haze (Barrie, 1986). These events are possible because particles that are transported to

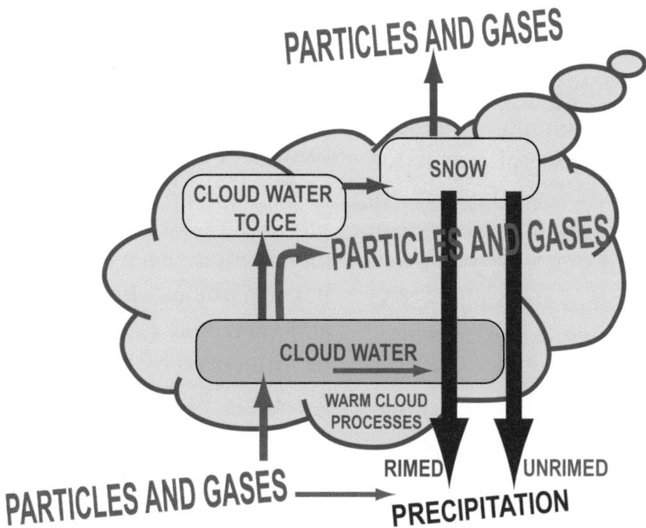

Figure 3.14. Schematic diagram of the processes of in-cloud scavenging of PM and gases from the atmosphere by a precipitating cloud. (adapted from Barrie, 1991).

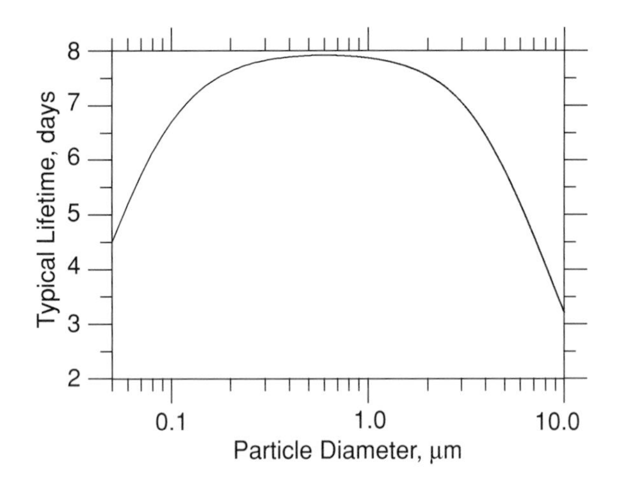

Figure 3.15. Typical lifetime of atmospheric particles in the lower troposphere accounting for losses due to coagulation, dry deposition, and wet removal (Jaenicke, 1993). The actual lifetime can be quite variable depending on the rain frequency, wind speed, concentrations of other particles, point of emission, etc. For areas (e.g., eastern United States) and periods with significant rainfall the lifetimes are shorter.

the free troposphere (above the cloud layer) can live substantially longer than those that remain in the boundary layer. Typical lifetimes of particles in the $PM_{2.5}$ range in the free troposphere are of the order of three weeks; thus intercontinental transport is possible even for large dust particles.

3.8 PM AND OTHER POLLUTANTS

PM, ozone, and other pollutants are related through a complex web of common emissions and precursors, common photochemical production pathways, and meteorological processes (Figure 3.16, Table 3.2). Integrated, multi-pollutant abatement strategies are possible. NO_x and VOCs are precursors of both ozone and of a fraction of atmospheric PM (NO_3^- and secondary organics) while they influence indirectly the formation of the rest of the secondary PM components like $SO_4^=$.

A number of modeling studies, once more focusing on Southern California, have quantified these effects (for example Meng et al., 1997). These results underline the possibility of counterintuitive interactions, like the increase of $PM_{2.5}$ for reductions of the VOC emissions in that area. As VOC levels are decreased, the concentration of RCO_3 radicals decreases, and the available NO_2 instead of being converted to PAN is converted to HNO_3, which reacts with available NH_3 to form NH_4NO_3. The increase in NH_4NO_3 in that area can exceed the decrease in the SOA concentration, and consequently the total $PM_{2.5}$ may increase even if one of the precursors decreases in concentration. At the same time the ozone concentration is predicted to decrease significantly (Figure 3.17). For example, this study shows that a 25 percent reduction in VOC emissions resulted in a 19 percent reduction of the peak ozone in Riverside California, but an 18 percent increase of the $PM_{2.5}$ concentration. In another test for the same receptor, a 25 percent decrease in NO_x emissions resulted in 3 percent reduction of the peak ozone and a 9 percent decrease of the $PM_{2.5}$ concentrations.

The above illustrates both the potential and also the pitfalls for the design of control strategies caused by the intricate coupling between ozone and PM. Control strategies for VOCs or NO_x that are optimal for ozone controls may even increase $PM_{2.5}$ concentrations. At the same time, careful choices can lead to synergism and reductions of the concentrations of both pollutants.

3.9 POLICY IMPLICATIONS

Particles remain in the atmosphere for days to a few weeks, depending on particle size, the rate at which they are removed by precipitation, etc. It follows that particles detected at a site may have been formed or emitted locally or may have originated at a site located hundreds or even thousands of kilometers away. Particles in the 0.1 to 1.0 µm diameter range have longer lifetimes than smaller or larger particles. *Implication:* Very small (<0.1 µm) and very large (>1 µm) particles are more likely to be influenced by local emissions or production than particles in the intermediate size range. This phenomenon is important when evaluating the relative roles of local and upwind sources in the design of compliance strategies.

Primary particle emissions from both anthropogenic and biogenic sources can contribute to PM

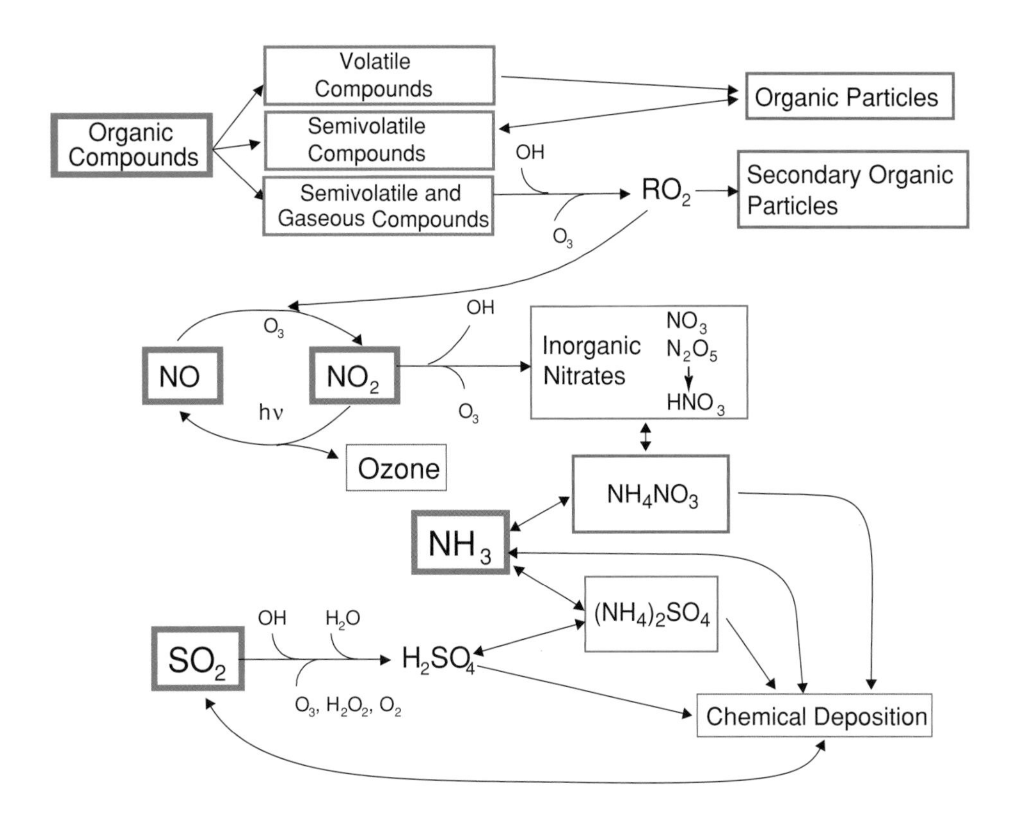

Figure 3.16. Chemical links between ozone and PM formation processes. The major precursors are shown in green squares. The VOC can be gaseous (always in the gas phase), non-volatile (always in the condensed phase), and semivolatile (partitioned between the gas and condensed phases (adapted from MSC, 2001)).

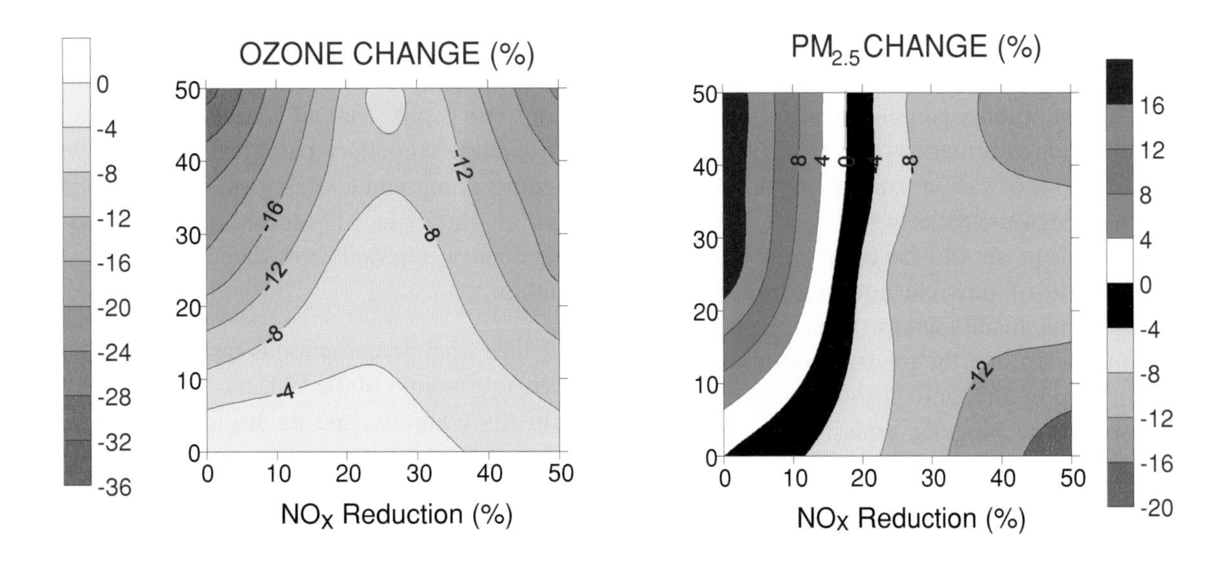

Figure 3.17. Percentage changes of maximum 1-hour average concentrations of ozone and $PM_{2.5}$ simulated at Riverside, California on 28 August 1987 for various combinations of VOC and NO_x emission reductions from base estimated 1987 basin-wide emissions (Meng et al., 1997).

Table 3.2. Typical pollutant / atmospheric issue relationships.[a]

Reduction in pollutant emissions	Change in associated pollutant or atmospheric issue					
	Ozone	PM Composition			$PM_{2.5}$[(k)]	Acid Deposition
		$SO_4^=$	NO_3^-	Organic compounds		
SO_2		↓	↑(e)		↓	↓
NO_x	↓(a)↑	↑↓(c)	↓(f)	↓↑(h)	↓↑	↓↑
VOC	↓	↑↓	↓↑ (g)	↓(i)	↓↑	↓↑
NH_3		↓(d)	↓		↓	↑(l)
Black Carbon	! (b)			↓(j)	↓	
Primary Organic Compounds	↑(b)			↓	↓	
Other primary PM (crustal, metals, etc.)	↑(b)				↓	↑(l)

[a] Arrow direction denotes increase (↑) or decrease (↓); arrow color denotes undesirable (red) or desirable (blue) response; arrow size signifies magnitude of change. Small arrows signify possible small increase or decrease. Blank entry indicates negligible response.
[b] In and downwind of some urban areas that are VOC limited.
[c] Effect on daytime ozone due to increase in solar flux and decrease in radical scavenging; effect on nighttime ozone unknown.
[d] Due to effect of NO_x on oxidant levels (OH, H_2O_2 and ozone); e.g., see SAMI modeling results.
[e] Due to effect of NH_3 on cloud/fog pH.
[f] Decrease in sulfate may make more NH_3 available for reaction with HNO_3 to form NH_4NO_3, more important when NH_4NO_3 is NH_3 limited.
[g] Decrease except special cases (e.g., SJV); decrease in NO_x may lead to increase in ozone with associated increase in HNO_3 formation.
[h] Increase due to less organic nitrate formation and more OH available for reaction with NO_2; decrease due to decrease in oxidant levels.
[i] Related to effect of NO_x on oxidant levels (OH, ozone, and NO_3^-).
[j] Decrease of secondary component; magnitude depends on OC fraction that is secondary anthropogenic.
[k] Reduction of OC adsorbed or emitted with black carbon.
[l] Refers to net acidity atmospheric deposition, not to acidification potential to ecosystem.

concentrations. Furthermore, the size and composition of primary particle emissions vary with the source. Such information can be used to estimate the contributions of various sources to ambient PM. *Implication:* Measurements of size-resolved ambient PM composition should be complemented with measurements of particle composition at both biogenic and manmade sources. Such measurements can help to determine the relative contributions of local anthropogenic emissions, which may be controllable, from biogenic emissions, which may not.

Secondary PM formed in the atmosphere by chemical transformations of reactive gaseous emissions typically constitutes a significant fraction of the PM mass. Most $SO_4^=$ and NO_3^- are secondary, and a portion of organic PM may be secondary. Both anthropogenic and biogenic sources emit gases that contribute to the formation of secondary PM, and gaseous emissions from all sources interact as they react to form secondary particles. *Implication:* In designing compliance strategies, it is essential to consider the role of secondary PM and the interactions of emissions from diverse sources in their formation.

Particulate organic compounds typically make up a significant fraction of the PM mass. Organic PM is chemically complex, and its origins and properties are not as well understood those of the inorganic fraction. It is likely that primary sources always contribute significantly to OC, although secondary sources may contribute up to 50 percent in some locations, especially where photochemical transformations of biogenic precursor gases are significant. *Implications:* More research on the sources, composition, and properties of the OC fraction is needed. The extent to which emission controls of OC and its gas-phase precursors can reduce ambient mass concentrations will likely vary with season and region, and is highly uncertain. Nevertheless, because OC composes such a significant fraction of PM mass, reducing its contributions should be considered when establishing strategies for meeting PM standards.

In some cases it may be possible to define "limiting reagents" for selected components of secondary PM. The mass concentrations of PM will respond readily to changes in the availability of this limiting reagent. *Implication:* Reducing the availability of limiting reagents can be an effective way to control the contributions of that species to PM.

Ammonium nitrate is a semivolatile compound formed by the reaction between HNO_3 and NH_3. Ammonium nitrate concentrations tend to increase with increasing emissions of NH_3, photochemical production of HNO_3 (from NO_x), and decreasing temperature. Control measures that lead to decreases in $SO_4^=$ have the potential to lead to increases in NH_4NO_3. Furthermore, either HNO_3 or NH_3 may be a "limiting reagent" in the formation of NH_4NO_3. *Implications:* Available modeling tools can be used to estimate the sensitivity of NH_4NO_3 concentrations to local emissions of NH_3 and NO_x, as well as its sensitivity to changes in $SO_4^=$ concentrations. The availability of suitable data on emissions, particle composition, and gas concentrations (especially NH_3 and HNO_3) will help to ensure that model predictions are accurate.

PM, ozone and other pollutants are related through a complex web of common emissions and precursors, photochemical production pathways, and meteorological processes. Therefore, changes in emissions of one pollutant can lead to changes in the concentrations of other pollutants. A variety of approaches can be used to explore such relationships including thermodynamic models, chemical-transport models, and empirical observations. For processes that are well understood, it may be possible to predict these interrelationships with confidence while in other cases one may be limited to "educated guesses." *Implications:* Interactions between pollutants need to be considered when making policy decisions. These interactions can be evaluated using available modeling tools.

3.10 REFERENCES

Ansari, A., Pandis, S.N., 1998. On the response of atmospheric particulate matter concentrations to precursor concentrations, Environmental Science and Technology 32, 2706-2714.

Ansari, A., Pandis, S.N., 1999. An improved method for the thermodynamic modeling of multicomponent inorganic atmospheric aerosols, Atmospheric Environment 33, 745-757.

Barrie, L.A., 1986. Arctic air pollution: an overview of current knowledge, Atmospheric Environment 20, 643-663.

Barrie 1991. Snow formation and processes in the atmosphere that influence its chemical composition, NATO ASI Series, Vol. G28, Seasonal Snowpacks, Davies, T., et al., eds.,1-20.

Bertram, A.K., Ivanov, A.V.. Hunter, M.. Molina, L.T., Molina, M.J., 2001. The reaction probability of OH on organic surfaces of tropospheric interest. Journal of Physical Chemistry 105, 9415-9421.

Blanchard, C.L., Roth, P.M., Tanenbaum, S.J., Ziman, S.D., Seinfeld, J. H., 2000. The use of ambient measurements to identify which precursor species limit aerosol nitrate formation. Journal of the Air & Waste Management Association 50, 2073-2084.

Dennis, R.L., Tonnesen, G.S., Mathur, R., 2001. Nonlinearities in the sulfate secondary fine particulate response to NOx emissions reductions as modeled by the Regional Acid Deposition Model. In: Air Pollution Modeling and its Application XIV, Gryning, S.E., Schiermeier, F.A., eds., Kluwer Academic/Plenum Publishers, New York, pp. 193-202.

Griffin R.J., Cocker, D.R., Flagan, R.C., Seinfeld, J.H., 1999. Organic aerosol formation from the oxidation of biogenic hydrocarbons. Journal of Geophysical Research 104, 3555-3567.

Grosjean, D., Seinfeld, J.H., 1989. Parameterization of the formation potential of secondary organic aerosols. Atmospheric Environment 23, 1733-1747.

Hildemann, L.M., Cass, G.R., Mazurek, M.A., Simoneit, B.R.T., 1993. Mathematical modeling of urban organic aerosol properties measured by high resolution gas-chromatography. Environmental Science and Technology 27, 2045-2055.

Honrath, R.E., Peterson, M.C., Dziobak, M.P., Dibb, J.E., Arsenault, M.A., Green, S.A., 2000. Release of NOx from sunlight-irradiated midlatitude snow. Geophysical Research Letters 27, 2237-2240.

Intergovermental Panel on Climate Change (IPCC), 2002. Climate Change 2001: The Scientific Basis. World Meteorological Office, United Nations Environmental Program

Jaenicke, R., 1993. Tropospheric Aerosols. In: Aerosol-Cloud-Climate Interactions. Hobbs, P.V., ed., Academic Press, San Diego, CA; pp. 1-31.

John, W., Wall, S.M., Ondo, J.L., Winklmayr, W., 1990. Modes in the size distributions of atmospheric inorganic aerosol. Atmospheric Environment–24A, 2349-2359.

Kavouras, I., Mihalopoulos, G.N., Stephanou, E.G., 1998. Formation of atmospheric particles from organic acids produced by forests. Nature 395, 683-686.

Kittelson, D.B., Arnold, M., Watts, W.F., 1999. Review of diesel particulate matter sampling methods. Final Report to the EPA, Center for Diesel Research, Department of Mechanical Engineering, University of Minnesota, Minneapolis, MN.

Langner, J., Rodhe, H., 1991. A global three-dimensional model for the tropospheric sulfur cycle. Journal of Atmospheric Chemistry 13, 225-263.

Lurmann, F.W., Wexler, A.S., Pandis, S.N., Musarra, S., Kumar, N., Seinfeld, J.H., 1997. Modelling urban and regional aerosols – II: Application to California's South Coast Air Basin. Atmospheric Environment 31, 2695-2715.

McHenry, J.N., Dennis, R.L., 1994. The relative importance of oxidation pathways and clouds to atmospheric ambient sulfate production as predicted by the Regional Acid Deposition Model. Journal of Applied Meteorology 33, 890-905.

McMurry, P.H., Woo, K.S., Weber, R., Chen, D.R., Pui, D.Y.H., 2000. Size distributions of 3-10 nm atmospheric particles: implications for nucleation mechanisms, Philosophical Transactions of the Royal Society, London, Series A-Mathematical Physical and Engineering Sciences 358, 2625-2642.

McMurry, P.H., Litchy, M., Huang, P.F., Cai, X.P., Turpin, B.J., Dick, W.D., Hanson, A., 1996, Elemental composition and morphology of individual particles separated by size and hygroscopicity with the TDMA. Atmospheric Environment 30,101-108.

Meng, Z., Dabdub, D., Seinfeld, J.H., 1997. Chemical coupling between atmospheric ozone and particulate matter. Science 277, 116-119.

Odum, J.R., Jungkamp, T.P.W., Griffin, R.J., Flagan, R.C., Seinfeld, J.H., 1997. The atmospheric aerosol-forming potential of whole gasoline vapor. Science 276, 97-99.

Pai, P., Vijayaraghavan, K., Seigneur, C., 2000. Particulate matter modeling in the Los Angeles Basin using SAQM-AERO. Journal of the Air and Waste Management Association 50, 23-42.

Pandis, S.N., Seinfeld, J.H., Pilinis, C., 1992. Heterogeneous Sulfate Production in an Urban Fog. Atmospheric Environment 26, 2509 2522.

Pun, B., Seigneur C., 1999. Understanding particulate matter formation in the California San Joaquin Valley: Conceptual model and data needs. Atmospheric Environment 33, 4865-4875.

Saxena, P., Hildemann, L.M., McMurry, P., Seinfeld, J.H., 1995. Organics alter the hygroscopic behavior of atmospheric particles. Journal of Geophysical Research 100, 18755-18770.

Schutz, L.W., Prospero, J.M., Buat-Menard, P., Carvalho, R.A., Cruzado, A., Harriss, R., Heidam, N.Z., Jaenicke, R., 1990. The Long-Range Atmospheric Transport of Natural and Contaminant Substances. NATO ASI Series C: Mathematical and Physical Sciences, 297, 197-230, Kluwer Acad. Publ.

Seigneur, C., Tonne, C, Vijayaraghavan, K., Pai, P., Levin, L., 2000. The sensitivity of PM2.5 source-receptor relationships to atmospheric chemistry and transport in a three-dimensional air quality model. Journal of the Air and Waste Management Association 50, 428-435.

Seinfeld, J.H., Pandis, S.N., 1998. Atmospheric Chemistry and Physics: From Air Pollution to Climate Change. J. Wiley, New York.

Sievering, H., Boatman, J., Gorman, E., Kim, Y., Anderson, L., Ennis, G., Luria, M., Pandis, S.N., 1992. Removal of sulfur from the marine boundary layer by ozone oxidation in sea-salt. Nature 360, 571-573.

Stein, A.F., Lamb, D., 2000. The sensitivity of sulfur wet deposition to atmospheric oxidants. Atmospheric Environment 34, 1681-1690.

Stonehouse, B., 1986. Arctic Air Pollution. Cambridge University Press.

Tang, I.N., 1997. Thermodynamic and optical properties of mixed-salt aerosols of atmospheric importance. Journal of Geophysical Research 102, 1883-1893.

Turpin, B.J., Lim, H.J., 2001. Species contributions to PM2.5 mass concentrations: Revisiting common assumptions for estimating organic mass, Aerosol Science and Technology 35, 602-610.

Turpin, B.J., Saxena, P., Andrews, E., 2000. Measuring and simulating particulate organics in the atmosphere: problems and prospects. Atmospheric Environment 34, 2983-3013.

U.S. National Acid Precipitation Assessment Program, 1991. The U.S. National Acid Precipitation Assessment Program, Vol. 1. U.S. Government Printing Office, Washington DC.

Wayne, R.P., 1991. The nitrate radical: Physics, chemistry and the atmosphere. Atmospheric Environment 25A, 1-203.

Watson, J.G., Chow, J.C., Lurmann, F.W., Musarra, S.P., 1994. Ammonium nitrate, nitric acid, ammonia equilibrium in wintertime Phoenix, Arizona. Journal of the Air and Waste Management Association 44, 405-412.

West, J., Ansari, A., Pandis, S.N., 1999. Marginal $PM_{2.5}$ - Nonlinear aerosol mass response to sulfate reductions. Journal of the Air and Waste Management Association 49, 1415-1424.

Zhang, X.Q., McMurry, P.H., Hering, S.V., Casuccio, G.S., 1993. Mixing characteristics and water content of submicron aerosols in Los Angeles and the Grand Canyon. Atmospheric Environment 27A, 1593-1607.

CHAPTER 4

Emission Characterization

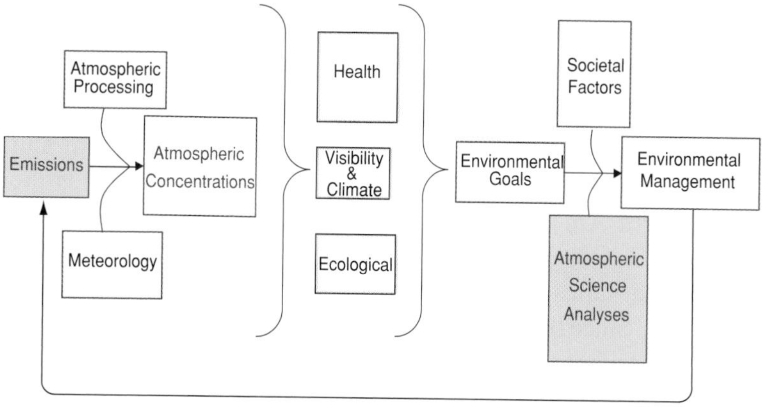

Principal Authors: George Hidy, David Niemi, and Thompson Pace

Emission characterization, including inventories, provides a critical starting point for strategies employed in the management of airborne PM in North America. As indicated in the framework illustration, Figure 1.1, inventories are used to identify anthropogenic sources and emissions of primary and secondary PM. The resultant information is used to track progress in reducing emissions, support development of control strategies, and provide input to air-quality models. In this chapter, emission information available in Canada, Mexico, and the United States is summarized; the discussion covers inventory characterization, quality control and assurance, and inventories or emission-processing systems or emission "models" used in chemical-transport modeling. The chapter also covers the nature of variables that affect or limit the reliability of emission estimates, including their spatial, temporal, and compositional resolution and accuracy. These uncertainties may impact policy decisions; measures that minimize the impact of these uncertainties are discussed. Additional details, including a summary of the methodology for developing emission data, are included in Appendix A.

Fine-particle emissions are generated both by man and by nature, as direct (or primary) PM emissions and as secondary PM produced from chemical reactions of gases in the atmosphere. Primary-particle and precursor emissions derive from manageable, unmanageable and fugitive sources. Manageable emissions can be identified readily with human enterprise: industrial sources, commercial operations, power plants, residential dwellings, and vehicle transportation. Natural emissions – volcanic eruptions, soil drifting, wind-blown sea spray, dust storms from remote arid areas, forest or brush fires initiated by lightning strikes or spontaneous combustion, and suspended waxy particles from vegetation – are considered unmanageable. Other unmanageable emissions include natural emissions of gases and vapors that react in the atmosphere to produce particles, including sulfur gases from terrestrial and marine sources, nitrogen oxides from soil respiration and lightning strikes, and organic compounds from vegetation. Fugitive particle sources that are related directly or indirectly to human activities compose an additional emission category. These include controlled or slash burning, vegetation clearing for agriculture, man-made dust from construction, agricultural operations and roadways, and windblown dust from surface mining, agriculture, roadways and construction. While conventional emission inventories include most of the categories noted here, they generally omit some fugitive particle sources.

Since the 1970s, considerable effort has been devoted to characterizing PM sources. Among other positive developments, this has resulted in an emerging broad-scale picture of natural emissions relative to anthropogenic PM contributions. This is illustrated by the PM_{10} emission data listed in Table 4.1 which suggests that, from a global perspective, primary natural PM_{10} emissions far outweigh their anthropogenic counterparts, while secondary natural and anthropogenic contributions are roughly equal in magnitude.

Similar global estimates for $PM_{2.5}$ suggest somewhat different behavior for finer particles. The fraction of soil dust and sea salt in $PM_{2.5}$ is perhaps 10 to 20 percent of that existing in PM_{10}, thus reducing these natural contributions substantially for fine particles. Considering all sources suggests that, exclusive of accidental vegetation fires, global $PM_{2.5}$ contributions from natural and anthropogenic sources are about the same.

Over North America, both the natural and the anthropogenic contributions make up a tropospheric burden that varies in concentration with proximity to sources. Since the residence times of fine particles in the troposphere typically range from 3 to 7 days or longer, a "natural" background of PM, much of which is unmanageable (Chapter 6), is expected to be present in varying ambient concentrations.

Caution should be exercised when applying relative natural and anthropogenic global-scale emissions, such as those shown in Table 4.1, to infer human exposure, which is more heavily dependent on localized or at least regional emissions, especially in and around population centers. In such situations anthropogenic emissions are generally of greater concern, because they are both manageable and are thought to be responsible for many of the (hypothesized) toxic components in particles.

Table 4.1. Global sources of airborne particles roughly less than 10 μm in diameter (after Andreae (1995)).

Source	Annual Emissions (Tg/yr)	Comments
Natural Particles		
Soil & rock debris [a, b]	1500	Principally coarse particles; contains an anthropogenic component.
Forest fires & slash burning [a]	50	Principally $PM_{2.5}$; contains an anthropogenic component.
Sea salt [b]	1300	Principally coarse particles
Volcanic debris	33	Highly intermittent source.
Gas to particle conversion		
Sulfate from sulfur gases	102	Principally $PM_{2.5}$
Nitrate from NO_x	22	Principally $PM_{2.5}$
VOC from plant exhalation [a, c] & fires	55	Principally $PM_{2.5}$
Subtotal	3060	
Anthropogenic Particles		
Primary particles - industrial transportation, etc.	120	Primarily $PM_{2.5}$
Gas to particle conversion		
Sulfate from SO_2 (& H_2S)	120	Primarily $PM_{2.5}$
Nitrate from NO_x	36	Primarily $PM_{2.5}$
VOC conversion	90	Primarily $PM_{2.5}$
Subtotal	366	
Total	3430	

[a] These categories are ambiguous in that they may include anthropogenic disturbances from land use, and prescriptive or accidental burning from man's activities.

[b] The low initial release height of soil and sea salt makes their raw emission estimates difficult to interpret; their magnitude biases their actual airborne contribution too high, because much suspended material is potentially deposited close to the source, or in the case of soil, removed from the air by collection on vegetation or structures.

[c] There is evidence that trees produce wind blown primary organic emissions as well in the form of waxy alkanes. (e.g., Diamond et al., 2000). The magnitude of these emissions is essentially not known.

4.1 TYPES OF EMISSION INVENTORIES AND THEIR USES

Emission inventories generally classify emissions in terms of point, area, or mobile sources. Identifiable point sources include power plants and industrial facilities. Area sources include diffuse non-point sources characteristic of a given area, such as commercial and residential operations. Mobile or transportation sources are represented by light-duty vehicles, heavy-duty trucking, railroads, aircraft, shipping, off-road construction, and agricultural and other movable equipment. The PM inventories are particularly complex because they include primary emission estimates for different PM size fractions. Inventories also include gaseous precursors (SO_2, NO_x, certain VOCs[1], and NH_3) for particle production, and the precursors that lead to oxidant formation in the air (NO_x, VOC, and in some cases, CO and CH_4). The methods conventionally used for estimating emissions are summarized in Appendix A.

Inventories are categorized according to their application, for example, the four levels used in the description of the U.S. Emissions Inventory Improvement Program (EIIP; U.S. EPA, 1997). In the context of implementing PM standards, a fifth level associated with air-quality modeling is added.

Level 1—Source specific, used for permitting and regulatory compliance programs.

Level 2—Urban area, used for local population center, state, regional or provincial planning (e.g., state implementation plans [SIPs]), and tracking (trends).

Level 3—Industry-wide, used for applications such as large utility-boiler characterization, control, or technology trends that do not necessarily drive regulatory concerns.

Level 4—Country-wide, used for national and international issues, including trends, such as acid rain, or climate alteration from greenhouse gases and suspended particles.

Level 5—Support for air-quality modeling, used for evaluating the response to emissions and their changes in future scenarios.

The fifth category concerns development and evaluation of chemical-transport models (CTMs), discussed in Chapter 8, and subsequent analysis for source attribution, as described in Chapter 7. This application of emission characteristics normally requires highly resolved spatial, temporal, and compositional information on model-specific areal grids that typically cross geo-political boundaries normally used for inventory compilation.

Emission inventories are commonly prepared as annual averages by source category, based on a methodology prescribed at the national level (e.g., U.S. EPA, 1997, 2002; EPS, 1999). Inventories are then developed using some combination of "bottom-up" aggregation from small geopolitical units like census districts in Canada, or counties in the United States, and "top-down" methodology prescribed at the national level (e.g., U.S. EPA, 1997, 2002). Local inventories through the provinces, states, or regions are prepared representing a "base year" for developing management strategies and for projecting future conditions. Projections are based on engineering estimates of technology deployment, and on population or infrastructure growth over a period of years.

Applications for the different levels have a broad range of requirements that place significant demands not only on accuracy of the estimates, but on the flexibility of the data base for access and use. For example, estimation of emission reductions needed to attain objectives or standards and the evaluation of planned emission controls for point sources require specific inventories with process-level resolution. Local PM management programs are developed using emission data by geographical (or geopolitical) regime for base years and future years. Applications require estimates of primary and secondary contributions, representative temporal distributions and spatial allocation of point and area sources, and chemical speciation profiles. Detailed emission

[1] The categories of VOC that are relevant to secondary organic aerosol particle formation (SOA precursors are an ill-defined subset of VOCs of carbon number 2-12 listed for oxidant chemistry.) The SOA precursors generally are believed to be aromatic compounds and species of carbon number greater than 6. As noted in Table 4.1, some evidence of airborne particles from waxy high molecular-weight alkanes from vegetation have been reported (Diamond et al., 2000).

estimates by source are used for reconciliation of ambient observations with expectations from inventories, analysis of ambient concentration trends, control prioritization, enforcement rule development, benefit-cost analysis, and compliance monitoring. In contrast, aggregated inventories of coarser resolution can be used to track national emission trends or record specific reductions from sources that have been controlled. Emission statements are also prepared at the specific industry-level resolution, which may not contain chemical speciation or temporal resolution needed for air-quality modeling. In the international setting, national governments support continental or global outlook analysis and strategy development. Examples of activities at the broad level include the Canada-U.S. Air Quality Agreement, the North American Free Trade Agreement (NAFTA) analysis and its associated environmental protocols, the United Nations Economic Commission for Europe, and the Intergovernmental Panel on Climate Change.

The long-standing conventions for preparing emission inventories clearly support the existing regulatory structure in North America, including response to the form of national goals or standards. For air-quality modeling to address concentration distributions for estimating human-health and visibility impacts, improved spatial and temporal resolution of inventories is often demanded beyond that required for other applications. The linkage to human exposure to PM, for example, is problematic; the spatial scale of individual or population exposure may require emission estimates and air-quality calculations within less than a kilometer spacing. Further, the cumulative effects of environmental exposure may not be capable of resolution since consistent historical inventories are not available. In the case of visibility impairment, the temporal scale is observationally defined, either in terms of daylight hours for aesthetic reasons, or both day and night for transportation safety. These are not compatible with conventional inventory development.

Comparing the inventory information with the particle characteristics hypothesized to be responsible for health impacts, examples of which are listed in Textbox 2.5, reveals some disparities (Table 4.2). If emissions are to be identified with example hypothetical mechanisms for health effects, the inventories are not sufficiently detailed chemically for such characterization. Recent preliminary attempts to more closely match the emission inventories to the needs of the health community have been made, such as the U.S. National Air Toxics Assessment (NATA; EPRI, 1994) and the risk analysis of hazardous air pollutants (HAPs) from large electric utility boilers. Implementation of residual risk and urban area source requirements of the U.S. Clean Air Act Amendments has been considered in the risk analysis of fossil-fueled power plant emissions (e.g., Levy et al., 1999).

As discussed in Chapter 3, emission data for SO_2 and NO_x associate strongly with observed ambient $SO_4^=$ and NO_3^-. Primary PM emission information can be at least partially associated with observed ambient mass concentration. Linkages also can be rationalized for BC and generic organic constituents. However, speciated carbon material, peroxides, and ultrafine particles really are not represented in current inventory information. Further, there is virtually no means of inventorying biological materials, such as airborne spores, algae, viruses, or bacteria. Improved knowledge of the composition of primary and secondary organic species associated with anthropogenic and biogenic sources is needed.

In the case of visibility impairment, the major components of $PM_{2.5}$ on relevant spatial scales, including carbon and the secondary particles from sulfur oxides and nitrogen oxides, are estimated in the conventional regulatory inventories. For appropriate temporal resolution, it is unclear whether the segregation into hourly estimates of primary emissions or secondary contributions is sufficiently precise for calculating optical effects or their response to emission changes. At least one study, however, has found that emission estimates with 5 km and hourly resolution have sufficient detail to evaluate optical effects over a city (Kleeman et al., 2001).

4.2 THE NATURE AND CHARACTERISTICS OF EMISSION INVENTORIES

The collection, codification, and recording of data and quality control by detailed source category is so labor-intensive that inventories are generally

Table 4.2. Illustrative linkages between current emission information used for estimating ambient concentrations relevant to hypothesized causal elements for PM health effects.

Hypothesis	Emission information available	Comments
1. Mass Concentration	Yes	Does not consider current multimodal mass distribution.
2. Surface Area	No [a]	Multimodal surface area distribution distinct from mass distribution.
3. Ultrafine PM	No [a]	Ultrafine particles are a distinct short-lived exposure phenomenon not covered in emission inventory approaches.
4. Metals and metal compounds	Yes (limited to crustal and HAPs; some transition metals reported in NPRI [b])	PM inventories generally do not include all relevant metals and especially valence states or compounds. HAPs and transition metals available from the NATA and NPRI.
5. Acids	Yes (acid sulfate and nitrate)	Generally excludes organic acids, but HCl is inventoried in the toxics release inventory (TRI); others including HCN,H_3PO_4 and HF are also inventoried in the NPRI.
6. Organic Compounds	Incomplete [a]	Highly limited inventories—little consideration of stable condensed material vs. semi-volatile material.
7. Biological or Biogenic Particles	No	These include biogenic species derived from vegetative VOC, and fungi, spores, bacteria, etc.
8. Sulfate and Nitrate Salts	Yes	Common salts included, but problematic for short lived species like sulfites, methane sulfonic acid, and organo- nitrates.
9. Peroxides		May be covered with VOC and NO_x emissions, but unproven.
10. BC (soot) Particles	No	Information on BC is available in modeling inventories but adsorbed material is uncharacterized.
11. Copollutant interactions	Yes	For internationally categorized "criteria" pollutants, and HAPs for the United States.

[a]In one study, emission inventories by source type and source composition profile data have been combined to provide some resolution of these factors (Cass et al., 2000).
[b] NPRI = National Pollutant Release Inventory.

compiled every few years rather than annually, and tend to lag two or more years behind present. The interval between detailed inventories for Canada is five years; an annual update begins in 2003. The detailed U.S. inventory is prepared every three years, with a less detailed update every year. These inventories form the basis for virtually all of the national and international considerations for air-quality management in these countries. A national particulate inventory is not yet available for Mexico, except for broad listings of SO_2, NO_x, and VOC prepared for international concerns for emissions affecting climate alteration.[2] The U.S. National Emission Inventory (NEI) is accessible in many ways through the U.S. EPA, including the Internet.[3] The Canadian national inventory for 1995 in summary form is available from the Environment Canada's Environmental Protection Service (EPS). A detailed annual emission inventory for the United States prepared by county for the latest compilation year, 1999, is accessible through the U.S. EPA.[4] Canadian provincial governments provide access to inventory summaries by province. Some Mexican emission data are available, which provide summaries by source category for certain large point sources. These cover several metropolitan areas, including Mexico City (CAM, 2001), and Monterrey, Guadalajara, Toluca, and Ciudad Juarez (INE, 1997). Recent inventory summaries for Mexico City are also accessible in the reports by Molina et al. (2000) and Molina and Molina (2002).

[2] e.g., SEMARNAT and INE (2001).
[3] www.epa.gov/ttn/chief/efig.
[4] ftp.epa.gov/EmisInventory.

Because of confidentiality restrictions in Canada and Mexico, detailed inventories by industry are not generally available to the public. However, they may be accessed for specific government-related studies, including regional air-quality management studies, by contacting the Canadian EPS, or the Mexican INE.

4.2.1 Characterization by Source Category

PM and precursor-gas emissions in the three nations show certain distinct characteristics that are related to the level of industrialization and other demographics. Canada maintains a PM inventory that includes PM[5], PM_{10}, and $PM_{2.5}$. The U.S. inventory includes only PM_{10} and $PM_{2.5}$. Both countries include the precursor gases: SO_2, NO_x, VOC, and NH_3. The VOC inventories categorize total emissions; in certain special circumstances, speciated VOC is available, mainly for low molecular-weight components that are unlikely SOA precursors.

Table 4.3 compares Canadian and U.S. emission estimates for similar aggregated emission categories for PM_{10} and $PM_{2.5}$ and precursor gases. A detailed breakdown of national emissions by source category for the two countries is included in Appendix A.

Table 4.3 illustrates important differences and similarities in the major source contributions for PM and for precursor gases. For example, the fugitive-dust PM_{10} component tends to be the largest contributor to primary PM_{10} emissions for both countries. But for primary $PM_{2.5}$ emissions, the U.S. combustion emissions, including carbonaceous material, tend to be important, along with fugitive dust, while in Canada fugitive dust and forest fires are predominant primary $PM_{2.5}$ source categories (see also Appendix A). In the case of SO_2 emissions, this gas is dominated in the United States by electric power production, whereas Canadian industrial sources such as smelters and petroleum-related production are the major contributors. In both countries NO_x is dominated by transportation sources and by electric power generation. Ammonia production is identified primarily with agricultural activities in both countries. However, NH_3 from other sources such as motor-vehicle exhaust, native vegetation, or native or domestic animals also may be of concern in certain circumstances.

Because of the large differences between land use, industrialization, and population in the United States and Canada, a national comparison like that in Table 4.3 does not tell the full story, especially in terms of emissions affecting large population segments. For a better "exposure related" comparison, emissions representing conditions in large urban centers are of interest. To illustrate annual emission patterns in different metropolitan areas, four cases have been selected: Atlanta, GA, Toronto, Ont., Mexico City, and Los Angeles, CA[6]. Some demographics describing these metropolitan areas are listed in Table 4.4. The cities range in population from about 2.4 million to 18 million over developed areas of about 600 km^2 to 88,000 km^2. The cities are economically diverse per capita. Their climates are varied from warm and arid to cool and continental. All of the cities are typical of North America with an industrial complex in the surroundings and a large vehicle population. The vehicle population varies substantially in terms of allocation per developed area. Improved and standardized approaches for developing emission inventories for Canada, Mexico, and the United States that provide compatible and accessible data need to be developed.

Atlanta typifies a large, relatively isolated city in the southeastern United States, with a high summer photochemical smog potential, accompanied by severely reduced visibility, whose ambient $PM_{2.5}$ is influenced strongly by $SO_4^=$ and carbon components, along with a contribution from soil dust. Atlanta's ozone chemistry is influenced strongly by biogenic VOC, which may influence secondary PM production in summer. Toronto is the largest city in Canada, imbedded in the heavily populated, industrial Great Lakes region just north of the Canada-U.S. border: the Windsor-Quebec City Corridor. Both Atlanta's and Toronto's pollution combine a local component with an externally transported component. Toronto's ambient $PM_{2.5}$ has a major $SO_4^=$ component with

[5] In Canada, PM is defined as particles less than 100 µm aerodynamic diameter. A definition used in the United States is total suspended particles (TSP), a measure of PM concentration based on filtration and gravimetric analysis for collections nominally <35 µm aerodynamic diameter.

[6] See also Chapter 10 for conceptual models of Los Angeles, Mexico City, and the Windsor-Quebec City Corridor.

Table 4.3. Summary of nationwide 1995 Canadian and 1999 U.S. emissions by similar categories (kTonnes/yr)[a]. Based on agglomeration of detailed breakdown by major source category shown in Appendix A[b].

Source	Primary PM_{10}		Primary $PM_{2.5}$		SO_2		NO_x		VOC		NH_3	
	CDN	US	CDN	US	CDN	US	CDN	US	CDN	US	CDN	US
Industrial	287	906	172	502	**1950**	**4227**	620	**3968**	**941**	1523	35	234
Non-Indust. Fuel	144	568	138	487	32	588	77	1175	404	670	2	7
Elect. Util.	35	225	19	128	534	**12698**	255	**5715**	3	56	1	7
Trans-portation	96	753	83	670	136	1299	**1290**	**14105**	734	**8529**	20	**270**
Incinera-tion	1.5	92	1.1	47	1.3	30	2.6	58	6.0	57	0.4	0
Open Sources	**4793**	**21126**	**1097**	**4930**	0.6	18	217	350	**937**	1072	**308**	**4322**
(Fugitive Dusts) [c]	(3380)	(14726)	(451)	(2631)	(0)	(0)	(0)	(0)	(0)	(0)	(0)	(0)
Miscel-laneous [d]	14	9	9	37	0	7	7	22	550	6244	188	124
Biog. [e]	--	0	--	0	--	0	(137)	0	(12770)	0	--	0
Total	5371	23679	1519	6800	2654	18867	2469 [f]	25393	3575 [f]	18150	554	4964

[a] Bold numbers are emphasized for large contributions by pollutant.
[b] There are some differences between Canada (EPS, 2000) and the United States (U.S. EPA, 1997) by detailed source category as noted in Appendix A. These differences do not affect the qualitative comparisons emphasized in this table.
[c] Fugitive dust is listed in this row for comparison but are part of open sources.
[d] The miscellaneous category includes solvent utilization, transport and storage of fuels and chemicals, and waste management and utilization.
[e] In the United States, biogenic emissions are estimated in the emission processor for the particular situation being simulated in the chemical transformation model.
[f] Excludes biogenic emission estimate.

contributions from dust and carbon. Mexico City is one of the world's largest metropolitan areas by population, but is distinct in its characteristics from other North American cities, both in climate and industrialization. Its pollution has a strong photochemical component with optically dense haze observed much of the year. Its ambient $PM_{2.5}$ is characterized by dominance of carbon and soil dust. Los Angeles is another of the world's largest cities, and is known for its long history of intense photochemical smog and poor visibility. Los Angeles $PM_{2.5}$ is characterized by unusually high NO_3^- levels, combined with carbon and soil dust.

The emission inventory for each of the cities is summarized in Tables 4.5 and 4.6. The inventories are difficult to compare with one-another in detail because the aggregated source categories differ somewhat from country to country. Nevertheless, qualitative insight can be gained from comparing their common features and differing source categories. Qualitative comparison of the emissions

Table 4.4. Demographics of example North American cities.

	Atlanta, GA [a]	**Los Angeles, CA** [a]	**Mexico City** [b]	**Toronto, Ont.**
Population (millions)	3.8 [a]	15.6 [a]	18	2.4
Nominal Metropolitan Area (km^2) (MSA in US [a])	13,300	88,000	5300 [c]	630 [d]
Climate/Regional Characteristics	Isolated continental; warm, moist-humid; regional air stagnation in summer; moderate photochemical potential	Isolated coastal; warm, semi-arid, Mediterranean-like; high photochemical potential	Isolated continental; arid, high altitude, low latitude; high photochemical potential	Variable: North Central States and southern Great Lakes continental; cool winter, warm summer; Regional influence in summer, early fall.
GDP per Capita ($US)	32,500	28,100 [b]	7800	21,100 [e]
No. of Vehicles (Millions)	3.35[f]	9.3	3	1.1
1999 Peak hourly ozone (ppb)	140	240	320	113 [e]
1999 Max. 24-hr PM_{10} ($\mu g/m^3$)	70	120	200	74.0 [e]
Dominant PM_{10} Composition	$SO_4^{=}$, carbon, dust	NO_3^-, carbon, dusts	Dust, carbon	$SO_4^{=}$, dust

[a] Based on designated metropolitan statistical area (MSA).
[b] From data given in Molina and Molina (2002).
[c] Nominal value of the metropolitan areas of Mexico City.
[d] Area for metropolitan Toronto; the area equivalent to an MSA is much larger.
[e] Data for 1999. Source: http://www.bea.doc.gov/bea/regional/reis/City of Toronto, November 1, 1999,: http://www.americascanada.org/eventabf/documents/mediakits/english/pf-provinceofontario-e.pdf
[f] Source: NET96 version 4; http://www.epa.gov/ttn/rto/areas/msa/msaindex.htm

of precursor gases adds insight into the different roles of the secondary component of PM.

Even though Mexico City has a very large population, its primary PM_{10} emission estimate is less than those in Atlanta or Toronto, perhaps amounting to only 15 percent of Toronto's PM_{10} emissions, and even less than Los Angeles. INE workers in Mexico have noted emissions from the open-source category as one their largest uncertainties. For all four cities, open sources, including fugitive dusts along with transportation sources tend to strongly influence the primary PM_{10} and $PM_{2.5}$ emissions. However, as will be discussed later and in Chapter 7, the inventory likely overstates the importance of these sources to regional air quality because much of these emissions are re-deposited close to the source. The PM from transportation sources is believed to be primarily carbonaceous material, either as BC or organic compounds. Thus there is no surprise in the large carbonaceous and soil-dust contributions to ambient PM composition observed in all the cities. The ratio of $PM_{2.5}$ to PM_{10} emissions is about 36 percent for Los Angeles, 34 percent for Atlanta, and 25 percent for Toronto.

The inventories of reactive gases also show differences between the cities. According to the inventory estimates, Toronto has the largest SO_2 emissions of the four cities, followed by Atlanta, Los Angeles, and Mexico City. The dominant SO_2 contributor in Toronto is industrial and to a lesser extent, electricity generation; in Atlanta, electricity production is dominant followed by transportation sources. In Mexico City, industrial sources of SO_2 tend to dominate, followed by transportation. The largest contributor to SO_2 emissions in Los Angeles is transportation, even though a very low sulfur concentration is mandated for gasoline in California. For NO_x, Los Angeles with its large vehicle population has by far the largest emissions, followed by Mexico City, and Atlanta, with Toronto the lowest. NO_x from transportation, industrial operations, electricity production, and non-industrial fuel burning are important in the four cities. Los Angeles represents a unique situation with very large NH_3 emissions and low SO_2 emissions. Thus with the large NO_x emissions, conditions are favorable for production of particulate NH_4NO_3. Estimates suggest that Mexico City has lower NH_3 levels than Toronto, with most attributable to animal wastes. The calculated NH_3 emissions in the Toronto area exceed those of the Atlanta area. In Atlanta, estimated transportation sources dominate NH_3 emissions, while open sources presumably primarily agricultural practices, are most important in Toronto and Mexico City. In Mexico City anthropogenic VOC emissions are larger than in the other three cities. VOC in Mexico City is dominated by transportation and miscellaneous source categories, as in the other three cities. Hypothetically, perhaps as much as 10 percent of the VOC emissions reported might be available

Table 4.5. Comparison of PM_{10} and $PM_{2.5}$ emissions by similar source categories between Atlanta [a], Toronto [b], Mexico City (MXC) [c], and Los Angeles (LA) (Tonnes/yr).

Source	**Primary PM_{10}**				**Primary $PM_{2.5}$**			
	Atlanta	**LA [d]**	**MXC**	**Toronto**	**Atlanta**	**LA [e]**	**MXC**	**Toronto**
Industrial	762	2360	2890	6930	470	4930	NA	3640
Non-Ind. Fuel	714	10,100	946	11700 [h]	701	584	NA	11,600 [h]
Electric Utilities	249	252	138	245	113	270	NA	194
Transportation	7600	13700	7140	4950	6360	13,600	NA	4220
On-road	3130	7800	7140	2210	2390	9200		1810
Off-road	4470	6080	NA	2740	3970	4400		2410
Incineration	--	27	--	58	--	26	NA	33
Open sources (incl. road and wind blown dust; farming)	84,800 [f]	84,700	8760 [i]	100,000	18,800 [f]	24,800	NA	10,200
Miscellaneous [g]	9220	12,900	167	1070	9120	365	NA	972
Total	103,000	124,000	20,000	125,000	35,600	44,600	NA	30,900

[a] Based on EPA 1999 MSA inventory, including counties making up Atlanta and neighboring suburban areas; i.e., Clayton, Cobb, DeKalb, Douglas, Fayette, Fulton, Gwinnett and Rockdale counties.
[b] Extracted from 1995 EC Ontario inventory estimates.
[c] Based on the 1998 Metropolitan Area of the Valley of Mexico (MAVM) Inventory [Comision Ambiental Metropolitana (CAM), 2001].
[d] PM_{10} from 2000 SCAB estimated emission summary; California Air Resources Board (www.arb/emssumcat)
[e] $PM_{2.5}$ from 1996 South Coast Air Quality Management District (SCAQMD) Inventory. (Note the apparent inconsistency in $PM_{2.5}$ and PM_{10} estimates between 1996 and 2000).
[f] Based on EPA Tier 1 category "Misc."
[g] Includes solvent utilization, storage and transportation, and waste disposal.
[h] Estimate includes residential wood burning (91 percent of non-industrial emissions). Note that this estimate of wood burning is believed to be in error, biased perhaps 2 to 3 times too high based on inconsistencies in ambient measurements and differences in spatial allocation between cities (Makar et al., 2001).
[i] Estimate is low compared with other cities; the estimate includes unpaved road contribution, and may underestimate windblown dust and other fugitive sources (noted, e.g., in Table 5.8 of Molina and Molina, 2002).

Table 4.6. Comparison of precursor gas emissions for Atlanta [a], Los Angeles [b], Mexico City [c], and Toronto [d] (Tonnes/yr).

Source	SO_2				NO_X				NH_3				VOC			
	Atlanta	LA	MXC	Toronto	Atlanta	LA	MXC	Toronto	Atlanta	LA [e]	MXC [h]	Toronto	Atlanta	LA	MXC	Toronto
Industrial	4,960	6,270	**12,100**	**23,300**	8,620	28,200	17,500	15,900	26	1,330	196	**2,630**	3,060	2,490	23,900	19,900
Non-Ind. Fuel	1,023	830	5,280	2,950	4,750	11,100	7,220	12,100	34	145	NA	195	10,200	5,180	315	44,000
Elect. Util.	**22,000**	86	16	**12,800**	4,900	4,750	9,540	12,700	5	570	NA	41	42	428	48	739
Transpor-tation	11,900	**21,400**	4,720	**11,600**	**135,300**	**353,000**	**168,000**	**71,700**	**3,381**	11,600	473	1,140	**94,400**	**226,000**	**188,000**	**427,000**
On-road	4,110	4,280	4,670	3,540	91,700	252,000	165,800	34,200	3,190	--	NA	1,090	70,100	187,000	187,600	249,000
Off-road	7,830	17,100	54	8,050	43,600	101,000	2,010	37,500	191	--	NA	57	24,300	39,500	419	178,000
Incinera-tion	--	27	0	234	--	488	0	272	--	--	NA	24	--	53	0	2,110
Open sources [g]	21	232	287	0	192	5,180	3,830	0	661	**31,300** [h]	8,560 [h]	**2,100**	562	6,840	3,770	0
Misc. [i]	32 [j]	1,060	100	0	299 [j]	0	0	23	817 [j]	8,500	2,050	1,190	37,500 [j]	**122,000**	**243,000**	1,990
Biogenic (veg.)		--	0			--	1160	650	44	--	NA			NA	15,700	6,450
Total	39,900	29,900	22,500	50,900	154,000	403,000	207,000	113,000	4,970	53,400	11,300	7,320	146,000	363,000	475,000	502,000

[a] Based on 1999 MSA estimates for counties representing metropolitan Atlanta and neighboring suburban areas; i.e., Clayton, Cobb, DeKalb, Douglas, Fayette, Fulton, Gwinnett, and Rockdale Counties.
[b] 2000 SCAB estimated emission summary California Air Resources Board (www.arb/emssumcat)
[c] Based on 1998 MAVM inventory (Comision Ambiental Metroplitana-CAM [www.sma.df.gob.mx/menu]).)
[d] Extracted from 1995 Ontario Inventory estimates.
[e] Sept 4, 1996 estimate of Kleeman et al. (1999)
[f] Estimates for 1995, Osnaya and Gasca (1998).
[g] Includes accidental fires, construction, and agricultural operations.
[h] Includes a large contribution from dairy farms and agricultural operations in LA; cattle, domestic animals, sewage, and other animals in MXC.
[i] Miscellaneous category by difference;
[j] Includes waste management practice, solvent use and petroleum transportation.

to produce secondary PM if significant amounts of aromatic or high-carbon-number components are present.

Qualitative insight about the nature of PM emissions can be further deduced from the emission inventory by inspecting relative amounts of different source categories, assuming that the PM_{10} observed in the cities is largely local in origin. For example, in the case of primary PM_{10}, the transportation emissions (mostly carbon) are much smaller than the open-source emissions, except for Mexico City. For primary $PM_{2.5}$, on the other hand, emissions from transportation sources are a larger fraction relative to soil dust. Generally the annually averaged ambient PM samples have a much larger ratio of carbon to soil, owing to the near-source removal of soil particles. The ratio of open sources for $PM_{2.5}$ to total SO_2 emissions is about 0.5 for Atlanta and Toronto, but 1.4 for Los Angeles, which is reflected in the ambient composition as well. Such insights are severely limited by inventory uncertainty, especially in light of the INE comment about the open-source inventory for Mexico City.

Another way to gain insight about similarities and differences in emissions is to use a normalization of emissions by GDP, by the number of motor vehicles present, or by vehicle kilometers traveled (VKT). For example, normalization of primary PM_{10} emissions in Table 4.5 by nominal population gives a value (kg/yr-person) of 27 for Atlanta, 52 for Toronto, 1.1 for Mexico City, and 7.9 for Los Angeles. Not surprisingly, this suggests a wide range of per capita pollution depending on climate, industrialization and level of pollution control. Another example considers the normalization with respect to vehicle kilometers traveled. Comparing Mexico City and Los Angeles shows motor vehicle PM_{10} emissions from vehicles for the former is 47 g/yr-VKT and for the latter, 27 g/yr-VKT. For VOC, the comparisons gives 1.2 kg/yr-VKT for Mexico City, and 0.44 kg/yr-VKT for Los Angeles. These numbers presumably reflect the differences in the vehicle fleet age and maintenance between the cities. While the application of this technique can add insight into the effectiveness of pollution control and the density of emissions, investigators need to use caution about the veracity of the demographic emission data available to ensure that comparisons are made on a reasonably consistent or common basis.

4.2.2 Geographical Distribution of Emissions

The geographical distribution of pollutants results from the location of emissions combined with atmospheric dispersion and chemical reactions of pollutants. Mapping the geographical distributions of emissions thus requires comparison and harmonization of estimates across neighboring jurisdictions, both within countries and across national boundaries. None of the three countries has yet to achieve an internally consistent application of emission estimation methodologies, though they are approaching this goal.

The distribution of specific pollutants such as SO_2 and NO_x is generally expected to be very similar and tied to population centers. However, different distributions for primary PM emissions might be observed depending on their origins. In general the link with population centers holds true for both PM and precursor gases, except for NH_3. The emission distributions of primary PM and example precursor gases, SO_2 and NH_3, are indicated in Figures 4.1 through 4.4 for Canada and the United States. The estimated emissions of primary PM_{10} and $PM_{2.5}$ for the two nations are shown respectively in Figures 4.1 and 4.2. The patterns of the two measures of PM indicate generally high emissions near population centers or agricultural areas in both countries. The emission distributions by county are concentrated east of the Mississippi River, in the industrialized and populated eastern United States, and in certain industrialized centers across the West. In Canada, PM emissions are concentrated along the southern half of the country, from the farming regions in the west through industrialized Windsor, Ontario, to Quebec City in the East, and in the Vancouver-Fraser Valley area and Edmonton-Calgary region of the West. There are large regions of Canada and the United States throughout the Midwest and West where PM sources are isolated or small. These general patterns are consistent with ambient PM concentration distributions throughout North America.

The geographical distributions of SO_2 and NH_3 are shown for Canada and the United States in Figures 4.3 and 4.4 for comparison with the primary PM distributions. As might be expected, the SO_2 distribution in Figure 4.3 is distinct from the primary $PM_{2.5}$ distribution, SO_2 reflecting a linkage with industrialization and population centers, along with fuel-use patterns. One noteworthy factor contributing to higher SO_2 emissions in eastern North America is the relative abundance and use of coal in that region. However, the $PM_{2.5}$ distribution identifies broad mobile and area sources associated with land use and open sources. The distributions of NO_x and VOC, not shown here, are basically similar to the SO_2 distribution. The SO_2, NO_x, and VOC distributions show high emission densities throughout most of the eastern United States and southeastern Canada, particularly from Windsor, Ontario to Quebec City, Quebec. Emissions east of the Mississippi River tend to be concentrated in and around urban centers and navigable rivers. This distribution, combined with regional-scale meteorological conditions, yields a complex picture of ambient PM distributions having multiscale emission contributions throughout most of eastern North America. Ozone concentrations have a similar multiscale character in eastern North America that has created the need for management strategies crossing geopolitical boundaries. Analogous strategies may need to be considered for $PM_{2.5}$ management in North America.

In contrast to the SO_2 emissions, the NH_3 distribution, shown in Figure 4.4 indicates that NH_3 is linked with agricultural regions of the Midwest and East, as well as certain western areas like southern and central California. Together with fugitive dusts, the differences in NH_3 and SO_2-NO_x emissions are believed to have some effect on acidity of particles and on the acidity of precipitation. The former, of course, would have an influence on the acidity exposure hypotheses for some populations in Canada and the United States.

Based on satellite imagery of Mexico, its national PM_{10} inventory is expected to be dominated by fugitive soil dust and burning vegetation, which have both affected broadly Mexico itself and on occasion the United States. Neither remote sensing nor direct measurements of PM and its precursors has characterized the regional extent of emission distributions in Mexico beyond certain isolated point sources and a few major population centers. While it is not possible to show the geographical distribution of PM from sources in Mexico, sources have been characterized at least qualitatively for PM_{10} and TSP as well as precursor gases in several cities, including Mexico City, Monterrey, Guadalajara, Tijuana, Ciudad Juarez, and Toluca (e.g., INE, 1998). Further, some large point sources have been characterized, especially in the northern parts of the country.

4.2.3 Temporal Variations, Trends and Forecasts in Emissions

4.2.3.1 Temporal Variations

The annual emission reports describe an average condition as a baseline for developing specific strategies for target emission reductions. Since emissions are not evenly distributed temporally within a given day, week, or year and their magnitude can vary significantly from year to year for certain source categories; thus the description of temporal variation is an important corollary to emission descriptions. These extremes in temporal variation can be quite large. Estimations of regular patterns of short-term variation such as the diurnal or daily changes in traffic frequency, or patterns of industrial operations, are important for the description of extremes in daily PM concentration patterns. These events are often characterized in considerable observational detail during short-duration field campaigns. In the absence of long-term observational data, these "snap-shots" are used for analyzing source receptor relationships. Short-term variations can be quite large in individual source categories. These need to be considered for characterizing short-term daily variations in ambient concentrations, as well as for air-quality modeling. They are important in that they introduce a random error or uncertainty in analyses that describe current conditions or project future conditions. They also reflect accidental releases of pollutants, or upset or (routine and non-routine) conditions that may occur over a day to a month. Short-term variations as estimated in emission-processing systems are integral to contemporary air-quality models. Currently these systems rely on "temporal profiles" (monthly,

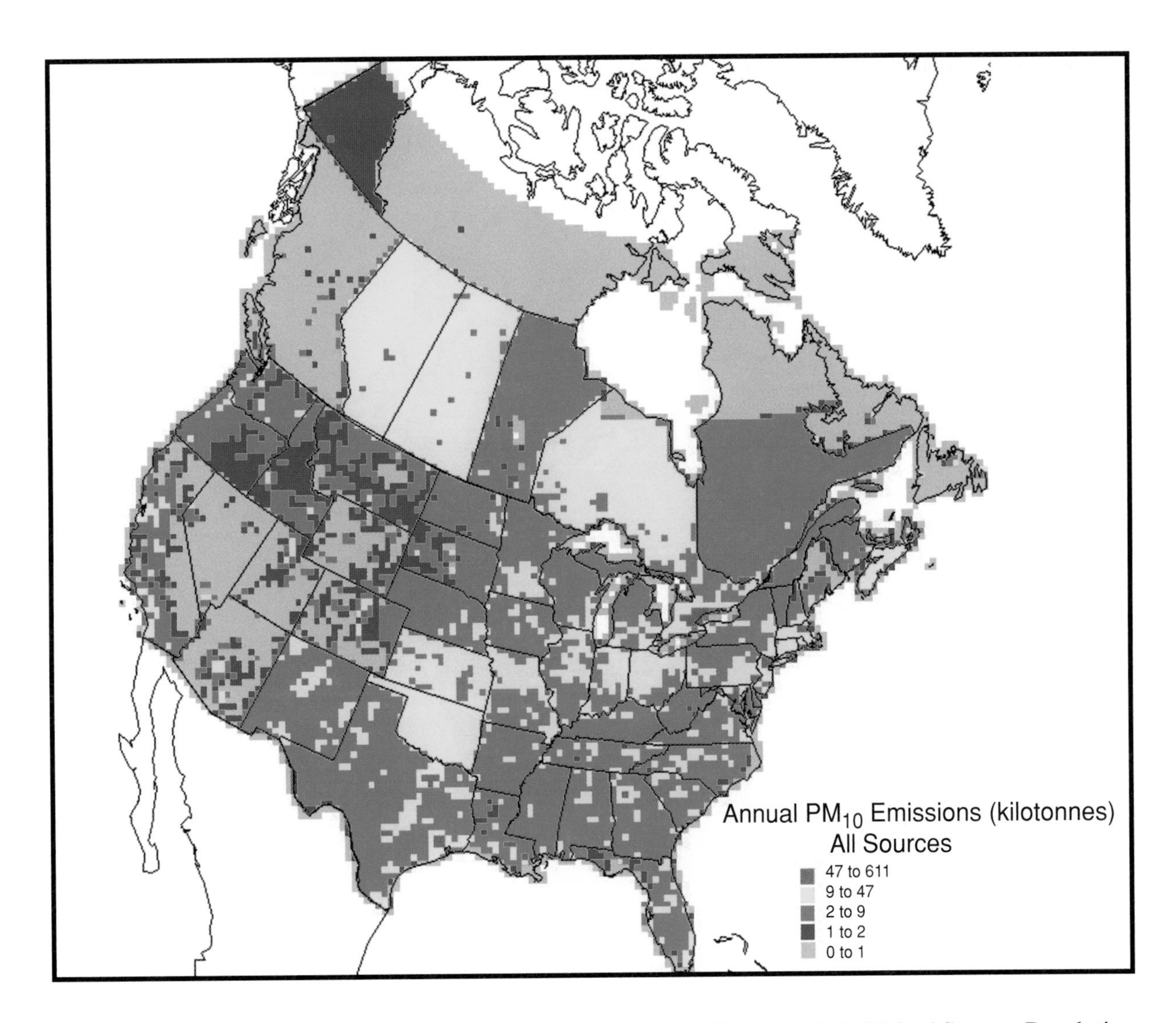

Figure 4.1. Distribution of sources of primary PM_{10} in southern Canada and the United States. Resolution is based on areas (squares) 40 km on a side (MSC, 2001).

weekend/weekday, and diurnal) to account for normal operational patterns of specific source categories.

Random short-term variations in emissions are difficult to estimate, but can be documented from records in relatively few cases. They represent a "stochastic" element in the uncertainty in emission descriptions, which may affect the estimation of short-term averages but probably not long-term (annual) averages. Assessments of these random, short-term variations in emissions generally are not available. However, one analysis of SO_2 emissions in the 1980s has been reported (Hidy, 1994), illustrating the range of variability in this precursor that might be expected. The range in variation, assumed to be random, was estimated to be 5 to 18 percent of the average rate; daily emission estimates could deviate from 70 to 120 percent of the monthly average. Seasonal average rates were found to vary about 20 percent from the annual average in the greater Ohio River Valley area. Similar analyses of other PM sources have not been reported, but one could expect as large or even larger deviations from an average for NO_x, and for primary PM emissions, which involve a wide variety of sources. Little is known about agricultural dust and NH_3 emission variations, which are dependent on meteorological conditions as well as daily and seasonal agricultural

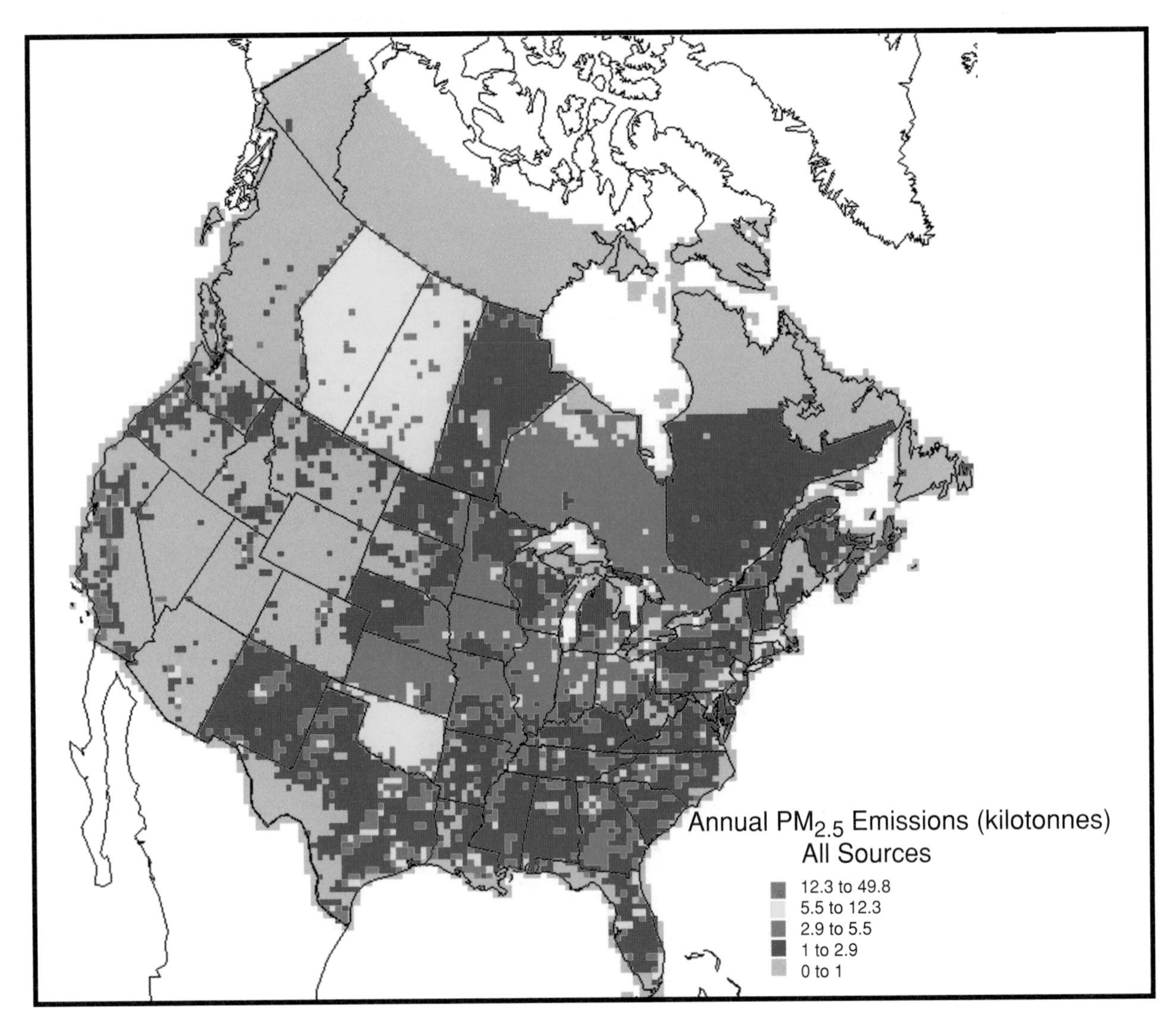

Figure 4.2. Distribution of sources of primary $PM_{2.5}$ in southern Canada and the United States resolved to 1600 km^2 areas (MSC, 2001).

practices. Work is currently underway at the U.S. EPA to link these emission estimates to crop planting calendars and seasonal influences.

4.2.3.2 Trends and Projections

Interannual changes in emission inventories reflect not only improvements in estimation methodology but, perhaps more importantly, they offer a means of measuring progress or trends in forecast emission reductions. Long-term changes in average annual emissions need to be estimated to establish a basis for determining the expected response of ambient concentrations to emission reductions. The record of U.S. emission estimates for PM_{10} and the Canadian and U.S. inventories for the precursor gases SO_2, NO_x, and VOC (e.g., MSC, 2001; U. S. EPA, 2000) are long enough to make some qualitative statements about recent changes. Historical estimates of emissions from 1980 and beyond are shown in Table 4.7 for Canada and the United States. These indicate that there have been changes in precursor-gas emissions, notably in SO_2 reductions, in both countries. Reductions in VOC have been recorded in the United States, but perhaps not in Canada. NO_x emissions have shown apparent slight increases in both countries up to 1995. PM_{10} emissions have declined in the United States between 1980 and 1999.

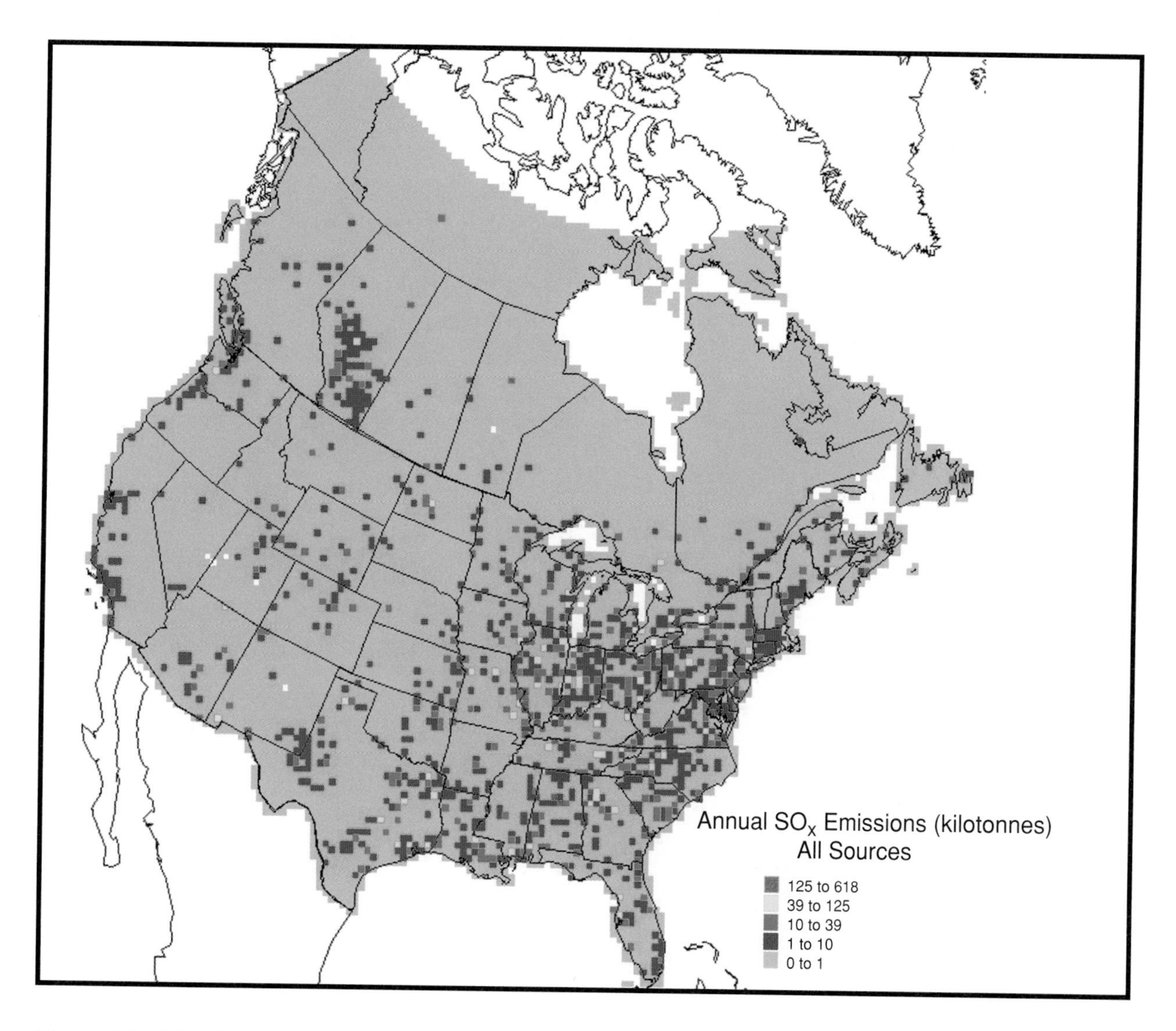

Figure 4.3. Distribution of SO_2 sources in Canada and the United States, resolved to 1600 km^2 areas. (This distribution would be similar to the NO_x and VOC emission distributions, as well (MSC, 2001).)

In general, historical emission trends for the major urban areas of the three nations are poorly documented owing to the difficulty in "backcasting" changes in emission-estimation methodologies. Perhaps the longest self-consistent record for PM precursors (as TSP or PM) can be found in the Los Angeles area, dating back to the 1960s. For example, SO_2 emissions in the metropolitan area have declined substantially since the 1960s, along with VOC and, to a lesser degree, NO_x emissions, consistent with the national picture. Other cities such as New York have a similar picture of precursor emission changes.

The written record in Canadian cities is less clear, as is the case for Mexico, in particular Mexico City. In Canada, estimates suggest that SO_2 emissions have declined since 1980, while a modest increase in NO_x emissions has occurred. Molina and Molina (2002) give some perspective on the trends in Mexico City emissions of TSP, SO_2, NO_x and VOC (see also Chapter 10). They note the ambiguities introduced in comparing the estimates because of changes in methods used for making estimates.

An important element of emission estimates desired for strategy development is projection of future

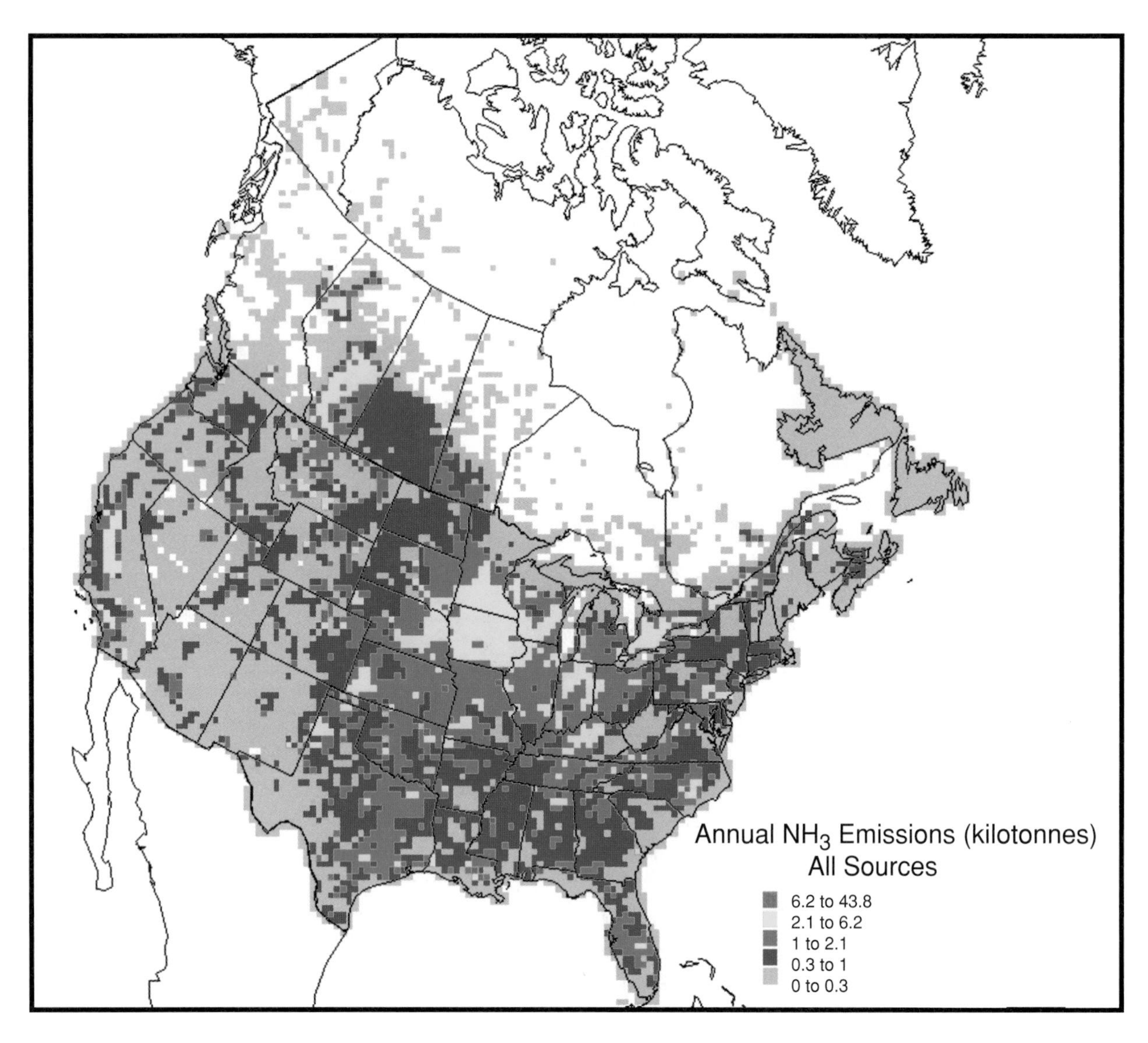

Figure 4.4. Distribution of NH_3 sources resolved to 1600 km^2 areas in the United States and Canada (MSC, 2001).

emissions, both on local and national levels. These projections attempt to account for changes in population, demography, technologies, and resource (e.g., fuel) consumption, while integrating future constraints imposed by environmental regulatory requirements. Such projections require a variety of information from planning agencies, resource and technology experts, and "futurists". The Canadian national projections are derived from consensus estimates, supported by the EGAS economic model. The U.S. national forecasts are calculated by the EPA. Projections in the United States also arise out of other federal agencies and DOE national laboratories, as well as private and academic sectors, including such cooperative entities as the Stanford University *Energy Forum* and the Washington, D.C.-based *Resources for the Future*.

Forecast national emission estimates for Canada and the United States are summarized in Table 4.7. Changes in primary PM emissions are highly uncertain because of questions about the magnitude and reduction of open sources, including fugitive dusts, agricultural operations, and vegetation fires. According to projections in Table 4.7, Canadian national emissions of primary PM_{10} are expected to rise by about 30 percent between 1995 and 2010, while $PM_{2.5}$ emissions will rise by about 20 percent. At the same time, Canadian SO_2 emissions are projected to rise about 8 percent. In the United States, estimates of PM suggest an apparent decline through 1999 followed by increases in 2007 and 2020. The requirements of Title IV of the U.S. Clean Air Act Amendments mandates that by 2007, SO_2 emissions from power plants will decline to 9000 ktonnes/yr, contributing substantially to the decrease in total SO_2 emissions. Total NO_x emissions are expected to decrease to about 16,000 ktonnes/yr by 2020. The U.S. gaseous precursor emissions are projected to continue in decline through 2020. Total U.S. SO_2 emissions are projected to remain above 15,000 ktonnes beyond 2007. Additional NO_x reductions may occur in the United States as a result of regional responses to the U.S. EPA SIP call for achieving the ozone NAAQS. For spatial and temporal resolution of projections, the CTMs are used in conjunction with future emission inventories and compared with a "base" year. The results are analyzed in terms of the air-quality goals imposed on future conditions to examine strategic options for meeting those goals.

4.3 ESTIMATING UNCERTAINTY IN EMISSION INVENTORIES

Emission inventories of the three NARSTO countries represent developmental stages that are characterized by different levels of uncertainty. At this stage, Mexico is relying heavily on the experience of the United States in developing its national and local inventories (e.g., U.S. EPA, 2002). In doing this, Mexican investigators are currently applying

Table 4.7. Summary of historical and projected national emissions for Canada [a] and the United States [b] excluding biogenics and NH_3 (kTonnes/yr).

Year	Primary PM_{10}	Primary $PM_{2.5}$	SO_2	NO_x	VOC
Canada					
1980	NA	NA	4640	1960	NA
1985	NA	NA	3700	1890	1850
1990	NA	NA	3310	2200	3030
1995	4640	907	2680	2030	3580
2000	5230	996	2830	2060	2690
2005	5670	1060	2910	2060	2790
2010	6140	1130	2910	2080	2930
United States					
1980	NA	NA	25,900	24,400	26,300
1985	NA	NA	23,700	23,200	24,400
1990	27,200	7600	23,700	24,200	21,000
1995	25,000	6900	18,800	26,000	18,100
1999	23,700	6770	17,100	23,000	18,100
2007	24,800	6980	17,000	20,500	15,000
2020	25,300	7240	15,800	16,000	14,800

[a] Source: MSC (2001); Environment Canada, 2001.
[b] Source: All current and past U.S. emissions are from U. S. EPA (2001) report, EPA 454/R-01-004; future projections are based on the methodology used in the U. S. EPA heavy duty diesel rule making (2001).

emission information contained in AP-42 (U.S. EPA, 2000) as well as models such as MOBILE V to account for emissions from vehicle fleets[7]. Strictly speaking, emission estimates obtained from these resources are based on U.S. process technology and transportation and may not be relevant to conditions in Mexico. The effort to obtain this relevance is large and expensive, so that progress on inventory development will take some time, even for current methodologies. In Canada, much of the U.S. experience has already assisted in development of the Canadian national inventory, but again Canadian investigators recognize the need to continue improving their own national versions of emission factors and activity patterns. Concerns common to all three nations are the reliability of estimates of fugitive emissions, open-source combustion, and to a lesser degree, NH_3. Also common to the development of the inventories are procedural matters that include receipt of information on plant emissions without supporting process data, limited information on current effluent data, or emission-control hardware updates. Emission inventories should be extended to include all available information from particle sources typically viewed as unmanageable, e.g., wildfires and blowing dust.

A significant element characterizing uncertainty is application of quality control and assurance (QC/QA) measures to establishing the voracity of the emission inventories. The quality of the input from state and local authorities supplying emission data varies considerably with expertise and resources available for this effort.

Relatively little has been reported about the validation procedures used by Mexican authorities. However, some discussion of their efforts in the Valley of Mexico is contained in CAM (2001) and Molina and Molina (2002). Canadian compilation and validation methods, on the other hand, are more accessible and documented. In Canada, industrial emissions are collated by each province, then are checked again by Environment Canada (EC) to ensure that all sources within a facility have been estimated using appropriate emission factors and that the emissions lie with a reasonable range. The information is then checked to verify that the facilities have been placed in the appropriate industrial sector, and if there have been any major changes from the previous year's estimates. Facility estimates are further checked to ensure that there are no major gaps in the estimates. During the QC/QA process gaps are noted, but little can be done to fill them. Because of the voluntary basis for reporting to the provinces and EC, none of the information is required, only requested, from local government. Area and transportation emission estimates rely on the emission models adopted. QC/QA of these results relies mainly on review of the results in terms of estimates made for similar conditions in Canada and the United States. Since few traffic data are available in some areas of Canada, for example, the mobile source estimates are likely to be highly variable in quality, but are believed to be reasonably reliable for the large metropolitan areas experiencing the more severe air-pollution problems.

Perhaps the most formal QC/QA process of the three countries is that for the United States. The United States recently has invested in a major emission improvement program with the states, the Emission Inventory Improvement Program (U.S. EPA, 1997). In this program a coordinated, sustained effort has been established with state and federal resources to systematically go through all local emission inventories to harmonize their data, and verify to the greatest degree possible their reliability. For most area-source categories, locally available information on source activity, and spatial and temporal patterns can enhance the information available to the federal agencies. Figure 4.5 shows a flow diagram indicating in general how the NEI is developed.

Generally the U.S. quality-assurance procedures involve three types of checking: a) initial format checks, b) data content checks, and c) public review. An initial set of format checks on the data will ensure that a file can be processed. The logic of these checks is based on the NEI input-format specifications. The second of the checks focuses on the integrity of the data content. Third, the inventory is published in draft format for public review. Comments that are received are addressed in consultation with the agency that originally submitted the data.

The NEI procedure tracks errors or missing information in a standard format for emission

[7] Mexican investigators have used MOBILE version 5.3 MCMA for VOC, NO_x and CO, but have used the US methodology for PM_{10}.

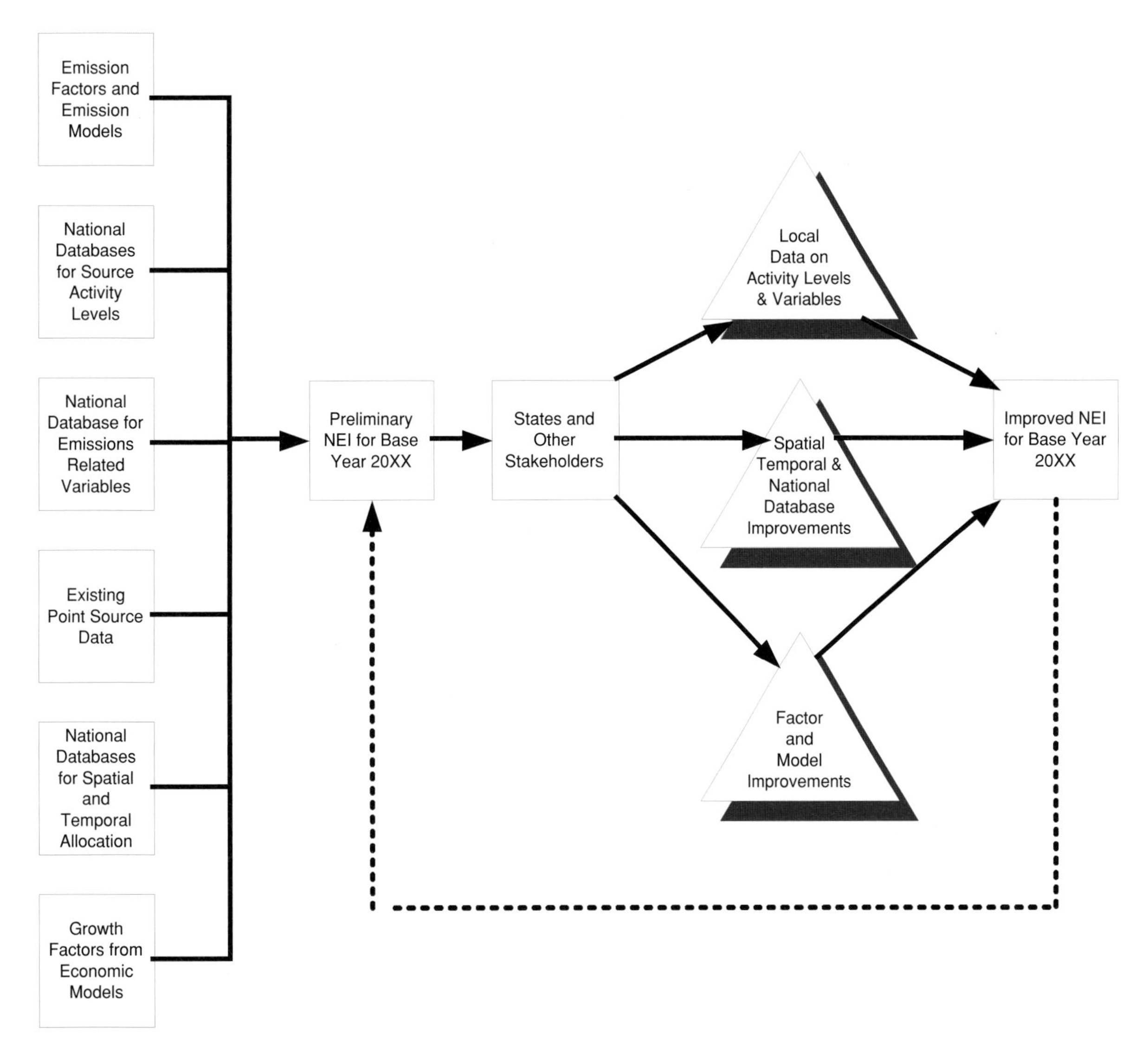

Figure 4.5. Flow diagram for development of the U.S. National Emission Inventory.

estimates based on prescribed emission factors and activity patterns. Ideally, the emission estimates should then be evaluated for their accuracy against independent measurements or observations of specific sources. Reconciliation of emission estimates then must depend on an independent assessment. This is usually accomplished through the use of "top down" consistency checks (e.g., macro-material balances, including comparison with mobile EFs based on mass/fuel used in an area, nationally vs. mass/vehicle mile traveled [VMT] or mass/VKT), comparisons between air-quality modeling and ambient observations, or on in-situ measurements near sources or at roadside sites, for example.

One of the critical elements in verifying the reliability of emission inventories is reconciliation with ambient data. Past work evaluating inventories has involved ad hoc studies that have adopted different approaches for intercomparisons. The use of ambient data for verifying emissions could be improved substantially by developing and using formal protocols for such intercomparisons. Improved protocols could include requirements for using receptor-based models, the length of time and location of sampling, and multiple geographic or source-diverse locations. The ambient data-reconciliation process is crucial if inventories continue improvement to provide a solid foundation for control-strategy decisions. Equally important, this type of "ground-truthing" assembles a substantial body of evidence ensuring that uncertainties in the steps of analyses (e.g., inventory, climatology, chemistry, and transport simulation) are constrained by collective insights for each analytical component. A major limitation that cannot be overlooked in using emission information to support management strategies is the age of the inventory data. Inventories take 2 to 3 years to prepare for a given base year; the processing to translate them into input for CTMs can be lengthy (12 months or more). Yet they form the basis for scenario development and predictions of future air quality. Ultimately control decisions may be made using information that at best may be 3 to 4 years old, and more typically 5 to 10 years old.

A growing body of knowledge of concentration and composition of particles in ambient samples can be used along with source-composition profiles to reconcile or "ground truth" the inventory. This is done through intercomparisons of the source emission speciation profiles with ambient measurements, using receptor-based data analysis tools such as the Chemical Mass Balance (CMB) Model, Positive Matrix Factorization, back-trajectory analyses, and other methods (e.g., Chapter 7). These results are then reconciled with ambient measurements and CTM estimates. This reconciliation process gives important feedback to validate (and modify, if necessary) specific parts of the inventory, and is essential to ensuring the reliability of the emission-estimation methods. Perhaps more importantly, this "ground-truthing" assembles a substantial body of evidence that collectively ensures that uncertainties in any of the steps in the process (e.g., inventory, climatology, atmospheric chemistry, transport simulations) are bounded by the collective insights of each approach. In Chapter 7, these so-called receptor-based methods are discussed, and some important examples of emission-inventory reconciliation are described to illustrate this means of validation.

The need for this kind of reconciliation and combination of various tools to apportion sources and identify inconsistencies is reflected by the relatively sparse quantitative knowledge concerning confidence levels in emission estimates. For most source categories, confidence limits can be based only subjectively on a combination of the record of emission testing, the source records of operations, the QC/QA records, the experience of the engineers maintaining the inventory, and feedback from the bringing together of receptor-and source-oriented approaches as described above. *Additional approaches that can be used to reconcile emission inventories with measurements of species concentrations observed in the atmosphere need to be developed.*

In cases where point sources employ continuous emission monitoring systems (CEMs), the confidence limits can be quantified readily using estimated errors in the monitoring measurements. There is a high confidence level for SO_2 from point sources as an example, but a low confidence level in estimates of fugitive dust emissions from roads or from agricultural operations. Different confidence levels for emission estimates for the same source category are probably acceptable for purposes of managing air quality in the different countries, depending on the approaches to emission reduction and the goals of the national and local programs. The higher levels of reliability are generally required in the United States where enforcement of the national standards has been in place since 1970. Fortunately, the combined use of receptor- and source-oriented approaches, coupled with a focus on regional impacts and strategies, places emphasis on the totality of evidence and reduces the need to evaluate the inventory "in a vacuum."

Table 4.8 provides a general, qualitative look at the relative confidence in the various components of the inventory of PM and its precursors.

4.4 IMPROVING ESTIMATION METHODS

As a practical matter, emission inventories cannot be pushed beyond a certain limit of accuracy, which is basically unknown. Even the specifications for accuracy for estimating ambient concentrations at receptor locations are undefined. The discussion above, however, describes the important uncertainties in current PM inventories that require further work, aside from the need to provide updated estimates at regular intervals to establish long-term emission trends. These are discussed below using two examples: a) improvement of

Table 4.8. Estimated confidence level of emission estimates.

Pollutant	Source	Method of Estimation	Estimated Confidence Of Category in Overall Inventory		
			Canada	USA	Mexico City
SO_2	Electric Utility	CEM/AP-42	H	H	H
	Ind. / Comm. Fuel Comb.	AP-42	M	M	M/L
	Other Fuel Comb.	AP-42	M	M	L
	Transportation	NR [c], Mobile, Mass Balance	M	M	L
	Industrial Processes	AP-42	M	M	N/A
	Other Man-made (Non Comb.)	AP-42	L	L	L
	Natural	Literature	L	L	L
NO_x	Electric Utility	CEM; AP-42	M-H	H	M
	Ind. / Comm. Fuel Comb.	AP-42	M	M	M
	Other Fuel Comb.	AP-42	M	M	M
	Transportation	MOBILE, NR [c]	H	H	M
	Industrial Processes	AP-42	M	M	L
	Other Man-made (Non Comb.)	AP-42	L	L	L
	Natural	BEIS	M	M	L
VOC [a]	Electric Utility	AP-42	M-H	M-H	M
	Ind. / Comm. Fuel Comb.	AP-42	M	M	M
	Other Fuel Comb.	AP-42	L	L	L
	Transportation	MOBILE, NR [c], Literature	M	H	L
	Industrial Processes	AP-42	M	M	L
	Other Man-made (Non Comb.)	AP-42	L	L	L
	Natural	BEIS	M	M	L
NH_3	Electric Utility	AP-42; literature	M	M	L
	Ind. / Comm. Fuel Comb.	AP-42; literature	L	L	L
	Other Fuel Comb.	AP-42;literature	L	L	L
	Transportation	Literature	M	M	L
	Industrial Processes	AP-42; literature	L	L	L
	Other Man-made (Non Comb.)	AP-42; literature	L-M	L	L
	Natural	Literature	L	L	NA
PM_{10} [b]	Electric Utility	AP-42	M	M	M
	Ind. / Comm. Fuel Comb.	AP-42	M	M	M
	Other Fuel Comb.	AP-42	L	L	L
	Transportation	NR [c], Mobile, Literature	M	M	L
	Industrial Processes	AP-42	M	M	L
	Other Man-made (Non Comb.)	AP-42	L	L	L
	Natural	AP-42/Literature	L	L	L
$PM_{2.5}$ [b]	Electric Utility	AP-42	M/L	M/L	NA
	Ind. / Comm. Fuel Comb	AP-42	M/L	M/L	NA
	Other Fuel Comb.	AP-42	L	L	NA
	Transportation	NR [c], Mobile, Literature	L	M	NA
	Industrial Processes	AP-42	L	L	NA
	Other Man-made (Non Comb.)	AP-42	L	L	NA
	Natural	AP-42/literature	L	L	NA

[a] For total VOC; speciation estimates rated low confidence level.
[b] For total PM; composition profiles rated low to medium confidence level.
[c] EPA's Non-Road Model.

conventional methodologies, and b) resolution of emission definitions taking into account the source/ ambient-air interface.

4.4.1 Methodological Improvements

Because of the regulatory setting, historical methods focus mainly on accurate emission estimates of total PM mass released to the atmosphere. Given the increasing knowledge about health effects associated with PM, a potential for a mismatch between the current regulatory management tools and health-based specifications, such as the current hypotheses, may hinder future improvements in evaluating air-quality health impacts.

Emphasis has increased on at least a preliminary resolution of particle size in emissions beyond PM_{10} and $PM_{2.5}$, as required by the changing air-quality standards. However, increased attention is likely to be given to more detailed size-distribution and composition information, especially for fine particles and ultrafine particles. Historically research on size distributions for different source categories began in the 1970s based on specialized source testing of stationary and mobile sources, as well as ambient measurements near sources. This research has expanded recently to encompass a broader range of source categories and operating conditions, but the costs are very high per source tested. *Emission inventories and inventory tools to include size distributions and size-resolved composition, organic speciation, and source profiles, need to be updated and expanded.*

As requirements for chemical speciation of primary particle emissions evolve, increased attention to this detail will be needed in each of the NARSTO nations. Presently, Canada and the United States use the U.S. EPA PMCALC or SPECIATE models. Other information derives from specific chemical-profile determinations, which support source-attribution analysis (Chapter 7). Today's speciation by chemical composition in the emission-inventory profiles is somewhat rudimentary, based largely on a highly simplified conceptual picture of ambient PM mixtures. The breakdown for $PM_{2.5}$ composition that is most often used in regulatory analyses segregates the chemical composition into (ammonium) sulfate, (ammonium) nitrate, BC, OC and "other" (including crustal material and other inorganic material). Emission processors then apportion the $PM_{2.5}$ direct emissions into these component "bins" before they are used in CTMs. These component "bins" are then compared to similar component groups obtained from analysis of ambient samples, as discussed in Chapters 7 and 8. Within the current CTM approach, a number of assumptions should be tested. One such assumption is the assignment of sulfates and nitrates as ammonium, where, for example, at least four species of sulfur oxide acid-salts are known to exist in the troposphere. Also, assigning organic matter a weighting factor of 1.4 (OC) is a major generalization. Recently Turpin and Lim (2001) have indicated that this assumption is not adequate to characterize organics in different environments. Organic species in PM emissions are not identified routinely, and metals are normally assumed to be crustal equivalents as oxides, without consideration for other species or valence state. Ambiguities such as semivolatile components and "bound" water content are not currently considered. Increasingly accurate and chemically detailed speciation of emitted particles will be needed for future considerations of human exposure, and source-attribution analysis. As noted above, this implies that considerably more effort will be required in profiling even the major sources of PM, and especially SOA precursors.

The breakdown into five or six generic PM components for CTM use is helpful in the interpretation of PM concentrations and CTM estimates, but has limited relationship to the evolving hypotheses about health effects. Modeling eventually will address exposure and may need to account for "minor" or trace components embodied in PM surface estimates, ultrafine particle concentrations, and trace-species composition.

In the more recent CTMs, coupled oxidant- and particle-chemistry computations require not only particle-source composition profiling, but speciation of VOC to be employed in oxidant-formation mechanisms. Since there are literally hundreds of VOC compounds present in emissions, the models apply classes of these compounds aggregated into alkanes, alkenes, aromatics, and oxygenates, for example. The U.S. inventory maintains a database of 400 VOC speciation profiles for emissions, as well as a PM database of about 400 profiles. The U.S. inventory also includes a listing of some 188 HAPs, and additional information for many specific sources is available through the U.S. National Toxics Inventory. Yet a SOA precursor inventory has not been called for specifically.

As an example of regulatory analysis requirements, some interactive CTMs use an aggregated composite of 66 PM profiles to represent the chemical composition of the national inventory. Each of the 5,000 Source Classification Codes in the U.S. inventory has one of the 66 PM profiles and one of the 400 VOC profiles assigned to it. Some of the profiles in U.S. use were developed in the mid-1970s from air-quality studies in the Los Angeles area, and are outdated. Also, if one or more of the health-effect hypotheses becomes a driver for regulation, these compositional profiles may need substantial modification to support PM regulation.

The use of the emission inventories could be facilitated by employing across North America a consistent set of source categories and disaggregations for complex species embodied in VOC and PM composition. Further, a standardized library of source profiles for chemical composition, and particle-size distributions accessible to the public would be a useful and welcome addition to the inventory literature. Work is underway at the U.S. EPA to update the profile database with the most recent data, but ultimately the quality of the data set will be limited by the resources available for additional testing.

Inspection of the current inventories indicates that the quality of differentiation by chemical composition and particle size is quite limited by the lack of direct testing or measurements of sources. This situation is particularly acute for determination of the carbonaceous material in aerosols from combustion of fuels and other sources. This is especially true of area and transportation sources that display a high level of annual seasonal and daily variation. A study of seasonal or shorter variations would be a major step forward.

While the U.S. EPA mobile-source emission inventories for VOC, CO, and NO_x include high emitters, the PM inventories presently include only normal emitters. Obviously, high emitting gasoline and diesel vehicles, including those consuming excess oil and those emitting visible smoke, contribute substantially higher PM emissions than normal emitters. However, there are no data on how many of these vehicles exist in the fleet. Some progress is being made in characterizing transportation sources through laboratory and field observations of on-road and off-road components. Yet, there is a practical limit to how much testing can be done under real world, or changing "fleet" conditions in different cities of North America. The U.S. EPA is now implementing a large emission-measurement program along with other organizations to determine the PM emission distribution for the light-duty vehicle fleet, and also more information on cold-start emissions and the specific organic compounds emitted. Similar work is underway for diesel trucks. It is hoped that these results can be extrapolated to off-road engines/vehicles which also have a significant PM contribution. When this program is completed, the emissions from high PM emitters will be incorporated into U. S. EPA emission models and inventories.

4.4.2 The Source/Ambient-Air Interface

Certain inherent limitations in the practice of inventory development are "definitional" in character. These relate to the boundary or interface between the emitting source and the ambient air. Historically, this boundary has been defined simplistically, but cleanly, in terms of the physical limits of the stack, the tailpipe of the source, or the areal surface. However, it is known that this boundary is not a simple one. Two cases in point include in–stack/in–plume chemistry, and the air–surface processes that affect fugitive emissions.

Careful observations of emissions from combustion processes and other sources have indicated that the physical and chemical characteristics of emitted material change in the plume forming a few meters between the stack or effluent outlet (e.g., Wei et al., 2001). The transition in this near-stack environment is exemplified in observations of changes in catalyst-equipped internal-combustion engine exhaust, diesel engine exhaust, and power-plant plumes. The transition conditions often vary with ambient temperature and humidity, as well as intensity of dilution of the effluent. If the composition, or more generally the thermodynamic state, of the emitted aerosol becomes important, then this transition regime may need characterization to ensure that the material injected into the ambient air after transition is fully accounted. The sophistication in emission characterization at this level is well beyond the present state of the art. For example, "condensable PM" is estimated to form outside the stack independently of any transformation processes that occur near the stack. Management of the PM problem may require more detailed consideration of near-stack processes in the long term.

A related problem is the potential for "double-counting" semi-volatile and non-volatile VOC emissions at the source/ambient-air interface as a result of inconsistent measurement methods. If measurements of low volatility, high carbon-number organics are collected at high temperature from the source, they may be in the gas phase. Measurements of gas-phase emissions that are condensable on cooling may include these species in the gas-phase profiles. If particle measurements are made concurrently but slightly downstream to estimate "particle" emissions at ambient temperature, then the emitted semi- or non-volatile mass may be counted twice. This is a significant concern, given that the current trend in organic particle formation parameterization in CTMs attempts to provide more details of speciation (to calculate partitioning between the gas and condensed phases). For example, the organic profile for light-duty diesel vehicles includes "vapor" carbon number emissions up to C_{45}. It is unlikely that constituents above C_{30} or so could exist as gases in the ambient air. Accounting for mass emitted of these compounds should be attempted in future work, and combined gas- and particle-OC measurements need to be balanced for consistency to avoid double counting. *Emission models that account for condensation or nucleation of semivolatile compounds as hot emissions mixing with cool air should be developed.*

Emission-rate estimates for fugitive sources at the earth's surface are sometimes inconsistent with resulting ambient concentrations. The causes of these discrepancies are not fully understood, but are believed to be associated in part with interpretation of ground-level measurements relative to a "mixed" volume of air above the surface. Typically, the emission rate is estimated in terms of a microscale "two-dimensional" measure of a horizontal flux at steady state through a vertical plane immediately downwind of a source activity. This measurement is compared with a similar one just upwind of the activity and the emission rate is determined by difference. This technique is believed to provide a reasonably accurate estimate of the net emission rate from the surface source. However, the approach does not account for other influences like deposition and removal that effectively remove airborne material before entrainment into the surface-air volume, depleting the surface boundary-layer prior to transport away from the source area. This emission-estimation method also does not account for microscale surface heterogeneity, including loss from re-deposition on nearby vegetation, other obstacles, or on near-surface collection of fine particles on falling larger particles. One hypothesis accounting for such discrepancies defines the surface source only in terms of the mass of suspended material that can rise above the surface features and be entrained into the local and/or larger scale wind patterns. This hypothetical treatment is "imbedded" into air-quality models as well, where the fugitive dust entrainment rate has to be linked with the surface winds and vertical mixing rate.

Refinements in the treatment of this ground-air interface are currently underway. Workers are attempting to resolve this issue in a consistent way with CTM deposition algorithms, to develop a defining framework for this problem and account for interaction of the dust plume with vegetation. Note that an analogous ambiguity in removal mechanism may also affect near-surface releases of NH_3. Ammonia exchange with vegetation is known to be capable of switching between deposition (e.g., leaf absorption) and emissions. This is identified with the "compensation point" for concentration. This point is a neutral condition where neither deposition nor emission occurs. It is estimated to be in the range of 2 to 6 ppb NH_3. At ambient concentrations below the compensation point, emissions of NH_3 take place and above, absorption occurs (e.g., Dabney and

Bouldin, 1990). *Emission models for dust need to be improved, and models for estimating NH_3 emissions need to be developed further.*

4.5 HOW WELL DO THE EMISSION INVENTORIES ADDRESS APPLICATION NEEDS?

The requirements for today's decision making differ across North America, depending on the nature of the practices, priorities, and laws or guidelines governing environmental protection. While each of the countries needs technical support in the five levels of inventory application, this support varies in quality and reliability. The response to this question has an important geopolitical context quite distinct from the technical aspects of the emission inventories. The current inventories generally provide the necessary general knowledge, but perhaps not sufficient detail for speciation and size-distribution information for insightful PM management in the next decade.

The five application levels for emission inventories listed earlier require somewhat different levels of output from the national inventories, with different confidence levels for spatial and temporal estimates. At the first level, estimation of source impact and compliance has important regulatory enforcement significance in each of three nations.

At Level 1,

- The inventories available in Canada and the United States generally are suitable for identification of key sources of particle pollutants, sulfur oxides, and nitrogen oxides for their management, both in a local and a state or provincial context.
- The inventory for Mexico City, and perhaps those for the other six Mexican cities documented to date, are probably sufficiently precise for the present emission strategies adopted by the Mexican authorities at their stage of development. Mexico is progressing toward a truly national inventory, at least for the northern half of the country, in the next few years. However, in the future, Mexico will likely require broader local-inventory development to meet its "Level 1" needs.
- In all three nations, there is a need to continue studies of individual sources or categories to ensure that the inventories are up-to-date, especially for area sources and transportation sources, where uncertainties in Mexico appear to be high relative to other categories as well.

At Level 2,

- For urban and population-center strategy planning and tracking, the U.S. and Canadian inventories are sufficiently accurate for identifying and managing sources of primary PM and gaseous precursors. Again they are not sufficiently precise to address the components of carbon chemistry and this may lead to a weakened capability to simulate the secondary interactions of NO_x (or NO_y), NH_3, and VOC with sulfur-oxide chemistry believed to involved in $PM_{2.5}$ pollution over many spatial scales.
- The characterization of the carbon-containing fraction of particles is important for its leverage on $PM_{2.5}$ reductions. This component presents a particularly difficult task since products of combustion derive from both primary and secondary contributions.
- The sources of carbon-containing components can be characterized better if their constituents can be identified chemically. Current analyses of organic particles have not been able to identify the majority of the components in this fraction.
- If the organic component turns out to be critical to health effects, considerably more in-depth effort will be required to trace its (primary and secondary) sources.
- Likewise, if components not currently tracked, such as some of the valence states of metals or ultrafine particles, turn out to be critical to health effects, the current inventories may not be sufficiently comprehensive for management practices.
- The inventories as currently developed need to be repeated over several years with a self-

consistent methodology to ensure that progress toward goals can be tracked for comparison with ambient concentration trends, to verify progress expected from management actions. At the present time, it is not possible to trace the response of ambient concentrations of some PM source categories to emission changes over the past several years. This is a highly desirable component to managing PM in accordance with national goals (e.g. NARSTO, 2000). *National emission inventories should be prepared over a period of years in parallel with long-term monitoring of health effects and ambient concentrations of PM and its precursors to facilitate analyses of trends and benefits.*

At Level 3,

- The present cooperation in the United States between industry and government has established a satisfactory means of applying emission data.

- The Canadian and Mexican legal requirements for confidentiality inhibit the open access to inventory information that could be useful to other experts than in government. This limitation is largely illusory on an international scale, where broad technology improvements or changes are readily described in the open literature.

At Level 4,

- The current national inventories of Canada and the United States serve the international community well for providing data input relevant to global-pollution issues, including assessment of hemispheric PM and precursor transport.

- Mexico provides a limited inventory for international programs, but its reliability is unknown.

At Level 5,

- Perhaps the most stringent requirements for emission inventories are those driven by the CTM community. For application of emissions to CTMs, demanding requirements on resolution of emission inventories are emerging, especially those requiring finer resolution in space and time.

- Depending on the averaging time and the spatial resolution of the output concentration fields from the model, the requirements for the emission inventory can range from an hour to annual average, and from about 4 km to more than 50 km grids.

- In addition, for a chemically specific computation including oxidant interactions for secondary particle formation and for regional-haze analyses, resolution with respect to particle size and chemical composition of particles and reactive gases is required.

- Composition of carbonaceous compounds is particularly important for both calculation of oxidant and secondary-particle concentrations. CTM input for both primary particle carbon and for high molecular-weight VOC is problematic.

- Precise estimates of emissions, including emission models for CTM applications, are very demanding on detailed spatially gridded data, to an extent beyond the present day capabilities of the inventories for quantification. While CTM research may specify such detail in emissions, the practical requirements of human-exposure estimation have not been defined well by the regulatory community.

- Experience has given an unclear picture of whether or not the three nations are achieving sufficiently precise emission estimates for their regulatory purposes (e.g., Table 8.1).

- In theory, the decision-making community has used or can use all of the detail in projections that the model analyst can provide. In practice, it is unclear if this is the case in all three nations, especially at the sub-national level.

Two elements for the use of inventories are important. These address the timeliness of inventory preparation and publication, and the documentation of inventories. In the first case, the use of inventories for evaluating current ambient data, and for estimating improvements in air quality is inhibited by the relatively long intervals between the updating and reporting of inventories. If inventories could be prepared on time schedules more frequent than presently considered, with consistent methodology, analysis of contemporary air-monitoring data could be facilitated, and investigation of apparent changes or lack of change ambient air quality with emissions could be evaluated for more progressive management opportunities. *Scientists should continue to improve the timeliness of emission inventories. In addition, approaches for obtaining emission information with adequate resolution in space and time, and composition for atmospheric modeling and exposure assessments, should be developed. These approaches will involve combinations of emission models and data with higher resolution than are currently included in the inventories.*

Making informed use of emission data at all application levels often requires some investigation of data sources, calculation methods, and the assumptions made to derive emission values. The national and local inventories available today have a range of documentation that is accessible to the public. The U.S. inventory is perhaps best documented with Canada following closely, except for confidentiality considerations. The Mexican inventories are perhaps least well documented, but if proposed protocols are followed, this potential problem should be alleviated.

In the future, improvements in the three national emission inventories will undoubtedly continue in response to the five levels of application. These will include:

- Completion of a national inventory for Mexico, at least for the northern half of the country.
- Development of a tri-national emission trend analysis by major category of contribution to PM, to document changes and progress toward reduction goals.
- Compilation of comprehensive, annual emission inventories in Canada with mandatory reporting of emissions from major facilities for 2002.
- Incorporation of spatially resolved estimates from the MOBILE model into a GIS positioning system to facilitate gridded emissions using local traffic and weather information.
- Incorporation of traffic volume data and road network data to estimate VKT on finer spatial and temporal resolutions within the Canadian provinces. Modification of the MOBILE VI model for Canadian and Mexican conditions.
- Advancements in open source estimation to improve input information, including fuel characteristics, and temporal and spatial resolution of fires, and fugitive dusts.
- Resolution of the fugitive-dust estimation techniques to account for their low release height and interaction with their surroundings.
- Resolution of spatial and temporal issues associated with agricultural NH_3 releases and the issues of NH_3 uptake and release by vegetation and soils.
- Overall improvements in the emission models and source profiles for both PM and SOA precursors, including additional specificity in particle size and chemical composition of primary particle emissions.

4.6 SUMMARY

Emission inventories represent an important component for managing PM air quality. $PM_{2.5}$ in particular is known to come from both natural and anthropogenic sources on the continent. With the exception of NH_3 and VOC, and perhaps locally blown soil dust or forest fires, the highest emission rates are estimated to be in urban-industrial areas, illustrating the dominating anthropogenic influence in populated areas. Ammonia emissions are linked with agricultural practices, and vegetation emits considerable VOC. The apportioning of VOC between natural and anthropogenic material responsible for secondary organic carbon generally is not well understood. The precise makeup of the PM depends on the geographical and climatic conditions in a specific location.

Currently, national emission inventories identify sources of PM and precursor gases, including SO_2, NO_x, VOC, and NH_3. In Canada and the United

States, the inventories are updated and revised in detail over periods of two to five years, with annual review of certain aspects. Mexico has completed an inventory for Mexico City, and has limited inventories for six other cities and certain large industrial sources. Mexico has initiated efforts to prepare a national inventory in the near future. The emission inventories are most accurate and comprehensive for aggregated emission estimates of annual averages segregated by geo-political jurisdictions. This form serves well for most applications, but additional calculations are required to translate inventory data into a spatially and temporally specific form for CTMs. This is generally done using emission-modeling systems.

Emission estimation involves calculations based on emission factors (average emission rates per unit process input over a range of operating conditions) and activity factors (from estimates of consumption or use rates and source operating characteristics). Both emission factors and activity estimates vary within source categories and across geographical regions. Since it is impractical to measure every source to derive these factors, source-category averages are derived based on small numbers of samples taken from specific source types. The accuracy and representativeness of these factors are often limited due to the small sampling on which they are based. This leads to differing levels of uncertainty depending on the nature of the source and its activity patterns. Uncertainties in emissions are difficult to estimate, but generally the highest uncertainties are associated with windblown dust, fugitive dust removal, NH_3 (emissions, uptake and removal) and carbon sources, including VOC and semivolatiles, motor-vehicle emissions (because of unknown in-use vehicle conditions), and open fires. The most certain estimates are associated with point sources such as continuous industrial processes and electric power generation.

Comparison of national emission inventories and CTM outputs indicates similarities and differences between Canada and the United States that follow lines of industrial capacity, transportation, and open or fugitive sources associated with land-use practices. For Mexico, these differences may be minimized with the current deployment of unified estimation methods for the United States and Mexico. Spatial distributions of $PM_{2.5}$ sources and their gaseous precursors largely follow population centers and areas of intensive energy use or industrial activity. PM_{10} and NH_3 patterns differ somewhat because of the contribution of airborne-dust sources and emissions from agricultural operations. Comparison of major urban emissions in Atlanta, Los Angeles, Mexico City, and Toronto suggests generally similar patterns from common source categories except for Los Angeles, whose metropolitan emissions of NH_3, NO_x, and VOC are large compared with other cities. The four major source categories based on magnitude of PM and precursor emissions are electric power generation, transportation, open sources, and industrial processes.

PM emissions vary substantially with time both in the short and long term. Short-term variations are difficult to take into account but necessarily must be estimated for precise CTM calculations. Long-term changes or trends in average emissions need to be estimated to establish that changes are actually taking place for determining the expected response of ambient concentrations to emission reductions. The record of national emission estimates for precursor gases is long enough in Canada and the United States to make some qualitative statements about recent changes. SO_2 emissions are estimated to have declined by a third or a half. NO_x emissions have risen in Canada and have fallen slightly in the United States. National forecasts in Canada estimate that SO_2 and NO_x emissions will rise slightly or remain stable in the next decade, while VOC emissions will fall substantially. In the United States, SO_2, NO_x, and VOC emissions are expected to decline by 25 to 40 percent.

Achievement of PM air-quality standards depends on reduction of the manageable component of PM sources. Sources derived from natural phenomena (e.g., volcanic activity, wildfires, biogenic emissions from vegetation, sea spray, and dust storms in remote areas) are generally unmanageable. Clearly most industrial and transportation sources are manageable using process modifications or end-of-pipe technology. There remains a largely uncertain area of emissions involving suspension of dusts, agricultural operations, and forestry practices, including prescribed burning, that also need to be considered in strategies. These sources are intermittent, but generally are straightforward to identify. They are difficult to characterize quantitatively, but can be potentially important in many locations at certain times of the year and can be readily controlled.

Reconciliation of emissions from inventory estimates using CTMs, ambient data, in-use activity, and source-apportionment tools is an important check on the inventory estimates, but is seldom done. Verification by these methods is important to ensure the reliability of the inventories and subsequently the confidence in the CTM projections. Examples of significant bias in current inventories have been identified with fugitive-dust and motor-vehicle emissions. Fugitive-dust emissions can be overestimated using existing methods; motor-vehicle particle emissions can be both underestimated or overestimated depending on category and location.

4.7 POLICY IMPLICATIONS

The following is a summary of the inferred policy implications from this chapter.

Emission inventories indicate that a relatively few common source categories generally make up the large majority of PM (combustion-related) and precursor emissions (transportation, electric power stations, industrial activities and processes, and open sources). *Implication:* Characterizing the relative contributions of these sources to local and regional PM problems will assist CTM applications and identification of effective management of these sources to improve air quality.

Annual emission information in all three countries continues to improve. Complete inventories for primary PM_{10} and $PM_{2.5}$ along with precursor-gas emissions have only just become available in Canada and the United States. It is anticipated that Mexico will have this information soon. Due to the trans(national)-boundary nature of the PM problem in many areas, it is necessary for emission information to be reported uniformly by country, using the same estimation methods supporting analysis of source-region contributions. *Implication:* The lack of standardized emission reporting methods and uniform emission information has constrained precise source characterization of local vs. upwind source contributions, particularly in trans(national)-boundary situations.

Because of the complex composition of PM, and both the primary and secondary origins of various size fractions, quantitative estimates of primary emissions and precursor gases are needed with suitable spatial and temporal resolution. The distinct contributions of natural and man-made sources also must be identified, particularly for selected VOC species, primary carbon, and NH_3. *Implication:* At present, the management of primary carbon and secondary carbon sources is based on limited quantitative information of high uncertainty.

Natural and fugitive emissions of PM (earth crustal material and carbon), particularly blowing dust and forest fires, and SOA precursors, especially high molecular-weight VOC, are poorly characterized but intermittently are major contributors to PM mass concentrations in some locations. These estimates need to be improved to precisely assess their contributions to exceedances of 24-hr and annual-average national standards. *Implication:* The contribution of natural and fugitive emissions to exceedances of national standards is considered to be modest. Yet some analyses suggest otherwise; however, the emission information to confirm this conclusion does not yet exist.

Carbon is a major component, 30 to 60 percent, of $PM_{2.5}$ composition. Yet the emissions of both direct and precursor carbon compounds are the most poorly characterized of the relevant contributing emissions. *Implication:* Until the emissions of primary carbon and selected VOC emissions are better estimated, management of PM through the sources of these emissions will be based on strongly limited information relative to other sources.

The estimates of point-source emissions for the precursor gases, SO_2 and NO_x, are of relatively high quality with little uncertainty. Uncertainty increases for sources with greater spatial and temporal variability. Uncertainty also increases as emissions become more chemically complex, for example, SOA precursor and primary particle emissions. *Implication:* Though the contribution of point sources responsible for the inorganic, secondary fraction of PM are the best characterized and are major components of the PM mass, there are other less well characterized sources, including carbon, that will be significant in some areas.

The compilation of emission inventories is a painstaking and detailed task that is often considered routine within the atmospheric-science and management communities. Recently, policy makers have begun to recognize the importance of this information, and have provided additional resources for improvements in methodologies and estimates for all three countries. Emission information is an important foundation of all planning. Limitations and uncertainties in this area propagate themselves through all aspects of the analyses. *Implication:* Investment in improving emission information will have far-reaching benefits for most air-quality improvement analyses. This investment will result in better overall technical guidance for policy makers.

4.8 REFERENCES

Andreae, M.O., 1995. Climatic effects of changing atmospheric aerosol levels. In World Survey of Climatology, Vol. 16, Future Climates of the World. pp. 347-398. Henderson-Sellers A., ed., Elsevier, Amsterdam.

Comision Ambiental Metropolitana (CAM), 2001. Inventorio de emisiones a la atmosfera, Zona Metripolitana del Valle Mexico, 1998. Mexico City.

Cass, G., Hughes, L., Bhare, P., Kleeman, M., Allen, J., Salmon, L., 2000. The chemical composition of ultrafine particles. Philosophical Transactions of the Royal Society of London 358, 2581-2592.

Dabney, S.M., Bouldin, D.R., 1990. Apparent deposition velocity and compensation point of ammonia inferred from gradient measurements above and through alfalfa. Atmospheric Environment 24A, 2655-2666.

Diamond, M.L., Gingrich, S.E., Fertuck, K., McCarry, B.E., Stern, G.A., Billeck, B., Grift, B., Booker, D., Yager, T.D., 2000. Evidence for organic film in an impervious urban surface. Environmental Science and Technology 34, 290-2908.

Environmental Protection Service (Canada) (EPS), 1999. 1995 Criteria Air Contaminants Emissions Inventory Guidebook. Environment Canada, Preliminary Draft. Hull QC.

EPS, 2000. Canadian Emissions Inventory of Air Contaminants (1995). (Draft). Pollution Data Branch, Environment Canada, Hull, QC.

EPRI, 1994. Electric Utility Trace Substances Synthesis Report. Report TR-104614, Electric Power Research Institute, Palo Alto, CA.

Fuentes, J.D., Wang, D., Neumann, H.H., Gillespie, T.J., Den Hartog, G., Dann, T.F., 1996. Ambient biogenic hydrocarbons and isoprene emissions from a mixed deciduous forest. Journal of Atmospheric Chemistry 25, 67-95.

Geron, C., Guenther, A., Pierce, T., 2000. An improved model for estimating emissions of VOC from forests in the eastern United States. Journal of Geophysical Research 99, 12,773-12,792.

Hidy, G.M., 1994. Atmospheric Sulfur and Nitrogen Oxides: Eastern North American Source-Receptor Relationships. p. 176-181, Academic Press, San Diego, CA.

Hildemann, L., Kleindinst, D., Klouda, G., Currie, L., Cass, G., 1994. Sources of urban contemporary aerosol. Environmental Science and Technology 28, 1565-1576.

Instituto Nacional de Ecologia (INE), 1998. Segundo Informe Sobre la Calidad del Air en Ciudades Mexicanas, 1997. Instituto Nacional de Ecologia, Mexico City, Mexico.

Irving, P., ed., 1991. Acidic Deposition: State of Science and Technology, Vol.1, Report No. 1 - Emissions. p. 1-111, et seq., U.S. National Acid Precipitation Assessment Program, Washington, DC.

Kleeman, M., Hughes, J., Allen J., Cass, G., 1999. Some contributions to the size and composition distribution of atmospheric particles: Southern California in September 1996. Environmental Science and Technology 33, 4331-4341.

Kleeman, M.J., Elderig, A., Hall, J.Rl, Cass, G., 2001. Evaluating the effectiveness of alternate emissions contributions strategies on visibility. Environmental Science and Technology 35, 4668-4674.

Levy, J.I., Hammitt, J.K., Yanagisawa, I., Spengler, J.D., 1999. Development of a new damage function for power plants: Methodology and applications. Environmental Science and Technology 33, 4364-4372.

Makar, P.A., Moran, M.D., Scholtz, M.T., Taylor, A., 2003. Speciation of volatile organic compound emissions for regional air quality modeling of particulate matter and ozone. J. Geophys. Res 108, in press.

Meteorological Service (Canada) (MSC), 2001. Precursor Contributions to Ambient Fine Particulate Matter in Canada. Environment Canada, Toronto, Ont.

Molina, M.J., Molina L.T., (eds.), 2002. Air Quality in the Mexico Megacity. Kluwer Academic Publishers, Boston, 384 pp.

Molina, M.J., Molina, L.T., Sosa, G. Gasca, J., West, J.J., 2000. Analysis and Diagnosis of the Inventory of Emissions of the Metropolitan Area of the Valley of Mexico. MIT Integrated Program on Urban, Regional and Global Air Pollution, Report No. 5, Massachusetts Institute of Technology, Cambridge, MA.

NARSTO, 2000. An Assessment of Tropospheric Ozone Pollution: A North American Perspective. NARSTO report available from NARSTO Management Coordination Office, Pasco, WA, or EPRI, Palo Alto, CA, Chapter 5. Also, NARSTO Assessment Team, 2000. "Assessing policy relevant science for managing O_3 air quality." Environmental Manager (Nov.), 11-15.

Osnaya, R.P., Gasca, J.R., 1998. Inventoriote amoniaco para la ZMCM, revision de diciembre de 1998. Instituto Mexicano del Petroleo, Project DOE-7238, IMP, Mexico City, 23pp.

SEMARNAT and INE (Secretaria de Medio Ambiente y Recursos Naturales and Instituto Nacional de Ecologia), 2001. Mexico: Second National Communication to the United Nations Framework Convention on Climate Change, Mexico.

Turpin, B., Lim, H.J., 2001. Species contribution to $PM_{2.5}$ mass concentrations: Revisiting common assumptions for estimating organic mass. Aerosol Science and Technology 35, 602-610.

U.S. Environmental Protection Agency (U.S. EPA), 1997. Introduction and Use of the EIIP Guidance for Emission Inventory Development. Emission Inventory Improvement Program (EIIP), U.S. Environmental Protection Agency, Research Triangle Park, NC.

U.S. EPA, 2000. Air Quality and Emission Trends Report, 1998. Report EP 450/R-98-00-003, U.S. Environmental Protection Agency, Research Triangle Park, NC.

U.S. EPA, 2002. Guidance for estimation of emissions from various sources in Mexico. (Series of ten working documents) www.epa.gov/ttn/catc/cica (In English and Spanish). Office of Air Quality Planning and Standards, Research Triangle Park, NC.

Wei, Q., Kittelson, D., Watts, W., 2001. Single stage dilution tunnel performance. in In-cylinder Diesel Particulate and NO_x Control 2001. Paper SP-1592. 2001 World Congress, Society of Automotive Engineers, Warrenburg, PA.

CHAPTER 5

Particle and Gas Measurements

Principal Authors: Fred Fehsenfeld, Don Hastie, Judith Chow, and Paul Solomon

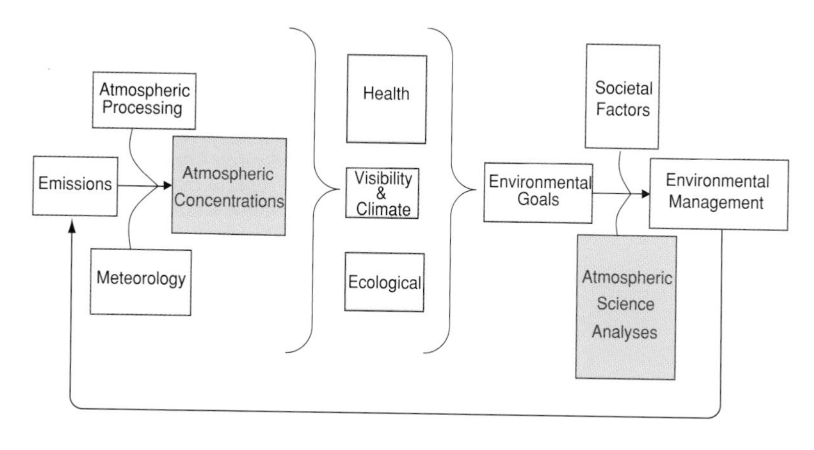

Measurements provide fundamental information for evaluating and managing the impact of PM on air quality. As illustrated in Figure 5.1, data obtained through measurements form the foundation of all approaches used to define and mitigate PM problems. Diverse measurements are required by different users to meet a variety of needs:

1) To monitor **exposure.** *(Who has the problem?)* Exposure monitors document actual human exposure to indoor and outdoor pollution.

2) To determine specific causes of **health effects.** *(What is the problem?)* Initial emphasis focuses on testing related to the eleven hypotheses listed in Textbox 2.5. The wide array of measurements that are needed to determine the connection between health effects and specific PM properties are enumerated in Appendix B, Table B.1. *Measurements of novel PM properties are needed to test hypotheses regarding causal agents for human health effects.*

3) To monitor **visibility impairment.** *(Is the problem visible?)* Fine particles in the atmosphere reduce local visibility, increase regional haze, and affect climate change.

4) To monitor **compliance.** *(Where is the problem?)* Measurements document locations where standards are exceeded.

5) To develop reliable **source attribution.** *(Who or what is causing the problem?)* Measurement techniques, modeling, and data analysis use measurement data to identify the sources of PM pollution.

6) To establish **program evaluation.** (accountability) *(Are air-quality goals being met? If not, where and when are shortfalls occurring?)* Measurements monitor the atmospheric distributions of regulated pollutants and pollutant precursors, and the trends in those distributions.

7) To provide information for **scientific understanding** required for analysis and forecasting. *(Is the problem understood?)* Measurements are used to establish if the sources and atmospheric chemical and physical processes that shape PM concentrations are sufficiently well understood; to explain failures to attain desired air quality; and to suggest alternative management strategies. Measurements also support research to develop an air-quality forecast system that can predict air-quality problems and issue appropriate guidance, warnings, or alerts.

To provide the measurement data for these and other applications, different types and configurations of instruments have been developed and deployed. However, data acquisition, analysis, and interpretation are interrelated activities. A full understanding of available sampling and analysis strategies, and a determination of their uncertainties and costs, are required before acceptable measurement techniques can be specified to meet PM measurement user-community needs. This chapter briefly describes the measurement techniques available, explains how the information obtained through measurements is applied, and discusses the confidence that can be placed in the data based on current uncertainty estimates. More detailed

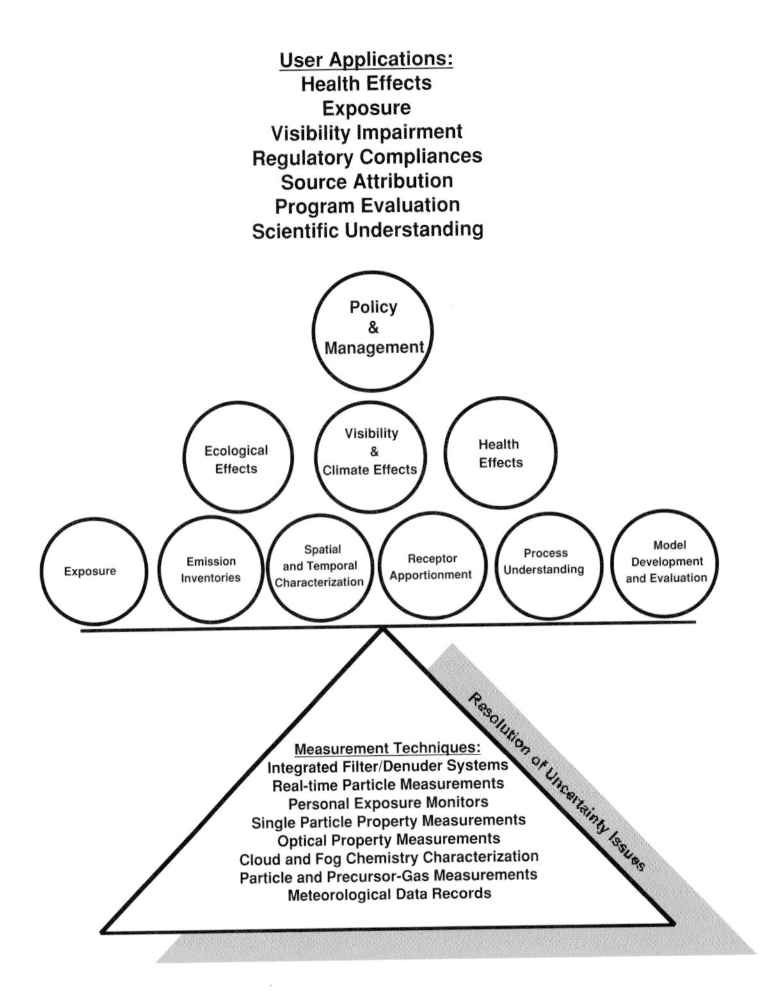

Figure 5.1. Relationships of data-measurement techniques, air-quality management tools, and user applications.

descriptions of measurement techniques and applications are given in Appendix B.

Data obtained through measurements quantify the observable properties of particles: size and size distribution, mass, chemical composition, and optical characteristics. Measurements of other variables – for example, cloud-water content, atmospheric chemistry, and meteorological conditions – also characterize the behavior of particles in the atmosphere, document particle sources and relationships to other pollutants, and provide data on meteorology related to PM formation, removal, and transport. Information obtained on particle characteristics is applied according to the user's needs. Examples of particle characteristics are as follows:

Size/Number - Investigators concerned about health effects have focused on particles with diameters less than 10 μm and less than 2.5 μm, including ultrafine particles with diameters ranging from 0.001 μm to 0.1 μm. Monitors that determine human exposure to fine particles use particle inertia to partition particles into different size ranges or measure size directly using methods such as optical sensing. Investigators examining visibility degradation, regional haze, or long-range transport need measurements of particles in a range of sizes from about 0.1 μm to 1 μm, since particles in this range scatter and adsorb light efficiently and have the longest atmospheric residence times.

Mass - Several techniques measure the mass of particles in a given size range in sampled air. Particle inertia is used to separate particles into different size fractions. After separation, particles are usually collected on a filter and the mass is determined gravimetrically. Frequently such mass-based selection methods inadvertently alter the observed PM mass by allowing desorption of volatile compounds from the particles or sorption of ambient gases on the collected particles and/or the collection substrate.

Chemical composition - Many of the hypothesized causes for health effects are related to the chemical composition of PM. Chemical composition also determines the scattering and light-absorbing properties of particles that contribute to visibility reduction. The chemical components include major ions (NO_3^-, $SO_4^=$, NH_4^+); alkali (Li^+, Na^+, K^+) and alkali earth (Mg^{++}, Ca^{++}) metals; BC; OC; sea salt; trace elements, including transition metals; other minerals; and biological particles.

Three groups of techniques developed to determine particle composition are 1) methods, which collect particles on suitable substrates for subsequent laboratory analysis; 2) real-time techniques, which collect particles, detect certain chemical constituents, and give on-line information concerning the amount of the detected compounds; and 3) single-particle measurements, which give semi-quantitative, real-time information regarding the chemical composition of each particle detected.

Optical properties - Measurements of the optical properties of particles characterize temporal trends and spatial distributions of particles that impair visibility, and link the visual effects of those particles to their sources through particle composition or other properties. The fundamental particle properties related to visibility are light scattering and absorption. Particles with size ranges from 0.1 μm to 1 μm are most efficient at scattering light, while BC contained in aerosol is the most effective light-absorbing PM component.

Aside from the measurement of particle size, mass, composition, and optical properties of aerosols, additional measurements are required to explain the fate of aerosols in the atmosphere, their sources, and their relation to other pollutants. Since PM concentrations are strongly affected by the presence of fogs and clouds, measurements of fog- and cloud-water content and composition are important for characterizing the formation and accumulation of PM in air. Measurements that quantify gas-phase compounds, such as SO_2, NH_3, NO_x or NO_y, and VOCs, which are precursors to PM formation, are needed. Other gas-phase measurements are required to determine the relation between atmospheric co-pollutants and PM. Finally, meteorology affects all of the processes related to PM formation, loss, and transport. The understanding of PM and PM relationships to precursor compounds and to copollutants requires measurements of meteorological parameters: wind speed, wind direction, temperature, and relative humidity.

5.1 CURRENTLY AVAILABLE TECHNOLOGY AND INSTRUMENT CAPABILITIES

Eight measurement techniques are described: 1) integrated denuder and gravimetric filter-based systems; 2) real-time particle measurement techniques; 3) personal exposure monitors; 4) single-particle property measurements; 5) optical properties of aerosols and long-path optical measurements; 6) fog- and cloud-water measurements; 7) gas-phase fine-particle and copollutant precursor measurements; and 8) meteorological measurements.

5.1.1 Size-Selective Inlets

Before particles can be collected and measured they must enter the sampler through an inlet, where size-fractionation may occur. Size-selective inlets define the particle-size fraction being sampled and are typically the inlets for many of the measurement techniques described in the following sections. These inlets usually rely on inertial separation of large (heavy) particles from small (light) particles to obtain size fractionization. Five approaches for size separation are in use currently and are illustrated in Figure 5.2.

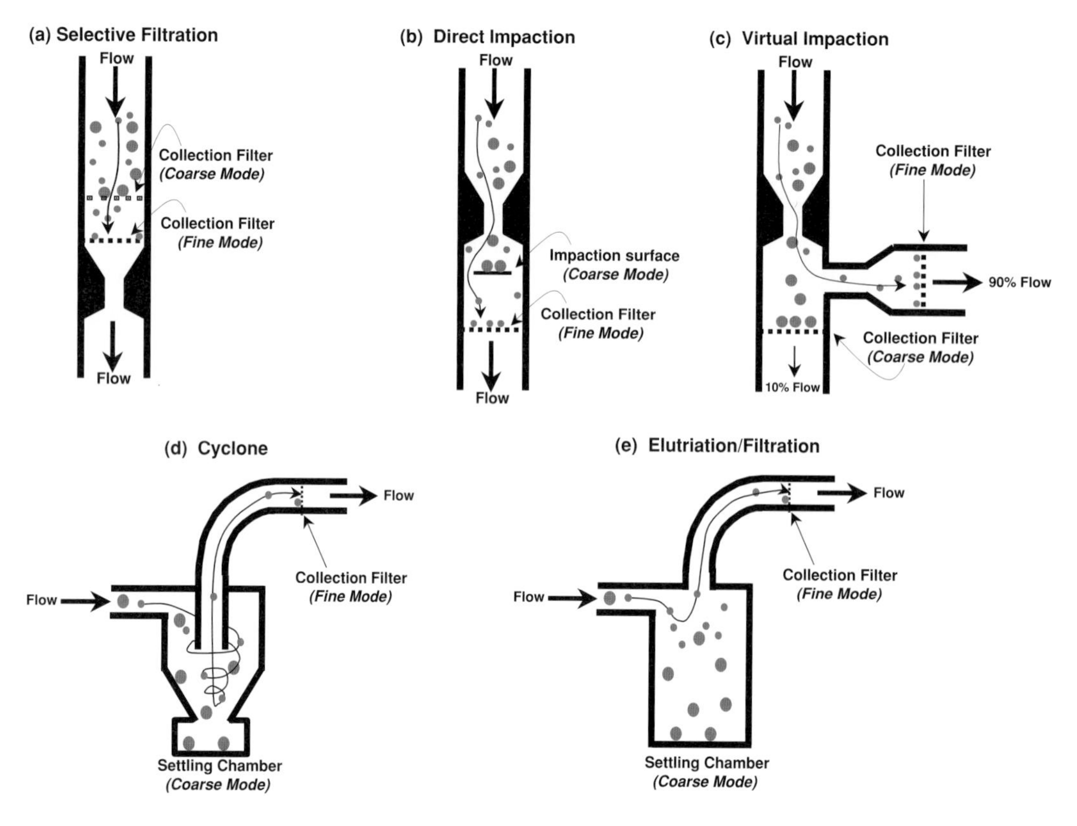

Figure 5.2. Inlets used for particle size separation, illustrating the separation and collection of the smaller (fine mode) particles from larger (coarse mode) particles. The fine mode collection filter is often replaced by a variety of measurement devices.

1) Selective filtration inlets use a filter as the particle fractionator (see Figure 5.2a). As air is sampled through the filter, light particles follow the streamlines of the flow through small pores in the filter into the device where they are collected or measured. The larger particles are collected on the surface of the filter.

2) Direct impaction uses a collecting surface in the flow stream to collect large particles (see Figure 5.2b). This surface is placed orthogonal to the direction of flow. The momentum of heavy particles carries them to this surface, across a stable boundary layer, where they are collected; particles below the cut-point follow the streamlines of the flow around the surface into the device where they are collected or measured. Multiple stages may be employed to collect PM in several size ranges.

3) Virtual impaction uses a collection nozzle rather than a collection surface as in direct impaction (see Figure 5.2c). Small particles, below the cut-point, follow the major flow, which is diverted orthogonally to the inlet flow and into the major flow channel. Heavier particles, those above the cut-point, continue straight into the collection probe, across a stable boundary layer near its top (the virtual surface) following the minor flow (typically 5-10%), and are then collected or measured downstream.

4) Cyclones use a conical chamber into which air enters from the side along the inside circumference to impart centrifugal motion to particles in the chamber (see Figure 5.2d). Flow out of the chamber is tangential to the rotating gas in the chamber. The radial momentum imparted to the heavier particles allows them to impact the walls of the cyclone and be collected by gravitational settling, while lighter particles follow the tangential flow out of the cyclone where they can be collected or measured.

5) Elutriation/ impaction inlets place a large chamber in the flow path (see Figure 5.2e). The

flow exits the chamber through the top. Gravitational settling of the heavier particles carries them to the bottom of the chamber where they are collected; the lighter particles migrate to the top of the chamber from which they exit into the device in which they are measured.

All routine monitoring techniques use these methods to sort particles by size, with impactors and cyclones used most widely. These devices must operate with minimal maintenance over extended periods in all types of weather. Since a large particle may have as much material as hundreds or thousands of smaller particles, a likely source of error in these measurements is the mobilization or re-entrainment of the collected larger particles followed by their migration into the measurement device. These problems are reduced by surface design (e.g., virtual impactors) or surface coatings, which effectively retain the larger particles, and by careful maintenance of the inlet system. However, coatings often interfere with chemical analysis. Low-flow impactors and cyclones are used most commonly for ambient sampling (1 m^3/hr) in urban and rural areas and for personal monitoring (0.1 m^3/hr). High-volume samplers are used to improve detection limits and/or time resolution.

5.1.2 Integrated Denuder and Gravimetric Filter-Based Systems (substrate- and absorbent-based measurements) for Mass and Composition Sampling

Time-integrated denuder and gravimetric filter-based systems are used to collect PM to determine particle mass and composition[1]. Such systems, in use for decades, have undergone steady progress in reducing interferences and artifacts in mass and composition measurements, thereby improving precision, accuracy, and detection limits. Measurements made using integrated denuder and filter systems have long sampling times (typically several hours to 24 hours), and laboratory analysis of these samples is labor-intensive and time-consuming. In addition, the time delay and sample handling between collection and analysis provide the opportunity for added error (e.g., loss of semivolatile species or contamination by absorbing gas-phase compounds), and contribute to delays in data reporting.

Integrated systems allow for the determination of a large set of PM properties. Sampling systems typically include size-selective inlets, denuders, and filters, along with the necessary connecting lines, flow controllers, and flow meters. The simple diagram shown in Figure 5.3 illustrates the principles of this sampling approach. The air sample is introduced through a size-selective inlet that limits the PM size distribution to the range of interest. The sample then passes into separate channels where PM is collected on different filter types, which are subsequently used to determine properties of sampled PM. For example, in Figure 5.3, four channels collect PM for analysis of: 1) the organic compounds present in the sampled particles; 2) the PM mass and elemental composition; 3) the ionic composition; and 4) the semivolatile NO_3^- components. Because semivolitile VOC (SVOC) and NH_4NO_3 may be present in the sample, and because the gaseous fractions can interfere with the collection of PM on filters, denuders are used to remove the gas-phase fraction from the sampled air before the PM filter. The PM is then collected on a suitable substrate or filter. Since the SVOC and NH_4NO_3 may evaporate from the collected particles, an adsorbing or reactive backing-filter may be placed after the primary filter to capture the compounds that evolve from the primary filter. After exposure, the filters and denuder are removed for chemical analysis. Atmospheric concentrations are obtained by dividing the mass loading measured on the filter by the total volume of air that passes though the filter. The principal analysis components for these systems are given in Appendix B, in Table B.2 and the accompanying text.

Specialized versions of these samplers have been successfully developed for use as personal exposure monitors (c.f., Section 5.1.4). In addition, integrated samplers also have been adapted for use aboard aircraft. However, because of the long sampling times required by time-integrated denuder and filter-based systems, they are not able to resolve the valuable short-duration information concerning the

[1] Integrated denuder and gravimetric filter-based systems include the U.S. Federal Reference Method and other gravimetric methods commonly in use in regulatory or compliance networks to determine size-resolved bulk particle mass.

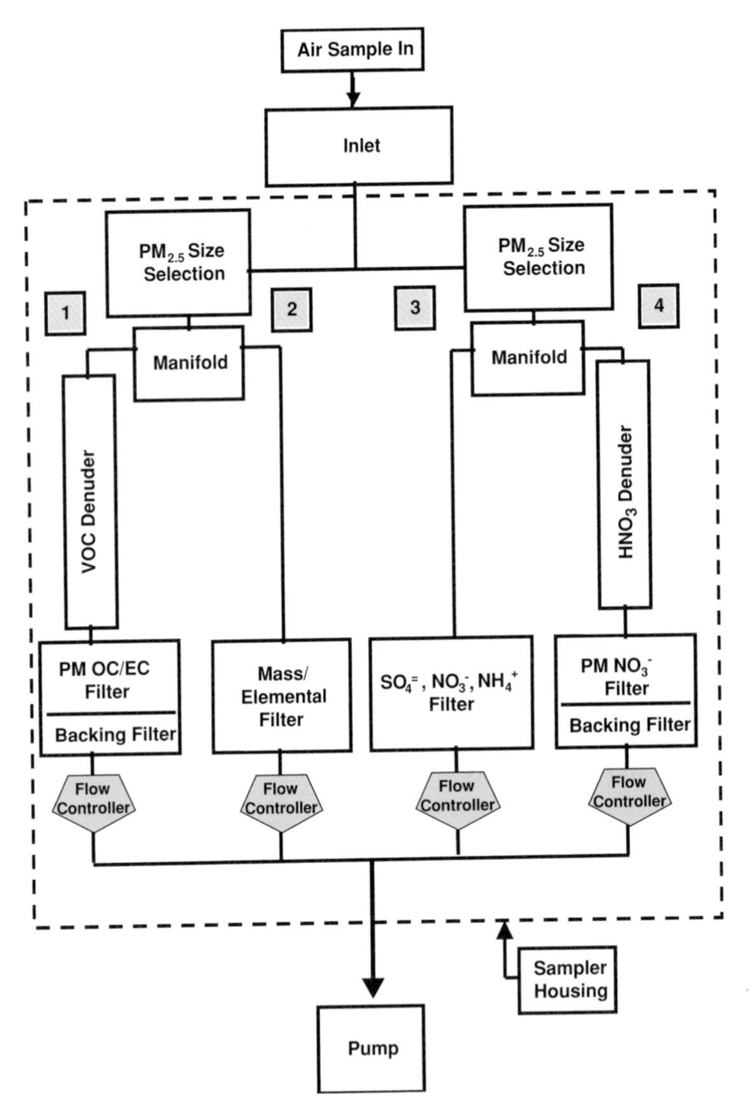

Figure 5.3. Representative denuder and filter sampler for collecting $PM_{2.5}$.

chemical and physical properties of the atmosphere encountered by fast-moving aircraft.

Denuders and Their Gas-Phase Measurements. Denuders are placed in the sampling system to remove selected gases while transmitting particles. Typically denuders are cylinders coated with a substance that absorbs and retains the selected gases. Molecules in the gas phase diffuse rapidly to the denuder surface where they are absorbed, while particles diffuse much less rapidly and pass through the denuder to the filter where they are collected. The absorbed gas-phase compound concentrations can be obtained for laboratory analysis by extracting denuders in a solvent. Denuders can be used as part of, or immediately behind, size-selective inlets to remove gases that might interfere with the PM measurement (positive interference), or to quantify the concentrations of gases that are precursors to secondary PM formation. In addition, denuders may be placed after the filter to capture semivolatile gases that may evaporate from PM collected on the filters (negative interference). The removal and collection of SO_2, HNO_3, and NH_3 by denuders are well developed and characterized. However, the collection and analysis of organic gases using denuders is particularly difficult due to the wide range of physical and chemical properties exhibited by this family of compounds. This has been the focus of much recent study.

Substrate- and Absorbent-Based Measurements of Particle Chemical Composition. The principal method used in integrated systems to determine size-selected PM properties is filtration. Filters for collecting PM are selected based on the gravimetric and chemical analyses to be undertaken. Most systems use a Teflon-membrane or quartz-fiber filter followed by an absorbent or a chemically treated filter (i.e., a filter impregnated with a reactive material). Particle mass is usually obtained by determining the weight of particles collected on the front filter. The second filter collects vapors that pass through the first filter or that are released by the particles collected on the first filter. Chemical analysis to determine particle composition includes bulk analysis and particle-scanning analysis.

Stable species (e.g., $SO_4^=$) and trace elements (e.g., Al, Si, Ca, Fe, Mn, Ti) can be collected with minimal sampling bias on filters. A wide variety of compounds that contain trace elements can be detected in PM, including metals such as Cr, Mn, Co, Ni, Cd, Hg, Be, As, Se, Sb, and Pb, which may be related to the potential health effects noted in Textbox 2.5.

Labile species, such as NH_4NO_3 and SVOC existing in gas and particle phases in the atmosphere, require specialized filter-sampling protocols that typically use denuders and reactive collection substrates. Methods that use only filters for collecting these species may suffer from sampling artifacts associated with the capture of gas-phase species during sampling (often referred to as a "positive artifact") or the loss of compounds from collected PM (often referred to as a "negative artifact"). Combined denuder-filter methods are required to provide the least biased measurements of semivolatile components in aerosols: for example, NO_3^-, HNO_3, NH_4^+, NH_3, and SVOC.

Overall, measurements of carbonaceous material are most challenging and prone to errors owing to PM volatility, interferences from gaseous organic species, limitations with collection and analytical methods, and lack of accepted primary and transfer standards (c.f., Turpin, et al., 2001). The most common method used to quantity the bulk carbon/organic content of collected PM is thermal oxidation. In this method the sample is heated and sequentially less volatile organics are evolved and oxidized. Eventually, graphitic carbon in the sample is oxidized. In each case the carbon in the collected sample is evolved from the particles and detected as CO_2. The result is a measure of EC, often described as BC or soot, and OC. Both of these terms are operationally defined (Chow et al., 2001). In this chapter we distinguish between EC as determined by thermal methods and BC as determined by optical methods but continue, in accordance with Textbox 1.2, to apply the term BC when no distinction is required. The organic compounds that compose OC contain carbon plus other elements such as oxygen and hydrogen. Because these other elements are typically not measured, it is commonly assumed that organic mass concentrations can be inferred by multiplying the measured carbon mass by a factor that accounts for other associated elements. This factor is typically assumed to be 1.4, but may be as high as 2.2, depending on particle composition (Turpin, et al., 2001).

Improved methods are needed to measure OC and BC concentrations and to identify and quantify organic PM species, especially those important as source tracers or health effects markers. More information on the relationship between the carbon content of OC and OC mass concentrations is also needed. Effort should be made to characterize the composition of primary and secondary organic species associated with anthropogenic and biogenic source emissions.

Beyond the determination of the BC and OC fractions, the identification of the exact chemical composition of these carbonaceous fractions is desirable. Organic aerosols emitted by different sources have distinct chemical signatures. Therefore, measurements of organic speciation at a receptor can provide valuable insights into the origins of ambient OC. However, the measurement problem becomes much more challenging. The carbonaceous material in PM can include thousands of compounds covering a vast range of properties. The most sophisticated laboratory methods available can identify only 10 to 20 percent of the total organic-carbon mass by compound; such measurements are expensive and require highly trained personnel. Furthermore, only a few laboratories in North America are equipped to carry out such analyses.

Two strategies have been applied to the determination of organic species. In the first, the organic compounds that are evolved by heating the collection filter can be analyzed by mass spectrometry. Although the organic compounds contained on the filter may be altered or partially decomposed during the heating process, the evolved material contains information concerning the identity of the substances that are contained in the OC fraction.

A second, more commonly used approach involves the selective and quantitative removal of various classes of organic fractions or species from the filter by solvent extraction. Separation often is based on the chemical affinity of different organic-compound classes (e.g., polar vs non-polar). The fractions can be analyzed by GC-MS or LC-MS to quantify individual compounds, a number of which are important source tracer species (Schauer et al., 1996).

Impactors. Cascade impactors represent a special class of substrate- and absorbent-based techniques for composition sampling. These instruments are used to obtain size-resolved measurements in ambient chemistry and visibility studies. These devices use a series of impactor stages as described in the inlet section above, with each subsequent stage designed to collect smaller particles, thus, providing a series of size-resolved particle samples over a range of particle sizes (e.g., Marple and Olsen, 2001). The substrate on the impactor stage can be removed and analyzed for the mass and composition of the collected particles. By having the collector surfaces move continuously in a stepwise fashion, time-resolved measurements can be obtained.

A special type of impactor, the dichotomous sampler, uses virtual impaction to separate the particles into two size fractions, typically into fine ($<PM_{2.5}$) and coarse ($PM_{10-2.5}$). In this device, which has been widely used in routine monitoring, particles are drawn in through a single inlet. The flow splits inside the impactor. A smaller portion of the flow continues along a straight path, while the larger flow is diverted into a separate channel (Figure 5.2c). The particle-size distribution is determined by the flow rate and the configuration of the flow splitter or virtual impactor. The particles in each channel are collected on appropriate filters and analyzed for mass and composition.

5.1.3 Continuous and Semi-continuous Real-time Measurements

For specific PM properties, continuous and semi-continuous real-time measurements have a faster time response (seconds to an hour) than the integrated methods (6 to 24 hours) and often afford real-time quantification of the specific property that is measured. Thus, continuous and semi-continuous measurements give much more information concerning the sources and atmospheric processes that shape PM formation and transport because they add temporal variability to these analyses. In addition, information on short-term variations may be useful in assessing health effects and validating models. The suite of chemical species that can be measured is continually evolving with the incorporation of new methods of detection. Significant progress has been made in the past several years on developing instrumentation for real-time measurements of PM mass and composition, and some of these instruments have been commercialized. Many continuous methods use size-selective inlets that restrict sampling (c.f., Section 5.1.1) to particles smaller than a certain size.

McMurry (2000) describes measurements of aerosol properties (e.g., cloud-condensation nuclei, particle-number concentration, refractive index, particle-size distribution, size-resolved composition). Available devices to measure physical properties over short durations (< 1 hour) and under conditions where continuous measurements may or may not be correlated with, or predictive of, integrated denuder and filter system measurements are discussed by Watson et al. (1998). Many of these measurements require collocated comparisons with well-established filter-based measurements to define their comparability.

Where reliable, low-cost and effective methods exist, filter samplers should be replaced with instruments that provide continuous, real-time information on PM mass and composition.

Continuous in-situ PM measurements can be classified, based on the properties they measure, into three categories: 1) mass (inertial mass, beta-ray attenuation, pressure drop); 2) size distribution and mobility (electrical mobility, aerodynamic mobility);

and 3) chemical components including carbonaceous compounds, (BC and OC), ions (NO_3^-, $SO_4^=$, NH_4^+) and elements (such as Si, Fe, Ca, Mg, and Ti). Examples of continuous and semi-continuous real-time measurements are reviewed in Appendix B, Table B.3 and the accompanying text. Discussions of recent developments are provided in the following subsection.

Mass and Mass Equivalent. Williams et al. (1993) have described a variety of dynamic principles used in continuous mass sampling techniques. These continuous measurements determine particle mass by measuring changes associated with 1) the inertia imparted to a dynamic element by collected particles; 2) electron attenuation through a surface by particles collected on the surface, and 3) pressure drop across small pores in a filter due to collected particles. The methods are reviewed in Appendix B, Table B.3a.

The continuous inertial methods in use include the widely used TEOM (tapered element oscillating microbalance) and the PEMB (piezoelectric microbalance). The TEOM element oscillates at a characteristic frequency when an electric field is applied and optically measures frequency shifts due to the increasing inertial mass of particles accumulating on the collector. The PEMB measures the shift in the resonant frequency of a piezoelectric crystal as mass is added to the surface of the crystal by the accumulating particles. Electron attenuation devices (beta gauges) measure the increasing attenuation of electrons emitted from a radioactive source by particles accumulating on a filter. Pressure-drop systems measure the pressure drop across a filter as the filter loads and its pores become clogged. Significant progress has been made in the past several years on developing instrumentation for real-time and/or continuous measurements of PM mass and composition, and some of these instruments have been commercialized. The availability and capabilities of commercialized real-time instrumentation will evolve substantially in the next few years, and efforts should be made to replace filter samplers as these methodologies mature.

Size Distribution and Mobility. A wide range of techniques and protocols are used to measure particle sizes ranging from <0.003 µm to >10 µm. Examples of continuous and semi-continuous real-time measurements of PM size-distribution and mobility are reviewed in Appendix B, Table B.3b and the accompanying text. Particle size is typically reported as "diameter." However, the different measurement techniques often define even this most fundamental parameter in different ways, as shown in the box (McMurry, 2000).

Aerodynamic diameter, which is measured by inertial methods such as impactors and cyclones, depends on particle shape, density, and size.

Electrical mobility diameter, which is obtained by electrostatic mobility analyzers, depends on particle shape and size, and is obtained by the rate of migration of charged particles in an electrostatic field.

Optical diameter, which is obtained by light scattering detectors, depends on particle refractive index, shape, and size.

Quantitative determination of the particle-size distribution is difficult to achieve. Sampling losses of particles >1 µm by inertial impaction, and of particles <0.1 µm (ultrafine) by diffusion deposition, present a major challenge in particle-size measurements. The effects of humidity and particle volatility on the determined size also must be considered. Calibration aerosols of known size distribution are often difficult to create and methodologies have not been standardized, making comparison between different measurements difficult.

To measure particle sizes down to approximately 0.1 µm, optical and aerodynamic particle counters are commercially available and can be used for continuous monitoring. Unfortunately, most current real-time instruments for measuring particle-size distributions below 0.1 µm are expensive and require continuous oversight, frequent calibration and maintenance, and considerable expertise to run with good accuracy and precision. Airborne measurements also require attention to the effects of high-velocity sampling inlets, rapid pressure changes, and spatially varying particle characteristics. Despite these difficulties, recent developments in particle-size sampling show promise for continuous or semi-continuous measurements.

For monitoring purposes, considerable effort needs to be made to ensure that particle-size distributions can be acquired on a long-term basis at urban and non-urban locations. Such measurements allow the role of particle size to be associated with health effects, and provide valuable information concerning the optical effects and likely residence time of the particles in the atmosphere. Even the most developed of available techniques require fairly complex devices and specific training to calibrate and to perform comprehensive data processing. Efforts have been made to develop standardized calibration devices and to generate calibration aerosol for collocated comparisons of different techniques.

Chemical Components. Continuous measurements of carbonaceous compounds, ions, and metals are described below.

Carbonaceous compounds: BC, EC, and OC. As mentioned earlier, the most challenging aspect of PM composition determination involves the carbonaceous fraction, which includes hundreds of organic compounds. Examples of continuous and semi-continuous real-time measurements of PM composition are reviewed in Appendix B, Table B.3c and the accompanying text. No real-time technique is available that can adequately speciate OC. Hence, measurements usually are restricted to major carbon fractions and labeled accordingly, OC or BC. As for time-integrated denuder and filter based measurements (see Section 5.1.2), OC and BC/EC measurements involve collecting particles on a substrate (e.g., a filter or impaction plate) for enough time to achieve the desired sensitivity. This is typically several minutes to an hour depending on the concentration of carbon-containing compounds in the collected particles. The substrate or plate is then heated in a controlled atmosphere to first volatilize the OC, then (in the case of EC measurements) further heated to burn off the EC. The OC and EC evolve to CO_2 gas, which can be measured directly by a non-disperse infrared detector, or converted to CH_4 and measured by a flame-ionization detector. As with the time-integrated denuder and filter-based systems, the mass of organic material is approximated by adjusting the evolved carbon mass by an appropriate factor.

Cost-effective techniques for obtaining more detailed spatial and temporal information on carbonaceous PM species should be developed.

Other approaches also are under development. In one such technique total organic carbon is measured by collecting particles into purified water and analyzing the aqueous medium. Although such techniques are only under development, they have the potential for time resolution on the order of seconds. However, they may be restricted only to the analysis of soluble organic components. The PAS2000 real-time polycyclic aromatic hydrocarbon (PAH) monitor measures particle-bound PAH based on the ionization of particles and their detection with an electrometer. The photoionization method for particle-bound PAH is reproducible, but at present can be related only to absolute concentrations of particle-bound PAH via collocated filter samples (Watson and Chow, 2002). More information regarding these systems can be found in Appendix B.

Identifying the origins of organic aerosols, understanding their transformations in the atmosphere, and understanding their effects will require measurements of speciated OC, and such measurements are needed with better spatial and temporal resolution than is currently possible. Although significant advances are expected in the near future, permitting more widespread measurements of OC and BC, instrumentation for continuous measurement of speciated OC is likely further off in the future.

More effort should be made to characterize the composition of primary and secondary organic species associated with anthropogenic and biogenic source emissions.

Ionic component of PM. The ionic component (particularly $SO_4^{=}$, NO_3^-, and NH_4^+) of PM is often a significant fraction of the total PM mass. Ion chromatography permits speciation and quantitative measurement of these and other ionic species (Chow and Watson, 2001). Filter extractions are analyzed with a dual channel ion chromatography for a range of ions, including cations (e.g., NH_4^+, Na^+, K^+), inorganic anions (e.g., Cl^-, $SO_4^{=}$, and NO_3^-), and organic acids (e.g., formate, acetate, and oxalate).

Efforts have been made to collect ambient particles directly into purified water for on-line ion-chromatography analysis. In all cases, diffusion denuders need to be placed upstream of the instrument to remove potentially interfering gases such as SO_2, NH_3, and HNO_3. The newer techniques can have high sampling rates (six samples per hour) but generally result in lower sensitivities. Continued developmental work is required for reliable real-time inorganic NO_3^- and $SO_4^=$ measurements. Additional emphasis also is needed on NH_4^+ and other anions and cations.

Other technology is being developed to measure the ionic components of PM. These alternative techniques are based on the choice of detectors and the constraints that these detectors place on the device. Real-time $SO_4^=$ detectors have been developed based on the use of a flame-photometric detector widely used to measure gaseous sulfur compounds. Commercially available continuous NO_3^- and $SO_4^=$ analyzers have been developed which involve the flash vaporization of particles collected by impaction onto a filament, followed by chemiluminescence detection of nitrogen oxides for NO_3^- or by ultraviolet absorption of SO_2 for $SO_4^=$. Finally, online PM mass spectrometry using thermal volatilization followed by electron-impact or chemical ionization (Section 5.1.5) is found to give quantitative size-resolved real-time measurements of the ionic species contained in PM.

PM Metals. Many of the techniques used to identify the metals contained in PM collected on filters are being adapted for continuous time-resolved measurement using moving filter strips. Detection methods include X-ray fluorescence and proton-induced X-ray emission. In each case, the metals are identified by the emission of X-rays from the metals that are excited in the sample. In addition, collection and flash vaporization followed by detection using atomic-absorption spectrometry, emission spectrometry and mass spectrometry is being developed to measure the metals in PM. These same detectors are used to distinguish the soluble metals contained in PM sampled into a solvent stream. Online laser desorption/ionization mass spectrometry (Section 5.1.5) has been used to detect the presence of metals, but the measurement has not been made quantitative.

5.1.4 Personal Exposure Monitors

The devices described in the previous two sections are used for monitoring PM and its chemical properties in ambient air. However, individuals are subject to non-ambient exposures in their homes, indoor workplaces, and stores (Williams et al., 2000a,b,c). With this in mind, some of the techniques that were described in the two previous sections have been adapted to monitor personal exposure. The sampling has focused on mass as well as specific pollutants that could have potentially significant human-health effects, such as combustion emissions, tobacco smoke, acid aerosols, and vapor-phase organics.

Personal-exposure monitors are designed to be carried by an individual and to record the exposure of that individual to pollution. The typical PM monitor is intended to obtain time-integrated measurement of PM over a 24-hour sampling period. To make these devices sufficiently portable, the methods used for ambient-air monitoring are altered to reduce the size, weight, and power consumption. Current technology is able to produce a measurement device that weighs no more than one kilogram. Recent comparisons have achieved agreement with FRM collectors of ± 20 percent for $PM_{2.5}$ mass concentration (Evans et al., 2000). In addition, measurements of the principal ionic components of PM have comparability within a factor of two of fixed ambient monitors, which is often adequate for their intended purposes.

Although most personal monitoring has focused on 24-hour collection, devices are being developed to give real-time continuous monitoring to estimate individual exposure to short-term (one minute) variations in PM concentrations. This is useful not only to determine the impact of brief high exposures but also to directly identify and characterize PM sources. Because of the sensitive nature of personal-exposure monitoring and the engineering required to miniaturize these devices, operation of the personal-exposure monitors requires highly skilled research personnel, but allows subjects to wear the monitor without extensive training.

5.1.5 Single-Particle Measurement Capabilities

The ability to measure properties of a single particle, rather than of a bulk collection of particles, has opened the way for more detailed understanding of the characteristics of particles and the chemical processes responsible for their formation and transformation. Development of automated scanning electron-microscope and X-ray fluorescence analyses (McMurry, 2000) has made it possible to perform elemental analysis on statistically significant numbers of particles for a reasonable number of samples.

More recently, single-particle mass-spectrometers have been developed that allow real-time detection of the chemical composition of individual particles. These approaches analyze the basic chemical and physical properties of individual particles and can be used to assess the sources of the particles and the history of atmospheric processes that determine their composition. Of particular value is their ability to provide information on the internal vs. external mixing of different aerosol components. Additional details concerning these techniques are given in Appendix B and listed in Table B.4. Also, Middlebrook et al. (2003) have discussed the operating principles of three of these instruments and compared the results obtained from their measurements. However, the use of single-particle measurement methods is limited because they are technologically sophisticated, costly, and the data thus far are only semi-quantitative.

A technique related to single-particle spectrometers, the aerosol mass-spectrometer, provides size-resolved, quantitative measurements of those PM components that are volatile at temperatures less than about 700°C. This technique operates in two modes: 1) to obtain size distributions associated with a limited number of PM components, and 2) to determine bulk concentrations of a larger number of PM components.

5.1.6 Optical Properties of Aerosols and Long-Path Optical Measurements

Issues associated with visibility, radiative transfer, and climate have motivated the use of optical long-path measurements, and in combination with other instruments also can contribute to other aspects of particle monitoring. Measuring optical properties of fine particles and visibility requires a combination of in-situ and long-path techniques. The principal techniques that are used are listed in Appendix B in Table B.5 and the accompanying text.

In-situ measurements of light scattering and light absorption. The in-situ properties of primary interest are aerosol light scattering and absorption. The light-scattering coefficient is measured directly by integrating nephelometers. In this device, a collimated beam of light passes through a chamber containing the sampled air. Scattering by the particles reduces the intensity of the light. The reduction in the intensity of the light by aerosol scattering is influenced by a combination of particle size and density. A number of vendors produce robust, field-ready versions of this instrument suitable for monitoring applications.

The most commonly used technique for monitoring light absorption measures the attenuation of light transmitted through the particles that are deposited on a filter. Time resolution can be achieved by depositing the particles on a filter-paper tape. The commonly used aethalometer and the particle soot absorption photometer convert optical attenuation to BC; however, conversion to mass concentrations requires use of an absorption coefficient, which is empirically determined.

These methods are adequate for routine measurements of the integral optical properties of aerosols related to visibility reduction. In most cases, sample air must be drawn from a sample inlet to the instrument, which means that care must be taken to avoid particle losses. Changes in relative humidity also affect light-scattering measurements, requiring measurement and, in general, control of sample relative humidity. Particles larger than a few micrometers in diameter are difficult to characterize quantitatively, due to inlet effects and optical truncation in the scattering measurement. More work is needed to characterize measurement artifacts of the filter-based methods for light absorption, to develop methods to determine the RH-dependence of aerosol light absorption, and to develop calibration and reference standards. Several new in-situ techniques, including cavity ring down cells, photo acoustic cells and direct extinction cells, are under development. These show promise for fast-response online measurements of aerosol scattering and absorption.

Long-Path Measurement Techniques: Remote sensing and visibility. The in-situ measurements described above give the visibility-relevant properties of air at the location of the sampling inlet. The long-path measurements described below indicate the burden and, in some cases, the particle distribution in a column of air.

Sun-tracking photometers measure the vertical optical depth by observing the transmission loss between the sun and the instrument. They obtain information about the average size-distribution of the particles in the column. Interpretation of the size distributions should be limited to general rather than detailed features, because the results are typically quite sensitive to measurement errors and data retrieval algorithms. Shadow-band radiometers measure both direct and diffuse radiation and provide information similar to that obtained from a sun photometer, but with less accuracy. Automated spectral sun photometers and shadow-band radiometers are commercially available, and are standard instruments in some radiation-monitoring networks. Radiative-transfer and climate issues usually motivate the use of these instruments, but they also can contribute to pollution research, especially in combination with other instruments.

Telephotometers or teleradiometers measure the contrast reduction of a distant reference target due to the intervening haze. The results are closely applicable to human perception, but depend on lighting conditions. Scene cameras are even more attuned to visual perception and less quantitative than telephotometers in terms of quantifying the optical properties of PM. Transmissometers measure the one-way transmission loss, and hence the average extinction coefficient, along a path. Transmissometers designed for long-term operational use are commercially available, and are used as a standard instrument in the IMPROVE network. Transmissometers often have been used at airports, where the design is optimized for low visibility.

Long-path measurements are becoming available that can provide spatial resolution. The lidar (light detection and ranging) uses a laser to transmit a pulse of light into the atmosphere. The small portion that is scattered back from the air is detected as a function of time, or equivalently, of distance from the device. The information provided by current instruments is qualitative, indicating the approximate distribution of particles in the atmosphere. This information is extremely valuable for locating aerosols and determining the thickness of the layers. Lidar can operate from the surface or from aircraft. Automated, eye-safe lidars are commercially available along with new emerging research prototypes.

Recent research has shown the value of data from various lidar systems for inferring profiles of the physical properties of particles, such as $PM_{2.5}$ and even the size distribution. The influence of relative humidity on the distribution and optical properties of aerosols and their behavior with height in the atmosphere can be studied using simultaneous lidar measurements of water vapor and aerosol backscatter or extinction. In addition, the lidar is capable of determining the depolarization ratio of backscattered light, a source of information relatively untapped in aerosol studies. The depolarization ratio indicates the morphology of particles by specifying the degree to which particles are non-spherical. Finally, coupling lidar measurements with other measurements (for example, a spectral sun photometer) can greatly enhance the available information. Additional development and evaluation of such methods should yield valuable dividends for future aerosol studies.

Network applications of light scattering and absorption. Prior to 1994, visual range was recorded hourly by human observations. In the United States and Canada, human observations were replaced with automated light scattering instruments in automated surface observing (or weather) systems (ASOS/AWOS). The automated sensors measure the extinction coefficient as one-minute averages. The visibility sensor operates on a forward-scatter principle, since forward-scatter measurements were found to correlate better with transmissometer extinction coefficients than backscatter measurements. Currently, the ASOS/AWOS data are provided as 18 specified visual ranges with a visual range upper bound of 10 miles, even though the instrument can provide meaningful data up to 20 to

30 miles. The raw visibility data need to be filtered to eliminate and correct for weather influences (fog, precipitation), high humidity, and visual range threshold. However, ASOS/AWOS data have great potential in aiding $PM_{2.5}$ mapping if adequate algorithms are developed to relate the forward scattering measured by these devices to fine-particle concentrations.

Satellite measurements. Several techniques exist for detecting aerosols from space using passive remote-sensing methods. Each method has its own advantages and disadvantages. Satellite techniques are at present most suited to define patterns of aerosol distributions and may be useful in evaluating total aerosol optical depth for comparison with total aerosol optical depth as computed by models. They usually provide the only information available in remote areas and can offer global coverage. However, current satellite sensors measure only relative concentrations of PM in the lower atmosphere.

Visible and near-infrared techniques also have been tested for detecting dust aerosols. The total ozone mapping spectrometer (TOMS) uses the ultraviolet part of the electromagnetic spectrum and has the ability to distinguish between absorbing and non-absorbing PM. A 20-year record is available globally, both over land and ocean. A major advantage of this technique is the availability of aerosol coverage, since the surface ultraviolet reflectivity is low and nearly constant over both land and water. However, the TOMS aerosol product is not sensitive to aerosols below 1 km, and the spatial resolution of the instrument is on the order of 40 by 40 km^2.

Although the accuracy of direct concentration measurements of the boundary layer from satellites is problematic, other satellite data can be used to characterize the physical atmosphere in which aerosols are produced. For example, atmospheric water is critical to the evolution and growth of particles and their ultimate optical and chemical properties. Visible and infrared measurements from polar orbiting satellites can be used to characterize cloud distributions and liquid water properties. Clouds and their physical properties are highly parameterized in models and satellite data potentially have the temporal and spatial resolution to replace the uncertain cloud characterizations produced in models. Satellites also have the potential to provide characteristics of the surface (moisture availability, albedo, heat capacity) that will ultimately improve the fidelity of PM modeling (e.g., Jones et al., 1998).

5.1.7 Chemical Analysis of Cloud and Fog Chemical Composition

Fog and cloud droplets collect local particles and soluble gaseous compounds. The amount of particulate matter in the air after the droplets evaporate can either increase or decrease (c.f., Pandis et al., 1998). While fog and cloud water have been sampled for a number of years, in the last few years multiple size-fractionating droplet collectors have been employed and the collected water analyzed for its chemical composition (Collett et al., 1998). These measurements allow for a better understanding of the chemistry within fogs and clouds, and of its variation with droplet size as a function of location, elevation, and time of day.

Fog- and cloud-water collectors are designed to collect water droplets ranging from 2 to 50 μm. Droplets typically are collected by impaction, either passively or actively. Passive collectors use ambient air movement to bring the droplets to the collector. Active collectors either draw air across the collection surface or move the surface through the air.

Fog- and cloud-water collectors are either bulk collectors, which collect all droplet sizes, or size-selective collectors, which collect the droplets in several discrete size ranges. Depending on the flow rates and liquid-water content of the fogs, sampling durations as short as 5 minutes can be achieved using active strand collectors. Size-selective collectors have been developed to collect cloud droplets in as many as five droplet size ranges. In addition, these samplers employ impaction either through multiple nozzle impactor plates or several impaction strings of different size.

Counterflow virtual impactors (Figure 5.2c) have been applied to inertially separate cloud droplets and ice crystals from ambient air, evaporate the water from them, and chemically examine the residue particles and evaporated gases. Such studies show promise in understanding the role of particles as ice nuclei and cloud-condensation nuclei, and for studying the relationship between cloud chemistry and PM processed through clouds

Bulk and size-fractionated fog and cloud-water samples can be analyzed for liquid-water content and chemical components, including inorganic and organic anions and cations. In general, the analytical methods for determining concentrations of these species in fogs and clouds are well established and summarized in Seigneur et al. (1998). Analytical limits of detection and analytical uncertainties are given in Collett et al., (1998). Intercomparisons to estimate the liquid-water content in fogs indicate agreement among the methods of about 10 percent. Important chemical species analyzed in fog and cloud water include inorganic anions (Cl^-, NO_3^-, and $SO_4^=$), inorganic cations (H^+ [i.e., pH], Na^+, K^+, NH_4^+, Ca^{++}, and Mg^{++}), total iron (FeX) and manganate (MnX), low molecular-weight organic anions (e.g., formate, acetate, and others), formaldehyde, S(IV), hydroxymethanesulfonic acid, hydrogen peroxide, and organic peroxides.

5.1.8 Gas-Phase PM Precursors, Ozone, Ozone Precursors, and Oxidants

The salient features of techniques currently available to measure the ambient concentrations of gas-phase PM precursors, ozone, ozone-precursors (including the odd-hydrogen free radicals), and oxidation products of these compounds are listed in Appendix B, Tables B.6.

Many PM precursors are also those that lead to tropospheric ozone. Several essential measurement methods are well developed (e.g., those for ozone, NO, NO_2, NO_y, CO, SO_2, and C_1-C_8 hydrocarbons). Routine measurements made in urban monitoring networks can reliably use low-cost commercially available instruments that measure these species, especially for compliance purposes, although higher sensitivity and improved selectivity are desired for measurements outside urban areas. However, aircraft measurements needed for understanding chemical processes and model development and evaluation require performance found only in the most sophisticated, research-grade instruments. Parrish and Fehsenfeld (2000) recently reviewed most of these techniques, their capabilities, and their limitations.

5.1.9 Meteorological Measurements

Accurate measurements of meteorological parameters at ground level and aloft are essential for understanding and modeling PM sources, chemistry, transport, and deposition. Major field programs have stimulated new insight in the complexity of the meteorological processes that control air quality and the need for better observations and models. Central to these studies has been the need for observations of horizontal and vertical transport and mixing, temperature, moisture, radiation, and the surface-energy budget. The meteorological measurements required for this research were reviewed recently (Seigneur et al., 1998). Ground-level methods for most measurements are routine, precise, accurate, and inexpensive. They are deployed in all monitoring networks and are well understood.

In contrast, measurements aloft are more complicated, difficult to operate and maintain, expensive, of lesser known accuracy (calibration methods are not straightforward), and relatively new. Unique insights into boundary-layer mixing and mesoscale or synoptic-scale transport processes can be made only with the application of these techniques. In addition, measurements aloft provide essential information about the distribution of PM and PM precursors, particularly the long-range transport of particles with sufficiently long atmospheric lifetimes to be transported thousands of kilometers from their sources. Much of the long-range transport may occur in transport layers in or above the planetary boundary layer. These measurements often are made in combination with routine observing networks. Indeed, profiling measurements provide important information that can be used to determine the best location for surface-based monitoring stations. Information concerning the measurement of meteorological parameters for critical parameters such as temperature, humidity, wind speed and direction, mixing height and solar radiation is given in Appendix B in Table B.7 and the accompanying text.

5.2 MEASUREMENT UNCERTAINTY AND VALIDATION

Estimated uncertainties of measurements made by instruments deployed in fine-particle monitoring networks are listed in Tables 5.1, 5.2, and 5.3.

The multitude of measurable parameters that may be necessary to define PM (e.g., size, mass, number, and composition) makes determining the uncertainty inherent in most particle measurements considerably more challenging than is the case for gas-phase measurements. In general, the uncertainty of any measurement consists of two parts: precision (random error) and accuracy (systematic error). Precision or reproducibility allows for estimates of random errors. These errors usually are determined for combined field and laboratory operations by collecting collocated replicate samples or by propagating individual instrument errors through all components of the measurement process. Accuracy or systematic error usually results in the largest uncertainty, since it incorporates instrument or method accuracy and includes the coupled effects of measurement interferences, artifacts, and detection limit. Interferences are constituents in the atmosphere that are detected in the measurement as the compound of interest in the case of gas-phase compounds or as the target property of interest in the case of PM. By contrast, artifacts are unintended alterations (gain or loss) in the compound of interest in the case of gas-phase compounds, or as the target property of interest in the case of PM in the course of the measurement or during collection and analysis of the sample. Detection limit is determined by the amount of a constituent that is required to be detectable above the intrinsic noise of the measurement technique and establishes the volume of air needed and the time required to make a measurement.

5.2.1 Estimation of Particle Measurement Uncertainty

With the possible exception of mass, it is difficult to determine the absolute accuracy in the measurement of PM properties. Mass can be considered an exception, but only by definition in a regulatory context as described later, and not in terms of certified Standard Reference Material (SRM, e.g., those prepared by NIST). Appropriate reference materials or standards for many particle-phase components that can account for the entire sampling and analysis process, and which are representative of many of the complex properties of atmospheric particles, are not routinely available or suitable for use with many PM measurement techniques. In general, the chemical properties of PM are referenced to a known level of that chemical component supplied to the system during the final stage of chemical analysis. It is this aspect of the determination of PM properties that is given as "accuracy" in Tables 5.1, 5.2, and 5.3. Although progress is being made in the development of calibration sources of known size with known physical-chemical properties (McMurry, 2000), many challenges remain in making a routine, absolute determination of true atmospheric PM properties and composition. Since SRMs are not available for most PM properties, the accuracy of the method relative to ambient is unknown and is estimated only by the convergence of several methods, also typically of unknown accuracy. Thus, while the gravimetric methods for mass may be regulatory-standard methods, they are not standard methods to judge accuracy or bias in the more classical analytical sense. As well, there are SRMs for the chemical analysis of sampled inorganic species, such as $SO_4^{=}$, NO_3^{-}, and NH_4^{+}, which provide confidence that the sampled material is appropriately analyzed. As with mass, the absolute determination of these properties in the atmosphere cannot be verified, since standards to test the complete sampling and analysis system are unavailable.

For this reason, measurement uncertainty (or perhaps more appropriately, consistency or comparability) is usually determined by comparisons among several methods that are aimed at determining similar properties of the sampled particles. Hence, for PM, convergence of methods through intercomparison studies, rather than calibration to a known standard as a guide to measurement uncertainty, is relied upon. Several such studies have been conducted over the past twenty years to determine the equivalence and comparability of PM mass and its chemical components (Seigneur, 1998). These studies provide equivalency or comparability among the methods tested, but not bias relative to a known standard or

Table 5.1. Estimated uncertainty in measurements of the physical properties of PM.

Compared Property	Analytic Method Accuracy [a]	Precision [b]	Range of Comparability [c]	Time Resolution	Comments[d]
Mass ($PM_{2.5}$)	±5 percent	±10 percent	within 20 percent	24-hr	Accuracy refers to gravimetric analysis only. There is no consideration of sampling artifacts. Adequate collocated precisions and comparability can be achieved among filter mass measurements if care is taken. (Watson and Chow, 2002; Solomon et al., 2003.)
Mass (coarse)	±5 percent	±10 percent	within 20 percent	24-hr	Accuracy refers to gravimetric analysis only. There is no consideration of sampling artifacts.
Size distribution *(coarse, fine, ultrafine)*	25 percent		within 25 percent	5 min	Accuracy based on comparison of integrated size distributions with integral measurements (e.g. total number).

[a] Accuracy is defined as the ability of the laboratory methods to correctly measure the samples of a standard reference material.
[b] Precision is defined as the coefficient of variation of repeated measurements obtained by identical collocated samplers in the field.
[c] The range of observed levels of comparability across techniques involved in field comparisons designed to test comparability.
[d] Estimates based on authors' experience unless a citation is given.

absolute particle metric for that property. The consistencies among evaluated techniques for measurements of specific properties are listed in Tables 5.1 through 5.3. In addition, references to the studies that have determined consistency among the instruments used for PM measurement are listed in Appendix B, Section B.3.

Tables 5.1, 5.2, and 5.3 indicate the level of confidence that has evolved in these measurements; that is, the comparability of the measurements of the physical properties and chemical composition for measurements made using integrated and, in some cases, continuous samplers operated side-by-side. It should be noted that the values listed in these tables represent current, state-of-the-art methods and that historical measurements are probably less accurate and precise. The particle transmission and collection properties of the inlet and particle volatilization in these integrated samplers have been identified as important factors that influence comparability between the different measurement methods (Chow, 1995; Solomon et al., 2003). Careful maintenance of the sampler and handing, storage, and analysis of the samples can minimize the influence of these parameters on uncertainty; however, they often are found to be as important as sampler design in determining the differences among systems. Differences in analysis methods also may result in significant differences in reported concentrations, which are important when, for example, data are compared across networks (e.g., National $PM_{2.5}$ Chemical Speciation Network and IMPROVE).

5.2.2 Mass and Size Distribution

The standard gravimetric techniques used to determine PM mass for many regulatory purposes involve filter collection and weighing. While such methods are relatively simple and inexpensive to implement, they require manual operation, provide limited time resolution, are prone to sampling interferences, and, due to human intervention required for weighing, are subject to other sampling errors that cannot be quantified. The gravimetric filter-based methods, such as the FRM for $PM_{2.5}$, have been shown to be biased low, under certain conditions, due to the loss of semivolatile compounds (NH_4^+, NO_3^-, SVOC) during sampling. Thus, while the $PM_{2.5}$ FRM is a reference method for compliance purposes, it is not a suitable reference method to

Table 5.2. Uncertainty in measurements of the chemical composition of PM: acids and inorganics.

Compared Property	Accuracy [a]	Precision [b]	Range of Comparability [c]	Time Resolution	Comments[d]
Sulfate ($SO_4^=$)	±5 percent	±10 percent	within 10 percent	3- to 24-hr	Generally good agreement among many methods (Solomon et al, 2003).
Nitrate (NO_3^-)	±5 percent	±10 percent	within 35 percent	3- to 24-hr	Significant differences among methods. Comparison involved measurements from both integrated systems and continuous measurements. (Solomon et al., 2003)
Chloride (Cl^-)	±10 percent	±30 percent	50 percent	3- to 24-hr	
Organic acids	±5 percent	±10 percent	unknown	3- to 24-hr	
Ammonium (NH_4^+)	±5 percent	±10 percent	within 30 percent	3- to 24-hr	Ammonium comparability similar to sulfate in PM dominated by ammonium sulfate or bi-sulfate (Solomon et al., 2003)
Alkali Metals (Li^+, Na^+, K^+)	±5 percent	±10 percent	within 30 percent	3- to 24-hr	
Alkali earth metals (Mg^{++}, Ca^{++})	±5 percent	±10 percent	within 30 percent	3- to 24-hr	

[a] Accuracy is defined as the ability of the analytic method to correctly measure the samples of a standard reference material.
[b] Precision is defined as the coefficient of variation of repeated measurements obtained by identical collocated samplers in the field.
[c] The range of observed levels of comparability across techniques involved in field comparisons designed to test comparability.
[d] Estimates based on the authors' experience unless a citation is given.

determine the accuracy of mass concentrations relative to what is actually in the air.

Continuous techniques for accurate measurements of mass that avoid such sampling errors are needed. These methods are being developed but will require considerable evaluation before they are ready for operation in routine networks. *SRM are not available for most PM properties, so accuracy of the method relative to actual ambient PM is unknown and only estimated by convergence of several methods also typically of unknown accuracy. Thus, the FRM methods for mass are regulatory standard methods, but are not standards to judge accuracy or bias in a more classical sense.*

5.2.3 Aerosol Chemical Composition

Non-volatile inorganic aerosol components such as $SO_4^=$ can be measured with acceptable precision and comparability among methods (i.e., within 10 percent). More labile compounds (e.g., NO_3^-, NH_4^+, and organic compounds) as well as BC are measured considerably less well. Organic-PM measurements (OC or OC species) are especially prone to error due to sampling artifacts, which include both the volatilization of semivolatile compounds from collected deposits of particles and the adsorption or absorption of organic gases on sampling substrates or collected particle deposits during sampling. It is likely that positive and negative artifacts and interferences occur during sampling, handling, and storage, and the relative magnitudes of these effects vary with environmental variables and sampling parameters (e.g., filter type, flow rate, sampling time). It is important to understand the magnitude of these artifacts and interferences so as not to bias the ability to manage air quality.

More effort is needed to understand properties of organic aerosols and their gas-phase precursors,

Table 5.3. Uncertainty in measurements of the chemical composition of PM: carbon and organics.

Compared Property	Accuracy [a]	Precision [b]	Range of Comparability [c]	Time Resolution	Comments[d]
Organic carbon (OC)	±10 percent	±20 percent	within 20-50 percent	3- to 24-hr	Serious problems have been identified using evolved gas analysis to determine total organic carbon in PM. Net positive artifacts (1-4 $\mu g/m^3$) with use of quartz-fiber filters also has been observed consistently. (Huebert and Charlson, 2000; Solomon et al., 2003)
Black Carbon (BC) or Elemental carbon (EC)	±10 percent	±20 percent	within 20 percent to 200 percent	3- to 24-hr	Serious problems have been identified using evolved gas analysis to determine elemental carbon in PM. Wide range in comparability is due to use of different analysis methods. (Huebert and Charlson, 2000; Solomon et al., 2003)
Total carbon	±10 percent	±10 percent	within 20 percent	3- to 24-hr	Adequate collocated precisions and comparability can be achieved among filter carbon measurements if care is taken. (Watson and Chow, 2002.)
Organic (speciated)	unknown	unknown	unknown		Involves extraction of organic fractions, based on polarity or other properties followed by GC-MS analysis (Schauer et al., 1996). To date a maximum of only 20 percent of the organic mass has been identified as individual species.
Trace elements	±5 percent	±10 percent	10-30 percent depending on species	3- to 24-hr	Comparability based on XRF analysis for PM collected on Teflon filters and species observed above their detection limit, primarily sulfur and crustal related species. (Solomon et al., 2003)
Transition metals	±5 percent	±10 percent	10-30 percent depending on species	3- to 24-hr	Comparability based on XRF analysis for PM collected on Teflon filters and species observed above their detection limit, primarily sulfur and crustal related species. (Solomon et al., 2003)
Biological PM	unknown	unknown	unknown		

[a] Accuracy is defined as the ability of the analytical method to correctly measure the samples of a standard reference material.
[b] Precision is defined as the coefficient of variation of repeated measurements obtained by identical collocated samplers in the field.
[c] The range of observed levels of comparability across techniques involved in field comparisons designed to test comparability.
[d] Estimates based on authors' experience unless a citation is given.

including factors that govern the gas-particle partitioning of semivolatile organic compounds, the hygroscopicity of particulate organic compounds, and the proclivity of gas-phase organic precursors to form secondary organic PM.

5.2.4 Uncertainties in Routine Gas-phase Measurements Used for Network Monitoring

The instruments and techniques available for the measurement of atmospheric concentrations of ozone and PM precursors, and other related trace gases are briefly described in Section 5.1.8 with additional detail given in Appendix B, Table B.6a. Uncertainty estimates are given in Table 5.4. There is a long evaluation history for instruments used to measure these gas-phase compounds. These instrument validations were in response to the need for reliable measurements that could be used to understand and monitor acid deposition, urban and regional photochemical air quality, climate change, and stratospheric-ozone depletion. As new techniques were devised, they were field tested to demonstrate worthiness and to determine the necessary housekeeping procedures (calibration, zeroing, routine maintenance). Several studies have been conduced over the past twenty years to determine the accuracy of these measurements (Parrish and Fehsenfeld, 2000).

With regard to uncertainties in the measurement of gas-phase compounds, two issues require serious attention. The first pertains to calibration procedures used for routine measurements. In particular, the approaches used to calibrate important measurements, such as ozone, NO, and NO_2, can lead to serious errors under some circumstances and in certain locations. Problems include the ability to reliably generate standards at low concentrations (ppt range) and deliver them through the inlet to the instrument. A careful review of calibration procedures used in routine monitoring networks should be undertaken.

The second issue involves measurement uncertainties associated with VOC sampling. Apel et al. (1999) identified large uncertainties for many of these VOCs. The approaches used to measure VOCs are discussed briefly in Appendix B (Section B.2.6). Serious problems have been observed in routine measurements of high molecular-weight VOCs, which have a high probability of participating in particle formation. Clearly one of the most critical needs of the ambient VOC measurement community is a rigorous field intercomparison of measurement techniques. Such an intercomparison will define more clearly measurement capabilities and identify prevalent problems that must be addressed.

5.3 MEASUREMENT STRATEGIES AND NETWORK ISSUES

The instruments described in the preceding sections eventually must be assimilated into a capable and cost-effective measurement system and strategy. An effective measurement strategy requires a complement of long-term routine measurements, short-term or special studies, and intensive field studies. Every measurement is not needed everywhere all the time, but a wide variety of measurements may be needed at some locations some of the time.

The design and implementation of the strategy must account for differences in measurement objectives: 1) to understand the composition and distribution of PM in the atmosphere, 2) to determine pollution effects, and 3) to provide the information needed for effective management. The system must identify when and where PM problems exist and the severity of these problems. The measurements must provide information concerning the sources, transport, and transformation of atmospheric PM. They must identify the sources, and monitor the distributions of, precursors that are involved in the generation of the PM pollution. Finally, they must provide data for developing and evaluating "receptor-based" and "emission-based" chemical-transport models (and the modules contained within these models) used to determine and test possible abatement strategies for PM pollution.

5.3.1 Deployment of Measurement Technology

Routine Surface-Network Observations. Monitoring networks at selected surface sites are vested with the primary responsibility for characterizing PM concentration and distribution for the purposes of air-quality management. The objectives, measurements, and design criteria for existing air-quality networks in Mexico, Canada, and the United States have been discussed extensively elsewhere (Demerjian, 2000; U.S. EPA, 2002). The information provided by the existing networks is

Table 5.4. Uncertainty in routine measurements of gas-phase compounds. References for the intercomparisons to establish limits can be found in Parrish and Fehsenfeld (2000).

Compared Property	**Accuracy**	**Time Resolution**	**Comments**
Sulfur dioxide (SO_2)	±10 percent	1 min.	Pulsed-fluorescence at 190-230 nm and fluorescence at 220-400 nm. Instrument currently in wide use for routine monitoring. Standard reference material available for through-the-inlet calibration. Recent studies indicate considerable advances in measurement of sulfur compounds. Similar results with longer-time resolution by filter pack measurements with potassium-carbonate-impregnated filters followed by ion chromatography analysis.
Ammonia (NH_3)	±10 percent	1-hr	Based on citric acid coated denuder sampling followed by ion chromatography analysis. Method currently used in research studies and networks. Careful handling of denuders required for optimum results. Newer techniques for ammonia measurement are being developed that promise significant improvements in ammonia measurement. Development of fast-response techniques must be accompanied by inlet designs that reduce ammonia uptake on inlet.
Ozone	±1 ppbv	1 min.	UV absorption at 254 nm, based on intercomparison of ozone measurements. Instrument in wide use currently for routine monitoring. Absolute calibration base on Beers' Law. Potential interferences in the standard UV absorption technique from VOCs such as the aromatic compounds benzene and toluene don't appear to be a problem. However, a clean, properly operating UV absorption instrument provides an absolute measurement of ozone. Additional calibration procedures only should identify field instruments needing cleaning and/or repair, never to alter instrument measurement results.
Carbon monoxide (CO)		1 min	There are presently a variety of instruments that are capable of measuring CO. The choice of instrument depends on the accuracy, precision, and frequency demanded of the measurements. Non-dispersive infrared absorption has proven to be satisfactory for most ground-based measurements. Instrument currently in wide use for routine monitoring. Standard reference material available for through-the-inlet calibration. The modified commercial NDIR instrument is capable of continuous, unattended measurements at 1-ppbv precision for one hour averaging times of the slowly varying ambient CO levels.
Nitric oxide (NO)	±30 percent	1 min	NO measurements are made in current networks and high-quality, commercially available instruments exist to make these measurements. The reliability of chemiluminescence technique to measure NO has been well established through several instrument intercomparisons. Standard reference material available for through-the-inlet calibration. Instrument currently in wide use for routine monitoring. `
Nitrogen dioxide (NO_2)	±30 percent	1 min	The reliability of chemiluminescence/photolysis technique to measure NO_2 has been well established through several instrument intercomparisons. Standard reference material available for through the inlet calibration. A high time resolution converter for NO_2 measurements by photolysis-chemiluminescence promises a significant improvement in the measurement of this compound.
Nitric acid (HNO_3)	±10 percent	1 hr	Routine monitoring usually done with denuders directly or denuder difference method and ion chromatographic analysis. The evaporation of nitric acid from ammonium nitrate containing aerosols can interfere with gas phase measurement. Calibration is an issue. Method is used currently in research studies and networks. The development of fast-response sampling techniques must be accompanied by improved inlet design to reduce nitric acid uptake on sampling lines. Several semi-continuous methods likely suitable for routine networks are emerging.
Reactive nitrogen oxides (NO_y)	±30 percent	1 min	NO_y measurements are made in current networks and high-quality, commercial instruments exist to make these measurements. Converter reduces NO_y to NO and detects NO by chemiluminescence. Technique has been well established through several instrument intercomparisons and is in wide use for routine monitoring. Calibration at very low levels is an issue. The development of fast-response sampling techniques must be accompanied by improved inlet design to reduce nitric acid uptake on sampling lines.

reviewed in Chapter 6. In meeting the air-quality information needs over all of North America, it is clearly in the interest of the three countries that the designs of these monitoring networks provide compatible information and meet diverse needs with maximum efficiency.

Monitoring networks that provide compatible information for Canada, the United States, and Mexico, and that meet diverse needs with maximum efficiency, need to be designed.

Several networks are in operation. They are aimed at various PM-related problems: harm to human health, ecosystem damage, degradation of visibility, and influence on global climate. The primary responsibilities of the existing networks are to document the severity and extent of the problem and to measure progress in its control. As described in Chapter 6, the existing networks have significantly advanced the understanding of the spatial and temporal patterns of PM concentration and composition. The sites in these networks are chosen and the instrumentation is selected to ensure that their primary goals are met. The network measurements are typically made on a nearly continuous basis throughout the year; although for PM a one-day-in-three or a one-day-in-six, 24-hr average sampling schedule often is used. Although the measured data typically vary according to the objectives of each network, there are significant elements of commonality in the instrumentation used in these networks.

Some networks measure a suite of PM components, including $SO_4^=$, NO_3^-, NH_4^+, OC, BC, and a variety of elements that represent crustal contributions, as well as gas-phase precursors (NO_x, VOC, SO_2 and NH_3) that are responsible for the formation and evolution of PM in the atmosphere. In some cases the network measurements also include ozone and ozone precursors (NO_x, VOC and CO). The measurements made in these networks, apart from their specified monitoring objectives, are useful for diagnostic air-quality applications: tracking effectiveness of control strategies, source identification and apportionment, and evaluation of chemical and physical process understanding. Additionally, information to support these objectives will be obtained by continuous methods, which are rapidly emerging to measure specific PM components, such as mass, $SO_4^=$, NO_3^-, NH_4^+, and trace elements. Continuous methods for carbonaceous compounds, however, are still research activities.

Current instruments, their temporal resolution, and spatial deployment are sometimes inadequate for these additional tasks, specifically for tracking trends and feeding receptor-modeling applications. Moreover, daily operation of the instruments in the network requires significant commitment of personnel resources for maintenance, quality assurance, data archiving, and reporting. Routine monitoring requires reliability, operational simplicity, and low cost. However, this monitoring also requires significant human intervention to reduce and analyze the measurements (Chow, 1995). These factors often result in insufficient time resolution to meet desired objectives.

Clearly, the length of monitoring needed depends upon monitoring objectives. An adequate baseline period may require several years (three or more) of data to specify current conditions and to provide sufficient data for use in designing the most efficient and effective monitoring programs. Most of the regulatory standards require at least three years of data for compliance monitoring. For the assessment of control programs by means of trend analyses, monitoring information is typically needed over a decade or more (see Chapter 6). In this respect, the evaluation of control measures imposes different requirements than does the characterization of population PM exposure over shorter periods.

Requirements for sampling duration and frequency for regulatory monitoring are dictated in part by the forms of the ambient air-quality standards. Twenty-four-hour measurements made once every three to six days have been judged adequate for determining compliance with the 24-hr and annual PM standards. However, 24-hr measurements typically provide little scientific insight into either the formation of secondary PM species or the linkages between emission sources and the eventual PM distribution.

Sampling frequency also affects statistical robustness. The collection of 24-hr PM samples once every six days provides representative but incomplete information for PM frequency distributions. The use of one-day-in-six sampling frequency is thought to introduce potential errors in estimating annual means (Chapter 6).

In addition, characterization of population PM exposure within acceptable uncertainty imposes significant requirements on the number and location of monitors. As discussed in Chapter 6, average-annual urban PM_{10} and $PM_{2.5}$ concentrations may show factor-of-two variations over distances of approximately 10 to 20 km and 50 to 100 km, respectively. A variety of statistical methods is available for spatially allocating monitoring stations. The design criteria to accomplish this have recently been presented (U. S. EPA, 2002)

Efforts to establish scientific bases for network design have not yet been incorporated into redesign of the existing networks. The new U.S. National Air Monitoring Strategy (http://www.epa.gov/ttn/amtic/stratdoc.html) appears to offer the prospect of integrating current single-pollutant monitoring approaches to better address the management of linked, multi-pollutant air-quality issues [2].

Some important monitoring requirements are not being met, or cannot be met until technology evolves. Virtually no speciated carbon measurements exist except for limited, special studies. Measurement of individual PM carbon compounds holds the potential for substantially improving knowledge of source contributions to ambient OC and BC concentrations. Precursor and copollutant gas measurements are necessary for fully understanding the production and evolution of PM in the atmosphere. This full suite of measurements is not available in as many locations as is required. Demerjian (2000) concluded that the monitoring information needed to provide scientific understanding compared with that needed strictly for air-quality management becomes critical for secondary pollutants. In the case of secondary pollutants, understanding the relative effectiveness of emission control strategies requires analyses of data that are capable of revealing the relations between the production of secondary species and their precursor concentrations.

Finally, the measurements made at surface monitoring sites have significant inherent limitations in providing information required for source apportionment and chemical-transport model evaluation. The concentrations of chemically active constituents can be influenced strongly by surface exchange and lead to vertical concentration gradients. The measurement requirements including the ancillary meteorological measurements needed to account for these effects are extremely demanding and cannot be accomplished with presently available network measurements. In addition, the measurements made at ground sites in populated areas are sensitive to nearby sources. Some of these limitations can be overcome with the addition of ground-based instruments that can provide vertical-profile measurements of chemical composition and meteorology.

Special studies and intensive field studies. Routine surface-based monitoring networks alone are neither sufficient nor cost-effective approaches to providing all of the information required by users with different needs (Figure 5.1). To overcome some of the limitations of routine monitoring networks, other approaches, (e.g., special and intensive field studies), are necessary. Three general types of special studies can play a critical role in improving the overall capability of atmospheric-science tools.

1) *Intercomparison studies* are aimed at determining the capability, reliability and/or consistency of the measurement techniques used to characterize the atmospheric concentration of individual gas-phase compounds or specific chemical and physical properties of sampled particles. As discussed earlier, comparisons are critical for PM and related species since field standards are not available for through-the-inlet calibration. These studies also provide an arena to test the proficiency of new measurement techniques and to estimate the confidence, comparability, and reliability of existing techniques (c.f., Section 3.2). The atmospheric-science community addresses measurement uncertainties in an arduous but effective way: formal, rigorous, and unbiased inter- and intra-method instrument comparisons. There have been many such special

[2] This monitoring strategy is specifically focused on the National Air Monitoring Stations (NAMS), the State and Local Air Monitoring Stations (SLAMS), the Photochemical Air Monitoring Stations (PAMS), and IMPROVE.

studies, carried out at ground sites and in aircraft, devoted specifically to the assessment of instrument capability, reliability, and comparability.

2) *Source-characterization studies* are used to pinpoint sources, to identify their emissions, and to evaluate emission inventories using receptor-modeling methods (c.f., Chapter 7).

3) *Major field intensives* use a comprehensive collection of measurements, including research techniques, to characterize source emissions and to better understand chemical and meteorological processes. These studies are especially useful for evaluating chemical-transport models (c.f., Chapter 8). These two types of studies are often carried out in a region that has routine monitoring sites. The studies involve altering the sampling frequency of the routine measurements and/or deploying additional measurement techniques at selected locations for the duration of the special study.

Intensive field programs have been used effectively for several years, to fill gaps in the understanding of processes and sources that shape the distribution of gas-phase and PM pollution, and to provide information needed to critically evaluate model performance. These studies use an array of advanced instruments operated by teams of scientists that cooperate to carry out these large field campaigns. The intensive measurement periods are followed by an extended period of analysis and interpretation by the scientists involved in the study as well as others in the scientific community.

To achieve greater cost effectiveness, the goals of all three special-study types can, and often are, merged into one integrated intensive field study involving ground sites, aircraft, and specific source measurements. The studies are carried out over periods ranging from several weeks to more than a year, and typically involve one or more highly instrumented ground sites along with satellite ground sites that are less highly instrumented and distributed over the region of interest. The major ground sites need sufficient measurements in a highly variable atmosphere to reduce the degrees of freedom in interpretation and quantification of the important processes and sources. The measurement sites are chosen to best identify important sources or processes that shape atmospheric composition and pollution accumulation, and may be collocated at previously established network monitoring sites. The distributed ground sites and aircraft permit the study to be placed in a regional context. In addition, aircraft measurements allow investigation through the boundary layer and above. The studies are supported by detailed estimates of the local and regional emissions of compounds of interest, which often have much higher temporal resolution than the information supplied to national emission-inventory programs. Specific source-type measurements may include, for example, tunnel studies or continuous emission monitoring of major point sources. These studies have provided the modeling community with the range of data that is required to test current models (see Chapter 7 and Chapter 8). In addition, receptor models (see Chapter 7) have been used with these data to uncover problems with emission inventories. Because these studies are expensive and require elaborate planning, they occur infrequently (for a given region, perhaps one or twice a decade).

5.3.2 Future Requirements for Measurement Strategies

PM represents a large collection of components, whose identity and source are poorly understood. Thus the current situation with regard to developing PM measurement strategies and monitoring networks is substantially different from that encountered with gas-phase pollution monitoring. The characteristics of the particles that cause harmful effects are not yet well established. For this reason, the measurement strategies and networks required to monitor these properties are not yet fully defined and implemented. There is a need for a basic suite of chemical measurements at spatially representative, or unique, sites to provide the full picture of PM formation, transport, and trends. For example, a minimum suite includes PM mass, speciation of the six major components ($SO_4^{=}$, NO_3^-, NH_4^+, OC, BC, crustal species), ground-level ozone, and precursor species (NO_x, SO_2, VOCs and NH_3).

The source-receptor relationships that determine PM concentration and visibility are similar; thus networks and field studies should be designed to permit particle concentrations and visibility relationships to be studied simultaneously. As new measurement technology emerges, it must be adapted to maximize the information-gathering capacity and utility of the existing and planned measurement platforms and sites. The development of accurate, real-time (measurement time of a few seconds or less) instruments opens opportunities for both airborne instruments (on manned and unmanned aircraft and balloons) and ground-level instruments (on vans, boats, and trains) that can make measurements with better spatial resolution and coverage. It will be necessary to increase the reliability and operational simplicity of the techniques so that they can be more easily adapted to operational needs. This includes the development of methods to calibrate these measurements, along with the development and maintenance of calibration and certified reference standards. The measurement strategies that use these resources must be aimed at providing data products and data-product formats that fulfill specific needs and can be readily and easily used.

Where networks are responsible for compliance, the principle of accountability should be incorporated into network design and the interpretation of network data. This means that the changes in emissions should be reflected in the data, and those changes should be compatible with expected changes based on a theoretical understanding of atmospheric processes and CTM predictions.

There is a natural synergism between PM research and research involving other air pollutants. The most effective management of these copollutants requires the combined understanding of the sources and processes that are responsible for them and the relationships between them. Measurement programs aimed at the study of PM should include sufficient scope to address the questions associated with these linkages.

5.4 SUMMARY

Measurement Uncertainty

The large number of properties needed to fully define PM (e.g., size, mass, number, and composition) makes the determination of the uncertainty inherent in any of the particle measurements more challenging than for gas-phase measurements. Appropriate reference materials or standards for many particle-phase components are limited and virtually unavailable for in-the-field through-the-inlet calibration. For this reason, measurement uncertainty (or perhaps more appropriately, consistency or equivalency) is determined by comparisons among several methods that determine similar properties of sampled particles. Hence, for PM, convergence of methods is used rather than calibration to a known standard, as a guide to measurement uncertainty. However, measurement accuracy relative to what actually exists in air is undefined by current methods. Considerable research is required to develop in-field certified calibration methods and reference standards, and may not be possible or feasible for all particle properties.

Comparisons of PM mass measurement using filter-based methods indicate that $PM_{2.5}$ mass determinations are comparable to within 20 percent. The measurements of the stable ionic components, such as $SO_4^=$, are comparable within 10 percent. The measurement of important volatile and semivolatile PM components, including NH_4NO_3 and SVOCs is much less certain, even with the use of denuders and reactive filters.

The measurement of particulate carbon and carbon-containing species by filter/denuder-based methods is especially problematic. Measurements of organic compounds in PM are prone to errors due to particle volatility, interferences from gaseous organic species, limitations with analytical methods, and lack of calibration and certified reference standards. It is likely that both positive and negative artifacts and interferences occur during sampling, and there is no consensus as to the net effect of these interferences and artifacts. Recent studies using quartz-fiber filters indicate a net positive artifact for OC in the range of 1-4 $\mu g/m^3$ under the variety of conditions examined. It is likely that the relative magnitudes of these effects vary as a function of a number of variables. For these reasons, current organic speciation explains only 10 percent to 20 percent of total organics in PM. A list of all the estimated uncertainties in integrated denuder and filter systems is given in Tables 5.1 through 5.3.

Integrated Denuder and Gravimetric Filter-Based Systems

Integrated denuder and filter-based systems have been widely used to measure PM mass, PM chemical composition and PM precursor gases (e.g., HNO_3 and NH_3). These systems typically consist of an inlet and subsequent particle-size separating device to allow only particles of interest to be collected; a denuder to remove potential gas-phase interferences if the species of interest is semivolatile (e.g., NH_4NO_3); an inert or reactive collection substrate, often composed of Teflon, quartz-fiber, or nylon; followed by a vacuum pump to pull air through the system and a device to monitor and regulate the flow. Several different size-fractionating modules are available, as well as denuders and collection media. Denuders and reactive filters to measure NO_3^- and/or HNO_3 are fairly well developed, although new designs require significant testing. The collection of organic aerosols is still quite challenging. However, the chemical measurements using these systems in combination with well-established analytical methods are capable of identifying and quantifying the major components ($SO_4^=$, NO_3^-, NH_4^+, OC, BC, and crustal-related species) of the collected mass. While filter/denuder based systems are robust and relatively easy to use, their application in national PM monitoring networks is limited because they are labor intensive and costly. As described below, continuous measurements would be expected to reduce cost and to provide unique data in terms of temporal resolution.

Continuous and Semi-continuous Measurements

Instrumentation is needed to measure aerosol composition semi-continuously or continuously in real-time. While suitable real-time bulk mass instruments are available now, measurement techniques for PM chemical components are emerging rapidly. Real-time instruments also can be less expensive to operate than filter-based sampler systems and can provide valuable information on diurnal trends. While filter measurements are relatively simple and inexpensive to implement, they require manual operation, provide poor time resolution, and are subject to sampling errors that cannot be quantified. Continuous and semi-continuous real-time methods supply high-frequency, short-duration measurements needed to improve understanding of the formation, redistribution, and loss of atmospheric PM, as well as the sources of this material and the relation between PM and human health. They also facilitate timely communication to the public. Significant progress has been made in the past several years to develop instrumentation for real-time measurements of PM mass and composition, and some of these instruments have been commercialized. However, sampling errors and uncertainty in continuous and semi-continuous composition measurements still need to be better identified and quantified before wide use in routine monitoring networks. The availability and capabilities of commercialized real-time instrumentation will evolve substantially in the next few years, and as greater confidence in these methods is achieved, efforts should be made to replace filter samplers.

Personal Exposure Monitors

Techniques have been adapted to monitor actual personal exposure. These monitors are designed to be carried by an individual and record the actual exposure of that individual to particulate matter. The sampling has focused on mass and collection of environmental pollutants having potentially significant human-health effects such as combustion emissions, tobacco smoke, acid aerosols, and vapor-phase organics. Although most personal monitoring has focused on 24-hr collection, devices are being developed for real-time continuous monitoring to determine individual exposure to short-term (one minute) variations in PM concentrations to determine the impact of brief high exposures and to identify and characterize PM sources. These small, lightweight devices have uncertainties of about 20 percent for mass and about 20 percent for stable particle-phase components (e.g., $SO_4^=$).

Single-Particle Measurements

During the past ten years, the ability to measure properties of a single particle in near-real time, rather than just the properties of a bulk collection of particles over an integrated time period, has opened the way for more detailed understanding of particle morphology and chemical composition. Of particular value is the information provided on the mixing of different aerosol components. Currently, the techniques are only semi-quantitative. As a consequence, these methods are not used in routine monitoring applications. It is possible that particle mass-spectrometry will evolve a capability to provide accurate real-time measurements of size-resolved composition, but much more research will be required to develop and evaluate the potential for such measurements. Single-particle analysis by scanning electron microscopy (SEM) also has evolved, allowing for simultaneous detailed observations of particle morphology, size, and chemical composition, providing valuable insights for source apportionment of ambient, micro-environmental, and human-exposure studies.

Optical Properties of Aerosols and Long-Path Optical Attenuation

Issues associated with visibility, radiative transfer, and climate have motivated the use of optical long-path measurements. In combination with other instruments these techniques also can contribute to other aspects of PM monitoring. Satellite techniques are currently most suited to define global aerosol-distribution patterns. Satellite techniques usually provide the only information available in remote areas and can offer global coverage, but typically measure only relative concentrations of a PM in the atmosphere. Satellites also have the potential to monitor cloud distributions and characteristics of the surface such as moisture availability, albedo, and heat capacity. Such measurements can ultimately improve the fidelity of PM modeling.

Routine surface-based visibility measurements provide data on total extinction and the major components of extinction – light scattering and light adsorption. Along with the chemical composition of PM from filter-based or continuous methods the contributions to light extinction can be apportioned to emission sources. The ASOS/AWOS visibility measurements and the ancillary meteorological data represent a valuable resource for air-quality investigations as well. Resources should be made available to the agencies that operate and evaluate the ASOS/AWOS network to make the data more widely available and in a form that represents the full dynamic range of the instruments.

Analysis of Cloud and Fog Chemical Composition

Fog and cloud droplets collect local aerosols and soluble gaseous compounds. Cloud-chemistry measurements allow for a better understanding of the chemistry within fogs and clouds and the variation with droplet size as a function of location, elevation, and time of day. Along with ambient PM measurements, fog and cloud measurements allow for an improved knowledge of the impact of water droplets on PM concentrations during and after fog and cloud occurence. Studies show that clouds and fog can play a significant role in PM formation, either increasing or decreasing particle loading after the event. Intercomparisons to estimate liquid-water content in fogs indicate agreement among the methods of about 10 percent. In addition, important chemical species can be analyzed in fog and cloud water within several droplet size ranges. The non-volatile fraction of the chemical compounds in cloud water can be determined to within 10 percent.

Gas-Phase PM and Copollutant Precursors

Techniques are available to measure the ambient concentrations of gas-phase particle precursors, ozone, ozone precursors, and oxidation products of these compounds in the atmosphere. In general, the measurement methods and calibration procedures are well developed and adequate for most studies. Several intercomparisons have been conducted over the past twenty years to determine the equivalence and comparability of measurement methods for these species. The measurement methods for routine monitoring application are well developed for ozone, NO, NO_2, NO_y, CO, and SO_2. These measurements can be made routinely with uncertainties of less than 30 percent. A list of all the estimated uncertainties in the relevant gas-phase measurements was given in Table 5.4.

The VOCs are of particular interest in connection with measurement uncertainties. They constitute a very large class of compounds generally defined as volatile at standard temperature and pressure. VOC measurements are difficult to perform because of the extreme complexity of the organic mixtures that may be present in the atmosphere. Although the C_1-C_6 hydrocarbons usually can be determined with uncertainties of less than 20 percent, difficulties associated with VOC measurements increase with complexity of the compound, and depend on whether carbon bonds in the VOC are saturated and whether the VOC is oxygenated. Serious problems have been observed for higher molecular weights, ($\geq C_8$) particularly NMHCs, which have a high probability of participating in particle formation. Clearly, one of the most critical needs of the ambient VOC measurement community is a rigorous field intercomparison of measurement techniques. Such an intercomparison will more clearly define measurement capabilities and identify prevalent problems that must be addressed.

Meteorological Measurements

Ground-level methods for meteorological observations are routine, precise, accurate, and inexpensive. They are deployed in all monitoring networks and are well understood and appreciated by most scientists who use them. In addition, accurate measurements of meteorological parameters aloft are essential for understanding and modeling PM sources, chemistry, transport, and deposition. New insight has been gained from measurements aloft regarding the complexity of the meteorological processes that control air quality and the need for better observations and models. However, these measurements are more complicated, difficult to operate and maintain, expensive, difficult to calibrate, and relatively new. NOAA maintains a sparse national network of radar profilers to measure wind speed and direction aloft in the U.S. Midwest and along both coasts for tracking severe weather. This network also can provide valuable information for air quality management.

5.5 POLICY IMPLICATIONS

The ultimate deployment of measurements must consider the level of confidence needed in the data, the uncertainly associated with the measurements, the resolution of the data, and the intended use of the information. Multiple measurement strategies – routine, special study, intensive periods – are required to estimate human exposure, support epidemiological studies, identify the sources and understand the evolution of aerosols in the atmosphere, and track the success of implementation. *Implication:* The intended purpose of the data should drive selection of the measurement method, and the spatial and temporal resolution of the measurement. The most effective measurement strategy requires a complement of long-term routine measurements, focused special studies and intensive field studies.

There is a need for a basic suite of chemical measurements at spatially representative sites to provide the full picture of PM formation, transport, trends, and impacts. For example, a minimum suite would be: PM mass (fine and coarse), speciation of the six major components ($SO_4^=$, NO_3^-, NH_4^+, OC, BC, crustal material), ground-level ozone, the precursor gases (NO_x, SO_2, $>C_7$ VOCs, and NH_3), HNO_3 and CO, with meteorological measurements. *Implication:* A focused suite of measurements is necessary to characterize the limiting processes and chemical reagents leading to particle formation and growth, to link ambient concentrations and exposure to sources, and to illustrate the short- and long-term impact of management strategies.

The large number of properties needed to fully define PM (e.g., size, mass, number, and composition) as well as the impact of related parameters (e.g., gases, meteorological variables) makes it challenging to determine the uncertainty inherent in any of the particle measurements. At present, PM measurement uncertainty for many properties is estimated by instrument comparisons among different methods which determine similar properties of sampled particles. *Implication:* The science community relies on convergence of methods, not absolute calibration, as a guide to measurement certainty. Currently, measurement confidence varies according to the specific particle characteristic that is being measured and even the conditions under which it is being measured.

Significant progress has been made in the past several years to develop instrumentation for real-time, continuous measurements of particle mass and composition. Comparability studies indicate that filter-based methods and continuous measurements each have useful roles to play in managing and tracking PM. However, continuous real-time methods can supply additional information that can improve understanding of the formation, redistribution and loss of PM in the atmosphere, the sources of that PM, and the relation between PM and human health. *Implication:* Over the next few years the availability and capability of routine real-time, continuous instrumentation will improve substantially. Efforts should be made to augment and/or replace the filter methods with these continuous methods.

Measurement confidence and ability to characterize the carbon fraction, speciated organics, the semivolatile species, the relative amounts of OC and BC, and the relative contributions of primary and secondary carbon is evolving, but limited. *Implication:* The carbon fraction is typically 1/5 to 1/2 of the total particle mass, and the limited information about this major component handicaps understanding of PM formation and evolution, and inhibits the development of management strategies.

The health-research community (including epidemiological, toxicological and clinical researchers) is restricted to investigations of the impacts of PM characteristics that can be measured, and require sufficient data from these measurements to assess health impacts. Of the many hypothesized causal elements of PM impacts, there are only sufficient ambient data to test about half. Measurements of many of the potential causal agents also require highly trained investigators. *Implication:* Progress in elucidating causal mechanisms and the sources that may be most responsible for PM health impacts are contingent upon the development and monitoring of the hypothesized elements. Population-level studies typically require relatively long-term data sets to achieve statistically significant results.

Investigation and assessment of health impacts have been based upon PM data obtained from gravimetric methods and are the basis of current ambient PM standards. Numerous advanced measurement methods are emerging that can measure additional PM properties, e.g., single-particle mass spectrometry, particle density, ultrafine particle size and number distribution, and size-resolved composition. *Implication:* New PM mass and composition measurements related to health impacts need to be linked to the historical context, and will provide insights into the fundamental properties and relative source contributions of particles providing the ability to differentiate between sources, local versus regional contributions, primary versus secondary aerosol constituents, and natural versus anthropogenic contributions.

The gravimetric filter-based methods (for example, the U.S. Federal Reference Method) may underestimate or overestimate the concentration of $PM_{2.5}$ due to loss of volatile components or the adsorption of gases. These losses or gains may be significant under certain circumstances and the parameters that affect these variables and related amounts are only now being quantified. *Implication:* Uncertainty in the gravimetric characterization of PM mass can impact the status of compliance within a given region. A better understanding of the uncertainty of these gravimetric methods for $PM_{2.5}$ is needed to determine if national standards are being achieved.

5.6 REFERENCES

Apel, E.C., Calvert, J.C., Gilpin, T.M., Fehsenfeld, F. C., Parrish,D.D. , Lonneman, W.A., 1999. The Non-Methane Hydrocarbon Intercomparison Experiment (NOMHICE): Task 3. Journal of Geophysical Research 104, 26,069-26,086.

Chow, J.C., 1995. Critical review: measurement methods to determine compliance with ambient air quality standards for suspended particles. Journal of the Air and Waste Management Association 45, 320-382.

Chow, J.C., Watson, J.G., 2001. Ion chromatography. In Elemental Analysis of Airborne Particles, Landsberger, S. and Creatchman, M., eds. Gordon and Breach, Newark, NJ, pp. 97-137.

Chow, J.C., Watson, J.G., Crow, D., Lowenthal, D.H., Merrifield, T., 2001. Comparison of IMPROVE and NIOSH carbon measurements. Aerosol Science and Technology 34, 23-34.

Collett, J.L., Hoag, K.J., Pandis, S.N., 1998. The influence of drop size-dependent fog chemistry and aerosol production and deposition in San Joaquin Valley fogs. Prepared for the California Regional PM10/PM2.5 Air Quality Study c/o the

California Air Resources Board, Technical Support Division, Sacramento, CA.

Demerjian, K.L., 2000. A review of national monitoring networks in North America. Atmospheric Environment 34, 1861-1884.

Evans, G.F., Highsmith, R.V., Sheldon, L.S., Suggs, J.C., Williams, R.W., Zweidinger, R.B., Creason, J.P., Walsh, D., Rodes, C.E., Lawless, P.A., 2000. The 1999 Fresno particulate matter exposure studies: comparison of community, outdoor, and residential PM mass measurements. Journal of the Air and Waste Management Association 50, 1887-1896.

Huebert, B.J., Charlson, R.J., 2000. Uncertainties in data on organic aerosols. Tellus 52B, 1249-1255.

Jones, A.S., Guch, I.C., Vonder Haar, T. H., 1998. Data assimilation of satellite-derived heating rates as proxy surface wetness data into a regional atmospheric mesoscale model. Part I: methodology. Monthly Weather Review 126(3), 634-645.

McMurry, P.H., 2000. A review of atmospheric aerosol measurements. Atmospheric Environment 34, 1959-1999

Marple, V.A., Olson, B.A., 2001. Interial, gravitational, centrifugal, and thermal collection techniques. In Aerosol Measurement: Principles, Techniques and Applications, Second Edition, Baron, P.A. and Willeke, K., eds. Van Nostrand Reinhold, New York, NY.

Middlebrook, A.M., Murphy, D.M., Lee, S.H., Thomson, D.S., Prather, K.A., Wenzel, R.J., Liu, D.Y., Phares, D.J., Rhoads, K.P., Wexler, A.S., Johnston, M.V., Jimenez, J.L., Jayne, J.T., Worsnop, D.R., Yourshaw, I., Seinfeld, J.H., Flagan, R.C., Hering, S.V., Weber, R.J., Jongejan, P., Slanina, J., and Dasgupta, P.K., 2003. A comparison of particle mass spectrometers during the 1999 Atlanta Supersite Experiment. Journal of Geophysical Research, in press.

Pandis, S., Lillis, D., Collett, J., Richards, R.L., 1998. Fog effects on PM concentration and composition in the SJV. Prepared for the California Regional PM10/PM2.5 Air Quality Study c/o the California Air Resources Board, Technical Support Division, Sacramento, CA.

Parrish, D.D., Fehsenfeld, F.C., 2000. Methods for gas-phase measurements of ozone, ozone precursors and aerosol precursors. Journal of Geophysical Research 34, 1921-1957.

Schauer, J.J., Rogge, W.F., Hildemann, L.M., Mazurek, M.A., Cass, G.R., Simoneit, B.R.T., 1996. Source apportionment of airborne particulate matter using organic compounds as tracers. Atmospheric Environment 30, 3837-3855.

Seigneur, C., Pun, B., Prasad, P., Louis, J-F., Solomon, P.A., Koutrakis, P., Emery, C., Morris, R., White, W., Zahniser, M., Worshop, D., Tombach, I., 1998. Guidance for the performance evaluation of three-dimensional air quality modeling systems for particulate matter and visibility, Appendix A: Measurement methods. Prepared for the American Petroleum Institute, Washington, DC by Atmospheric Environmental Research, Inc., San Ramon, CA.

Solomon, P.A.,Baumann, K., Edgerton, E., Tanner, R., Eatough, D., Modey, W., Marin, H., Savoie D., Natarajan, S., Meyer, M.B., Norris, G., 2003. Comparison of integrated samplers for mass and composition during the 1999 Atlanta Supersites project. Journal of Geophysical Research, in press.

Turpin, B.J., Saxena, P., Andrews, E., 2001. Measuring and simulating particulate organics in the atmosphere: Problems and prospects. Atmospheric Environment, 34, 2983-3013.

U.S. EPA, 2002. National Air Quality Monitoring Strategy. Available at www.epa.gov/ttn/amtic/stratdoc.html.

Watson, J.G., Chow, J.C., Moosmüller, H., Green, M.C., Frank, N.H., Pitchford, M.L., 1998. Guidance for using continuous monitors in $PM_{2.5}$ monitoring networks . Report No. EPA-454/R-98-012. Prepared by U.S. Environmental Protection Agency, Research Triangle Park, NC.

Watson, J.G., Chow, J.C., 2002. Comparison and evaluation of in-situ and filter carbon measurements at the Fresno Supersite. Journal of Geophysical Research, Submitted for publication.

Williams, K.R., Fairchild, C.I., Jaklevic, J.M., 1993. Dynamic mass measurement techniques. In Aerosol Measurement: Principles, Techniques and Applications, Willeke, K. and Baron, P.A., eds. Van Nostrand Reinhold, New York, NY, pp. 296-312.

Williams, R., Suggs, J., Zweidinger, R., Evans, G., Creason, J., Kwok, R., Rodes, C., Lawless, P., and Sheldon, L., 2000a. The 1998 Baltimore particulate matter epidemiology-exposure study: Part 1. Comparison of ambient, residential outdoor, indoor and apartment particulate matter monitoring. Journal of Exposure Analysis and Environmental Epidemiology 10, 518-532.

Williams, R., Suggs, J., Creason, J., Rodes, C., Lawless, P., Kwok, R., Zweidinger, R., and Sheldon, L., 2000b. The 1998 Baltimore particulate matter epidemiology-exposure study: Part 2. Personal exposure assessment associated with an elderly study population. Journal of Exposure Analysis and Environmental Epidemiology 10, 533-543.

Williams, R., Creason, J., Zweidinger, R., Watts, R., Sheldon, L., and Shy, C., 2000c. Indoor, outdoor, and personal exposure monitoring of particulate air pollution: the Baltimore elderly epidemiology-exposure pilot study. Atmospheric Environment 34, 4193-4204.

CHAPTER 6

Spatial and Temporal Characterization of Particulate Matter

Principal Author:
Charles Blanchard

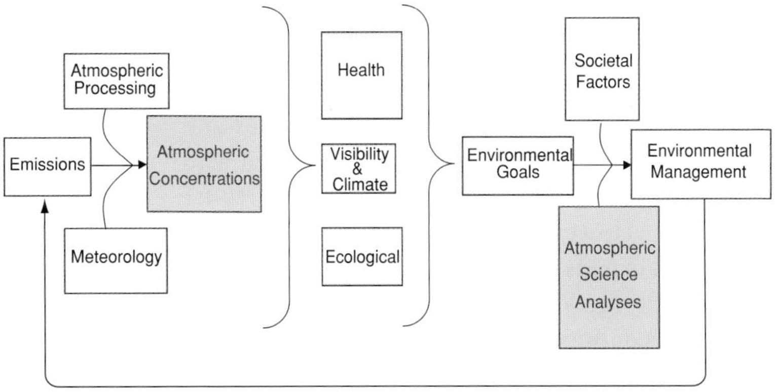

6.1 INTRODUCTION

Geographical and temporal variations of the concentration and composition of $PM_{2.5}$ influence PM exposure (Chapter 2) and provide important insights into the processes that influence particle formation and distribution (Chapter 3). Characterizing the temporal, spatial, and chemical variations of particles and their precursor gases, as well as the size properties of particles, enhances understanding of emission sources (Chapters 4 and 7), secondary PM formation, and transport. This information establishes key features of the data that should be reproduced by modeling studies and allows air-quality models to be tested against ambient measurement data (Chapter 8). Measurements thereby support the use of models as an important tool for assessing proposed emission-management strategies or evaluating source contributions to existing aerosol concentrations. Where a data record exists that is long enough to support trend assessments (typically decadal), characterization of trends provides an important technique for evaluating the success of ongoing management programs.

In this chapter, the spatial and temporal patterns of PM mass concentration and composition are presented. The topics covered are:

- Section 6.2: spatial variations of PM across North America
- Section 6.3: regional and urban contributions to PM
- Section 6.4: influence of intercontinental aerosol on North America
- Section 6.5: trends, along with their implications for source attribution and management
- Section 6.6: covariation of PM and ozone concentrations
- Section 6.7: summary
- Section 6.8: policy implications.

Two appendices provide further information on data availability and monitoring networks (Appendix C) and the global distribution of aerosols (Appendix D). Chapters 7 (Receptor Modeling) and 10 (Conceptual Models) extend upon the present chapter through discussion of methods for analyzing observations (Chapter 7) and presentation of regional case studies (Chapter 10).

The terminology presented in Chapter 1 is used throughout this chapter: $PM_{2.5}$ is defined as particles of aerodynamic diameter 2.5 $\mu g/m^3$ and less, PM_{10} is particulate matter of aerodynamic diameter 10 $\mu g/m^3$ and less, and $PM_{10\text{-}2.5}$ consists of particles between 2.5 and 10 $\mu g/m^3$ aerodynamic diameter. Because PM regulations are expressed in terms of mass, rather than particle number or surface area, the data in this chapter are presented in mass units $\mu g/m^3$.

6.1.1 General Features Affecting Particulate Levels in North America

Regional and local PM concentrations in North America are affected by the location and magnitudes of anthropogenic and natural emissions, topography, land cover, winds, and a variety of processes affecting the rates of conversion of gases to particles. Human population densities are greatest in central Mexico, the northeastern United States, southeastern Canada, and portions of the Pacific Coast, leading to higher anthropogenic emissions of primary particles and particulate precursors in those areas compared with other parts of North America (Chapter 4, Figures 4.1 and 4.2). However, the relative sparseness of vegetation in parts of the southwestern United States and Mexico favors airborne suspension of dust in those areas during the dry seasons.

Topography, through its influence on circulation, strongly affects daily variations of PM mass concentrations in some locations. Mexico City (Edgerton et al., 1999) and Los Angeles are two notable examples where terrain substantially influences the circulation of air masses, affecting the concentrations of PM mass from day to day.

The general pattern of air movement across North America influences the prevailing transport directions for particles. During summer, a generally south-to-north-or-northeast transport direction is expected for eastern North America, whereas a west-to-east pattern is more typical of western North America (Figure 6.1). Prevailing air-mass patterns make it possible for particle-laden air masses to reach the Gulf Coast states in the United States from northern Africa, or for air pollutants to travel from Asia to western North America (see Appendix D).

The potential transport distances of particles depend strongly upon particle size (see Chapter 3) and wind speed. Typical residence times for particles of 0.1-1.0 µm aerodynamic diameter are on the order of ten days. However, this residence time can vary with altitude. Within the atmospheric boundary layer (the lowest 1 to 2 km), the typical residence time is about 3 to 5 days. If aerosols are lifted to ~1 to 10 km in the troposphere (e.g., by deep convection at fronts or convergence zones), they can be transported for weeks and many thousand kilometers before removal. Particles at higher levels of the atmosphere may or may not affect ground-level concentrations, so information about particle plumes derived from satellite measurements typically needs to be compared with data from surface monitors.

6.1.2 Spatial and Time Scales of Interest

The spatial and temporal distributions of aerosols are determined by the patterns of their causal factors: emissions, transport, deposition, and chemical reactions. Consequently, temporal and spatial pattern analysis of PM data is particularly useful for source identification and characterization.

In the atmosphere, larger spatial scales are associated with longer time scales (Chapter 1, Figure 1.3). However, regulatory approaches for managing PM exposure use both 24-hr and annual-average standards, applicable at all spatial scales. Spatial scales of interest for health impacts include the neighborhood scale (1 to 10 or more km) for characterizing individual and community exposure, the urban scale (10 to 100 or more km) for characterizing population exposure, and regional

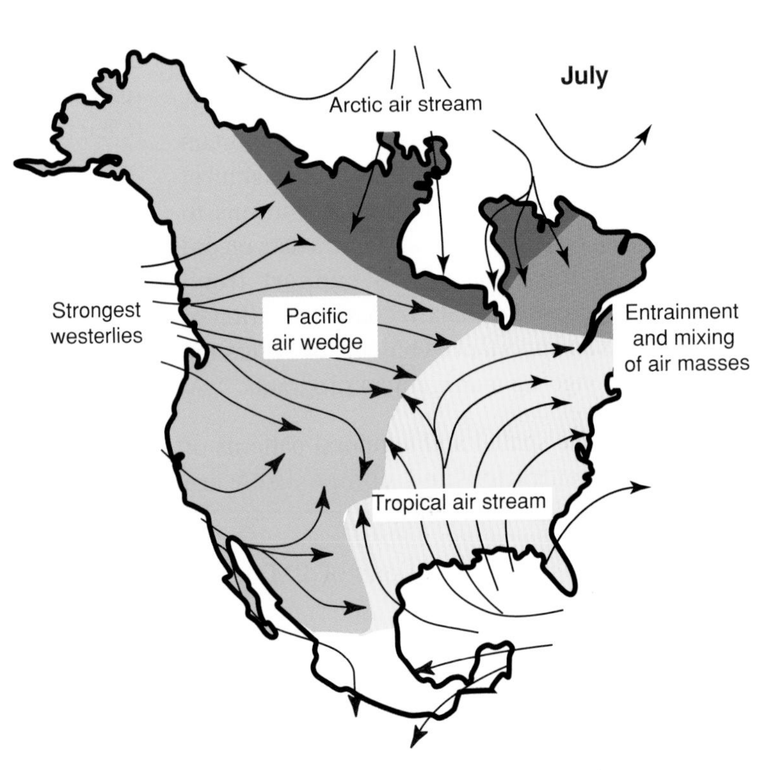

Figure 6.1. Average air mass source regions and transport during July. (Source: adapted from Bryson and Hare, 1974).

(100 to 1000 or more km), continental, and intercontinental scales for characterizing large-scale contributions to PM mass upon which urban aerosol concentrations are superimposed. Management of PM mass concentrations focuses on the full range of spatial and temporal scales to address a set of linked problems: sorting out the regional, urban, and neighborhood contributions to measured PM mass concentrations, establishing the spatial representativeness of monitoring locations, apportioning PM observed at a receptor back to its sources, and designing monitoring networks capable of characterizing population exposure.

Full characterization of the spatial variations of PM is exceedingly difficult. Virtually the entire record of PM mass and composition underestimates the true variability occurring in the atmosphere, because monitors operated at fixed locations typically do not capture concentration variations occurring over distances of tens to hundreds of meters. Indeed, monitors are generally located where the measurements may be considered representative of conditions over tens to hundreds of kilometers (urban to regional scales). Yet, health effects may occur on finer spatial scales than typical PM networks can resolve, such as adjacent to roadways (Brunekreef et al., 1997).

Visibility concerns focus primarily on the regional scale, especially in the scenic vistas of western North America, but also include urban- and neighborhood-scale concerns. As noted, time scales of interest expand along with spatial scales, so that multi-day episodes of impaired visibility are typically regional in spatial extent. Visibility impairment is strongly related to $PM_{2.5}$ levels, but linkages to gas-phase pollutant issues exist (Chapters 3 and 9).

6.1.3 Monitoring Capabilities

The data used in this chapter were obtained primarily from multiple surface-based monitoring networks originally designed for differing purposes (Appendix C). As discussed in Chapter 5, measurements of PM mass obtained by different methods typically agree within 20 percent. Comparability across techniques for specific PM components can be better or worse than comparability for mass, depending on the difficulty of the measurement: about 10 percent for $SO_4^=$, 35 percent for NO_3^-, 10 percent to 30 percent for trace elements, 20 percent to 50 percent for OC, and 20 percent to 200 percent for BC (see Chapter 5). Spatial patterns and time trends compiled across monitoring networks may reflect some of the variability among sampling and measurement methods. However, analyses of the spatial and temporal variations using data derived from consistent measurement techniques need not reflect the full range of variability across measurement methods, though the absolute magnitudes of any set of measurements will reflect whatever biases may exist in the techniques.

6.2 CONTINENTAL AND REGIONAL VARIATIONS OF PM CONCENTRATIONS

In this section, the spatial patterns of aerosol concentrations and composition are examined across North America and by region. Spatial and temporal variations, and their significance, are discussed using comparisons and contrasts of measurements from nine different regions within North America (Figure 6.2):

- Southeastern Canada (Windsor-to-Quebec City corridor)
- Northeastern United States
- Southeastern United States
- Ohio River Valley
- Mexico City
- Southwestern United States
- Los Angeles, California
- San Joaquin Valley, California
- Pacific Northwest (United States and Canada).

The selected regions illustrate specific characteristics of PM for a variety of situations that represent a wide range of emission sources, terrain, and meteorological characteristics. Both urban and regional measurements are considered, with the aim of

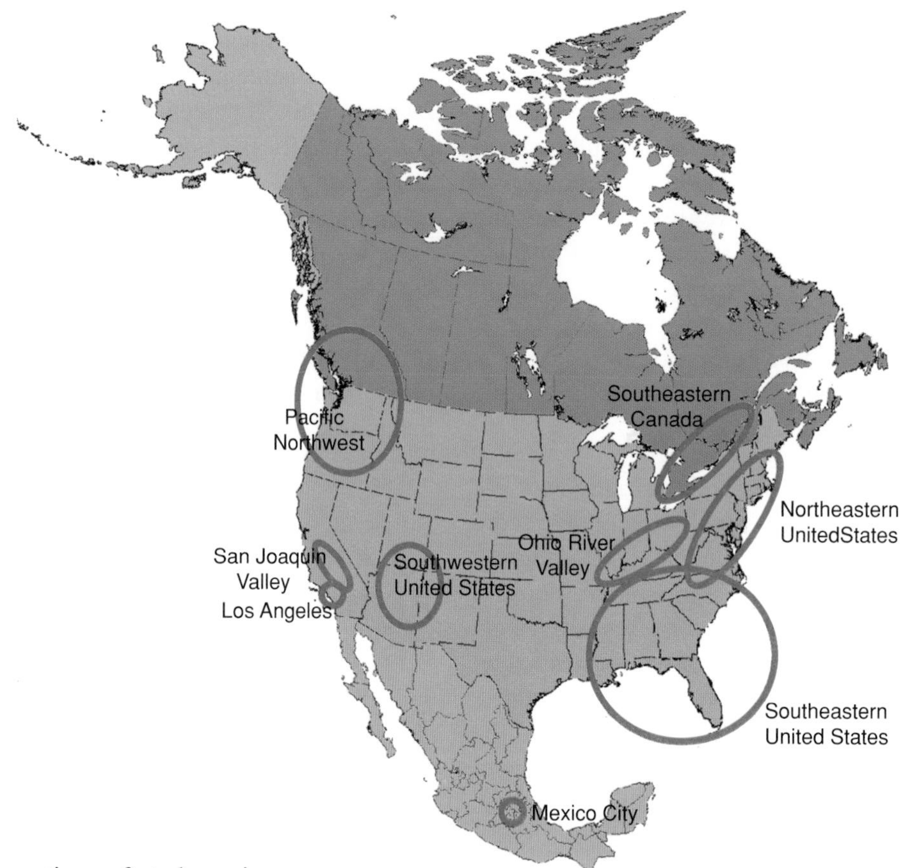

Figure 6.2. Locations of study regions.

showing the relation of urban and regional PM in each area. Much of the material from this chapter will be applied further in Chapter 10, where key infomation is condensed in terms of region-specific "conceptual models" of PM behavior. It should be noted in this context, however, that the areas designated on Figure 6.2 correspond only loosely to the conceptual-model regions in Chapter 10.

The eastern regions illustrate conditions in which $SO_4^=$ is a major component of $PM_{2.5}$ during much or most of the year. SO_2 emission densities are greater in the Ohio Valley than in any other region, and SO_2 emissions in the other three eastern regions exceed those in western North America (Chapter 4).

Distinguishing features of the southeastern United States include the widespread, persistent haze observed over much of the region, especially in summer, the amount of rural power generation, and rapidly growing urban areas (Chapter 9). Organic carbon and $SO_4^=$ are usually major constituents of that haze. More needs to be understood about the sources of the carbonaceous component of the particle samples obtained in the southeastern United States, including primary particle emissions and secondary contributions from atmospheric chemical reactions involving VOCs.

With about 20 million inhabitants, Mexico City is one of the largest urbanized areas in the world. Unlike U.S. and Canadian cities, it is situated in a tropical latitude (19 degrees north). Its elevation of approximately 2240 m provides it with a temperate climate. Mexico City occupies an area of about 1300 km^2 within the Valle de Mexico, a basin with mountain ranges rising about 1000 m above the valley floor to the east and west. The terrain limits ventilation, and pollutant concentrations are affected by thermal and topographic circulation patterns. Of particular significance to PM is the presence of urban-rural transition zones with unpaved areas, agriculture, livestock, and deforested areas.

For many years, the southwestern United States, particularly the Grand Canyon region, has been the focus of research on visibility, as well as regulatory efforts aimed at eliminating regional haze and thereby maintaining the scenic vistas of the southwestern United States. Typically, the best visibility in the conterminous United States occurs in a region including the Grand Canyon and other national parks in northern Arizona, southern Utah, as well as portions of Colorado, Nevada, and Wyoming (Malm, 2000).

The California examples illustrate two different sets of conditions, each of which results in some of the highest PM mass concentrations recorded in North America. Whereas the Los Angeles area is primarily urban, with a population density of about 2100 people per square mile, the San Joaquin Valley is primarily rural with several cities of moderate size (~50,000 to 400,000) and an average population density of about 120 people per square mile. Los Angeles borders the Pacific Ocean and exhibits a marine influence with fog commonly occurring during late spring and summer. The San Joaquin Valley is inland, bordered by the Pacific Coast Range to the west and the Sierra Nevada to the east; winter inversion fog is common. In both cases, the presence of fog plays a key role in the conversion of gas-phase species to particles, the growth of hygroscopic particles, and particle volatility.

Sites in the Pacific Northwest exhibit PM mass concentrations lower than in eastern North America, and PM composition differing from that typical of either other portions of western North America or of eastern North America.

6.2.1 Spatial Variations of PM_{10} Mass: Where and When are PM_{10} Concentrations Highest?

Substantial spatial (Figure 6.3) and seasonal (Figure 6.4) variations occur in the concentrations of PM_{10} mass. Annual PM_{10} averages in Mexico City, Toluca, Ciudad Juarez, Monterrey, and Guadalajara all exceed 50 $\mu g/m^3$ (Figure 6.3). High concentrations predominate during the months of December through May (Figure 6.4). This period falls within the dry season – cold and dry from November through February and warm and dry from March through June. In Mexico, PM_{10} mass concentrations drop substantially during the wet season, from July through October.

In the United States, the highest annual PM_{10} averages are recorded at sites in California, especially in the San Joaquin Valley and the Imperial Valley. During the period 1995 through 1999, maximum 24-hr PM_{10} mass concentrations ranged from 153 to 279 $\mu g/m^3$ in the San Joaquin Valley and from 229 to 1342 $\mu g/m^3$ in the Imperial Valley, based on sampling once every six days (CARB, 2001). Occasional high daily PM_{10} concentrations occur at many California sites, especially during autumn (e.g., Figure 6.4).

High PM_{10} mass concentrations occur along much of the U.S.-Mexico border, and represent a transboundary pollution issue, with transport likely occurring in both directions. Mexicali (Baja California), Calexico (California), El Paso (Texas), and Ciudad Juarez (Chihuahua) exhibit annual PM_{10} averages in excess of 50 $\mu g/m^3$. The U.S.-Mexico border extends for 3141 km, with over 10 million people living within 100 km on either side of the border (Mukerjee, 2001). This area, which is subject to the 1983 Agreement for the Protection and Improvement of the Environment in the Border Area (La Paz Agreement), is experiencing rapid growth in population and development, especially since the 1993 North American Free Trade Agreement. Ongoing environmental studies are described in Mukerjee (2001) and companion papers. The 1992-93 Imperial/Mexicali Valley Cross-Border PM_{10} Transport Study documented significant (factor of two) PM_{10} mass-concentration differences within an 80 km by 20 km study area (Chow and Watson, 2001).

From 1984 to 1993, mean PM_{10} mass concentrations varied from 11 to 44 $\mu g/m^3$ at the 19 sites of the Canadian National Air Pollution Surveillance (NAPS) Network (Brook and Dann, 1997). Average $PM_{2.5}$ mass was 36 to 68 percent of the PM_{10} mass concentrations, with mean coarse ($PM_{10-2.5}$) mass ranging from 4 to 24 $\mu g/m^3$. Differences in PM_{10} mass concentrations at paired urban-nonurban sites exceeded the differences between regions (Brook and Dann, 1997).

PM_{10} remains a management issue, as exceedances of the U.S. 24-hr PM_{10} standard of 150 $\mu g/m^3$ and the U.S. and Mexican annual-average PM_{10} standards of 50 μg/m^3 occur in both Mexico and the United States (Canada does not have a standard for PM_{10} but is considering a standard for $PM_{10-2.5}$).

PM_{10} mass concentrations can show substantial variations over urban-scale (10 to 100 km) distances. Five-fold variations of average annual PM_{10} mass concentrations are evident over distances of about 100 km or less (Figure 6.3). In urban areas having multiple monitoring sites, average PM_{10} mass may vary by up to roughly a factor of two over distances as small as approximately 10 to 20 km. For example, recent data from southern California, shown in Figure 6.5, reveal two-fold or greater differences in average PM_{10} mass within the metropolitan area.

Mexico City shows twofold variations of annual average PM_{10} concentrations (Figure 6.6). The northeastern monitoring site (Xalostoc; XAL in Figure 6.6) is located in an industrialized area and has historically shown the highest concentrations of total suspended particulate (Vega et al., 2002). The eastern site (Netzahualcoyotl; NET in Figure 6.6) is surrounded by paved and unpaved roads, loose dirt surfaces, and open landfills (Chow et al., 2002). PM_{10} concentrations in the eastern and northeastern areas may be influenced also by resuspended dust from the bed of dry Lake Texcoco, which is 5 km east and 10 km north, respectively, of the monitors at Xalostoc and Netzahualcoyotl (Vega et al., 2002; Chow et al., 2002). Such variations of ambient outdoor PM mass concentrations suggest that people living in different portions of the Mexico City urban area are experiencing substantially different levels of exposure to PM.

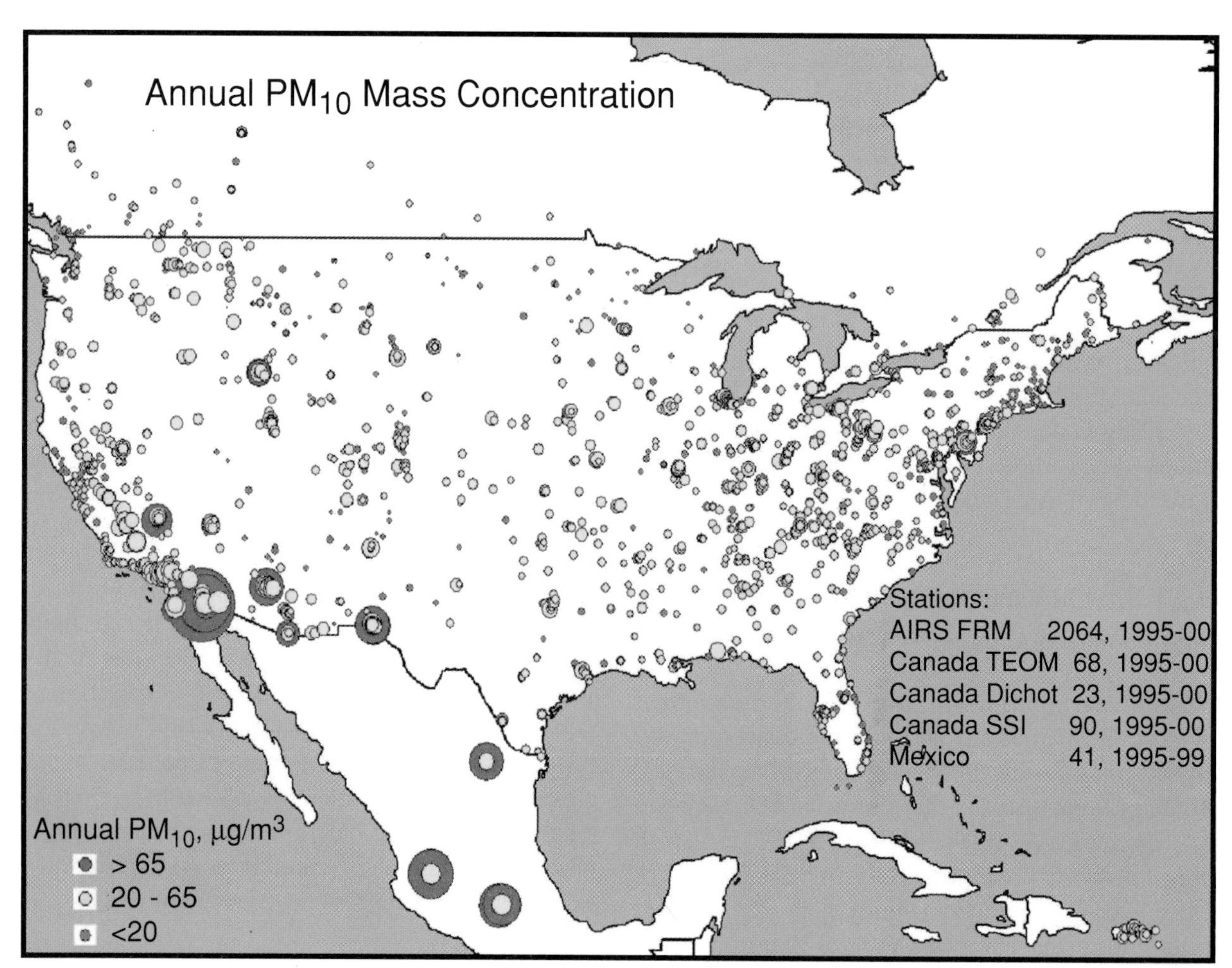

Figure 6.3. Average annual PM_{10} mass concentrations. The U.S. data are from sites in the EPA AIRS database. Canadian data were provided by Environment Canada. PM_{10} data were available for five cities in Mexico. Spot diameter varies in proportion to concentration. (Source: R. Husar, pers. comm.).

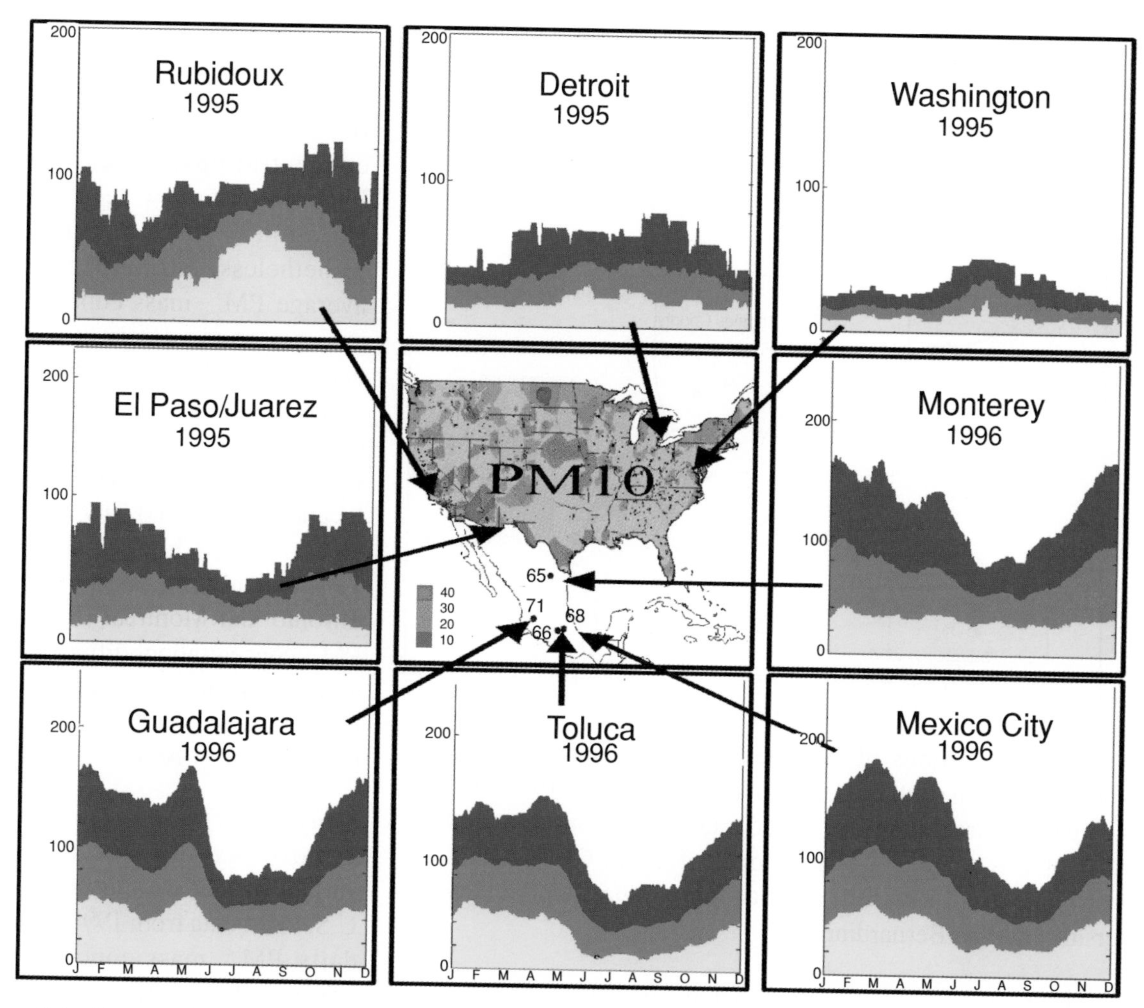

Figure 6.4. Seasonal variation of PM_{10} mass concentrations at selected monitors in the United States and Mexico. The graphs indicate the estimated 20th percentile (yellow), average (green), and 90th percentile (blue) on each date beginning in 1995 or 1996. Data for U.S. locations represent one site; data for Mexican locations include five to eight sites per location. Statistical smoothing procedures were applied to each time series in estimating the percentiles. (Source: R. Husar, pers. comm.).

Spatial variations of PM_{10} mass concentrations are also evident in other urban areas. In Montreal, two long-term monitoring locations have shown mean PM_{10} concentrations differing by 14 μg/m³ (28 compared with 42 μg/m³). A short-term (34-day) study that placed an additional seven monitors at 1 or 3 km distances around the two primary locations revealed daily (24-hr) concentration differences ranging from ~5 to a maximum of 20 μg/m³ (Brook et al., 1999). A special study in California's San Joaquin Valley during December 1995 and January 1996, using 25 monitors in Fresno (population ~480,000) and 12 in Bakersfield (population ~300,000), showed that mean concentrations of PM_{10} and its constituents varied by 10 to 15 μg/m³ (~20 percent of mean values) over distances of 4 to 14 km (Blanchard et al., 1999).

The observed intra-urban variations of PM_{10} and its constituents imply that local (within about 10 km) emission sources often contribute much (i.e., greater than half) of the PM_{10} mass on an annual-average basis; it is possible for local contributions to be greater on shorter time scales (daily or hourly). These variations also imply that annual PM_{10} exposure may vary by a factor of two or more across an urban area, with potentially larger variations occurring in daily exposure. As a result, epidemiological studies run the risk of misclassifying PM_{10} exposures if the

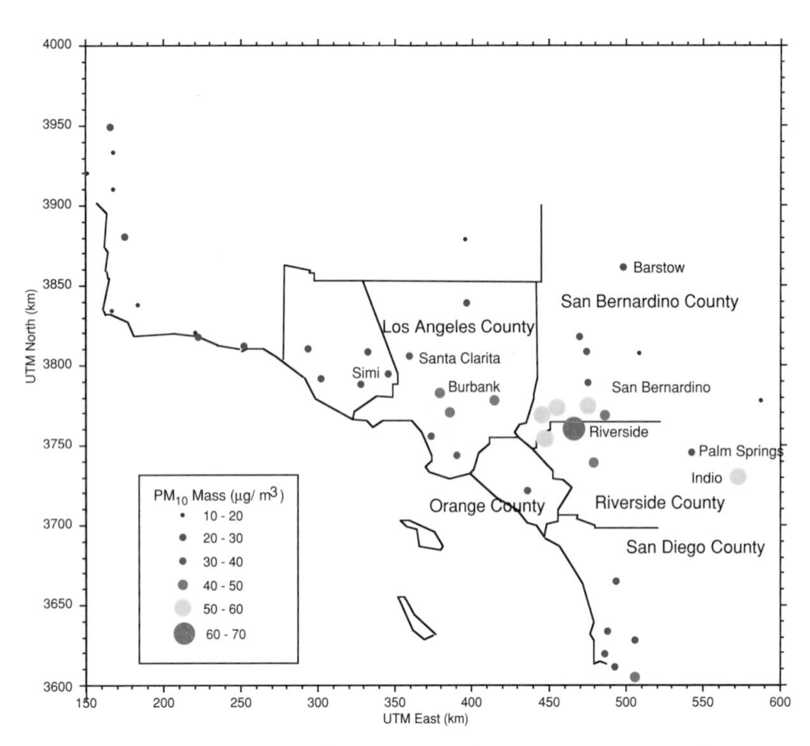

Figure 6.5. Variations of average PM_{10} mass concentrations at monitoring sites in southern California over the years 1997-99. The Los Angeles metropolitan area includes Orange County, the southern portion of Los Angeles County (south of Santa Clarita), and the western portions of Riverside and San Bernardino counties (west of Riverside and San Bernardino).

number of monitors within an urban area is insufficient to represent the spatial variations of outdoor population exposures.

6.2.2 Spatial Patterns of $PM_{2.5}$ Mass

The highest mean annual $PM_{2.5}$ concentrations in North America occur at sites in California and over a broad portion of the eastern, and especially the southeastern, United States (Figure 6.7) (short-term $PM_{2.5}$ data for Mexico City indicate the occurrence of high fine-particle mass concentrations there as well). Throughout North America, both urban and rural monitoring locations may exhibit $PM_{2.5}$ mass concentrations exceeding the national standards of Canada or the United States (Mexico does not currently have a fine-particle standard). Although determination of compliance with PM standards is the responsibility of various regulatory agencies, the ambient air-quality standards established by those agencies provide a useful benchmark for delineating the dimensions of the $PM_{2.5}$ management task.

In comparison with PM_{10} concentrations, annual-average $PM_{2.5}$ levels show less regional variation (Figures 6.7 and 6.3). Nonetheless, intra-urban annual-average $PM_{2.5}$ mass concentration can vary up to a factor of two over distances of about 50 to 100 km, as is the case in Los Angeles (Figure 6.8). Factor of two inter-site differences also occur among locations in Mexico City (Edgerton et al., 1999; Chow et al., 2001). Paired monitoring locations within Toronto and Montreal show mean $PM_{2.5}$ concentrations differing by 2.3 and 4.0 $\mu g/m^3$, respectively (for mean values of ~15 to 20 $\mu g/m^3$) (Brook et al., 1999).

Daily (24-hr) $PM_{2.5}$ concentrations usually exhibit more spatial variability than do annual averages. U.S. FRM data from 1999 show that daily $PM_{2.5}$ mass concentrations were strongly correlated ($r>0.9$) at all pairs of multiple monitoring locations within only 5 of 28 metropolitan areas (Lefohn, 2001). Statistically significant differences between daily $PM_{2.5}$ concentrations occurred up to 50 percent of the time; median intra-urban daily differences ranged from zero to 8.4 $\mu g/m^3$, and maximum daily differences exceeding 30 $\mu g/m^3$ were frequent.

The typically smaller spatial variations of $PM_{2.5}$ mass than PM_{10} mass are consistent with the well-known long residence time of fine PM, which permits transport over distances of 10 to 1000 km and tends to homogenize spatial variations in mass concentrations. As is the case for PM_{10}, epidemiological studies run the risk of misclassifying $PM_{2.5}$ exposures if the number of monitors within an urban area is insufficient to represent the spatial variations of outdoor population exposures. Further attention to network-design requirements is needed to ensure that future monitoring data are adequate to properly characterize exposure.

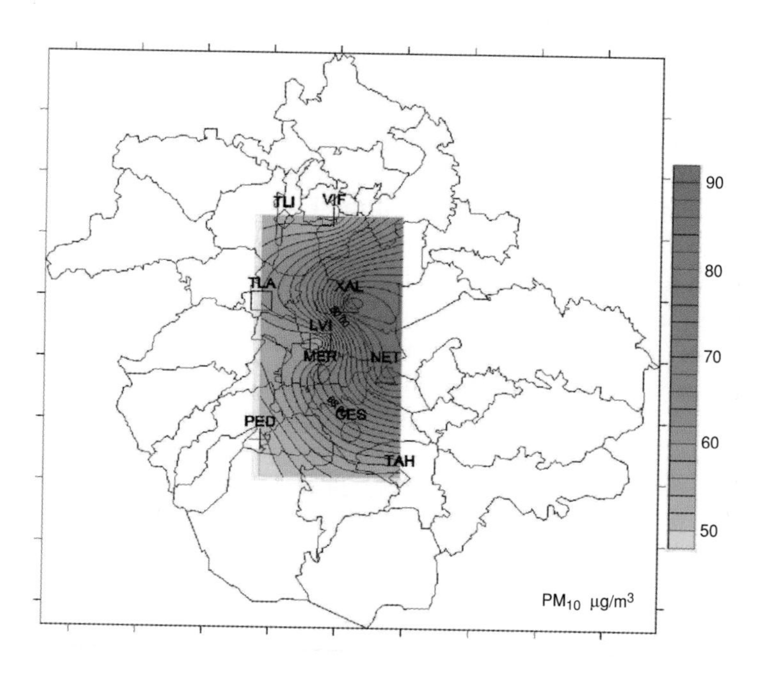

Figure 6.6. Variations of average PM_{10} mass concentrations at monitoring sites in Mexico City over the years 1995 to 2000. The area covered by the monitoring sites is approximately 20 km in the east-west direction by 45 km in the north-south direction. (Source: E. Vega, pers. comm. Summary data are given in Cicero-Fernandez et al., 2001. Spatial distributions of PM_{10} mass in the Mexico City metropolitan area are also presented in Cicero-Fernandez et al., 2001, and Molina and Molina, 2002).

For 1998 and 1999, urban locations from Chicago to Mississippi and Ohio to Georgia showed annual-average $PM_{2.5}$ mass concentrations exceeding the U.S. federal standard of 15 µg/m³, with sites in Tennessee and in northern Alabama, Mississippi, and Georgia exhibiting mean annual $PM_{2.5}$ levels exceeding 18 µg/m³ (Figure 6.7). A substantial inland-to-coast gradient of mean PM concentrations exists in the southeastern United States and monitoring locations near the Gulf or Atlantic coasts exhibit mean annual $PM_{2.5}$ levels less than the U.S. federal standard of 15 µg/m³. The spatial variations indicated for FRM sites in Figure 6.7 are also characteristic of the concentrations reported by the IMPROVE and Clean Air Status and Trends Network (CASTNet) networks, which consist almost exclusively of rural sites.

For the first year of FRM monitoring in 1999, 68 of the 73 FRM samplers in California had data and 33 sites showed annual-average $PM_{2.5}$ mass concentrations exceeding the U.S. federal standard of 15 µg m⁻³ (the exceeding locations recorded annual means ranging from 15.1 to 30.2 µg/m³). All eight FRM monitors with one year of data in the San Joaquin Valley had annual $PM_{2.5}$ averages over 15 µg/m³ (range 19.6 to 27.0 µg/m³), as did 14 of 16 monitors in the Los Angeles metropolitan area (exceeding means ranged from 17.3 to 30.2 µg/m³). Fine-particulate measurements made since 1987 at 8 IMPROVE (or IMPROVE-protocol) sites and 32 sites in three networks operated by the California Air Resources Board provide a long-term record confirming that mean $PM_{2.5}$ mass concentrations in southern California and the San Joaquin Valley typically exceed the annual-average $PM_{2.5}$ standard (Motallebi et al., 2003a; 2003b).

A longer monitoring record is available for Canada, where 11 of 22 monitoring sites in the NAPS exhibited 98th percentile 24-hr $PM_{2.5}$ mass concentrations exceeding the Canada Wide Standard (CWS) during the period 1997 through 1999 (the CWS is 30 µg/m³ expressed as a three-year average of the 98th percentile 24-hr values). All eleven were located in the provinces of Ontario and Quebec, and included nine urban and two rural sites. From Windsor, Ontario, eastward, the three-year mean annual $PM_{2.5}$ concentrations ranged from 4.9 µg/m³ to 13.6 µg/m³ at Toronto-Evans. The four monitoring locations in British Columbia (Vancouver and Victoria) all had three-year mean annual $PM_{2.5}$ concentrations less than 10 µg/m³, and Victoria recorded the minimum 98th percentile 24-hr value of 16.5 µg/m³.

Long-term monitoring data, from which 98th percentile 24-hr values or mean annual concentrations may be computed, are not available for Mexico City. However, data from many short-term studies do exist and indicate the occurrence of some of the highest

$PM_{2.5}$ and PM_{10} concentrations in North America. Recent measurements from an intensive field study during February and March 1997 yielded monthly average $PM_{2.5}$ concentrations ranging from 21.6 to 50.0 μg/m^3 at six monitoring locations within Mexico City (Edgerton et al., 1999). The maximum 24-hr $PM_{2.5}$ concentrations ranged from 33.9 to 183.7 μg/m^3. While Mexico has no fine PM standard, Mexico's 24-hr PM_{10} standard of 150 μg/m^3 was exceeded at one or more sites on 7 of the 28 days sampled. The $PM_{2.5}$ mass concentration was typically about half the PM_{10} mass concentration.

In five of the seven geographical regions shown in Figure 6.9, the 98th percentile 24-hr $PM_{2.5}$ mass concentrations at the majority of urban U.S. and Canadian monitoring sites exceeded 30 μg/m^3 (the level of the Canada Wide Standard). In three regions, California, the southeastern United States, and the Ohio Valley – Great Lakes states, annual-average $PM_{2.5}$ mass concentrations at about half the urban sites exceeded the U.S. three-year average annual mean standard of 15 μg/m^3 (Figure 6.9). With the exception of California, few sites had 98th percentile 24-hr $PM_{2.5}$ exceeding the level specified by the U.S. 24-hr standard (65 μg/m^3). Thus, in the United States, achieving the annual $PM_{2.5}$ standard will be the focus of the regulatory effort at most sites, whereas in Canada the regulatory effort will necessarily be oriented to the 24-hr Canada Wide Standard. In California, PM management needs to address both the U.S. annual and 24-hr standards.

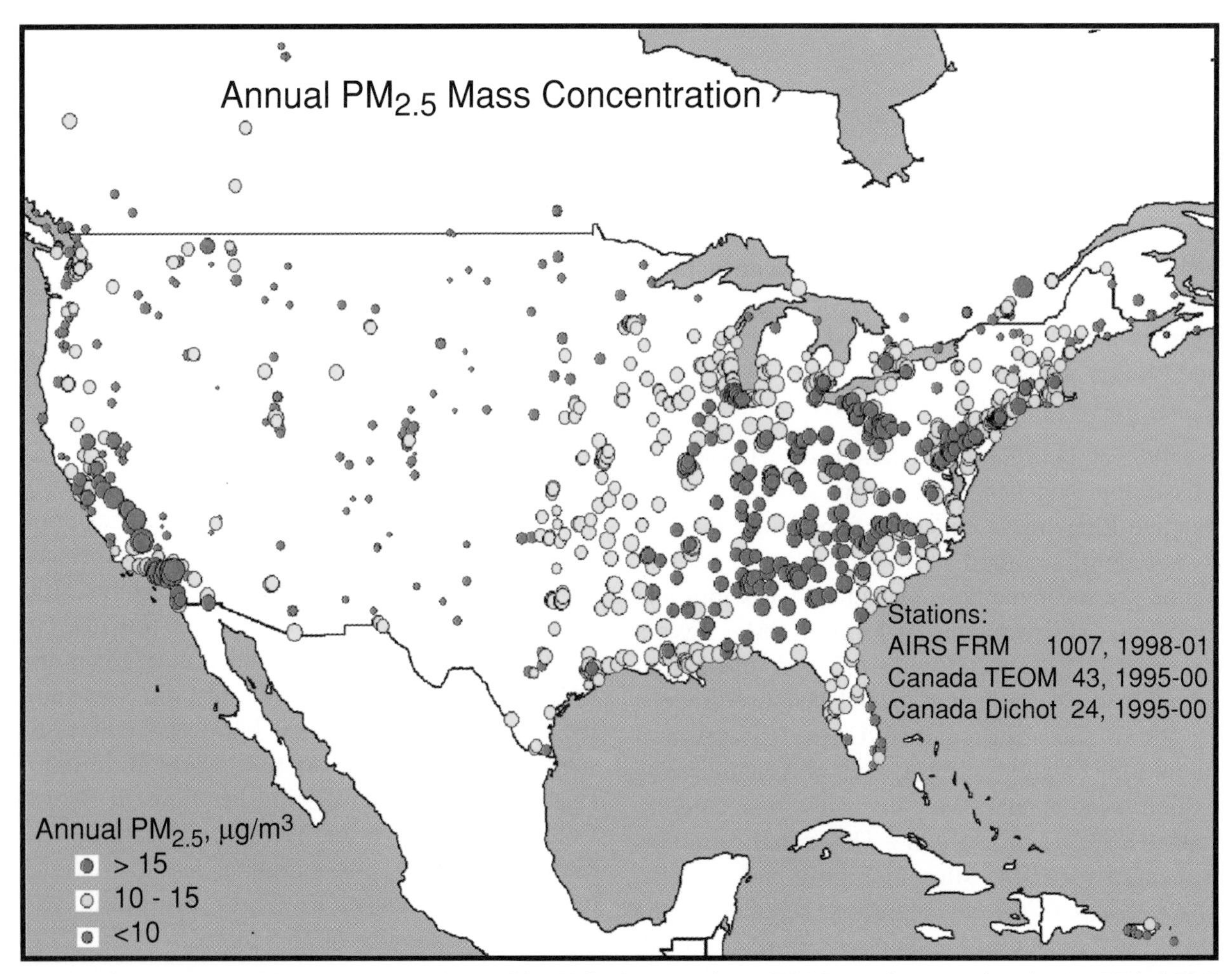

Figure 6.7. Average $PM_{2.5}$ concentrations. The U.S. data are from FRM monitors at sites in the EPA AIRS database for July 1998 through July 2000. Canadian data are from TEOM and dichotomous samplers operating from 1995 through 2000. The currently available data from sites in Mexico represented less than one year of sampling and were excluded from the computation of annual averages. Spot diameter varies in proportion to concentration. (Source: R. Husar, pers. comm.).

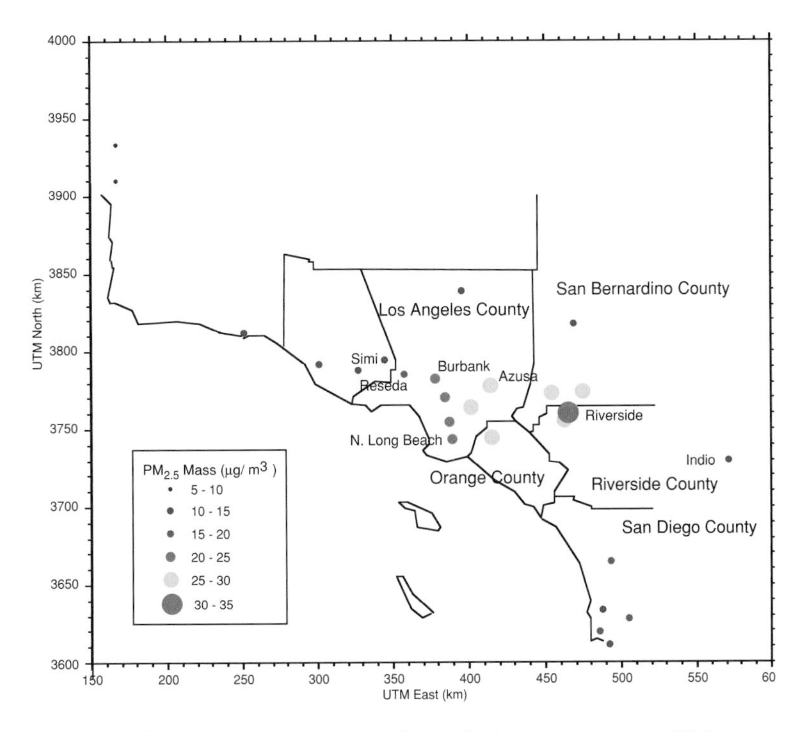

Figure 6.8. Example urban-scale variations of average $PM_{2.5}$ mass concentrations in southern California. The data are the 1999 average concentrations from all monitoring locations having at least 11 months data. (Source: data provided by California Air Resources Board).

$PM_{2.5}$ concentration measurements typically exhibit strongly skewed frequency distributions, dominated by a large number of low values and including a smaller number of high concentrations (Figure 6.10). At locations where exceedances of national 24-hr or annual-average $PM_{2.5}$ standards occur, numerous episodes of elevated PM mass concentrations contribute. High concentrations do not occur on the majority of days, but neither are they rare events: as shown in Figure 6.10, the ten-year record at 15 California sites shows 2.7 percent of the samples exceeding the U.S. 24-hr standard of 65 $\mu g/m^3$. Similarly, during the months of June through September at NAPS sites in Ontario, approximately 5 to 15 percent of the days exhibit 24-hr $PM_{2.5}$ concentrations exceeding 30 $\mu g/m^3$. For skewed pollutant distributions, such as shown in Figure 6.10, the arithmetic average and the median are not identical. The difference between these two measures of the central tendency of the distribution, with the mean being greater than the median, indicates the influence of the relatively fewer high-concentration samples on the average. Particle-management strategies that reduce the frequency and magnitude of high PM mass concentrations will help reduce the annual mean, though in many instances (e.g., California sites), elimination of the highest concentrations will not suffice to bring the annual averages below the U.S. annual standard.

6.2.3 Seasonal Variations of $PM_{2.5}$ Mass

Most monitoring locations exhibit seasonal variations of $PM_{2.5}$ mass, as well as some or all of its major constituents. In much of eastern North America, higher $PM_{2.5}$ concentrations tend to occur during the summer (Figure 6.11). Winter stagnation episodes also occur in eastern North America, but less often than during summer. For example, during the period 1994-98, 6 percent of February days and 16 percent of August days exceeded the Canada Wide Standard at sites in Ontario (Vet et al., 2001). The seasonal variation of the average PM values may differ from the seasonal variations of the highest concentrations (Figure 6.11).

In eastern North America, SO_2 emissions are greater than elsewhere (Chapter 4), and $SO_4^=$ is a substantial component of $PM_{2.5}$, especially during summer. Higher insolation and humidity during summer months enhance both homogeneous and heterogeneous reactions that produce secondary $SO_4^=$ particles. Photochemical reactions that produce particulate organic carbon from gas-phase precursors also proceed at faster rates during summer months, though evidence for significant secondary organic PM formation is inconclusive at present.

Historical analyses of visual range indicate that haze levels in the eastern United States changed from seasonal maxima in winter during the period 1948 through 1954 to seasonal maxima in summer during 1975 through 1983 (Husar and Wilson, 1989). These patterns, with different rates of change occurring in

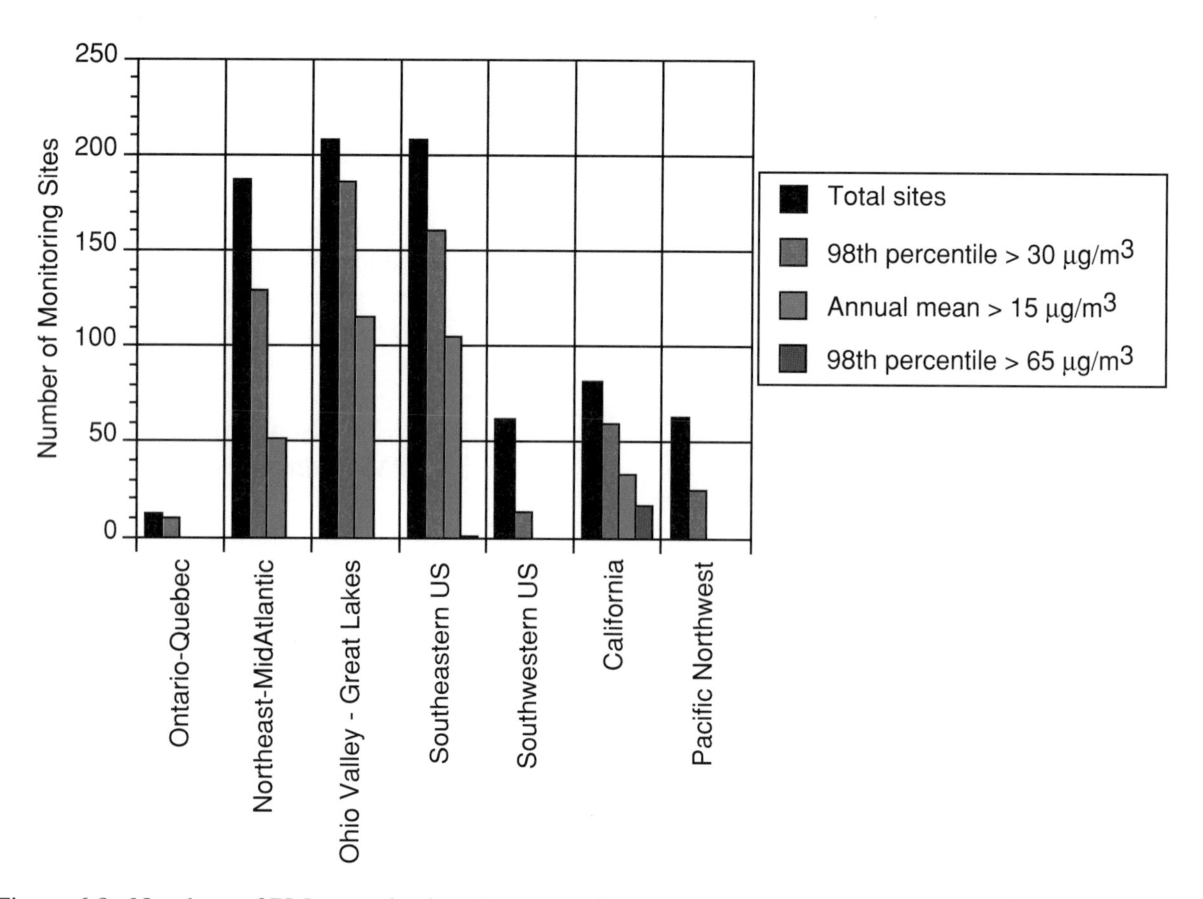

Figure 6.9. Numbers of $PM_{2.5}$ monitoring sites exceeding three benchmark levels. The reference levels are the CWS of 30 µg/m³ expressed as a three-year average of the 98th percentile 24-hr values, the U.S. three-year average annual mean standard of 15 µg/m³, and the U.S. 24-hr standard of 65 µg/m³ expressed as a three-year average of the 98th percentile 24-hr values. Data from Canada are for the period 1997 through 1999 (1995 through 1997 at two sites) and are measurements made using dichotomous samplers operated for 24 hours once every six days. Data from the United States are for 1999 and 2000 and are measurements from FRM samplers operated for 24 hours once every one, three, or six days. Site averages and completeness levels were compiled by the U.S. EPA. Sites were included if completeness levels indicated at least 30 days per year (50 percent sampling completeness for sampling schedules of once every six days). The standard U.S. EPA reporting protocol of 75 percent completeness eliminates many of the sites. The Northeast-MidAtlantic region includes all monitors in the states from Maine through Maryland and the District of Columbia. The Ohio Valley - Great Lakes region includes all monitors in West Virginia, Kentucky, Ohio, Michigan, Indiana, Illinois, and Wisconsin. The Southeastern U.S. region includes monitors located in Virginia, North Carolina, South Carolina, Tennessee, Georgia, Florida, Alabama, Mississippi, and Louisiana. The Southwestern U.S. region includes monitors in New Mexico, Arizona, Colorado, Utah, and Nevada. The Pacific Northwest region includes monitors in British Columbia, Washington, and Oregon.

the northeastern and southeastern states, corresponded to a historical shift from higher emissions of SO_2 during winter to roughly comparable emission levels during winter and summer (Husar and Wilson, 1989).

PM mass concentrations tend to be higher during winter in many parts of California (basins with poor ventilation), though this seasonal pattern may not occur in areas removed from population centers (Figure 6.11). In Los Angeles and the San Joaquin Valley, $PM_{2.5}$ concentrations are highest during the autumn or winter months of October through February, with 50 to 100 percent of the highest PM mass concentrations occurring during those months.

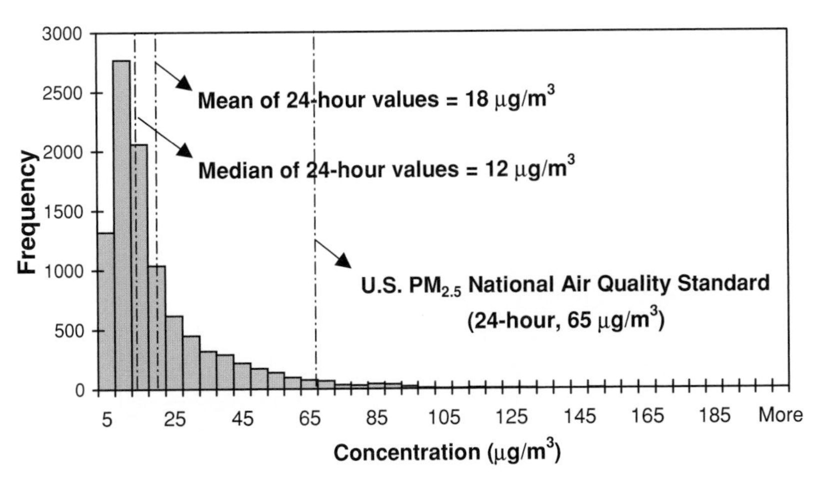

Figure 6.10. Statistical distribution of 24-hr $PM_{2.5}$ concentrations at 20 sites in California over the period 1988 through 1998. The measurements were made using dichotomous samplers. The mean and median of the data are shown. The frequency counts are labelled using the upper value of each concentration interval. The maximum value was 195 µg/m^3 and 25 measurements exceeded 100 µg/m^3. (Source: California Air Resources Board; Motallebi et al., 2003a; 2003b).

Measurements of $PM_{2.5}$ made by the California Air Resources Board from 1988 through 1994 at Yosemite and Sequoia National Parks in the Sierra Nevada, east of the San Joaquin Valley, show that the park locations experienced higher summer levels of $PM_{2.5}$ and lower winter concentrations, a pattern opposite to that of all the valley sites. Since the summer $PM_{2.5}$ mass concentrations at the two parks were of comparable magnitude to summer concentrations at urban locations such as Sacramento and Bakersfield, the data indicate different degrees of coupling of the San Joaquin Valley and the western slope of the Sierra Nevada during summer and winter. The lower winter inversion height inhibits dispersion within the valley, resulting in higher PM mass concentrations there. Similarly, the IMPROVE monitoring site at San Gorgonio National Monument, east of Los Angeles, also exhibits summer PM maxima, when dispersion within and from the Los Angeles basin is more limited than during summer. In addition, NH_3 concentrations at mountain sites may be lower than in the San Joaquin Valley or the Los Angeles area, affecting NO_3^- concentrations.

Throughout California, $SO_4^=$ is a lesser fraction of total PM and NO_3^- is greater than at eastern locations. Nitrate exists in a quasi-equilibrium with its gas-phase precursors, and the equilibrium favors the particulate species under cool, moist conditions, such as prevail during winter in California. Meteorological factors also play a role in favoring winter PM maxima. Winter is the rainy season in California, but between frontal passages, dispersion is weaker during winter than summer. Cooler temperatures, lower wind speeds, and lower inversion heights result in weaker mixing, and pollutants tend to remain within the confines of enclosed air basins such as the Los Angeles area and the San Joaquin Valley. In the San Joaquin Valley, winter PM mass concentrations typically build up over several days and valley-wide episodes occur. Such episodes are associated with high pressure over the southwestern United States, restricted vertical mixing, weak or offshore pressure gradients, flow reversals, and net transport speeds of about 1 m/s. Fog also plays a role in PM accumulation in the San Joaquin Valley.

Many other urban locations (e.g., Phoenix, Tucson, Salt Lake City, Denver) in the western United States exhibit seasonal patterns with higher winter PM concentrations. As for urban California sites, limited dispersion during winter months is a contributing factor. In some locations residential wood combustion may also play a role.

Mexico lacks a long-term record of fine PM mass, but as previously noted PM_{10} mass concentrations are lowest during summer, the rainy season (Figure 6.4). The seasonal variations of PM_{10} mass are related to rainfall.

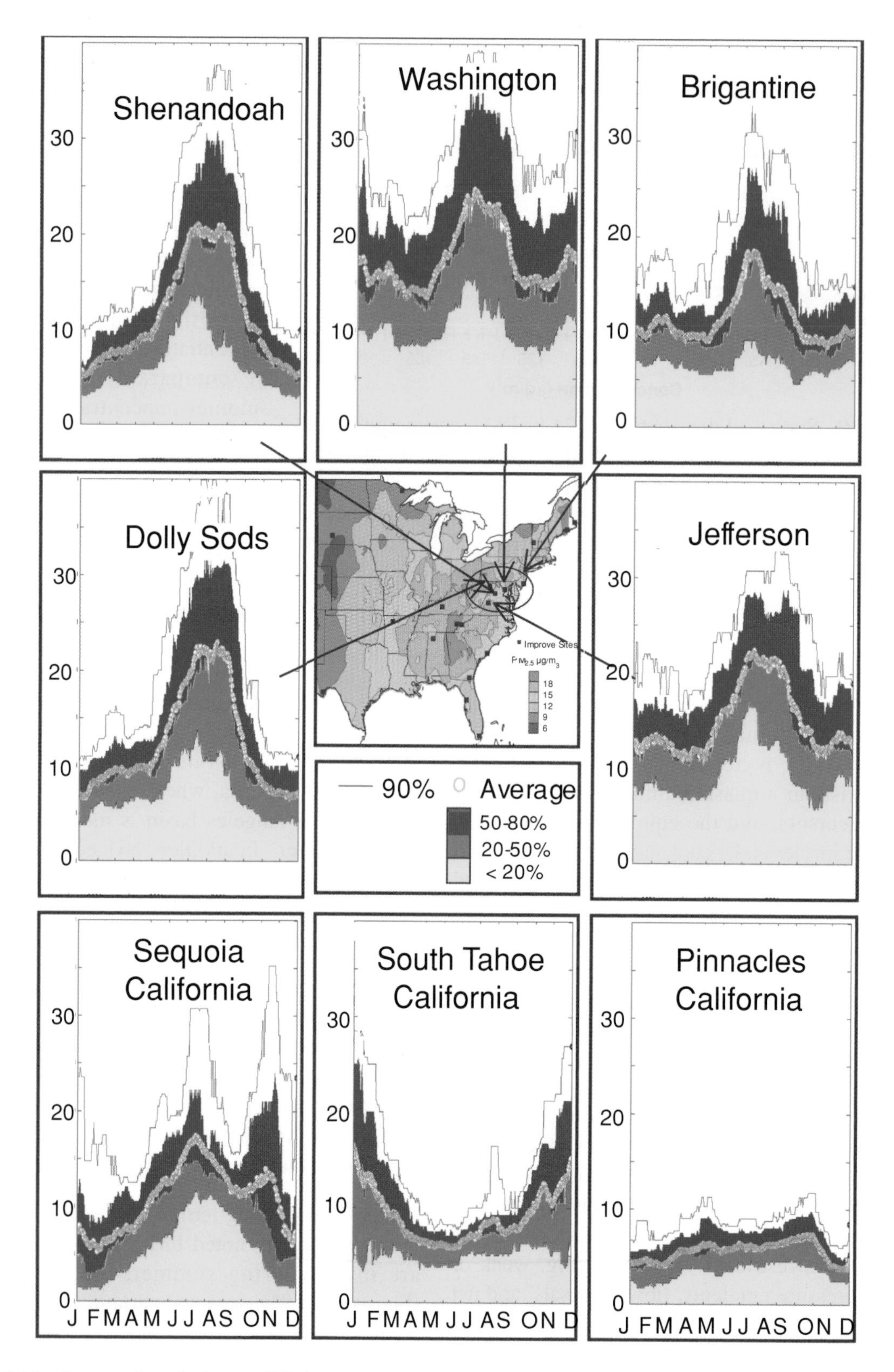

Figure 6.11. Seasonal variations of $PM_{2.5}$ mass concentrations at selected IMPROVE monitoring locations (1992-1999) for the cleanest (20 percent), moderate (20-50 percent), and highest PM (50-80 percent and 90 percent) days. The units of measurement are $\mu g/m^3$. (Source: R. Husar, pers. comm.).

6.2.4 The Composition of $PM_{2.5}$ and Its Geographical Variation

The average composition of $PM_{2.5}$ varies considerably among geographical regions. In eastern North America, $SO_4^=$ typically constitutes about one-fifth to one-half the average annual $PM_{2.5}$ mass concentration (Figure 6.12). Along with the associated NH_4^+, roughly one-quarter to over one-half the average fine mass is $(NH_4)_2SO_4$ in eastern North America. Maximum North-American $SO_4^=$ levels occur in the Ohio River Valley, mid-Atlantic states, and the southeastern United States (Figure 6.13). This area is largely coincident with the area of maximum anthropogenic SO_2 emissions (Chapter 4, Figure 4.3).

Besides $SO_4^=$, the components most affecting geographical variations in $PM_{2.5}$ mass concentration are carbon and NO_3^-. Carbon and NO_3^- concentrations are, in turn, affected by geographical variations in emissions and in the influences of atmospheric processes on secondary particle formation. Organic carbon mass, which includes carbon and associated elements such as oxygen and hydrogen, constitutes roughly one-fifth to one-half the average annual $PM_{2.5}$ mass at monitoring locations throughout North America (Figure 6.12). Mexico City shows higher OC concentrations than do other North American cities, with nearly a factor of two difference in $PM_{2.5}$ mass concentration among different locations within the city (Figure 6.12). Although the data shown are from a single month, other short-term studies confirm the importance of organic species as contributors to total $PM_{2.5}$ in Mexico City. For example, organics were 22 ± 2 percent of the mean $PM_{2.5}$ mass concentration of 42 ± 4 µg/m³ during September 1990 and 47 ± 2 percent of the mean $PM_{2.5}$ mass concentration of 39 ± 6 µg/m³ during February 1991 (Miranda et al., 1994).

Fires, both as wildfires and controlled biomass combustion, are an important and geographically dispersed source of OC during summer, especially in Mexico, the western United States, and western Canada. Fireplace burning contributes significantly to PM in the winter in western North America. The relative importance of secondary OC formation from gas-phase VOC precursors is not well understood and may be more significant in the southeastern United States than in other areas.

Nitrate concentrations are greatest at California sites, representing over one-quarter of the average annual $PM_{2.5}$ mass there (Figure 6.12). At sites in California's San Joaquin Valley, NO_3^- accounts for ~30 percent of the annual-average $PM_{2.5}$ particle mass (Figure 6.12). In Los Angeles, NO_3^- also accounts for ~30 percent of the annual-average $PM_{2.5}$ particle mass at sites in the western and central portions of the basin (Figure 6.12), but this fraction increases to ~40 percent in the eastern basin. Annual-average NO_3^- concentrations in the two areas range from about 6 to 16 µg/m³, with daily-average NO_3^- levels reaching as high as 100 µg/m³ on some days in the eastern Los Angeles basin. The NH_4^+ associated with NO_3^- typically accounts for ~10 to 20 percent of the annual-average $PM_{2.5}$ particle mass in the San Joaquin Valley and southern California.

Nitrate is also a significant component of the annual-average $PM_{2.5}$ mass concentration at some locations in southeastern Canada and the northeastern United States (Figure 6.12). Nitrate levels in eastern North America tend to be higher in urban than in rural locations (section 6.3).

Nitrate derives partially or predominantly from the equilibrium reaction between two gas-phase species, HNO_3 and NH_3 (Chapter 3). The equilibrium favors the condensed phase at lower temperatures and higher humidities, so NO_3^- concentrations are generally higher during winter months.

Ammonia reacts preferentially with H_2SO_4, and, if sufficient NH_3 is available, it then combines with HNO_3 to form NO_3^-. In California, $SO_4^=$ concentrations are typically lower than in many other parts of the country (Figure 6.12), and sufficient NH_3 is available at most times and locations to allow the formation of NH_4NO_3 (Blanchard et al., 2000). In contrast, the availability of NH_3 limits NO_3^- formation during most times at sites in the southeastern United States (Blanchard and Hidy, 2003).

Since the availability of NH_3 typically does not limit the amount of NO_3^- occurring at most California locations, the amount of NO_3^- that forms depends upon the amount of HNO_3. This amount in turn depends upon the rate of conversion of NO_2 to HNO_3.

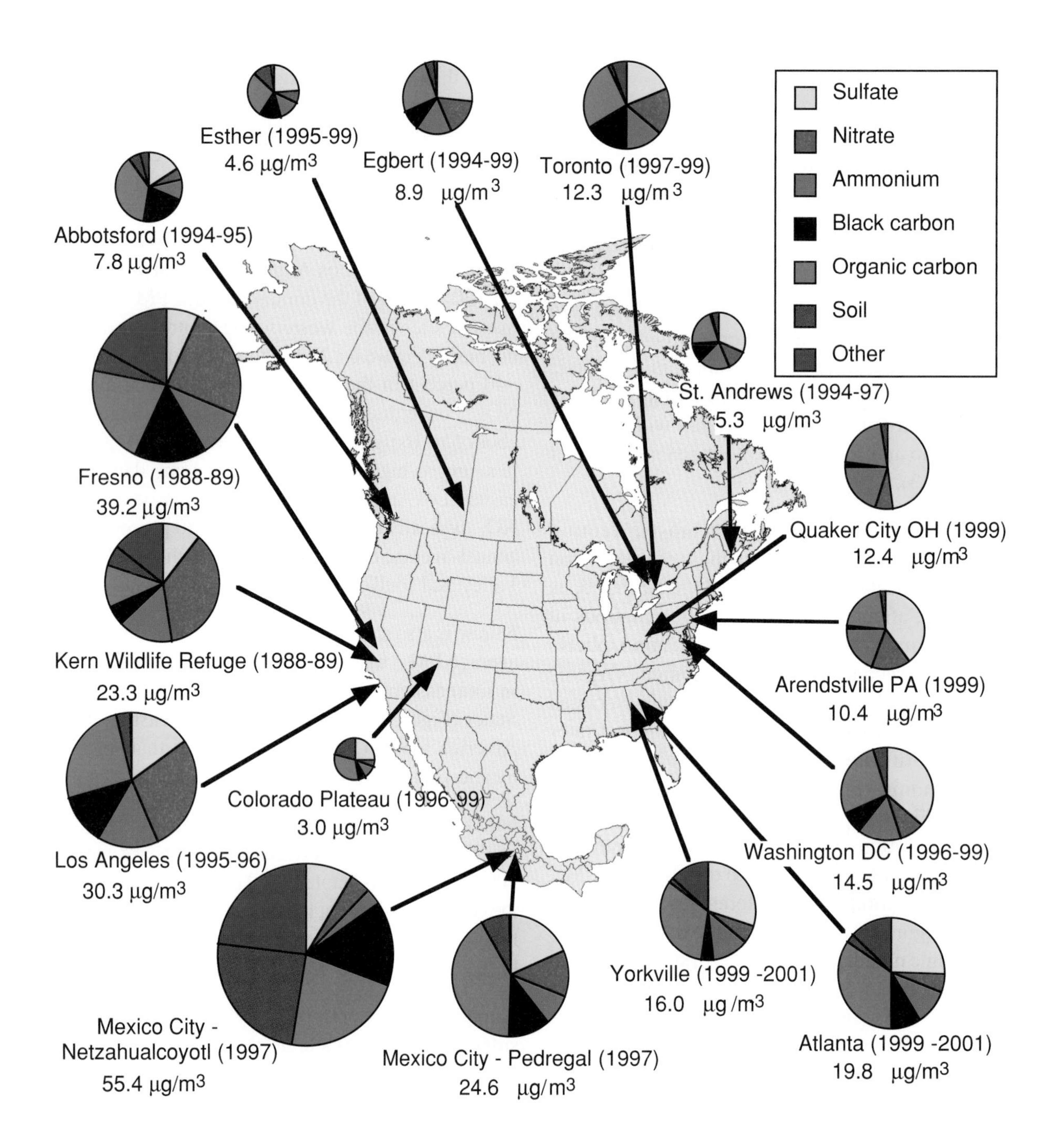

Figure 6.12. Composition of $PM_{2.5}$ at representative urban and rural locations. The urban sites are Toronto, Washington DC, Atlanta, Mexico City, Los Angeles, and Fresno. Averaging periods and average $PM_{2.5}$ mass are indicated. All sites have at least one year of sampling except Mexico City, for which the average was determined from one month (14 days). More recent short-term measurements from December 1995 and January 1996 at Fresno and Kern Wildlife refuge show lower $PM_{2.5}$ mass concentrations but similar composition to the data displayed here. The Colorado Plateau data are the averages of the IMPROVE sites located at Bryce Canyon, Canyonlands, Grand Canyon, Petrified Forest, Mesa Verde, and Zion National Parks. (Source: Vet et al., 2001; Malm, 2000; Hansen et al., 2003; Chow et al., 1993; U.S. EPA, 2000; Kim et al., 2000).

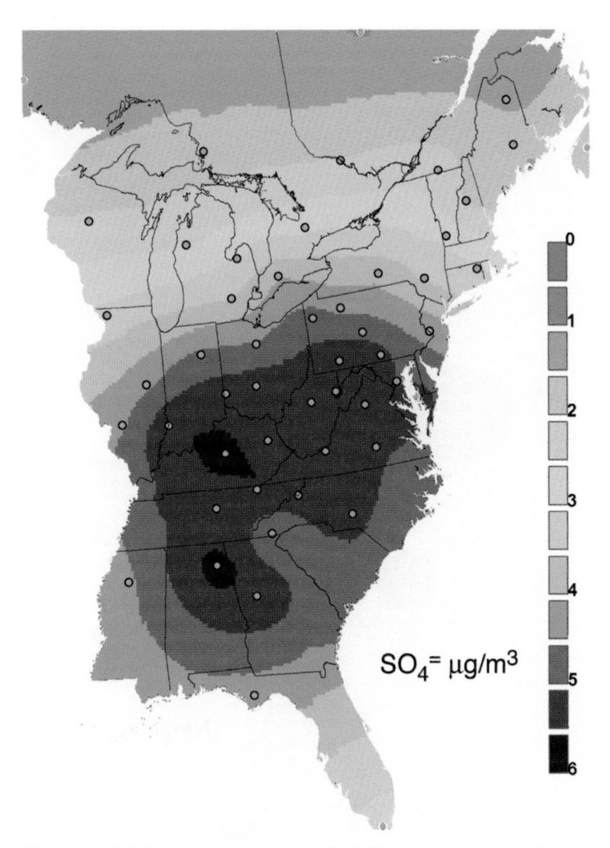

Figure 6.13. Mean annual $SO_4^=$ concentration in eastern North America during 1997-98. Measurements are from CASTNet and CAPMoN sites. (Source: J. Narayan, Environment Canada).

During winter in the San Joaquin Valley, NO_2 concentrations typically exceed concentrations of HNO_3 (Kumar et al., 1998), implying that the rate of conversion of NO_2 to HNO_3 is a key determinant of ambient concentrations of HNO_3, and, hence, of NO_3^- levels. Initial modeling efforts suggest that in one of the urban areas of the San Joaquin Valley, HNO_3 formation is limited by the availability of radical species, and NO_3^- formation may be more effectively reduced through reductions of VOC than NO_x emissions (Pun and Seigneur, 2001). Outside the urban areas, however, HNO_3 formation in the San Joaquin Valley may be responsive to NO_x emissions (Stockwell et al., 2000), especially over multiday stagnation episodes. *These findings have significant implications for the management of PM, and additional research efforts are needed for fully understanding the implications.*

While summer NO_3^- concentrations in eastern North America are low in comparison with $SO_4^=$ levels, higher winter NO_3^- concentrations occur in some areas. On an annual basis, NO_3^- levels in eastern North America show a geographical pattern that closely parallels the NH_4^+ pattern (Figure 6.14), except that NO_3^- levels are lower where $SO_4^=$ concentrations are highest (Figure 6.13). Some analyses suggest that the availability of NH_3 may limit NO_3^- levels in much of eastern North America during summer months, with more NH_3 available during winter months when $SO_4^=$ levels are lower (Vet et al., 2001). The response of NO_3^- concentrations to reduced SO_2 emissions and lower $SO_4^=$ concentrations is of importance for the management of PM concentrations in eastern North America. *Further research efforts are needed for fully characterizing the relations between $SO_4^=$ and NO_3^- concentrations in eastern North America.*

Trace elements, such as antimony, arsenic, beryllium, cadmium, cobalt, chromium, iron, lead, manganese, mercury, nickel, selenium, and zinc, typically represent a portion of the $PM_{2.5}$ mass that is too small to be apparent in illustrations such as Figure 6.12. While they do not contribute much to mass, such species are of special interest because current research has not yet determined which, if any, PM components are related to health effects, and because such trace elements are designated as hazardous pollutants by one or more North American governments and are useful in PM studies for apportionment.

The principal anthropogenic sources of most of the trace elements are industrial operations. Antimony, arsenic, cadmium, chromium, lead, nickel, and zinc may be present in the plumes of smelters. Zinc is also present in emissions from waste combustion. Iron is typically associated with foundry emissions. Manganese, mercury, and selenium are found in emissions from coal-fired power plants; manganese and selenium are also crustal elements. Mercury is emitted as a vapor (elemental mercury) or as particles (mercury compounds). Selenium tends to follow transport paths similar to sulfur, being emitted as gas-phase SeO_2 in hot stack gases; SeO_2 is condensable at ambient temperatures.

Monitoring data typically show larger intersite differences in trace-element concentrations than in $PM_{2.5}$ mass concentrations. Median concentrations of the most abundant trace elements vary by factors

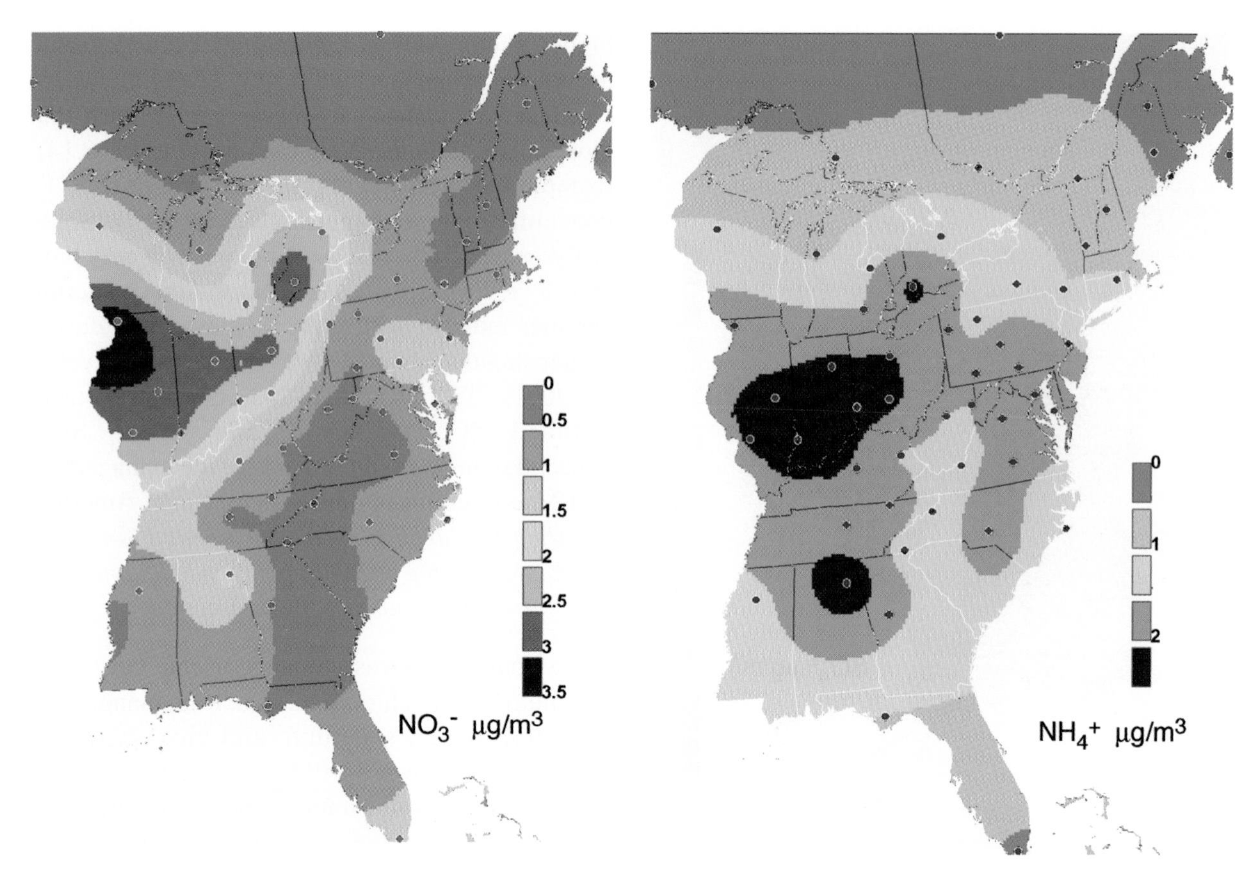

Figure 6.14. Mean NO_3^- and NH_4^+ concentrations during 1997-98. Measurements are from the CASTNet and CAPMoN networks. (Source: J. Narayan, Environment Canada).

of two to four among Canada's NAPS sites, reflecting differing source influences (Brook et al., 1997). Within Montreal, median iron, lead, and zinc concentrations show twofold differences. Proximity to oil-combustion sources affects vanadium and nickel concentrations at the Halifax site, and the higher concentrations of selenium in Toronto and Montreal than at other NAPS sites appear linked to coal-combustion sources (Brook et al., 1997).

In California, factor-of-two or greater variations occur among air-toxic monitoring locations within single metropolitan areas. The five sites located in the San Francisco Bay area show mean multi-year (1990-99) averages of 0.5 to 0.9 ng/m^3 for arsenic, 10.5 to 18.8 ng/m^3 for lead, and 10.1 to 19.1 ng/m^3 for manganese. The six sites located in Los Angeles area show mean multi-year averages of 0.8 to 5.3 ng/m^3 for arsenic, 20.9 to 47.0 ng/m^3 for lead, and 26.1 to 60.5 ng/m^3 for manganese. The two monitoring sites in Fresno (San Joaquin Valley) recorded 1.4 to 7.2 ng/m^3 for arsenic, 14.0 to 106 ng/m^3 for lead, and 28.2 to 54.0 ng/m^3 for manganese. These geographical variations within and between urban areas have not been thoroughly examined, particularly with respect to their implications for spatial variations of personal exposure to hazardous air pollutants.

Two other networks presently supplying trace-element monitoring data from locations in the United States and Canada are the Integrated Atmospheric Deposition Network (IADN) and the Mercury Deposition Network (MDN), which is a component of the National Atmospheric Deposition Program (NADP). These networks provide estimates of wet and dry deposition of lead, cadmium, and mercury, as well as organic compounds. The average wet deposition of mercury reported by the MDN is about 10 µg/(m^{2}y), with higher precipitation concentrations and deposition amounts during summer than winter, and a low-to-high deposition gradient from New England and eastern Canada (lowest) to southern Florida (highest) (Sweet et al., 1999).

With the exception of lead, virtually no long-term monitoring measurements exist with sampling frequencies sufficient for providing good comparisons over time. The reduction of ambient lead levels between 1970 and the present, in large part due to removing lead from gasoline, has been one of the principal successes of North American air-pollution management programs. Since about 1990, for example, virtually all U.S. monitoring locations have reported ambient lead levels below the EPA ambient air-quality standard for lead (U.S. EPA, 2000b).

6.3 REGIONAL AND URBAN CONTRIBUTIONS TO PM

6.3.1 Comparisons Between Rural and Urban Sites

Differences between PM mass concentrations at urban and rural monitoring sites help indicate the relative magnitudes of local and regional contributions. In general, the regional contribution includes PM transported from various urban areas or generated in nonurban settings, as well as nonanthropogenic background. Quantitative comparisons usually require pairing of an urban and a nearby rural monitor operating with the same sampling schedule over the same time period; the nonurban monitor should be upwind of the urban area for the most commonly occurring wind directions. In eastern North America, urban-rural differences imply that typical average local urban contributions are roughly 25 percent of the mean urban $PM_{2.5}$ mass concentrations, with regional PM contributing the remaining, and larger, portion (Figure 6.15). On average, $SO_4^=$ and OC are strongly regional in eastern North America, with ~75 to 95 percent of the urban $SO_4^=$ concentrations and ~60 to 75 percent of the urban OC concentrations occurring at the nonurban sites (Figure 6.15). In Los Angeles, PM mass, $SO_4^=$, NO_3^-, and BC contributions are strongly local; the mean $SO_4^=$ concentrations at the upwind (offshore) site were about one-third of the urban concentrations (a recirculation pattern that moves air from the mainland offshore and back onshore at a later time is known to transport $SO_4^=$ from the city to the upwind location at times).

The magnitude of the regional component of PM varies among regions. Data from the IMPROVE network indicate that mean annual $PM_{2.5}$ levels exceed 10 μg/m^3 at rural locations in much of the southeastern United States. Rural locations in other portions of the eastern United States, California, Ontario, and Quebec exhibit mean annual $PM_{2.5}$ levels of 5 to 10 μg/m^3. It will be difficult to meet PM standards in urban areas where the regional concentration is at present close to the standards, and regional PM management programs may be required.

Nitrate concentrations in eastern North America are higher at urban than at rural locations (Figure 6.16). This comparison indicates the potential importance of NO_3^- within urban areas, and implies that one or both of the gas-phase precursors of NO_3^- (NH_3 and HNO_3) are present at higher concentrations in urban areas (Chapter 3 and section 6.2.3). Since the sampling devices used for collecting the urban samples are known to underestimate NO_3^- concentrations, due to losses caused by volatilization, the actual ambient NO_3^- levels at the urban locations are even greater than those shown in Figure 6.16. PM management strategies in eastern North America may therefore need to address urban-scale NO_3^- formation.

6.3.2 Evidence for Local PM Sources: Temporal Variations

Significant variations in PM mass concentrations can occur on time scales of a few days, a few hours, or even a few minutes. Such short time-scale variations result from meteorological or emission changes occurring over relatively short periods or distances, and provide additional insights into the relative magnitudes of local and regional PM contributions.

At most urban monitoring locations, a weekly cycle is apparent in either PM_{10} or $PM_{10\text{-}2.5}$ mass concentrations (Figure 6.17). The weekly periodicity is attributable to weekly variations of anthropogenic emissions, and provides one indication of the magnitudes of PM_{10} or $PM_{10\text{-}2.5}$ mass contributions

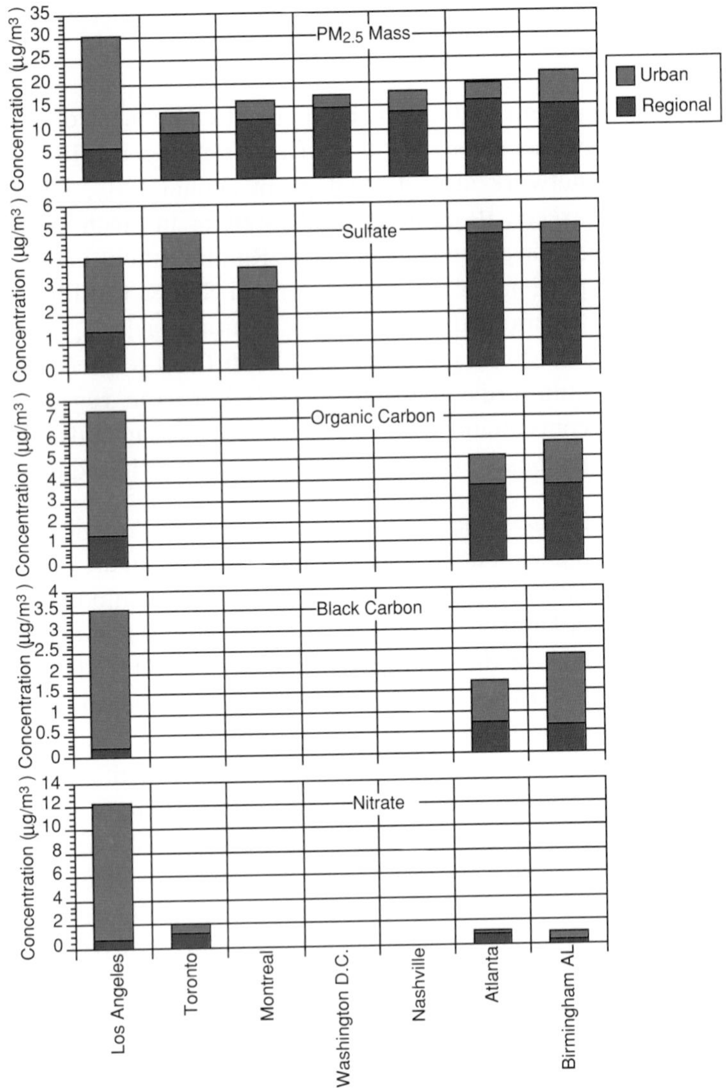

Figure 6.15. Comparisons of average $PM_{2.5}$ mass and species concentrations at paired urban and rural locations. The mean urban concentration is given by the sum of the estimated urban and rural contributions. Not all species were measured at all sites. (Source: Brook et al., 1999; Kim et al., 2000; Hansen et al., 2003).

from relatively local emission sources. Nonurban monitoring sites generally do not show prominent weekly PM_{10} or $PM_{10-2.5}$ mass cycles (Figure 6.17). Weekly cycles of $PM_{2.5}$ concentrations tend to be less evident than weekly cycles of PM_{10} concentrations, reflecting the larger fractions of secondary species and regional contributions to the $PM_{2.5}$ mass.

Diurnal patterns of PM mass concentrations provide even stronger evidence of the influence of local PM sources on PM_{10} concentrations in some areas. For example, sites in Mexico City show diurnal concentration changes consistent with the patterns of human activities (Figure 6.18). In contrast, diurnal $PM_{2.5}$ variations do not always provide evidence for local source contributions. However, local source influences on $PM_{2.5}$ mass concentrations were evident during a special study from December 1995 through January 1996 in California's San Joaquin Valley. $PM_{2.5}$ concentrations were lowest from about 3:00 a.m. to 3:00 p.m. at two urban sites, increased during late afternoon, and persisted at their highest levels through midnight, whereas two rural sites showed the reverse pattern (Chow et al., 1999). The diurnal profiles of $SO_4^{=}$, NO_3^{-}, and organic and elemental carbon showed that the variation of the PM mass at the urban sites was driven by the variation of organic and elemental carbon, which appeared to be strongly influenced by local (~20 km scale) residential wood combustion.

6.3.3 PM Mass Concentrations at Remote Locations

The magnitude of nonanthropogenic background PM mass concentrations is difficult to quantify because anthropogenic influences extend even to remote monitoring sites. Moreover, background levels vary among locations and seasons. Surface monitoring at global background locations provides a long-term record of surface-level PM concentrations. Data are available from several monitoring locations within the Global Atmosphere Watch (GAW) network (http://www.wmo.ch with links to GAW). These locations may show elevated $PM_{2.5}$ mass concentrations at times, influenced by anthropogenic emissions (e.g., at Arctic sites) or strong natural sources (e.g., Sahara desert dust at Canary Islands and Bermuda). Crustal material is the largest component at many such sites, such as Izana in the Canary Islands (e.g., Maring et al., 2000).

Annual Average $PM_{2.5}$ Concentrations ($\mu g/m^3$)
and Particle Type in Rural Areas, 1999

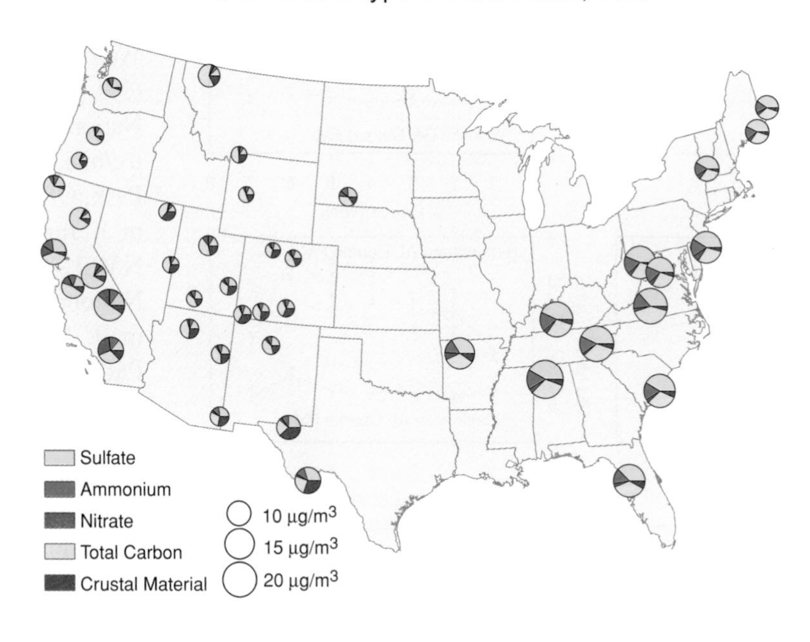

Annual Average $PM_{2.5}$ Concentrations ($\mu g/m^3$)
and Particle Type in Urban Areas, 2001

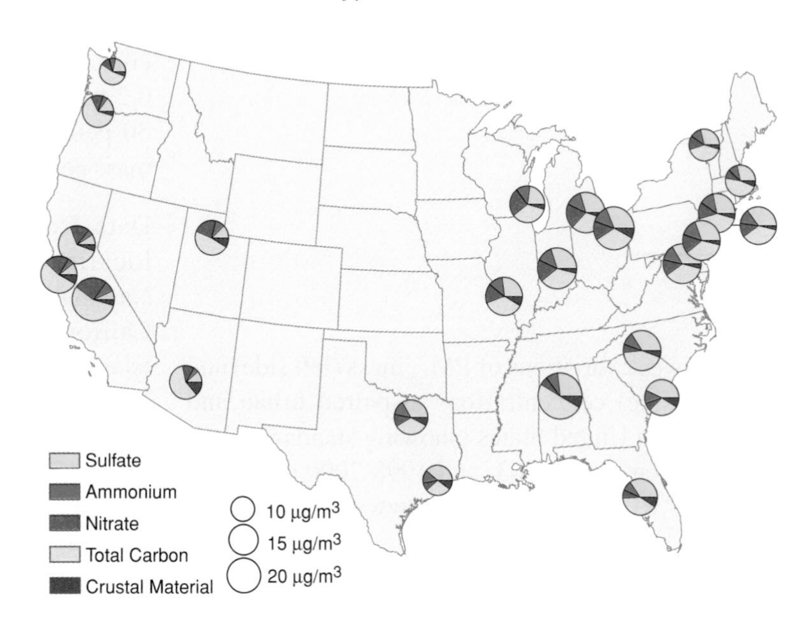

Figure 6.16. Comparisons of average $PM_{2.5}$ mass and species concentrations at urban and rural locations. The rural data are from the IMPROVE network, while the urban data are from the EPA Speciation Trends Network. Note that urban nitrate concentrations are distinctly higher than rural levels. When directly comparing the information in these two maps, consider 1) they represent different years; 2) one is an urban network and the other is a rural network; and 3) instruments and measurement methods differ. (Source: U.S. EPA, 2002).

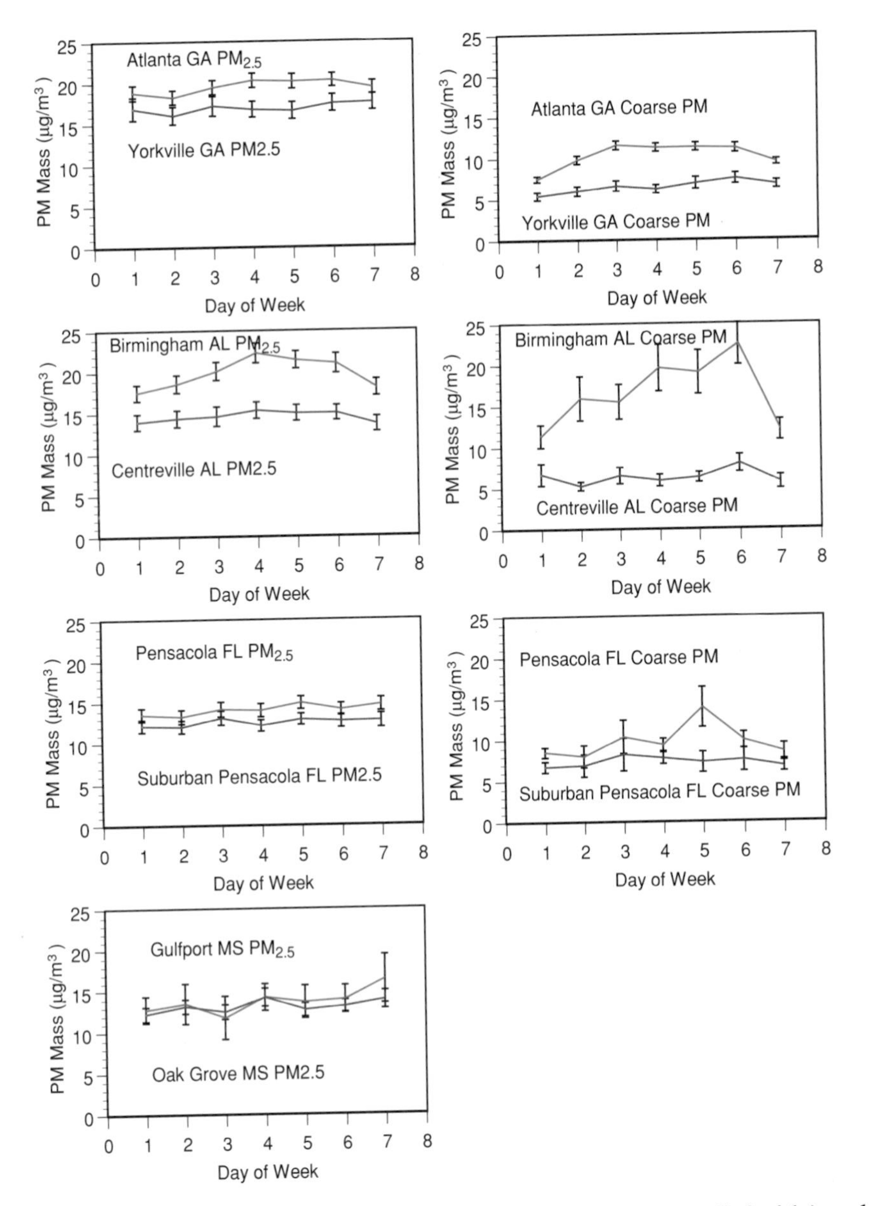

Figure 6.17. Example day-of-week variations of $PM_{2.5}$ mass (left side) and coarse mass ($PM_{10-2.5}$, right side) concentrations at paired urban and nonurban sites in the southeastern United States (showing standard errors of the means). (Source: Edgerton et al., 2003, and 1998-2000 data from SEARCH network, from SEARCH public archive www.atmospheric-research.co/searchhome.htm).

The mean annual $SO_4^=$ concentrations at GAW sites are a few tenths up to less than 1 μg/m³. Carbon concentrations exceed $SO_4^=$ concentrations at some sites.

Thirty-one sites in the IMPROVE network reported monitoring data from 1988 through 2002. Of these sites, the lowest mean fine PM mass concentrations occurred at Denali National Park, AK (2.0μg/m³), Bridger Wilderness, WY (2.7 μg/m³), Weminuche Wilderness, CO (2.9 μg/m³), Bryce Canyon National Park, UT (3.1 μg/m³), Crater Lake National Park, OR (3.3 μg/m³), Jarbridge Wilderness, NV (3.3 μg/m³), Lassen National Park, CA (3.3 μg/m³), Mesa Verde National Park, CO (3.5 μg/m³), and Great Sand Dunes National Monument, CO (3.5 μg/m³). The mean OC and $SO_4^=$ concentrations at these sites ranged from 0.7 to 1.4 μg/m³ (about 30 to 50 percent of mean fine PM mass concentration) and 0.4 to 1.1 μg/m³ (about 20 to 40 percent of mean fine PM mass concentration), respectively. Mean concentrations of crustal material accounted for 0.2 to 0.9 μg/m³ (about 10 to 30 percent of mean fine PM mass concentration).

Data from two monitoring locations, Point Reyes National Seashore in northern California and San Nicolas Island in southern California, have been used by the California Air Resources Board (CARB) to characterize coastal California background $PM_{2.5}$ mass levels. During 1995, the annual-average $PM_{2.5}$ mass concentrations were 6.5 μg/m³ at Point Reyes and 5.6 μg/m³ at San Nicolas Island (Motallebi et al., 2003a; 2003b). Sea salt, as sodium chloride, was 27 percent of the average mass at Point Reyes and 22 percent at San Nicolas Island. The average $SO_4^=$ concentrations were 1.2 and 1.4 μg/m³, while the average OC concentrations were 1.0 and 1.2 μg/m³. However, wind recirculation patterns are

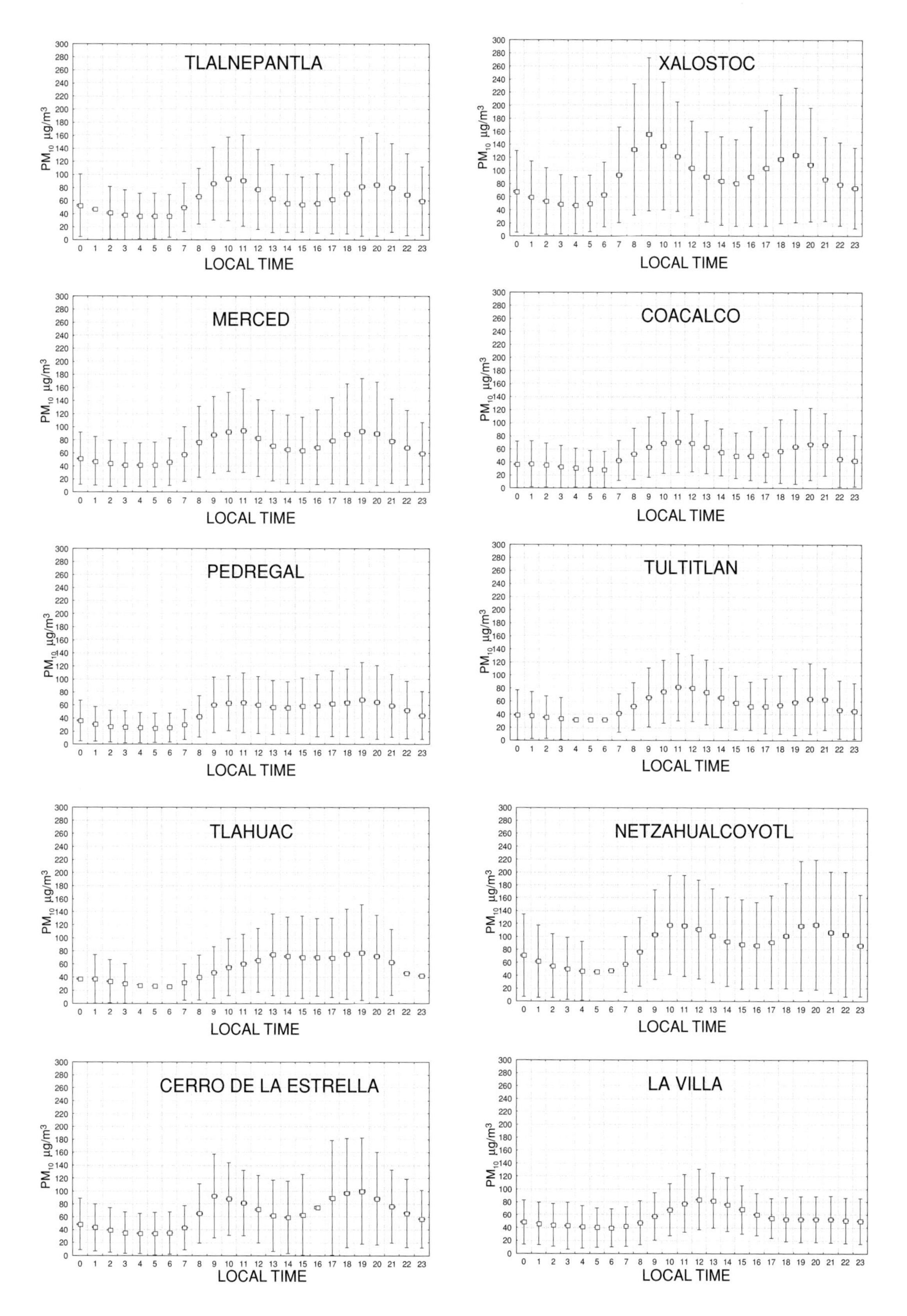

Figure 6.18. Example diurnal variations of PM_{10} mass concentration at ten sites in Mexico City. (Source: E. Vega, data from Manual Atmospheric Monitoring Network, 1995-99).

known to transport anthropogenic pollutants from coastal, urban locations to these monitoring sites at times.

6.3.4 Regional Transport

The Sulfate Regional Experiment (SURE), a pioneering field study conducted during 1977 and 1978, provided observational evidence for regional transport of $SO_4^=$ in eastern North America and linked transport distances to specific weather patterns (for information on SURE, see Hidy, 1984 and references therein). Regionally elevated $SO_4^=$ concentrations occurred during summer most often under two types of meteorological conditions. When a high-pressure system dominated the area along the Atlantic coast, the resulting stagnation on the lee (western) side of the system caused $SO_4^=$ concentrations to increase at monitoring locations across eastern North America. Transport distances under such stagnation conditions were usually within roughly 100 km, not exceeding over about 300 km. However, in a typical summer pattern, a frontal system would then advance slowly south and east from Canada into the United States across the Great Lakes, channeling windflow between the advancing front and the Atlantic high-pressure system along the line of the Appalachians. More rapid transport then developed from south to northeast, along the western side of the Atlantic high-pressure system, leading to observable transport distances of about 500 to 700 km. Higher concentrations of $SO_4^=$ at locations in southern Ontario were linked to transport from the midwestern and southern United States.

During October of the SURE study, the stagnation conditions became more important than transport (Hidy, 1984). By January, high $SO_4^=$ concentrations were associated almost entirely with regional stagnation or with nonregional events. Winter stagnation events indicated that maximum $SO_4^=$ concentrations were centered within about 100 km of areas of highest SO_2 emissions.

In contrast with the evidence for $SO_4^=$ transport developed by the SURE study, PM mass (TSP) showed only local (less than approximately 100 km) areas of source influence (Hidy, 1984). The spatial patterns of TSP were identifiable with a west-east industrialized region south of Lake Erie and with specific areas of higher particulate and SO_2 emission densities. At that time, 24-hr $SO_4^=$ concentrations during summer events were generally in the range of 10 to 40 µg/m^3, whereas TSP concentrations often exceeded 100 µg/m^3.

Continuing study of meteorological influences on PM mass concentrations has led to several methods for classifying meteorological conditions and characterizing their association with high PM mass concentrations in eastern North America. In eastern Canada, PM mass concentrations have been shown to vary among as many as nineteen types of meteorological conditions (Vet et al., 2001). As for the SURE study, regional PM episodes were characterized by the presence of a quasi-stationary high-pressure system sitting over eastern North America, with a high-pressure ridge oriented in an east-west direction across Virginia and North Carolina. (Regional episodes were defined as periods when mass concentrations exceeded 30 µg/m^3 as a 24-hr average for two or more consecutive days at sites in two of the three regions of eastern Canada, i.e., Ontario, Quebec, and Atlantic Canada, and, on one of those days, exceeded 30 µg/m^3 at sites in all three regions). The episodes typically lasted five to six days and ended when a trough of low pressure moved rapidly (typically in one day) from the Great Lakes to the eastern seaboard, displacing the high-pressure ridge.

Back trajectories (e.g., Chapter 7, Figure 7.4) calculated from Canadian cities during regional episodes indicate the potential for transport from the United States. Trajectories do not in themselves establish the actual occurrence of transport, since they do not indicate mass removal nor do they explicitly treat the relative contributions of local and distant emission sources. However, in combination with the SURE study findings and other analyses, the trajectory analysis indicates that emission sources in the midwestern United States and southwestern Ontario sometimes play a role in the formation of large-scale $PM_{2.5}$ episodes in Ontario, Quebec, and the Atlantic provinces. In other cases, locally generated emissions or a combination of local and distant sources may be equally important.

As discussed in Chapter 9, visibility impairment in national parks, particularly those of the southwestern United States whose scenic vistas are highly prized, has been a subject of concern for over twenty years. In its final report to the U.S. EPA in 1996, the Grand Canyon Visibility Transport Commission (GCVTC) recommended a variety of PM-management measures for maintaining and improving visibility in the Grand Canyon and other national parks in the southwestern United States (Grand Canyon Visibility Transport Commission, 1996). The recommendations of the GCVTC were based upon many years of scientific studies that indicated the importance of contributions of PM from near (< 100 km) and distant (> 500 km) emission sources or source areas. No single source or group of sources was the dominant contributor.

In an area such as the Grand Canyon National Park, where local emission levels are low, the relative amount of transported PM as a fraction of the total measured ambient PM mass concentrations is typically greater than would be the case in areas where local emissions densities are greater. Thus, while transport distances may not be greater than in eastern North America, the proportions of transported PM may be greater, and PM management needs to focus comprehensively on regional and long-distance (> 500 km) management options, though not to the exclusion of local management measures.

Key contributors to visibility impairment in the southwestern United States include $SO_4^=$, derived from SO_2 emissions, fine crustal material, and smoke (carbon from fires, including controlled burns and wildfires). Sulfate plays a particularly important role (Chapter 9), and has therefore been the focus of much study. Inert tracers have been used to investigate $SO_4^=$ transport (Pitchford et al., 2000). CMB analyses apportioned $SO_4^=$ concentrations at two locations within the Grand Canyon among four coal-fired power plants and eight regional sources (Eatough et al., 2000). The results of the CMB source apportionment showed that the relative contributions to $SO_4^=$ were different at the two locations within Grand Canyon National Park, so source proximity was a factor. Sources as far as Los Angeles and Baja California (about 600 km) contributed to $SO_4^=$ concentrations. The contributions of power generating facilities to SO_2 were greater than to $SO_4^=$ concentrations, thus indicating that full conversion of primary emissions to $SO_4^=$ required more time than elapsed during transport.

According to the report of the GCVTC, clean-air conditions were related to good ventilation and the occurrence of precipitation along the paths followed by air masses before they arrived at the Grand Canyon. Such conditions occurred more often for air masses arriving from the northwest than from other directions, and these differences, rather than the presence or absence of specific emission sources, accounted for differences in air quality on different days.

6.4 THE INFLUENCE OF INTERCONTINENTAL AEROSOL TRANSPORT ON PM MASS CONCENTRATIONS IN NORTH AMERICA

From the point of view of air-quality management in North America, the transport of aerosols around the world represents a largely uncontrollable contribution to North American particle levels, and is of uncertain magnitude. Appendix D provides an overview of global aerosol patterns, and that information is used here to estimate preliminary bounds for the potential contributions of intercontinental transport to North American $PM_{2.5}$ mass concentrations.

Evidence to date suggests that intercontinental aerosol transport typically does not in itself produce surface-level concentrations that violate North American PM standards, but has the potential to contribute to exceedances at some locations when high concentrations from local PM emissions are superimposed upon this long-range background contribution. To date, no systematic quantification of the frequency of occurrence of such transport events, or their average contributions to 24-hr or annual average concentrations, has been completed. However, a number of individual episodes have been studied.

Satellite imagery shows a PM plume from North Africa reaching Central America, Mexico, and the

U.S. Gulf Coast states, with the time of maximum impact occurring in July (Appendix D). Measurements at surface sites along the transport path (e.g., Canary Islands, Bermuda) reveal that the dominant component of the North African PM plume is crustal material, also known as soil or dust. In the southeastern United States, sites in the Interagency Monitoring of Protected Visual Environments (IMPROVE) network show seasonal maximum dust concentrations in July. However, on an annual basis, soil, or dust, constitutes only about 5 to 10 percent of the average fine-particle concentrations at IMPROVE (rural) monitoring locations in the southern United States, representing about 0.4 to 1.1 $\mu g/m^3$ (annual average). The transported dust therefore is a minor (probably less than 5 percent) contribution to the annual-average concentrations.

Analyses by the Texas Commission on Environmental Quality indicate that the North African plume contributed as much as 15 to 20 $\mu g/m^3$ at sites in the Houston area on two days in June and August 1997 (Price et al., 1998). On those days, the maximum 24-hr $PM_{2.5}$ concentrations were in the range of 24 to 33 $\mu g/m^3$ and no sites exceeded the U.S. 24-hr $PM_{2.5}$ standard. For further comparison, during 1999 only one FRM monitor in the southeastern United States showed a value exceeding the U.S. 24-hr 98th percentile standard of 65 $\mu g/m^3$, and 74 of the 91 sites showed 98th percentile 24-hr $PM_{2.5}$ concentrations between 25 and 45 $\mu g/m^3$. These simple comparisons suggest that the North African plume does not contribute enough $PM_{2.5}$ mass to affect the regulatory status of any monitoring locations; however, more precise estimates of the impact of the dust plume on 24-hr and annual-average $PM_{2.5}$ particle concentrations are needed.

Satellite data provide evidence of an aerosol plume reaching into North America from Asia during April 1998 (Appendix D). Elevated surface-level $PM_{2.5}$ mass concentrations were also observed at IMPROVE sites in the Pacific Northwest then, with the most prominent effect being a simultaneous sharp rise of $PM_{2.5}$ mass to 3 to 11 $\mu g/m^3$ on April 29, 1998 from more typical peak values of 1 to 3 $\mu g/m^3$. Transport episodes of the magnitude of the April 1998 event appear to occur less frequently than once per year and perhaps no more often than the order of once per ten years (Appendix D), so their potential regulatory consequences are likely limited to their effects on 24-hr concentrations. During 1999, no FRM monitors in the Pacific Northwest (Oregon, Washington, and Idaho) showed a 98th percentile value exceeding the U.S. 24-hr 98th percentile standard of 65 $\mu g/m^3$, and 28 of the 30 sites showed 98th percentile 24-hr $PM_{2.5}$ concentrations between 25 and 45 $\mu g/m^3$. During 1997-99, most sites in British Columbia showed a 98th percentile value less than 20 $\mu g/m^3$, well below the level specified by the Canada Wide Standard (30 $\mu g/m^3$). These statistical summaries suggest that the Asian plume would rarely, if ever, contribute enough $PM_{2.5}$ mass to affect the regulatory status of any monitoring locations. However, more precise estimates of the impact of the dust plume on 24-hr and annual-average $PM_{2.5}$ concentrations are needed to support this preliminary conclusion.

Fires generate fine particles, consisting especially of organic carbon, which are then lifted by the heat of the fire. PM from fires may affect large areas of North America on a seasonal basis, from northwestern or northeastern Canada and the Pacific Northwest to Central America, Mexico, and the southeastern United States. Satellite data show the presence of large numbers of fires in Central America and prominent aerosol clouds at times (Appendix D). Such fires represent a significant source of internationally transported $PM_{2.5}$ affecting considerable portions of North America. Analyses of satellite and surface data by the Texas Commission on Environmental Quality, for example, indicate that smoke from agricultural burning in the Yucatan Peninsula likely contributed about 2 to 5 $\mu g/m^3$ to the daily average $PM_{2.5}$ mass concentration at Corpus Christi during an episode in May 1997 (Price et al., 1998) (the 24-hr $PM_{2.5}$ concentration was 21.4 $\mu g/m^3$ and no sites exceeded the U.S. 24-hr $PM_{2.5}$ standard). Similarly, fires in Northern Quebec during July 2002 caused extensive haze in regions of Canada and the United States to the south.

As each of these examples of long-range transport indicates, the combined use of satellite and surface measurements provides evidence linking changes in surface PM mass concentrations to visible, transported aerosol plumes. The satellite data provide the means for identifying when long-range plumes are impacting an area, while the surface measurements provide the data needed for quantifying the excess PM levels occurring during those impact periods. While a number of analyses of transport episodes have been documented in the literature, what is now needed is a more systematic and precise quantification of the plume impacts in terms of their potential contributions to 24-hr and annual-average PM mass concentrations, including assessment of their potential for leading to violations of standards when local emissions are superimposed upon the transport component. Since transport episodes occur at irregular intervals, full characterization of the frequency and magnitude of transport requires daily sampling at a number of regionally representative surface sites; however, the larger number of compliance sites can operate on a more practical schedule of one sampling day every six days. Currently, global plumes may be tracked with satellite data so transport days can be identified and removed from statistics as rare events.

6.5 TRENDS AND THEIR IMPLICATIONS

Comparing emission changes with changes in ambient particle concentrations potentially provides information on the effectiveness of emission management programs. However, emission-related trends (the signal of interest) in ambient concentrations are difficult to quantify because the majority of the day-to-day, and even year-to-year, variation in concentrations is attributable to variations in weather (noise). This variation cannot be removed by simply adding more monitoring locations, because all locations within a particular area tend to experience the same weather patterns and to therefore show highly correlated variations in concentrations.

The detectability and quantifiability of trends in ambient pollutant concentrations depends upon the magnitudes of emission reductions or increases, the quality and length of record of the monitoring data, and the relative magnitudes of the emission-related (signal) and weather-driven (noise) variations in ambient pollutant concentrations. For a signal-to-noise ratio of 1:1, about two to four years of monthly data are typically needed to detect a linear trend with high probability (90 percent) at a 95 percent confidence level, whereas 10 to 20 years of data are generally needed when the signal-to-noise ratio falls to 0.1:1 (Weatherhead et al., 1998). The latter, low signal-to-noise ratio, is usually characteristic of ambient pollutant measurements. No single method for detecting trends in precipitation chemistry and deposition is considered optimal, or even appropriate, for all purposes (Holland and Sirois, 2001), and a similar conclusion likely applies to aerosol concentrations and composition.

Many sites in the IMPROVE network have shown declining $PM_{2.5}$ concentrations over the 11-year period from 1988 through 1998 (Figure 6.19). Statistical significance ($p<0.05$) of the $PM_{2.5}$ mass trends was associated with changes of about 0.05 µg/m per year or greater, or roughly 0.5 µg/m or greater over 10 years (Malm, 2000). For many of the IMPROVE sites, that level of change corresponded to about 10 percent of the median $PM_{2.5}$ concentrations, or roughly 5 percent of the mean concentration of the upper 20 percent of the measurements.

Related work on the detection of trends in precipitation $SO_4^=$ is relevant, as such trends also involve low signal-to-noise ratios. The minimum emission decrease detectable as a trend in precipitation $SO_4^=$ over 11 years has been estimated as a 10 percent reduction occurring across eastern North America (Schreffler and Barnes,1996). Reliable detection of changes in precipitation $SO_4^=$ concentrations resulting from SO_2 emission reductions of ~20 percent occurring over 3 years (1994-96) was predicted to require ~10 years of monitoring before the reduction and 2 to 3 years after (Blanchard, 1999). When emission changes of that magnitude occurred in much of the eastern United States within a single year, with new controls in place by the end of 1994, statistically significant changes in precipitation $SO_4^=$ were reported 1 to 2 years later (Lynch et al., 1996; Shannon, 1999). The error limits of the regression equations implied uncertainties of about 20 to 40 percent in the estimated magnitudes of the changes in $SO_4^=$ concentrations, and 10 or more years post-reduction measurements may be required to reduce the uncertainties to less than 20 percent (Blanchard, 1999).

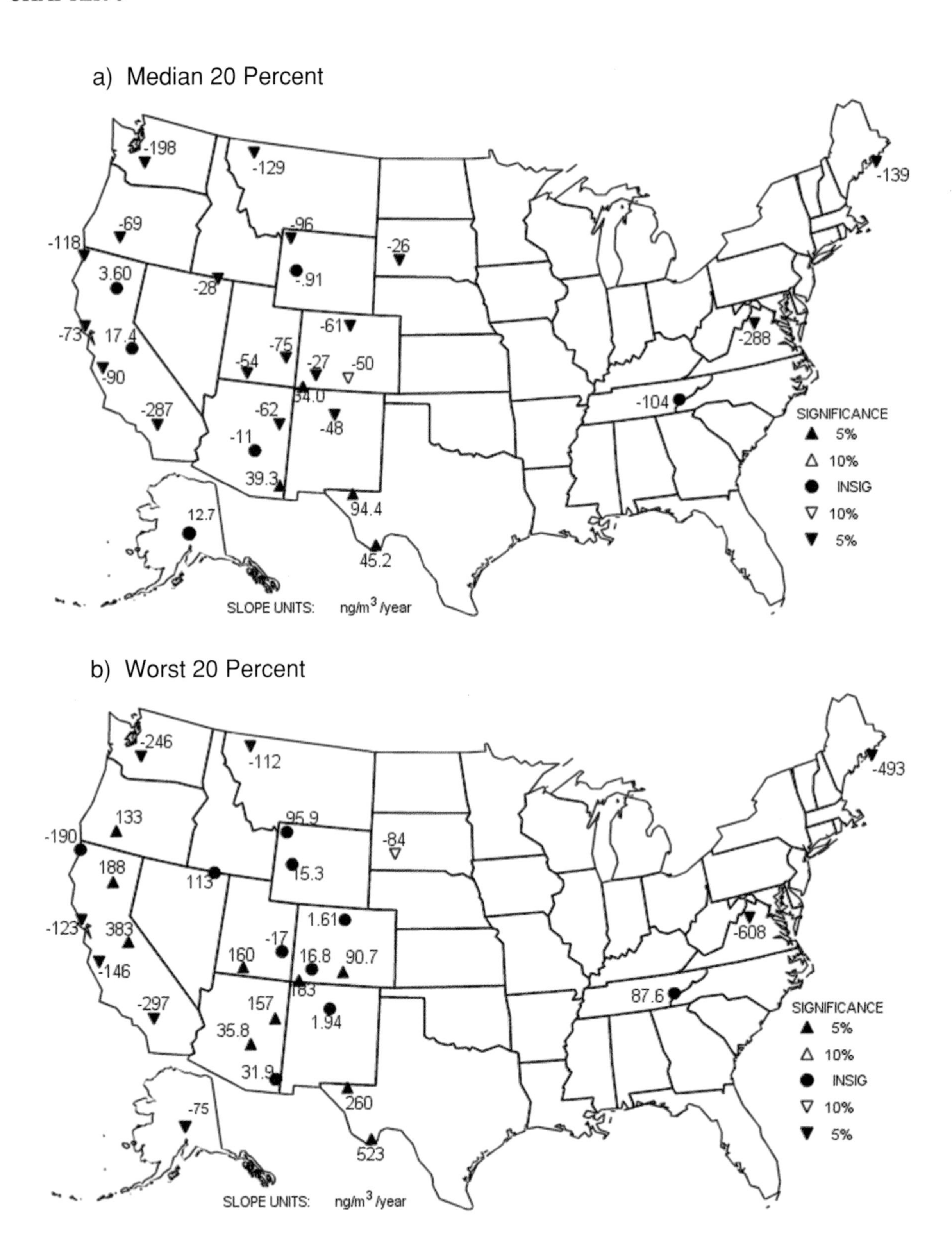

Figure 6.19. Trends in $PM_{2.5}$ mass at IMPROVE sites, 1988-98. The top panel shows trends in the median reconstructed $PM_{2.5}$ mass concentrations, while the bottom panel shows trends for the worst 20 percent of the days (highest daily fine mass concentrations). Arrows denote upward or downward trends. Solid arrows indicate significance at $p<0.05$ while open arrows indicate significance in the range of $0.05<p<0.10$. (Source: Malm, 2000).

Assessing the effects of emission-control strategies on ambient concentrations of pollutants typically requires about a decade. Measurements across North American networks should be designed to enable the determination of spatial and temporal trends of a multipollutant mixture. Issues including reference standards, measurement methodologies and sampling strategies need to be considered in assessing the compatibility of sampling networks. Measurement uncertainties should be as small as possible and should be reported routinely.

Since the length of monitoring required for detecting trends depends upon the magnitude and shape (e.g., step change versus gradual change) of the trend, as well as the quality of the monitoring data and the type of statistical procedure used, no simple statement can accurately characterize all situations. As described above, many of the existing trend analyses suggest that a 10-year record is typically necessary for detecting changes on the order of 10 to 20 percent of the measured concentrations. Indeed, many of the more sophisticated trend techniques have shown the existence of 3- to 5-year cycles in precipitation $SO_4^=$ concentrations (Sirois, 1993) and in particulate $SO_4^=$, NO_3^-, and NH_4^+ concentrations (Vet et al., 2001). The presence of such multi-year cycles implies that a record of 10 years or more is necessary for establishing the existence and significance of long-term trends.

From a regulatory perspective, it is significant that a pronounced decrease of $SO_4^=$ concentrations occurred in the eastern United States and Canada during the 1990s, related in part to the U.S. Phase I SO_2 emission controls that were implemented as of the end of 1994 (Figure 6.20). From 1989 to 1998, SO_2 emissions in the states east of and including Minnesota to Louisiana declined by about 25 percent. Average SO_2 and $SO_4^=$ concentrations at CASTNet monitoring sites in the same region declined by about 40 percent. This decline was not statistically different from the 25 percent reduction in emissions. Furthermore, ambient concentrations of SO_2 and $SO_4^=$ exhibited a statistically significant correlation with the SO_2 emission trend (U.S. EPA, 2000b). These associations support the utility of regional reductions of SO_2 emissions for effecting near-proportional reductions of $SO_4^=$.

In Canada, $PM_{2.5}$ concentrations from the NAPS network (11 cities nationwide or 6 eastern cities) showed interannual variability from 1986 through 1992, a 40 percent decline in median and 90th percentile values from 1992 through 1996, and a 14 percent increase from 1996 through 1998 (Vet et al., 2001). Although the temporal pattern (a decline from 1992 through 1996) was consistent with the timing of the U.S. Phase I SO_2 emission reductions, $SO_4^=$ concentrations did not exhibit a corresponding decline during the 1992-96 time period. However, the rural Canadian Aerosol and Precipitation Monitoring Network (CAPMoN) sites did exhibit declining concentrations of SO_2 and $SO_4^=$ between 1990 and 1999 (Figures 6.20 and 6.21). At both the CAPMoN (Canada) and CASTNet (U.S.) sites, the ambient SO_2 declines generally exceeded the decreases in $SO_4^=$ concentrations.

In Los Angeles, SO_2 emissions varied from about 250 to 350 tons per day during the years from 1970 to 1977, and decreased to about 265 tons per day in 1979 and 150 tons per day by 1983 as power plants switched from oil to natural gas beginning in 1978 (Hidy, 1994; Alexis et al., 2001). SO_2 emissions further declined to about 120 tons per day in 1987, following the introduction of low-sulfur gasoline in 1983, and to about 77 tons per day by 1996 with continuing adoption of other control measures. During the period from 1976 to 1996, annual-average SO_2 concentrations in the Los Angeles area declined by over 70 percent, which corresponds closely with the 71 percent decrease in SO_2 emissions from 1979 to 1996. In contrast, annual-average $SO_4^=$ concentrations declined by about 50 percent from 1976 to 1996 (Hidy, 1994; Christoforou et al., 2000). The decline in $SO_4^=$ concentrations lagged the SO_2 concentration decline by about five years, and did not become apparent until the ambient SO_2 concentrations had been reduced by over 40 percent. Current scientific understanding indicates that the time lag and nonproportional decline of $SO_4^=$ concentrations occurred because the conversion of SO_2 to $SO_4^=$ was oxidant-limited; no other explanations are known. This historical example illustrates the nonproportional response between SO_2 emission changes and ambient SO_2 or $SO_4^=$ concentrations that may occur in some situations (see also Chapter 3), and the need to maintain a

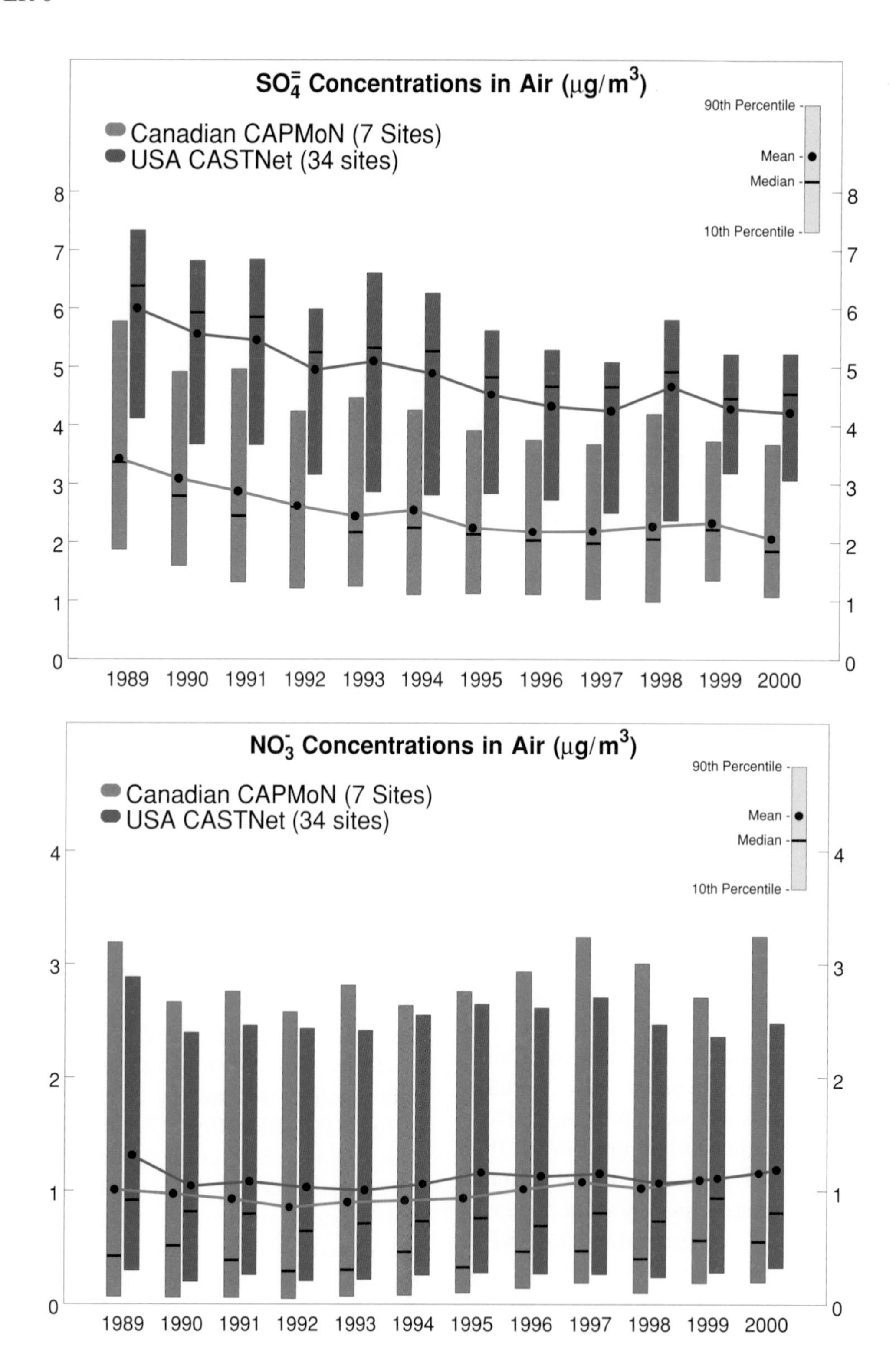

Figure 6.20. Trends in annual particulate $SO_4^=$ (top) and NO_3^- (bottom) concentrations based on 34 U.S. CASTNet and 7 Canadian CAPMoN sites. Shown are boxplots of annual mean, median and 10th and 90th percentile concentrations for the combined sites. The calculation of annual means at the sites is as described in US EPA (1999). All concentrations are expressed in μg/m³ at 25 degrees C and 1 atmosphere. CAPMoN data were converted from daily to weekly values to match CASTNet sampling periods before producing the mean, median and percentile values. (Source: M. Shaw, Environment Canada).

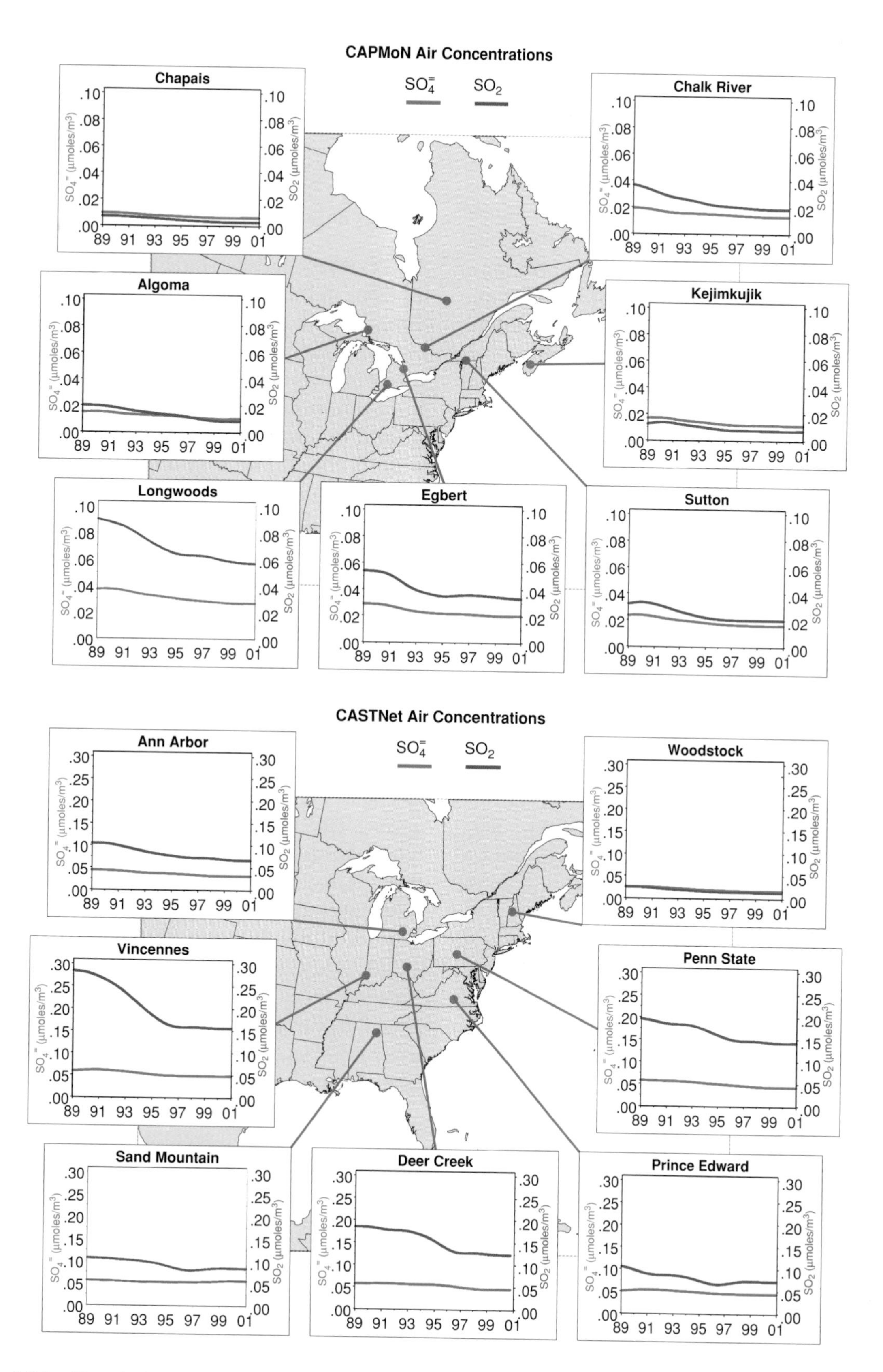

Figure 6.21. Trends in concentrations of SO_2 and $SO_4^=$ at CAPMoN and CASTNet sites, 1980 - 2000. (Source: Vet et al., 2001; M. Shaw, Environment Canada).

long-term monitoring program over many years to detect and quantify the effects of emission changes.

IMPROVE monitoring sites in the Four Corners states (Arizona, Utah, Colorado, and New Mexico) showed mixed trends in $PM_{2.5}$ concentrations over the period 1988 through 1998 (Figure 6.19). Sulfate trends tended to parallel those shown for $PM_{2.5}$, and trends in the median concentrations did not always track trends in the upper 20 percent of the measurements (Malm, 2000). At these and other IMPROVE sites, $PM_{2.5}$, $SO_4^=$, and carbon concentrations trended downward from 1988 until approximately early 1996, then leveled off or increased (White, 2001). The linkages to either meteorological variations or emission changes remain to be established.

California monitoring data show downward trends in annual-average $PM_{2.5}$ mass as well as in the concentrations of some components over the period 1988 through 1998 (Figure 6.19 and Motallebi et al, 2003b). Annual-average $PM_{2.5}$ mass levels declined from a range of ~25 to 30 μg/m³ to ~ 10 to 25 μg/m³ at sites in southern California and the San Joaquin Valley. Considerable year-to-year variability exists, and these trends are not statistically significant at all locations. However, all California IMPROVE sites showed statistically significant trends in $PM_{2.5}$, $SO_4^=$, NO_3^-, and OC concentrations. Meteorological differences may be a contributing factor, as the average precipitation during the latter half of the period covered by the trend analyses was up to twice the average during the earlier years. The influence of meteorological cycles is also suggested by the presence of statistically significant trends at Point Reyes National Seashore and Redwood National Park, two locations on the Pacific Coast that are minimally influenced by emissions from sources within California.

Ambient concentrations of $SO_4^=$, NO_3^-, and NH_4^+ are linked through equilibrium reactions. Declining $SO_4^=$ levels have the potential to exert an increasing influence on NO_3^- concentrations where NH_3 concentrations currently limit the formation of NO_3^-. Existing measurements are not definitive. In the eastern United States, particulate NO_3^- levels generally did not decline at CASTNet sites during the 1990s (U.S. EPA, 2000) (Figure 6.20). Of the three eastern IMPROVE sites, NO_3^- concentrations showed statistically significant declines from 1988 through 1998 at Acadia and Great Smoky Mountains National Parks, but not at Shenandoah National Park (Malm, 2000). Thus, pronounced NO_3^- trends are not evident in the way that $SO_4^=$ shows substantial declines during the 1990s.

As previously noted, at rural CAPMoN monitoring locations in Canada, $SO_4^=$ and NH_4^+ concentrations decreased from the early to late 1990s (Vet et al., 2001). During this same time period, NO_3^- concentrations increased. Either, or both, increasing NO_x emissions or increasing rates of NO_3^- formation (caused by increasing availability of NH_3 as $SO_4^=$ levels declined) may have contributed; both causes are considered probable (Vet et al., 2001). Seasonal data from the CAPMoN sites shows that NO_3^- concentrations were highest when $SO_4^=$ concentrations were lowest and more free NH_3 (Chapter 4) was available.

Longer-term records of PM mass concentrations show that PM levels have declined in many or most U.S. cities over the past 40 to 50 years (Lipfert, 1998). Primary PM emissions decreased by 50 to 60 percent from 1940 to 1990. Trends reconstructed from available TSP or PM_{10} measurements indicate that average TSP concentrations declined by two- to three-fold in urban areas between 1950 and 1980, while $PM_{2.5}$ concentrations (estimated from proxy information on emissions) likely declined by about a factor of two between 1960 and 1990 (Lipfert, 1998). Despite the uncertainties in analyses that of necessity relied on differing types of measurements from different years, the available data indicate that average PM levels in U.S. cities are roughly half those occurring in the 1950s. Thus, present-day exposures of individuals to PM are substantially lower in U.S. cities than those of 50 years ago. These changes are of significance for the characterization of lifetime PM exposures in epidemiological analyses. The compositional changes in PM over this time are unknown, however.

Fifty-year trends in ambient concentrations of PM in Mexico City are unavailable. Population in Mexico City increased six-fold from 1950 to 2000 and monitoring data indicate that PM levels have declined since 1988 (Molina and Molina, 2002).

6.6 COVARIATION OF PM WITH OZONE

Many air pollutants are related either through commonality of emission sources or atmospheric processes. Considering how the concentrations of different pollutants covary helps indicate if changes in the emissions of one pollutant, or its precursors, may lead to changes in the ambient concentrations of another.

PM is a complex mixture of chemical compounds, varying with particle size, location (source influence), and time of the year or even the day (meteorological factors). Primary PM emissions related to soil-derived species (windblown dust, construction dust, road dust, sea salt, and other mechanically derived materials, perhaps associated with industrial processes) typically have aerodynamic diameters greater than 2.5 µm, and PM_{10} concentrations can be dominated by such primary emissions. Seasonal variations of PM_{10} generally are not strongly related to ozone concentrations (Figure 6.22), and usually have little relation with regional haze and acid deposition. Ozone formation is most rapid during summer when radiation levels are greatest; summer ozone concentrations tend to be highest during conditions of stagnation, when windblown dust would typically be at a minimum.

Secondary particles are formed through photochemical processes, nighttime radical chemistry (such as NO_3^- via the nitrate radical), and aqueous mechanisms (for example, during foggy conditions where SO_2 can be absorbed into water droplets and oxidized efficiently by H_2O_2 or metal catalysts). Ozone has precursor species (NO_x and VOC) in common with secondary PM. In eastern North America, secondary particles and $PM_{2.5}$ in general tend to accumulate in air during stagnant conditions in the summer, often when ozone forms (Figure 6.22).

At most urban and many rural sites in California, the autumn or winter fine PM is often dominated by NH_4NO_3 and carbon compounds. During summer, equilibrium favors the precursors of NH_4^+ and NO_3^- (HNO_3 and NH_3); as well, semivolatile organic compounds are also found preferentially in the gas phase in the summer. In the fall and winter, with cooler temperatures, the NO_3^- equilibrium is shifted to the particle phase. Thus, the winter $PM_{2.5}$ maxima observed in California are associated with low ozone concentrations (Figure 6.22). Changes in NO_x or VOC emissions may have various effects on PM mass concentrations (Chapter 3), depending on whether the NO_3^- is limited by the amount of HNO_3 or NH_3 present in air, and on whether the conversion of NO_2 to HNO_3 is radical- or NO_x-limited. It is possible that winter NO_3^- concentrations in some locations may not respond to NO_x reductions even though summer ozone concentrations do.

At many locations, seasonally higher concentrations of PM persist throughout the autumn, when ozone levels are declining toward their winter minima (Figure 6.22). Higher PM mass concentrations may occur during periods of stagnation, whereas ozone levels decline with declining seasonal radiation.

In the eastern half of the United States, $(NH_4)_2SO_4$ and carbonaceous material dominate during summer. Like ozone, $SO_4^=$ and some fraction of the carbon are derived from photochemical processes, and higher concentrations of both ozone and $PM_{2.5}$ occur during periods of stagnation. Thus, $PM_{2.5}$ mass tends to correlate with ozone concentrations on a daily time scale (Figure 6.23), although high PM and ozone concentrations do not necessarily occur on the same day at the same location.

In California, summer ozone and PM mass concentrations typically show statistically significant (but not strong) correlations on a daily time scale in the San Joaquin Valley, but not necessarily in Los Angeles (Figure 6.23). The $PM_{2.5}$ and ozone peaks occasionally coincide, but often do not (Figure 6.23). These summer $PM_{2.5}$ mass concentrations are much lower than those occurring during winter (Figure 6.22). PM_{10} concentrations are often highest during windy conditions, with wind blown dust being a major contributor, and under conditions when ozone does not accumulate, so peak ozone and peak PM_{10} days do not usually coincide.

On an hourly time scale, summer PM and ozone concentrations at sites in eastern North America do not necessarily track each other, indicative of variations in the relative importance of the different chemical and physical processes that drive ozone and

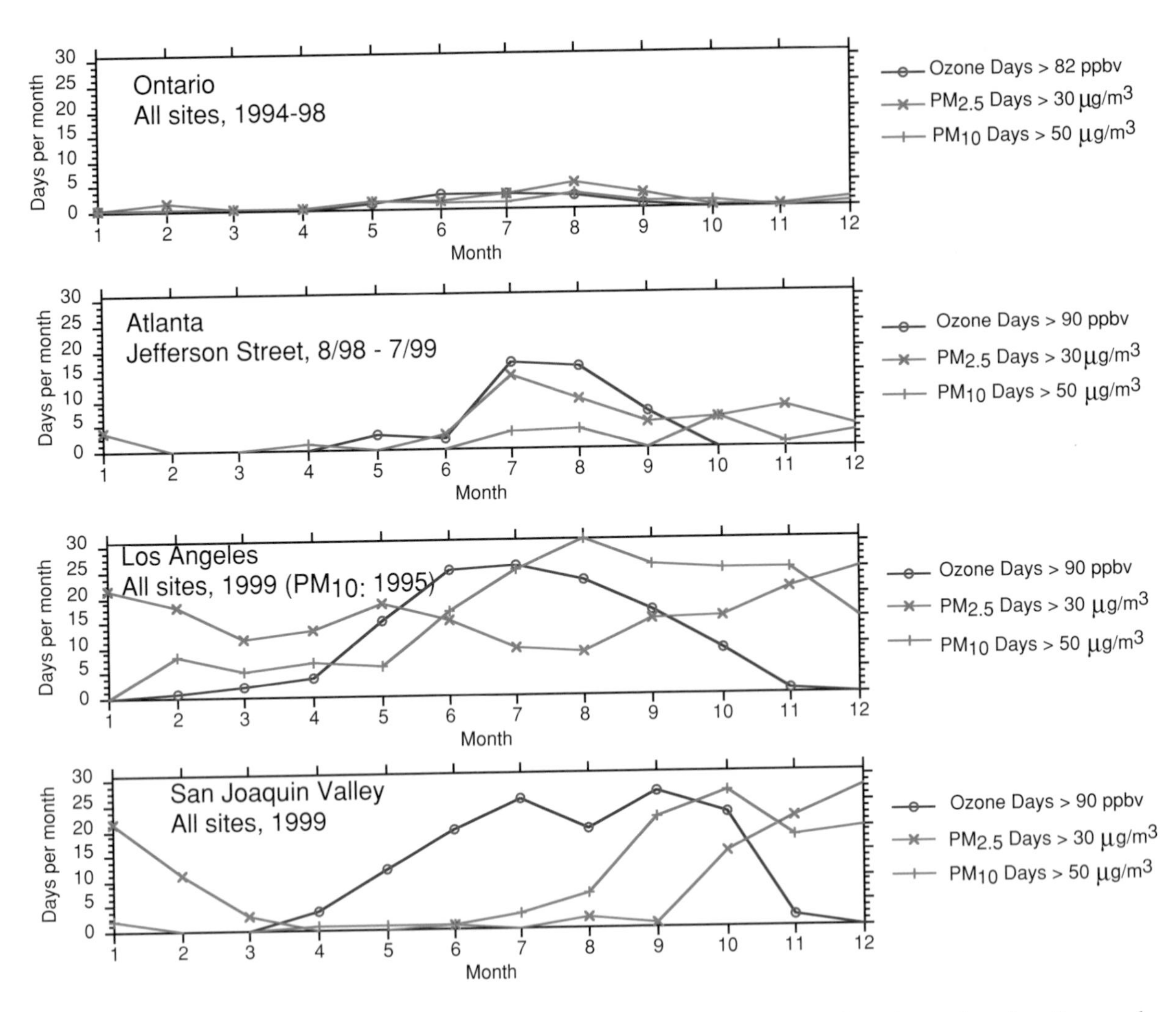

Figure 6.22. Seasonal variation of ozone, $PM_{2.5}$ mass, and PM_{10} mass at four locations, showing the number of days per month exceeding specified threshold levels. For ozone, the thresholds refer to the peak hourly value. For $PM_{2.5}$ and PM_{10}, the thresholds refer to 24-hr concentrations. $PM_{2.5}$ and PM_{10} frequencies were scaled according to the number of days with missing data each month. (Source: data provided by Environment Canada; Vet et al., 2001; Hansen et al., 2003; Alexis et al., 2001).

PM formation and accumulation (Figure 6.24). For example, summer data from Atlanta and Ontario show that the typical diurnal profile observed in ozone levels is not seen in $PM_{2.5}$ mass concentrations (Figure 6.24). In the summer Atlanta data, periods of both correlation and anticorrelation occur; these periods also appear as time lags between hourly ozone and $PM_{2.5}$ maxima. The Ontario data clearly reveal the dependence of both ozone and PM mass concentrations on multi-day meteorological variations. During winter, ozone continues to show daytime maxima, whereas $PM_{2.5}$ mass concentrations are typically greatest at night and appear to respond more to the presence or absence of the nocturnal inversion or to emissions such as wood smoke (Figure 6.24).

6.7 SUMMARY

The highest mean annual $PM_{2.5}$ concentrations in North America occur at sites in California and over a broad portion of the eastern, and especially the southeastern, United States. Shorter-term monitoring data from Mexico City suggest that mean annual $PM_{2.5}$ concentrations there could be among the highest recorded in North America.

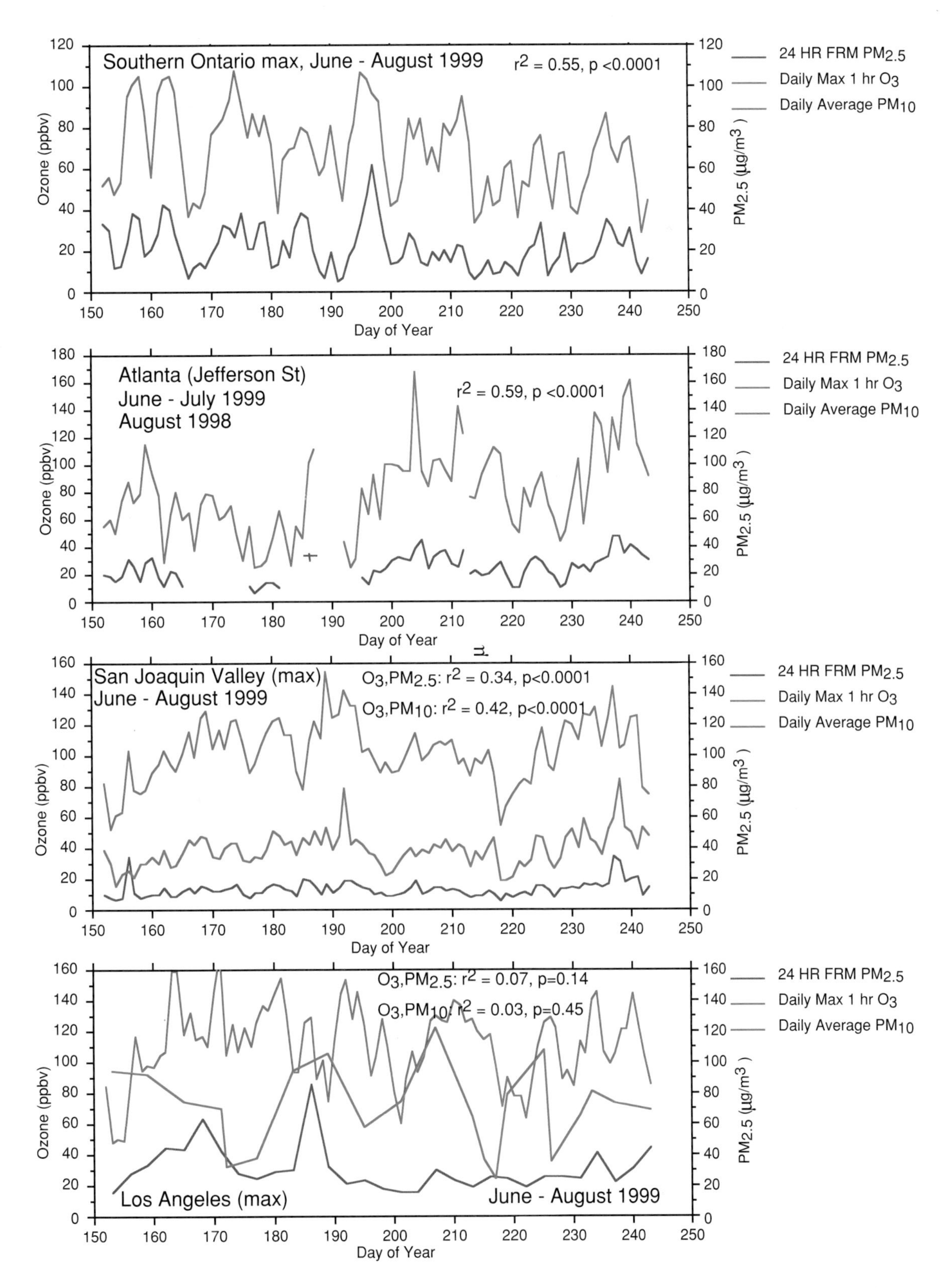

Figure 6.23. Daily variations of ozone, $PM_{2.5}$ mass, and PM_{10} mass during June through August 1999. In Los Angeles, the sampling frequency for $PM_{2.5}$ was once each three days and for PM_{10} was once each six days. (Source: data provided by Environment Canada; Vet et al., 2001; Hansen et al., 2003; Alexis et al., 2001).

Although determination of compliance with PM standards is the responsibility of various regulatory agencies, the ambient air-quality standards established by those agencies provide a useful benchmark for delineating the dimensions of the PM management task. Throughout North America, both urban and rural monitoring locations may exhibit $PM_{2.5}$ concentrations exceeding the levels specified by the national standards of Canada, the United States, or Mexico. The 98th percentile 24-hr $PM_{2.5}$ mass concentrations at the majority of urban U.S. and Canadian monitoring sites exceeded 30 µg/m³ (the level of the Canada Wide Standard; U.S. FRM data are presently limited to two years of measurements). With the exception of California, few sites had 98th percentile 24-hr $PM_{2.5}$ mass concentrations exceeding the level specified by the U.S. 24-hr standard (65 µg/m³). In California, the southeastern United States, and the Ohio Valley – Great Lakes states, annual average $PM_{2.5}$ mass concentrations at about half the urban sites exceeded the U.S. three-year average annual mean $PM_{2.5}$ mass standard of 15 µg/m³ in 1999 and 2000. Thus, in the United States, achieving the annual $PM_{2.5}$ standard will be the focus of the regulatory effort at most sites, whereas in Canada the regulatory effort will necessarily be oriented to the 24-hr Canada Wide Standard. In California, PM management needs to address both the U.S. annual and 24-hr standards.

$PM_{2.5}$ concentration measurements typically exhibit strongly skewed frequency distributions, dominated by a large number of low values and including a smaller number of high concentrations. At locations where exceedances of national 24-hr or annual-average $PM_{2.5}$ standards occur, numerous episodes of elevated PM mass concentrations contribute. High

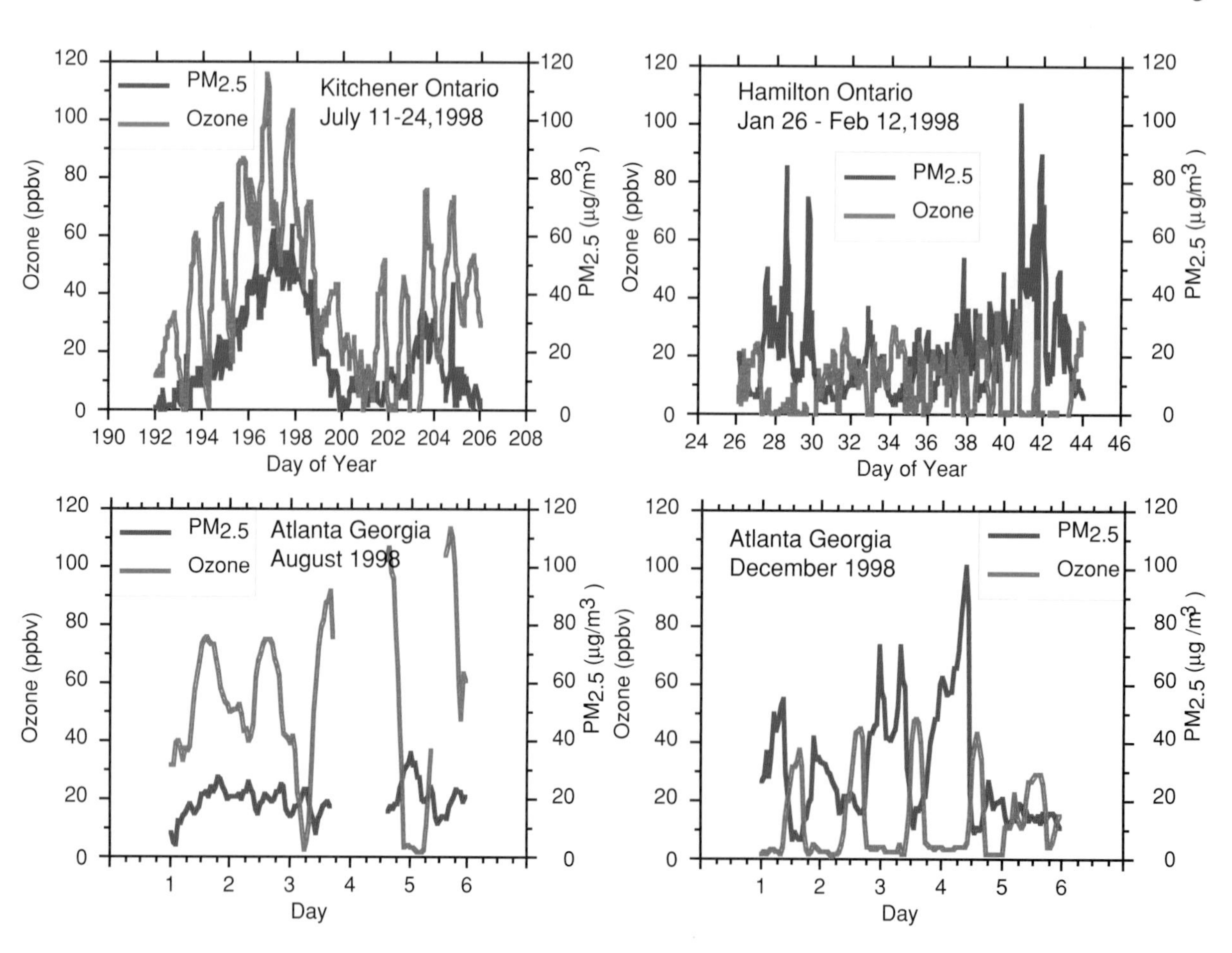

Figure 6.24. Hourly variations of ozone and $PM_{2.5}$ mass during two time periods at two monitoring sites in Ontario (top) and one site (Jefferson Street) in Atlanta. (Source: data provided by Environment Canada; Vet et al., 2001; Hansen et al., 2003; Alexis et al., 2001).

concentrations do not occur on the majority of days, but neither are they rare events.

Fine particles are produced from primary emissions, including windblown dust, fires and other combustion sources of carbon compounds, and as secondary particles from precursor emissions of sulfur, NO_x, NH_3, and VOC. Thus, in virtually all locations the major chemical components of PM are $SO_4^=$, NO_3^- ,NH_4^+, OC, BC, and crustal-related compounds (e.g., oxides of Si, Fe, Ca, and Mg, collectively known as soil or dust). Trace elements, such as Sb, As, Be, Cd, Co, Cr, Fe, Pb, Mn, Hg, Ni, Se, and Zn, do not contribute much to $PM_{2.5}$, but are of interest because of their possible adverse effects on health and their usefulness as tracers of the sources of fine particulate emissions. Some trace elements are identified as hazardous air pollutants by one or more North American governments.

Of the major components of fine particles, OC is found at all urban and rural monitoring locations, composing typically 25 to 70 percent of the mass in urban locations. Sulfate is a significant component of PM at rural and urban sites in eastern North America during the summer, but in winter $SO_4^=$ concentrations are lower and NO_3^- concentrations are greater than in summer. Nitrate is a major constituent at rural and urban sites in California and dust is a contributor at some locations in western North America.

Differences between PM mass concentrations at urban and rural monitoring sites provide indications of the relative magnitudes of local and regional contributions. Regional aerosol includes inter-urban or longer-range transport as well as nonanthropogenic background particulate mass. Urban areas show mean $PM_{2.5}$ levels exceeding those at paired rural sites. In eastern North America, the differences imply that local urban contributions are roughly 25 to 50 percent of the mean urban concentrations, with regional PM contributing the remaining, and larger, portion. Sulfate and OC appear to be strongly regional (~50 to 90 percent of the urban values may be due to the regional component). The regional component is large enough that both regional and local PM and PM-precursor emission reductions will be required to meet the U.S. and Canadian $PM_{2.5}$ standards in many urban areas.

The regional contribution to PM varies among regions. Data from the IMPROVE network indicate that mean annual $PM_{2.5}$ levels exceed 10 $\mu g/m^3$ at rural locations in much of the southeastern United States. Rural locations in other portions of the eastern United States, California, Ontario, and Quebec exhibit mean annual $PM_{2.5}$ levels of 5 to 10 $\mu g/m^3$. It will be difficult to meet the U.S. annual PM standard in urban areas where the regional component is at present close to the level of the standard, and regional PM management strategies may be required.

Specific studies, such as the Sulfate Regional Experiment, the Grand Canyon visibility studies, and the San Joaquin Valley Air Quality Study have examined the long-range transport of pollutants (PM and PM precursors) within North America. These studies have shown that transport of anthropogenic $PM_{2.5}$ can occur over distances of 500 to 1000 km or more. For example, pollutants from Los Angeles have been shown to impact visibility in Class 1 areas in and around the Grand Canyon. While this transport may not cause an exceedance of the U.S. $PM_{2.5}$ standard, it is sufficient to affect visibility in Class 1 areas in the southwestern United States.

Global-scale long-range transport can also affect PM mass concentrations. Both satellite and surface monitoring show that long-range transport to or within North America occurs, from east Asia across the North Pacific, from North Africa to the Caribbean and Gulf Coast states, and from Central America northward into the United States. Evidence to date suggests that such transport typically does not in itself produce surface-level concentrations that violate PM standards, but has the potential to contribute to exceedances when high concentrations from local PM emissions are superimposed upon this long-range background. To date, no systematic quantification of the frequency of occurrence of such transport events, or their average contributions to 24-hr or annual average concentrations, has been completed. However, a number of individual episodes have been studied. Since transport episodes occur at irregular intervals, full characterization of the frequency and magnitude of transport requires daily sampling at designated monitoring sites; however, the larger number of compliance sites can operate on a more practical schedule of one sampling day every six days. Currently, global plumes may be tracked with satellite data so transport days can be identified.

A pronounced decrease of $SO_4^=$ concentrations occurred in the eastern United States during the 1990s, and its timing indicates that it is a reflection of the U.S. Phase I SO_2 emission controls that were implemented as of the end of 1994. From 1989 to 1998, SO_2 emissions in the states east of and including Minnesota to Louisiana declined by about 25 percent. Average SO_2 and $SO_4^=$ concentrations at CASTNet monitoring sites in the same region declined by about 40 percent and exhibited a statistically significant correlation with the SO_2 emissions trend. The magnitudes of the emission and concentration changes were not statistically different, supporting the utility of regional reductions of SO_2 emissions for effecting near-proportional reductions of $SO_4^=$.

In Canada, $PM_{2.5}$ concentrations in six eastern cities showed about a 40 percent decline from 1992 through 1996, and a 14 percent increase from 1996 through 1999. Although the temporal pattern was consistent with the timing of the U.S. Phase I SO_2 emission reductions, the $SO_4^=$ concentrations did not exhibit a corresponding decline during the 1992-96 time period. However, the rural CAPMoN sites did exhibit declining concentrations of SO_2 and $SO_4^=$ between 1990 and 1999, suggesting that the effects of the U.S. management program may have been masked in the urban data by more local influences.

In Los Angeles, a 70 percent reduction of SO_2 emissions between 1977 and 1995 led to a halving of $SO_4^=$ concentrations, and a time-lag of about 5 years occurred between the onset of emission reductions in 1978 and the first observable declines in $SO_4^=$ concentrations. This historical example illustrates the potentially nonproportional response between $SO_4^=$ concentrations and SO_2 emissions, as well as the need to maintain a long-term monitoring program to characterize the effects of the emission reductions.

Ammonia reacts preferentially with H_2SO_4, and, if sufficient NH_3 is available, it also combines with HNO_3 to form NO_3^-. Declining $SO_4^=$ levels therefore have the potential to cause increasing NO_3^- concentrations where NO_3^- formation is limited by the availability of NH_3. Existing measurements are not definitive. At rural (CAPMoN) monitoring locations in Canada, $SO_4^=$ and NH_4^+ concentrations decreased from the early to late 1990s. During this same time period, NO_3^- concentrations increased. Either, or both, increasing NO_x emissions or increasing rates of NO_3^- formation (caused by increasing availability of NH_3 as $SO_4^=$ levels declined) may have contributed; both causes are considered probable. Seasonal data from the CAPMoN sites shows that NO_3^- concentrations were highest when $SO_4^=$ concentrations were lowest and the levels of NH_3 were highest.

Data from California locations indicate the importance of NO_3^- as a component of $PM_{2.5}$ concentrations there, accounting for ~30 to 40 percent of the annual-average $PM_{2.5}$ mass; annual-average NO_3^- concentrations ranged from about 6 to 16 $\mu g/m^3$. Nitrate is a prominent component of $PM_{2.5}$ at other urban locations in the western United States as well. In California, $SO_4^=$ concentrations are typically lower than in eastern North America, and sufficient NH_3 is available at most times and locations to allow the formation of NH_4NO_3. The amount of NO_3^- that forms depends upon the amount of HNO_3. This amount in turn depends upon the rate of conversion of NO_2 to HNO_3. Initial modeling efforts suggest that HNO_3 formation may be limited by the availability of radical species in one urban location of the San Joaquin Valley, and NO_3^- formation may be more effectively reduced through reductions of VOC than NO_x. However, NO_3^- concentrations in nearby rural areas may respond to reductions in NO_x emissions. These findings have significant implications for the management of both PM and ozone, and additional research efforts are needed for fully understanding these implications.

Long-term records of PM show that PM emissions and ambient levels have declined in many or most U.S. cities over the past 40 to 50 years. Primary PM emissions decreased by 50 to 60 percent from 1940 to 1990. Trends reconstructed from available TSP or PM_{10} measurements indicate that average TSP concentrations declined by two- to three-fold in urban areas between 1950 and 1980, while $PM_{2.5}$ concentrations (estimated from proxy information on emissions) likely declined by about a factor of two between 1960 and 1990. Despite the uncertainties in analyses that of necessity relied on differing types of measurements from different years, the available data therefore indicate that average PM levels in U.S. cities are roughly half those occurring in the 1950s. Thus, present-day PM exposures are substantially lower than those of 50 years ago. The compositional changes in PM are unknown.

In urban areas having multiple monitoring sites, average PM_{10} mass may vary by up to roughly a factor of two over distances as small as approximately 10 to 20 km. In some cities, average $PM_{2.5}$ mass may also show factor-of-two variations, though usually over distances of 50 to 100 km or more. The typically smaller spatial variations of $PM_{2.5}$ mass than PM_{10} mass are consistent with the well-known long residence time of fine PM, which permits transport over distances of 10 to 1000 km and tends to homogenize spatial variations in mass concentrations. Monitoring data also indicate that significant short-term (24-hr) variations of $PM_{2.5}$ mass may occur within urban areas. Monitoring data show typically larger intersite differences in trace-element concentrations than in $PM_{2.5}$ mass concentrations. Epidemiological studies run the risk of misclassifying PM exposures if the number of monitors within an urban area is insufficient to represent the spatial variations of outdoor population exposures. Further attention to network-design requirements is needed to ensure that future monitoring data are adequate to properly characterize exposure.

PM management requires strong ambient monitoring programs. Many key dimensions of PM monitoring programs must be specified, including size fractions, chemical species, sampling duration, sampling frequency, number and location of monitors, length of monitoring program, and instrumentation required to accomplish each of the preceding accurately and affordably. Specification of these choices requires clear identification of program objectives. In practice, monitoring networks are resource-limited and cost considerations typically limit their ability to address more than a limited number of objectives. Variables are typically optimized for the purpose of each network, and these differences add to uncertainty in the use of the data when applied for different purposes. Therefore, an ongoing need is to provide complete documentation on network methods and comparability of data across networks. Although the data required typically vary according to the objectives of each network, an effort should be made to ensure that existing monitoring networks meet diverse needs with maximum efficiency. Some useful monitoring activities are not now well supported. Apart from limited special studies, virtually no speciated carbon measurements exist. Measurement of a variety of PM carbon compounds at emission sources and receptor locations holds the potential for substantially improving current knowledge of source contributions to ambient OC and BC concentrations. Similarly, gas-phase measurements of semivolatile compounds are usually not included within most PM networks, yet they are necessary for fully understanding the behavior of such compounds (e.g., NH_4NO_3 and semivolatile organics). Conversely, with many networks making the same types of measurements, it is likely that duplication of effort exists in some specific locales. With appropriate harmonization of operating schedules and methods, more data could be shared across networks, freeing resources for measurements that are not now made by any network.

6.8 POLICY IMPLICATIONS

Long-term measurements (typically 10 years) are required to assess trends. Commitment should be made to routinely monitor at representative sites gases and particulate properties of interest for studies of population exposures, effects, model performance, and efficacy of emission-control measures. These measurements should extend to properties other than mass, e.g., size, composition. *Implication:* Comprehensive measurements of the multipollutant mixture over a period of at least 10 years are required to understand relationships between emissions, ambient concentrations, and effects.

There is a relationship between the spatial scale over which particulate pollutants are distributed and the spatial and temporal scales of measurements needed to assess their impacts. Relatively coarse spatial and temporal resolution is typically adequate for regional pollutants. For locally emitted pollutants or pollutants that vary diurnally due to photochemical production or variations in temperature and relative humidity, measurements must be made with much higher spatial and temporal resolution. *Implication:* It is essential to carry out measurements with adequate spatial and temporal resolution to document population exposure patterns and to obtain data that can be used to evaluate the performance of chemical-transport models.

$PM_{2.5}$ and PM_{10} exhibit seasonal dependencies that vary with location. For example, $PM_{2.5}$ concentrations in eastern North America tend to reach their highest values during summer, when high relative humidities and solar radiation favor the formation of $SO_4^=$ from regional SO_2 emissions, while in Mexico PM_{10} concentrations reach their maxima during the dry period from November through May. *Implications:* Management strategies, especially those for short-term standards, need to take account of seasonal patterns and to focus on species and processes that lead to the elevated concentrations.

The composition of $PM_{2.5}$ varies with region and locale. For example, $SO_4^=$ and OC are relatively abundant in eastern North America; the contribution of NO_3^- increases in winter, when low temperatures favor the condensed phase. Organic carbon and NO_3^- are abundant in much of California. Within a given region, local emissions lead to urban concentrations that are ~25 percent higher than concentrations at nearby nonurban sites. *Implication:* PM management strategies should take account of local, regional, and seasonal differences in composition.

In areas where local emissions are low, PM concentrations are likely to be determined by small contributions from many distant sources. For example, regional haze is affected by emissions that occur in regions that are hundreds or thousands of kilometers in extent. *Implications:* When local concentrations are dominated by long-range transport, PM-management strategies need to account for the impact of multiple distant sources on PM concentrations.

Many air pollutants are related either through commonality of emission sources or atmospheric processes. Considering how the concentrations of different pollutants covary helps indicate if changes in the emissions of one pollutant, or its precursors, may lead to changes in the ambient concentrations of another. For example, between 1989 and 1998, $SO_4^=$ concentrations in the eastern United States and Canada have declined approximately in proportion to SO_2 emissions, which decreased by about one third, while in Los Angeles a 70 percent reduction in SO_2 emissions between 1976 and 1996 led to a halving of ambient $SO_4^=$ concentrations. Such empirical observations, however, need to be examined in light of what is known about the behavior of the multipollutant mixture. Assuming that historical trends will continue in the future can lead to errors. *Implication:* Empirical relationships between emissions and ambient concentrations from historical data can provide valuable insights into changes in ambient concentrations that might be anticipated in the future if emissions change. Confidence in such extrapolations will be improved if the extrapolated trends can be confirmed using available modeling tools.

Intercontinental transport of dust from Asia or Africa occurs but does not contribute significantly to annual-average concentrations of PM concentrations in North America. It is possible, however, that it may occasionally contribute significantly to 24-hr average concentrations. *Implication:* Intercontinental transport of dust typically does not significantly impact PM concentrations in North America. PM management in North America should continue to focus on manageable local and regional PM and precursor emissions. Intercontinental transport of dust should be tracked, allowing for exclusion of rare and unmanageable events from PM management.

Forest fires or biomass burning can contribute significantly to local PM concentrations. Satellite images and measurements of composition can be used to determine periods when such fires are significantly affecting concentrations. *Implications:* Air-quality forecasting can be used to determine periods when the effects of smoke emissions from controlled burns can be minimized. Data that are routinely collected (including satellite data) can be used to identify periods when smoke emissions from distant fires are affecting local PM concentrations.

6.9 REFERENCES

Alexis, A., Delao, A., Garcia, C., Nystrom, M., Rosenkranz, K., 2001. The 2001 California Almanac of Emissions and Air Quality. California Air Resources Board, Sacramento, California.

Blanchard, C.L., 1999. Methods for attributing ambient air pollutants to emission sources. Annual Review of Energy and the Environment 24, 329-365.

Blanchard, C.L., Carr, E.L., Collins, J.F., Smith, T.B., Lehrman, D.E., Michaels, H.M., 1999. Spatial representativeness and scales of transport during the 1995 integrated monitoring study in California's San Joaquin Valley. Atmospheric Environment 33, 4775-4786.

Blanchard, C.L., Roth, P.M., Tanenbaum, S.J., Ziman, S.D., Seinfeld, J.H., 2000. The use of ambient measurements to identify which precursor species limit aerosol nitrate formation. Journal of the Air and Waste Management Association 50, 2073-2084.

Blanchard, C.L., Hidy, G.M., 2003. Effects of changes in sulfate, ammonia, and nitric acid on particulate nitrate concentrations in the southeastern United States. Journal of the Air and Waste Management Association. In press.

Brunekreef, B., Jannsen, N.A., de Hartog, J., Harssema, H., Knape, M., van Vliet, P., 1997. Air pollution from truck traffic and lung function in children living near motorways. Epidemiology 8, 298-303.

Brook, J.R., Dann, T.F., Bonvalot, Y., 1999. Observations and interpretations from the Canadian fine particle monitoring program. Journal of the Air and Waste Management Association 49, PM35 – PM44.

Bryson, R., Hare, F.K., 1974. Climate of North America, Vol 11. Bryson, R., Hare, F.K., eds. World Survey of Climatology. Elsevier, New York.

California Air Resources Board (CARB), 2001. The 2001 California Almanac of Emissions and Air Quality. California Air Resources Board, Sacramento CA.

Christoforou, C.S., Salmon, L.G., Hannigan, M.P., Solomon, P.A., Cass, G.R., 2000. Trends in fine particle concentration and chemical composition in southern California. Journal of the Air and Waste Management Association 50, 43-53.

Chow, J.C., Watson, J.G., Lowenthal, D.H., Solomon, P.A., Magliano, K.L., Ziman, S.D., Richards, L.W., 1993. PM_{10} and $PM_{2.5}$ compositions in California's San Joaquin Valley. Aerosol Science and Technology 18, 105-128.

Chow J.C., Watson, J.G., Lowenthal, D.H., Hackney, R., Magliano, K., Lehrman, D., Smith, T., 1999. Temporal variations of $PM_{2.5}$, PM_{10}, and gaseous precursors during the 1995 Integrated Monitoring Study in central California. Journal of the Air and Waste Management Association 49, PM16 – PM24.

Chow, J.C., Watson, J.G., 2001. Zones of representation for PM_{10} measurements along the US/Mexico border. The Science of the Total Environment 276, 33-47.

Chow, J.C., Watson, J.G., Edgerton, S.A., Vega, E., 2002. Chemical composition of $PM_{2.5}$ and PM_{10} in Mexico City during winter 1997. The Science of the Total Environment 287: 177-201.

Cicero-Fernandez, P., Torres, V., Rosales, A., Cesar, H., Dorland, K., Munoz, R., Uribe, R., Martinez, A.P., 2001. Evaluation of human exposure to ambient PM_{10} in the metropolitan area of Mexico City using a GIS-based methodology. Journal of the Air and Waste Management Association 51, 1586-1593.

Dolislager, L.J., Motallebi, N., 1999. Characterization of particulate matter in California. Journal of the Air and Waste Management Association 49, PM45 – PM56.

Eatough, D.J., Farber, R.J., Watson, J.G., 2000. Second generation chemical mass balance source apportionment of sulfur oxides and sulfate at the Grand Canyon during the Project MOHAVE summer intensive. Journal of the Air and Waste Management Association 50, 759-774.

Edgerton, S.A., Bian, X., Doran, J.C., Fast, J.D., Hubbe, J.M., Malone, E.L., Shaw, W.J., Whiteman, C.D., Zhong, S., Arriaga, J.L., Ortiz, E., Ruiz, M., Sosa, G., Vega, E., Limon, T., Guzman, F., Archuleta, J., Bossert, J.E., Elliot, S.M., Lee, J.T., McNair, L.A., Chow, J.C., Watson, J.G., Coulter, R.L., Doskey, P.V., Gaffney, J.S., Marley, N.A., Neff, W., Petty, E., 1999. Particulate air pollution in Mexico City: a collaborative research project. Journal of the Air and Waste Management Association 49, 1221-1229.

Edgerton, E.S., Hartsell, B.E., Jansen, J.J., Hansen D.A., 2003. The Southeastern Aerosol Research and Characterization Study (SEARCH): 2. Filter-based measurements of $PM_{2.5}$ mass concentration and composition. Journal of the Air and Waste Management Association. In press. http://www.atmospheric-research.com/searchhome.htm

Grand Canyon Visibility Transport Commission, 1996. Recommendations for Improving Western Vista: Report of the Grand Canyon Visibility Transport Commission to the U.S. Environmental Protection Agency.

Hansen, D.A., Edgerton, E.S., Hartsell, B.E., Jansen, J.J., Hidy, G.M., Kandasamy, K., 2003. The Southeastern Aerosol Research and Characterization Study (SEARCH): 1. Overview. Journal of the Air and Waste Management Association. In press. http://www.atmospheric-research.com/searchhome.htm

Hidy, G.M., 1994. Atmospheric Sulfur and Nitrogen Oxides: Eastern North America Source-Receptor Relationships. Academic Press, San Diego.

Hidy, G.M., 1984. Aerosols: An Industrial and Environmental Science. Academic Press, San Diego.

Holland D., Sirois A., 2001. Acid rain modeling. In: Encyclopedia of Environmetrics. J. Wiley & Sons.

Husar, R.B., Wilson, W.E., 1989. Trends of seasonal haziness and sulfur emissions over the eastern United States. In: Mathai, C.V., ed. Transactions: Visibility and Fine Particles, An AWMA/EPA Specialty Conference. Pittsburgh PA: Air & Waste Management Association. pp. 318-327.

Kim B.M., Teffera, S., Zeldin, M.D., 2000. Characterization of $PM_{2.5}$ and PM_{10} in the South Coast air basin of southern California: Part 1 n–spatial variations. Journal of the Air and Waste Management Association 50, 2034-2044.

Kumar, N., Lurmann, F.W., Pandis, S., 1998. Analysis of Atmospheric Chemistry During 1995 Integrated Monitoring Study. Sacramento CA: California Air Resources Board.

Lefohn, A.S., 2001. Assessing the existence of gradients for $PM_{2.5}$ Ambient Concentrations Within MSAs and CMSAs Across the United States. In: American Petroleum Institute Comments on the March 2001 Draft Particulate Matter Criteria Document.

Lynch, J.A., Bowersox, V.C., Grimm, J.W., 1996. Trends in Precipitation Chemistry in the United States, 1983-1994: An Analysis of the Effects in 1995 of Phase I of the Clean Air Act Amendments of 1990, Title IV. USGS 96-0346. United States Geological Survey, Washington DC. http://water.usgs.gov/public/pubs/acidrain

Malm, W.C., 2000. Spatial and seasonal patterns and temporal variability of haze and its constituents in the United States. National Park Service.

Maring, H., Savoie, D.L., Izaguirre, M.A., McCormick, C., Arimoto, R., Prospero, J.M., Pilinis, C., 2000. Aerosol physical and optical properties and their relationship to aerosol composition in the free troposphere at Izana, Tenerife, Canary Islands, during July 1995. Journal of Geophysical Research 105(D11), 14677-14700.

Mejia-Velaquez, G.M., Rodriguez-Gallejos, M., 1997. Characteristics and estimated air pollutant emissions from fuel burning by the industry and vehicles in the Matamoros-Reynosa border region. Environment International 23, 733-744.

Miranda, J., Cahill, T., Morales, R., Aldape, F., Flores, J., Diaz, R., 1994. Determination of elemental concentrations in atmospheric aerosols in Mexico City using Proton Induced X-ray Emission, Proton Elastic Scattering and Laser Absorption. Atmospheric Environment 28, 2299-2306.

Molina, L.T., Molina, M.J., Air Quality in the Mexico Megacity: An Integrated Assessment. Kluwer Academic Publishers, Dordrecht, The Netherlands, 2002.

Motallebi, N., Croes, B.E., Taylor Jr., C.A., Turkiewicz, K., 2003a. Intercomparison of selected $PM_{2.5}$ samplers in California. Journal of the Air and Waste Management Association, In press.

Motallebi, N., Croes, B.E., Taylor Jr., C.A., Turkiewicz, K., 2003b. Spatial and temporal patterns of $PM_{2.5}$, coarse PM, and PM_{10} in California. Journal of the Air and Waste Management Association, In press.

Mukerjee, S., 2001. Selected air quality trends and recent air pollutant investigations in the US-Mexico border region. The Science of the Total Environment 276, 1-18.

Pitchford, M., Green, M., Kuhns, H., Farber, R., 2000. Characterization of regional transport and dispersion using Project MOHAVE tracer data. Journal of the Air and Waste Management Association 50, 733-745.

Price, J., Dattner, S., Lambeth, B., Kamrath, J., Aguirre, M., McMullen, G., Loos, K., Crow, W., Tropp, R., Chow, J., 1998. Preliminary results of early $PM_{2.5}$ monitoring in Texas: separating the impacts of transport and local contributions. In: J Chow and P Koutrakis, eds. $PM_{2.5}$: A Fine Particle Standard. Proceedings of a Specialty Conference, January 28-30, 1998. Pittsburgh PA: Air & Waste Management Association.

Pun, B.K., Seigneur, C., 2001. Sensitivity of Particulate Matter Nitrate Formation to Precursor Emissions in the California San Joaquin Valley. Environ. Sci. Technol. 35, 2979-2987.

Schreffler, J.H., Barnes Jr., H.M., 1996. Estimation of trends in atmospheric concentrations of sulfate in the northeastern United States. Journal of the Air and Waste Management Association 46, 621-630.

Shannon, J.D., 1999. Regional trends in wet deposition of sulfate in the United States and SO_2 emissions from 1980 through 1995. Atmospheric Environment 35, 807-816.

Sirois, A., 1993. Temporal variation of sulphate and nitrate concentrations in precipitation in Eastern North America: 1979-1990. Atmospheric Environment 27A, 945-963.

Stockwell, W.R., Watson, J.G., Robinson, N.F., Steiner, W., Sylte, W.W., 2000. The ammonium nitrate particle equivalent of NO_x in Central California's San Joaquin Valley. Atmospheric Environment 34, 4711-4717.

U.S. EPA, 1999. Clean Air Status and Trends Network (CASTNet), 1998 Annual Report. Research Triangle Park NC: U.S. Environmental Protection Agency.

U.S. EPA, 2000. Clean Air Status and Trends Network (CASTNet), 1999 Annual Report. Research Triangle Park NC. U.S. Environmental Protection Agency.

U.S. EPA, 2000b. National Air Quality and Emissions Trends Report, 1998. Research Triangle Park NC. U.S. Environmental Protection Agency.

U.S. EPA, 2002. Latest Findings on National Air Quality: 2001 Status and Trends Report. EPA 454/K-02-001. Research Triangle Park NC. U.S. Environmental Protection Agency. www.epa.gov/airtrends

Vega, E., Reyes, E., Oritz, E., Ruiz, E., Chow, J.C., Watson, J.G., Edgerton, S.A., 2002. Basic statistics of $PM_{2.5}$ and PM_{10} in the atmosphere of Mexico City. The Science of the Total Environment 287: 167-176.

Vet, R., Brook, J., Dann, T., Dion, J., 2001. The Nature of $PM_{2.5}$ Mass, Composition and Precursors in Canada. In: JR Brook, PA Makar, MD Moran, MF Shepherd, RJ Vet, TF Dann, and J Dion. Precursor Contributions to Ambient Fine Particulate Matter in Canada. Toronto, Ontario: Meteorological Service of Canada.

Weatherhead, E.C., Reinsel, G.C., Tiao, G.C., Meng, X.L., Choi, D., Cheang, W.K., Keller, T., DeLuisi, J., Wuebbles, D.J., Kerr, J.B., Miller, A.J., Oltmans, S.J., Frederick, J.E., 1998. Factors affecting the detection of trends: Statistical considerations and applications to environmental data. Journal of Geophysical Research 103, 17149-17161.

White, W., 2001. Inter-annual variations in the composition of North American haze. Washington University, St. Louis MO.

Zheng, M., Cass, G.R., Schauer, J., Edgerton, E., 2002. Source apportionment of fine particle air pollutants in the southeastern United States using solvent extractable organic compounds as tracers. Environmental Science and Technology 36, 2361-2371.

CHAPTER 7

Receptor Methods

Principal Authors: Jeffrey Brook, Elizabeth Vega, and John Watson

7.1 INTRODUCTION AND OVERVIEW

Receptor methods for source attribution or apportionment include the interpretation of physical and chemical measurements of ambient particles and their precursors to infer possible or probable sources and to quantify contributions from these sources. Source attribution implies that individual PM emitters are identified (e.g., a specific power plant), while apportionment identifies general source types or categories (e.g., vehicle exhaust, fugitive dust) responsible for the observed PM.

Receptor methods play a role at several points within the overall PM-management framework used in this Assessment (Figure 1.1). Application of the receptor methods described in this chapter helps to develop and refine conceptual models for explaining ambient PM concentrations and their precursors. When combined with information on emissions, receptor methods can provide insight on aspects of the inventory that may need improvement or on key atmospheric processes governing the observed PM. Receptor methods lead to a better understanding of the atmospheric environment. In areas with excessive $PM_{2.5}$ or PM_{10} concentrations, receptor methods can identify possible solutions, especially when applied systematically in combination with examination of the local emission inventory. Receptor methods also are useful for analysis and development of policy and/or specific PM management strategies.

Receptor methods encompass a variety of observationally based techniques that use differences in chemical composition, particle size, and concentration patterns in space and time, as well as ratios among specific compounds (particles and/or gases), to identify source types and limiting precursors, and to quantify source contributions affecting particle mass concentrations, light extinction, or deposition. "Receptor models" are receptor methods that provide the theoretical and mathematical framework for quantifying source contributions.

Receptor models contrast with "source-oriented" CTMs, discussed in Chapter 8. Receptor models start with observations at given locations (receptors) and work backward, using as much information as is practical and available to determine the sources contributing to the observations and to quantify their contributions. CTMs combine estimated emission rates with meteorological transport, chemical changes and deposition rates to estimate concentrations and their temporal variations at different receptors, often an array of geographically-distributed grid points. CTMs can be used to predict how atmospheric concentrations or deposition fluxes could change if emission rates are changed. Receptor models explain events that have already occurred and are not used for predictive purposes, except for establishing current conditions for a source-category specific linear rollback (deNevers and Morris, 1975).

Complete understanding of the atmospheric environment and a thorough assessment of possible PM-management strategies are possible only when information from receptor methods, emission inventories, and CTMs is combined. This corroborative, or "weight-of-evidence," approach (U.S. EPA, 2001) provides the strongest scientific basis for specific emission controls. Through this approach decisions can be made with greater confidence when considering air-quality management options and costs in relation to other societal needs.

Several reviews and evaluations of receptor models have been published (Chow and Watson, 2002; Cooper and Watson, 1980; Gordon, 1980, 1988; Gordon et al., 1984; Henry et al., 1984; Henry, 1997a, 2002; Hopke, 1985, 1991, 1999, 2001; Javitz et al., 1988; Watson, 1984; Watson et al., 1981, 1989, 2001, 2002a; Watson and Chow, 2002). This chapter summarizes and complements these efforts, expanding them to a broader concept of receptor methods.

Receptor methods are most useful for: 1) identifying potential sources, especially those that are not included in emission inventories; 2) quantifying contributions from source types to ambient concentrations, initially to target emission reductions, and over the long-term to evaluate control-strategy effectiveness; 3) developing and refining conceptual models for source-receptor relationships; 4) evaluating and improving emission inventories; 5) guiding the application of CTMs; and 6) evaluating and improving CTM results. At present, receptor models have limited capability to distinguish sources of secondary PM compounds except when combined with elements of CTMs. CTMs play an important role in linking secondary PM to its sources while receptor methods can be used to refine the questions asked of the CTM. Successful application of all receptor methods depends on the quality and relevance of the measurements, the availability and specificity of source markers, and the interpretations made by the scientist and air-quality manager.

This chapter describes several receptor methods that have been applied to better explain $PM_{2.5}$ and PM_{10} source contributions and to reconcile emission inventories and source-model estimates. It provides examples of how PM, gas, and meteorological measurements are analyzed to identify and quantify contributions from PM and precursor gas emitters to $PM_{2.5}$ and PM_{10} concentrations measured at receptors.

Source apportionment by receptor methods begins with plausible hypotheses about the causes of high concentrations or effects. These initial hypotheses, which generally focus on various components of the overall conceptual model (Figure 1.1), are essential. They should provide reasonable, although not necessarily accurate, explanations of: 1) potential sources; 2) size, chemical, and temporal characteristics of particle and precursor-gas emissions from these sources; 3) particle and precursor-gas emission rates; 4) meteorological conditions that affect emissions, transport, and transformation; and 5) frequency and intensity of the effect.

The initial hypotheses, as well as the conceptual model, are derived from previous experience (e.g., tests on similar sources, particle movement under similar meteorological conditions), the nature of the problem (neighborhood complaints, consistently poor visibility over a local area or large region, exceedance of air-quality standards), and available measurements (from existing air-quality and meteorological sensors). Sometimes evaluation of the hypotheses immediately quantifies contributions and implies appropriate remedial actions. More often the evaluation directs the application of appropriate receptor methods. The conceptual model, regardless of its level of sophistication, guides the location of monitoring sites, sampling periods, sampling frequencies, sample durations, selection of samples for laboratory analysis, and the particle properties that are quantified in those samples to test and discard hypotheses until, ideally, only one remains.

Several semi-quantitative, observationally based (receptor-oriented) data analyses should be performed prior to quantitative receptor modeling. These simple observational techniques, which are also a type of receptor method, can provide information about sources or source regions influencing the PM concentrations in an area. This helps to develop or refine the conceptual model. These methods often can be completed with data that are already available from existing networks. An overview of these methods and examples of the information they provide are listed below.

Time series plots of single-day, average, or median hourly $PM_{2.5}$ or PM_{10} concentrations. Continuous monitors, described in Chapter 5, acquire hourly (or shorter duration) PM measurements. Diurnal averages of weekend and weekday measurements may demonstrate morning traffic peaks that are accentuated during weekday rush hours. Summertime averages may show only a weekday morning peak, and the evening rush-hour peak may not occur due to deeper atmospheric mixing layers. An evening peak is more likely in the wintertime because an earlier sunset results in a pollutant-trapping surface inversion before rush-hour traffic diminishes. Systematic evening peaks may also reveal other behavior, such as an increase in combustion (wood or fossil fuels) for home heating, which may occur with cooling after sunset and/or when people return to their homes at the end of the day. Hourly measurements from a single day may reveal that a relatively brief event dominates the 24-hr average concentration (e.g., windblown dust, nearby fire, truck idling near sampling inlet). Using short-duration measurements (e.g., 5 to 30 minutes, possible with TEOMs, nephelometers and aethelometers), concentration spikes caused by nearby emitters can be integrated and subtracted from hourly or 24-hr averages to estimate the effects of intermittent local sources (see Figure 7.1). High temporal-resolution measurements may help test the hypothesis that very high concentrations are caused by local sources and that these values do not represent human exposure over larger areas such as a neighborhood or a city.

Averaging by wind speed. When hourly or more frequent measurements of PM_{10} or coarse particles and wind speed are available, concentration averages can be calculated for different wind-speed categories to determine the extent to which wind speed affects dust suspension and the threshold for suspension. The hypothesis that "windblown dust" is a natural and uncontrollable PM_{10} contributor can be tested with measurements near natural surfaces, such as a crusted desert, and near surfaces with manmade disturbances, such as traffic or construction (see Figure 7.2).

Comparisons among source-oriented, neighborhood-scale and urban-scale PM mass and chemical concentrations. Spatial differences in chemical composition and/or mass concentration can indicate influence zones of different sources, which might arise from widespread emitters that affect an entire area, or from middle-scale emitters that affect a smaller population (see Figure 7.3). Large-scale urban/rural differences illustrated in Figure 6.21 can provide an estimate of the importance of upwind sources relative to the emissions within urban areas. PM_{10} and $PM_{2.5}$ mass, ion, element, and carbon measurements are becoming more commonly available in long-term networks and from special studies. These provide a first estimate of the potential for different source types to affect ambient concentrations. High $SO_4^=$ levels imply that precursor SO_2 emissions must be reduced. Large geological contributions determined from Al, Si, Ca, and Fe concentrations imply that fugitive-dust sources require control. High proportions of BC imply primary emissions from incomplete combustion sources such as diesel and gasoline engine exhaust and vegetative burning (e.g., wood fireplaces). High OC levels can result from cooking, small vegetation fragments, and SOA formation (some of which is from biogenic hydrocarbon emissions), as well as vegetative burning and vehicle exhaust. Many trace metals (Fe, As, Se, Cr, Pb, Cu, Zn, Ni, V, Mn) result from industrial emissions. PM enriched in these elements identifies the presence of industrial contributions and the potential for secondary $SO_4^=$ and NO_3^- from SO_2 and NO_x gases that are co-emitted with the primary particles.

Concentration directionality. Particles move with wind-flow patterns. Averaging concentrations associated with wind-direction sectors or multi-day transport pathways narrows down the locations of influential sources. For nearby sources, a simple pollution rose that averages hourly concentrations by wind-direction sector may reveal higher concentrations from sectors that contain point sources or major roadways. This receptor method requires hourly resolution (or better) for PM and wind-direction measurements. For regional contributions, concentration averages may be classified by trajectories that pass over areas with high or low emissions to test the hypothesis that higher concentrations correspond with longer and more frequent transport over these source areas (see Figure 7.4). Transport or wind-direction sectors can be defined according to points on the compass, locations of suspected sources, or applications of cluster analysis or principal-component analysis to identify multivariate meteorological associations (e.g., Green et al., 1992). *There is a need for more observation-based analyses including the use of multiple data sources (e.g., meteorological, emissions, gaseous pollutant measurements). These analyses can provide valuable insight and will*

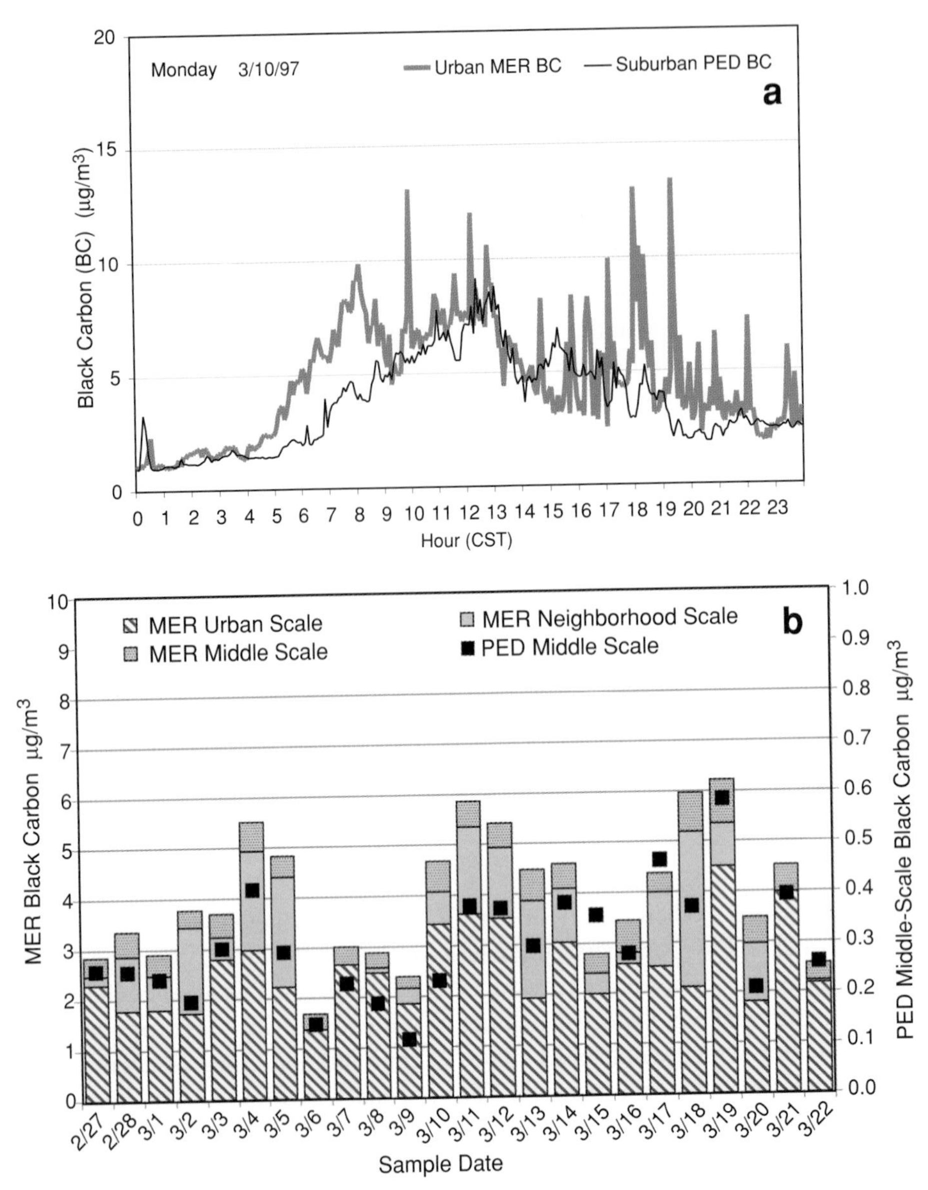

Figure 7.1. (a) Five-minute averages of BC measured with an aethalometer at a downtown (MER, La Merced) and a suburban site (PED, Pedregal) in Mexico City. The sharp spikes at MER are from middle-scale sources (within 0.5 km from the site). The morning traffic buildup (5:00 a.m. to 9:00 a.m.) around the downtown site is evident and is not seen at the suburban site. (b) To estimate the overall urban baseline, the short-term peaks in 7.1a are subtracted from BC at the suburban site (PED) before computing the daily, 24-hr average (peak-adjusted average). The difference between the peak-adjusted and non-adjusted 24-hr average at PED (black squares) is the suburban middle-scale BC. The difference between the peak-adjusted average at MER and PED yields an estimate of the neighborhood-scale (1 to 5 km) contribution at the downtown site (MER). The middle-scale contribution at the downtown site is obtained by subtracting the peak-adjusted 24-hr average at the downtown site from the unadjusted 24-hr average at this site. The magnitudes of the contributions from these 3 different spatial scales are shown for a series of days in 7.1b. This analysis indicates that only ~10 percent of the BC downtown at MER is of very local origin (middle-scale), and thus the site reasonably represents neighborhood exposures. Similar analyses can be performed on 5-minute PM_{10} and $PM_{2.5}$ mass measurements from TEOMs and light-scattering devices (See Chapter 5). (After Watson and Chow, 2001).

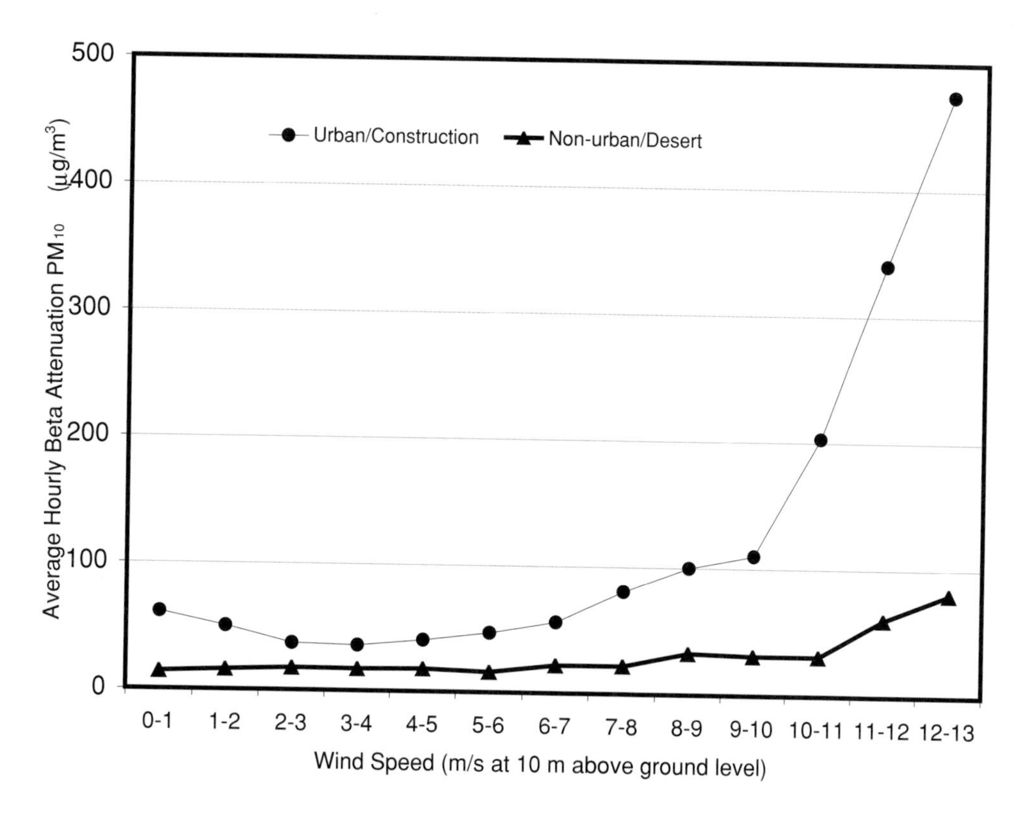

Figure 7.2. Average PM_{10} associated with different wind-speed categories at urban and non-urban locations in the Las Vegas, NV area, (Chow et al., 1999). Dust suspension at the construction site occurs at a lower wind speed, ~6 m/s, than at the desert site where PM_{10} increases are not observed until ~11 m/s. The concentrations are much higher near the disturbed land compared to the crusted desert surface at high wind speeds. This analysis convinced decision-makers that fugitive dust contributions were not just from, and were not dominated by, natural windblown dust. Methods to minimize land disturbances and to stabilize disturbed land during construction were needed to reduce excessive PM_{10} concentrations in Las Vegas. (After Watson et al., 2000).

increase in value as the size of the datasets with PM information grows. These semi-quantitative receptor-oriented data analyses (observational methods) should be performed prior to quantitative receptor modeling.

7.2 RECEPTOR MODEL TYPES

Table 7.1 summarizes receptor models that have been used for source apportionment/attribution along with their data requirements, strengths, and weakness. Watson et al. (2002a) provide more detailed references to their theory and application. Receptor models, such as the Chemical Mass Balance (CMB) model and its derivatives, are derived from the same physical principles as source models. Others, such as eigenvector analysis, are statistically based, showing associations between variables from which causative factors may be inferred.

Receptor models are measurement-intensive, requiring many measurements of physical and chemical properties taken at monitoring sites that represent different spatial scales. Filter samples used for PM_{10}- or $PM_{2.5}$-compliance measurements are not entirely suitable for the type and amount of chemical and physical analysis needed, so special sampling programs must be planned. Some techniques (e.g., CMB) require PM chemical-composition measurements of primary emissions for the sources influencing the ambient PM. Once these data are available, however, receptor methods do not require large manpower or computational resources. Costs for receptor-oriented studies range from ~$US 50K

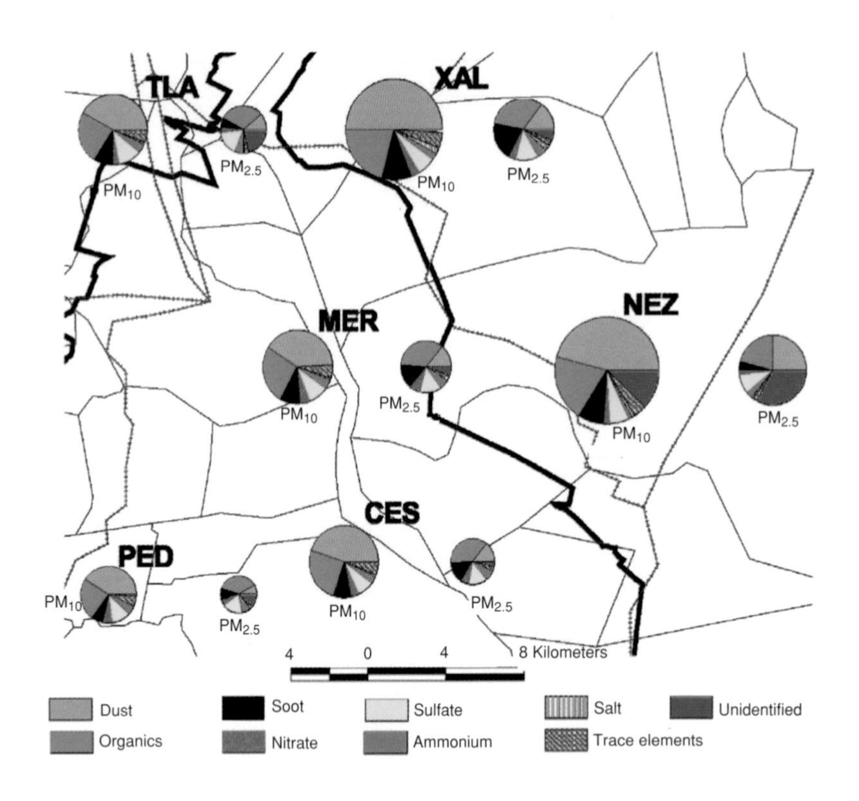

Figure 7.3. Material balance for PM_{10} (larger circles) and $PM_{2.5}$ (smaller circles) at six base sites in Mexico City averaged for the period of March 2 through 19, 1997. Most of the material is of primary origin composed of fugitive dust and carbon. NO_3^- and $SO_4^=$ levels are relatively low. The effect of salt from dry Lake Texcoco, located in the upper right-hand corner, is evident at the XAL and NEZ sites, but not at the other sites. The XAL and NEZ sites appear to have larger local influences than the other sites, especially for the dust portion of PM_{10}. The MER, TLA, and CES locations have similar PM concentration and compositions. Differences in soot (BC), measured by the IMPROVE protocol (Chow et al., 1993a), between the MER and PED sites are similar to those derived independently in Figure 7.1. This plot is being used by the air-quality district to evaluate and improve the compliance and alert monitoring network operated in Mexico City. (After Chow et al., 2002a).

for a small community with an established air-quality network (e.g., Chow et al., 1992) to ~$US 10M for a large region with a complex mixture of sources requiring establishment of a network and survey of potential sources (e.g., Pitchford et al., 1999). Many of the receptor-model measurements are the same as those that should be acquired to support a well-rounded source-model application.

Measurements of elements, ions, OC, and BC were usually sufficient in the past for successful source apportionment using receptor models. However, emission controls (e.g., precipitators and baghouses on stacks) and changes in processes and fuels (e.g., removal of lead from gasoline) have depleted many of the marker species, or have made their abundances more variable, in source emissions. Properties of single particles, organic compounds, and operationally-defined carbon fractions are being measured to compensate. *There is a need to identify a greater variety of unique markers or tracers of specific sources and processes, particularly organic compounds, to improve the ability of receptor-modeling techniques to identify and quantify source contributions. This requires development of source and ambient sampling techniques and chemical extraction and analysis methods, and then testing of these new tools under a variety of situations.*

Multivariate receptor models require a large number of PM samples with chemical characterization. These models derive information from the temporal and spatial variability in the chemical concentrations and inter-relationships between the different chemical

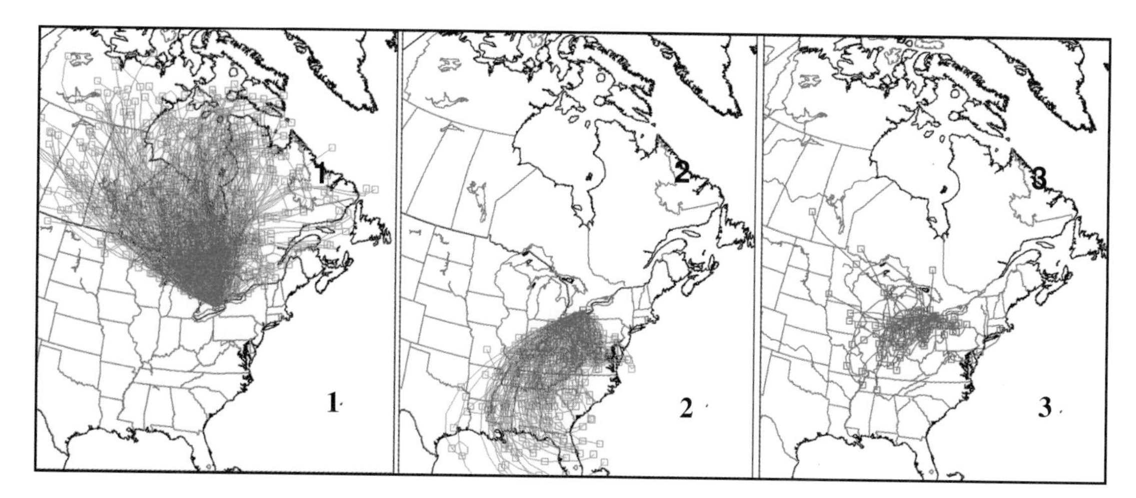

Figure 7.4. Three-day back-trajectories arriving at Simcoe, ON, during May-September of 1998 and 1999 were sorted by transport sector. Back-trajectories represent the most probable path that the air mass followed en route to Simcoe. Sectors 1 and 2 represent: 1) northerly flow over predominantly Canadian source regions, and 2) southerly flow over U.S. source regions. Six-hr average $PM_{2.5}$ loadings measured by a TEOM were 6.7 ±6.1 µg/m^3 (±1SD) for sector 1 and 22.4 ±11.7 µg/m^3 for sector 2. Sector 3 includes trajectories corresponding to $PM_{2.5}$ >30 µg/m^3, which could not be classified into either of the other sectors because they cross over regions. The cut-off value of 30 was used because it approximates the Canadian standard for 24-hr average $PM_{2.5}$ (see Chapter 1 for details). The unclassifiable high-concentration cases (sector 3) were associated with very short transport distances, indicating stagnant conditions in the Midwest and Great Lakes region. There were 5 six-hr concentrations >30 µg/m^3 in sector 1 and 51 in sector 2. This information is being used to estimate the portion of the $PM_{2.5}$ in the Simcoe and Toronto areas that is caused by local emissions vs. transport from other parts of Canada and the United States. (Brook et al., 2002).

concentrations. At least 50 samples are needed, and more than 100 samples from many locations and/or one location over a long time period are preferred. Samples should include high and low contributions from the suspected sources, usually requiring <24-hr averages. Sampling sites should be most densely located along strong concentration gradients for a pollutant or source. Network adequacy (Chow et al., 2002b) is not usually evident until after the measurements have been acquired.

7.2.1 Chemical Mass Balance

The CMB model (Friedlander, 1973) expresses ambient chemical concentrations as the sum of products of species abundances and source contributions. These equations are solved for the source contributions when ambient concentrations and source profiles are supplied as model input. Source profiles consist of the mass fractions of selected particle properties in source emissions; examples are given below. The CMB is the only receptor air-quality model that calculates quantitative uncertainty values for source-contribution estimates. It provides the basic structure for other receptor models that can be derived from physical principles. The CMB model has been applied to evaluate PM control effectiveness (e.g., Chow et al., 1990; Engelbrecht et al., 2000), determine causes of high concentrations (Lowenthal et al., 1997; Schauer and Cass, 2000; Vega et al., 1997), quantify contributions to light extinction (e.g., Watson et al., 1988, 1996), and to estimate sources of particle deposition (Feeley and Liljestrand, 1983).

Several different CMB solution methods have been applied, but the multivariate, effective-variance, least-squares estimation method (Watson et al., 1984) is most commonly used because it incorporates precision estimates for all of the input data into the solution and propagates these errors to the model outputs. The tracer solution, in which a single chemical component or particle property is used to represent a single source or source type, is also in common use, but source-contribution estimates are

Table 7.1. Summary of receptor model source-apportionment models.

Receptor Model (Example Applications)	Data Requirements	Strengths	Weaknesses
7.2.1 Chemical Mass Balance	-Source and receptor measurements of stable aerosol properties that can distinguish source types. -Source profiles (mass abundances of physical and chemical properties) that represent emissions pertinent to the study location and time. -Uncertainties that reflect measurement error in ambient concentrations and profile variability in source emissions. -Sampling periods and locations that represent the effect (e.g., high PM, poor visibility) and different spatial scales (e.g., source dominated, local, regional).	-Simple to use, software available. -Quantifies major primary source contributions with element, ion, and carbon measurements. -Quantifies contributions from source sub-types with single particle and organic compound measurements. -Provides quantitative uncertainties on source contribution estimates based on input data, uncertainties and colinearity of source profiles. -Has potential to quantify secondary SO_4 contributions from single sources with gas and particle profiles when profiles can be "aged" by chemical transformation models.	-Completely compatible source and receptor measurements are not commonly available. -Assumes all observed mass is due to the sources selected in advance, which involves some subjectivity. -Does not directly identify the presence of new or unknown sources. - Typically does not apportion secondary particle constituents to sources. Must be combined with profile aging model to estimate secondary aerosol. -Much colinearity among source contributions without more specific markers than elements, ions, and carbon.
7.2.1 Injected Marker CMB Tracer Solution	-Non-reactive marker(s) added to a single source or set of sources in a well-characterized quantity in relative to other emissions. Sulfur hexafluoride, perfluorocarbons, and rare earth elements have been used.	-Simple, no software needed. -Definitively identifies presence or absence of material from release source(s). -Quantifies primary emissions contributions from release source(s).	-Highly sensitive to ratio of marker to PM in source profile; this ratio can have high uncertainty. -Marker does not change with secondary aerosol formation—needs profile aging model to fully account for mass due to "spiked" source. -Apportions only sources with injected marker. -Costly and logistically challenging.

Table 7.1. Summary of receptor model source-apportionment models. (continued)

Receptor Model (Example Applications)	Data Requirements	Strengths	Weaknesses
7.2.2 Enrichment Factor (EF)	-Inorganic or organic components or elemental ratios in a reference source (e.g., fugitive dust, sea salt, primary carbon). -Ambient measurements of same species.	-Simple, no software needed. -Indicates presence or absence of emitters. -Inexpensive. -Provides evidence of secondary PM formation and changes in source profiles between source and receptor.	-Semi-quantitative. More useful for source/process identification than for quantification.
7.2.3 Multiple Linear Regression (MLR)	-100 or more samples with marker species measurements at a receptor. -Minimal covariation among marker species due to common dispersion and transport.	-Operates without source profiles. -Abundance of marker species in source is determined by inverse of regression coefficient. -Apportions secondary PM to primary emitters when primary markers are independent variables and secondary component (e.g. $SO_4^{=}$) is dependent variable -Implemented by many statistical software packages.	-Marker species must be from only the sources or source types examined. -Abundance of marker species in emissions is assumed constant with no variability -Limited to sources or source areas with markers. -Requires a large number of measurements.
7.2.4 Eigenvector Analysis [a]	-50 to 100 samples in space or time with source marker species measurements. -Knowledge of which species relate to which sources or source types. -Minimal covariation among marker species due to common dispersion and transport. -Some samples with and without contributing sources.	-Intends to derive source profiles from ambient measurements and as they would appear at the receptor. -Intends to relate secondary components to source via correlations with primary emissions in profiles. -Sensitive to the influence of unknown and/or minor sources.	-Most are based on statistical associations rather than a derivation from physical and chemical principles. -Many subjective rather than objective decisions and interpretations -Vectors or components are usually related to broad source types as opposed to specific categories or sources.

[a] Includes Principal Component Analysis [PCA], Factor Analysis {FA], Empirical Orthogonal Functions [EOF], Positive Matrix Factorization [PMF], and UNMIX).

Table 7.1. Summary of receptor model source-apportionment models. (continued)

Receptor Model (Example Applications)	Data Requirements	Strengths	Weaknesses
7.2.5 Time Series	-Sequential measurements of one or more chemical markers. -100s to 1000s of individual measurements.	-Shows spikes related to nearby source contributions. -Can be associated with highly variable wind directions. -Depending on sample duration, shows diurnal, day-to day, seasonal, and inter-annual changes in the presence of a source.	-Does not quantify source contributions. -Requires continuous monitors. Filter methods are impractical.
7.2.6 Aerosol Evolution	-Emissions locations and rates. -Meteorological transport times and directions. -Meteorological conditions (e.g., wet, dry) along transport pathways.	-Can be used parametrically to generate several profiles for typical transport/ meteorological situations that can be used in a CMB.	-Very data intensive. Input measurements are often unavailable. -Derives relative, rather than absolute, concentrations. -Level of complexity may not adequately represent profile transformations.
7.2.7 Aerosol Equilibrium	-Total (gas plus particle) $SO_4^=$, NO_3^-, NH_4^+ and possibly other alkaline or acidic species over periods with low temperature and relative humidity variability. -Temperature and relative humidity.	-Estimates partitioning between gas and particle phases for NH_3, HNO_3, and NH_4NO_3. -Allows evaluation of effects of precursor gas reductions on ammonium nitrate levels.	-Highly sensitive to T and RH. Short duration samples are not usually available. -Gas-phase equilibrium depends on particle size, which is not usually known in great detail. -Sensitivity to aerosol mixing state not understood/quantified.

more uncertain than with the multivariate solution because they are highly dependent on the variability of the tracer-to-mass ratio. Artificial tracers (injected markers), such as rare-earth elements, sulfur hexafluoride, and halocarbon gases have been released with normal emissions and used in a CMB tracer solution to estimate contributions from many sources using the same fuel and from individual emitters (Horvath et al., 1988; Ondov et al., 1992; Pitchford et al., 1999).

The accuracy and precision of CMB source-contribution estimates depend on the variability of chemical abundances in the source profiles. Source profiles should represent source compositions as they would be perceived at a receptor. Profiles are most often measured at the source, and changes that take place during transport are not realized. These "fresh" profiles do not account for secondary PM formation and particle volatilization that may take place from source to receptor. Source profiles also must contain several chemical components, which differ greatly in their abundances among different source types. Profiles often have not been measured on representative sources, or on enough different sources, to obtain reasonable estimates of profile variability. The propagated standard errors associated with CMB source-contribution estimates underestimate the true uncertainties of the apportionment in these cases.

The CMB attempts to explain the ambient chemical concentrations selected for its calculation with the source profiles provided. Excessive differences between calculated and measured concentrations usually indicate that profiles are insufficient, or that a contributing profile is missing. It is the responsibility of the modeler, not the CMB model, to evaluate the CMB uncertainties and performance measures, along with other data such as emission inventories, before accepting the results. This is true of all receptor- and source-oriented models.

Size-specific source profiles, including a variety of chemical species, need to be obtained. Consistent and well-characterized measurement techniques should be applied and a wide variety of sources, particularly those that are suspected to impact many geographic areas, should be surveyed. Enough source measurements from a series of representative sources to determine reasonable estimates of profile variability should be obtained. Gaseous pollutant measurements and extensive VOC and SVOC profiles should be obtained in conjunction with the PM source profiles. Not only will source-profile information benefit receptor modeling, leading to more confident results, but significant improvements in emission inventories and emission models will also result. This will improve the results from source-oriented models.

7.2.2 Enrichment Factors

The EF model (Sturges, 1989; Lawson and Winchester, 1979; Mason, 1990) compares ratios of chemical concentrations in the atmosphere to the same ratios in a reference material associated with a stable and well-characterized source, such as average crustal composition for a region, sea salt, or OC to BC in combustion sources. These ratios are often normalized to the Si concentrations in air and in soil, as Si is the most abundant element in suspended soil. Reference ratios vary substantially between bulk soil and suspendable particles, and among different regions (Rahn, 1976). Differences are explained in terms of other sources, but not with quantitative apportionment as in the CMB.

Heavy-metal enrichments are usually attributed to industrial emitters. Sulfur enrichment is attributed to secondary $SO_4^{=}$-containing particles formed from gaseous SO_2 emissions. Organic carbon enrichment is attributed to SOA and non-reference primary emitters. Potassium, and sometimes Cl^-, enrichments are attributed to burning and cooking. Vanadium and Ni enrichments indicate contributions from residual oil combustion or refinery catalyst crackers, while enriched Se is usually attributed to coal-fired power stations. Iron, Mn, Cu, Zn, and Pb enrichments indicate steel-mill, smelting, or plating contributions. Calcium is often enriched near cement manufacture or the use of cement products in construction. Sodium and Cl^- are enriched near the coast and dry lake beds, and after de-icing materials are applied to streets.

For many applications, EFs can be determined by elemental analysis of Teflon filter samples, typical of those obtained with $PM_{2.5}$ compliance monitors. Enrichment factors are useful for forming a conceptual model of potential sources that might affect higher loadings, especially industrial emitters. The EF method also has been used to examine different geological strata to determine how specific elements, such as Pb, may have deposited over long time periods (Weiss et al., 1999).

7.2.3 Multiple Linear Regression on Marker Species

The MLR model expresses particle mass, a component of mass such as $SO_4^{=}$, light extinction, or another observable, as a linear sum of unknown regression coefficients multiplied by source-marker concentrations measured at a receptor. The markers, which may be chemical elements or compounds, must originate only in the source type being apportioned. The regression coefficients represent the inverse of the source-profile chemical abundance of the marker species in the source emissions. The product of the regression coefficient and the marker concentration for a specific sample is the tracer solution to the CMB equations and yields the source-contribution estimate. MLR has been used to estimate source contributions to high PM concentrations (Morandi et al., 1991), to relate light extinction and $SO_4^=$ to power-station and smelter emissions in the western United States (Malm et al., 1990), and to apportion $SO_4^=$ to different source regions in the eastern United States (Lowenthal and Rahn, 1989).

7.2.4 Temporal and Spatial Correlation Eigenvectors

Principal Component Analysis (PCA), Factor Analysis (FA), Empirical Orthogonal Functions (EOF), Positive Matrix Factorization (PMF), and UNMIX are all variations on temporal and spatial eigenvectors (e.g., Henry, 1991; Hopke 1988; Paatero, 1997). There is a large variability in how these methods are applied and interpreted. The patterns from most of these methods are empirically derived, as a basic-principles derivation has not yet been formulated. PCA, FA, and EOF are applied to correlations between variables over time or space. Temporal correlations are calculated from a long time series (at least fifty, but preferably several hundred samples) of chemical concentrations at one or more locations. Eigenvectors of this correlation matrix (a smaller matrix of coordinates that can reproduce the ambient concentrations when added with appropriate coefficients) are determined and a subset is rotated in multi-dimensional space to maximize and minimize correlations of each vector with each measured species. These rotated vectors (also called "factors," "principal components," or "orthogonal functions") are interpreted as source profiles by comparison of factor loadings with source measurements. Several different normalization and rotation schemes have been used, but this is where the physical derivation of the model is lost; the rotations depend on statistical or judgmental rather than physical criteria. The PMF method (Paatero, 1997) provides a physical basis for estimating eigenvectors that can be related directly to sources when a fairly stringent set of assumptions is met. This method is also reported to provide source profiles from the ambient data, but these have not been verified against actual profile measurements in most cases. The UNMIX method (Henry, 2000) also has a physical justification by searching for "edges" in multi-dimensional space to determine ratios among ambient marker species.

For a large spatial network (25 to 50 locations), spatial correlations are calculated from chemical measurements taken on simultaneous samples across the monitoring locations (Henry, 1997b; White, 1999). Eigenvectors of this correlation matrix represent a spatial distribution of source influences over the area, provided that the samplers have been located to represent the gradients in source contributions. As with temporal correlation models, several normalization and rotation schemes have been applied and the physical derivation is still lacking.

Eigenvector receptor methods have been used to evaluate contributions to high particle concentrations that might affect health (Thurston and Spengler, 1985), distinguish regional contributors to $PM_{2.5}$ and $SO_4^=$ concentrations (Poirot et al., 2001), and to determine $SO_4^=$ contributions to light extinction over large regions (Henry, 1997c). They are used in exploratory analyses to identify the potential for emissions that are not yet inventoried. For example, Magliano (1988) identified an eigenvector containing a common variation between Cu and $SO_4^=$ in California's San Joaquin Valley. Further investigation determined that $CuSO_4$ was sprayed on orchards during winter to lower the freezing point of the fruit.

7.2.5 Time Series

Spectral analysis (Hies et al., 2000), intervention analysis (Jorquera et al., 2000), lagged regression analysis (Hsu, 1997), and trend analysis (Somerville and Evans, 1995) models separate temporal patterns for a single variable and establish temporal relationships between different variables. These models have been used to identify sources, to forecast future pollutant concentrations based on past experience, to infer relationships between causes and effects, and to relate long-term trends in PM chemical concentrations to emission changes. It is especially important to include meteorological indicators in time-series models (e.g., Shively and Sager, 1999; Weatherhead et al., 1998) and to use data sets with comparable measurement methods and sampling frequencies (Henry et al., 2000).

7.2.6 Neural Networks

The neural-network approach (Reich et al., 1999; Wienke et al., 1994) involves constructing a synthetic logic network, patterned after biological reasoning systems, by presenting the neural network with known inputs and outputs representing the source-receptor system. These known inputs and outputs are referred to as training sets, and once the linkages are established by the neural net, it recognizes similar patterns and relationships among ambient variables with unknown causes. During training, the network assigns weights to the inputs that reproduce the outputs. Neural networks can provide functional relationships that generally turn out to be solutions to the MLR and CMB equations (Song and Hopke, 1996). Their results are only as good as the known relationships within the training sets and the extent to which the unknown data sets share the same patterns within the training sets.

7.2.7 Aerosol Evolution and Equilibrium

Atmospheric "aging" of source profiles containing PM chemical components (with or without size-distribution information) and gaseous precursors is modeled using appropriate chemical-reaction and aerosol-process schemes (Lewis and Stevens, 1985; Watson et al., 2002b). Aging is assumed to occur in a "box" or "puff" moving along a trajectory path from source to receptor (Kleeman et al., 1999a; Gordon and Olmez, 1986). Several of the substances reaching a receptor, such as NH_3/NH_4^+ and NO_3^-, may be in both the vapor and particle phases, depending on temperature, humidity, and precursor-gas concentrations. Aged profiles have been used to estimate $SO_4^=$ contributions to light extinction (Latimer et al., 1990).

Aerosol-evolution models provide the most explicit and useful link between source-oriented chemical-transport and receptor models. A source-oriented air-quality model with a range of inputs could develop individual source or regional source profiles, which might be used with the CMB or other receptor samples. The results also could be used to evaluate source model inputs and representations of transformation processes. The profiles derived from specialized source-modeling experiments could be applied to future samples with similar meteorology but without a full source-modeling effort. *Methods for developing "aged" source profiles and for subsequent application of these profiles in receptor models, such as CMB, need to be evaluated. This could extend the capabilities of receptor models to include apportionment of secondary PM or to account for differential attrition of or addition to the primary emissions due to physical and chemical processes.*

Aerosol-equilibrium models (referred to in Chapter 8 as observation-based models) have been used to determine whether HNO_3 or NH_3 reductions are needed to reduce NH_4NO_3 (Ansari and Pandis, 1998; Blanchard et al., 2000; Watson et al., 1994a). Equilibrium models also have been used to determine the extent to which $SO_4^=$ reductions will free up NH_3, potentially leading to an increase in NH_4NO_3 in $PM_{2.5}$ (West et al., 1999). These models are simpler than the aerosol-evolution models because fewer processes are considered and transport is not taken into account.

Chapter 3 includes some examples of the application of an equilibrium model for eastern North America. Equilibrium models have been combined with aerosol evolution models to determine the extent to which HNO_3 formation is limited by NO_x or VOC emissions and to determine PM credits for precursor NO_x emission reductions (Stockwell et al., 2000).

7.3 RECEPTOR-MODEL INPUT MEASUREMENTS

As previously noted, receptor models depend on emission characteristics that remain stable, or change in predictable ways, during transport from source to receptor. These include particle size, chemical composition, and variability in any of these parameters in space and time. Chapter 5 describes the methods used to quantify these characteristics. Chemical abundances and spatial/temporal variations are indicative of the main processes and sources, manageable or unmanageable, contributing to PM concentrations. For example, the relative amount of $SO_4^=$ decreases East to West in Canada and the United States, while the amounts of other compounds, such as NO_3^- and OC increase, indicating that SO_2 sources are a more important focus for $PM_{2.5}$ reductions in the East than in the West.

7.3.1 Particle Size

As explained in Chapter 3, different particle-size fractions are dominated by different source types. $PM_{2.5}$ and PM_{10} size fractions are most commonly measured, often at the same locations, and most receptor-modeling studies have used temporal or spatial variations in the bulk chemical composition of these fractions. Variations in particle size and chemical composition can be further exploited for quantitative source apportionment.

The ultrafine fraction is usually measured in terms of particle number by condensation particle counters or as mass concentration on the final stages of low-pressure impactors. Ultrafine particles are abundant in fresh combustion emissions, but they rapidly coagulate with each other and with larger particles shortly after emission (Preining, 1998). As a result, high ultrafine number counts typically indicate the influence of fresh combustion emissions, and lower particle-number concentrations indicate that nearby sources are not large contributors. Buzorius et al. (1999) measured 10,000 to 50,000 particles/cm^3 near a heavily traveled street, while levels at a more pristine site were typically 2,000 to 3,000 particles/cm^3. Time series of the particle-number distribution, along with CO and NO_x concentrations, have been used together to apportion particle numbers to gasoline and diesel sources (Wahlin et al., 2001).

The accumulation mode usually contains most of the $PM_{2.5}$ mass and most of its chemical variability. Particle properties in the accumulation mode most often have been used for receptor modeling. These particles can be directly emitted "primary particles" or arise from rapid condensation of primary SVOC. Secondary particles that form in the atmosphere from gaseous emissions of SO_2, NO_x, NH_3, and some VOCs also occupy this mode. Within the accumulation mode there may be sub-modes indicating secondary particle-formation processes (John et al., 1990). The condensation mode (peaking at ~0.2 μm) consists of smaller particles, which form from condensation and coagulation in dry air. The droplet mode (peaking at ~0.7 μm) forms from activated particles and gases adsorbed by, and reacted in, these cloud and fog droplets; when the water in the droplet evaporates, a larger amount of $SO_4^=$ or NO_3^- remains than is in the condensation mode. The sub-mode occupied by $SO_4^=$ in the size spectrum allows inferences about the history of the gases and processes that formed it.

Coarse particles, often reported as TSP minus $PM_{2.5}$ or PM_{10} minus $PM_{2.5}$ mass, result from the disaggregation of liquid or solid material, mostly fugitive dust or sea salt. Concentrations are bounded at larger particle sizes by gravitational settling. This settling results in dustfall nuisances near coarse-particle emitters such as heavily-traveled unpaved roads, construction sites, and mineral-handling facilities. As a result of settling, the peak of the coarse distribution shifts to smaller particles with transport distance from the point of emissions. Coarse particles can travel over long distances when they are injected high into the atmosphere by severe wind storms (Perry et al., 1997), but the peak of this transported coarse mode is closer to 3 μm than 10 μm.

7.3.2 Chemical Composition

Chemical composition of particles provides the most information on their origins. The most abundant and commonly measured constituents of $PM_{2.5}$ or PM_{10} mass are OC, BC, geological material, $SO_4^=$, NO_3^-, and NH_4^+. Soluble salts such as NaCl are often found near the ocean, open playas, and after road de-icing. Analysis for these common components indicates the extent to which mass concentrations result from fugitive dust, secondary particles, or carbon-generating combustion activities such as vehicle exhaust, vegetative burning, and cooking. More specific chemical characterization then can be focused on the components with the largest mass fractions.

For quantitative source apportionment, source profiles must contain chemical abundances (fractions of emitted $PM_{2.5}$ or PM_{10} mass) for a range of components that differ among source types, that do not change appreciably during transport between source and receptor (or that allow changes to be simulated by measurement or modeling), and that are reasonably constant among different emitters of the same type (e.g., motor vehicles) and source operating conditions. Minor chemical components, constituting of less than 1 percent of particle mass, are needed for quantitative apportionment as they are more numerous and more likely to occur with patterns that allow differentiation among sources.

7.3.2.1 Soil, Dust, and Industrial Markers

Aluminum, Si, K, Ca, and Fe are most abundant in geological material, although the abundances vary by type and use of soil. Figure 7.5a is a profile from a paved road. Low levels of Pb deposited from exhaust and exhaust systems due to prior use of leaded gasoline is evident. Organic carbon from deposited exhaust, oil drippings, asphalt, tire wear and ground up plant detritus is often more abundant in road dust than in pristine desert soils. Trace metals such as Cu and Zn probably originate from brake and clutch linings as well as from metals used in vehicle construction. Water-soluble K is about one-tenth of the total K abundance in most fugitive dusts, indicating that other sources are responsible for water-soluble K in atmospheric samples when this ratio is exceeded.

Trace elements have been useful as markers for different industrial or combustion-source contributions (Biegalski et al., 1998). However, elements identified and quantified 1) in residual oil combustion and refinery catalyst crackers (V and Ni); 2) in leaded-gasoline combustion (Pb, Cl and Br); and 3) in metal smelting and refining processes (Cu, Fe, Mn, Cr, Zn, Zr, and As) have been removed from many fuels, processes, and industrial exhaust streams. While good for the environment, these modifications have decreased the utility of elemental abundances to distinguish among source contributions in receptor models. As highlighted above, additional chemical and physical properties beyond the commonly measured elements, ions, and carbon need to be examined to determine the ones that are useful in practical source-apportionment applications.

7.3.2.2 Combustion Markers

Figure 7.5b from a coal-fired power station contains many of the same elements found in fugitive dust, but in different proportions. Selenium, Sr, and Pb are enriched over the dust profile, and OC is depleted. The ratio of SO_2 to $PM_{2.5}$ emissions in this sample is large, and SO_2 is a good indicator of fresh emissions from this source type. This ratio changes with time, however, as SO_2 deposits more rapidly than $PM_{2.5}$ and augments $PM_{2.5}$ mass, generally within a day after emission, as it transforms into $SO_4^=$.

The gasoline-vehicle exhaust profile in Figure 7.5c shows OC and BC as its major components, with OC two to three times BC. A wide variety of different elements is found in these emissions, but their abundances are highly variable. This is consistent with many of the metals originating from exhaust-system deterioration and motor oil that differ substantially from vehicle to vehicle. The CO to $PM_{2.5}$ ratio is highest for gasoline-vehicle exhaust, but it is also highly variable.

Figure 7.5d shows that wood-burning emissions are also dominated by carbon, but the OC abundance is usually much higher than the BC fraction. Most of the K is water-soluble, in contrast to the insoluble K in fugitive dust. Other soluble salts containing $SO_4^=$, NO_3^-, NH_4^+, and Cl^- are evident. Trace metals from dust deposited on the burned wood and from stove or fireplace deterioration are often evident in these emissions.

7.3.2.3 Secondary Sulfate and Nitrate

Sulfate, NO_3^-, and NH_4^+ abundances are low in the primary emission profiles shown in Figure 7.5 because most of these are formed in the atmosphere from primary emissions of gaseous SO_2, NO_x, and NH_3. Receptor models can easily separate primary from secondary origins when stoichiometric abundances in compounds such as H_2SO_4, NH_4HSO_4, $(NH_4)_2SO_4$, NH_4NO_3, and $NaNO_3$ are included as source profiles. Although receptor models do not generally attribute these compounds to a specific source, the quantities of the different secondary compounds indicate the age of the material (e.g., $H_2SO_4$is usually from a nearby emitter that has not been exposed to sufficient NH_3 for neutralization).

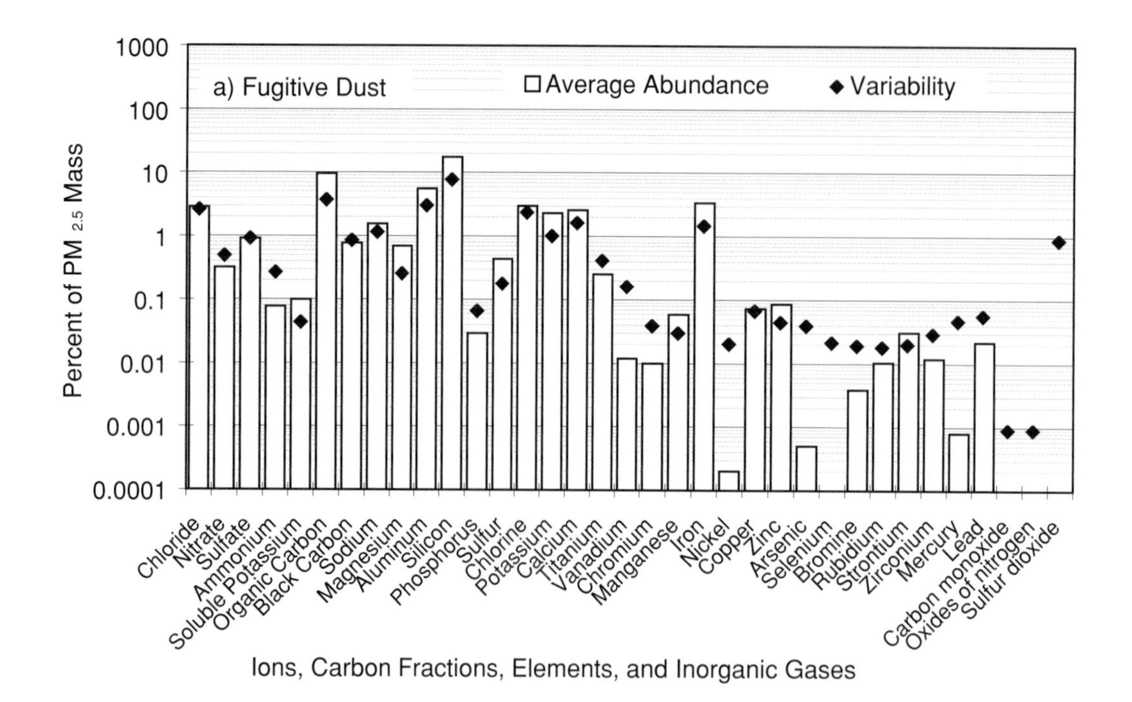

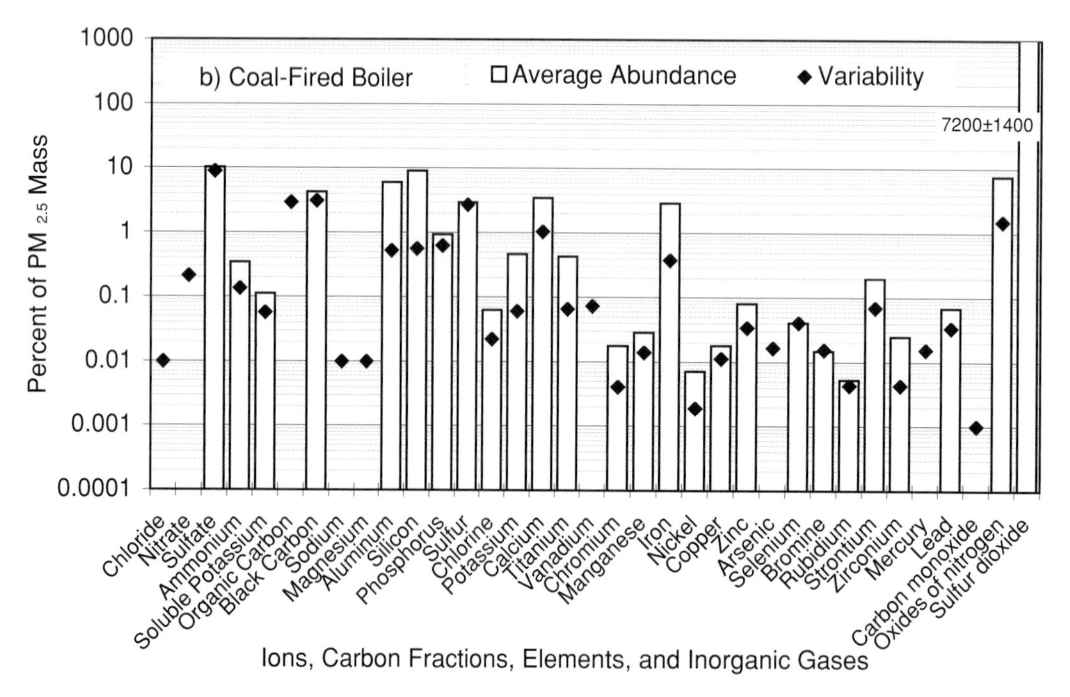

Figure 7.5. Examples of $PM_{2.5}$ source profiles determined from carbon, ion, and elemental analyses of diluted source samples from representative sources for a) fugitive dust, b) a coal fired power station, c) hot stabilized gasoline vehicle exhaust, and d) hardwood burning from emitters typical of Denver, CO, during winter, 1996 (Zielinska et al., 1998). The heights of the bars represent the average abundance of each species as determined by the average from several source tests and the diamonds represent the standard deviation of the average derived from the same tests. Ratios of CO, NO_x, and SO_2 to $PM_{2.5}$ mass emissions can also be included in source profiles and used in aerosol evolution models to determine how the profile might change during source/receptor transport. BC and OC were determined using the IMPROVE thermal/optical reflectance protocol (Chow et al., 1993a). (After Watson and Chow, 2002).

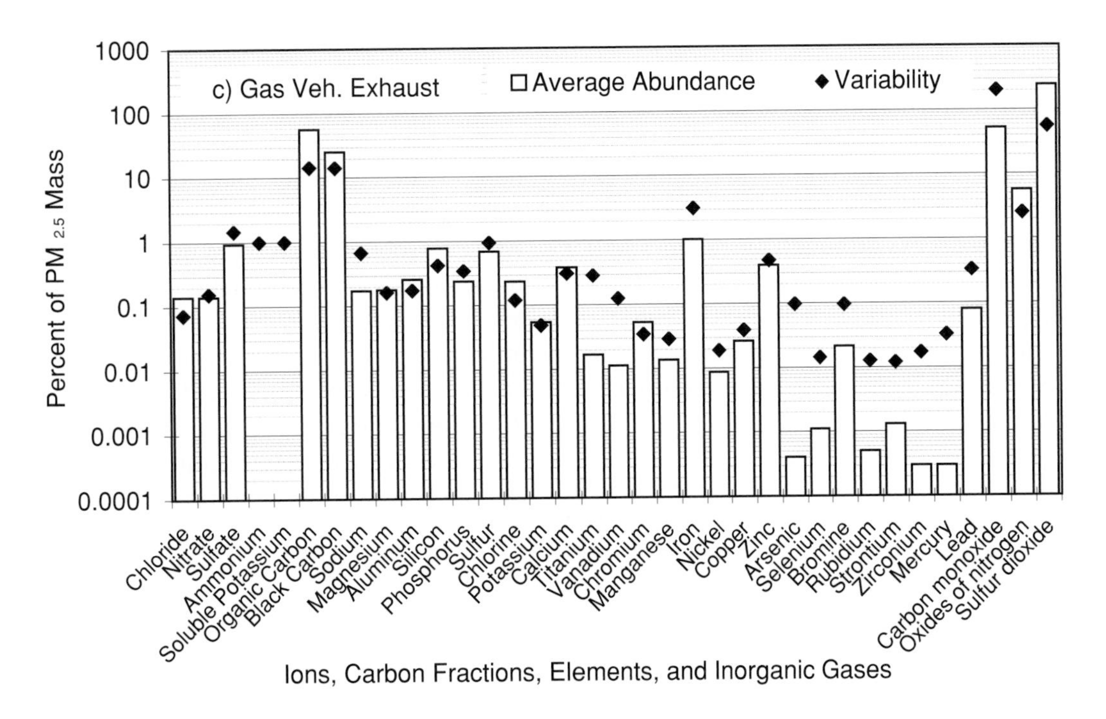

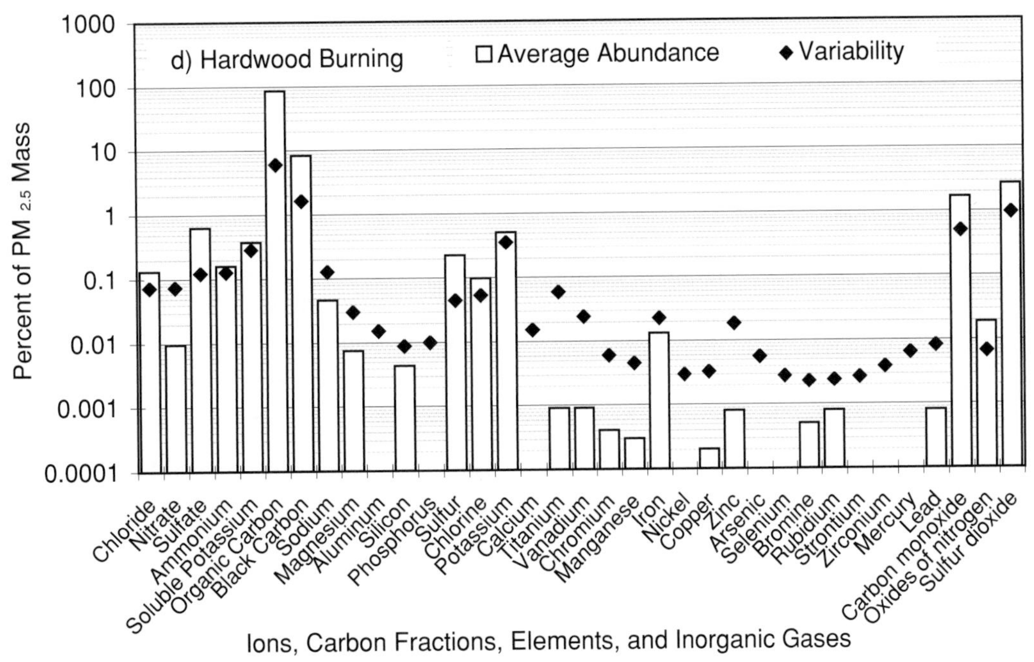

Figure 7.5. Examples of $PM_{2.5}$ source profiles. (continued)

Over North America, anthropogenic emissions, specifically those from coal and oil combustion for power generation and metal smelting and refining, are the largest SO_2 emitters. Diesel fuel and gasoline contain some residual sulfur, but this is being reduced in the United States and Canada because it interferes with the effectiveness of catalytic converters (Lloyd and Cackette, 2001). In remote areas, natural sources, such as DMS, H_2S, and OCS from marine and terrestrial biota, volcanoes, and biomass burning are important, but the resulting $SO_4^=$ levels are typically <1 μg/m³.

Attributing $SO_4^=$ to its original SO_2 source (e.g., smelters, power plants, steel manufacturing, natural emissions) is uncertain because SO_2 in the atmosphere behaves the same regardless of its origin. However, within a large (~1000 km) geographic region, the largest SO_2 emitters, on average, probably contribute the largest fraction of $SO_4^=$. Direct or primary $SO_4^=$ emissions also can arise from combustion sources, although in much smaller amounts than SO_2. Primary $SO_4^=$ is often part of a more complex molecule such as $FeSO_4$, $NiSO_4$, or organic sulfur. However, most (>90 percent) fine-particle $SO_4^=$ forms in the atmosphere and usually is associated with NH_4^+.

Small amounts of PM NO_3^- are produced during combustion or formed in hot exhaust. As described in Chapter 3, most NO_3^- (>95 percent) derives from atmospheric oxidation of NO_x into species such as gaseous HNO_3, N_2O_5 and NO_3 radical. Much of the N_2O_5 and NO_3 end up as HNO_3, which can combine with NH_3 to produce NH_4NO_3. NO_x is produced in the presence of high temperature (e.g., combustion), so fuel combustion is ultimately responsible for much of the NO_3^-. Similar to $SO_4^=$, attributing NO_3^- to a particular combustion source is uncertain. This is compounded by the interplay between NH_3 and HNO_3 and the fact that PM NO_3^- can convert back to these gaseous precursors.

NH_4^+ is usually associated with ammonium sulfate and ammonium nitrate, and derives from gaseous NH_3 emissions. Most NH_3 is of biogenic origin, associated with livestock waste products and vegetative decay. NH_3 is also emitted by some industries and in cities (due to people and pets, catalyst-equipped vehicles and wastewater). NH_3 rapidly neutralizes H_2SO_4. Once the $SO_4^=$-related acidity has been neutralized, the remaining NH_3 may react with HNO_3 to create NH_4NO_3 particles. In an NH_3-depleted atmosphere containing HNO_3, reducing $SO_4^=$ levels may free NH_3 that can then create NH_4NO_3 particles, as described in Chapter 3 and as exemplified later in this chapter.

7.3.2.4 Carbonaceous Particles

As defined in Chapter 1, carbonaceous particles refer to both BC and OC. Although total carbon is comparably measured by different methods, the observed OC and BC fractions depend on the analysis method. Several different methods are in common use (e.g., Schmid et al., 2001; Currie et al., 2002), and their operational definitions for OC and BC do not always coincide. For receptor modeling, the definition of carbon fractions, especially OC and BC, must be the same for ambient and source-profile measurements. As noted in earlier chapters the terms BC, EC, soot, and LAC are used loosely and often interchangeably by air-quality, atmospheric, health, and industrial researchers. EC is not found in the atmosphere in its purest forms of diamond (four carbon bonds) or graphite (three carbon bonds). Therefore, referring to EC on atmospheric particles is misleading and a more ambiguous term like BC, as used here, is more appropriate. Atmospheric BC particles are believed to be the product of incomplete combustion of carbon-containing fuels in an oxygen-starved environment. Chang et al. (1982) define BC as "...a complex three-dimensional polymer with the capability of transferring electrons..." Seinfeld and Pandis (1998) refer to BC as soot that forms when carbon-to-oxygen ratios during combustion are less than one. This soot contains randomly oriented crystals 2 to 3 nm in diameter with a graphite-like structure, interspersed with other elements. Ebert (1990) found that fresh BC generated from incomplete combustion of diesel fuel was ~92 percent carbon, ~6 percent oxygen, ~1 percent hydrogen, ~0.5 percent sulfur, and ~0.3 percent nitrogen by weight. It is crucial when reporting OC and BC concentrations, as done in Chapter 6 and in Figures 7.3 and 7.5, to indicate the specific measurement method(s) that were utilized.

The most common sources of BC are combustion of fossil fuels, such as coal, oil, and natural gas (old carbon) and combustion of biomass, such as wood in fireplaces or forests/shrubs/grasses (new carbon). The amount of BC that forms during combustion depends on the combustion conditions (temperature, fuel-oxygen ratio). BC is emitted directly in the ultrafine or accumulation size ranges and in carbon aggregate chains that can be geometrically larger than 2.5 µm. Since BC arises from most oxygen-starved combustion processes, and since thermal measurements of total BC do not reveal much detail about its structure, BC is not easily relatable to specific sources without more information.

As shown in Chapter 6, OC makes up a large fraction of $PM_{2.5}$ in all locations. It also contains hundreds to thousands of individual organic compounds. The main natural sources are: spores, bacteria, humic material and leaf litter in soils and abrasion of plant waxes; and heavy organic gases from biological processes that are oxidized and condense, partition, or react on particles (e.g., monoterpenes). Naturally-occurring forest and wildfires are also large contributors. Consequently, a fraction of the particle-phase OC is of natural origin, even in densely populated areas with manmade contributions. The main anthropogenic sources of OC are fossil-fuel combustion, burning of wood and other vegetation for heating or land clearance, cooking (both the fuel and the meat), wear/friction of carbon-containing material (e.g., tires and asphalt), and oxidation of some VOCs that condense or partition onto existing particles.

As described in previous chapters, OC may be emitted directly in the particulate phase or as an organic vapor (semivolatile OC or SVOC) that condenses on other particles shortly after being emitted. OC condensation that occurs quickly as the exhaust plume cools should be included as part of direct particle emissions, but this is not well-represented by hot-exhaust sampling (England et al., 2000). To realistically capture these condensables, source-profile tests should be made with cooling to ambient temperatures, dilution, and a short aging period prior to collection on filters (Hildemann et al., 1989). Considerable uncertainty exists regarding the true amount of direct particle-phase OC emissions because many source-profile tests have not followed this approach. Some SVOC compounds alternate between the gas and particle phase in the atmosphere, depending upon their concentrations and the ambient temperature. Typical diurnal, seasonal and geographic variations in temperature are large enough for some compounds to change phase several times during their atmospheric lifetime. Since gases often deposit more rapidly than particles, these phase changes alter the relative abundances among different components in a profile.

Past information on OC emissions needs to be examined to determine if appropriate sampling methods, capable of properly capturing SVOCs that condense rapidly, were utilized. The extent to which positive artifacts, due to VOC/SVOC adsorption onto the filter media, may have biased the OC emission information and also needs to be assessed. Appropriate and consistent techniques need to be applied for a wide variety of sources to update the current source profile or speciation information.

The IMPROVE OC/BC fractions (Chow et al., 1993a, 2001) in Figure 7.5 are one of several ways to separate carbon into functional and repeatable categories (see Appendix B) for which the exact composition is not generally known. Watson et al. (1994b) found differences in the quantities of carbon that evolved at different temperatures, with diesel exhaust yielding more carbon at higher temperatures than gasoline-vehicle exhaust. Greaves et al. (1987) and Jeon et al. (2001) have experimented with thermal desorption followed by gas-chromatographic and mass-spectrometric detection, techniques that yield many carbon fractions, some of which can be associated with organic compounds, in receptor samples. These profiles have not yet been quantitatively related to source emissions via source testing. Thermal-desorption and pyrolysis methods may be less costly and more convenient to implement than detailed organic-compound analyses, which require large samples and laborious solvent-extraction operations. Comparisons must be made between the simple and complex analysis methods on source and receptor samples to demonstrate their utility.

Solvent extraction followed by gas or liquid chromatography with different detectors (e.g., Mazurek et al., 1987), and comparison with standards, has identified hundreds of organic compounds; but these account for only 10 percent to 20 percent of the OC in most ambient samples. Some organic compounds are dominant in certain source types and remain stable in the atmosphere; these are good source markers for receptor models. Figure 7.6 compares several different organic-compound abundances in hardwood burning with those in meat cooking. Meat-cooking profiles for elements, ions, and OC are similar to those for wood burning (Figure 7.5d), and this precludes separation of wood burning and meat cooking by the CMB. However, with the additional organic compounds shown in Figures 7.6a and 7.6b, these two sources can be distinguished. Hardwood burning is rich in guaiacols and syringols, but low in sterols such as steroid-m and cholesterol. Just the opposite is true for meat cooking, where

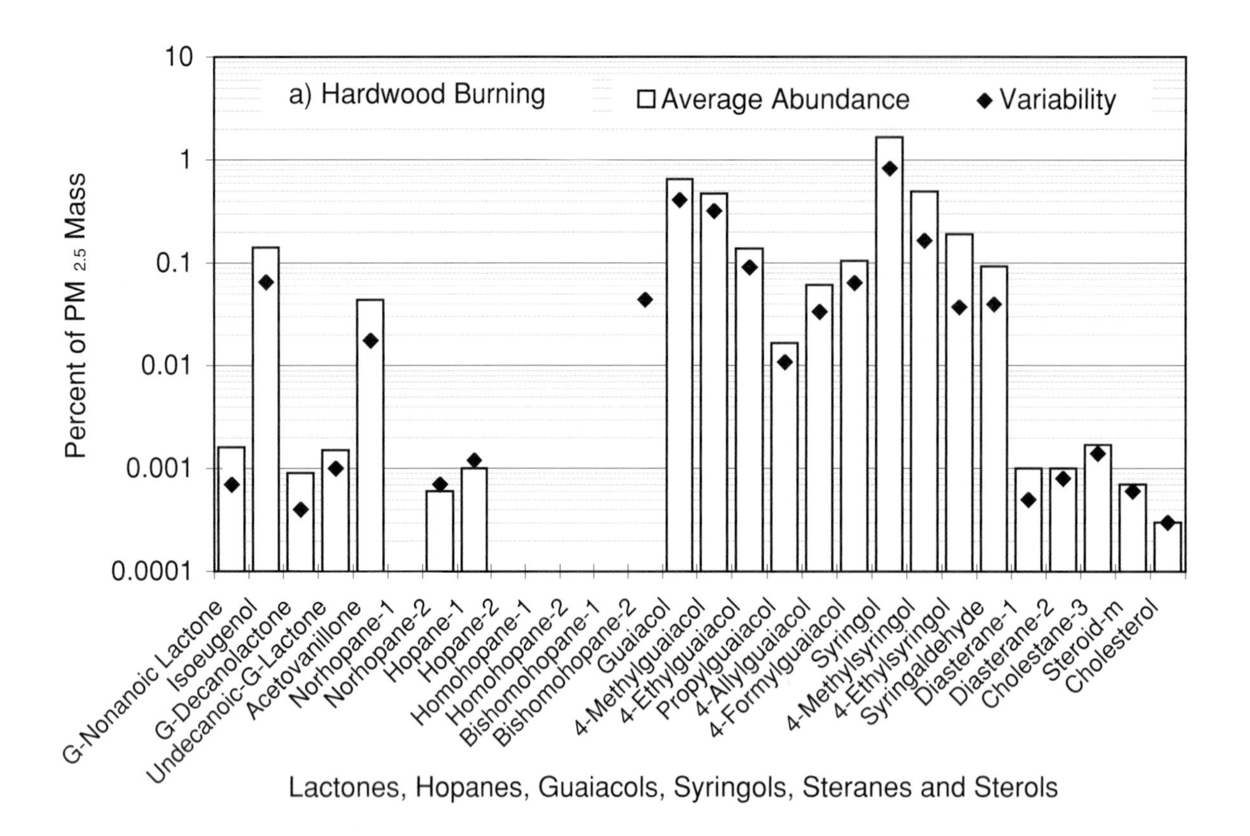

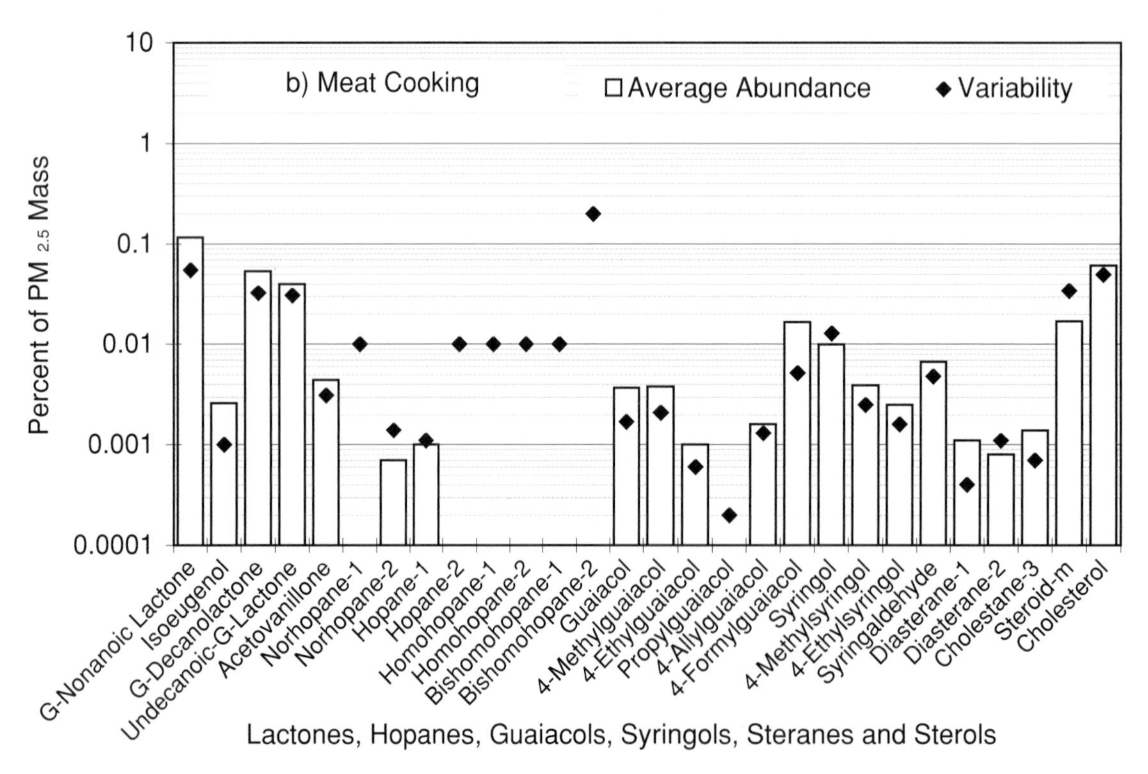

Figure 7.6. Examples of $PM_{2.5}$ source profiles for lactones, hopanes, guaiacols, syringols, steranes, and sterols for a) hardwood combustion and b) meat cooking (Zielinska et al., 1998). These are among many different categories of molecular organic markers measured by filter extraction and gas chromatography with mass-spectrometric detection. (After Watson and Chow, 2002).

cholesterol is among the most abundant species. Syringols are more abundant in hardwoods, such as oak or walnut, and they are depleted in softwoods such as pine, thereby allowing even greater differentiation to be achieved in source apportionment.

Some organic-compound source profiles have been measured for gasoline- and diesel-vehicle exhaust (Fraser et al., 1998; Kleeman et al., 2000; Schauer et al., 1999a, 2002a; Rogge et al., 1993a;), cooking (Nolte et al., 1999; Schauer et al., 1999b, 2002b), natural gas combustion (Rogge et al., 1993b), coal combustion (Oros and Simoneit, 2000), oil combustion (Rogge et al., 1997), wood burning and forest fires (Elias et al., 1999; Fine et al., 2002; Hays et al., 2002; McDonald et al., 2000; Rogge et al., 1998; Schauer et al., 1998; Simoneit et al., 1999; Standley and Simoneit, 1994), and fugitive dust (Rogge et al., 1993c). Most of these results are from emitters in southern California. Different investigators measure and use different markers for these sources, and these do not always correspond to the organic compounds measured at receptors. Organic-compound measurement methods are rapidly evolving and may soon be in widespread use for source profiles and receptor concentrations.

Simoneit (1999) cites several examples of odd- and even-numbered carbon molecules in the n-alkane series as indicating the presence or absence of OC from man-made sources along with ubiquitous contributions from natural sources. Biogenic materials, such as plant waxes, tend to have more molecules with odd numbers of carbon atoms, whereas n-alkanes from combustion processes have nearly equal quantities of even and odd carbon-numbered molecules. A Carbon Preference Index (CPI), the ratio of odd to even n-alkane masses, is used to assess the relative abundance of natural vs. anthropogenic contributions.

7.3.2.5 Secondary Organic Aerosol

SOAs are formed when some of the VOCs participate in photochemical reactions over short time periods under intense sunlight (Pandis et al., 1992), over longer time periods of stagnant conditions (Strader et al., 1999), or during long-range transport. The VOC classes that most commonly form SOAs are described in Chapter 3. The relative importance of SOA vs. directly emitted OC is highly variable in time and space. Enrichment of the OC/BC ratio over that expected from source profiles of primary emissions has been used to estimate potential SOA contributions to OC (Turpin and Huntzicker, 1995). This approach assumes that the OC/BC ratio from primary emissions falls within a reasonable range (typically 1 to 4, depending on dominant primary contributions and OC/BC measurement method), that primary OC and BC disperse and deposit in the same way, and that all BC is of primary origin. Ambient OC/BC ratios are higher (e.g., >6) when secondary formation contributes OC in excess of that from primary emissions. Other primary OC emitters (e.g., vegetation fragments, cooking, unburned fuel, smoldering fires) can interfere with this interpretation. Organic vapors adsorbed onto quartz fiber filters may inflate OC measurements in pristine areas where actual particulate carbon concentrations are low. The same carbon analysis method must also be applied to the ambient samples and source emissions from which the primary OC/BC reference is derived (Dreher and Harley, 1998).

SOA may be a large fraction of OC during high $PM_{2.5}$ episodes and/or at particular times of the day, such as the afternoon (Turpin and Huntzicker, 1995). Experiments in California's South Coast Air Basin (e.g., Hildemann et al., 1993; Schauer et al., 1996) and San Joaquin Valley (Strader et al., 1999; Schauer and Cass, 2000), which are discussed in Chapter 3, showed that, on average, SOA was a minor fraction of ambient OC. In the midwestern United States, Cabada et al., (2002) estimated 10 to 35 percent of annual average OC to be from SOA based on OC/BC enrichments. More specific PM end-products for SOA reactions would provide more definitive estimates of its contributions to ambient OC.

7.3.2.6 Other Chemical Markers

Isotopic abundances differ among source materials depending on formation processes and geologic origins. Radioactive carbon-14 (^{14}C) is formed by cosmic rays in the atmosphere and is incorporated into living things through respiration and ingestion. Fossil fuels formed millions of years ago are depleted in ^{14}C because its half life is ~5,000 years. ^{14}C abundances allow the separation of biogenic or "modern carbon" from carbon derived from fossil-fuel combustion (Currie et al., 1999). Estimates of fuel age must be made, as wood that grew before outdoor nuclear testing in the 1950s has much lower ^{14}C abundances than plant life that grew during and after that period. Other receptor-model applications using isotopic abundances include ^{4}He as a marker of continental dust (Patterson et al., 1999), ^{210}Pb to determine sources of lead in house dust (Adgate et al., 1998), and ^{34}S to determine sources of acid deposition (Turk et al., 1993) and to distinguish secondary $SO_4^=$ contributions from residual-oil combustion (Newmann et al., 1975). Isotopes such as non-radioactive ^{34}S or radioactive ^{35}S follow the transformations of SO_2 to $SO_4^=$ from a source type and offer a way to directly apportion the secondary PM in cases where there are large differences in profiles among contributing sources (Hidy, 1987).

Single particles can be identified in source emissions based on their shapes, sizes, mineralogy, optical properties, and elemental composition (Casuccio et al., 1983, 1989; Draftz, 1982; McCrone and Delly, 1973). Results from microscopic examination are semi-quantitative, as thousands of particles must be examined to obtain an adequate statistical representation of the millions deposited on a typical air-sampling filter. Computer-automated scanning systems with pattern-recognition methods are being developed to perform these analyses automatically (Hopke and Casuccio, 1991).

7.3.3 Temporal and Spatial Variability

Some emitters have source profiles so similar that their contributions cannot be distinguished from each other based on chemical and size characteristics. However, sources often can be discriminated by where and when the samples are taken. Although vegetative burning from wood stoves and forest fires cannot be differentiated chemically, high vegetative-burning contributions found in neighborhoods during winter logically result from residential wood combustion, while region-wide contributions during hot, dry summer months are more reasonably attributed to forest fires or prescribed burning. Bracketing the morning traffic peak shown in Figure 7.1 with samples before, during, and after gives higher confidence that the carbon is from traffic rather than from other sources. Comparing measurements from within cities to those taken in nearby rural areas, as shown in Chapter 6, helps discriminate locally generated particles from those potentially transported into the area from upwind sources (Brook et al., 1999, 2002).

7.3.4 Combining Size, Composition, Space, and Time

Chapter 5 discusses many of the new technologies that are being developed for continuous measurements of carbon, $SO_4^=$, NO_3^-, and metals. As illustrated in Figure 7.1, higher time and chemical resolution will assist the association of chemical components with different sources and spatial scales. A semi-continuous elemental-analysis system (Kidwell and Ondov, 2001) shows clear, short-duration impacts from plume touchdown of nearby point sources. Several single-particle or particle-ensemble, sizing, and chemical-characterization methods are now being applied (Noble and Prather, 1996; Ge et al., 1998; Tan et al., 2002). Suess and Prather (1999) describe laser ablation of individual particles followed by on-line time-of-flight mass-spectroscopy to characterize particles from individual sources, such as biomass burning, automobile emissions, and suspended soils. Aerosol mass-spectrometers are also beginning to be applied to analyze emissions such as diesel exhaust (Jayne et al., 2001). Biomass particles are characterized by K clusters associated with Cl^- and $SO_4^=$ (Silva et al., 1999). Vehicle-exhaust particles have been found to contain Pb, Ce, Pt, Mo, Ca, and Na (Noble and Prather, 1996; Silva and Prather, 1997). Soils are typified by individual particles containing Al, Fe, Na, Mg, K, Ca, Si, and Ti (Silva et al., 2000). Particle spectra containing Ca abundances are attributed to building materials, and K and Na have been associated with marine origins (Wiess et al., 1996). Particles containing C_2 and C_3 organic compounds with $SO_4^=$, NO_3^-, and Cl^- have been associated with combustion processes (Noble and Prather, 1996).

7.4 RECEPTOR MODELS AND DECISION-MAKING

Receptor methods have been widely used to justify pollution-control decisions. State Implementation Plans (SIPs) for several dozen PM_{10} non-attainment areas in the western United States were based almost entirely on receptor methods during the late 1980s and early 1990s. Many of these non-attainment areas were dominated by wood-smoke and fugitive-dust contributions under stagnant wintertime inversions (e.g., Chow et al., 1993b; Mathai et al., 1988), which were not amenable to accurate emission estimates or dispersion modeling. Watson (2002) details how receptor and source methods complemented each other to estimate contributions to haze in the Grand Canyon. Several recent examples are summarized below to demonstrate how receptor methods have been used in conjunction with other approaches in decision-making studies.

7.4.1 Sulfur Reductions in Canadian Gasoline

Removing sulfur from gasoline and diesel fuels helps decrease tailpipe emissions by improving catalyst performance and by direct reduction in the amount of primary $SO_4^=$ and SO_2 produced. Based on this information, Canadian policy-makers considered new sulfur-content regulations. Receptor methods provided quantitative estimates of the potential improvement in air quality. These estimated improvements enabled determination of the health and economic benefits of lower sulfur in fuel – important information in the decision-making process. Health researchers initially requested that the estimated changes focus on ambient PM and $SO_4^=$ because of the amount of quantitative information available on their health effects. Methods to estimate these changes included:

1. Source apportionment based on measurements from Canadian cities to establish that motor-vehicle exhaust was an important contributor to ambient PM levels.

2. A simple observation-based model, consisting of discrete and quantifiable sub-components describing the processes linking vehicle emissions to $PM_{2.5}$ in urban areas.

3. Description of aerosol-related processes derived using an aerosol evolution/equilibrium observation-driven box model.

4. Application of the model and estimation of the uncertainties.

Previous Canadian and U.S. CMB studies were reviewed for information on the likely contribution of motor vehicles to PM. Combining the results of multiple, independent source-contribution estimates provided an understanding of uncertainties and similarities for different urban areas. The weight of evidence indicated that motor-vehicle exhaust was an important contributor to $PM_{2.5}$ in urban areas, regardless of location. Average primary motor-vehicle contributions were 44±23 percent of $PM_{2.5}$

CMB source-contribution estimates from vehicle exhaust and other sources are shown for Toronto and the Lower Fraser Valley (e.g., Vancouver, BC, area) in Figure 7.7. These estimates, although uncertain because of the lack of locally-derived source profiles, were consistent with differences in carbon concentrations for different site locations and with estimates for U.S. cities with similar sources and meteorology.

An emission and air-quality modeling system, including appropriate measurements of emissions, meteorology, boundary conditions, and initial concentrations, was not available to credibly simulate the processes that would relate emission changes to ambient $PM_{2.5}$ in Canadian urban areas. A simpler, receptor-oriented conceptual model linking ambient $PM_{2.5}$ to vehicle emissions was constructed, as illustrated in Figure 7.8. The main assumptions, which were required in order to have quantifiable model sub-components, were: 1) particles in urban air can be separated into vehicle-related fractions that will and will not be affected by sulfur fuel reductions and that are independent of particles from other sources; 2) primary vehicular emissions of $SO_4^=$, and of SO_2, NO_x, and VOC (PM precursors), disperse throughout an urban area, reaching typical ambient concentrations before chemical reactions take place; 3) after "inert" dispersion of the precursors, they form some $SO_4^=$, NO_3^-, and SOA before moving out of the urban area; 4) the urban-air concentration of the vehicle-related $PM_{2.5}$ that is sensitive to fuel sulfur content is equal to the sum of the mass of secondary particles formed from the dispersed precursors and of the mass of the dispersed primary $SO_4^=$.

Although the conceptual model linking urban particles to tailpipe emissions in Figure 7.8 is simple, it identified the main processes and the chemical components that needed to be quantified (red and green boxes in Figure 7.8). The effects of sulfur fuel reductions on direct particle emissions of OC, BC, and other elements were unknown, and thus changes in their ambient concentration were assumed to be zero (gray boxes in Figure 7.8). Sulfate, SO_2, NO_x and VOCs from the primary vehicle emissions were assumed to disperse like CO to estimate their ambient concentrations (red box). Observations and vehicle-emission rates for CO were used to derive a dispersion coefficient, which was then used in a CMB tracer solution to estimate vehicle-related primary pollutant concentrations. CO was used because most of it is emitted from vehicles and it does not react or deposit appreciably during the time required for dispersion throughout an urban area. A simple dispersion model and use of NO_x data in the CMB tracer solution were also considered. The primary-emission dispersion rates using these two approaches were similar to the CO-based estimate, thus lending confidence to the results and providing additional information for uncertainty estimates.

After instantaneous dispersion, SO_2, NO_x and VOCs formed particles for a time-period equivalent to the length of time urban emissions typically remain over a city (green box in Figure 7.8). This residence time, accounting for the hours of possible photochemical transformations, was determined from available meteorological data using back-trajectories. Conversion factors allowing determination of the amount of secondary $SO_4^=$, NO_3^- and SOA formed during this time (green box) were derived using an aerosol-evolution model. This model was applied for several possible conditions (e.g., different seasons, different ambient NH_3 concentrations) to quantify changes in emission characteristics, and it was "tuned" to fit the observed conditions. This was done for Toronto, Montréal, and Vancouver (these conversion factors were also used for other cities where appropriate).

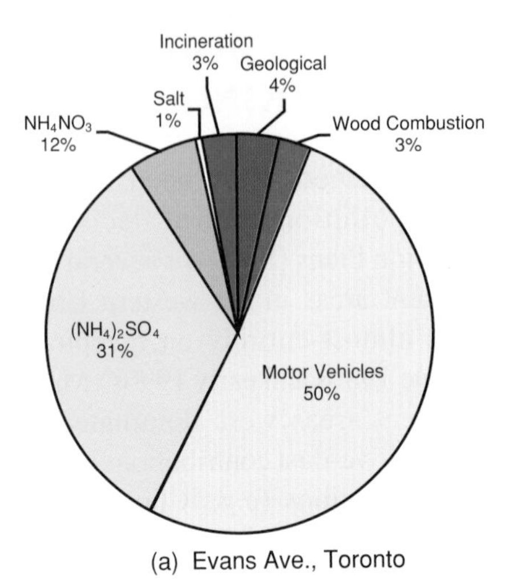

(a) Evans Ave., Toronto

Average $PM_{2.5}$ = 10.8 $\mu g/m^3$

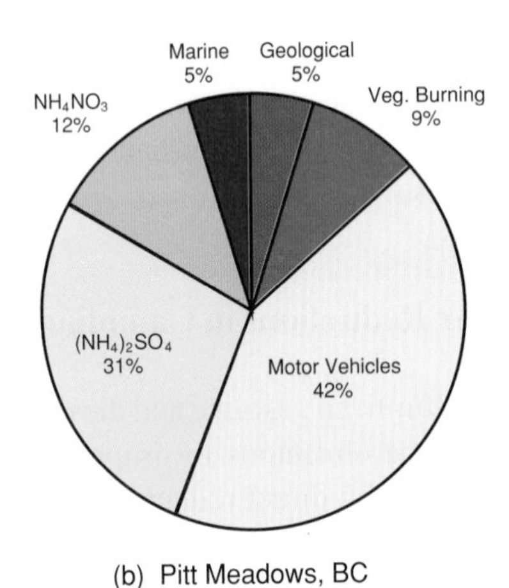

(b) Pitt Meadows, BC

Average $PM_{2.5}$ = 8.1 $\mu g/m^3$

Figure 7.7. CMB $PM_{2.5}$ source-contribution estimates for two regions of Canada. The analyses demonstrate that, given the range of sources assumed to be important, motor vehicles contribute a large fraction. This information was used to support policy development focusing on S in gasoline. (a) Average motor-vehicle exhaust CMB contributions were 50 percent at Evans Ave., Toronto (average of twelve 24-hr observations from September 1995), which is within the range estimated for U.S. urban areas, but above the average. The Evans Ave. monitor is near high traffic volumes. (b) Average vehicle-exhaust $PM_{2.5}$ contributions in the Lower Fraser Valley of BC, at a site more distant from dense traffic, constituted ~42 percent of $PM_{2.5}$ (average for twenty-six 24-hr observations from July-August 1993). (From Bloxam et al., 1997).

The quantities of secondary $SO_4^=$, NO_3^-, NH_4^+, and SOA formed from the dispersed precursors (Box 1) were added to the "dispersed primary $SO_4^=$ concentration" (Box 2) to determine the $PM_{2.5}$ mass associated with vehicle emissions that is sensitive to fuel-sulfur content. This 'fuel-sulfur-sensitive' $PM_{2.5}$ was calculated using current emissions, based on actual fuel sulfur content and projected emissions, based on several different options for sulfur content. Subtracting $PM_{2.5}^{projected}$ from $PM_{2.5}^{current}$ provided estimates of the changes in total ambient $PM_{2.5}$ mass due to the different fuel-sulfur scenarios. The structure of the conceptual model allowed uncertainties to be incorporated into the calculations. A range of probable values for emission changes, dispersion rates, and secondary-formation functions were defined and used to estimate uncertainty.

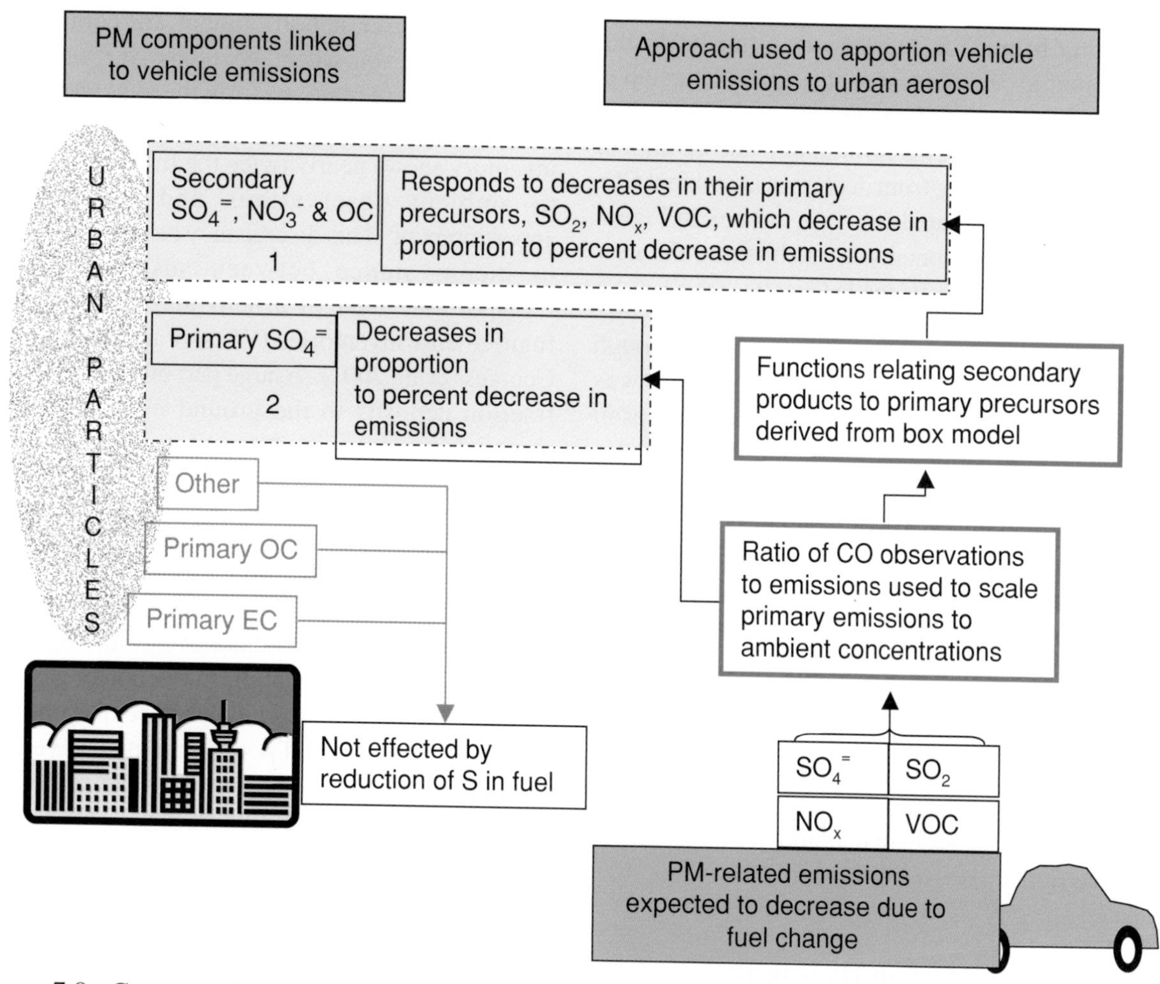

Figure 7.8. Conceptual model describing links between urban particles and tailpipe emissions of primary $SO_4^=$ and particle precursors (NO_x, SO_2, VOCs). Receptor methods were used to estimate the ambient concentrations of the precursors and primary $SO_4^=$, which were assumed to disperse like CO within an urban area (red box). After dispersion, a portion of the precursors were assumed to contribute to secondary $SO_4^=$, NO_3^- and SOA based on conversion rates determined from observational data and an aerosol evolution model (green box). The portion of the $PM_{2.5}$ from the tailpipe emissions that is expected to respond to reductions in fuel S was determined by summing the secondary products (box 1) and the 'dispersed' primary $SO_4^=$ (box 2). Primary PM, OC, and BC emissions and other (e.g., trace elements, other inorganics) primary particle emissions were assumed to be unaffected by changes in fuel S. From this simple combination of models, estimates of the changes in urban $PM_{2.5}$ concentration due to decreasing fuel S were determined and subsequently related to health and economic benefits. (From Bloxam et al., 1997).

A gasoline sulfur reduction from 579 ppmw to 25 ppmw was estimated to result in a 0.27 $\mu g/m^3$ decrease in the annual average $PM_{2.5}$ in Toronto, which represented a ~1.8 percent decrease from the current levels. Toronto was estimated to have the largest improvement, and hence the greatest potential health benefit, because its base-case sulfur levels in fuel were highest among the cities. Although a 1.8 percent reduction seems negligible, when combined with human mortality and morbidity concentration-response functions and summed over millions of people for 20 years, the estimated health and economic benefits were substantial (Thurston et al., 1998) and outweighed the estimated costs of the fuel reformulation. This information and additional health benefits arising from decreases in ambient CO, NO_2 and SO_2 (estimated from the red box in Figure 7.8) were key components in establishing the new Canadian regulations (Canada Gazette, 1998).

Although this Assessment was carried out through an open, multi-stakeholder process, and was supported through an external review, many gaps in knowledge became evident. For example, distinguishing diesel- and gasoline-powered vehicle contributions with the CMB was not possible with the available measurements. Emission rates and source profiles for Canadian vehicles using the current and reformulated fuels were not available, and it was not possible to assess how the proposed fuel changes might affect non-sulfur primary particle emissions, including those of OC and BC. The decision-making process helped identify knowledge gaps with strong relevance to policy. This provided guidance for research that could be expected to improve future assessment cycles.

7.4.2 $PM_{2.5}$ and Urban Haze in Denver, CO

The Denver Brown Cloud has been the subject of five major studies, and numerous smaller ones, since 1973. A 1987-88 study (Watson et al., 1988) identified secondary NH_4NO_3 and carbon as the largest chemical components, with wood smoke and vehicle exhaust being the largest contributors to the carbon as determined by the CMB model. With the available measurements, diesel contributions could not be distinguished from gasoline-vehicle exhaust, and wood burning could not be distinguished from cooking contributions. A 1996-97 study (Watson et al., 1998) refined these CMB results by measuring organic markers (see Figures 7.5 and 7.6), which were sufficient to separate these source contributions, as well as gasoline-exhaust contributions from cold-starts and poorly-maintained vehicles. This more recent study also used an aerosol-equilibrium model to examine the limiting precursors for NH_4NO_3.

Relative contributions from the emission inventory and CMB source apportionment are compared in Figure 7.9. Even when secondary $SO_4^=$ and NO_3^- contributions are removed from the total $PM_{2.5}$, there are substantial discrepancies. For fugitive dust, the inventory shows nearly twice the fraction compared to ambient observations and CMB model apportionment. This discrepancy has been attributed to inequivalence between suspendable and transportable dust in urban- and regional-scale fugitive dust inventories (Watson and Chow, 2000; Countess et al., 2001). A large part of the suspendable fraction deposits to the ground near its point of suspension and does not contribute ambient PM at larger distances.

The other discrepancy in Figure 7.9 is that between gasoline-vehicle exhaust and diesel exhaust contributions: the inventory indicates that diesel emissions are four times those of gasoline-vehicles. This is true for the hot-stabilized fraction of gasoline-vehicle exhaust. However, when the emissions from gasoline cold-starts and poorly-maintained vehicles are added, the ratio reverses. These results are confirmed from the emission rates measured during the source characterization tests (Cadle et al., 1998; Zielinska et al., 1998). Denver (Watson et al., 1998) and Los Angeles (Hildemann et al., 1994; Kleeman et al., 1999b) have included PM emissions from cooking in their inventories, but this and other potential sources have been neglected in most urban inventories. Through application of receptor models, the importance of these sources and others has been identified. This has led to improvements in emission inventories and has expanded the number sources that potentially could be controlled to reduce particle levels. *The discrepancy between the importance of dust to $PM_{2.5}$ emissions versus its importance to the observed ambient $PM_{2.5}$ observed in Denver is*

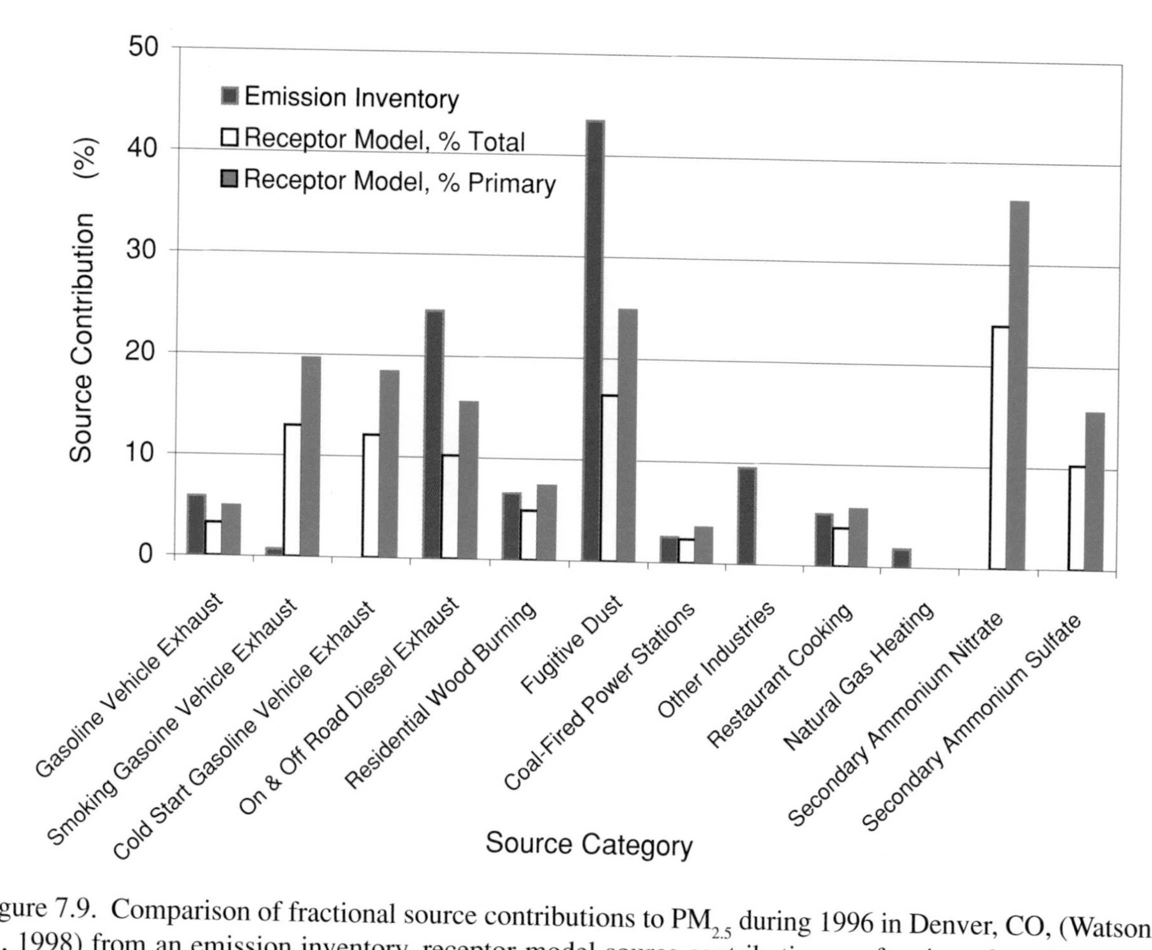

Figure 7.9. Comparison of fractional source contributions to $PM_{2.5}$ during 1996 in Denver, CO, (Watson et al., 1998) from an emission inventory, receptor-model source contributions as fraction of total $PM_{2.5}$, and receptor-model source contributions as fraction of primary $PM_{2.5}$ ($PM_{2.5}$ minus secondary NH_4NO_3 and $(NH_4)_2SO_4$ contributions). Fugitive dust emissions are overestimated by the inventory relative to quantities in ambient air. The cold-start gasoline category is not in the inventory because the standard Federal Test Procedure (FTP) for gasoline-vehicle exhaust emissions allows vehicles to equilibrate at summertime temperatures prior to testing. High emitters (poorly-maintained vehicles) are also underrepresented in most FTP testing. These discrepancies point toward improvements needed in the emission inventory. They also highlight the observation that inventories by themselves cannot be used to plan $PM_{2.5}$ reduction strategies. (After Watson and Chow, 2001b).

common throughout North America. This suggests that methods for estimating fine-particle dust emissions require significant improvement.

The 1996-97 Denver study showed that measures to reduce wood-burning emissions implemented as a result of the 1988 study had been effective. Approximately half of the $PM_{2.5}$ that would have been attributed to wood smoke without the organic markers was found to derive from meat cooking, as evidenced by the ambient sterol levels. Although diesel-emission reductions are still being considered, additional control measures are being investigated for the other sources.

Figures 7.10a through 7.10c show how changes in NH_3, HNO_3, and $SO_4^=$ affect ambient concentrations. This analysis was important for local decision-makers because NH_3 emission reductions were being considered (and were eventually implemented) by the South Coast Air Quality Management District (1996) to reduce PM_{10} concentrations in southern California. Logically, this was also being considered for Denver. Figure 7.10a shows that no increases in NH_4NO_3 result from increases in NH_3 concentrations where the present concentration (i.e., no change in NH_3) is given as 100 percent along the x-axis. A 25 percent reduction in NH_3 (from 100 percent to

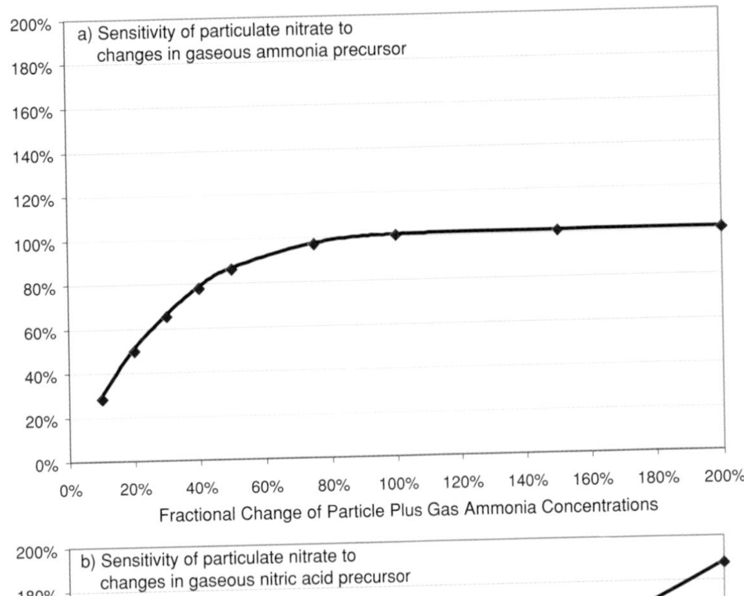

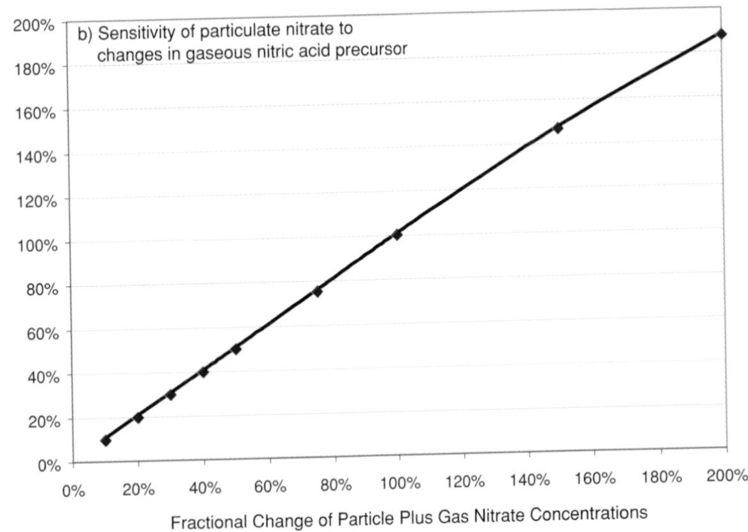

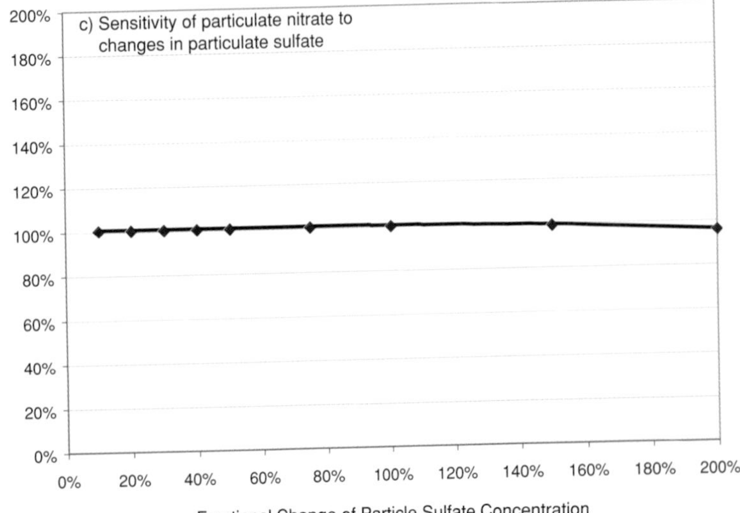

Figure 7.10. Effects of changes in (a) NH_3, (b) HNO_3, and (c) $SO_4^=$ levels that might result from emissions-control strategies on average NH_4NO_3 concentrations estimated using the Simulating Composition of Atmospheric Particles at Equilibrium (SCAPE) equilibrium model (Kim et al., 1993a, 1993b; Kim and Seinfeld, 1995). The aerosol-equilibrium receptor model apportions Na, NO_3^-, $SO_4^=$, NH_4^+, and Cl^- among gas, liquid, and solid phases using thermodynamic equilibrium theory, as described in Chapter 3. These averages were derived from thousands of individual SCAPE simulations applied to 3-hour periods over which PM NO_3^-, $SO_4^=$, and NH_4^+ and gaseous HNO_3 and NH_3 were measured at urban and non-urban receptor sites in the Denver area. Ambient temperature and relative humidity corresponding to the samples are the other SCAPE inputs. This analysis is being used by decision-makers to focus emission-reduction efforts on HNO_3 precursors rather than NH_3 emissions. The horizontal axis represents the fraction of the measured 1996 concentration. (From Watson et al., 1998).

75 percent) results in a minor reduction in particle NO_3^- and a reduction of 50 percent NH_3 reduces particle NO_3^- by only 15 percent. After a 50 percent NH_3 reduction, however, PM NO_3^- decreases are nearly proportional to NH_3 reductions. Ammonia levels must be reduced by more than half before large reductions of PM NO_3^- are realized. This contrasts with Figure 7.10b, which examines effects of HNO_3 reductions on PM NO_3^-. Changes in HNO_3, presumably resulting from changes in NO_x emissions that contribute to HNO_3, result in a direct and proportional reduction in $PM_{2.5}$ NH_4NO_3 concentrations. Figure 7.10c shows that there is no sensitivity to changes in $SO_4^=$ levels. This is because there is sufficient free NH_3 in the Denver area that the small amount of NH_3 freed by removing $SO_4^=$ from the atmosphere has no effect on NH_4NO_3 levels.

7.4.3 Haze in the Grand Canyon

Frequent and intense hazes in Grand Canyon National Park have obscured natural vistas to the extent that a Grand Canyon Visibility Transport Commission (GVTC) was established to determine how particle concentrations should be reduced. Earlier studies (Malm et al., 1989; Richards et al., 1991) used receptor-oriented methods to attribute portions of the $SO_4^=$ measured in the Grand Canyon to the Navajo Generating Station located at the northeastern edge of the Canyon. The source-attribution methodology and results were contested (Markowski, 1992, 1993; Malm et al., 1993; Richards, 1993), but they were qualitatively similar to results from a follow-up study that used independent measurements and models (Richards et al., 1991). Sulfur dioxide scrubbers were installed and recently commenced operation at that facility.

Another uncontrolled coal-fired power station located southwest of the Canyon, the Mohave Power Project (MPP), was not found to be a wintertime contributor in these studies, but it was believed to be a contributor during summer. Another study (Pitchford et al., 1999) was undertaken during summer of 1992 to determine the MPP contribution to $SO_4^=$ and its subsequent contribution to the haze. This ~$US 10M study included perfluorocarbon tracer gas releases from the MPP stack and two other source areas. Upper-air meteorological measurements, ground-based chemistry measurements at more than 30 locations, special studies of PM size distributions, fogs, and peroxides, and compilation of measurements into a common data base were included. Project MOHAVE data were also used for the GVTC assessment related to regional haze (Mathai, 1995, 1996). This large data base allowed many different receptor and source models to be applied, as summarized in Table 7.2.

Source-contribution estimates from the applied models generally showed that MPP $SO_4^=$ contributions were a small fraction of the measured $SO_4^=$. However, the inert tracer gas could not mimic gas-to-particle conversion. It was useful to determine when MPP emissions were present or absent, but it did not add much information on $SO_4^=$ contributions. Tracer levels were typically highest when a receptor was directly impacted by a coherent plume, but generally there was insufficient time for conversion of SO_2 to $SO_4^=$ under these circumstances. Source-contribution estimates from the many models applied were inconsistent with each other, as illustrated in Figure 7.11. The weight-of-evidence was that the MPP contribution was detectable on several occasions, but that it was not the major cause of the haze and that quantitative MPP source contribution estimates were very uncertain.

Regulatory negotiations subsequent to the study used its results with other considerations to agree that the MPP would install state-of-the-art SO_2 scrubbers or cease operation by 2006.

Grand Canyon haze studies of individual source and regional source contributions have provided an evolving understanding of the cause of haze in the desert Southwest. The involvement of stakeholders (including power-station owners and operators) in the acquisition of this knowledge has resulted in commitments for progress toward continual improvement (Mathai et al., 1996; Chow et al., 2002b). The direct influence of individual studies on regulatory policy regarding emission controls on power stations and other sources was limited. However, their results have had a major influence on subsequent policy development employing air-quality models. The timing of this impact is ambiguous because of the lengthy cycle of assimilation of scientific results into the regulatory hierarchy. The results added an important perspective on the magnitude of expected benefits to visibility improvement associated with single sources. This perspective reinforced the need for a corroborative approach to establish the basis for regulatory involvement in prescribing emission controls for regional haze improvement. At this time, there is no formal protocol for measuring the benefits of SO_2 emission reductions from power stations to visibility improvement in the Grand Canyon area. However, this appears to be an ideal location for tracking such progress, in the light of the influence of other regional sources (Watson, 2002).

Table 7.2. Receptor and source methods used to attribute $SO_4^=$ in the Grand Canyon to Mohave generating station and other source emissions (Pitchford et al., 1999).

Method	Description	Inputs	Outputs
Receptor Data Analyses			
Tracer Max (Tracer Scaling)	Multiplies perfluorocarbon tracer (PFT) concentration at receptor by ratio of total sulfur to PFT in stack emissions. Provides upper bound for MPP $SO_4^=$ contributions.	PFT, SO_2, and PM sulfur concentrations at receptors; emission ratio of SO_2/PFT.	Contribution of PFT source to ambient S (SO_2+ $SO_4^=$); upper bound estimate of contribution to PM sulfur.
Exploratory Data Analysis	Statistical analysis of SO_2, PM sulfur, and PFT measurements.	PFT, SO_2, and PM sulfur concentrations, and particle light scattering at receptors.	Spatial correlations of PM sulfur, temporal correlations of PFT, SO_2, and PM sulfur at specific sites.
Tracer Regression	Regression of total light extinction (b_{ext}) against PFT, industrial methylchloroform, and water vapor mixing ratio.	PFT, methylchloroform, and mixing ratio measurements at receptors.	Contributions to light extinction from emissions in source regions of the chosen tracers.
Tracer Aerosol Gradient Interpretative Technique (TAGIT)	PFT measurements identify nearby sites with and without MPP impact. MPP $SO_4^=$ contribution is the increment over unimpacted sites.	PFT, SO_2, and PM sulfur concentrations at multiple receptors.	SO_2 and PM sulfur concentrations attributable to sources/source regions where PFT was emitted .
Modified CMB	CMB regional profiles modified to account for conversion and deposition of SO_2 and $SO_4^=$ with aerosol evolution model.	Source/source-regions and receptor concentrations of SO_2, $SO_4^=$, and markers elements, spherical aluminosilicate, light absorption (b_{abs}), relative transport times; ROME transport model estimates of relative conversion rates for emissions from different sources/source-regions.	SO_x and $SO_4^=$ attributable to sources/source- regions, including region containing MPP emissions.
Tracer Mass Balance Regression (TMBR)	Multiple linear regression of SO_2 against PFTs and of PM sulfur against PFTs.	Concentrations at receptors of PFT, SO_2, and PM sulfur.	SO_2 and PM sulfur concentrations attributable to MPP
Differential Mass Balance Regression (DMBR)	Combination of tracer-based dilution calculation with parameterized deposition and conversion.	Concentrations at receptors of PFT and SO_2; times of travel from source to receptors; estimates of conversion rates; index of cloud cover.	SO_2 and PM sulfur concentrations attributable to MPP.

Table 7.2. Receptor and source methods used to attribute $SO_4^=$ in the Grand Canyon to Mohave generating station and other source emissions (Pitchford et al., 1999) (continued).

Method	Description	Inputs	Outputs
Source Emissions Simulations			
HAZEPUFF (Modified)	Lagrangian puff model; interpolated wind field; first order SO_2 to SO_4 conversion; modified dispersion classes.	Vertical wind measurements, PFT and SO_2 emissions from MPP, relative humidity.	Plume locations and concentrations of PFT, SO_2, SO_4, and light scattering attributable to MPP.
CALPUFF/ CALMET	Multi-layer Gaussian puff model with parameterized first order chemical conversion; diagnostic meteorological model.	Surface and upper air meteorological data, topography, PFT and SO_2 emissions from MPP, solar radiation, ambient O_3.	Distribution of concentrations of PFT, SO_2 and SO_4 attributable to MPP.
ROME/ RAPTAD/ HOTMAC	Lagrangian plume model with explicit reaction chemistry; three-dimensional Lagrangian random puff dispersion; primitive equation meteorological model.	Meteorological soundings, topography and land use, solar radiation; MPP emissions of PFT, SO_2, NO_x, and trace metals; background chemical concentrations; PFT concentrations at receptors.	Concentrations of PFT, SO_2 and SO_4 in MPP plume, at surface and aloft.

7.4.4 Understanding the Sources of PM_{10} and $PM_{2.5}$ in Mexico City

Much of the recent effort to understand the causes and control of PM in Mexico City has been directed at developing and using a technical infrastructure. A CMB source-apportionment study (Vega et al., 1997) was conducted at one site in 1989-1990. $PM_{2.5}$ was strongly influenced by primary carbon particles from motor-vehicle emissions, with substantial contributions from secondary particles ($SO_4^=$, NO_3^-, and NH_4^+), as well as dust. This study found refinery emissions to be a major contributor, and the large, but obsolete, refinery was eventually shut down. Infrastructure development for PM assessment then focused on an intensive wintertime field study in 1997 and a source-testing study in 1998. This involved several research groups from Mexico and the United States. This effort included application of several of the receptor-oriented methods described here. They are being used to develop a conceptual model of the causes of elevated concentrations (Molina and Molina, 2002).

Chemical analysis of the samples taken in the 1997 study provided a picture of spatial variation across Mexico City (Figure 7.3). The study showed that fugitive dust was one of the largest PM_{10} contributors and affected some parts of the city more than others. $PM_{2.5}$ was dominated by carbon, with significant additions from secondary particles ($SO_4^=$ and NO_3^-), and dust. These results were consistent with the earlier work of Vega et al. (1997) and demonstrated the spatial extent of the problem and likely contributors. Evidence of the high $SO_4^=$ and trace metals caused by the closed-down refinery were not detected in 1997, indicating the success of the previous decision. The results confirm a continuing need for reduction in carbon sources, including motor-vehicle emissions, and fugitive dust, as a part of the Mexico City clean-air strategy. Equilibrium modeling (Moya et al., 2001) showed that NH_4NO_3 was not limited by available NH_3, which was plentiful. Reductions in secondary $SO_4^=$ will require regional pollution controls, as 70 percent of $SO_4^=$ in the city was accounted for at the boundary monitors. Secondary PM reductions will require continuing efforts to reduce industrial emissions, combined with fuel sulfur reductions, and reductions in NO_x from stationary and mobile sources (e.g., Molina and Molina, 2002).

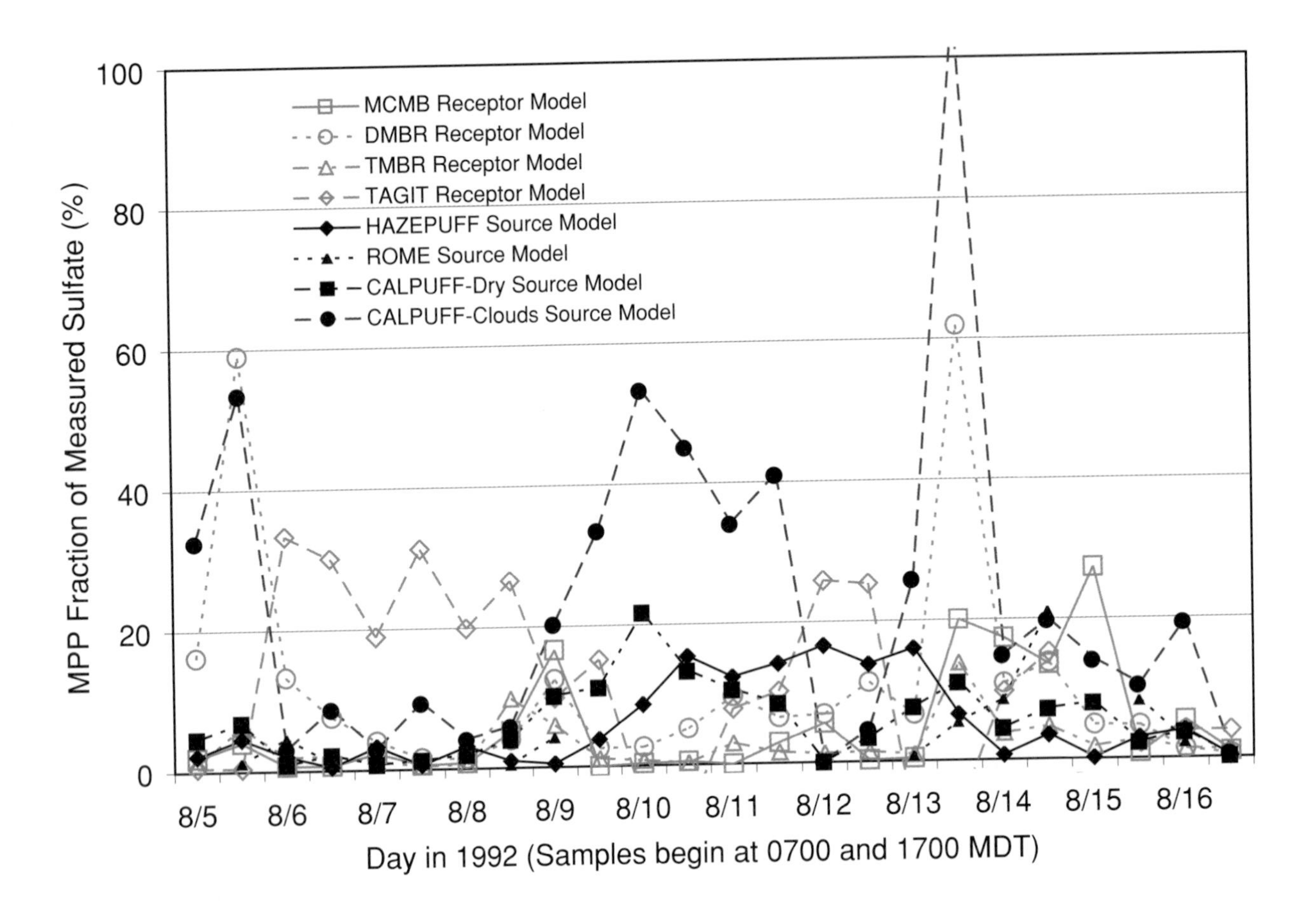

Figure 7.11. Comparison of fractions of $SO_4^=$ at Meadview, AZ, contributed by the MPP as estimated by different source and receptor models (Pitchford et al., 1999). This is a 12-day excerpt from the 7/12/92 through 8/31/92 results (Watson, 2002). Twelve-hour fractional contributions were estimated by a Modified Chemical Mass Balance (MCMB, Eatough et al., 2000), Differential Mass Balance Regression (DMBR, Malm et al., 1997), Tracer Mass Balance Regression (TMBR, Malm et al., 1997, the Tracer Aerosol Gradient Interpretive Technique (TAGIT, Kuhns et al., 1999,) U.S. EPA's HAZEPUFF plume model, Reactive and Optics Model of Emissions (ROME, Karamchandani et al., 2000), and the U.S. EPA's CALPUFF model with clear air (CALPUFF-Dry) and aqueous (CALPUFF-Clouds) $SO_4^=$ transformation. CALPUFF results were intended to represent low (dry) and high (clouds) transformation conditions to bound potential MGS contributions rather than estimate them. (After Watson, 2002).

7.5 DEVELOPING PM MANAGEMENT STRATEGIES

Receptor analyses can assist in determining how much of the observed PM is manageable and unmanageable, as well as the source types that contribute to undesirable concentrations. In some situations, receptor analyses may help identify specific sources of importance (attribution).

7.5.1 Manageable and Unmanageable Source Contributions

The broad distinctions between manageable and unmanageable sources were discussed in Chapters 1 and 3. Cases where unmanageable emissions cause exceedances are uncommon and usually identifiable as special events (e.g., visible smoke from wildfires or windblown dust plumes). These events need to be recorded so that they can be specially considered when determining compliance with air-quality standards. Unmanageable natural emissions usually account for a minor (<20 percent) fraction of $PM_{2.5}$ during normal situations. Confidence in the contributions from natural OC and natural SOA is low.

From a practical standpoint, manageable sources can be classified as being either managed or unmanaged. Managed emitters are those that arise from human activities subject to permits and/or to emission-reduction strategies within a given political jurisdiction, such as a city, county, state, province or air-quality management district. Conversely, unmanaged emitters arise also from human activity, but they are not subject to such measures. This may be due to the nature of the emissions (e.g., difficult to control) and possibly because they are assumed to be relatively small. In this context, measures can be undertaken to reduce unmanaged emissions, albeit at a cost that may not be considered acceptable. Agricultural operations, for example, are often exempt from on-road emission standards for vehicles and dust controls applied to construction sites. *The fraction of OC from natural sources is unknown, but is potentially large. Detailed organic measurements and subsequent receptor-model studies need to be carried out to determine the importance of natural primary and secondary OC sources to $PM_{2.5}$ and PM_{coarse}.*

At a local level, emissions from human activities outside of an air-quality management district's jurisdiction can be viewed as being unmanageable. They may or may not be managed under the auspices of another jurisdiction, but the downwind districts have no direct management authority. Regional transport of $PM_{2.5}$ or its precursors across city, county, state/provincial, and international boundaries is common, especially for $PM_{2.5}$ and $SO_4^=$. As the data in Chapter 6 show, this transport can cause high $PM_{2.5}$ concentrations in rural areas which are comparable to, and are part of, concentrations in nearby urban centers and which may affect downwind populations. Transported anthropogenic $PM_{2.5}$ is manageable, but management requires broader cooperation among jurisdictions and control strategies that are consistent among national and international governing bodies.

Receptor models that rely solely on chemical components cannot distinguish between manageable and unmanageable sources, or between managed and unmanaged sources because their chemical compositions are often similar (e.g., natural vs. prescribed fire, natural windblown vs. construction dust). The greater the distance between the source and the receptor, the more difficult the distinction becomes, owing to the greater potential for atmospheric processes to alter the characteristics of the $PM_{2.5}$. Source apportionment on samples taken at different times and locations that maximize the potential differences between these managed, unmanaged, and unmanageable contributions can be contrasted to estimate contributions from each category.

7.5.2 Main Contributors to Manageable PM

Receptor methods are necessary, but not sufficient, for identifying source types and quantifying their contributions to PM. These methods depend on the measurements available, and quantitative apportionment is less likely to be accurate when applied to measurements that have not been taken for this specific purpose. As with all approaches for studying the relationship between emissions and ambient concentrations, receptor methods have strengths and weaknesses. Therefore, multiple observation-based, receptor- and source-oriented approaches, including detailed examination of emission inventories, should be undertaken to compensate. This corroborative approach helps characterize uncertainties and assess the consistency and strength of the results, which leads to more well-rounded information for decision-making, even when there are disagreements. Since receptor methods are based on actual observations, they provide an independent check on the emission inventories and some aspects of the source model (CTM) results.

From the discussion above, it is apparent that a sustained, iterative effort is needed to continually refine the conceptual model for high concentrations in an area. Building upon current knowledge leads to the design of measurement programs that make receptor- and source-modeling methods more effective. The cycle of assessments and measurement/modeling studies provides clearer focus on uncertainties identified in previous studies so that understanding of a problem improves over time.

After undertaking case-studies employing simpler observational methods, which may be used directly to guide some decisions, the following steps, which fit into the overall framework for PM science and management (Figure 1.1), should be followed:

1. Formulate conceptual model. Use the conceptual model derived from analysis of existing data (observational methods) and studies in similar areas to guide the location of monitoring sites, sampling periods, sampling frequencies, sample durations, sampling methods, the selection of samples for laboratory analysis, and the species that are quantified in those samples. The conceptual model may also help identify which factors that affect PM concentrations in a given area (e.g., emissions, secondary formation, transport) are most important and/or most uncertain so that subsequent actions/activities are focused accordingly.

2. Compile emission inventory. Receptor models need to know which sources are potential contributors. The models then quantify those contributions and help focus resources on those emitters that are the most important contributors to PM concentrations. A receptor-model inventory requires only source categories, not the locations and rates of specific sources.

3. Characterize source emissions. Chemical or physical properties that are believed to distinguish among different source types are measured on a representative set of emitters. Source profiles are the mass abundances (fraction of total mass) of a chemical species in source emissions and the expected variability in the mass fraction. Source profiles are intended to represent a category of sources rather than individual emitters. Several compilations of particle profiles are available that might be applicable to an initial source apportionment, but these profiles will not necessarily represent the sources affecting the current area of interest if they were not measured from sources in this area.

4. Analyze ambient samples for mass, elements, ions, and carbon and other components from sources. Elements, ions ($SO_4^=$, NO_3^-, Cl^-, NH_4^+, Na^+, K^+), OC, and BC are sufficient to account for most of the particle mass, with reasonable assumptions about unmeasured oxides and hydrogen contents. Additional categories such as organic compounds, operationally-defined carbon fractions, isotopic abundances, and single-particle characteristics further distinguish source contributions from each other, even though they may not constitute large mass fractions.

5. Confirm source types with multivariate model. If a sufficient number of chemically characterized ambient samples is available (more than 50), multivariate analyses are helpful to determine the source types and profile characteristics that might be contributors.

6. Quantify source contributions. A CMB model estimates source contributions based on the degree to which source profiles can be combined to reproduce ambient concentrations. The CMB attributes primary particles to their source types and determines the chemical form of secondary PM when the appropriate chemical components have been measured. The PMF and UNMIX models are CMB applications that attempt to derive source profiles from the ambient measurements.

7. Estimate profile changes and limiting precursors. Source characteristics may change during transport to the receptor, the most common change being conversion of SO_2 and NO_x gaseous emissions to $SO_4^=$ and NO_3^- particles. These changes can be simulated with aerosol-evolution models under certain, but not all, conditions. Secondary $(NH_4)_2SO_4$ and NH_4NO_3 involve NH_3 from non-combustion sources that may be a limiting precursor. Chemical-equilibrium receptor models determine the extent to which one or the other precursor needs to be diminished to achieve reductions in NH_4NO_3 levels.

8. Apply source-oriented models. If available, source-oriented models can be used to examine the contribution of precursor-gas emissions from a variety of source categories to secondary particles and to contrast this contribution with the total particle mass. More advanced aerosol-evolution modeling, taking into account

horizontal/vertical transport, can be undertaken to test for limiting precursors. Depending on the spatial scale of the problem, it may be necessary to use the source-oriented model (CTM) to assess the role of upwind sources of primary and secondary particles relative to local sources.

9. Develop scenarios for source-oriented model (CTM) runs. Sources or source categories found to be significant contributors based upon the observational analyses or the receptor models should be targeted in emission-control scenarios set up for the CTM. This will help confirm the sensitivity of the PM levels to these sources and potentially identify optimum approaches for reducing PM.

10. Reconcile source contributions from the receptor models with other data analyses, with the inventories and with source models. Since no model, source, or receptor is a perfect representation of reality, the results must be independently challenged. Receptor-model source contributions should be consistent among locations in the current area of interest and across different sampling times and methods. If possible, discrepancies between source contributions suggested by the emission inventories and estimated by source and receptor models should be resolved. This will improve the models and/or the emission data. Remaining discrepancies should be made explicit to decision-makers.

7.6 SUMMARY

Receptor methods are useful for quantifying contributions from different sources to ambient concentrations, developing and refining conceptual models for source-receptor relationships, verifying emission inventories, and as an independent check on CTM or source model results. Given the complexity of the physical and chemical processes influencing PM-related emissions and atmospheric behavior, it is not surprising that receptor methods are not usually sufficient, by themselves, to characterize and quantify the source-receptor system. The approach to deal with the uncertainties and to obtain the best understanding of source-receptor relationships involves careful analysis of all data and the application of multiple techniques, from analysis of emission data to CTMs. Receptor methods play an important role in this corroborative approach and new techniques that combine receptor models, which tend to be empirical, with physically based models will lead to further insights. As is the case for all approaches, successful receptor modeling depends on the quality of the measurements, ambient and source, and the availability of source markers, as well as the experience of the air quality analyst.

Source apportionment by receptor or source models must begin with plausible hypotheses about the causes of high concentrations or effects. These are derived from previous experience and background information/data, which might immediately suggest appropriate remedial actions or, if not, will direct the application of appropriate receptor methods and/or help guide the development of targeted field studies. With the available background information/data, semi-quantitative receptor-oriented data analyses should be performed prior to quantitative receptor modeling. These simple observational methods of analysis can provide insights regarding the sources or source regions influencing the PM concentrations in an area and possible limiting precursors, which will help develop or refine a conceptual model describing the source-receptor relationships in the region of interest.

Receptor methods apply to the measurement periods and locations where the required measurements are available, and the conceptual model should be used to select these sites and times. Receptor methods are measurement-intensive, requiring many chemical components taken over a representative period at a number of monitoring sites. Once these data are available, however, receptor methods are inexpensive to apply. However, it is important to realize that the required measurements are seldom available in retrospect and thus, must be planned for in advance.

Receptor methods that rely solely on chemical components do not distinguish between manageable and unmanageable emissions or between managed and unmanaged emissions because the particles from these different source categories can have very similar chemical compositions. The greater the distance from the source to the receptor, the more difficult the distinction becomes due to the greater potential for atmospheric processes to alter the characteristics of the particles. Source apportionment on samples taken at different times and locations that maximize the differences between these managed, unmanaged, and unmanageable contributions can be contrasted to estimate contributions from each category. Multiple observation-based, receptor- and source-oriented approaches need to be applied to gain as complete of an understanding as is possible. Conclusions drawn from this corroborative approach will be more defensible for decision-making, even when there are disagreements, and agreement between approaches clearly increases confidence.

Specific answers to NARSTO science questions require specification of where and when they apply. Receptor methods have a role to play in obtaining answers that are accurate and precise enough to make well-informed decisions that are likely to result in successful emission-reduction strategies. The answers given below highlight the extent to which receptor methods have the potential to gain these answers with the appropriate data.

7.7 POLICY IMPLICATIONS

The contributions of different species to ambient PM concentrations can be assessed reasonably well. The major chemical components of fugitive dust, organics, BC, $(NH_4)_2SO_4$, and NH_4NO_3 can be separated by commonly used chemical analyses, although sampling and analysis methods give operational, rather than absolute, concentrations of OC and BC that are not always comparable (Chapter 4). The presence or absence of industrial contributions can be determined from trace-element measurements.

Anthropogenic and biogenic sources of PM species can be distinguished reasonably well. In areas with a PM problem, the large majority of the $SO_4^=$ and NO_3^- is of anthropogenic origin. The main uncertainty currently lies with the carbon component of the PM. Carbon resulting from fossil-fuel combustion can be separated from carbon of more recent (modern or contemporary) origin because ^{14}C is depleted in fossil-fuel carbon. However, within the group of contemporary carbon compounds there remain both natural and anthropogenic components (e.g., wildfires vs. fireplaces, vegetative emissions or decay vs. cooking). For primary particles, there is reasonable success in apportioning the modern or total OC to sources using specific organic markers or operationally-defined carbon fractions. In theory, this approach can distinguish between different types of vehicle exhaust, cooking, vegetative burning, vegetation fragments, etc. However, given the large fraction of OC on particles that cannot be identified through chemical analysis, there also may be more primary biogenic PM carbon compounds for which there are no clear marker species and which thus will not be apportioned properly. Carbon fractions that are not necessarily associated with specific compounds may yield different profiles for different source types. Unfortunately, at the present time there are insufficient data from source-characterization studies to fully test the use of organic markers to determine the extent to which they are successful at quantifying source contributions.

Relative contributions of primary and secondary PM can be distinguished. Secondary $SO_4^=$, NO_3^-, and NH_4^+ can be measured accurately and subtracted from total mass to estimate the fraction of PM composed of inorganic secondary material. A portion of $PM_{2.5}$ OC also results from the conversion of heavy hydrocarbons to condensable organic compounds. Polar oxygenated compounds are indicative of SOA formation, but they also result from primary PM emissions. Many of the SOA compounds are not quantified by existing measurement technologies. OC/BC ratios in excess of five or six may indicate an important contribution from SOA, but there are also primary OC sources that are depleted of BC. The OC/BC ratio is highly dependent on the thermal evolution protocol, and the method needs to be the same for primary emissions and receptor samples. However, this method may be not be as appropriate in eastern North America due a greater impact from long-range transport and to the more "aged" air mass that affects the region. It also is not a very reliable method when BC concentrations are low, owing to measurement uncertainties. The ability to separate OC both in terms of natural vs. anthropogenic and primary vs. secondary (i.e., four groups) is limited at this time. This is a major knowledge gap that requires considerable research. Some secondary organic products on particles formed from biogenic emissions (i.e., natural and secondary) can be chemically identified and apportioned to natural sources, but reliable separation of secondary organic carbon into natural and anthropogenic origin is not entirely possible at this time. An obvious complicating factor is that the distribution of OC into these four groups changes from place to place and over multiple time scales (i.e., daily and seasonally) at a given location. Consequently, a definitive answer may never be known, but through a corroborative approach enough can usually be learned to support specific policy decisions.

Relative contributions of local and long-range transport to $PM_{2.5}$ and coarse-particle concentrations can be distinguished. Source profiles are often similar for distant and nearby sources and thus, multiple techniques including detailed observation-based analyses need to be used to address this question. To gain insight into the relative importance of local and long-range transport, source-receptor models must be applied at locations and times that are expected to be more or less influenced by various scales of influence, nearby (<1 km of the monitor), neighborhood (1 to 5 km), urban (>10 km), and regional (100 to 1000 km). Conducting such an analysis clearly requires a well thought-out conceptual model and obviously, the situation varies geographically. Comparison of source-contribution estimates for short-term or long-term averages from the locations impacted over different time and space scales can be used to separate source contributions related to the different spatial scales. However, there has been very little work done on this issue at the present time and thus, new measurements and research are needed.

7.8 REFERENCES

Adgate, J.L., Willis, R.D., Buckley, T.J., Chow, J.C., Watson, J.G., Rhoads, G.G., Lioy, P.J., 1998. Chemical mass balance source apportionment of lead in house dust. Environmental Science and Technology 32, 108-114.

Ansari, A.S., Pandis, S.N., 1998. Response of inorganic PM to precursor concentrations. Environmental Science and Technology 32, 2706-2714.

Biegalski, S.R., Landsberger, S., Hoff, R.M., 1998. Source-receptor modeling using trace metals in aerosols collected at three rural Canadian Great Lakes sampling stations. Journal of the Air and Waste Management Association 48, 227-237.

Blanchard, C.L., Roth, P.M., Tanenbaum, S.J., Ziman, S.D., Seinfeld, J.H., 2000. The use of ambient measurements to identify which precursor species limit aerosol nitrate formation. Journal of the Air and Waste Management Association 50, 2073-2084.

Bloxam, R., Pandis, S.N., Brook, J., Dann, T., Rogak, S., Caton, R., Grappolini, G., 1997. Atmospheric Science Expert Panel Report Joint Industry/ Government Study Sulphur in Gasoline and Diesel Fuels. Environment Canada, Toronto, Ontario, Canada.

Brook, J.R., Dann, T.F., Bonvalot, Y., 1999. Observations and interpretations from the Canadian fine particle monitoring program. Journal of the Air and Waste Management Association 49, PM-35-44.

Brook, J.R., Lillyman, C.D., Mamedov, A. and Shepherd, M., 2002. Regional transport of and local urban contribution to $PM_{2.5}$ over eastern Canada. Journal of the Air and Waste Management Association 52, 874-885.

Buzorius, G., Hämeri, K., Pekkanen, J., and Kulmala, M., 1999. Spatial variation of aerosol number concentration in Helsinki City. Atmospheric Environment 33, 553-565.

Cabada, J.C., Pandis, S.N., Robinson, A.L., 2002 Sources of atmospheric carbonaceous particulate matter in Pittsburgh, Pennsylvania. Journal of the Air and Waste Management Association 52, 732-741.

Canada Gazette, 1998. Sulphur in gasoline regulations, 132,2988-3022, October 31.

Cadle, S.H., Mulawa, P., Hunsanger, E.C., Nelson, K., Ragazzi, R.A., Barrett, R., Gallagher, G.L., Lawson, D.R., Knapp, K.T., Snow, R., 1998. Light-duty motor vehicle exhaust particulate matter measurement in the Denver, Colorado area. In Proceedings, $PM_{2.5}$: A Fine Particle Standard, Chow, J.C., Koutrakis, P., Eds. Air & Waste Management Association, Pittsburgh, PA, pp. 539-558.

Casuccio, G.S., Janocko, P.B., Lee, R.J., Kelly, J.F., Dattner, S.L., Mgebroff, J.S., 1983. The use of computer controlled scanning electron microscopy in environmental studies. Journal of the Air Pollution Control Association 33, 937-943.

Casuccio, G.S., Schwoeble, A.J., Henderson, B.C., Lee, R.J., Hopke, P.K., Sverdrup, G.M., 1989. The use of CCSEM and microimaging to study source/receptor relationships. In Transactions, Receptor Models in Air Resources Management, Watson, J.G., Ed. Air & Waste Management Association, Pittsburgh, PA, pp. 39-58.

Chang, S.G., Brodzinsky, R., Gundel, L.A., Novakov, T., 1982. Chemical and catalytic properties of elemental carbon. In Particulate Carbon: Atmospheric Life Cycle, Wolff, G.T., Klimisch, R.L., Eds. Plenum Press, New York, NY, pp. 159-181.

Chow, J.C., Watson, J.G., Egami, R.T., Frazier, C.A., Lu, Z., Goodrich, A., Bird, A., 1990. Evaluation of regenerative-air vacuum street sweeping on geological contributions to PM_{10}. Journal of the Air and Waste Management Association 40, 1134-1142.

Chow, J.C., Liu, C.S., Cassmassi, J.C., Watson, J.G., Lu, Z., Pritchett, L.C., 1992. A neighborhood-scale study of PM_{10} source contributions in Rubidoux, California. Atmospheric Environment 26A, 693-706.

Chow, J.C., Watson, J.G., Pritchett, L.C., Pierson, W.R., Frazier, C.A., Purcell, R.G., 1993a. The DRI Thermal/Optical Reflectance carbon analysis system: Description, evaluation and applications in U.S. air quality studies. Atmospheric Environment 27A, 1185-1201.

Chow, J.C., Watson, J.G., Ono, D.M., Mathai, C.V., 1993b. PM_{10} standards and nontraditional particulate source controls: A summary of the A&WMA/EPA International Specialty Conference. Journal of the Air and Waste Management Association 43, 74-84.

Chow, J.C., Watson, J.G., Green, M.C., Lowenthal, D.H., DuBois, D.W., Kohl, S.D., Egami, R.T., Gillies, J.A., Rogers, .F., Frazier, C.A., Cates, W., 1999. Middle- and neighborhood-scale variations of PM_{10} source contributions in Las Vegas, Nevada. Journal of the Air and Waste Management Association 49, 641-654.

Chow, J.C., Watson, J.G., Crow, D., Lowenthal, D.H., Merrifield, T., 2001. Comparison of IMPROVE and NIOSH carbon measurements. Aerosol Science and Technology 34, 23-34.

Chow, J.C., Watson, J.G., 2002. Review of $PM_{2.5}$ and PM_{10} apportionment for fossil fuel combustion and other sources by the chemical mass balance receptor model. Energy & Fuels 16, 222-260.

Chow, J.C., Watson, J.G., Edgerton, S.A., Vega, E., 2002a. Chemical composition of PM_{10} and $PM_{2.5}$ in Mexico City during winter 1997. The Science of the Total Environment 287, 177-201.

Chow, J.C., Engelbrecht, J.P., Watson, J.G., Wilson, W.E., Frank, N.H., Zhu, T., 2002b. Designing monitoring networks to represent outdoor human exposure. Chemosphere 49, 961-978.

Chow, J.C., Bachmann, J.D., Wierman, S.S.G., Mathai, C.V., Malm, W.C., White, W.H., Mueller, P.K., Kuman, N., Watson, J.G., 2002c. 2002 Critical review discussion - Visibility: Science and regulation. Journal of the Air and Waste Management Association 52, 973-999.

Cooper, J.A., Watson, J.G., 1980. Receptor oriented methods of air particulate source apportionment. Journal of the Air Pollution Control Association 30, 1116-1125.

Countess, R.J., Barnard, W.R., Claiborn, C.S., Gillette, D.A., Latimer, D.A., Pace, T.G., Watson, J.G., 2001. Methodology for estimating fugitive windblown and mechanically resuspended road dust emissions applicable for regional scale air quality modeling. Report No. 30203-9, Western Regional Air Partnership, Denver, CO.

Currie, L.A., Klouda, G.A., Benner, B.A., Jr., Garrity, K., Eglinton, T.I., 1999. Isotopic and molecular fractionation in combustion: Three routes to molecular marker validation, including direct molecular 'dating' (GC/AMS). Atmospheric Environment 33, 2789-2806.

Currie, L.A., Benner, B.A., Jr., Cachier, H., Cary, R., Chow, J.C., Druffel, E.R.M., Eglinton, T.I., Gustafsson, Ö., Hartmann, P.C., Hedges, J.I., Kessler, J.D., Kirchstetter, T.W., Klinedinst, D.B., Klouda, G.A., Marolf, J.V., et al., 2002. A critical evaluation of interlaboratory data on total, elemental, and isotopic carbon in the carbonaceous particle reference material, NIST SRM 1649a. Journal of Research of the National Bureau Standards 107, 279-298.

deNevers, N., Morris, J.R., 1975. Rollback modeling: Basic and modified. Journal of the Air Pollution Control Association 25, 943-947.

Draftz, R.G., 1982. Distinguishing carbon aerosols by microscopy. In Particulate Carbon: Atmospheric Life Cycle, Wolff, G.T., Klimisch, R.L., Eds. Plenum Press, New York, NY, pp. 261-271.

Dreher, D.B., Harley, R.A., 1998. A fuel-based inventory for heavy-duty diesel truck emissions. Journal of the Air and Waste Management Association 48, 352-358.

Eatough, D.J., Farber, R.J., Watson, J.G., 2000. Second-generation chemical mass balance source apportionment of sulfur oxides and sulfate at the Grand Canyon during the Project MOHAVE summer intensive. Journal of the Air and Waste Management Association 50, 759-774.

Ebert, L.B., 1990. Is soot composed predominantly of carbon clusters? Science 247, 1468-1471.

Elias, V.O., Simoneit, B.R.T., Pereira, A.S., Cabral, J.A., Cardoso, J.N., 1999. Detection of high molecular weight organic tracers in vegetation smoke samples by high-temperature gas chromatography-mass spectrometry. Environmental Science and Technology 33, 2369-2376.

Engelbrecht, J.P., Swanepoel, L., Zunckel, M., Chow, J.C., Watson, J.G., Egami, R.T., 2000. Modelling PM_{10} aerosol data from the Qalabotjha low-smoke fuels macro-scale experiment in South Africa. Ecological Modelling 127, 235-244.

England, G.C., Zielinska, B., Loos, K., Crane, I., Ritter, K., 2000. Characterizing $PM_{2.5}$ emission profiles for stationary sources: Comparison of traditional and dilution sampling techniques. Fuel Processing Technology 65, 177-188.

Feeley, J.A., Liljestrand, H.M., 1983. Source contributions to acid precipitation in Texas. Atmospheric Environment 17, 807.

Fine, P.M., Cass, G.R., Simoneit, B.R.T., 2002. Chemical characterization of fine particle emissions from the fireplace combustion of woods grown in the southern United States. Environmental Science and Technology 36, 1442-1451.

Fraser, M.P., Cass, G.R., Simoneit, B.R.T., 1998. Gas-phase and particle-phase organic compounds emitted from motor vehicle traffic in a Los Angeles roadway tunnel. Environmental Science and Technology 32, 2051-2060.

Friedlander, S.K., 1973. Chemical element balances and identification of air pollution sources. Environmental Science and Technology 7, 235-240.

Ge, A., Wexler, A.S., Johnston, M.V., 1998. Laser desorption/ionization of single unltrafine multicomponent aerosols. Environmental Science and Technology 32, 3218-3223.

Gordon, G.E., 1980. Receptor models. Environ. Science and Technology 14, 792-800.

Gordon, G.E., Pierson, W.R., Daisey, J.M., Lioy, P.J., Cooper, J.A., Watson, J.G., Cass, G.R., 1984. Considerations for design of source apportionment studies. Atmospheric Environment 18, 1567-1582.

Gordon, G.E., Olmez, I., 1986. Hybrid receptor modeling with multiple sources and vertical mixing. In Transactions, Receptor Methods for Source Apportionment: Real World Issues and Applications. Pace, T.G., Ed. Air Pollution Control Association, Pittsburgh, PA, pp. 229-238.

Gordon, G.E., 1988. Receptor models. Environmental Science and Technology 22, 1132-1142.

Greaves, R.C., Barkley, R.M., Sievers, R.E., Meglen, R.R., 1987. Covariations in the concentrations of organic compounds associated with springtime atmospheric aerosols. Atmospheric Environment 21, 2549-2561.

Green, M.C., Flocchini, R.G., Myrup, L.O., 1992. The relationship of the extinction coefficient distribution to wind field patterns in Southern California. Atmospheric Environment 26A, 827-840.

Hays, M.D., Geron, C.D., Linna, K.J., Smith, N.D., Schauer, J.J., 2002. Speciation of gas-phase and fine particle emissions from burning of foliar fuels. Environmental Science and Technology 36, 2281-2295.

Henry, R.C., Lewis, C.W., Hopke, P.K., Williamson, H.J., 1984. Review of receptor model fundamentals. Atmospheric Environment 18, 1507-1515.

Henry, R.C., 1986. Fundamental limitations of receptor models using factor analysis. In Transactions, Receptor Methods for Source Apportionment: Real World Issues and Applications. Pace, T.G., Ed. Air Pollution Control Association, Pittsburgh, PA, pp. 68-77.

Henry, R.C., 1987. Current factor analysis receptor models are ill-posed. Atmospheric Environment 21, 1815-1820.

Henry, R.C., Wang, Y.J., Gebhart, K.A., 1991. The relationship between empirical orthogonal functions and sources of air pollution. Atmospheric Environment 25A, 503-509.

Henry, R.C., 1997a. History and fundamentals of multivariate air quality receptor models. Chemometrics and Intelligent Laboratory Systems 37, 37-42.

Henry, R.C., 1997b. Receptor model applied to patterns in space (RMAPS) Part I - Model description. Journal of the Air and Waste Management Association 47, 216-219.

Henry, R.C., 1997c. Receptor model applied to patterns in space (RMAPS) Part II - Apportionment of airborne particulate sulfur from Project MOHAVE. Journal of the Air and Waste Management Association 47, 220-225.

Henry, R.C., 2000. UNMIX Version 2 Manual. Report, Dept. Civil and Environmental Eng., U. of Southern California, Los Angeles.

Henry, R.C., 2002. Multivariate receptor models - Current practice and future trends. Chemometrics and Intelligent Laboratory Systems 60 (1-2), 43-48.

Henry, R.F., Rao, S.T., Zurbenko, I.G., Porter, P.S., 2000. Effects of changes in data reporting practices on trend assessments. Atmospheric Environment 34, 2659-2662.

Hidy, G.M., 1987. Conceptual design of a massive aerometric tracer experiment (MATEX). Journal of the Air Pollution Control Association 37, 1137-1157.

Hies, T., Treffeisen, R., Sebald, L., Reimer, E., 2000. Spectral analysis of air pollutants Part 1. Elemental carbon time series. Atmospheric Environment 34, 3495-3502.

Hildemann, L.M., Cass, G.R., Markowski, G.R., 1989. A dilution stack sampler for collection of organic aerosol emissions: Design, characterization and field tests. Aerosol Science and Technology 10, 193-204.

Hildemann, L.M., Cass, G.R., Mazurek, M.A., Simoneit, B.R.T., 1993. Mathematical modeling of urban organic aerosol: Properties measured by high-resolution gas chromatography. Environmental Science and Technology 27, 2045-2055.

Hildemann, L.M., Nazurek, M.A., Cass, G.R., Simoneit, B.R.T., 1994. Seasonal trends in Los Angeles ambient organic aerosol observed by high-resolution gas chromatography. Aerosol Science and Technology 20, 303-317.

Hopke, P.K., Alpert, D.J., Roscoe, B.A., 1983. FANTASIA - A program for target transformation factor analysis to apportion sources in environmental samples. Computers & Chemistry 7, 149-155.

Hopke, P.K., 1985. Receptor Modeling in Environmental Chemistry. John Wiley & Sons, Inc., New York.

Hopke, P.K., 1988. Target transformation factor analysis as an aerosol mass apportionment method: A review and sensitivity study. Atmospheric Environment 22, 1777-1792.

Hopke, P.K., 1991. Receptor Modeling for Air Quality Management. Hopke, P.K., Ed. Elsevier Press, Amsterdam, The Netherlands.

Hopke, P.K., Casuccio, G.S., 1991. Scanning electron microscopy. In Receptor Modeling for Air Quality Management, Hopke, P.K., Ed. Elsevier, Amsterdam, The Netherlands, pp. 149-212.

Hopke, P.K., 1999. An introduction to source receptor modeling. In Elemental Analysis of Airborne Particles, Landsberger, S., Creatchman, M., Eds. Gordon and Breach Science, Amsterdam, pp. 273-315.

Hopke, P.K., 2001. Advances in receptor modeling. Journal of Aerosol Science 32, 363-368.

Horvath, H., Kreiner, I., Norek, C., Preining, O., Georgi, B., 1988. Diesel emissions in Vienna. Atmospheric Environment 22, 1255-1269.

Hsu, K.J., 1997. Application of vector autoregressive time series analysis to aerosol studies. Tellus 49B, 327-342.

Javitz, H.S., Watson, J.G., Guertin, J.P., Mueller, P.K., 1988. Results of a receptor modeling feasibility study. Journal of the Air Pollution Control Association 38, 661-667.

Jayne, J.T., et al., 2001. In-situ measurement of diesel bus exhaust in New York City. AAAR Annual Meeting 2001, Abstract. Portland, Oct. 2001.

Jeon, S.J., Meuzelaar, H.L.C., Sheya, S.A.N., Lighty, J.S., Jarman, W.M., Kasteler, C., Sarofim, A.F., Simoneit, B.R.T., 2001. Exploratory studies of PM receptor and source profiling by GC/MS and principal component analysis of temporally and spatially resolved ambient samples. Journal of the Air and Waste Management Association 51, 766-784.

John, W., Wall, S.M., Ondo, J.L., Winklmayr, W., 1990. Modes in the size distributions of atmospheric inorganic aerosol. Atmospheric Environment 24A, 2349-2359.

Jorquera, H., Palma, W., Tapia, J., 2000. An intervention analysis of air quality data at Santiago, Chile. Atmospheric Environnment 34, 4073-4084.

Karamchandani, P., Santos, L., Sykes, I., Zhang, Y., Tonne, C., Seigneur, C., 2000. Development and evaluation of a state-of-the-science reactive plume model. Environmental Science and Technology 34, 870-880.

Kidwell, C.B., Ondov, J.M., 2001. Development and evaluation of a prototype system for collection of sub-hourly ambient aerosol for chemical analysis. Aerosol Science and Technology 35, 596-601.

Kim, Y.P., Seinfeld, J.H., Saxena, P., 1993a. Atmospheric gas-aerosol equilibrium I. Thermodynamic model. Aerosol Science and Technology 19, 157-181.

Kim, Y.P., Seinfeld, J.H., Saxena, P., 1993b. Atmospheric gas-aerosol equilibrium II. Analysis of common approximations and activity coefficient calculation methods. Aerosol Science and Technology 19, 182-198.

Kim, Y.P. and Seinfeld, J.H. (1995). Atmospheric gas-aerosol equilibrium III. Thermodynamics of crustal elements Ca^{2+}, K^{+}, and Mg^{2+}. Aerosol Sci. Technol. 22, 93-110.

Kleeman, M.J., Hughes, L.S., Allen, J.O., Cass, G.R., 1999a. Source contributions to the size and composition distribution of atmospheric particles: Southern California in September 1996. Environmental Science and Technology 33, 4331-4341.

Kleeman, M.J., Schauer, J.J., Cass, G.R., 1999b. Size and composition distribution of fine particulate matter emitted from wood burning, meat charbroiling, and cigarettes. Environmental Science and Technology 33, 3516-3523.

Kleeman, M.J., Schauer, J.J., Cass, G.R., 2000. Size and composition distribution of fine particulate matter emitted from motor vehicles. Environmental Science and Technology 34, 1132-1142.

Kuhns, H.D., Green, M.C., Pitchford, M.L., Vasconcelos, L., White, W.H., Mirabella, V., 1999. Attribution of particulate sulfur in the Grand Canyon to specific point sources using Tracer-Aerosol Gradient Interpretive Technique (TAGIT). Journal of the Air and Waste Management Association 49, 906-915.

Latimer, D.A., Iyer, H.K., Malm, W.C., 1990. Application of a differential mass balance model to attribute sulfate haze in the Southwest. In Transactions, Visibility and Fine Particles, Mathai,C.V., Ed. Air & Waste Management Association, Pittsburgh, PA, p. 819.

Lawson, D.R., Winchester, J.W., 1979. A standard crustal aerosol as a reference for elemental enrichment factors. Atmospheric Environment 13, 925-930.

Lewis, C.W., Stevens, R.U., 1985. Hybrid receptor model for secondary sulfate from an SO_2 point source. Atmospheric Environment 19, 917-924.

Lloyd, A.C., Cackette, T.A., 2001. 2001 Critical review - Diesel engines: Environmental impact and control. Journal of the Air and Waste Management Association 51, 809-847.

Lowenthal, D.H., Rahn, K.A., 1989. The relationship between secondary sulfate and primary regional signatures in northeastern aerosol and precipitation. Atmospheric Environment 23, 1511-1515.

Magliano, K.L., 1988. Level 1 PM_{10} assessment in a California air basin. In Transactions, PM10: Implementation of Standards, Mathai, C.V., Stonefield, D.H., Eds. Air Pollution Control Association, Pittsburgh, PA, pp. 508-517.

Malm, W.C., Gebhart, K.A., Cahill, T.A., Eldred, R.A., Pielke, R.A., Stocker, R.A., Watson, J.G., Latimer, D.A., 1989. The Winter Haze Intensive Tracer Experiment. by National Park Service, Ft. Collins, CO.

Malm, W.C., Iyer, H.K., Gebhart, K.A., 1990. Application of tracer mass balance regression to WHITEX data. In Transactions, Visibility and Fine Particles, Mathai, C.V., Ed. Air & Waste Management Association, Pittsburgh, PA, p. 806.

Malm, W.C., Gebhart, K.A., Iyer, H., Watson, J.G., Latimer, D., Pielke, R., 1993. Response to "The WHITEX Study and the role of the scientific community: A critique" by Gregory R. Markowski. Journal of the Air and Waste Management Association 43, 1128-1136.

Malm, W.C., Gebhart, K.A., 1997. Source apportionment of sulfur and light extinction using receptor modeling techniques. Journal of the Air and Waste Management Association 47, 250-268.

Markowski, G.R., 1992. The WHITEX Study and the role of the scientific community: A critique. Journal of the Air and Waste Management Association 42, 1453-1460.

Markowski, G.R., 1993. Reply to Malm, et al.'s Discussion of the WHITEX Critique. Journal of the Air and Waste Management Association 43, 1137-1142.

Mason, B., 1966. Principles of Geochemistry, Third Edition. John Wiley & Sons, Inc., New York, NY.

Mathai, C.V., Stonefield, D.H., and Watson, J.G., 1988. PM_{10}: Implementation of standards. A

Summary of the APCA/EPA international specialty conference. Journal of the Air Pollution Control Association 38, 888-894.

Mathai, C.V., 1995. The Grand Canyon Visibility Transport Commission and visibility in Class I areas. Environmental Management 1, 20-31.

Mathai, C.V., Kendall, S.B., Trexler, E., Carlson, A.J., Teague, M.L., Steele, D.S., 1996. Integrated assessment and recommendations of the Grand Canyon Visibility Transport Commission. Environmental Management 2, 16-24.

Mazurek, M.A., Simoneit, B.R.T., Cass, G.R., Gray, H.A., 1987. Quantitative high-resolution gas chromatography and high-resolution gas chromatography/mass spectrometry analyses of carbonaceous fine aerosol particles. International Journal of Environmental and Analytical Chemistry 29, 119-139.

McCrone, W.C., Delly, J.G. (1973). The Particle Atlas, Volume I: Principles and Techniques. 2nd ed. Ann Arbor Science Publishers, Inc., Ann Arbor, MI.

McDonald, J.D., Zielinska, B., Fujita, E.M., Sagebiel, J.C., Chow, J.C., Watson, J.G., 2000. Fine particle and gaseous emission rates from residential wood combustion. Environmental Science and Technology 34, 2080-2091.

Molina, L.T., Molina, M.J., 2002. Air Quality in the Mexico Megacity: An Integrated Assessment. Molina, L. T., Molina, M. J., Eds. Kluwer Academic Publishers, Dordrecht, The Netherlands.

Moya, M., Ansari, A.S., Pandis, S.N., 2001. Partitioning of nitrate and ammonium between the gas and particulate phases during the 1997 IMADA-AVER study in Mexico City. Atmospheric Environment 35, 1791-1804.

Morandi, M.T., Lioy, P.J., Daisey, J.M., 1991. Comparison of two multivariate modeling approaches for the source apportionment of inhalable particulate matter in Newark, NJ. Atmospheric Environment 25A, 927-937.

Newmann, L., Forrest, J., Manowitz, B., 1975. The application of an isotopic ratio technique to a study of the atmospheric oxidation of sulphur dioxide in the plume from a oil-fired power plant. Atmospheric Environment 9, 969-977.

Noble, C.A., Prather, K.A., 1996. Real-time measurement of correlated size and composition of individual atmospheric aerosol particles. Environmental Science and Technology 30, 2667-2680.

Nolte, C.G., Schauer, J.J., Cass, G.R., and Simoneit, B.R.T., 1999. Highly polar organic compounds present in meat smoke. Environmental Science and Technology 33, 3313-3316.

Ondov, J.M., Kelly, W.R., Holland, J.Z., Lin, Z.C., Wight, S.A., 1992. Tracing fly ash emitted from a coal-fired power plant with enriched rare-earth isotopes: An urban scale test. Atmospheric Environment 26B, 453-462.

Oros, D.R., Simoneit, B.R.T., 2000. Identification and emission rates of molecular tracers in coal smoke particulate matter. Fuel 79, 515-536.

Paatero, P., 1997. Least squares formulation of robust, non-negative factor analysis. Chemometrics and Intelligent Laboratory Systems 37, 23-35.

Pandis, S.N., Harley, R.A., Cass, G.R., Seinfeld, J.H., 1992. Secondary organic aerosol formation and transport. Atmospheric Environment 26A, 2269-2282.

Patterson, D.B., Farley, K.A., Norman, M.D., 1999. ^{04}He as a tracer of continental dust: A 1.9 million year record of aeolian flux to the west equatorial Pacific Ocean - The geologic history of wind. Geochimica et Cosmochimica Acta 63, 615.

Perez, P., Trier, A., Reyes, J., 2000. Prediction of $PM_{2.5}$ concentrations several hours in advance using neural networks in Santiago, Chile. Atmospheric Environment 34, 1189-1196.

Perry, K.D., Cahill, T.A., Eldred, R.A., Dutcher, D.D., 1997. Long-range transport of North African dust to the eastern United States. Journal of Geophysical Research 102, 11225-11238.

Pitchford, M.L., Green, M.C., Kuhns, H.D., Malm, W.C., Scruggs, M., Farber, R.J., Mirabella, V.A., White, W.H., McDade, C., Watson, J.G., Koracin, D., Hoffer, T.E., Lowenthal, D.H., Vimont, J.C., Gebhart, D.H., et al., 1999. Project MOHAVE Final Report. U.S. Environmental Protection Agency, Region IV, San Francisco, CA.

Poirot, R.L., Wishinski, P.R., Hopke, P.K., Polissar, A.V., 2001. Comparative application of multiple receptor methods to identify aerosol sources in northern Vermont. Environmental Science and Technology 35, 4622-4636.

Preining, O., 1998. The physical nature of very, very small particles and its impact on their behaviour. Journal of Aerosol Science 29, 481-495.

Rahn, K.A., 1976. Silicon and aluminum in atmospheric aerosols: Crust-air fractionation? Atmospheric Environment 10, 597-601.

Reich, S.L., Gomez, D.R., Dawidowski, L.E., 1999. Artificial neural network for the identification of unknown air pollution sources. Atmospheric Environment 33, 3045-3052.

Richards, L.W., Blanchard, C.L., Blumenthal, D.L., 1991. Navajo generating station visibility study. Report No. STI-90200-1124-FR. Prepared for Salt River Project, Phoenix, AZ, by Sonoma Technology, Inc., Santa Rosa, CA.

Richards, L.W., 1993. A comment on the WHITEX CD_4 tracer data. Journal of the Air and Waste Management Association 43, 1143-1144.

Rogge, W.F., Hildemann, L.M., Mazurek, M.A., Cass, G.R., Simoneit, B.R.T., 1993a. Sources of fine organic aerosol 2. Noncatalyst and catalyst-equipped automobiles and heavy-duty diesel trucks. Environmental Science and Technology 27, 636-651.

Rogge, W.F., Hildemann, L.M., Mazurek, M.A., Cass, G.R., Simoneit, B.R.T., 1993b. Sources of fine organic aerosol 5. Natural gas home appliances. Environmental Science and Technology 27, 2736-2744.

Rogge, W.F., Hildemann, L.M., Mazurek, M.A., Cass, G.R., Simoneit, B.R.T., 1993c. Sources of fine organic aerosol 3. Road dust, tire debris, and organometallic brake lining dust: Roads as sources and sinks. Environmental Science and Technology 27, 1892-1904.

Rogge, W.F., Hildemann, L.M., Mazurek, M.A., Cass, G.R., Simoneit, B.R.T., 1997. Sources of fine organic aerosol 8. Boilers burning No. 2 distillate fuel oil. Environmental Science and Technology 31, 2731-2737.

Rogge, W.F., Hildemann, L.M., Mazurek, M.A., Cass, G.R., Simoneit, B.R.T., 1998. Sources of fine organic aerosol 9. Pine, oak, and synthetic log combustion in residential fireplaces. Environmental Science and Technology 32, 13-22.

Schauer, J.J., Rogge, W.F., Mazurek, M.A., Hildemann, L.M., Cass, G.R., Simoneit, B.R.T., 1996. Source apportionment of airborne particulate matter using organic compounds as tracers. Atmospheric Environment 30, 3837-3855.

Schauer, J.J., Cass, G.R., Simoneit, B.R.T., 1998. Characterization of the emissions of individual organic compounds present in biomass aerosol. Journal of Aerosol Science 29, S223.

Schauer, J.J., Kleeman, M.J., Cass, G.R., Simoneit, B.R.T., 1999a. Measurement of emissions from air pollution sources 2. C_1 through C_{30} organic compounds from medium duty diesel trucks. Environmental Science and Technology 33, 1578-1587.

Schauer, J.J., Kleeman, M.J., Cass, G.R., Simoneit, B.R.T., 1999b. Measurement of emissions from air pollution sources. C_1 through C_{29} organic compounds from meat charbroiling. Environmental Science and Technology 33, 1566-1577.

Schauer, J.J., Cass, G.R., 2000. Source apportionment of wintertime gas-phase and particle-phase air pollutants using organic compounds as tracers. Environmental Science and Technology 34, 1821-1832.

Schauer, J.J., Kleeman, M.J., Cass, G.R., Simoneit, B.R.T., 2002a. Measurement of emissions from air pollution sources 5. C_1-C_{32} organic compounds from gasoline-powered motor vehicles. Environmental Science and Technology 36, 1169-1180.

Schauer, J.J., Kleeman, M.J., Cass, G.R., Simoneit, B.R.T., 2002b. Measurement of emissions from air pollution sources. 4. C_1-C_{27} organic compounds from cooking with seed oils. Environmental Science and Technology 36:567-75.

Schmid, H.P., Laskus, L., Abraham, H.J., Baltensperger, U., Lavanchy, V.M.H., Bizjak, M., Burba, P., Cachier, H., Crow, D.J., Chow, J.C., Gnauk, T., Even, A., ten Brink, H.M., Giesen, K.P., Hitzenberger, R., et al., 2001. Results of the "Carbon Conference" international aerosol carbon round robin test: Stage 1. Atmospheric Environment 35, 2111-2121.

Seinfeld, J.H., Pandis, S.N., 1998. Atmospheric Chemistry and Physics: From Air Pollution to Climate Change. John Wiley & Sons, New York, NY.

Shively, T.S., Sager, T.W., 1999. Semiparametric regression approach to adjusting for meteorological variables in air pollution trends. Environmental Science and Technology 33, 3873-3880.

Silva, P.J., Prather, K.A., 1997. On-line characterization of individual particles from automobile emissions. Environmental Science and Technology 31, 3074-3080.

Silva, P.J., Liu, D.Y., Noble, C.A., Prather, K.A., 1999. Size and chemical characterization of individual particles resulting from biomass burning of local southern California species. Environmental Science and Technology 33, 3068-3076.

Silva, P.J., Carlin, R.A., Prather, K.A., 2000. Single particle analysis of suspended soil dust from Southern California. Atmospheric Environment 34, 1811-1820.

Simoneit, B.R.T., 1999. A review of biomarker compounds as source indicators and tracers for air pollution. Environmental Science and Pollution Research 6, 159-169.

Simoneit, B.R.T., Schauer, J.J., Nolte, C.G., Oros, D.R., Elias, V.O., Fraser, M.P., Rogge, W.F., Cass, G.R., 1999. Levoglucosan, a tracer for cellulose in biomass burning and atmospheric particles. Atmospheric Environment 33, 173-182.

Somerville, M.C., Evans, E.G., 1995. Effect of sampling frequency on trend detection for atmospheric fine mass. Atmospheric Environment 29, 2429-2438.

Song, X.H., Hopke, P.K., 1996. Solving the chemical mass balance problem using an artificial neural network. Environmental Science and Technology 30, 531-535.

South Coast Air Quality Management District, 1996. 1997 air quality maintenance plan: Appendix V, Modeling and attainment demonstrations. South Coast Air Quality Management District, Diamond Bar, CA.

Standley, L.J., Simoneit, B.R.T., 1994. Resin diterpenoids as tracers for biomass combustion aerosols. Atmospheric Environment 28, 1-16.

Stockwell, W.R., Watson, J.G., Robinson, N.F., Steiner, W.E., Sylte,W.W., 2000. The ammonium nitrate particle equivalent of NO_x emissions for continental wintertime conditions. Atmospheric Environment 34, 4711-4717.

Strader, R., Lurmann, F.W., Pandis, S.N., 1999. Evaluation of secondary organic aerosol formation in winter. Atmospheric Environment 33, 4849-4863.

Sturges, W.T., 1989. Identification of pollution sources of anomalously enriched elements. Atmospheric Environment 23, 2067-2072.

Suess, D.T., Prather, K.A., 1999. Mass spectrometry of aerosols. Chemical Reviews 99, 3007-3035.

Tan P.V., Evans, G.J., Tsai, J., Owega, S. Fila, M., Malpica, O., Brook, J.R., 2002. On-line analysis of urban particulate matter focusing on elevated wintertime aerosol concentrations. Environmental Science and Technology 36, 3512-3518.

Thurston, G.D. and Spengler, J.D., 1985. A quantitative assessment of source contributions to inhalable particulate matter pollution in metropolitan Boston. Atmospheric Environment 19, 9-25.

Thurston, G.D., Bates, D., Burnett, R.T., Lipfert, F., Ostro, B., Hale, B., Krupnick, A., Rowe, A., and Ireland, D., 1998. Health and environmental impact assessment panel report: Joint industry/ government study of sulphur in gasoline and diesel fuels. Environment Canada, Toronto, Ontario, Canada.

Turk, J.T., Campbell, D.H., Spahr, N.E., 1993. Use of chemistry and stable sulfur isotopes to determine sources of trends in sulfate of Colorado lakes. Water Air and Soil Pollution 67, 415-431.

Turpin, B.J., Huntzicker, J.J., Larson, S.M., and Cass, G.R., 1991. Los Angeles summer midday particulate carbon: Primary and secondary aerosol. Environmental Science and Technology 25, 1788-1793.

Turpin, B.J., Huntzicker, J.J., 1995. Identification of secondary organic aerosol episodes and quantification of primary and secondary organic aerosol concentrations during SCAQS. Atmospheric Environment 29, 3527-3544.

U.S. EPA, 2001. Draft guidance for demonstrating attainment of air quality goals for $PM_{2.5}$ and regional haze. by U.S. Environmental Protection Agency, Research Triangle Park, NC.

Vega, E., García, I., Apam, D., Ruíz, M.E., Barbiaux, M., 1997. Application of a chemical mass balance receptor model to respirable particulate matter in Mexico City. Journal of the Air and Waste Management Association 47, 524-529.

Wahlin, P., Palmgren, F., van Dingenen, R., 2001. Experimental studies of ultrafine particles in streets and the relationship to traffic. Atmospheric Environment 35, S63-S69.

Watson, J.G., Henry, R.C., Cooper, J.A., Macias, E.S., 1981. The state of the art of receptor models relating ambient suspended particulate matter to sources. In Atmospheric Aerosol, Source/Air Quality Relationships, Macias, E.S. and Hopke, P.K., Eds. American Chemical Society, Washington, D.C., pp. 89-106.

Watson, J.G., 1984. Overview of receptor model principles. Journal of the Air Pollution Control Association 34, 619-623.

Watson,J.G., Cooper,J.A., and Huntzicker,J.J. (1984). The effective variance weighting for least squares calculations applied to the mass balance receptor model. Atmospheric Environment 18, 1347-1355.

Watson, J.G., Chow, J.C., Richards, L.W., Andersen, S.R., Houck, J.E., Dietrich, D.L., 1988. The 1987-88 Metro Denver Brown Cloud Air Pollution Study, Volume III: Data interpretation. Report No. DRI 8810.1. Prepared for 1987-88 Metro Denver Brown Cloud Study, Inc., Greater Denver Chamber of Commerce, Denver, CO, by Desert Research Institute, Reno, NV.

Watson, J.G., Chow, J.C., Mathai, C.V., 1989. Receptor models in air resources management: A summary of the APCA International Specialty Conference. Journal of the Air Pollution Control Association 39, 419-426.

Watson, J.G., Chow, J.C., Lurmann, F.W., Musarra, S., 1994a. Ammonium nitrate, nitric acid, and ammonia equilibrium in wintertime Phoenix, Arizona. Journal of the Air and Waste Management Association 44, 405-412.

Watson, J.G., Chow, J.C., Lowenthal, D.H., Pritchett, L.C., Frazier, C.A., Neuroth, G.R., Robbins, R., 1994b. Differences in the carbon composition of source profiles for diesel- and gasoline-powered vehicles. Atmospheric Environment 28, 2493-2505.

Watson, J.G., Blumenthal, D.L., Chow, J.C., Cahill, C.F., Richards, L.W., Dietrich, D., Morris, R., Houck, J.E., Dickson, R.J., Andersen, S.R., 1996. Mt. Zirkel Wilderness Area reasonable attribution study of visibility impairment - Vol. II: Results of data analysis and modeling. Prepared for Colorado Department of Public Health and Environment, Denver, CO, by Desert Research Institute, Reno, NV.

Watson, J.G., Fujita, E.M., Chow, J.C., Zielinska, B., Richards, L.W., Neff, W.D., Dietrich, D., 1998. Northern Front Range Air Quality Study. Final report. Prepared for Colorado State University, Fort Collins, CO, by Desert Research Institute, Reno, NV.

Watson, J.G., Chow, J.C., 2000. Reconciling urban fugitive dust emissions inventory and ambient source contribution estimates: Summary of current knowledge and needed research. Report No. 6110.4D2. Prepared for U.S. Environmental Protection Agency, Research Triangle Park, NC, by Desert Research Institute, Reno, NV.

Watson, J.G., Chow, J.C., Pace, T.G., 2000. Fugitive dust emissions. In Air Pollution Engineering Manual, Second Edition, 2nd ed., Davis,W.T., Ed. John Wiley & Sons, Inc., New York, pp. 117-135.

Watson, J.G., Chow, J.C., 2001. Estimating middle-neighborhood-, and urban-scale contributions to elemental carbon in Mexico City with a rapid response aethalometer. Journal of the Air and Waste Management Association 51, 1522-1528.

Watson, J.G., Chow, J.C., Fujita, E.M., 2001. Review of volatile organic compound source apportionment by chemical mass balance. Atmospheric Environment 35, 1567-1584.

Watson, J.G., Chow, J.C., 2002. Particulate pattern recognition. In Introduction to Environmental Forensics, Murphy, B.L. and Morrison, R., Eds. Academic Press, New York, NY, pp. 429-460.

Watson, J.G., Zhu, T., Chow, J.C., Engelbrecht, J.P., Fujita, E.M., Wilson, W.E., 2002a. Receptor modeling application framework for particle source apportionment. Chemosphere 49, 1093-1136.

Watson, J.G., Chow, J.C., Lowenthal, D.H., Robinson, N.F., Cahill, C.F., Blumenthal, D.L., 2002b. Simulating changes in source profiles from coal-fired power stations: Use in chemical mass balance of $PM_{2.5}$ in the Mt. Zirkel Wilderness. Energy & Fuels 16, 311-324.

Weatherhead, E.C., Reinsel, G.C., Tiao, G.C., Meng, X.L., Choi, D., Cheang, W.K., Keller, T., DeLuisi, J., Wuebbles, D.J., Kerr, J.B., Miller, A.J., Oltmans, S.J., Frederick, J.E., 1998. Factors affecting the detection of trends: Statistical considerations and applications to environmental data. Journal of Geophysical Research 103, 17149-17162.

Weiss, D., Shotyk, W., Appleby, P.G., Kramers, J.D., Cheburkin, A.K., 1999. Atmospheric Pb deposition since the Industrial Revolution recorded by five Swiss peat profiles: Enrichment factors, fluxes, isotopic composition, and sources. Environmental Science and Technology 33, 1340-1352.

West, J.J., Ansari, A.S., and Pandis, S.N., 1999. Marginal $PM_{2.5}$: Nonlinear aerosol mass response to sulfate reductions in the eastern United States. Journal of the Air and Waste Management Association 49, 1415-1424.

White, W.H., 1999. Phantom spatial factors: An example. Journal of the Air and Waste Management Association. 49, 345-349.

Wienke, D., Gao, N., Hopke, P.K., 1994. Multiple site receptor modeling with a minimal spanning tree combined with a neural network. Environmental Science and Technology 28, 1023-1030.

Zielinska, B., McDonald, J.D., Hayes, T., Chow, J.C., Fujita, E.M., Watson, J.G., 1998. Northern Front Range Air Quality Study. Volume B: Source measurements. Prepared for Colorado State University, Fort Collins, CO, by Desert Research Institute, Reno, NV.

CHAPTER 8

Chemical-Transport Models

Principal Authors:
Christian Seigneur and Michael Moran

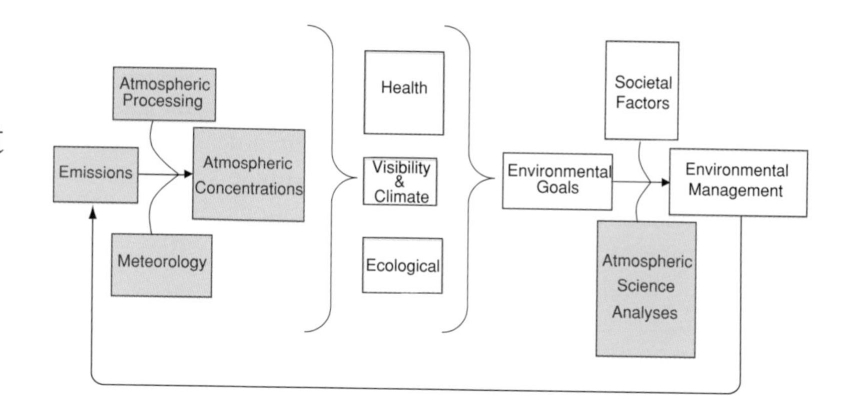

8.1 INTRODUCTION

A crucial issue in understanding and managing atmospheric PM is the ability to link emissions of primary PM and precursors of secondary PM quantitatively to ambient PM concentrations and other physiologically and optically important properties. Chemical-transport models (CTMs) for PM are an important quantitative tool with which to address this relationship. CTMs consist of mathematical representations of the relevant physical and chemical atmospheric processes, which are solved using numerical algorithms to obtain pollutant concentrations as a function of space and time for a given set of pollutant emissions and meteorological conditions (e.g., Peters et al., 1995; Seinfeld and Pandis, 1998; Jacobson, 1999; NARSTO Synthesis Team, 2000; Russell and Dennis, 2000). CTMs thus include mathematical descriptions of all aspects of the atmospheric environment shown in Figure 1.1 : atmospheric processes (Chapter 3), emissions (Chapter 4), and meteorology (Section 8.2.2), and are included as one of the analysis tools listed in the "Atmospheric Science Analyses" box in Figure 1.1.

CTMs for atmospheric pollutants often are referred to by other names, including air-quality models, air-quality simulation models, air-pollution models, emission-based models, source-based models, source-oriented models, source models, first-principles models, and comprehensive models. All of these names suggest at least one of the defining characteristics of this category of model: that is, they are prognostic models that, given the emission rates of selected pollutants and their precursors and prevailing meteorological conditions, predict the atmospheric concentrations of those pollutants based on a combination of fundamental and empirical representations of the relevant physicochemical atmospheric processes. Although most current CTMs tend to treat the same major physicochemical processes, there are significant differences among CTMs in their characterization of PM chemical composition and size distribution. Air-quality models that neglect atmospheric chemistry are often termed dispersion models, and such models are limited to the treatment of chemically inert species. An *air-quality modeling system* is the set of emission, meteorological, and air-quality models needed to simulate air quality.

In addition to PM, CTMs have been developed for other pollutants, including photochemical oxidants and chemical deposition (wet and dry). As a general rule, all of the atmospheric processes relevant to photochemical oxidants and chemical deposition are also relevant to PM, but some processes relevant to PM (for example, size-dependent droplet chemistry) are usually neglected in CTMs for photochemical oxidants or chemical deposition. Consequently, PM CTMs are normally more complex than CTMs developed for other air pollutants, and they are often able to predict these other pollutants as well as PM. For this reason, they may also be referred to as "one-atmosphere," "multi-pollutant," or "unified" air-quality models.

Two other categories of PM models are commonly applied: these are the receptor-based models and observation-based models described collectively in Chapter 7. Such models are independent of and complementary to PM CTMs because they use different methodologies and input data sets. Section 7.5 provides a guide to how these three different categories of PM models can be used corroboratively. PM receptor-based models are diagnostic models that use statistical techniques, ambient PM measurements, and, often, PM-emission composition (but not transformation rate) by source category to determine empirical relationships between PM concentrations at measurement sites and contributing source categories. Examples of some source categories include diesel- and gasoline-powered vehicles, industries such as power generation and petroleum refining, residential wood burning, and natural sources such as wind-blown dust and wildfires. PM receptor-based models have the advantage of not needing emission inventories[1], but their application to secondary PM has been limited. Also, they cannot provide source-apportionment information on a geographically resolved basis (unless coupled with meteorological analyses or dispersion models) or *predict* the complex relationships between ambient secondary PM concentrations and emissions, including the impact of changes in PM precursor emissions on PM concentration levels. PM observation-based models, like receptor-based models, are diagnostic models that primarily use ambient PM data as opposed to emission inventories to infer the possible responses of ambient measured PM to changes in its precursor compounds and/or the sources of these compounds. Unlike receptor-based models, observation-based models are non-statistical; instead, they use selected physicochemical theory incorporated in conceptual models or numerical modules to process and analyze the ambient PM data (e.g., West et al., 1999; Blanchard et al., 2001). Although they have been used successfully to address similar policy questions related to tropospheric ozone (e.g., Cardelino and Chameides, 2000; NARSTO Synthesis Team, 2000), such models are just beginning to be applied to the PM problem because speciated PM data are just now becoming available.

Two other types of "models" are also discussed elsewhere in this Assessment: thermodynamic models and conceptual models. Thermodynamic models of PM (or aerosol equilibrium models; cf. Section 7.2.7) correspond to the gas-particle partitioning module of a PM CTM; they can be used with atmospheric ambient data on speciated PM and related condensable gases to investigate the distribution of condensable chemical species between the gas phase and the condensed phase as discussed in Chapter 3. PM conceptual models are the qualitative "paper" or "mental" models of current understanding that are discussed in Chapters 1 and 10.

There are two subcategories of CTMs based on the reference frame employed: Eulerian (or grid-based) models and Lagrangian (or trajectory) models. Eulerian models use a reference frame that is fixed in space. Their simplest form is the 1-dimensional box model. However, most Eulerian CTMs are three-dimensional (3-D) grid-based models (see Figure 8.1a). Chemical and physical transformations are treated in situ within each grid cell, and transport and diffusion processes move chemical species between grid cells. Lagrangian trajectory models, on the other hand, use a reference frame that follows the movements of individual air parcels from sources to receptors (see Figure 8.1b). Transformations take place within the parcels and the transport processes that are not resolved by the mean advective transport (e.g., vertical turbulent diffusion) may move chemical species between parcels or increase the size of parcels. Lagrangian models are usually simpler in their formulation and less demanding in their computational requirements than Eulerian models. However, they are also limited by their simplicity in that they generally do not treat several atmospheric physical processes realistically, including differential advection due to vertical wind shear, vertical transport, and horizontal diffusion (some Lagrangian models can treat such processes but the computational costs increase accordingly). Since the CTMs that will be used for policy and regulatory purposes are likely to be 3-D Eulerian grid models due to the need for realistic representations of all relevant processes, the focus of this chapter is Eulerian CTMs. However, Lagrangian CTMs may still be useful for screening assessments, where their lower computational requirements are an asset, and for studies where realistic treatments of chemical and aerosol processes are required but simpler representations of transport and diffusion processes are acceptable (e.g., Barthelmie and Pryor, 1996; Kleeman and Cass, 1999).

[1] Although PM source speciation profiles are usually required.

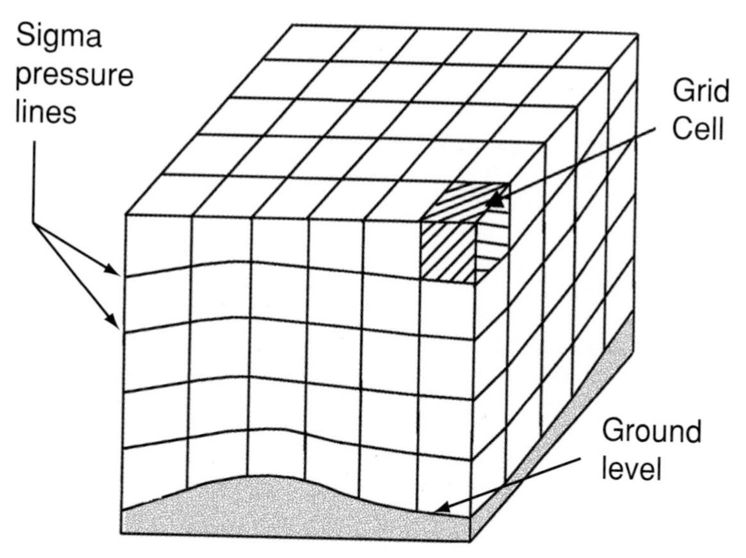

Figure 8.1a. Schematic of 3-D Eulerian framework for a chemical transport model.

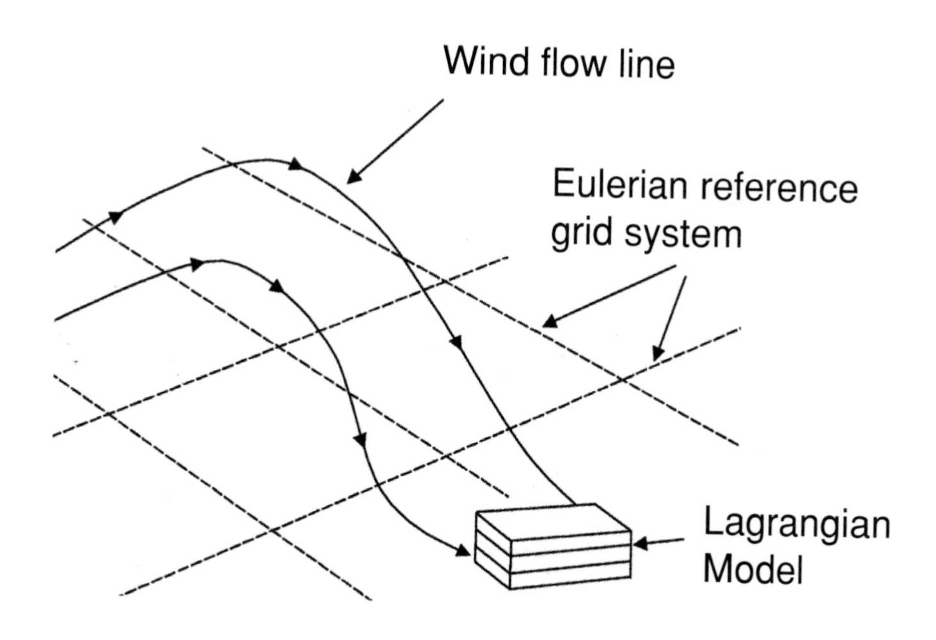

Figure 8.1b. Schematic of 1-D Lagrangian framework for a chemical transport model on an Eulerian reference grid system.

Even 3-D Eulerian CTMs can never be exact representations of the atmosphere. This is both because a perfect understanding of atmospheric processes and conditions is an unreachable goal and because simplifying assumptions must be made to maintain the required inputs and simulation times of CTMs to manageable levels so that they can be used effectively in a policy context (e.g., to investigate the potential effects of various emission scenarios on PM ambient concentrations). Nevertheless, using reasonable assumptions and, to the extent possible, reliable data for their inputs and formulation, CTMs should provide approximations of reality that are sufficiently accurate so as to be useful for policy applications. For cases where the accuracy of a CTM is not satisfactory (e.g., inability to reproduce observed ambient concentrations of some PM components) or uncertainties are high, other

Box 8.1. Characteristics of a typical 3-D PM CTM

Modeling framework	The model takes as inputs emissions of gases and particles, meteorological fields, initial concentrations of gases and particles, and concentrations of gases and particles at the boundaries of the modeling domain. It calculates the concentrations of gases and particles as a function of space and time, as well as their deposition fluxes to the earth's surface. CTMs are sometimes coupled with meteorological models so that some feedbacks from air quality to meteorology (e.g., effect of PM on the radiative budget) can be taken into account.
Spatial scales	Applications range from the urban scale (about 100 km range with a spatial resolution of a few km) to the regional/continental scale (a range of one to a few thousands km with a spatial resolution of a few tens of km) and the global scale (with a spatial resolution of a few hundreds of km). These various scales can be imbedded (nested modeling domains) to provide the desired spatial resolution where needed most.
Temporal scales	Most air-quality simulations have historically focused on air-pollution episodes of a few days, although longer simulation times (e.g., one year) have now become more common. The temporal resolution of CTM inputs and outputs is typically one hour (time steps used for the computations are usually considerably less, on the order of a few minutes).
Emissions	Natural and anthropogenic emissions enter the model domain from either ground-level area sources and elevated point sources for both gases and particles. Some sources are affected by meteorology.
Transport and diffusion	Transport and diffusion processes include horizontal and vertical advection by the resolved wind flow, vertical convection by winds not resolved by the model's spatial resolution (e.g., strong updrafts from cumulus clouds), and turbulent diffusion.
Gas-phase	Gas-phase transformations include the chemistry of the formation of ozone and other oxidants and the oxidation of SO_2, NO_x and some VOC to condensable $SO_4^=$, NO_3^-, and organic species, respectively. In such gas-phase chemical kinetic mechanisms, the inorganic chemistry is represented in detail but the organic chemistry is simplified by grouping VOC species according to their chemical characteristics (this is necessary because of the huge number of VOC species). The chemistry of ozone formation is coupled to that of PM formation in an intricate, and sometimes non-intuitive, manner.
Cloud dynamics	Clouds (and, to a lesser extent, fogs) must be represented because they modulate photolysis rates, lead to secondary PM formation via cloud chemistry, remove PM and precursor species from the atmosphere (or redistribute vertically) via precipitation (and evaporation), and transport PM vertically via updrafts and downdrafts. The chemistry of sulfur and nitrogen deposition is, therefore, closely coupled to that of PM formation.
Aerosol processes	Primary PM is considered mostly chemically inert and is only affected by emission, transport and diffusion, and deposition processes; however, alkaline dust and sea salt can react with H_2SO_4 and HNO_3. The processes leading to the formation of particulate $SO_4^=$, NO_3^- and NH_4^+ are known; they are governed by thermodynamic equilibrium between the particles and the gas phase and, for large particles (greater than 1 μm), mass transfer from the bulk gas phase to the particles' surfaces. There is currently considerable uncertainty regarding the formation of SOA and several approaches are being developed and evaluated. Currently, at least two particle size fractions (fine and coarse) are treated in most CTM applications. More detailed representations of the particle-size distribution can be implemented using either modal or sectional representations; however, they are not supported by current emission inventories. Nucleation, condensation, and coagulation all modify particle-size distributions. Aerosols are usually assumed to be internally mixed. Aerosol optical properties can be used to estimate visibility.
Deposition	Deposition of gases and particles via dry processes (to the ground, vegetation, and man-made structures, and including wetted surfaces) and wet processes (precipitation, cloud impaction, and fog settling) removes chemical species from the atmosphere. Sedimentation redistributes coarse particles in the vertical direction.

analytical techniques (see Chapter 7 for a discussion of receptor-modeling techniques) can be used to corroborate and/or complement the CTM results. Additional experimental research also can be conducted in both the laboratory and the field to provide the data that will allow the scientific community to improve the accuracy of the CTMs for their future applications.

What is the current status of these models for helping us to estimate the particle concentrations associated with any given set of PM primary and precursor emissions and meteorological conditions? The rest of this chapter first reviews the current status of CTM process representations and the application of CTMs to both episodic (lasting from a few days to a few weeks) and long-term (from seasons to year) simulations. Next, the discussion considers the suitability of CTMs for addressing policy-relevant questions, followed by the performance evaluation of current CTMs and the use of CTMs to support the design and evaluation of ambient monitoring networks and the estimation of PM exposure in support of epidemiological and human health studies. The chapter reviews policy-relevant results and guidance obtained to date from CTMs, then lists major uncertainties in current CTMs and recommendations on how to address those uncertainties. The chapter concludes with a summary of policy implications related to the use of CTMs for PM management.

8.2 CURRENT STATUS OF PM CHEMICAL-TRANSPORT MODELS

Recent reviews of CTMs for PM identified about a dozen 3-D grid models that have been applied to various regions or urban airsheds of North America (e.g., Seigneur et al., 1999). The status of these CTMs is rapidly evolving, as some existing CTMs are undergoing modification to include the latest state-of-the-science and new CTMs for simulating PM are being developed. Therefore, a compendium of existing CTMs is not presented here, since it would likely be quickly outdated. Instead, an overview of the general formulation of the atmospheric processes that need to be simulated in any CTM to estimate particle concentrations (see Figure 8.2) is presented; the discussion also covers the limitations associated with those formulations and identifies where new advances are currently taking place. Additional information is available in several recent review articles (see Peters et al., 1995; Seigneur et al., 1999; Russell and Dennis, 2000).

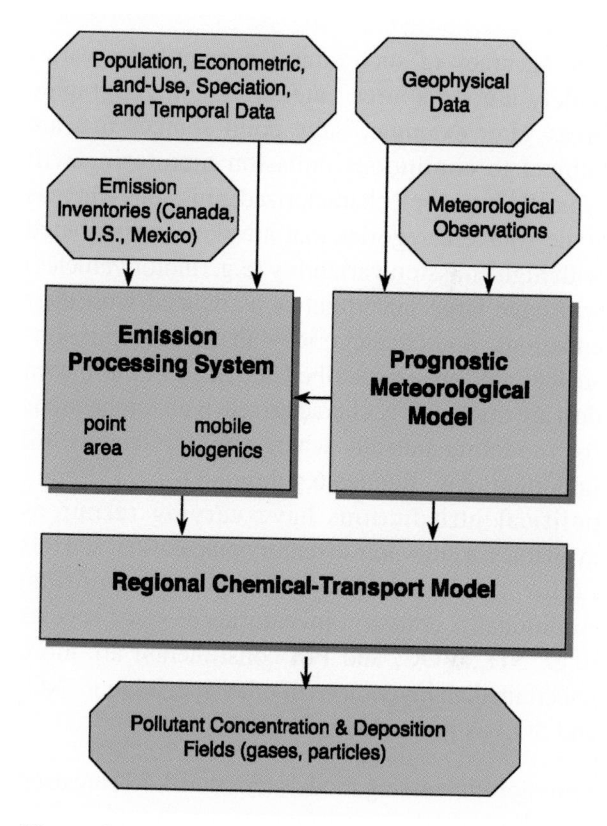

Figure 8.2. Schematic description of the components of a PM modeling system.

8.2.1 Emissions

All CTMs require quantitative information on the emissions of PM and its gaseous precursors. Emission data are required for primary PM with chemical composition and size resolution, and for NO_x, SO_2, VOC, CO, and NH_3.

As described in Chapter 4, emission inventories are typically developed for various source categories by combining emission factors with activity data. Emission factors represent the amount of pollutant emitted by a unit amount of activity, where the activity refers, for example, to the amount of electricity produced or fuel consumed by a power plant or the kilometers traveled by different classes of motor vehicles. Some types of emissions depend on meteorology (e.g., wind-blown dust emissions, evaporative losses of VOC, biogenic emissions of VOC from vegetation, sea-salt emissions from ocean surfaces, application of road salt) and on land use (e.g., biogenic emissions, NH_3 emissions from cattle and from fertilizer application).

The accuracy of such emission inventories varies widely among source categories and geographic areas. For example, large point sources that are subject to continuous emission monitoring will typically have well characterized emissions whereas some source categories that are widely distributed with high emission variability (e.g., motor vehicles) will have large uncertainties associated with their emissions. Source categories with episodic emissions such as wildfires, prescribed burns, and wind-blown dust are often poorly characterized, with implications for modeling periods where such emissions are important (e.g., Figures 6.4 through 6.7). Different political jurisdictions have varying resources available for emission-inventory preparation, and this factor can result in increased uncertainties. Additionally, emission inventories of some species (e.g., NH_3, VOC, and PM constituents) are more uncertain than inventories of other species (e.g., NO_x and SO_2) as illustrated in Table 4.8.

Emission-processing models (Section 4.4.1) are used to calculate the components of the emission inventories that depend on meteorology and land use. Emission-processing models are also used to distribute the raw emissions from the inventories spatially, temporally, and by chemical species for input to CTMs. For example, CTMs require 1) total VOC emissions to be separated into a more detailed set of individual VOC species and VOC classes and 2) primary PM emissions to be separated into its main chemical constituents (BC, OC, crustal material, $SO_4^=$, and possibly sea salt). Once a baseline emission inventory has been developed, the emission-processing models, when combined with predictions from socio-economic models, can be used effectively to develop a variety of quantitative future emission scenarios that can be tested as emission-management options.

But emission-processing models have limitations of their own. For example, they depend on the use of libraries of temporal profiles, speciation profiles for SO_x, NO_x, VOC, and PM, size-distribution profiles for PM, and spatial allocation surrogates, and there are far fewer such profiles and surrogates available than there are source types. This is a particular concern for the speciation and size-characterization of primary PM emissions. There is also confusion at present over whether condensable VOC emissions will be accounted for as primary particle OC emissions or as VOC emissions, especially if VOC- and PM-speciation profiles were developed separately for the same source type. In addition, the differentiation between primary BC and OC mass components made by both the emission community (Chapter 4) and the measurement community (Chapter 5) is an operational one. Thus, the quantity called "black carbon" in CTMs for PM will depend entirely on the operational definition employed in source testing and characterization by the emissions community, and may or may not correspond exactly to the BC definition implicit in ambient speciated PM measurements.

In some cases, CTMs can flag potential errors in emission inventories if significant discrepancies between simulated and observed PM concentrations are more likely to be due to inputs than to model formulation. One example is the analysis of CTM output over the northeastern United States, which suggested that the NH_3 emission inventory was likely in error. Introducing seasonal variability in the NH_3 emission inventory improved the CTM's performance. Such analyses, however, must be conducted with care because uncertainties in the model formulation may affect the analyses, and emission inventories ultimately should be revised from the bottom up.

8.2.2 Meteorology

All CTMs require meteorological information (e.g., winds, turbulence, temperature, relative humidity, clouds, and precipitation) to simulate the transport and removal of gases and particles in the atmosphere. Meteorological fields can be constructed by interpolation of available meteorological data (diagnostic approach) or by simulation of the meteorology using a computer model that solves the fundamental equations of atmospheric fluid dynamics (prognostic approach). The prognostic approach offers the advantage of providing a self-consistent description of the meteorology, because it is based on fundamental principles of atmospheric physics. Furthermore, it can be improved by "nudging" the simulation to the available meteorological data via data assimilation, thereby combining the advantage of the diagnostic approach (the use of real-world measurements) with the comprehensiveness and internal consistency of the prognostic approach. The prognostic models used to generate the meteorology for CTMs may be the same as those used for weather prediction, as is the approach adopted, for example, by Environment Canada.

In some cases, due to differences in grid specifications used by the prognostic meteorological model vs. the CTM, it is necessary to perform an intermediate interpolation step to transform the meteorological fields predicted by the meteorological model to equivalent fields on the CTM grid. In other cases, it may be necessary to diagnose a meteorological field that is needed by the CTM but not predicted by the meteorological model from related fields. Such "interfacing" steps may be necessary but are undesirable in general because they introduce errors and inconsistencies into the modeling system (e.g., Byun, 1999; Seaman, 2000). A few air-quality modeling systems avoid this issue by simulating the air-quality processes together with the meteorological processes in the same model (e.g., Jacobson, 1998). Such "on-line" or "in-line" models[2] have the additional advantage of allowing feedback between the air quality and meteorology (e.g., effect of PM on the radiative budget). However, they are both more complex and more demanding computationally when applied to simulate emission-management strategies because the meteorology must be computed repeatedly for each CTM simulation.

The accuracy of the meteorological predictions has a direct effect on the accuracy of CTM predictions because meteorology affects certain emission processes, the transport and dispersion of pollutants, the rate of chemical transformations and particle formation, and the removal of pollutants from the atmosphere (e.g., Pielke and Uliasz, 1998; Russell and Dennis, 2000). Meteorological models tend to be less accurate at smaller spatial scales, in complex terrain, for near-surface stable stratification (e.g., nocturnal jets, turbulent bursts), for calm conditions (i.e., conditions that may be conducive to PM pollution), and in predicting clouds and precipitation.

For example, at smaller spatial scales, the meteorological observing network whose measurements are used to specify either diagnostic meteorological fields or initial conditions for the meteorological model may not have sufficient spatial resolution to detect small-scale weather features. Another issue is that the closure assumptions used in some parameterizations of atmospheric physical processes are scale-dependent. If the grid spacing of a meteorological model is then reduced in order to consider smaller spatial scales, these closure assumptions may be violated if a different parameterization designed for smaller spatial scales is not used. Most parameterizations of deep convection, for example, are not valid for horizontal grid spacings smaller than 10 km because they assume that thunderstorm updrafts cover only a negligibly small fraction of a grid cell (e.g., Seaman, 2000). In complex terrain, on the other hand, topographic and land-surface features that are smaller in size than the grid spacing will not be represented in the model but can influence meteorological conditions locally. Episodic meteorological models are also likely to be less accurate for longer simulations because of the simple representations they usually employ for some processes such as atmospheric radiation and the hydrological cycle (e.g., soil moisture, snowpack). This may be a concern if not addressed by re-initialization or data assimilation when longer-term meteorological and air-quality simulations are needed related to seasonal or annual air-quality standards.

Clearly, the availability of meteorological data for assimilation by the model can greatly enhance model performance. Meteorological measurements (particularly aloft) are always needed to confirm the quality of the meteorological simulations, particularly

[2] As opposed to "off-line" air-quality models where the meteorological model and CTM are separate, the meteorological model must be run first but only once, the CTM may be run many times for the same meteorological episode, and the air-quality fields cannot influence the meteorological fields.

in mountainous and coastal areas where horizontal gradients and wind shear aloft can be significant. Here, as has been the case in the past, improvements to meteorological observational networks and technologies in support of numerical weather prediction will also benefit air-quality modeling.

8.2.3 Transport and Diffusion Processes

Transport and diffusion processes govern the movement and dilution of gases and particles in the atmosphere due to atmospheric circulations on many scales. Transport generally refers to the mean movement of air parcels resulting from model-resolved winds, whereas diffusion refers to the mixing and dilution of gases and particles by turbulent eddies that are not resolved by the mean wind flow (e.g., Moran, 2000). Transport can be further categorized as horizontal movement of air parcels parallel to the surface and vertical movement of air parcels up and down. Strong vertical movement may occur when cumulus clouds are present, in complex terrain when the wind flow meets a steep surface slope, and when there is considerable convergence of the horizontal wind field (e.g., a lake-breeze front).

Several approaches have been used to characterize turbulent mixing in CTMs. However, there is still considerable uncertainty in those parameterizations, especially under conditions of stable stratification (e.g., nocturnal boundary layer, free troposphere) where turbulence occurs sporadically in both space and time due to shear instabilities. A few CTMs consider "venting" to the free troposphere by subgrid-scale boundary-layer clouds or subgrid-scale vertical transport by deep convective clouds. Because turbulent diffusion is a random process, its parameterization in CTMs is approximate and this has some impact on model predictions. It should be noted also that vertical diffusion is treated in both the meteorological model and the CTM, but typically using different algorithms; reconciliation is warranted in such cases to provide better consistency to the modeling process.

The uncertainties that exist in the predictions of the meteorological model for the wind and turbulence fields translate directly into uncertainties for the transport and diffusion processes in the CTM. This is a particular problem when the grid size of the meteorological model and/or CTM is too coarse to resolve real-world horizontal and vertical gradients in the wind and turbulence fields.

8.2.4 Chemical Transformations

Gas-phase chemistry. Because the number of organic chemical species involved in the chemistry of the troposphere is very large (several thousands), it is not feasible to simulate all the chemical reactions of all atmospheric constituents. Therefore, some approximations must be made in the representation of VOC chemistry. Two major approaches have been used. In the first approach, VOCs are grouped by classes (e.g., short-chain olefins) and represented by one chemical species of each class (those species act as surrogates for all species within the class). In the second approach, VOC molecules are decomposed into their functional groups (e.g., double bond, carbonyl group) and the reactivity of the functional groups is simulated. Both approaches have successfully reproduced the dynamics of ozone formation in experiments conducted in laboratory smog chambers and have been successfully applied to actual photochemical smog episodes. However, there has been little evaluation of those mechanisms for fall/winter episodes, for which days are shorter, solar radiation is weaker, and temperatures are cooler as compared to summer episodes. Such evaluations are warranted, since high PM episodes can occur during fall/winter in some areas (for example, California's San Joaquin Valley and southeastern Canada: see Chapter 10).

Chemical kinetic mechanisms that are based on either of the two above approaches still involve on the order of 100 chemical species and 100 to 200 chemical reactions (e.g., Dodge, 2000). To date, such mechanisms have focused primarily on anthropogenic VOC and have not provided explicit treatment of biogenic compounds beside isoprene. Mechanisms are now being developed that include a detailed treatment of the oxidation of several biogenic VOC (e.g., monoterpenes, oxygenates) including their effect on PM formation. Reduced chemical mechanisms also have been developed, which use a minimum number of chemical species (say, three species to represent all VOC) in order to minimize the computational costs of simulating long periods of time (e.g., one year).

Aqueous–phase chemistry. Reactions in cloud and fog droplets and, to some extent, raindrops typically enhance the oxidation of SO_2 and NO_x to $SO_4^=$ and NO_3^-, respectively. They may also play a role in the formation of condensable organic compounds. Chemical mechanisms for the formation of $SO_4^=$ and NO_3^- are well established, and the more advanced CTMs include fairly comprehensive descriptions. On the other hand, little is known about the aqueous chemistry of organic compounds, and CTMs currently do not treat it. Chemical reactions in droplets may depend on droplet size (because pH may vary with droplet size); however, most aqueous-phase chemical mechanisms currently used in CTMs only treat a single average droplet size (i.e., "bulk" chemistry). The role of snow and ice in the oxidation of volatile precursors to condensable compounds is not well known except for those reactions that are of particular interest to Arctic chemistry (e.g., polar stratospheric clouds); however, this process is unlikely to affect CTM outputs significantly. Clearly, aqueous chemical processes need to be included in CTMs since a significant fraction of secondary fine PM may be formed in clouds or fogs.

Heterogeneous chemistry. Chemical reactions may also occur at the surface of particles (e.g., Jacob, 2000). Such reactions may lead to the formation of additional PM (e.g., N_2O_5 hydrolysis at the surface of aqueous particles will form NO_3^-) or may affect the chemistry of the gas phase (e.g., scavenging of radicals by particles will reduce ozone formation). The heterogeneous formation of PM typically is not a major reaction pathway under most circumstances. Consequently, most CTMs do not treat heterogeneous reactions, although these reactions may contribute significantly to NO_3^- formation under some conditions (e.g., high humidity).

8.2.5 Representation of PM

PM consists of a mixture of particles of different sizes and chemical composition. Therefore, it is necessary to provide some representation of particle size and chemical composition to properly simulate the evolution of PM in the atmosphere in order to calculate PM concentrations and other properties of interest. Although CTMs for PM vary in their level of detail for characterizing particles, there are some basic characteristics that are simulated by most CTMs.

As discussed in Chapter 3, the fine and coarse PM size fractions tend to have different origins and differ significantly in their chemical composition and atmospheric behavior. Therefore, it is imperative to differentiate between fine and coarse particles, and most CTMs treat at the minimum these two PM size fractions. Because some processes such as deposition, light scattering, and particle retention in the respiratory system are strong functions of particle size, some CTMs attempt to calculate the evolution of particle-size distributions within the fine and coarse size fractions. To that end, two approaches have been used. On one hand, the particle-size distribution can be approximated by several log-normal distributions. Typically, three log-normal distributions are used to represent the nucleus, accumulation, and coarse modes (this approach is often referred to as the "modal" approach, Figure 3.2). On the other hand, the particle-size distribution can be discretized into a conterminous set of size sections (this approach is generally referred to as the "sectional" approach). Both approaches compromise between the accuracy of the numerical solution and the associated computational costs (e.g., Zhang et al., 1999). Either approach will require compatible size characterization of the primary PM emissions, that is, into modes or sections. (Because emission inventories do not currently provide any level of detail about PM-emission size-distribution beyond separation into fine and coarse fractions, emission-processing systems must assume more detailed size distributions for various primary PM source types. This size-disaggregation step in emission processing thus introduces additional uncertainties, which will contribute to uncertainties in the predicted PM size distributions. Note that detailed size-distribution information can be aggregated into fine and coarse fractions corresponding to PM air-quality standards.)

CTMs that provide a detailed resolution of the particle-size distribution must in theory simulate all the relevant dynamic processes that may modify the distribution, including the emission of primary particles of different sizes, the nucleation of new ultrafine particles, the growth of existing particles via condensation of vapors and the shrinkage of particles resulting from volatilization of particulate species, the coagulation of particles, and the size-dependent removal of particles resulting from dry- and wet-deposition processes (see Chapter 3). Some, mostly urban-scale, CTMs neglect nucleation (thereby assuming that condensation prevails under conditions with moderate to high PM concentrations) and coagulation (which is typically slow compared to other processes), but regional-scale CTMs with their larger domains and longer transport times may consider these two processes.

Most CTMs assume that particles in a given size range have the same chemical composition. This is not realistic for fresh emissions or ambient particles, but becomes an increasingly better approximation for aged aerosols owing to condensation and coagulation (e.g., Chapter 3; Winkler, 1973). Also, the vast majority of PM monitoring instruments measure features of aerosol populations rather than of individual ambient particles (Chapter 5). Recent work, though, has shown that it is feasible (with limitations) to remove this assumption and allow the treatment of particles of different chemical compositions in the same size range. This can be valuable for estimating optical properties or for "tagging" particles for purposes of source attribution or apportionment (e.g., Kleeman and Cass, 1999, 2001; Jacobson, 2001).

Most CTMs calculate at least the concentrations of particulate $SO_4^=$, NO_3^-, NH_4^+, OC, BC, and water, in addition to total PM mass (the remaining mass is generally assigned to soil dust or "other"). Because sea salt and soil dust can interact chemically with strong acids such as HNO_3 and H_2SO_4 (see Chapter 3), some models explicitly treat chemical species such as Na, Cl, Ca, K, Mg, and $CO_3^=$. The various algorithms (thermodynamic models) used to calculate the partitioning of inorganic species between the gas phase and the particles tend to agree within about 10 percent on average (e.g., Ansari and Pandis, 1999; Zhang et al., 2000). However, treatment of the effects of coarse sea-salt or soil-dust particles on PM chemical composition varies widely among CTMs. A correct treatment must take into account the fact that coarse particles are generally not in equilibrium with the gas phase (Wexler and Seinfeld, 1990, 1991). Because such non-equilibrium calculations are computationally demanding, some CTMs either assume equilibrium (which can lead to incorrect PM chemical composition in some cases) or use other simplifying assumptions. Finally, the formation of SOA from gas-phase VOC is an area of ongoing research. As a result, the treatment of particulate organic compounds in CTMs keeps evolving and currently there are large differences in the formulation and simulation results of SOA formation in existing CTMs.

The calculation of water existing in the condensed phase is also important because 1) some residual water is present in the gravimetric measurements conducted for the PM ambient standards and 2) condensed water associated with the ambient PM affects the scattering of light by particles. Currently, the amount of water associated with the inorganic PM fraction can be predicted (albeit with some uncertainty); however, the complex mix of organic compounds (some hydrophilic and some hydrophobic) limits reliable predictions in the case of organic PM.

As source- or emission-based models, the performance of CTMs will depend directly on the comprehensiveness and accuracy of their input emission data. Improvements to emission inventories and emission modeling, including the description of natural and often intermittent sources such as wildfires, windblown dust, and sea salt, of anthropogenic PM, NH_3, and semivolatile VOC emission levels, and of the size distributon and chemical composition of primary PM emissions, should result in improved CTM performance and reduced uncertainty.

8.2.6 Deposition Processes

Deposition processes include both dry- and wet-removal phenomena. Dry deposition of gases and particles occurs when a gas molecule or a particle comes into contact with, and adheres to or is absorbed at a surface (e.g., soil, vegetation, structures). In addition, large particles are sufficiently affected by gravity to deposit via sedimentation (also referred to as gravitational settling). Sedimentation is particularly relevant to the development of emission inventories for dust, because a large fraction of coarse dust particles will deposit close to their point of emission. Thus, dust emissions will not be correctly estimated for CTM grid-cell scales if one ignores the fraction that is rapidly deposited or intercepted immediately after suspension. CTMs treat dry deposition as a function of meteorological conditions, land-use type (e.g., urban area, vegetation, water), and the nature of the depositing species (e.g., size in the case of particles). At present there is considerable uncertainty about parameterizing particle dry-deposition, and few measurements exist to evaluate the modeled dry-deposition rates of many gas-phase species considered by CTMs (e.g., Wesely and Hicks, 2000; Zhang et al., 2002). This uncertainty translates directly into an uncertainty in PM ambient concentrations.

Wet deposition includes in-cloud scavenging ("rainout") and below-cloud scavenging by precipitation ("washout"). In clouds or fogs, particles may act as condensation nuclei (thereby leading to the formation of droplets) or be scavenged by existing droplets via collision. Some CTMs treat all forms of precipitation similarly (e.g., rain, snow, and hail), which may not be appropriate for many cases. Fog-droplet deposition, also known as occult deposition, is typically not treated in CTMs although it may be an important removal mechanism in foggy coastal areas or mountainous areas (e.g., Harvey and McArthur, 1989; Schemenauer et al., 1995; Lillis et al., 1999). Overall, wet deposition is not represented well in CTMs, both because precipitation-scavenging processes are not well understood and because clouds and precipitation are difficult for meteorological models to predict. However, because wet deposition is the major removal mechanism for $SO_4^{=}$ and other nonvolatile fine PM constituents as well as a significant removal mechanism for NO_3^-, it is essential that it be simulated correctly in CTMs, especially those being used for regional or long-term simulations.

8.2.7 Computational Aspects

The processes discussed above must be described in terms of mathematical equations, which are then solved numerically. These solutions introduce additional uncertainties, which must be minimized to the extent possible. In most cases, a balance needs to be achieved between numerical accuracy[3] and computational speed so that a CTM simulation does not require excessive computer time. Therefore, the numerical solvers used in CTMs are not the most accurate but sacrifice some accuracy (say, 10 percent) to gain significant speed. Similar compromises must be made for the number of VOC species considered, the grid spacing, and the domain size considered (see next section for further discussion of the design of the modeling domain).

A typical simulation using a CTM with comprehensive treatment of gas-phase and aqueous-phase chemistry and an empirical treatment for SOA formation currently takes about two hours of computing per simulation day on a typical desktop computer for 100,000 grid cells. If a detailed representation of the formation of SOA is used, the computing time may increase by a factor of 4. On the other hand, if a reduced gas-phase chemistry mechanism and a parameterized approach to particle formation are used, the computing time of the former simulation can be lowered.

Another important computational aspect of CTMs is the large size of the input and output files (for example, hourly data on emissions, winds, temperature, and relative humidity for one entire year). Large files can present challenges not only for their efficient processing but also for their quality assurance/quality control (QA/QC). Because of the importance of QA/QC procedures to ensure that CTMs are properly applied, protocols are needed to verify that the input files, especially the emission and meteorological files, contain valid and reliable data, (e.g., internally documented files, verification of units, processing steps, file version, and temporal period).

[3] Dependent upon the accuracy of the scientific modules of the CTM and the temporal and spatial resolutions used by the CTM.

8.3 APPLICATIONS OF CHEMICAL-TRANSPORT MODELS TO THE SIMULATION OF EPISODIC AND LONG-TERM PM CONCENTRATIONS

8.3.1 Episodic Simulations

An episodic simulation covers only a few days, typically less than a week. Since PM concentrations may be regulated by both short-term and long-term standards, episodic simulations are appropriate to address a short-term PM standard in areas where it may be violated. Episodic simulations are conducted routinely for assessing ozone pollution since the ozone air-quality standards are defined for short periods of time (i.e., 1 to 8 hours). (Episodic simulations are also relevant to air-quality forecasting as discussed in Section 8.4.7.) However, episodic simulations for PM may differ from simulations conducted for ozone assessments for several reasons.

First, since the atmospheric lifetime of fine particles is several days (and possibly 1000 or 2000 kilometers in terms of transport distance) in the absence of precipitation (Chapter 3), it may be necessary to use a large modeling domain to ensure that the calculated PM concentrations are governed by the emissions within the domain rather than by the chemical concentrations entering at its upwind boundary. Alternatively, it will be essential to characterize the lateral boundary concentrations accurately, and these are likely to vary with time. Typically, a "spin-up" period that corresponds to the time to flush out the initial concentrations is used to minimize the effect of the initial concentrations. For cases where a large domain is used, this "spin-up" period may cover several days.

Clearly, computational costs increase as the size of the domain (i.e., the number of grid cells being simulated) increases. There are several ways to minimize these costs. For example, the number of grid cells can be minimized by using a coarse grid over the entire modeling domain and one or several grids of finer spatial resolution nested within the coarse grid. This "nested-grid" approach minimizes computations over the coarse domain while focusing most computational resources on the areas of interest, where fine spatial resolution is used. A complementary approach consists of using a subgrid-scale treatment for major point sources that would otherwise have their emissions diluted over large air volumes. This "plume-in-grid" treatment allows one to maintain a more realistic representation of the dynamics and chemistry of the major point-source plumes even when a coarse grid resolution is used.

Second, the vertical extent of the domain may also differ in many cases from that used in ozone modeling. As discussed above, cloud processes can contribute significantly to the formation of secondary particles. For regional-scale modeling, it may be necessary to provide sufficient vertical resolution to properly represent both boundary-layer and cloud processes, including deep convective systems.

Finally, one must keep in mind that PM episodes may not coincide with ozone episodes, which mostly occur during summer. For example, some areas experience high PM episodes during late fall and winter (see Chapter 6). Meteorological models have undergone performance evaluation for a variety of seasons and locations. However, the chemical mechanisms that simulate the formation of oxidants and secondary PM typically have not been evaluated for non-summer conditions, and their performance may be poorer than for summer episodes (see Section 8.4.3).

8.3.2 Long-Term Simulations

In many cases, the long-term PM air-quality standard may be the governing standard, and long-term simulations of PM will be required to address source-receptor relationships and control-strategy impacts properly (i.e., commensurate with the time scale of the standard). There are two approaches to simulate long-term PM concentrations, as depicted in Figure 8.3.

One approach consists in simulating the actual long-term period. However, one must note that the long-term PM standard may represent a period of several years (e.g., the U.S. annual standard is a three-year average standard). Consequently, one must either simulate several years or assume that the year being

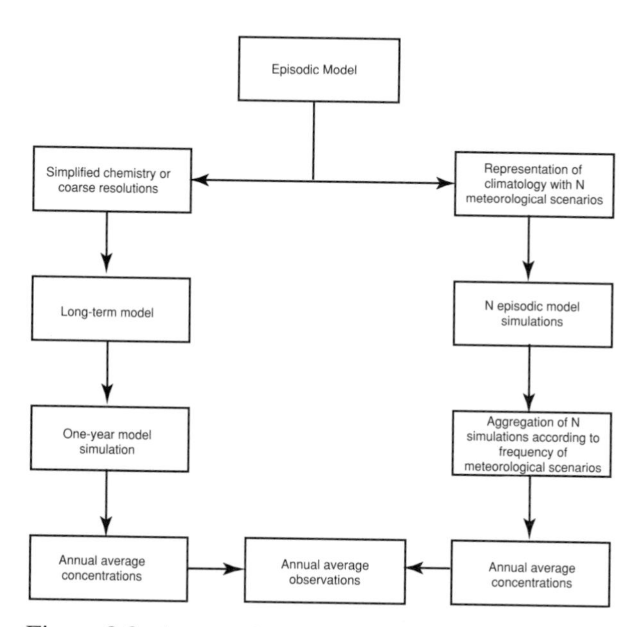

Figure 8.3. Approaches to calculating long-term PM concentrations (after Seigneur et al., 1999).

simulated is representative of multi-year periods. Note that the year being simulated does not need to be a calendar year nor does it need to be a continuous year (e.g., it could consist of quarters selected among various years). Nevertheless, some sampling bias must be expected if only one year is simulated to represent a period of several years (e.g., Brook and Johnson, 2000). Moreover, some approximations are typically made to manage the large computational costs associated with long-term simulations. For example, a reduced chemical mechanism may be used to minimize the chemistry calculations or a coarser spatial resolution may be selected to reduce the number of grid cells being simulated. Those approximations will also affect model accuracy and need to be thoroughly evaluated before the model is used to develop source-receptor relationships.

The other approach aggregates individual episodic simulations to construct a synthetic long-term period (e.g., one or several years). The aggregation is typically based on a statistical analysis of the meteorological regimes that occur in the area of interest (e.g., Brook et al., 1995a,b; Cohn et al., 2001). By assuming that only a limited number of meteorological episodes may be used to represent the long-term meteorology of an area, one can limit the air-quality simulations to those representative episodes[4]. The results of the air-quality simulations then can be weighted according to the relative frequency of the meteorological regimes to construct long-term simulation results. This more climatological approach does not require any simplification of the CTM or selection of a nominally representative year. However, some approximation is involved in the use of a limited number of meteorological episodes.

Both approaches offer advantages and shortcomings. To date, there has been no formal comparative evaluation of these two approaches and, until such an evaluation is conducted, a combination of both approaches is recommended.

8.4 WHAT QUESTIONS CAN CHEMICAL-TRANSPORT MODELS ADDRESS AND HOW WELL?

Policy-related questions that will be typically addressed in most applications of PM CTMs are reviewed here. For each question, the best assessment of how well it can be answered by current CTMs is provided.

8.4.1 Can the Contributions of Various Precursors and Source Types to PM Be Quantified?

Clearly, the development of effective emission-control strategies will require relating observed PM concentrations to their precursors. Such precursors must then be identified and attributed to their sources of emission.

The processes that lead to the formation of PM $SO_4^=$, NO_3^-, and NH_4^+ are currently sufficiently well understood to permit their adequate representation in a CTM. It is slightly more difficult to calculate NO_3^- than $SO_4^=$ because NO_3^- involves the calculation of a gas/particle equilibrium, which requires knowledge of the temperature, relative humidity, and

[4] Bearing in mind (e.g., Section 6.5.2) that long-term averages can be significantly influenced by relatively infrequent episodes.

NH_3 concentration. Therefore, uncertainties in any of these variables will be compounded in the NO_3^- concentration.

In contrast, understanding of the processes that lead to SOA formation is still incomplete. As a consequence, current CTM parameterizations of SOA formation have large uncertainties. Moreover the representation of the SOA-formation processes requires some simplifications, because the large number of chemical species involved needs to be reduced to a manageable number in a CTM. Primary OC is problematic too, because emissions of particulate organics and condensable organic gases are also not well characterized at present. Consequently, CTMs cannot currently provide an accurate representation of the contribution of primary or secondary particulate organic compounds to PM.

Other factors that affect the accuracy of the PM concentration predictions include the meteorological fields, the description of the transport, dispersion, and removal processes, and the accuracy of the emission inventories for individual source categories. These processes also directly affect the primary PM components, including the less volatile constituents such as dust, BC, and primary OC.

Therefore, CTMs currently have mixed abilities for predicting the contribution of various precursors and source types to PM. Overall, the formulation of CTMs is satisfactory for the inorganic component of PM (with the caveats noted in the preceding sections) but the treatment of organic PM must be considered to be in a developmental stage. The most advanced CTMs therefore should be able to provide reasonable estimates of $SO_4^=$, NO_3^-, and NH_4^+ concentrations (roughly half of $PM_{2.5}$ mass and less than half of PM_{10} mass on average: cf. Section 6.5.4), but are likely to give uncertain results for carbonaceous PM. In any case, the ability of a given CTM to predict PM concentrations must be carefully examined through performance evaluations (see Section 8.5) before its application for policy analysis.

There are currently large uncertainties associated with the treatment of carbonaceous PM. As a result, the relative amounts of BC versus OC, primary versus secondary OC, and anthropogenic versus biogenic OC cannot be estimated with good accuracy. Additional laboratory and field studies will help improve the formulation of CTMs for carbonaceous PM.

8.4.2 Can the Relative Contributions of Long-Range Transport and Local Emissions Be Quantified?

The simple answer is "yes." CTMs can be used to quantify the relative contributions of local sources vs. long-range sources at a receptor by carrying out a suite of emission sensitivity tests, in which emissions in different parts of the model domain are scaled upward or downward in different combinations (e.g., Dennis et al., 1990; Seigneur et al., 2000b). However, there are three main sources of uncertainty in such calculations. (Note that independent tools such as receptor models and observation-based models (Chapter 7) can be used corroboratively so as to minimize the potential effect of CTM-related uncertainties.)

As mentioned above, $PM_{2.5}$ can be transported over long distances because of its long atmospheric lifetime. Therefore, PM concentrations will include in most cases a significant component due to the long-range transport of PM and its precursors, in addition to the component arising from local emissions. Quantifying the relative contributions of long-range transport and local emissions requires a proper characterization by the CTM of 1) both local and upwind sources, 2) long-range transport processes, including deposition, and 3) upwind boundary concentrations of chemical species.

Since fine particles can persist for several days, the upwind boundary concentrations may significantly affect the predicted PM concentrations even for large domains (e.g., Brost, 1988). Such boundary concentrations must be specified using available data (e.g., aircraft data for gases, PM speciation and size distribution, and satellite and other real-time sensors to characterize plumes from fires and large dust events). However, data on concentrations of PM and precursors aloft are typically non-existent; thus large uncertainties will be associated with the upwind chemical boundary conditions. To minimize the importance of boundary contributions, one should attempt to include the great majority of the contributing sources within the modeling domain. Identifying and quantifying emission sources is best achieved if all sources are subject to the same emission-inventory criteria. In some cases, however, emission-inventory data for some upwind areas within the domain may be sparse and unreliable.

One must note that the accuracy of computed long-range transport trajectories will decrease as transport distance increases, even when using the most advanced meteorological models, making it harder to identify source regions with increasing distance from a receptor. Uncertainties related to chemical transformations and removal during long-range transport also will contribute to uncertainties in quantifying the magnitude of the long-range-transport contribution at a receptor.

Therefore, our inability to accurately quantify 1) the contribution of upwind boundaries, 2) the emissions of PM and precursors within the limited-area modeling domain, and 3) changes in PM concentrations during transport can be a limiting factor in the design of effective emission-control strategies.

8.4.3 Can the Relative Magnitude of Seasonal Contributions to PM Concentrations Be Represented?

The simple answer is again "yes". CTMs for PM can be run for different seasons and then the magnitudes of the predicted seasonal concentrations compared. But again, there are caveats.

CTM performance may vary from season to season because 1) the seasonal variations of some emissions (e.g., wood smoke, fugitive dust) may not be represented reliably by emission models, 2) gas-phase chemical mechanisms have not been fully evaluated for wintertime conditions, 3) gas-particle partitioning of NO_3^- and organic compounds may favor the particle phase under cold and humid conditions but predicting this partitioning adds uncertainty in the model calculation, and 4) removal processes, especially by snow and at wetted surfaces, are not as well characterized for wintertime conditions. Based on these concerns, model performance is anticipated to be worse for winter months than for summer months (see also Section 8.3.1).

Performance evaluation of existing PM models for different seasons is ongoing or pending, and measurements will become available in most cases to quantify seasonal behavior of PM concentrations (for total PM as well as for its components). Therefore, it should be possible in due course to evaluate the CTM performance by season and to assess how well such seasonal variations can be simulated.

In order to evaluate the skill of PM CTMs for predicting seasonal or annual PM concentrations, it is necessary to have access to measurements from long-term monitoring networks, including collocated and concurrent measurements of gas and particle phases of semivolatile species, and of particle-size distributions and size-resolved chemical composition.

8.4.4 Can the Response of PM Levels to Changes in Emissions and Upwind Concentrations Be Predicted?

Predicting the response of PM components to changes in precursor emissions is needed for the development of emission-control strategies. Since CTMs are emission-based models, they are by their nature capable of estimating the change in PM levels due to changes in emissions and upwind concentrations. But in order for such estimates to be accurate, a CTM must be able both 1) to predict properly the concentrations of the various components of PM for present emission levels (see Section 8.4.1) and 2) to represent accurately how atmospheric processes will propagate a change in emissions to give the resulting change in atmospheric PM concentrations. The second requirement is extremely important because it is conceivable that a CTM may by chance show adequate performance in terms of predicting current observed PM levels and chemical composition due to compensating errors, but may lack the correct scientific formulations necessary to simulate the complex nonlinear processes that lead to secondary PM formation.

As discussed in Chapter 3, previous CTM applications have exemplified the non-intuitive aspects of PM/precursor relationships. Evaluating the ability of a CTM to predict the response of the various PM components to changes in precursor levels is difficult because the necessary air-quality databases have not typically been available. There are, however, some alternative approaches that can shed some light on the expected performance of CTMs for such emission changes (e.g., ambient responses to weekday vs. weekend variations in emissions). In addition, the use of diagnostic information based on concentrations of chemical species that indicate the chemical regime of the atmosphere can support or challenge CTM results (see Chapters 3 and 7 and Section 8.5.4). Such analyses are consistent with a weight-of-evidence approach.

In summary, CTMs are capable of quantifying the impact of emission changes, but additional measurement data and time are required to evaluate the reliability of such predictions. Current CTMs are more likely to predict the direction of the atmospheric response to emission changes correctly, as opposed to the magnitude of that response.

8.4.5 Can the Relationships Between PM and Other Air-Pollution Problems Be Quantified?

As discussed in Chapter 3, there are some intricate relationships between the formation of secondary PM and that of ozone (see Table 3.1). Since particles scatter light and some PM components such as BC absorb light, there is a direct relationship between PM and visibility/regional haze (see Chapter 9 for further discussion). In addition, PM affects the atmosphere's radiative budget and therefore photolysis rates, and therefore ozone formation. Surface chemistry on particles can also affect gas-phase concentrations, and therefore ozone (Section 8.2.4). Two important PM components are $SO_4^=$ and NO_3^-; these two species are also primary contributors to acid deposition, and changes in the ambient concentrations of these two PM components will be reflected in qualitatively similar changes in their deposition fluxes. Finally, some PM compounds (e.g., some heavy metals and persistent organic pollutants) are known or suspected to be toxic. Thus, changes in PM levels also will be reflected in the levels of other air-pollution problems. Among the problems mentioned above, all of them but one will respond in the same direction as PM. An improvement in PM air quality is expected to result in some improvement in visibility, some reduction in acid deposition and, depending on the PM components that were reduced, some reduction in particulate air-toxic levels. However, similar direct relationships cannot always be expected between PM and ozone.

A CTM that is able to predict the composition and size distribution of PM as well as gaseous co-pollutant concentrations and the dry and wet deposition of these species should be capable in principle of quantifying the relationship between PM and these other air pollution problems. Current PM CTMs are more likely to be successful in predicting the direction, as opposed to the magnitude, of such relationships.

8.4.6 Can Other PM Properties that are Potentially Relevant to Health Effects Be Calculated?

As discussed in detail in Chapter 2, the exact causes of the adverse health effects of PM are still being investigated. Potential surrogate metrics for PM health effects that are being considered (see Table 1.3) include the number concentration of ultrafine particles, the acidity of particles, the concentrations of individual elements (e.g., metals), and gaseous co-pollutant interactions.

To date, PM simulations have focused on PM mass concentration because of its significance to air-quality regulations. It is therefore of particular interest to assess whether CTMs can predict other PM properties in addition to PM mass concentrations.

Those CTMs that do not calculate detailed particle-size distributions cannot provide information on the number concentration of ultrafine particles. CTMs that calculate particle-size distributions using either a modal or a sectional approach (see Section 8.2.5) can provide the number concentration of ultrafine particles either directly or with minor modifications.

PM acidity is either calculated or may be deduced from calculations of PM chemical composition. A comparative evaluation of several models for PM chemical composition showed that in most cases particle-acidity calculations agreed within a factor of two; however, there were some cases where significant discrepancies were obtained for low humidities (Zhang et al., 2000). Therefore, although most current CTMs can predict the acidity of particles, some careful evaluation will be required if CTMs are to be applied for that particular metric.

Calculation of individual PM classes is conducted in current CTMs only for $SO_4^=$, NO_3^-, NH_4^+, and carbon (some CTMs also calculate the components of sea salt and soil dust). Metals, biological matter, and specific organic toxics are not calculated. Therefore, those features will need to be added if they are to be addressed by CTMs.

A complementary discussion of the ability of CTMs to predict PM properties that are potentially relevant to visibility is provided in Section 9.3.2.

8.4.7 Can PM Episodes Be Forecast in Real Time?

It is now feasible to apply CTMs in real time for the prediction of air pollution, opening the possibility of advance public-health advisories and warnings. The limiting factors include the availability of a reliable numerical meteorological forecast, an emission inventory that is up-to-date and day-specific, and computational resources that allow a rapid turn-around of the simulation. Clearly, the limitations discussed above for the application of CTMs to policy and science issues will also apply to real-time forecasting. As our ability to simulate atmospheric PM improves, real-time forecasting of PM episodes may prove to be a valuable tool for air-quality management.

8.5 EVALUATION PROCESS FOR CHEMICAL-TRANSPORT MODELS

As discussed in Section 8.4.1, current CTMs can simulate some atmospheric processes rather well but are still deficient in their treatment of other processes. Thus, before a CTM is applied to analyze the effectiveness of various emission-management options, it is essential to ensure that it provides a credible representation of the processes that govern ambient PM concentrations. The main aspects of the performance-evaluation process are discussed here.

8.5.1 Model Simulations versus Ambient Measurements

Model output does not represent the same quantity as ambient measurements. Some of the differences between simulation results and measurements can be resolved, while others cannot. Comparisons should consider three key differences, as follows.

First, ambient measurements used to evaluate attainment of ambient air-quality standards for PM are based on filter-based techniques or continuous indirect methods with size-selective inlets (see Chapter 5). Therefore, the size cut between fine and coarse particles is based on the aerodynamic diameter of the particle (the aerodynamic diameter is defined as the diameter of a spherical particle with a density of 1 g/cm^3 that behaves similarly to the particle of interest). On the other hand, CTMs use the Stokes diameter of the particle and post-processing of the model results is needed to convert to an aerodynamic diameter (the Stokes diameter is the diameter of a spherical particle that behaves similarly to the particle of interest; it is equal to the aerodynamic diameter divided by the square root of the particle density). Also, the gravimetric measurements of PM mass concentrations are conducted under controlled conditions with a temperature between 15 and 30 °C and a relative humidity between 20 and 40 percent. CTM results are typically reported for the simulated ambient temperature and relative humidity. In addition, the filter measurements may involve artifacts for the volatile particulate species such as NO_3^-, NO_4^+, some organic compounds, and PM-bound water (see Chapter 5). The CTM simulation results can be made consistent with the PM gravimetric measurements by converting Stokes diameters to aerodynamic diameters and calculating the PM concentrations, including PM-bound water, for temperature and relative-humidity values representative of the gravimetric measurements. The potential artifact problem must be resolved by using additional measurement techniques that minimize or eliminate such artifacts.

Second, the comparison of measured and simulated PM carbonaceous-species concentrations is also challenging. CTMs typically simulate BC and the total organic compound mass. However, ambient measurements provide only total particulate carbon and some assumptions are necessary to 1) assign that carbon between BC and OC and 2) scale the estimated OC mass to organic compound mass (i.e., take into account the hydrogen, oxygen and nitrogen that were not included in the measurements: see Section 3.5.3). Note that this uncertainty in the particle mass and density will be present in the conversion from Stokes to aerodynamic diameter.

Third, the measurements are conducted at a specific location whereas the CTM-simulation results represent a volume-average value (corresponding to the volume of individual CTM grid cells). This discrepancy between the spatial representations of measurements and simulation results, sometimes referred to as incommensurability, can be resolved only if sufficient measurements are conducted within the volume of a grid cell to approximate a volume-averaged value. Such a density of measurements typically will not be available, and the discrepancy in spatial representations is a fundamental source of uncertainty when evaluating CTM simulation results against measurements. Incommensurability is likely to be more of an issue for primary PM components than for secondary PM components and in urban areas as opposed to rural areas, i.e., where subgrid-scale gradients are more important.

8.5.2 Overview of the Performance-Evaluation Process

A comprehensive evaluation of CTM for PM can be seen as consisting of the following four steps (e.g., Seigneur et al., 2000a):

1. An operational evaluation that tests the ability of the model to estimate PM mass concentrations ($PM_{2.5}$ and PM_{10}) and quantities required to address the regulations (i.e., the mass concentrations of major chemical components of PM, including $SO_4^=$, NO_3^- OC, BC, and crustal material).
2. A diagnostic evaluation that tests the ability of the model to predict the components of PM (or visibility/regional haze): At the minimum, PM chemical composition and associated gas-phase species (e.g., HNO_3, NH_3), PM precursors (e.g., SO_2, NO, NO_2, VOC), and some key atmospheric oxidants and products should be evaluated. More specific evaluation should be conducted to the extent possible for PM detailed chemical composition (e.g., individual organic compounds), PM size distribution (if warranted), temporal and spatial variation of PM and precursors, mass fluxes (emissions, transport, transformations, and deposition), and, for visibility, components of light extinction (i.e., scattering and absorption).
3. A mechanistic evaluation that tests the ability of the model to predict the response of PM concentrations or visibility to changes in 1) meteorology and 2) emissions.
4. A probabilistic evaluation that takes into account the uncertainties associated with the model predictions and observations of PM and visibility.

Because PM concentrations are a function of meteorology, emissions, and background concentrations of PM and its precursors, it is essential to test the ability of a model to predict PM concentrations over a wide range of meteorological conditions, emissions, and background concentrations. A single geographical area is unlikely to cover all of these variables. Therefore, comprehensive performance evaluations should be conducted in several different regional areas and for several seasons.

Once a model has undergone a comprehensive performance evaluation and the results have met accepted performance criteria, it can be applied to other geographic areas or to other time periods within a same area. Nevertheless, some performance evaluation should be conducted for each new application of the model to ensure that no major errors in the model input data or model design are associated with the new application. Because the model design should have been thoroughly tested during the comprehensive model evaluation, it is feasible to streamline the performance-evaluation process for new applications. A streamlined performance evaluation should address at a minimum concentrations of total PM and major PM components, PM precursors and key oxidants. In addition, the aspects of meteorological conditions and atmospheric chemistry that are relevant to PM concentrations in a new application should be reviewed and any conditions that were not tested in the comprehensive performance evaluations should be the subject of a diagnostic evaluation.

8.5.3 Data Needs for CTM Performance Evaluation

The databases required for CTM performance evaluation are extensive, and the field programs needed to acquire such databases are costly. For example, the California Regional PM Air Quality Study (CRPAQS), which was conducted in the California San Joaquin Valley during 2000 and 2001, cost about US $27M. Field-program design must take into account local geography, meteorology, emissions, and atmospheric chemistry, as well as the resources and the time frame available. It is recommended that, prior to the implementation of a field program for a comprehensive model-performance evaluation, a screening field program be conducted to identify the major characteristics of the meteorology and atmospheric chemistry of the area. Alternatively, if sufficient routine data have been collected in the area, an analysis of those data may suffice. Information obtained from this screening field program (or analysis of existing data) will be critical in determining which measurements should be given priority for the comprehensive field program. Many of the advanced measurement techniques are resource-intensive, and it is essential to determine which of those measurements are the most critical to the success of the field program. As discussed next, measurements of indicator species should be considered because they can provide qualitative information on the various chemical regimes of the atmosphere and are useful for the mechanistic evaluation of CTMs.

8.5.4 Corroboration of CTM Results with Indicator-Species Methods

Since CTMs are used eventually to predict the response of PM concentrations to changes in precursor emissions, it is imperative to evaluate to the extent possible the ability of a CTM to predict the correct response. One possible approach is to compare, where possible, the response predicted by a CTM with the corresponding response predicted by an independent technique such as an observation-based model. The various indicator-species methods are examples of observation-based models. Such methods use observations of indicator-species concentrations that can provide information on the chemical regimes of the atmosphere and consequently indicate how some chemical concentrations will evolve as others change. Presented below are indicator-species methods that can be used to identify chemical regimes of ozone, $SO_4^=$, and NH_4NO_3 formation.

Ozone formation results from the reactions of VOC and NO_x in the presence of sunlight; thus, it can be VOC-sensitive, NO_x-sensitive, or in a transition regime between those two main regimes. Several indicator species have been identified to characterize whether ozone formation is VOC- or NO_x-sensitive. These include H_2O_2/(HNO_3 + particulate nitrate), NO_y, HCHO/NO_y, and ozone/NO_x, where NO_y is the total oxidized nitrogen (excluding N_2O). They represent dominant products under VOC- or NO_x-sensitive regimes, ratios of these products, or chain lengths in the radical reactions that produce ozone (e.g., Sillman et al., 1997; Tonnesen and Dennis, 2000a,b). For example, the ratio H_2O_2/(HNO_3 + PM NO_3^-) represents the competition of the HO_2 radical termination product (H_2O_2 formation is dominant in a NO_x-sensitive regime) and the OH + NO_2 termination product (HNO_3 formation is dominant in a VOC-sensitive regime). Other indicators are related to ozone production efficiency (e.g., ozone/(NO_y – NO_x)) and to air-mass age (e.g., (NO_y – NO_x)/NO_y).

As discussed in Section 3.5, $SO_4^=$ can be formed in the gas phase or in cloud or fog droplets. Measurements of the size-distribution of $SO_4^=$-containing particles can help identify whether $SO_4^=$

was formed in the gas phase or the aqueous phase, since the median diameter of the $SO_4^=$-containing particles will differ noticeably depending on which process prevailed. The oxidation of SO_2 by H_2O_2 in cloud or fog droplets can be a major pathway for $SO_4^=$ formation. If the pathway is H_2O_2 limited, a small change in SO_2 concentrations will have little effect on $SO_4^=$ concentrations. Therefore, measurements of SO_2 and H_2O_2 are very valuable for identifying whether this $SO_4^=$ formation pathway is SO_2 limited or H_2O_2 limited.

Formation of NH_4NO_3 can be HNO_3-sensitive or NH_3-sensitive. Measurements of $SO_4^=$, total NO_3^- (HNO_3 plus particulate NO_3^-) and total NH_4^+ (NH_3 plus particulate NH_4^+) can be used effectively to determine whether changes in HNO_3 or NH_3 concentrations will affect NH_3 concentrations most (see Chapter 3). In addition, continuous measurements of total NO_3^- will shed some light on whether the daytime oxidation of NO_2 by OH radicals or the nighttime oxidation of NO_2 by ozone govern HNO_3 formation. As discussed in Chapters 3 and 10, these different pathways can lead to totally opposite responses of NO_3^- concentrations to NO_x concentration changes.

Such indicator-species methods provide quantitative or semi-quantitative information on the potential response of chemical species to changes in their precursors. Such information can be used to test whether a CTM predicts the correct chemical regimes, thereby corroborating or challenging the CTM results in terms of whether the responses are predicted for the correct reasons. Furthermore, such indicator-species methods are valuable in their own right and can be used effectively, for example, to develop conceptual descriptions of PM formation for specific airsheds (see Chapter 10).

8.6 CURRENT STATUS OF CTM PERFORMANCE AND INTERCOMPARISONS

This section presents a sample of PM CTM performance-evaluation results to illustrate different statistical and graphical presentations of model skill and CTM applications to different regions. Note that at this stage, relatively few results of individual PM CTMs and no results of formal PM CTM intercomparisons have been published in the peer-reviewed literature.

Table 8.1 presents a summary of published statistical performance evaluations of Eulerian CTMs for PM. All these evaluations were conducted with data from the 1987 Southern California Air Quality Study. Model applications to more recent Los Angeles episodes, e.g., September 1993 (Griffin et al., 2001) and September 1996 (Kleeman and Cass, 2001), are now becoming available but these do not report performance statistics. Since different meteorological inputs, VOC emissions, and performance statistics were used, no direct comparison between the performances of these models for this common case should be attempted based on this table. (CTM performance comparisons using identical or at least similar inputs for the same episode are strongly encouraged, with a focus on the same performance statistics for each model.) Nevertheless, these early results provide some indication of the type of performance that one can currently expect in PM modeling. All these performance evaluations have been conducted for California, and it is imperative that similar evaluations be conducted in other geographical areas with a variety of meteorological conditions and atmospheric chemical regimes.

Figures 8.4 and 8.5 present results of a simulation conducted with CMAQ-MADRID 1 (a modified version of CMAQ) to simulate regional haze in Big Bend National Park, Texas, near the U.S./Mexican border (Pun et al., 2002). The simulation used two nested grids with horizontal grid spacings of 12 km and 4 km, respectively. The coarse grid covered Texas and northern Mexico whereas the fine grid covered Big Bend National Park (BBNP) and surrounding areas. The results are presented for an episode of the BRAVO study during October 1999. Figure 8.4 shows time series of 24-hour average PM $SO_4^=$ concentrations during this October episode. The maximum 24-hour average $SO_4^=$ concentrations were observed on October 12. The simulated concentrations in the 12-km-grid case do not show any significant temporal evolution over the episode, but the simulated concentrations in the 4-km-grid case show a peak on October 12, consistent with the

Table 8.1. Performance evaluations of PM grid models for $PM_{2.5}$ and components with the SCAQS data base in the Los Angeles basin. Two episodes were used: 24-25 June 1987 and 27-28 August 1987. (Note that different statistics were used for the various model evaluations; therefore, comparisons should not be made among the various models using this summary.) (Source: Seigneur, 2001).

MODEL		UAM-AERO	GATOR	CIT	UAM-AIM	SAQM-AERO
Period		June 25	August 27-28	August 28	June 24-25	August 28
Statistics		Normalized statistics[a,b] (%)	Normalized statistics[a,c] (%)	Normalized statistics[a,d] (%)	Normalized statistics[e,f] (%)	Normalized statistics[e,g] (%)
$PM_{2.5}$ mass	error	32	44	46	NA[h]	NA
	bias	+24	-3	+46	NA	+10
$SO_4^=$	error	48	28	34	NA	NA
	bias	-10	+4	-30	-21	-33
NO_3^-	error	18	68	61	NA	NA
	bias	+11	-21	+47	+52	-14
EC	error	15	57	50	NA	NA
	bias	-10	+30	+35	NA	NA
OC	error	38	49	40	NA	NA
	bias	-38	-44	+14	NA	+38

[a] Normalized error = $\frac{1}{N}\sum_{i=1}^{N}\left|\frac{P_i - O_i}{O_i}\right|$; normalized bias = $\frac{1}{N}\sum_{i=-1}^{N}\left(\frac{P_i - O_i}{O_i}\right)$; P_i = prediction, O_i = observation; N: number of samples.

[b] Mean over all sites and sampling periods of the normalized errors of sampling-period average concentrations.

[c] Mean over all sites and hours of the normalized errors of 1-hour averaged concentrations (note that sampling periods exceeded 1 hour).

[d] Mean over all sites of the normalized errors of the 24-hour average concentrations

[e] Normalized bias of means = $\sum P_i - \sum O_i / \sum O_i$.

[f] Normalized bias of the means over all sampling periods and sites of the sampling-period average concentrations.

[g] Normalized bias of the means over all sites of the 24-hour average concentrations.

[h] Not available.

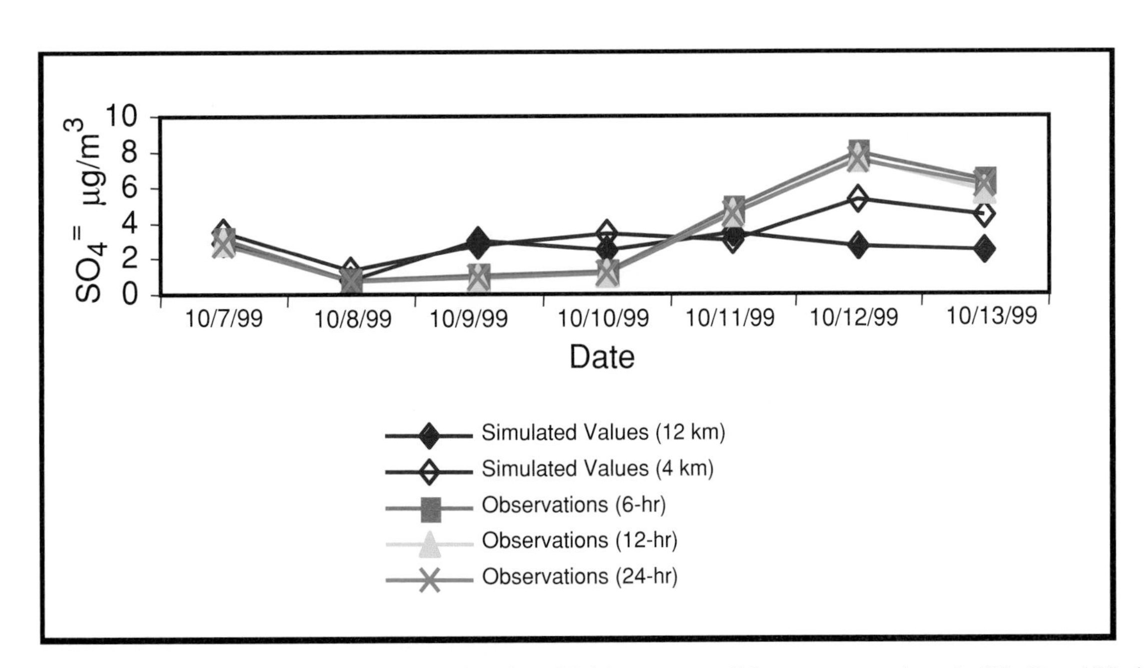

Figure 8.4. Comparison of measured and simulated 24-hr average $SO_4^=$ concentrations in Big Bend National Park, Texas.

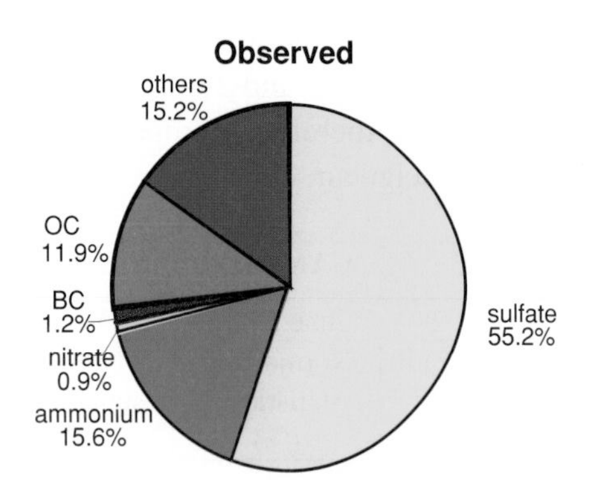

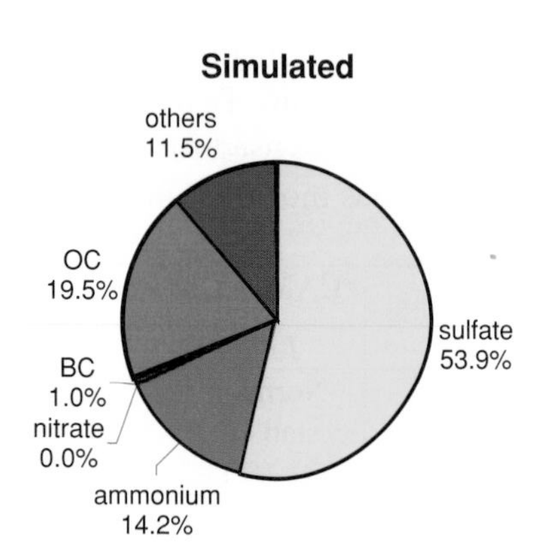

Figure 8.5. Comparison of measured and simulated chemical compositions of 24-hr average $PM_{2.5}$ in Big Bend National Park, Texas on 12 October 1999.

observations. In general, the 4-km-grid simulation reproduced the daily fluctuations in $SO_4^=$ concentrations, although with lesser amplitude than observed. A regression between the observed and 4-km-grid simulated data shows a coefficient of determination (R^2) of 0.69.

On October 12, the observed 24-hour average $PM_{2.5}$ concentration was 13.5 $\mu g/m^3$ at BBNP, whereas 9.7 $\mu g/m^3$ was predicted by the 4-km-grid CMAQ-MADRID simulation (i.e., 28 percent underprediction). A comparison between the observed and predicted fractional chemical compositions of $PM_{2.5}$ is shown in Figure 8.5. The largest component of $PM_{2.5}$ was $SO_4^=$ for both the observed (55 percent) and predicted (54 percent) PM. The other components in the measured data ranked as follows: the second most abundant component was NH_4^+, representing 16 percent of the fine PM mass, the third was the "other" component (15 percent), including small concentrations of sea salt. Organic compounds (12 percent) ranked fourth. Primary and secondary organic compounds (OC) ranked second in the simulated composition and accounted for 20 percent of the simulated PM mass. The contribution of SOA exceeded that of primary OC, with biogenic SOA dominating the simulated SOA concentrations. The third most abundant component in the simulated PM was NH_4^+, (14 percent). The "other" PM component constituted about 12 percent of the predicted PM. These components ranked second and third in the observed PM and were slightly under-represented in the model predictions. Black carbon and NO_3^-, which were minor components of $PM_{2.5}$, were under-represented by the model.

Figure 8.6 presents results from a regional PM simulation carried out with AURAMS (e.g., Zhang et al., 2002) over eastern North America for an ozone and PM episode that occurred during the 1988-90 Eulerian Model Evaluation Field Study (EMEFS). The EMEFS sites were generally located in rural or remote areas (e.g., McNaughton and Vet, 1996). Figure 8.6a shows observed 24-hr ambient $SO_4^=$ measurements stratified into seven color-coded concentration ranges while Figure 8.6b shows the corresponding AURAMS predictions for the 40-km model grid cells containing the EMEFS stations. The highest $SO_4^=$ values are found in the Ohio Valley and the lowest are found in northern Canada for both the observations and model, but the model predictions of $SO_4^=$ are generally too high. Note that the AURAMS domain did not extend far enough south to include the EMEFS sites in Alabama, Florida, and Mississippi.

In addition to performance evaluation of individual CTMs against data, it is valuable to conduct intercomparisons of CTMs. When conducted with consistent inputs, such intercomparisons can provide useful information on the uncertainties associated with the modeling process alone. For example, they can reveal how assumptions made independently by different model users regarding missing inputs can affect the model simulation results. A number of such comparative studies are currently taking place but peer-reviewed results are not yet available

Model intercomparisons are particularly useful to test the reliability of CTMs to predict the response of secondary pollutants such as ozone and PM to

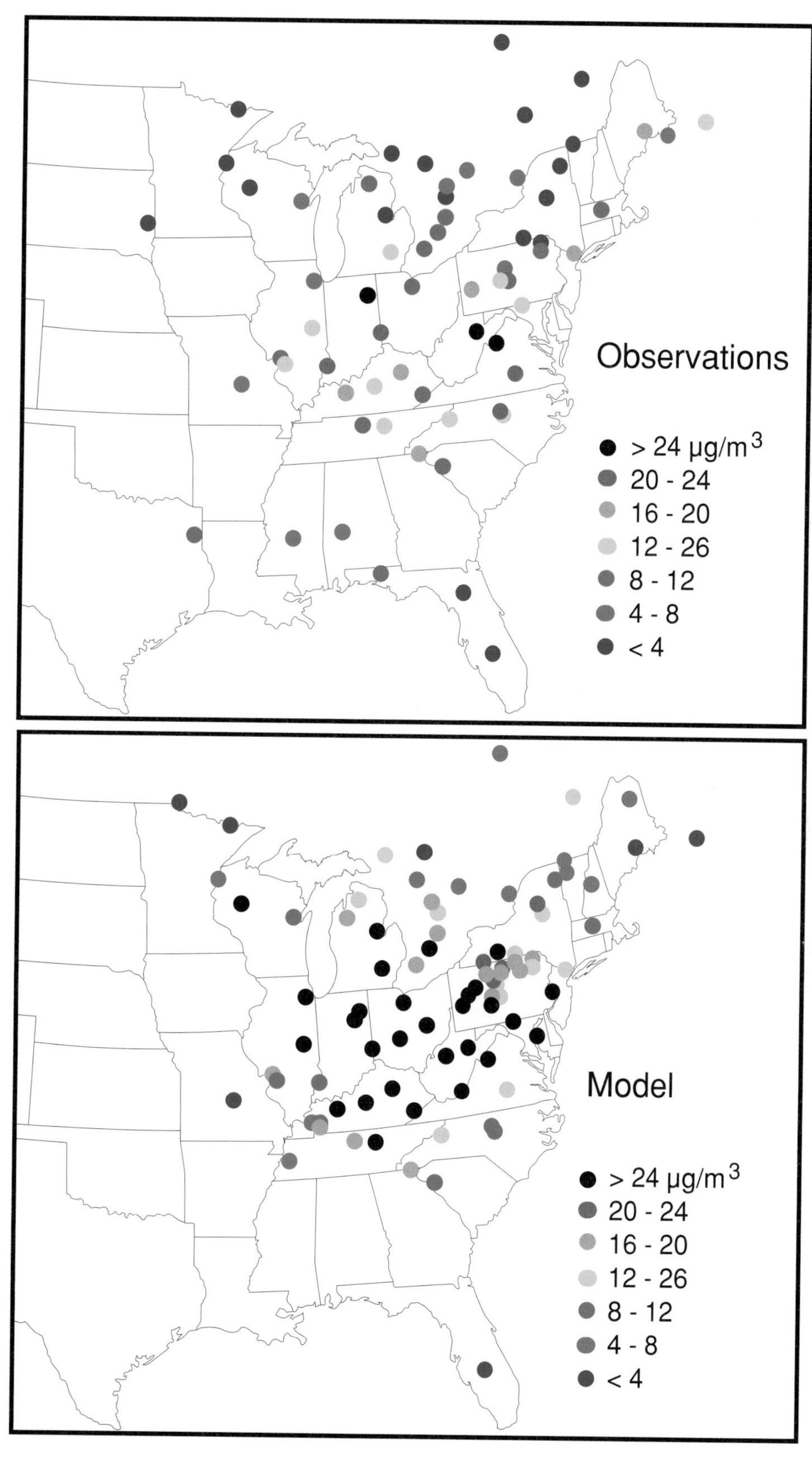

Figure 8.6. Comparison of measured (top) and simulated (bottom) 24-hr average $PM_{2.5}$ $SO_4^=$ concentrations over eastern North America for an EMEFS episode, 1 Aug. 1988.

changes in precursor emission levels. For example, two different CTMs were used to simulate the response of ozone values and PM to 50 percent reductions in NO_x and VOC emissions for an air-pollution episode (27-28 August 1987) in the Los Angeles basin, California (Meng et al., 1997; Pai et al., 2000). Both CTMs gave similar results for the effect of NO_x- and VOC-emission reductions on ozone and for the effect of NO_x-emission reductions on PM. However, the two CTMs differed in the magnitude of the effect of VOC-emission reductions on PM. Such model intercomparisons need to be performed. In particular, they definitely should be conducted prior to the application of long-term models that use simplified chemistry to ensure that the simplified chemical mechanism behaves similarly to the well-established comprehensive mechanisms.

CTMs must be satisfactorily evaluated against ambient data prior to their applications in a policy context. To that end, there is an urgent need for comprehensive field measurement programs as well as the development of sound modeling protocols to ensure that those model-performance evaluations meet appropriate scientific standards. Work must also continue to improve model treatments of some atmospheric processes.

8.7 USE OF CTMS TO COMPLEMENT MONITORING NETWORKS

Monitoring networks provide "ground truth," that is, ambient measurements of the real atmosphere of known accuracy, precision, and representativeness, whereas CTMs can only make independent estimates of what our best knowledge and understanding would lead us to expect in the real atmosphere. However, one significant difference between Eulerian CTMs and monitoring networks is that in a model, values are predicted everywhere in the domain at each time step for a full range of species for whichever set of meteorological conditions and emissions that has been specified. In laboratory terms, a CTM can be viewed as a fully instrumented, fully controllable testbed capable of performing a wide range of experiments, including observing system-simulation experiments. Thus, measurement networks and CTMs are very much complementary.

One obvious application of such experiments is to assist in the design and appraisal of PM monitoring networks, since the CTM will predict spatial patterns of PM, including the locations of maximum and minimum concentrations as well as strong gradients (see Section 6.10). Different network designs then can be tested before (or after) deployment to see how well the model-predicted spatial patterns are represented (e.g., Kuo et al., 1985; Trujillo-Ventura and Ellis, 1991). The model-predicted spatial patterns also can help analysts trying to interpolate either qualitatively or quantitatively between measurement sites, since the modeled patterns provide insights into the behavior of the concentration field between sites and can be used to calculate species-specific spatial covariance, a quantity required by some interpolation schemes such as kriging.

CTM-predicted fields also can be used to either supplement or supplant the use of measurements in calculating PM mass budgets, that is, domain-scale totals of the amount of PM mass that enters or leaves the modeling domain at the lateral boundaries due to advection and at the bottom boundary due to emissions and wet/dry deposition (e.g., Kuo et al., 1984; Wojcik and Chang, 1997). Mass budgets are useful to help understand the relative importance of these processes and to check mass consistency in the CTM. In addition, CTMs are very useful for planning intensive field campaigns, including measurements above the surface, particularly since one of the major objectives of such campaigns is to collect data for the evaluation of CTMs (e.g., Sheih et al., 1978; Hidy, 1994).

8.8 USE OF CTMS TO SUPPORT ESTIMATIONS OF EXPOSURE

While CTMs for PM can predict spatial and temporal variations in ambient PM mass concentration, size distribution, and chemical composition, the PM-related quantity of most interest to the health-effects community is PM dose, the total amount of PM that is deposited in an individual's pulmonary tract over a specified time period (e.g., Mage et al., 1999). PM dose depends upon respiration frequency, inspiration volume, mode of breathing (nasal vs. oral), and individual airway characteristics as well as on the PM concentration, size distribution, and hygroscopicity. To sidestep the need for such detailed respiratory information, a commonly used surrogate for dose is PM exposure, the time-averaged PM concentration entering an individual's breathing zone over a specified time period. Personal exposure is difficult to characterize properly because it depends on people's movements through different microenvironments and is a function of many factors including occupation, lifestyle, and place of residence.

Besides exposure to ambient outdoor PM concentrations, most people are exposed to PM in motor vehicles and in varied indoor microenvironments such as residences, workplaces, schools, stores, and restaurants. Within a given community, there are typically larger concentration differences between microenvironments than in the ambient outdoor atmosphere (e.g., Monn, 2001). Ambient outdoor PM contributes to indoor PM levels both by infiltration or penetration of outdoor air and by the resuspension indoors of settled outdoor PM due to such activities as walking on carpets, sitting on stuffed furniture, or vacuuming. Direct indoor PM sources, such as occupational activities, cooking, fireplaces, kerosene heaters, and smoking, may also contribute significantly to indoor PM exposure (Abt et al., 2000). PM concentrations inside motor vehicles depend upon vehicle type and speed, traffic density, types of other vehicles and their speeds, roadway type, and meteorological conditions as well as on background ambient levels. As a consequence, indoor PM concentrations are sometimes higher than outdoor levels and can vary enormously amongst indoor microenvironments.

This means that in order to quantify personal exposure, PM concentrations in each type of indoor microenvironment, including vehicle interiors, must be known in addition to outdoor PM concentrations, as must the amount of time a person spends outdoors and in each microenvironment. Estimates of indoor PM concentrations can be obtained through extensive series of measurements or through the use of indoor air-quality models. The latter account quantitatively for indoor/outdoor air exchange, indoor sources, transport and mixing, and deposition and sedimentation, with outdoor PM levels required as a time-dependent boundary condition (e.g., Nazaroff and Cass, 1989). Information about time spent outdoors and in each indoor microenvironment by an individual may be obtained from daily activity diaries.

For epidemiological studies, estimates of individual exposure must be scaled up to estimates of population exposure. This requires the additional availability of geographically resolved demographic data and population-specific human activity-pattern data resolved by subpopulation (e.g., by age, gender and socio-economic category). A number of models exist to estimate population exposure to PM (e.g., Seigneur, 1994). Such exposure models can use as input either outdoor PM measurements or the PM concentrations predicted by a CTM. The simplest population-exposure models calculate only the potential population exposure to outdoor PM and thus only require knowledge of outdoor PM concentrations and population density. However, outdoor PM population exposure is in general only a lower limit for total PM population exposure. The more comprehensive population-exposure models also treat exposure in indoor environments and, in some cases, take into account the movement of population groups within the region of interest. In the future, it is conceivable that CTMs for PM will include a population-exposure module, but it is more likely that CTMs for PM will continue to be used to provide outdoor PM concentration fields as one of a number of inputs required by separate population-exposure models. An approach where a CTM and population-exposure model are run in sequence rather than in a fully integrated fashion seems appropriate because one does not expect indoor pollution to have significant effect on outdoor PM concentrations.

One important issue related to the application of CTMs to exposure estimation is the horizontal spatial resolution needed for the predicted outdoor PM concentration fields, given the significant gradients in population density that occur within and near major urban centers. Ultimately, the resolution required will depend on the PM component or characteristic that is eventually identified as causal for human-health effects, because different PM components have been observed to vary differently on the urban scale. For example, PM mass can vary significantly on the sub-kilometer scale due to the presence or absence of local primary PM sources such as roadways, whereas some PM components such as $SO_4^=$ are quite uniform across a city. It may also prove feasible (and more affordable computationally) to superimpose subgrid-scale variations on grid-scale PM concentration values after the fact through knowledge of the location of emission sources, rather than applying the CTM at such fine grid spacings. (To date, simulations with CTMs have been performed down to 4 km horizontal grid spacing).

8.9 POLICY-RELEVANT RESULTS FROM CTM APPLICATIONS

Although CTMs for PM are still very new (at least outside of the Los Angeles basin) compared to those CTMs developed for some other air-quality issues such as chemical deposition and ground-level ozone, they already are beginning to provide policy-relevant guidance. Some results from CTMs developed for other air issues also may be useful to the PM policy community. This section summarizes some CTM results obtained to date that are relevant to the policy questions discussed in Section 8.4. Given the current state of the science, much more can be expected from CTMs for PM over the next decade.

8.9.1 PM CTMs

Section 8.4.1 addressed quantification of the contribution of various precursors and source types to PM. As described by Seigneur et al. (1999) and in Section 8.6, a number of CTMs for PM have been applied to simulate PM mass and composition at one or more sites in southern California (e.g., Jacobson, 1997; Meng et al., 1998; Kleeman and Cass, 1999; Pai et al., 2000). All of these models appear to possess some skill for predicting PM mass and chemical composition, but all of the models that have been applied to the same (August 1987) PM episode give different answers (see Table 8.1), and to date these models have been subjected to only limited performance evaluations based on only a few days of measurements[5] at small numbers of measurement stations. This suggests 1) that much more work is required to identify the more robust and better performing PM models and to characterize their performance, strengths, and weaknesses, and 2) that it is premature at this stage to formulate detailed PM-control strategies based on the results of a single PM model. One implication of the evaluation of CTM performance for only a few episodes is that demonstrated skill for these short periods is no guarantee of comparable performance at other times of year.

Elsewhere, few successful CTM simulations have been reported to date for size- and/or chemically resolved PM at other locations in North America. A Lagrangian PM model has been applied to the Lower Fraser Valley of southwestern British Columbia (Barthelmie and Pryor, 1996). Also, several groups have applied 3-D CTMs at larger scales over various areas in North America, including the entire contiguous United States (e.g., Jang et al., 2001; Pagowski et al., 2002). Figure 8.7 shows some sample results from regional simulations for eastern North America. These types of results for other regions are just starting to be reported.

Some approaches to the quantification of the relative contributions of regional and local emissions (cf. Section 8.4.2) may be impractical when applied to ambient PM. Any CTM for PM able to quantify the contribution of various precursors and source types to PM embodies the atmospheric source-ambient relationships implicitly in its formulation, code, and inputs. Quantifying these relationships explicitly, however, in the form of a source-receptor matrix (SRM) is computationally intensive, as one full CTM simulation is required per source region, precursor species, and period of interest. For example, a simple SRM was developed for the Los Angeles basin for $PM_{2.5}$ mass for one 2-day period for five source regions and four precursors (SO_2, NO_x, VOC, and primary $PM_{2.5}$). In total, twenty (= 5 x 4) full CTM simulations were required (Seigneur et al., 2000b). If more source regions or emitted species (e.g., NH_3, BC) were to be considered, more simulations would be required. If longer time periods were to be considered, more days would have to be simulated.

[5] Griffin et al. (2001) have considered a Sept. 1993 time period and Kleeman and Cass (2001) have considered a Sept. 1996 time period for Los Angeles.

Moreover, due to the nonlinear chemistry associated with secondary PM formation, the matrix elements of a PM SRM are likely to be dependent on the assumed base emission levels (consider the example of the EKMA ozone response surface to NO_x and VOC emissions). Some diagnostic techniques (e.g., sensitivity analysis techniques, forward and backward process-analysis techniques) also can be used to that end. However, they all have their limitations and involve additional computations.

Concerning the response of PM levels to changes in emissions and upwind concentrations (Section 8.4.4) and the relationships between PM and other air-quality problems (Section 8.4.5), a few (mostly simple) emission-change scenarios already have been carried out for southern California. Simulations with "across-the-board" 50 percent reductions in NO_x and VOC emissions, both individually and combined, were conducted for the Los Angeles Basin with one CTM for PM (Pai et al., 2000), then compared to those reported from a different CTM for the same period and scenarios (Meng et al., 1997). Although the magnitudes of the responses in ozone, particulate NO_3^-, and $PM_{2.5}$ mass were all different between the two models for the three emission-change scenarios, the directions (i.e., sign) of the responses were the same, providing corroborative support for the general conclusions (that is, two independent models rather than a single model). This directional consistency is particularly important in the case of the 50 percent VOC reduction scenario, in which both models predicted about a 30 percent decrease in ozone levels (at one station) but an increase in both NO_3^- and $PM_{2.5}$ mass levels, that is, a $PM_{2.5}$ disbenefit. Recognizing

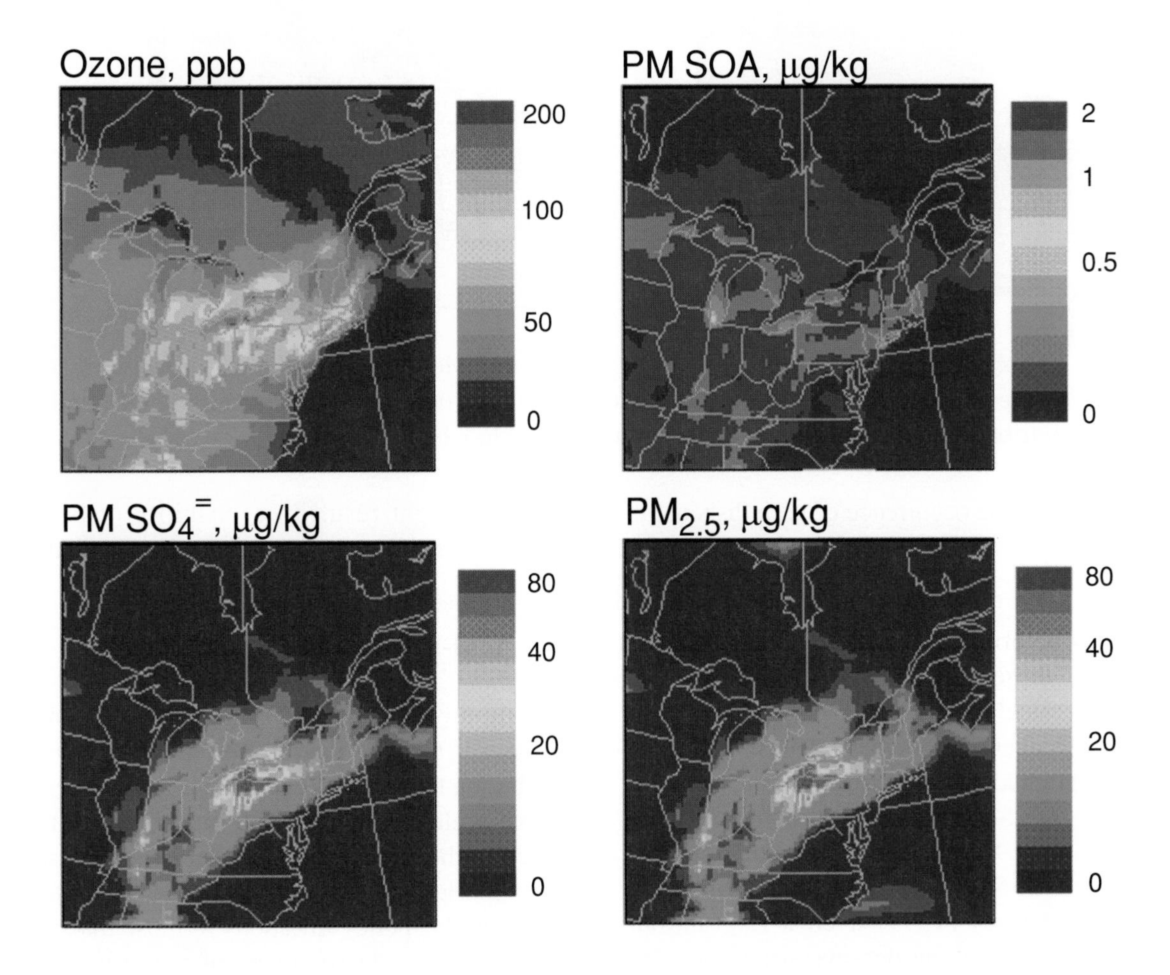

Figure 8.7a. Application of two distinct CTMs (AURAMS and Models-3/CMAQ) to northeastern North America: Instantaneous surface-level ozone, SOA, $SO_4^=$, and $PM_{2.5}$ mass concentration fields (µg/kg) at 2000 UTC, 3 August 1988, simulated by the "dry"-prototype version of AURAMS.

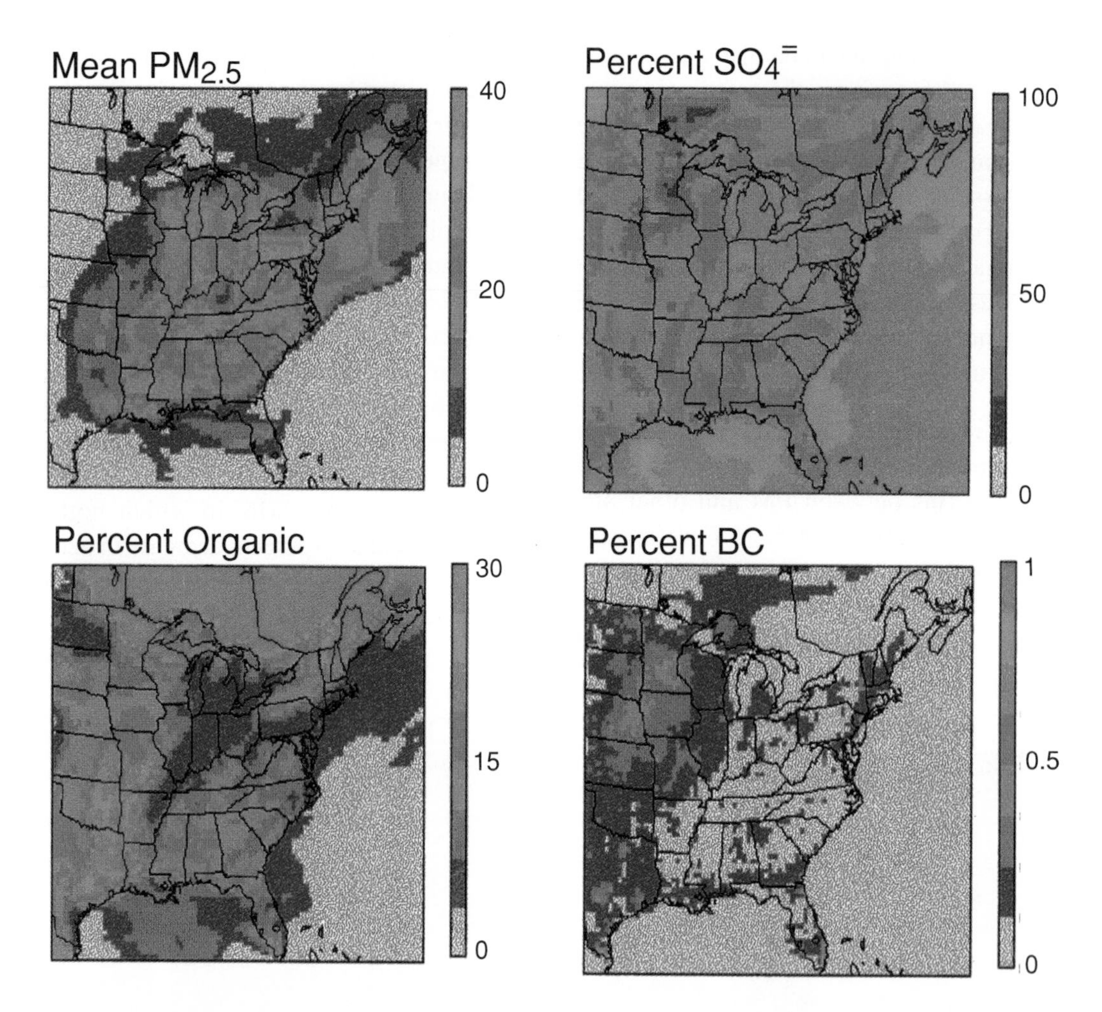

Figure 8.7b. Application of two distinct CTMs (AURAMS and Models-3/CMAQ) to northeastern North America: 24-hr average surface-level $PM_{2.5}$ mass concentration field (µg/m^3) and sulfate, OC, and BC percentage contribution fields on 13 July 1995, simulated by Models-3/CMAQ.

the possibility of the occurrence of such disbenefits is an important message for the PM and ozone policy communities.

In order to determine source-ambient "transfer coefficients," the response of $PM_{2.5}$ mass to emission changes was calculated with a CTM using 50 percent proportional reductions (Seigneur et al., 2000b). For several geographically-targeted precursor emission reductions, no consistent $PM_{2.5}$ disbenefit was associated with VOC reductions, but one was associated with SO_2 reductions (cf. Table 3.2). Kleeman and Cass (1999) used a Lagrangian sectional PM model to consider several quite detailed emission scenarios for different source types, again for the Los Angeles basin. One interesting result was the suggestion that a reduction in direct $PM_{2.5}$ emissions might result in a visibility disbenefit because of the change in the particle-size distribution.

8.9.2 Acid-Deposition CTMs

Not only are $SO_4^=$ and NO_3^- important PM mass constituents, they are also the most important species contributing to chemical deposition. A great deal of regional-scale, source-oriented modeling has been devoted to the chemical-deposition issue in eastern North America over the past two decades in order to answer similar policy questions to those discussed in Section 8.4. Clearly, many of the results from chemical-deposition modeling regarding the major inorganic ions will also be relevant to the PM issue.

One important result from chemical-deposition CTMs are the findings relating to the importance of aqueous-phase conversion of SO_2 to $SO_4^=$. A number of modeling studies have suggested that on a seasonal or annual basis, this conversion pathway is equal in importance to the gas-phase oxidation of SO_2 (e.g., Dennis et al., 1990, 1993; Karamchandani and Venkatram, 1992). The implication for PM modeling is that successful PM CTMs will need to have a reasonable representation of cloud/fog processes and aqueous-phase chemistry.

Chemical-deposition CTMs have been used to quantify the contribution of distant vs. local sources to $SO_4^=$ and NO_3^- deposition in eastern North America. Lagrangian models for the long-range transport of sulfur and nitrogen have been used to calculate annual source-receptor matrices for these substances for eastern North America. The U.S. EPA developed the Eulerian Tagged Species Engineering Model (TSEM) to calculate source-receptor relationships for $SO_4^=$ deposition in eastern North America. Sample maps produced by the TSEM showing the area of influence of several small (sub-state) SO_2 source subregions on annual total sulfur deposition are provided in Dennis et al. (1990) and McHenry et al. (1992). U.S. EPA (1995) presents area-of-influence plots for 53 subregions covering much of eastern North America. Figure 8.8a shows one such plot for the source subregion covering the Ohio/West Virginia/Pennsylvania border subregion. The TSEM also was used to estimate the area of influence of eastern Canadian SO_2 emissions on both Canada and the U.S. by tagging all Canadian SO_2 sources. Figure 8.8b shows the predicted percentage contribution of Canadian SO_2 sources to total dry sulfur deposition in eastern North America, with U.S. sources contributing the remainder.

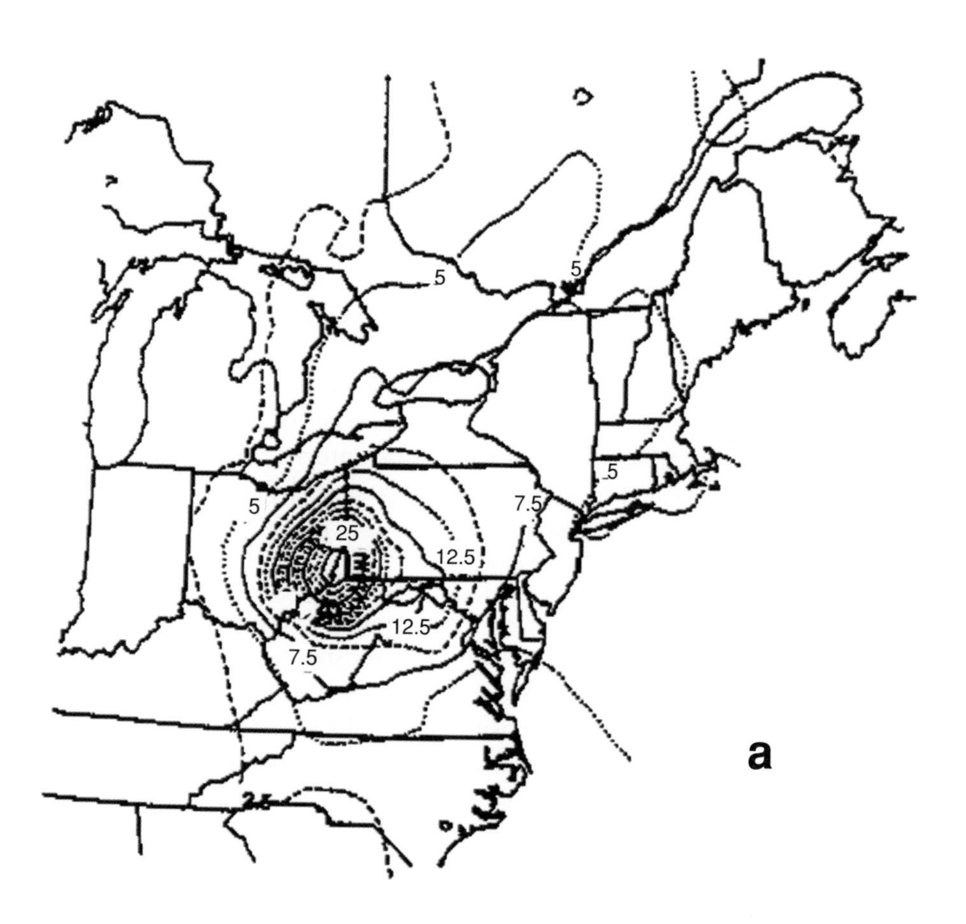

Figure 8.8a. Annual predicted percentage contribution of the Ohio/West Virginia/Pennsylvania border SO_2 source subregion to total sulfur deposition in eastern North America (from U.S. EPA (1995)).

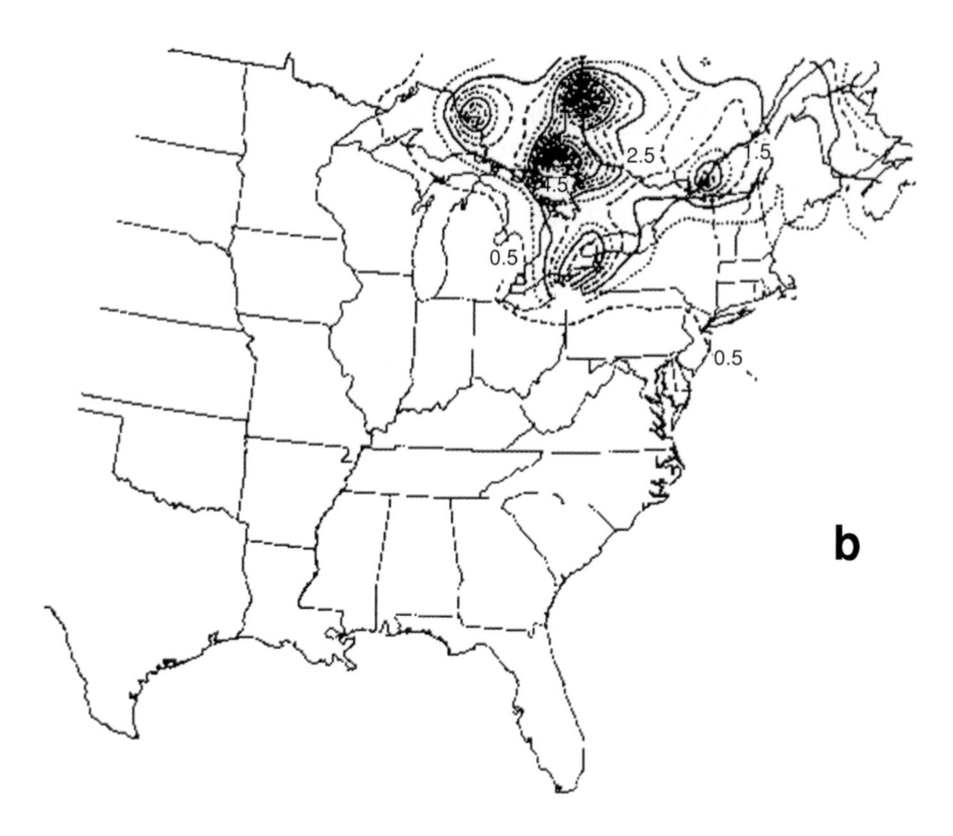

Figure 8.8b. Annual predicted percentage contribution of the (b) Canadian SO_2 sources to total dry sulfur deposition in eastern North America (from Dennis et al. (1990)).

The Eulerian Regional Acid Deposition Model (RADM), from which the TSEM is derived, was applied to derive similar area-of-influence plots for NO_x emissions on integrated nitrogen deposition for nine of the same 53 source subregions. The range of influence of NO_x emissions was found to be similar to that of SO_2 emissions, on the order of 800 km (Dennis, 1997). The area-of-influence results were then used to estimate the size and location of the "airshed" whose NO_x emissions are significant contributors to nitrogen deposition in the Chesapeake Bay watershed. This airshed was found to be about five times the size of the watershed.

Because chemical deposition is of most concern on time scales of a year or more, chemical-deposition CTMs have had to predict annual concentration and deposition fields using the two approaches discussed in Section 8.3.2. Accordingly, performance evaluations for these CTMs have looked at all seasons, and model performance has been found to vary with season (e.g., Clark et al., 1989; Dennis et al., 1990). Poorer model performance for a particular season may point to problems with the model representation of one or more of those processes known to vary with season (e.g., gas-phase chemistry, dry deposition, wet scavenging). For example, SO_2-to-$SO_4^=$ conversion via in-cloud oxidation is known to be more oxidant-limited in the winter than in other seasons (e.g., Dennis et al., 1990). Emission inventories that do not properly reflect seasonal variability may also be a source of error.

Much of the impetus for the development of Eulerian chemical-deposition CTMs came from concerns over nonlinearities or nonproportionalities in the atmospheric chemistry of sulfur (e.g., Venkatram and Karamchandani, 1986). Based on Eulerian CTM results, on an annual basis such atmospheric process nonlinearities are now expected to be small in eastern North America, on the order of 10 to 15 percent, for total sulfur deposition (e.g., Dennis et al., 1990; Dennis, 1997). However, chemical-deposition CTM results have also demonstrated another important source of nonlinear responses to SO_2 emission changes, namely nonuniformity in the spatial pattern of emission reductions. For example, in one CTM simulation in which domain-wide SO_2 emissions were reduced by 50 percent but all reductions were made to coal-fired utility sources, state-by-state emission reductions ranged from 30 to 70 percent and the resulting total annual sulfur deposition was reduced by a range of values from less than 20 to 55 percent (Dennis et al., 1990). This distinction is very relevant to PM-control strategies.

In summary, the experience gained to date from regional chemical-deposition modeling suggests that the inorganic component of PM will have to be managed as a regional issue involving transport scales of 1000 to 3000 km.

8.9.3 Photochemical CTMs

Previous applications of photochemical CTMs to policy issues have been addressed recently by the NARSTO Synthesis Team (2000) and Russell and Dennis (2000). Therefore, we only provide here a summary of the more salient points of these reviews.

The NARSTO Ozone Assessment noted that the application of a given CTM to the assessment of emission-control strategies could give different (and sometimes conflicting) results depending on the inputs selected. Two examples were given, one in California and the other in New York. In those two modeling studies, different (but basically acceptable) inputs for the meteorology and boundary conditions were used resulting in simulations that could have led to vastly different strategies for reducing ozone precursors. Clearly, these results reinforce the point made in Section 8.8.1 that emission-control strategies should not be based on the use of a single model and that it is essential to obtain corroborative information to confirm any modeling results. They also emphasize that modeling includes not only the model but also the modeling process, that is, all of the steps involved in applying the model, from domain specification and input preparation to the interpretation and analysis of results.

Russell and Dennis (2000) emphasize that one must distinguish between the error in the CTM predictions and the error in the CTM response to emission changes. Clearly, the ability of the CTM to predict the correct response is critical to the development of effective emission-control strategies. The uncertainties in the inputs (emissions, meteorology) and CTM formulation that lead to uncertainties in model response therefore should be identified and quantified. Somewhat disturbingly, despite several decades of photochemical modeling, Russell and Dennis were not able to identify a single example of such a comprehensive CTM sensitivity and uncertainty analysis, perhaps suggesting implicitly the level of effort required.

These considerations on the application of photochemical CTMs to the development of policy decisions clearly apply also to CTMs for PM and are consistent with several issues that were discussed in previous sections.

8.10 CRITICAL UNCERTAINTIES

Table 8.2 summarizes qualitatively the best estimates, of the authors of this chapter, of the degree of confidence in various aspects of current PM CTM simulations. Uncertainties in the ability to predict the concentrations of PM arise from a variety of sources, including the input data supplied to the CTMs (meteorological predictions, emission inventories for PM and its precursors, chemical boundary conditions), the formulation of the CTMs for PM themselves, and the lack of suitable evaluation data sets. Listed below are the major sources of uncertainties at the moment.

Pielke and Uliasz (1998) and Seaman (2000) have provided excellent reviews of the uncertainties associated with meteorological model simulations. Those uncertainties will affect CTM predictions. For PM, the prediction of atmospheric liquid and solid water (i.e., clouds, fogs, and precipitation) is particularly important because droplets tend to enhance the formation of $SO_4^=$, NO_3^- and, possibly, condensable organic compounds. Also, precipitation is a major removal mechanism for PM. The treatment of fog is typically ignored or highly simplified in most current CTMs; it should be included since it has been shown to be important for wintertime PM episodes in some areas. The treatment of convective transport at the subgrid scale also needs to be improved in both

Table 8.2. Present qualitative levels of confidence in various aspect of PM CTM simulations.

Field	Confidence Level[(a)]	Comments
PM Mass Components		
PM ultrafine	VL	Number distribution and nucleation and coagulation are not well represented
PM fine	M	Can be handled by existing size-resolved models, some chemical components not well represented
PM coarse	M	Can be handled by existing size-resolved models, some chemical components not well represented
PM Composition		
$SO_4^=$	M-H	Extensive chemical-deposition modeling experience, precursor emissions well characterized, relatively simple gas-phase chemistry but aqueous-phase chemistry requires good knowledge of presence of fog or clouds
NO_3^+	M	Extensive chemical-deposition modeling experience, precursor emissions well characterized, some uncertainties related to gas/particle partitioning
NH_4^+	M	Extensive chemical-deposition modeling experience, precursor emissions not well characterized, some uncertainties related to gas/particle partitioning
OC primary	L	Significant emission uncertainties, natural sources may be important
OC secondary	VL	Emissions uncertainties for VOC precursors, natural sources may be important, incomplete understanding of formation chemistry and gas/particle partitioning
BC	L	Chemistry unimportant but emissions not well known
Crustal material	L	Chemistry unimportant but emissions not well known and natural sources are important but sporadic & intermittent
Sea salt	L	Chemistry may be important; emissions are function of wind speed and fetch and may be uncertain
Water	L	Gas/particle partitioning is complex and not well described for multi-constituent aerosols, especially organic component
Metals, biologicals, peroxides	VL	Emissions are poorly known; to date have been modeled separately if at all
Gases		
SO_2	H	Emissions and chemistry well characterized
NO_x	H	Emissions reasonably well characterized, chemistry well characterized
NH_3	M	Emissions poorly characterized, bi-directional surface fluxes
VOC	M	Emissions less well characterized; speciation uncertainties
HNO_3	M	Secondary species: extensive previous modeling experience but complex gas-phase chemistry and gas/particle partitioning
O_3	M	Secondary species: extensive previous modeling experience but only for summer conditions; complex gas-phase chemistry

[(a)] H: high, M: medium, L: low, VL: very low

Table 8.2. Present qualitative levels of confidence in various aspect of PM CTM simulations. (continued)

Field	Confidence Level[a]	Comments
Spatial Scale		
Continental	L	Must consider all processes, computationally demanding, spatial resolution limitations, more emission uncertainties
Regional	M	Must consider all processes, computationally demanding, emissions may be multi-national, impact of upwind boundary conditions
Urban	L-M	May be possible to neglect some aerosol processes sometimes, difficulty of simulating small-scale meteorology, upwind boundary conditions may be very important and even dominant
Temporal Scale		
Annual	L	Computationally demanding, must handle all seasons, additional uncertainties associated with long-term predictions (cf. Sec. 8.3.2)
Seasonal	L	Computationally demanding, knowledge of emissions may vary by season, accuracy may vary by season
Episodic	M	Can use existing comprehensive PM models
Forecast	VL	Input emission data always out of date, have to forecast meteorology well, computationally very demanding (i.e., simulation time must be considerably less than real time)

(a) H: high, M: medium, L: low, VL: very low

meteorological models and CTMs (Sections 8.2.2, 8.2.3, 8.2.4).

Emission inventories need to be improved, particularly for PM, VOC, and NH_3. PM emissions need to be resolved by size (at a minimum, fine and coarse) and by species (at a minimum, $SO_4^=$, crustal material, OC, and BC). Fugitive emissions of crustal material (soil dust) and biomass burning (wildfires and prescribed burning) need to be better characterized. Emissions of VOC and SVOC need to be refined and provided with chemical speciation. Biogenic VOC inventories must include individual terpene species in addition to isoprene and oxygenated compounds (Section 8.2.1).

In the CTMs themselves, there are several areas that require further improvement. The formation of gas-phase condensable organic compounds and their partitioning between the gas phase and the particulate phase is currently a major source of uncertainty. Atmospheric data with particulate organic speciation are needed to evaluate the various models that are currently available. As mentioned above, cloud and fog processes can be a major source of PM formation, and their treatment (i.e., cloud dynamics and associated transport processes, precipitation, fog dynamics and deposition, aqueous-phase chemistry, wet scavenging) in CTMs must be improved. In particular, most CTMs only treat $SO_4^=$ formation in droplets; heterogeneous and homogeneous processes leading to NO_3^- formation must be included. Size-resolution of droplets may be required for accurate simulations of aqueous chemistry. Research is also needed to assess whether cloud droplets are a significant source of condensable organic compounds (Sections 8.2.4, 8.2.6, 8.5.3).

Dry deposition of particles is a strong function of particle size, atmospheric conditions, and land use. Particle dry-deposition algorithms need to be evaluated with reliable atmospheric data and, if warranted, improved. Dry deposition of PM precursors (e.g., organic compounds, HNO_3, NH_3) also needs to be carefully evaluated with reliable data (Section 8.2.6).

The formation of $SO_4^=$ and NO_3^- in the plumes of large point sources needs to be treated with subgrid-scale representations in CTMs (plume-in-grid modeling). Plume-in-grid models for PM need to be developed and evaluated (Section 8.2.3).

Existing gas-phase chemical kinetic mechanisms have been developed and evaluated primarily for summer daytime conditions. Such mechanisms need to be evaluated for nighttime conditions, and for winter conditions in areas where PM episodes occur during winter. The need for surface chemistry and heterogeneous reactions to be added to these mechanisms should also be investigated (Section 8.2.4).

Currently, only a few PM modeling systems include feedbacks between aerosol formation in the CTM and radiative forcing in the meteorological model. Although such feedbacks may not be necessary for PM air-quality applications, they may need to be incorporated for global climate studies (Sections 8.2.4, 8.2.5).

For CTM applications, specification of the upwind boundary conditions (particularly, fine PM and precursor concentrations) is critical. Measurements both at the ground and aloft are needed to specify associated upwind concentrations (Section 8.4.2).

More work is required to evaluate and compare the two main approaches to long-term CTM simulations shown in Figure 8.3: reduced-form versions of comprehensive CTMs and episode aggregation (Sections 8.3.2, 8.4.3).

CTMs that have been developed for PM will need to be evaluated thoroughly with comprehensive data bases so that their major sources of uncertainties can be identified and, if warranted, addressed. To that end, there is a dire need for large, multi-seasonal regional PM field programs that can provide comprehensive data bases for model evaluation with good spatial (including measurements aloft) and temporal resolutions. CTMs should be used for the design of such field programs to ensure that all the necessary data are collected. The CTM evaluation should then include operational, diagnostic, and, to the extent possible, mechanistic and probabilistic evaluations. Such evaluations need to be conducted in several areas that present a variety of geographical, meteorological, and atmospheric-chemistry conditions (Sections 8.5, 8.6).

As policy-makers' needs for guidance from PM CTMs increase, so does the need for trained air-quality modelers. As Figure 8.2 suggests, the successful use of a CTM and interpretation of results requires an understanding of emission inventories, meteorological and air-quality measurements, emissions, meteorological, and chemical-transport modeling, and the processes represented in those models. Such knowledge and understanding requires years of study and experience to acquire. There is an urgent need to train a generation of new air-quality modelers to contribute to CTM development, evaluation, and application.

Finally, CTMs may need to be improved to treat PM characteristics besides PM mass concentrations that are relevant to health effects, such as PM number concentrations, acidity, and metal concentrations (Section 8.4.6).

As discussed above, current CTM predictive uncertainties are of the order of ±50 percent for $PM_{2.5}$ mass. Among secondary $PM_{2.5}$ components, performance is best for $SO_4^=$, and typically more uncertain for NO_3^- and NH_4^+. Large uncertainties are associated with the organic compounds. Once the issues discussed above are addressed, improved performance from CTMs is anticipated.

8.11 SUMMARY

Source-oriented chemical-transport models exist to simulate PM concentrations using natural and man-made emissions of PM and PM precursors as inputs. These CTMs can be applied to calculate the fraction of PM that is transported into the area of interest from upwind areas and the fraction that is emitted or generated locally. The CTMs also can be applied to assess the impact of changes in emissions, to perform source attributions, and to quantify the relationships between PM and other air-pollution problems.

There are, however, uncertainties associated with the results of CTM simulations. Those uncertainties arise from the scientific formulation and numerical solution of the CTM as well as from the input data (i.e., meteorology, emissions, upwind air pollutant concentrations). Some major sources of uncertainties include emissions of primary PM, VOC, and NH_3 (and the chemical speciation of the first two); meteorology in complex terrain, under calm conditions, and for clouds/fog/precipitation; and upwind air-pollution concentrations (particularly aloft) at model boundaries.

The most advanced CTMs for PM can currently predict the formation of $SO_4^=$ and HNO_3 satisfactorily (e.g., within 50 percent). The accuracy of NO_3^- predictions is likely to be limited by the reliability of NH_3 emissions. CTMs can predict concentrations of primary PM (e.g., BC, crustal material) satisfactorily, provided the emissions are well characterized; however, CTMs cannot predict with certainty the concentrations of primary PM from sources whose emissions are currently highly uncertain (e.g., fugitive dust, biomass burning).

At the moment, CTMs have large uncertainties associated with OC predictions because of uncertainties in 1) the emissions of primary OC and condensable organic gases, 2) the emissions of precursors of secondary OC (particularly biogenic VOC), 3) the chemical reactions leading to condensable organic compounds, and 4) the partitioning of those compounds between the gas and condensed phases.

As a consequence, CTMs are likely to be most useful initially to address the $SO_4^=/NO_3^-/NH_4^+$ component of PM. CTM predictions of the organic fraction of PM (e.g., contributions of primary vs. secondary and man-made vs. natural sources) will be much less certain. Other complementary techniques (receptor modeling, indicator-species approaches) should be used to corroborate CTM results, particularly for those PM components known to be highly uncertain in CTM simulations. One must note too that because PM is comprised of many components, there will likely be a number of possible emission-control strategies to attain PM air-quality standards.

CTMs will typically perform better for long-term periods (e.g., one year) than for short-term periods (e.g., 24 hours or less) due to "averaging out" the impact of short-term, small-scale meteorological fluctuations. However, the better performance obtained for long-term average PM concentrations also may result from offsetting errors, and thorough model-performance evaluations must be conducted to identify such errors, if any. For example, the ability of CTMs to reproduce seasonal variations must be tested. Longer-term CTM simulations also require additional approximations (e.g., simplified parameterizations or episode aggregation) to be made at present due to computer-resource limitations.

The performance evaluation of CTMs requires extensive data sets, including speciated and size-resolved PM measurements and measurements of associated gas-phase species for different seasons. As a result, CTM evaluations have been limited to date outside of the Los Angeles airshed due to a lack of suitable evaluation data sets. A comprehensive performance evaluation consists of four primary stages: operational; diagnostic; mechanistic; and probabilistic. At all stages, care must be taken to match the form of CTM predictions of various physical and chemical quantities to the form and methodology of the measurements, though incommensurability can never be avoided entirely.

CTMs can be used effectively to identify the major sources contributing to PM locally. However, the impact of specific sources at distances of 100 km or more cannot be assigned as reliably. Instead, source attribution in regional-scale studies should focus on source regions and source categories rather than on individual sources. This uncertainty is due in part to our inability to predict wind flow and precipitation accurately over long distances and periods. It also results from compounding uncertainties in the model formulation, which will increase as the simulation proceeds. Uncertainties in meteorological aspects can be reduced to some extent by using data assimilation when conducting the meteorological simulations. Further advances in the current formulation of CTMs (e.g., by carefully evaluating specific components such as dry deposition against ambient data) will help reduce the latter source of uncertainty.

There is ample evidence that emission-control strategies for PM will affect, positively or negatively, other atmospheric problems such as ozone, regional haze, chemical deposition, air toxics and visibility. Because an emission-control strategy may have a positive effect on one pollutant but a negative effect on another, it is imperative that all pollution problems be analyzed in an integrated manner before the emission-control strategy becomes implemented. CTMs that can simulate several air-pollution problems jointly are therefore preferable for regulatory applications. However, those various air-pollution problems may have different temporal and spatial scales, thereby complicating their joint analysis.

CTMs currently treat PM mass concentrations resolved for different size ranges and chemical components. If other surrogates for PM health effects are selected (e.g., number of ultrafine particles, particle acidity, trace-metal concentrations), CTMs will need to be adapted accordingly.

CTMs can be used as a complement to monitoring networks to help understand what is being observed, to help define what needs to be measured, and to estimate progress in areas where air-monitoring data are sparse. For such applications, the CTMs should be tested to verify that the observed progress (or lack of progress) corresponds to their predictions. They should be corrected if significant discrepancies are found.

CTMs can be used to support estimation of population exposure to outdoor air pollutants. However, they need to be coupled with indoor and vehicle-interior air pollution models to allow realistic predictions of population exposure.

CTMs also can be used for daily forecasting, although applications to date have been limited to ozone. Such forecasts will require real-time meteorological forecasts and up-to-date emission inventories. Forecasting for PM will also require evaluation of forecast accuracy, especially at the proof-of-concept stage.

Because CTMs for PM are complex analytical tools, their application requires highly trained professionals to ensure that the proper inputs are being used, potential problems are promptly identified, and the CTM results are correctly interpreted. There is an urgent need for new modelers to be trained. Also, establishing best modeling practices via networks of modelers at key public and private institutions and via the establishment of regional or national modeling centers will improve future modeling endeavors.

8.12 POLICY IMPLICATIONS

Performance evaluation of the CTMs against appropriate ambient data is required before they are used to guide policy development. *Implication:* Appropriate ambient data sets, adding to what is now available, need to be collected via intensive field studies designed to meet model-performance evaluation needs. Current CTM outputs can be used as part of the collective scientific analysis to guide policy.

At this stage of scientific understanding of PM formation and transport there is variable confidence in specific CTM outputs (chemically, spatially, temporally). *Implication:* If the output of interest has relatively low confidence, then it is necessary to acquire supporting information from other approaches: e.g., ambient characterization, receptor modeling, and/or another CTM platform.

At their current proficiency levels, PM CTMs have served mainly as experimental research tools, and policy applications are just starting. Currently, each scenario takes from a few months to a year (contingent upon having appropriate meteorological and emission information at hand) to prepare, run, and evaluate. This timing is also dependent upon the resources available, both in terms of expert modelers and computational facilities. The performance evaluation will indicate the level of confidence that can be placed in the current generation of CTMs. It is expected that within the next few years CTMs for PM will be available for some applications to guide policy development, being used in conjunction with other analyses. *Implication:* Within the next few years CTM uncertainties will be better characterized, and in some cases quantified, making their outputs more useful to policy development. The better-performing CTMs will also be identified. Experience with ozone modeling, though, has shown that building confidence in CTM predictions will take longer.

CTMs are currently being developed and operated on different geographic scales (urban to global) for different time periods (few days to entire year). Results of performance evaluations suggest that there are still some large uncertainties, particularly for simulations that cover long time periods and larger domains. *Implication:* Application of CTM outputs to long time-periods and/or larger geographic regions to provide input to development of long-term management approaches is not yet appropriate. Their application may be appropriate for episodic conditions over limited isolated areas, contingent upon satisfactory performance evaluation.

CTM capabilities and uncertainties, as well as being bound by current understanding of atmospheric processes, are limited by the type and quality of the ambient and meteorological data and emission information used to evaluate and run them. *Implication:* Higher CTM resolution (with respect to chemical species and particle size) and confidence can only be achieved as inputs improve. Current spatial and temporal resolution of CTMs does not yet match the requirements for estimating human exposure; additional models such as local dispersion and indoor models need to be used in combination with CTMs to obtain the appropriate spatial and temporal resolutions. This also indicates that, as the health community focuses on specific particle characteristics, the models will need to be adapted to provide relevant predictions.

Currently, the measurements of some species, most notably NO_3^- and semivolatile organic species, are not accurate. Thus the use of these data in CTM performance evaluation is problematic. The measurement technology needs to evolve in order for CTMs to be thoroughly evaluated. *Implication:* Evaluated, policy-ready CTMs will become available gradually, with growing confidence, over the next decade in step with other improvements in PM science.

Current CTM outputs can simulate the effect of SO_2 emission reductions on $PM_{2.5}$ mass and chemical deposition with some confidence. However, the relative contributions of primary and secondary and anthropogenic and natural contributions (most notably for the organic fraction but also for NO_3^-) to $PM_{2.5}$ and PM_{10} mass are much less certain. *Implication:* There is moderate confidence in the CTM outputs for estimating changes in the inorganic PM fraction over monthly to seasonal periods on urban to regional scales, and chemical-deposition CTM results suggest that the inorganic component of PM will have to be managed as a regional issue.

In some cases, CTMs can flag potential errors in emission inventories if significant discrepancies between simulated and observed PM concentrations are more likely to be due to inputs than to model formulation. *Implication:* Application of CTMs will result in guidance for improving emission-inventory information, which can then in turn improve CTM performance.

8.13 REFERENCES

Abt, E., Suh, H.H., Catalano, P., Koutrakis, P., 2000. Relative contribution of outdoor and indoor particle sources to indoor concentrations. Environmental Science and Technology 34, 3579-3587.

Ansari, A.S., Pandis, S.N., 1999. An analysis of four models predicting the partitioning of semi-volatile inorganic aerosol components. Aerosol Science and Technology 31, 129-153.

Barthelmie, R.J., Pryor, S.C., 1996. Assessing limits on inorganic aerosol formation. Proc. 85th AWMA Annual Meeting, Paper 96-MP1A.02, June, Nashville, Air & Waste Management Association, Pittsburgh.

Blanchard, C.L., Roth, P.M., Tannenbaum, S.J., Ziman, S.D., Seinfeld, J.H., 2001. The use of ambient measurements to identify which precursor species limit aerosol nitrate formation. Journal of the Air and Waste Management Association 50, 2073-2084.

Brook, J.R., Johnson, D., 2000. Identification of representative warm season periods for regional air quality (ozone) model simulations. Atmospheric Environment 34, 1591-1599.

Brook, J.R., Samson, P.J., Sillman, S., 1995a. Aggregation of selected three-day periods to estimate annual and seasonal wet deposition totals for sulfate, nitrate, and acidity. Part I: A synoptic and chemical climatology for eastern North America. Journal of Applied Meteorology 34, 297-325.

Brook, J.R., Samson, P.J., Sillman, S., 1995b. Aggregation of selected three-day periods to estimate annual and seasonal wet deposition totals for sulfate, nitrate, and acidity. Part II: Selection of events, deposition totals, and source-receptor relationships. Journal of Applied Meteorology 34, 326-339.

Brost, R.A., 1988. The sensitivity of input parameters of atmospheric concentrations simulated by a regional chemical model. Journal of Geophysical Research 93, 2371-2387.

Byun, D.W., 1999. Dynamically consistent formulations in meteorological and air quality models for multiscale atmospheric issues. Part II: Mass conservation issues. Journal of the Atmospheric Sciences 56, 3808-3820.

Cardelino, C.A., Chameides, W.L., 2000. The application of data from photochemical assessment monitoring stations to the observation-based model. Atmospheric Environment 34, 2325-2332.

Clark, T.L., Voldner, E.C., Dennis, R.L., Seilkop, S.K., Alvo, M., Olson, M.P., 1989. The evaluation of long-term sulfur deposition models. Atmospheric Environment 23, 2267-2288.

Cohn, R.D., Eder, B.K., LeDuc, S.K., Dennis, R.L., 2001. Development of an aggregation and episode scheme to support the Models-3 Community Multiscale Air Quality Model, Journal of Applied Meteorology 40, 201-228.

Dennis, R.L., 1997. Using the Regional Acid Deposition Model to determine the nitrogen deposition airshed of the Chesapeake Bay watershed. In: Atmospheric Deposition of Contaminants to the Great Lakes and Coastal Waters. Baker, J.E., Editor, Society for Environmental Toxicology and Chemistry (SETAC) Press, Pensacola, Florida, 393-413.

Dennis, R.L., Binkowski, F.S., Clark, T.L., McHenry, J.N., Reynolds, S., Seilkop, S.K., 1990. Selected applications of RADM (Part II). App. 5F, NAPAP SOS/T Report 5, In: National Acid Precipitation Assessment Program: State of Science and Technology, Volume 1, National Acid Precipitation Assessment Program, Washington, D.C., 37 pp.

Dennis, R.L., McHenry, J.N., Barchet, W.R., Binkowski, F.S., Byun, D., 1993. Correcting RADM's sulfate underprediction: discovery and correction of model errors and testing the corrections through comparisons against field data. Atmospheric Environment 27A, 975-997.

Dodge, M.C., 2000. Chemical oxidant mechanisms for air quality modeling: critical review. Atmospheric Environment 34, 2103-2130.

Griffin, R.J., Dabdub, D., Kleeman, M.J., Fraser, M.P., Cass, G.R., Seinfeld, J.H., 2002. Secondary organic aerosol: III. Urban/regional scale model of size- and composition resolved aerosols. Journal of Geophysical Research 107.

Harvey, M.J., McArthur, A.J., 1989. Pollution transfer to moor by occult deposition. Atmospheric Environment 23, 1073-1082.

Hidy, G.H., 1994. Atmospheric Sulfur and Nitrogen Oxides: Eastern North American Source-Receptor Relationships, Academic Press, San Diego.

Jacob, D.J., 2000. Heterogeneous chemistry and tropospheric ozone. Atmospheric Environment 34, 2131-2159.

Jacobson, M.Z., 1997. Development and application of a new air pollution modeling system - Part III. Aerosol-phase simulations. Atmospheric Environment 31, 2829-2839.

Jacobson, M.Z., 1999. Fundamentals of Atmospheric Modeling, Cambridge University Press, Cambridge, UK.

Jacobson, M.Z., 1998. Studying the effects of aerosols on vertical photolysis rate coefficient and temperature profiles over an urban airshed. Journal of Geophysical Research 103, 10593-10604.

Jacobson, M.Z., 2001. Strong radiative heating due to the mixing state of black carbon in atmospheric aerosols. Nature 409, 695-697.

Jang, C., Dolwick, P., Possiel, N., Timin, B., Tikvart, J., 2001. Annual application of U.S. EPA's third-generation air quality modeling system over the continental United States. Paper 379, 94th Annual Air & Waste Management Conference, June 24-28, Orlando, Florida.

Karamchandani, P., Venkatram, A., 1992. The role of non-precipitating clouds in producing ambient sulfate during summer: results from simulations with the Acid Deposition and Oxidant Model (ADOM). Atmospheric Environment 26A, 1041-1052.

Kleeman, M.J., Cass, G.R., 1999. Effects of emissions control strategies on the size and composition distribution of urban particulate air pollution. Environmental Science and Technology 33, 177-189.

Kleeman, M.J., Cass, G.R., 2001. A 3D Eulerian source-oriented model for an externally mixed aerosol. Environmental Science and Technology 35, 4834-4848 .

Kuo, Y.-H., Anthes, R.A., 1984. Accuracy of diagnostic heat and moisture budgets using SESAME-79 field data as reveled by observing system simulation experiments. Monthly Weather Review 112, 1465-1481.

Kuo, Y.-H., Skumanich, M., Haagenson, P.L., Chang, J.S., 1985. The accuracy of trajectory models as revealed by the observing system simulation experiments. Monthly Weather Review 113, 1852-1867.

Lillis, D., Cruz, C.N., Collett, Jr., J., Richards, L.W., Pandis, S.N., 1999. Production and removal of aerosol in a polluted fog layer: Model evaluation and fog effect on PM. Atmospheric Environment 33, 4797-4916.

Mage, D., Wilson, W., Hasselblad, V., Grant, L., 1999. Assessment of human exposure to ambient particulate matter. Journal of the Air and Waste Management Association 49, 1280-1291.

McHenry, J.N., Binkowski, F.S., Dennis, R.L., Chang, J.S., Hopkins, D., 1992. The Tagged Species Engineering Model (TSEM). Atmospheric Environment 26A, 1041-1052.

McNaughton, D.J., Vet, R.J., 1996. Eulerian Model Evaluation Field Study (EMEFS): a summary of surface network measurements and data quality. Atmospheric Environment 30, 227-238.

Meng, Z., Dabdub, D., Seinfeld, J.H., 1997. Chemical coupling between atmospheric ozone and particulate matter. Science 277, 116-119.

Meng, Z., Dabdub, D., Seinfeld, J.H., 1998. Size-resolved and chemically resolved model of atmospheric aerosol dynamics. Journal of Geophysical Research 103, 3419-3435.

Monn, C., 2001. Exposure assessment of air pollutants: a review on spatial heterogeneity and indoor/outdoor/personal exposure to suspended particulate matter, nitrogen dioxide and ozone. Atmospheric Environment 35, 1-32.

Moran, M.D., 2000. Basic aspects of mesoscale atmospheric dispersion. In Mesoscale Atmospheric Dispersion, Boybeyi, Z., Editor, WIT Press, Southampton, 27-119.

NARSTO, 2000. An Assessment of Tropospheric Ozone Pollution: A North American Perspective. North American Research Strategy for Tropospheric Ozone, [Available from NARSTO Management Coordinator Office, Pasco, Washington, or Electric Power Research Institute, 1-800-313-3774 or www.epri.com]

Nazaroff, W.W. Cass, G.R., 1989. Mathematical modeling of indoor aerosol dynamics. Environmental Science and Technology 23, 157-166.

Pagowski, M., Chtcherbakov, A., Bloxam, R., Wong, S., Lin, X., Sloan, J., Soldatenko, S., 2002. Evaluation of Models-3/CMAQ performance during a winter high particulate matter episode in February 1998. Fourth AMS Conf. on Atmospheric Chemistry, Jan. 13-17, Orlando, Florida, American Meteorological Society.

Pai, P., Vijayaraghavan, K., Seigneur, C., 2000. Particulate matter modeling in the Los Angeles Basin using SAQM-AERO. Journal of the Air and Waste Management Association 50, 23-42.

Pandis, S.N., Wexler, A.S., Seinfeld, J.H., 1995. Dynamics of tropospheric aerosols. Journal of Physical Chemistry 99, 9646-9659.

Peters, L.K., Berkowitz, C.M., Carmichael, G.R., Easter, R.C., Fairweather, G., Ghan, S.J., Hales,

J.M., Leung, L.R., Pennell, W.R., Potra, F.A., Saylor, R.D., Tsang, T.T., 1995. The current state and future direction of Eulerian models in simulating the tropospheric chemistry and transport of trace species: a review. Atmospheric Environment 29, 189-222.

Pielke, R.A., Uliasz, M., 1998. Use of meteorological models as input to regional and mesoscale air quality models – limitations and strengths. Atmospheric Environment 32, 1455-1466.

Pun, B., Seigneur, C., Wu, S.-Y., 2002. Modeling regional haze in Big Bend National Park with CMAQ. CMAQ Workshop, Research Triangle Park, NC, 21-23 October 2002.

Russell, A., Dennis, R., 2000. NARSTO critical review of photochemical models and modeling. Atmospheric Environment 34, 2283-2324.

Schemenauer, R.S., Banic, C.M., Urquizo, N., 1995. High elevation fog and precipitation chemistry in southern Quebec, Canada. Atmospheric Environment 29, 2235-2252.

Seaman, N.L., 2000. Meteorological modeling for air-quality assessments. Atmospheric Environment 34, 2231-2259.

Seigneur, C., 1994. Review of mathematical models for health risk assessment. VI. Population exposure, Environmental Software 9, 133-145.

Seigneur, C., 2001. Current status of air quality modeling for particulate matter. Journal of the Air and Waste Management Association 51, 1508-1521.

Seigneur, C., Pai, P., Hopke, P., Grosjean, D., 1999. Modeling atmospheric particulate matter. Environmental Science and Technology 33, 80A-86A.

Seigneur, C., Pun, B., Pai, P., Louis, J.-F., Solomon, P., Emery, C., Morris, R., Zahniser, M., Worsnop, D., Koutrakis, P., White, W., Tombach, I., 2000a. Guidance for the performance evaluation of three-dimensional air quality modeling systems for particulate matter and visibility. Journal of the Air and Waste Management Association 50, 588-599.

Seigneur, C., Tonne, C., Vijayaraghavan, K., Pai, P., 2000b. The sensitivity of PM2.5 source-receptor relationships to atmospheric chemistry and transport in a three-dimensional air quality model. Journal of the Air and Waste Management Association 50, 428-435.

Seinfeld, J.H., Pandis, S.N., 1998. Atmospheric Chemistry and Physics – From Air Pollution to Climate Change. John Wiley & Sons, New York, NY.

Sheih, C.M., Hess, G.D., Hicks, B.B., 1978. Design of network experiments for regional-scale atmospheric pollutant transport and transformation. Atmospheric Environment 12, 1745-1753.

Sillman, S., He, D., Cardelino, C., Imhoff, R.E., 1997. The use of photochemical indicators to evaluate ozone-NO_x-hydrocarbon sensitivity: Case studies from Atlanta, New York, and Los Angeles. Journal of the Air and Waste Management Association 47, 1030-1040.

Tonnesen, G.S., Dennis, R.L., 2000a. Analysis of radical propagation efficiency to assess ozone sensitivity to hydrocarbons and NOx. Part 1: Local indicators of instantaneous odd oxygen production sensitivity. Journal of Geophysical. Research 105, 9213-9225.

Tonnesen, G.S., Dennis, R.L., 2000b. Analysis of radical propagation efficiency to assess ozone sensitivity to hydrocarbons and NOx. Part 2: Long-lived species as indicators of ozone concentration sensitivity. Journal of Geophysical Research 105, 9227-9241.

Trujillo-Ventura, A., Ellis, J.H., 1991. Multiobjective air pollution monitoring network design. Atmospheric Environment 25A, 469-479.

U.S. EPA, 1995. Acid Deposition Standard Feasibility Study: Report to Congress, Report EPA-430-R-95-001a, October, Office of Air and Radiation, U.S. Environmental Protection Agency, Washington, DC.

Venkatram, A., Karamchandani, P., 1986. Source-receptor relationships: a look at acid deposition modeling. Environmental Science and Technology 20, 1084-1091.

Wesely, M.L., Hicks, B.B., 2000. Review of current status of knowledge on dry deposition. Atmospheric Environment 34, 2261-2282.

West, J.J., Ansari, A.S., Pandis, S.N., 1999. Marginal PM2.5: nonlinear aerosol mass response to sulfate reductions in the Eastern United States. Journal of the Air and Waste Management Association 49, 1415-1424.

Wexler, A.S., Seinfeld, J.H., 1990. The distribution of ammonium salts among a size and composition dispersed aerosol. Atmospheric Environment 24A, 1231-1246.

Wexler, A.S., Seinfeld, J.H., 1991. Second-generation inorganic aerosol model. Atmospheric Environment 25A, 2731-2748.

Winkler, P., 1973. The growth of atmospheric aerosol particles as a function of the relative humidity. Part II: An improved concept of mixed nuclei. Aerosol Science 4, 373-387.

Wojcik, G.S., Chang, J.S., 1997. A re-evaluation of sulfur budgets, lifetimes, and scavenging ratios for eastern North America. Journal of Atmospheric Chemistry 26, 109-145.

Zhang, Y., C. Seigneur, J.H., Seinfeld, M.Z., Jacobson and F. Binkowski, 1999. Simulation of aerosol dynamics: A comparative review of algorithms used in air quality models, Aerosol Sci. Technol, 31, 487-514.

Zhang, Y., Seigneur, C., Seinfeld, J.H., Jacobson, M.Z., Clegg, S.L., Binkowski, F.S., 2000. A comparative review of inorganic aerosol thermodynamic equilibrium modules: Similarities, differences, and their likely causes. Atmospheric Environment 34, 117-137.

Zhang, L., Moran, M.D., Makar, P.A., Brook, J.R., Gong, S., 2002: Modelling gaseous dry deposition in AURAMS — A Unified Regional Air-quality Modelling System. Atmospheric Environment 36, 537-560.

CHAPTER 9

Visibility and Radiative Balance Effects

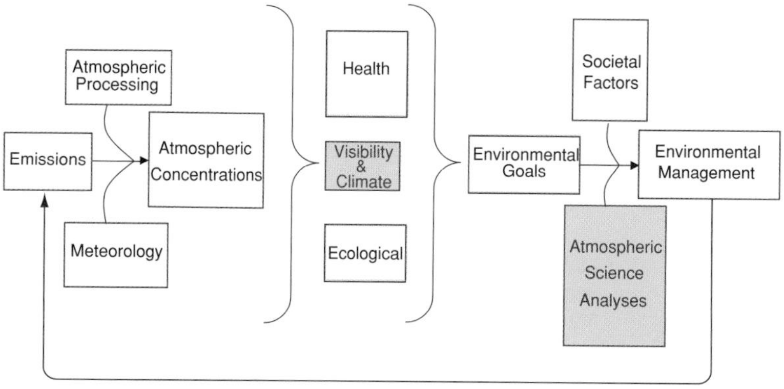

Principal Authors: Ivar Tombach and Karen McDonald

In addition to their effects on human health, particles in the atmosphere contribute to impairment of visibility and affect the global radiation balance (which, in turn, affects climate). These effects are governed by the atmospheric concentrations of particles, their spatial distribution (both geographically and vertically), and their size distribution and chemical composition. Thus all of the factors discussed in Chapters 3 through 8 are of relevance for understanding and managing visibility impairment and radiative-balance effects due to particles. This chapter addresses the linkage between PM and visibility and radiative-balance effects, as well as additional issues unique to visibility and radiative balance. The principal focus here is on visibility impairment, which is closely linked to PM concentrations and composition. The radiative-balance effects of PM, which are only one component of climate change due to air pollution, are discussed in less detail.

9.1 HOW IS VISIBILITY LINKED TO PM?

Particles suspended in the air interfere with the transmission of light through the atmosphere. The scattering and absorption of light by the particles result in a degradation of visibility, which is manifested as a reduction in the distance to which one can see and a decrease in the apparent contrast and color of distant objects. Furthermore, the particles may scatter sunlight into the image, causing a washed-out appearance, or the atmosphere itself may have a hazy appearance that has different color and brightness than a haze-free atmosphere. These processes are illustrated in Figure 9.1.

To the public, these optical effects are the most readily recognized indications of the presence of particulate air pollution. Visibility is also affected by natural clouds and fog and by precipitation, but the focus here is on visibility effects resulting from air-pollutant particles.

Visibility is concerned with the propagation of light in the horizontal direction. The same processes of atmospheric light scattering and absorption also occur in the vertical direction and affect the balance between the light and heat that reaches the earth from the sun and that which is propagated back into space. The impacts of particles on this radiation balance influence weather and climate, as discussed briefly later in this chapter.

Visibility effects due to particulate air pollution take place over two spatial scales: local and widespread. Localized plumes or clouds of pollution can obscure visibility, a process that is sometimes called plume blight. Because of the short travel time of a plume from a nearby source, local visibility effects are usually caused mostly by primary particles. Widespread pollution can produce a regional haze that extends over distances of tens or hundreds of kilometers. Many sources distributed over a wide area may contribute to regional haze and, because of the longer transport times (typically over many hours or a few days), regional haze is usually caused mostly by secondary particles.

Physical linkages exist between PM concentrations and visibility effects. Increasing concentrations of PM result in roughly proportional increases in the amount of light that is scattered and absorbed by the particles. Increasing relative humidity also increases the amount of light that is scattered relative to that of

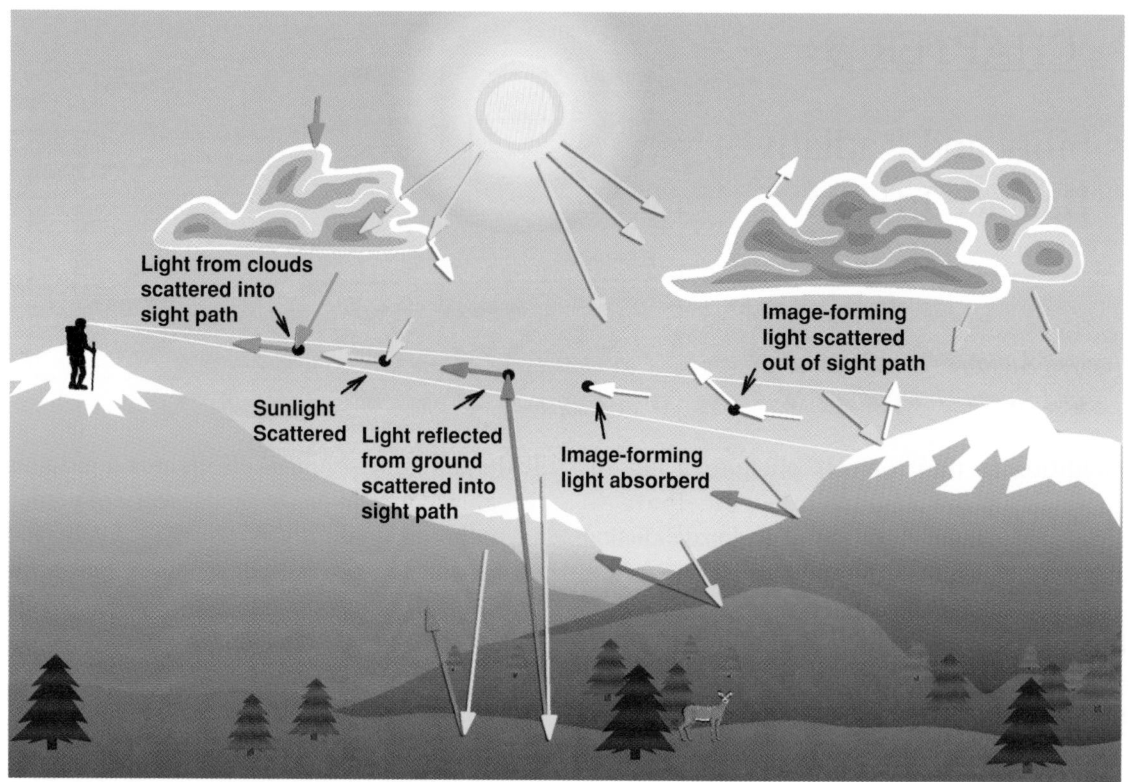

Figure 9.1. Illustration of the three processes by which particles and gases in the atmosphere affect visibility– the scattering of image-forming light, the absorption of image-forming light, and the scattering of light into the sight path. (Adapted from Malm, 2000a).

dry PM. In most cases in ambient air, the scattering of light by particles is much greater than the absorption of light by either particles or gases. In some urban areas and major point-source plumes, however, absorption by soot particles and/or by NO_2 gas can be important. The combined effect of light scattering and absorption, due to both particles and gases, is called light extinction and is characterized by the extinction coefficient. For any given scene and illumination condition, an increase in the extinction coefficient is associated with a decrease in visibility. Visibility can be defined in terms of the distance one can see (the visual range), which is inversely proportional to the extinction coefficient. (For additional discussion of the optical effects of particles, see Watson (2002), Malm (2000a), NRC (1993), and Trijonis et al. (1990).)

The extinction coefficient, denoted b_{ext}, is a measure of the fraction of light that is lost from a beam of light as it traverses a unit distance through a uniform atmosphere. Formally, for a beam of light with initial intensity I_0, its intensity after passing through a distance r of atmosphere with extinction coefficient b_{ext} is $I = I_0 \exp(-b_{ext}\, r)$.

For example, the extinction coefficient of particle-free air (the Rayleigh scattering coefficient) at sea level is 0.012 km^{-1} for light at the peak wavelength of human sensitivity. This means that 1.2 percent of the light in the beam is lost (scattered by air molecules) as the beam passes through one kilometer of a clean atmosphere. For convenience, a megameter, denoted Mm and equal to 1000 km, is often used as the unit of distance, in which case the sea level Rayleigh scattering coefficient is 12 Mm^{-1}. The corresponding visual range along a path that follows the earth's surface would be about 325 km, were it possible to see around the curvature of the earth.

This Rayleigh scattering by air molecules can make up the majority of the light extinction on the clearest days in arid, relatively unpopulated areas, such as the intermountain region of the western United States and Canada, and in the Arctic. On less clear days and elsewhere, scattering and absorption by PM dominate the light extinction.

The light-extinction effects of PM vary with particle size and chemical composition. The particles with the greatest influence on visibility are fine particles of the same scale as the wavelengths of visible light (approximately 0.3 to 1 µm in diameter), which are composed primarily of $SO_4^=$, NO_3^-, OC, and BC. Fine soil particles and coarse particles of all compositions usually play much smaller roles, except in areas where wind-blown dust is a factor. Because particle sizes and compositions depend on the natures of the sources that emit the particles or their precursors, emissions from different sources will have different effects on visibility. Thus, although reducing ambient concentrations of PM will also yield benefits in visibility, the most effective approach to managing visibility may differ from the optimum one for the management of PM mass concentrations to protect human health at the same location.

9.1.1 How Is Visibility Distributed and How Has It Varied over the Years?

As was shown in Chapter 6, PM due to man's emissions is present to some degree over all of North America most of the time. As a result, visibility impairment caused by air pollution is not limited to cities with substantial PM pollution, such as Los Angeles and Mexico City, but is also present in rural and remote areas (even the high Arctic). The U.S. National Research Council reports that visibility impairment caused by air pollution occurs in varying degrees at many U.S. national parks virtually all of the time (NRC, 1993).

Extensive measurements of the fine PM species that cause light extinction have been carried out in rural areas of the United States for the past decade. Figure 9.2 shows the distribution of average light extinction over the United States, calculated from particle measurements made by the IMPROVE (Interagency Monitoring of Protected Visual Environments) network at national parks and wilderness areas. (The

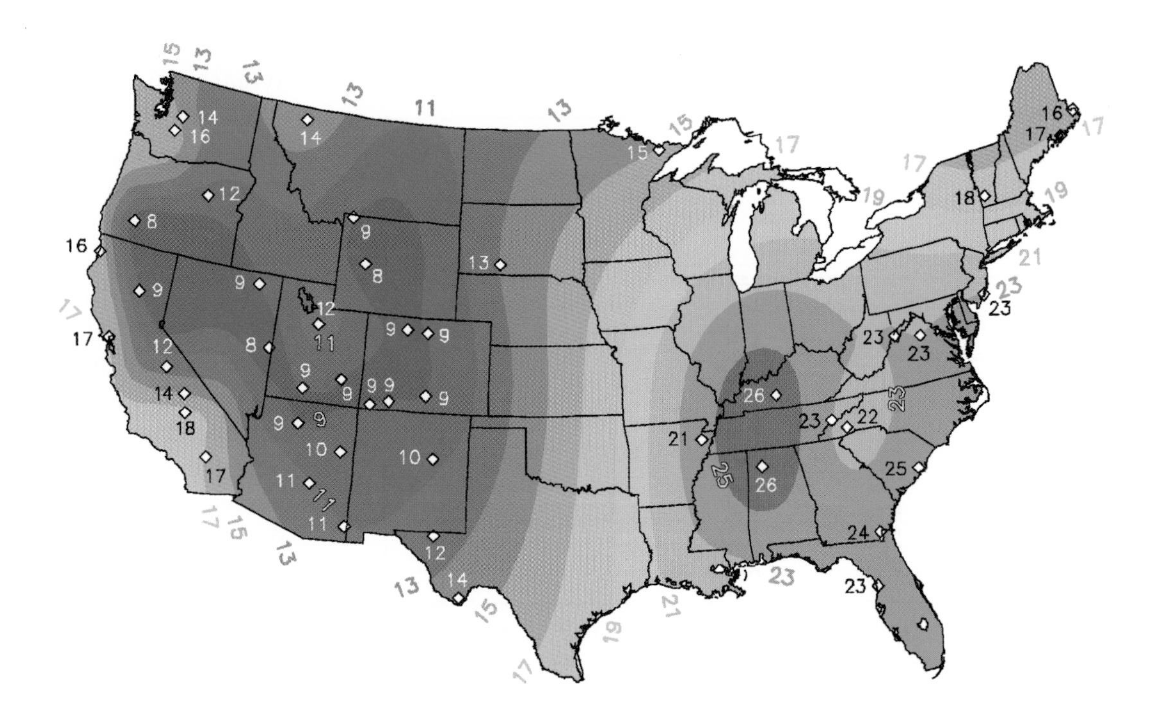

Figure 9.2. Distribution of average light extinction in national parks and wilderness areas of the United States, in units of deciviews, as calculated from particulate matter measurements in the IMPROVE network (Malm, 2000b).

methodology used for such calculations will be described below in Section 9.1.3.) This figure shows the dramatic difference between the relatively impaired visibility in the more humid and more populated East and the much clearer visibility in the arid and relatively unpopulated intermountain West. Visibility along the Pacific Coast is again poorer because of increased population density and, in the Northwest, higher humidity. Urban areas normally have greater PM concentrations and thus the average visibility there can be expected to be worse than that shown in Figure 9.2.

Figure 9.2 presents extinction in units of deciviews. The deciview (Pitchford and Malm, 1994) has come into vogue recently for representing extinction because a given increment in deciviews is assumed to be perceived approximately the same by a human observer, independent of the absolute level of extinction. For example, a change by 2 dV is assumed to be perceived similarly whether the original extinction is 5 dV or 25 dV. In analogy with the decibel scale for hearing, the formula for deciviews is $dV = 10 \ln(b_{ext}/10)$, where b_{ext} is in units of Mm^{-1}. Zero dV corresponds to Rayleigh scattering at an altitude of about 1500 m, and increasing values of the deciview index represent increasing light extinction, i.e., decreasing visibility. A difference of 1 dV corresponds to a difference of about 10 percent in the extinction coefficient.

Canada has a similar east-to-west spatial pattern of visibility as the United States. Figure 9.3 illustrates light extinction statistics averaged over Canadian regions, based on fine PM speciation measurements made in urban areas (McDonald, 2002). (Recall that Figure 9.2 represents rural areas.) As in the United States, the best visibility in Canada occurs in the mid-continental region, with a median for the prairie cities of about 13 dV. Urban centers of the west coast and of the maritime provinces of the east coast both have median visibility near 15 dV. Upper Canada, the region along the St. Lawrence River that includes Ottawa, Montreal and Quebec City, has a slightly hazier median level. Not surprisingly when compared with Figure 9.2, the Golden Triangle region of Canada, the populated region around the Great Lakes, shows the greatest light extinction with an average median near 20 dV. These urban values closely match the rural values shown south of the border in Figure 9.2.

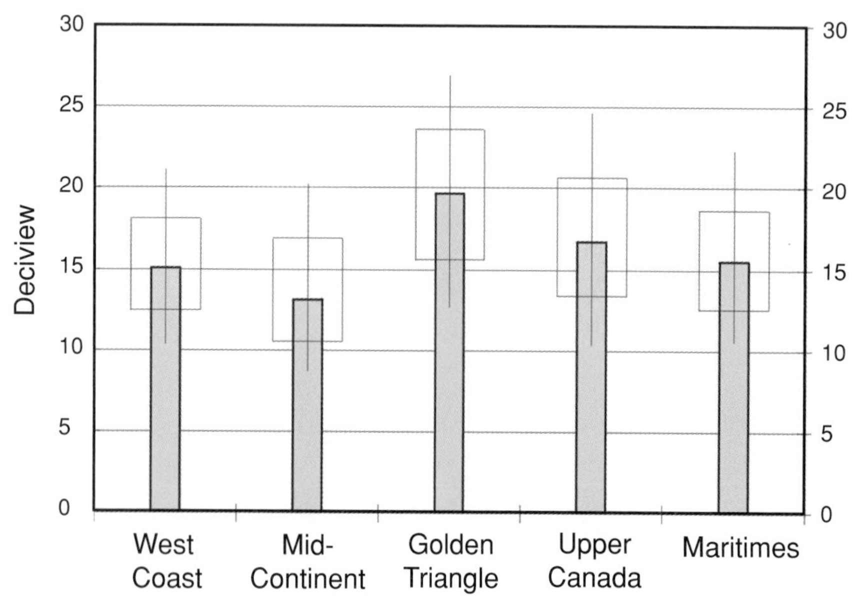

Figure 9.3. Light extinction statistics (west to east) in deciviews for Canadian urban regions, calculated from fine PM measurements. The height of the gray bar indicates the median. The 10th to 90th percentile range is described by the thin line and the 25th and 75th percentiles are defined by the box. (Based on McDonald, 2002).

Reported long-term visibility information for Mexico is limited to the Mexico City area. At two observation sites there, the mean visual range attributable to air pollution has been less than 10 km during the 1990s (Mora et al., 2001). In terms of the deciview index used in Figures 9.2 and 9.3, the corresponding extinction is greater than 37 dV. (As a comparison with another urban area, the light extinction in Washington, D.C., measured and calculated on the same basis as the information shown in Figure 9.2, is 24 dV.)

Visibility in much of the rural United States has generally improved in the past two decades, especially in the eastern half of the country. The widespread improvement in visibility reverses a trend of decreasing visibility during the preceding quarter century. This improvement is not uniform under all conditions, however. For example, although visibility generally has improved at most national parks and wilderness areas throughout the country between 1988 and 1998, the 20 percent of the days with the worst visibility are showing further degradation at some such locations (Sisler and Malm, 2000). The recent visibility improvement that has occurred is associated with a reduction in $PM_{2.5}$ concentrations throughout much of the country. It is largely attributable, in the eastern United States at least, to a decrease in $SO_4^{=}$-containing particle concentrations as a result of mandated reductions in SO_2 emissions from electric power plants in the mid 1990s.

Visibility-monitoring information in Canada is limited spatially and temporally, making an assessment of trends difficult. Airport visual-range observations have not been found useful for characterizing the whole range of visibility conditions in much of the country because visual ranges reported by the Meteorological Service of Canada often have a maximum of 15 miles (24 km).

Available data are not sufficient to evaluate visibility trends in most of Mexico. In Mexico City the visual range is reported to have decreased from 4 to 10 km in 1940 to 1 to 2 km in recent years, with 2 to 4 km of the decline occurring in the 1950s (Ciudad de México, 1999).

A full understanding of visibility throughout North America will require more extensive measurements in many areas. *Long-term monitoring of PM mass concentrations and composition, using consistent protocols, should be done throughout North America to assess spatial and temporal visibility trends.*

9.1.2 Factors Affecting the Relationship between PM and Visibility

The relationship between the light extinction coefficient and the PM mass concentration depends on the sizes and chemical compositions of the particles, and the ambient relative humidity. The roles of these factors and the relationships that result are discussed below.

Particle Size. The effectiveness of particles at scattering light depends on their sizes. Particles whose diameters are approximately the same as the wavelength of light are the most efficient, in terms of light scattering per unit mass (the scattering efficiency). As shown in Figure 9.4, particles whose diameters are between 0.3 and 1.0 µm tend to scatter the most light per unit mass. Since the human eye is most sensitive to light with wavelengths in the vicinity of 0.5 µm, these effects are very evident to human observers. Note that larger particles, such as those in the coarse mode with diameters between 2.5 and 10 µm, are not very efficient at scattering light.

Particles with diameters in the general vicinity of 0.5 µm also account for most of the mass in the fine-particle fraction ($PM_{2.5}$). As a result, the $PM_{2.5}$ mass-concentration corresponds to those particles that are most efficient at scattering light.

Chemical Composition. As has been described earlier in this report, the principal chemical constituents of particulate matter are $SO_4^{=}$, NO_3^{-}, BC, a broad variety of OC compounds, crustal (soil) compounds, water, and small amounts of some other species, such as potassium from vegetation burning, sea salt, and various trace metals. Figure 9.4 demonstrates that the scattering efficiency varies somewhat with chemical composition for a given particle size. Note that BC (labeled "Carbon-2" on the figure), does not scatter light as effectively as other species: it is very effective at absorbing light, however (which is not portrayed in Figure 9.4).

In actuality, the aerosol is not composed of particles of a single species, but rather is a mixture. It can be considered as externally mixed, in which each species is found in separate particles, or as internally mixed, in which several species are combined in a single particle. Actual ambient aerosol appears to be a combination of both types of mixtures. Particles formed at different times in different air masses may mix together to form an external mixture, while particles formed by secondary processes and condensation from a mixture of precursors may be internally mixed. This complex structure makes it difficult to calculate the light-extinction properties

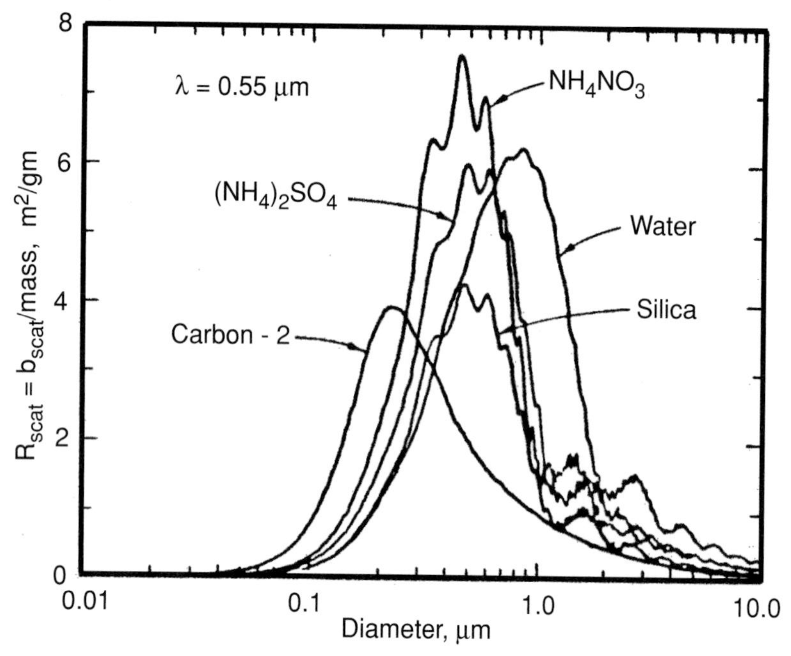

Figure 9.4. Calculated light scattering efficiency vs. particle diameter for several chemical species (Shah, 1981).

of individual real particles and, as discussed below, empirical estimates of extinction efficiencies are used instead.

As was described in Chapter 3, some of the material in ambient aerosol particles is semivolatile; that is, it can change phase from gas to particle, and back again, depending on ambient conditions such as temperature, relative humidity, and gas-phase concentration. The principal semivolatile species of interest for visibility are NH_4NO_3 (which is in equilibrium with gaseous HNO_3), the ammonium ion associated with some sulfates and nitrates (which is in equilibrium with NH_3 gas), and some organic compounds that are in equilibrium with volatile organic compounds (VOCs).

Humidity. An important factor that affects particle size, and hence the scattering efficiency of a species, is the water that is adsorbed by some hygroscopic species – sulfates, nitrates, and some organics. The amount of water uptake depends on the chemical species, the structure of the particle, and on the ambient relative humidity. At high relative humidity the size of a particle can be several times that of the dry particle, as was shown in Figure 3.4. This results in an even more dramatic effect on light extinction.

For example, Figure 9.5 illustrates the growth in light scattering with increasing relative humidity for pure NH_4NO_3, $(NH_4)_2SO_4$, and a mixture of the two, at two different particle sizes. The increase in scattering going from 40 percent relative humidity to 97 percent is nearly a factor of ten for the particles that are originally 0.3 μm in diameter. The relative increase in scattering for the 0.6 μm particles is a bit less because the initial scattering is already around the peak shown in Figure 9.4.

The influence of relative humidity on the growth of light scattering by particles that contain organic compounds is not as well quantified as it is for inorganic species. Furthermore, the physical location of the organic material in or on a particle may affect the particle's affinity for water. For example, some particles that originate in urban areas may have an organic film on their surfaces that inhibits the uptake of water by the hygroscopic salt inside. Thus, although a particle may be composed predominantly of a hygroscopic salt, the organic coating could reduce its effective hygroscopicity.

9.1.3 Empirical Relationships between PM and Visibility

Light extinction can be broken down into the sum of its scattering and absorption components as follows:

$$b_{ext} = b_{Ray} + b_{ag} + b_{sp} + b_{ap},$$

where

b_{ext} = light extinction coefficient

b_{Ray} = Rayleigh scattering (light scattering by molecules of air)

b_{ag} = light absorption due to gases (mainly NO_2)

b_{sp} = light scattering by particles

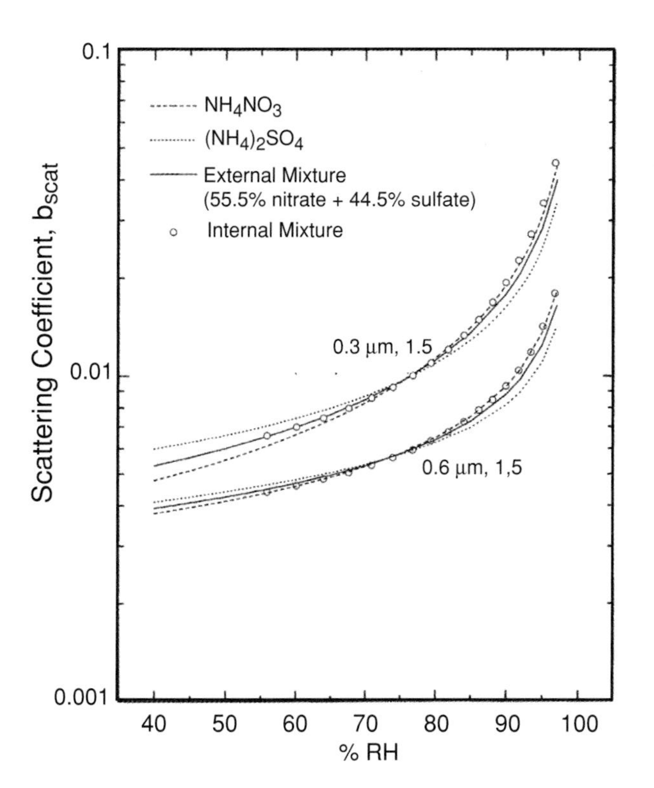

Figure 9.5. Effect of relative humidity on light scattering by mixtures of NH_4NO_3 and $(NH_4)_2SO_4$ (Tang, 1996). Reproduced by permission of the American Geophysical Union.

$$b_{ap} = \text{light absorption by particles.}$$

The scattering and absorption by gases (b_{Ray} and b_{ag}) can be calculated readily knowing the air pressure (altitude), temperature, and the concentration of NO_2. To deal with the particle-related extinction (b_{sp} and b_{ap}), one can allocate portions of the extinction to each species of the mixture and then sum the contributions to arrive at the total particle-caused extinction. This construction is sometimes called chemical extinction (Watson, 2002). Although there are complex theoretical considerations that limit the rigor of such an approach when assigning culpability to a particular species (White, 1986), the approach has been found useful for estimating the relative contributions of various species to the total light extinction.

The formulas typically used for chemical extinction are, for scattering of light by particles,

$$b_{sp} = e_{Sulfate}[\text{Sulfate}] + e_{Nitrate}[\text{Nitrate}] + e_{OMC}[\text{OMC}] + e_{FSoil}[\text{FSoil}] + e_{Other}[\text{Other}] + e_{Coarse}[\text{Coarse}]$$

and, for absorption of light by particles,

$$b_{ap} = e_{BC}[\text{BC}]$$

where

[Sulfate] is the concentration of fine sulfate-containing compounds

[Nitrate] is the concentration of fine nitrate-containing compounds

[OMC] is the concentration of fine organic carbon compounds

[FSoil] is the concentration of soil compounds in the fine fraction

[Other] is the concentration of other species in the fine fraction

[Coarse] is the concentration of all species in the coarse fraction

[BC] is the concentration of black carbon

e_X is the scattering efficiency (or, for BC, the absorption efficiency) of species X.

The scattering (or absorption) efficiency e_X is interpreted here as the scattering (or absorption) attributable to a unit concentration of species X. The value of e_{BC} sometimes also includes the contribution to scattering, which is small compared to the absorption. Here, fine particles are those that make up $PM_{2.5}$ and the coarse-particle concentration equals $[PM_{10}]$ - $[PM_{2.5}]$. Since the amount of material described as "Other" is typically very small, that term is often ignored.

Note that, in most cases, the concentrations used in the above formula are those of chemical compounds, not specific ions or elements. Since the full molecular form may not be measured, or even known, empirical factors are sometimes used to estimate the mass concentration of the compound based on ionic or elemental measurements. Examples include assuming $(NH_4)_2SO_4$ and NH_4NO_3 for $[SO_4^{=}]$ and $[NO_3^{-}]$, specific oxides for the soil compounds, and an empirical (and poorly quantified) factor to scale up the OC concentration to represent the contributions of hydrogen, oxygen, and other elements to [OMC]. (See Section 3.5.3 for a discussion of the carbon factor.)

The three equations above describe a light-extinction budget (LEB). This is the approach that was used to derive the light extinction values in Figures 9.2 and 9.3 from measured particle species concentrations.

Values for all the e_X have been developed by a variety of theoretical and experimental means. Extinction efficiencies are not fundamental constants, but rather vary with the size distribution, the properties of the particles, and the molecular composition of the sulfates, nitrates, and OC. Thus it is only a crude approximation to use a single value of e_X to represent the extinction efficiency at different locations or at different times. Also, as discussed, the e_X for $SO_4^=$, NO_3^-, and OC depend on the relative humidity. Various empirical formulas have been developed for sulfates and nitrates to represent the type of humidity-dependent behavior that was illustrated in Figure 9.4. There is little agreement on how to represent the humidity-dependent behavior of e_X for OC, however, although half the $SO_4^=$ or NO_3^- growth factor has been used as a compromise in some work (e.g., Malm et al., 1994). The current approach used by the U.S. IMPROVE network, which provided the information in Figure 9.2, is to assume that organics do not reflect any humidity-related growth. *Research is needed to develop a better understanding of the hygroscopicity of organic components of particulate matter.*

To illustrate the relative magnitudes of the various e_X, the predominant ranges of values of e_X deduced from various studies are presented below. The units are Mm^{-1} per $\mu g/m^3$, hence m^2/g. All values are for dry particles; recall that the contribution of water can multiply the given $SO_4^=$ and NO_3^- efficiencies many fold at high relative humidities, while the effect of water on the OC efficiency at any specific location is an open question. The ranges of prevailing dry extinction efficiencies listed below indicate the uncertainty that could result from using fixed extinction efficiencies (such as those used by IMPROVE, listed in parentheses) under varying circumstances.

$$e_{Sulfate} = 1.5 \text{ to } 4 \text{ m}^2/\text{g} \ (3 \text{ m}^2/\text{g})$$

$$e_{Nitrate} = 2.5 \text{ to } 3 \text{ m}^2/\text{g} \ (3 \text{ m}^2/\text{g})$$

$$e_{OMC} = 1.8 \text{ to } 4.7 \text{ m}^2/\text{g} \ (4 \text{ m}^2/\text{g})$$

$$e_{Fsoil} = 1 \text{ to } 1.25 \text{ m}^2/\text{g} \ (1 \text{ m}^2/\text{g})$$

$$e_{Coarse} = 0.3 \text{ to } 0.6 \text{ m}^2/\text{g} \ (0.6 \text{ m}^2/\text{g})$$

$$e_{BC} = 8 \text{ to } 12 \text{ m}^2/\text{g} \ (10 \text{ m}^2/\text{g}).$$

From the perspective of air-quality management, the above values show that the biggest visibility benefits per unit change in concentration are gained by reducing sulfates, nitrates, organics, and BC, because they have the largest efficiencies. In humid areas the visibility benefit of reducing sulfates, nitrates, and (perhaps) organics is even greater because of the effect of adsorbed moisture. On the other hand, it would take about five to fifteen times as great a reduction in coarse-particle mass concentration to equal the benefit of reducing the concentration of one of the four most efficient species by a given amount.

Based on three recent years of PM measurements made by the IMPROVE program (Malm, 2000b) and light-extinction calculations made using the formulas above, Figure 9.6 presents the proportion of rural particle extinction that is contributed by the individual chemical constituents (including associated water) in several regions of the United States. The positions of the pie diagrams are in a roughly geographic arrangement. The relative contributions of the various species in these diagrams illustrate implications for control options as a consequence of the combination of the relative magnitudes of the extinction efficiencies described above and the regional composition of ambient PM described in Chapter 6. (Figure 9.6 can be compared with a similar presentation for speciated mass concentrations in Figure 6.12.)

According to this figure, $SO_4^=$-containing particles and associated water dominate rural light extinction throughout much of the country, particularly in the eastern half, which implies that sulfur controls are a necessary part of most programs to improve regional visibility. The exception is southern California, where the proportion of extinction due to NO_3^--containing particles is greatly enhanced by NO_x emissions from nearby urban areas. (A similar enhancement of the NO_3^- contribution is observed at the IMPROVE sampling site in Washington, D.C., which is not shown here.) Organic-containing particles produce a greater fraction of the extinction in the West than in the East, offering another means for visibility improvement. Black carbon is a fairly consistent, but small, relative contributor across the continent, while the relative contributions due to fine soil and coarse particulate matter can be substantial in the more arid regions.

Although it is not shown in Figure 9.6, the composition for Denali National Park in Alaska looks very much like the one for the Colorado Plateau. The total extinction there, as represented by the sum of the species contributions, is a somewhat clearer 20 Mm^{-1}.

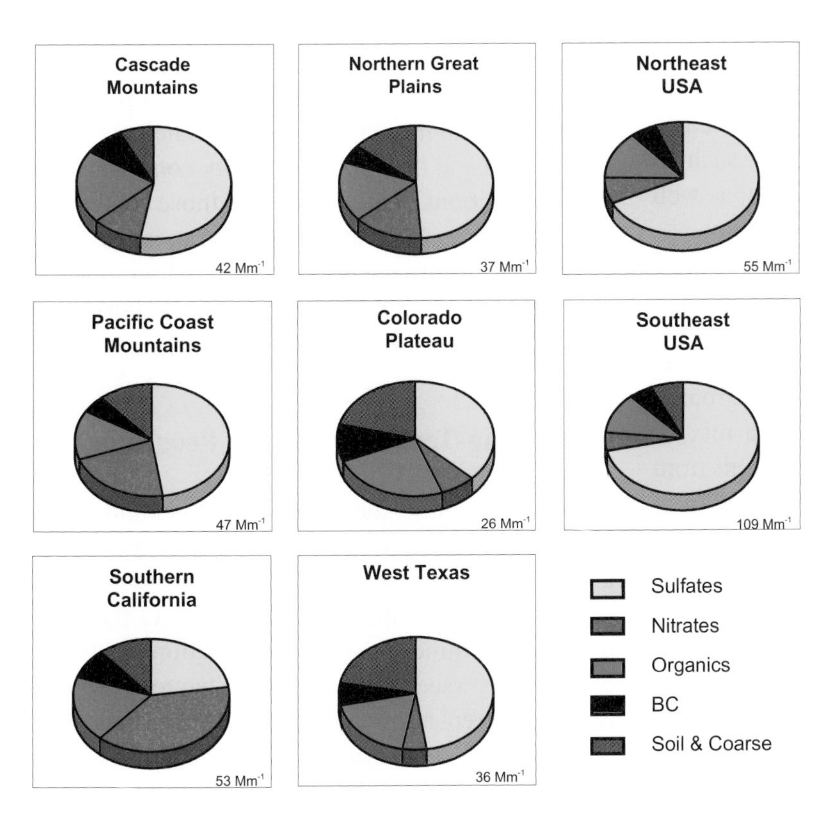

Figure 9.6. Relative contributions of PM constituents (including associated water) to particle-caused light extinction for several rural regions of the United States, calculated from IMPROVE PM measurements from March 1996 through February 1999. The numerals are the total chemical extinction (the sum of the component contributions) at each region. (Derived from data in Malm, 2000b).

9.1.4 What are Some Special Issues with Visibility?

The above technical discussion has shown that the linkage between visibility and particulate matter is a complex one. To set the stage for later discussion on visibility management, it is useful to lay out some special factors that have to be considered. They include the following:

- Under conditions that prevail in the atmosphere, some species (fine crustal particles and coarse particles of all compositions) are much less efficient, per unit mass, at scattering or absorbing light than are other species ($SO_4^=$, NO_3^-, OC, BC). This means that reductions in the concentrations of some species will be more effective at improving visibility than will reductions in others. Since the more-efficient species are mostly in the form of secondary particles, in many cases air-quality management strategies that are regional and may encompass many states or provinces are likely to be more important for regional haze than are local strategies.
- In humid areas, the light-scattering efficiency of $SO_4^=$, NO_3^-, and (perhaps) some organics is greatly enhanced. This puts even greater emphasis on control of these species for effective mitigation of visibility impairment.
- Visibility tends to vary seasonally by an even greater extent than PM concentrations, because the effect of particle growth under high humidity is superimposed on the factors of temperature, availability of oxidants, and presence of clouds that affect the formation of secondary PM.

- In the clearest, almost particle-free air, only a minute additional amount of fine PM (1 μg/m^3) can reduce visibility by 20 to 30 percent. Therefore, strategies to preserve visibility on the clearest days may require stringent limitations on emission growth, and dilute impacts from very distant sources can still be important.

- At the other extreme, in locations such as urban areas, where the concentrations of PM are higher and visibility is more limited, visibility is relatively insensitive to small changes in PM concentrations. Rather, substantial decreases in PM concentrations (typically greater than 20 percent) are required to achieve perceptible changes in visibility.

- Visibility is perceived instantaneously, so there is no averaging over time, as is appropriate when considering the health effects of PM. As a result, visibility-management strategies may need to consider mitigation of brief episodes, such as might be caused by passage of plumes, as well as the longer-term effects of regional hazes.

- The level of the natural background is important for those situations where, for example, the goal is to minimize the anthropogenic influences on scenic visibility. Determining that background, which varies daily and consists of a mix of particles of soil dust, products of emissions from vegetation and oceanic spray, and emissions from wildfires, is difficult since it is not feasible to separate out man's influence. In addition to the natural background, emissions from human activities throughout the globe and in neighboring countries also affect the overall visibility, but are beyond the reach of a single nation's visibility-management program.

- The actual appearance of a view depends not only on the extinction coefficient, but also on the color and contrast of the scene, the sun angle relative to the sight path (which changes during the day), and the presence of clouds in the sky. These factors mean that the human perception of different scenes, or of the same scene at different times, may differ even though the extinction coefficient and the PM concentration and composition are unchanged.

9.2 ROLES AND USES OF PM AND OPTICAL MEASUREMENTS IN VISIBILITY ASSESSMENT AND MANAGEMENT

Measurements and modeling are the two basic approaches for assessing visibility conditions and for developing strategies for managing visibility. Measurements serve to describe current situations and provide data for developing, testing, and applying air-quality models for visibility. Modeling, which is discussed in the next section, is used to forecast future situations and to estimate the effects of visibility management strategies. Modeling also can be used to fill in estimates of current data between measurement locations and times.

Measurements of pollutant-species concentrations, atmospheric optical parameters, meteorological conditions, and the appearances of scenes or objects all play roles in describing visibility conditions and the contributions of air pollution to those conditions. Methods for making many of these measurements have been discussed in Chapter 5. Their roles in characterizing visibility and the contributions of air pollutants are discussed here.

9.2.1 Long-Term Monitoring Programs

The earliest routine measurements of visibility were those of prevailing visibility by human observers at airports. A data base of more than five decades is now available for the United States, and has been used to determine the spatial distribution and trends of visibility (see, for example, Chapter 6 of this report). Current airport visibility data even have been used to characterize visibility over the entire globe (Husar et al., 2000). Automatic observations of airport visibility by the Automated Surface Observing System (ASOS) and the Automated Weather Observation System (AWOS) (see descriptions in Chapter 5) are replacing human observers in many locations, which will eventually bring an end to the human-observer data base in most jurisdictions

Since the focus of airport observations is for the safety of aircraft operations, there is sometimes an artificial maximum to the reported visibility. For example, the maximum prevailing visibility often reported at Canadian airports is 15 miles (24 km) and the new ASOS and AWOS light-scattering data at airports in the United States are routinely reported to a maximum visual range equivalent of 10 miles (16 km). These upper bounds in reported data limit the usefulness of these methods in areas with good visibility. However, some ASOS optical data are being archived at full instrument resolution. *ASOS and AWOS light extinction measurements from all locations, archived with a time resolution of one-hour average or shorter, should be made readily available for purposes of tracking air-pollution-related visibility.*

Regional long-term monitoring of aerosol and visibility began in the national parks and wilderness areas of the western United States in 1978-79 under the Western Fine Particle Network (WFPN) and the National Park Service's visibility monitoring network (Flocchini et al., 1981; Malm and Molenar, 1984). Information from these two networks provided a mapping of rural visibility and PM in the West.

The WFPN measurements, which did not include carbon, led initially to the conclusion that $SO_4^=$ played the dominant role in western U.S. visibility (Malm and Johnson, 1984). However, subsequent measurements of the PM carbon fraction, such as in the Western Regional Air Quality Studies (WRAQS; Tombach et al., 1987), indicated that these initial conclusions had overstated the relative contribution of $SO_4^=$ to extinction in the West (White, 1985; White and Macias, 1989).

Five years (1984-89) of detailed measurements were made in the U.S. Southwest, in the general vicinity of the Grand Canyon, by SCENES (Subregional Cooperative Electric Utility, Department of Defense, National Park Service, and Environmental Protection Agency Study; Mueller et al., 1986). In addition to providing much insight into the composition of the aerosol, SCENES also produced advances in optical and aerosol measurement technology, including improved means of quantifying atmospheric concentrations of semivolatile species (especially nitrates and organics).

A comprehensive characterization of Class-I area visibility and aerosol throughout the entire United States began with IMPROVE (Interagency Monitoring of Protected Visual Environments), which initiated operation in 1988 (Malm et al., 1994). Many of the advances in measurement technology developed by WRAQS and SCENES were incorporated into IMPROVE, which replaced the WFPN. IMPROVE was substantially expanded in 2000 (Pitchford et al., 2000) for the purpose of providing 5 years of baseline measurements for the U.S. national regional haze management program. One of the results of the IMPROVE measurements is a sufficiently long data base to infer trends, such as those that were discussed earlier in this chapter.

Because PM in the U.S. Southeast is different from that in the West, a multi-year program of measurements was initiated in the Southeast in 1998. The study, named SEARCH (Southeastern Aerosol Research and Characterization; Hansen et al., 2000; Jansen et al., 2001), is planned to run through 2005. SEARCH now includes optical measurements, coupled with in-depth, state of the art, particulate and gas-phase measurements, at eight urban and rural locations. For the purposes of understanding visibility in this section of the country, these measurements will supplement those of IMPROVE and will provide insight on the relationships between urban and rural conditions.

The U.S. EPA has recently initiated speciated fine-particle measurements, similar to those of IMPROVE, in both urban and rural areas throughout the country. These measurements will be useful for estimating visibility outside of the scenic areas that are the focus of IMPROVE. Some methodological differences mandate care when comparing measurements made by the two programs, however.

Measurements of PM and light extinction have also been made at a few locations in Canada, near the border with the United States (Hoff et al., 1997), under a program called GAViM (Guelph Aerosol and Visibility Monitoring). The PM measurements use part of the IMPROVE sampling system, and thus can be compared with U.S. measurements.

Such long-term measurement programs have provided insight into the causes of visibility impairment, its temporal and spatial variability, and its year-to-year trends over much of North America. Several of the studies have included, or currently include, research components that address specific issues concerning the physics and chemistry of the aerosol or of atmospheric optics. These research efforts have provided considerable insight concerning the challenging visibility-related issues of particle-size distributions, semivolatile species, hygroscopicity of particles, and estimation of particle light-extinction efficiencies. Until very recently, however, the visibility studies have focused almost entirely on rural areas, especially national parks and wilderness areas. Insight at the same level of detail is not yet available for urban visibility, except from a few of the short-term studies that are described below.

9.2.2 Short-Term Measurement Programs

A multitude of shorter-term studies (typically involving a year or less of measurements) has focused on specific visibility issues in the United States. Many of them have been concerned with determining the effects of specific emission sources or source regions on visual air quality, and thus demonstrate the application of measurement and modeling approaches to visibility management. There are too many studies of this type to describe them all here, but significant recent ones are summarized in Table 9.1.

Electric power-plant emissions were a major focus of interest of almost all of the short-term studies in Table 9.1. In fact, the last three studies in the table were followed by actions to reduce emissions from the targeted sources, and results of the research studies played a role in leading to these actions. One factor contributing to this emphasis may be that the point-source emissions from electric power plants are easier to identify than the dispersed emissions from the area sources that characterize urban areas and are the other major sources of visibility-impairing PM.

In addition to the studies noted in Table 9.1, major visibility-related urban studies have been carried out in Tucson, Arizona, in 1994; Phoenix, Arizona, in 1990; and Denver, in 1982 and 1987. The 1987 South Coast Air Quality Study (SCAQS) in the Los Angeles basin also had a visibility component.

Short-term studies focused on urban areas and their surroundings have also taken place in Canada and Mexico. In Canada, the area in the vicinity of Vancouver, on the Pacific Coast, has been the subject of several air-quality studies that have included a visibility component. The most recent such study took place in September 2001. In Mexico City, a short-term study in 1997 measured both the chemical composition of $PM_{2.5}$ (Chow et al., 2002) and light scattering and absorption (Eidels-Dubovoi, 2002).

9.2.3 An Example of a Scenic Visibility Setting – The Colorado Plateau

As an example of the understanding provided by studies such as those described above, a description of the aerosol and visibility situation in the Colorado Plateau area of the western United States is provided here. The Colorado Plateau, which lies west of the Rocky Mountains, contains Grand Canyon National Park and several other well-known scenic areas for which good visibility is considered a national asset. The description provided here for visibility can be compared with those for PM presented in Chapter 10. Much of the aerosol and visibility information presented here is found in Malm (2000b).

The average visibility on the Colorado Plateau, in terms of light extinction, is about 9 deciviews (25.6 Mm^{-1}), as was indicated in Figure 9.2. This corresponds to an average visual range of about 150 km. Although conditions vary over the wide geographic extent of the Colorado Plateau, the annual average is relatively uniform over the region. The 20 percent of the days with best visibility average in the range from 4 to 6 dV (about 15 to 19 Mm^{-1}), while the 20 percent of the days with the poorest visibility average in the range from 11 to 15 dV (30 to 45 Mm^{-1}).

The average $PM_{2.5}$ concentration in this region is 3.0 $\mu g/m^3$, a value that is well below the U.S. national annual $PM_{2.5}$ standard of 15 $\mu g/m^3$. The average concentration of coarse mass (PM_{10} – $PM_{2.5}$) is 4.3 $\mu g/m^3$. The principal components of $PM_{2.5}$ are $SO_4^=$ (35 percent) and OC (32 percent), with fine soil contributing 22 percent. Nitrates and BC each account for around 6 percent. The composition of the coarse

Table 9.1. Recent short-term visibility measurement programs.

Study Name, Year Of Observations, And Location	Description And Comments	References
CRPAQS (California Regional $PM_{10}/PM_{2/5}$ Air Quality Study), 1999-2000. Central California.	Regional characterization and modeling of PM and visibility, and the emission sources that contribute to them. Data analysis is ongoing.	Magliano and McDade (2001)
BRAVO Study (Big Bend Regional Aerosol and Visibility Observational Study), 1999. Texas.	Characterization of aerosol and visibility in the vicinity of Big Bend National Park, tracer study of atmospheric transport, and modeling of contributions of source regions in Texas and Mexico. Conclusions available in 2003.	Green et al. (2000)
Big Bend National Park Regional Visibility Preliminary Study, 1996. Texas and Mexico.	Bi-national cooperative measurement program in both Texas and Mexico in 1996. Precursor to BRAVO.	Big Bend Air Quality Work Group (1999); Gebhart et al. (2000)
SEAVS (Southeastern Aerosol and Visibility Study), 1995. Great Smoky Mountains National Park.	Research to characterize the chemical and optical properties of the aerosol in the Southeast. Concluded that more reliable techniques are needed for measuring OC in particulate matter.	Andrews et al. (2000)
Dallas-Ft. Worth Winter Haze Project, 1994-95. Northeastern Texas.	Study of wintertime visibility and aerosol in the Dallas-Ft. Worth area and assessment of the contributions of distant coal-burning power plants to the haze there. Concluded that the plant emissions contributed "subtly" to the haze	Tombach et al. (1996); McDade et al. (2000)
Mt. Zirkel Visibility Study, 1995. Yampa Valley, Colorado.	Study of the causes of visibility impairment at the Mt. Zirkel Wilderness Area with elevations above 3000m. Emphasis on the contributions of two coal-fired power plants some 50 km distant. Concluded, based largely on a videotaped observation, that the two power plants contributed at times to puffs of haze in the wilderness area, but that the haze was generally of more regional origin.	Watson et al. (1996)
Project MOHAVE (Measurement of Haze and Visual Effects), 1992. Southern Nevada and northern Arizona, including Grand Canyon National Park.	Field and modeling study of the effect of emissions from the coal-fired Mohave Generating Station and other sources on visibility in Grand Canyon National Park. Concluded that the generating station was generally an insignificant contributor to visibility impairment at the Grand Canyon, which is mostly attributable to other sources, but that the plant's contribution could occasionally be important.	Pitchford et al. (1999)

fraction has not been studied routinely, but is presumed to consist mainly of soil.

The contributions of these species to light extinction have been displayed in Figure 9.6. At the Colorado Plateau, $SO_4^=$ accounts for 37 percent of the annual extinction due to particles (i.e., excluding Rayleigh scattering), followed by OC at 24 percent. These fractions differ from those for mass because of the enhancement of $SO_4^=$-particle scattering by moisture. The combination of fine soil and coarse particles accounts for another 21 percent. Black carbon accounts for 11 percent and NO_3^- for 7 percent.

The clearest days in the region tend to occur when there is transport from the northwest. Those air masses traverse a region of low emissions, and the concentrations of pollutants are further reduced by the aggressive atmospheric mixing that is typically associated with air masses arriving from that direction. The more typical transport is from the southwest. Since that transport route passes over the urbanized areas of southern Nevada, southern California and northern Mexico, transport from the southwest can be associated with poor visibility whenever the atmospheric mixing is limited. Poor visibility conditions can also occur on the Colorado Plateau during the monsoons of late summer, when transport from the south, passing over urban areas, smelters and power plants in Arizona, New Mexico, and Mexico, is combined with high relative humidity.

In addition to the contributions due to transport of pollution from distant urban areas, sources of air pollution in and near the Colorado Plateau include emissions from several coal-fired power plants, and local generation of fugitive dust from disturbed soil and motor-vehicle travel in this arid region. Also, prescribed and wild vegetation fires cause substantial episodic air-pollution and haze events.

The power plants can produce plumes that are visibly brownish-yellow due to light absorption by NO_2, but their impact on regional visibility is due to the formation of $SO_4^=$ from SO_2. Although the rate of formation of $SO_4^=$ is slow in the clear skies that are typical of this region, several studies have indicated rapid formation of $SO_4^=$ and haze when the emissions interact with moisture in clouds. The major coal-fired power plants in the region have installed SO_2 control systems in recent years, or are scheduled to do so in the next few years, so one should expect decreases in the $SO_4^=$ contribution to visibility from the amounts that were described above.

As part of the U.S. regional-haze program, the Western Regional Air Partnership (WRAP) has developed a strategy to continue to reduce power-plant emissions and is currently developing strategies to mitigate emissions of fugitive dust and fires that affect the Colorado Plateau. Additional progress in protecting visibility will have to come from emission-control efforts in urban areas in the United States and Mexico.

9.2.4 Can One Use PM Studies for Visibility?

All of the studies described in this section have one common feature – they were specifically designed, at least in part, for the purpose of addressing visibility impairment. Since PM measurements for the purpose of complying with health-based standards or for research on the health effects of PM are more common than those for visibility, the question arises as to whether one can use the information from a PM measurement program for visibility purposes. The answer is a qualified "Yes" - provided that the measurements include the concentrations of the optically-important species in $PM_{2.5}$ and care has been taken to minimize sampling artifacts for semivolatile species. Measurements of the mass concentrations of $PM_{2.5}$ and PM_{10} without speciation, which are the primary measurements of most routine PM monitoring programs, are not of great use for visibility characterization.

If the species concentrations are available, they can be used to estimate the allocation of light-extinction to each species, following the light extinction budget method described in Section 9.1.3. Measurements of the relative humidity and of the coarse-fraction mass concentration are also needed to carry out this process. (These data are also useful for the air-quality modeling that is described in the next subsection.) More rigorous calculations of light-extinction contributions require measurements of the particle-size distributions in the fine fraction, however, which usually are not available in most PM studies.

9.3 HOW ARE MODELS USED IN VISIBILITY MANAGEMENT?

Visibility models produce estimates of the optical effects of simulated or measured distributions of PM and gases. The typical output is the spatial distribution of the light-extinction coefficient or a description of the color and opacity of a plume. Some visibility models also include modules that emulate aspects of the response of the human eye-brain system to these optical effects, and thus describe the effects of the PM on the appearance of an actual scene.

Visibility models are used for much the same purposes as other air-quality models. One application is to simulate present and past conditions where optical or visibility measurements were not made or are not feasible, usually for the purpose of better understanding the factors that affect current visibility. Such modeling can also fill in spatial and temporal gaps in measurements. A second application is to simulate future conditions and thus to project future trends in visibility, usually for the purpose of developing and evaluating air-quality management strategies. Another related purpose is the assessment of impacts of a proposed new source or benefits of applying emission controls to existing sources.

9.3.1 What Specific Features are Required when Modeling Visibility?

Visibility modeling commonly starts with application of a chemical transport model (CTM), such as described in Chapter 8. (The exception is when the optical or visual effects of a measured actual aerosol are to be calculated.) A plume model is the starting point for analyzing a plume-blight situation, while a three-dimensional CTM would be used for regional haze. The minimum outputs needed from the CTM are spatial distributions of PM species concentrations (both fine and coarse), NO_2 gas concentrations, and relative humidity. The chemical species of interest for visibility are described in Section 9.1.

The simplest visibility models use the species' concentrations, together with assumed scattering and absorption efficiencies, to arrive at local descriptions of extinction through the empirical formulas given in Section 9.1.3. More advanced models calculate particle scattering based on the physics of the interaction of light with particles. Such aerosol-optics models also require information concerning the particle-size distribution of the aerosol, which generally has to be estimated based on specialized measurements. Since not all of the above information is produced routinely by CTMs, the intended use of the CTM outputs for visibility modeling must be taken into account when setting up the modeling exercise. One should note that models developed for PM_{10} alone or those that deal only with primary PM do not provide an adequate foundation for visibility modeling.

Visibility is of concern at all time scales, from instantaneous observations to averages over periods of years. Because visibility management tends to depend on measurements of PM, and these are typically made over 24 hours in routine networks, modeling with hourly resolution (and averaging over 24 hours) is generally the shortest time frame required. At the other extreme, there is a need to characterize conditions over longer periods, such as the annual averages used in some Canadian analyses and those 20 percent of the days in a 5-year period that reflect the worst visibility that are used in the U.S. regional haze program. For practical reasons, simulations over such long periods of time typically involve compromises – either a simplified model with coarse spatial resolution and/or simplified chemistry has to be used, or the longer period is constructed as a combination of a limited number of shorter periods that reflect the range of atmospheric or emission conditions of interest.

Ideally, light extinction is calculated directly through mechanistic models that consider the effects of individual particles on the transmission of light. Such calculations are sensitive to the size distribution of the particles. Chapter 8 notes that the ability of models to construct size distributions of secondary particles or to use detailed size-distribution information explicitly in the simulation is currently embryonic, and that few successful CTM simulations have been reported with size-resolved and/or chemically resolved PM anywhere in North America. Therefore, the capability of using a CTM and then specifically modeling the light extinction caused by the predicted PM is currently quite limited.

However, as was noted in Section 9.1.3, light extinction can be estimated from speciated chemical concentrations and assumed extinction efficiencies. This simplified approach eliminates the need for detailed information about the particle-size distribution, and thus places fewer demands on the CTM. Such an estimating technique for determining light extinction is used in several air-quality models, as described below, and is the basis for determining light extinction from both modeled and measured PM concentrations under the U.S. EPA's Regional Haze Rule.

9.3.2 Are Current Models Able to Simulate Visibility Conditions?

Different modeling approaches are used for simulating the visibility effects of point sources and of multiple sources over a region.

For point sources, modeling the appearance of a coherent plume and its effect on visibility (typically within 50 km of the source) requires a different method than the modeling of the visibility impact of a more dispersed plume at greater distances. For regulatory applications, the U.S. EPA has established detailed procedures for modeling the dispersion and optical effects of a plume and the PLUVUE-2 model is recommended for treating the effects of a coherent plume on contrast and color.

For greater distances from the source, the U.S. Interagency Workgroup on Air Quality Modeling (IWAQM) has recommended that, for regulatory purposes, the CALPUFF model be used to simulate long-range transport from point sources. CALPUFF simulates the dispersion of emissions from point sources and includes simple representations of the formation of $SO_4^=$ and NO_3^- from SO_2 and NO_x. CALPUFF is being proposed by the U.S. EPA as a preferred model for simulating long-range (>50 km) PM impacts from point sources. Furthermore, CALPUFF is the recommended tool for managers of national parks and wilderness areas to use for assessing the long-range visibility impacts of proposed point sources (FLAG, 2000). Using CALPUFF for modeling the combined effects of many sources of different types becomes cumbersome, however, and its chemistry calculations are too simplistic to deal with interactions between emissions from different sources.

Other plume-aerosol and visibility models, applicable over a wide range of travel distances and with more advanced simulation of the diffusion and chemical processes than either PLUVUE-2 or CALPUFF, are also available. Examples include the Reactive and Optics Model of Emissions (ROME; Seigneur et al., 1997) and the second-order closure puff model SCICHEM (Karamchandani et al., 1999).

Simulating regional distributions of visibility requires a three-dimensional CTM. The set-up and input needs for determining PM concentrations with this kind of model are significant, as described in Section 8.2. The additional calculation of light extinction can increase this complexity substantially if detailed aerosol optical calculations are made, or it can require relatively little additional resources if empirical light-extinction efficiencies of the species are used. For visibility, as with the PM calculations, calculation of the inorganic PM contribution is significantly more straightforward and certain than the calculation of the secondary organic contribution.

A regional air-quality model system that has been developed with visibility applications in mind is the Community Multiscale Air Quality (CMAQ) Modeling System developed by the U.S. EPA. CMAQ can explicitly calculate light extinction from particle properties (Mie theory approach) or via the simplified approximate approach of using species concentrations and extinction efficiencies. CMAQ has been evaluated against measured visibility data for a summer season in eastern North America (Binkowski et al., 2001) and recent evaluations have been made by WRAP for the western United States and in BRAVO for the Texas area. These evaluations have shown that, at least with current versions of the model, the predictions of extinction tend to be less than the observations, though spatially- and temporally-averaged spatial patterns and increments of change are sometimes comparable to observations. Most other regional PM models (e.g., AURAMS, PM-CAMx, and the urban-scale model UAMV-PM) do not incorporate a module for the explicit calculation of light extinction, although some others (e.g., REMSAD) estimate extinction using the products of species concentrations and extinction efficiencies.

In addition to the physical and chemical modeling of source emissions, receptor-modeling and statistical techniques (such as factor-analysis methods) have been used to relate observed PM concentrations to source categories or regions of emissions, as described in Chapter 7. These methods can be extended to visibility. Such methods allow for easy identification of spurious events such as dust storms and forest fires. For the secondary PM that is important for visibility, however, assumptions may need to be made about conversion rates in order to apply such methods. If information is also available on meteorological trajectories, then the results of these techniques can be refined to give stronger estimates of the source apportionment for secondary PM and, hence, visibility.

A special form of visibility modeling uses the aerosol and optical information from a CTM or from measurements, calculates the effect of the aerosol on the contrasts and colors in an actual scene, and displays the results in a computer-modified photograph (Molenar et al., 1994). Such computer-generated images approximate how the scene will appear to the human eye, and thus provide a concrete visualization of the abstract concept of light extinction. Although the analyses and calculations involved for simulating one view are quite extensive, results can be demonstrated using simplified software that is available for use on a personal computer (WinHaze, 2002). Using slightly simplified algorithms, this software has been applied to demonstrate the effects of varying extinction levels on a number of views at several national parks and wilderness areas in the United States and in some other locations.

The simulation of the human perception of actual scenes by using photographs of computer images is not perfect, however. Based on color-matching experiments, Henry (1999) points out that such images are less colorful and bluer than the true scene, and therefore the artificial images overstate the visual effects of increasing haziness. Nevertheless, such images are useful for portraying the essential visual effects of extinction and the relationship to PM concentration, albeit only semi-quantitatively.

9.3.3 What Would Improve the Capacity to Model Visibility?

Models developed and tested for the purposes of estimating primary and secondary $PM_{2.5}$ concentrations are suitable for application to the visibility question, provided that there is also a means for calculating the optical effects of the PM, either within the model or in an external post-processor. Since visibility itself is an instantaneous response to atmospheric conditions, the basis for the model calculations should be short period (e.g., hourly) results, which then can be averaged for longer periods. Also, since current visibility regulatory programs are based on averages over many days (e.g., in the United States, over the haziest 20 percent of the days in a 5-year period) and the days of importance may be different for different locations in the modeling domain, the model's computing time needs to be taken into consideration.

Dealing with the critical uncertainties in CTMs for PM, which are described in Section 8.10, is equally relevant for modeling visibility. Particularly important concerns are descriptions of the meteorological field and shortcomings in emission inventories. In addition, there are some unique aspects to the needs for visibility modeling.

As one example, improvement in modeling of secondary organic PM will be important to visibility, both in urban areas and in those pristine areas that may have significant emissions of natural VOCs. One specific concern for regional visibility modeling in scenic areas is the validity of the gas-particle partitioning simulations, when applied in a clean environment rather than the polluted urban one for which they have typically been developed. Models that rely on condensibility through absorption will have difficulties due to the low PM concentrations found in pristine environments and also may not properly address increases in organic-particle growth with changes in humidity. Because of the necessary treatment of water chemistry, models whose approach addresses the water solubility of VOCs are most appropriate for visibility. Current models with sound treatment of water chemistry and ability to model condensational growth include, for example, GATOR, CMAQ and AURAMS.

Also, conditions on the 20 percent of the days with best visibility play an important role in the implementation of the U.S. regional-haze program. Many of those days have conditions very close to natural background. Natural background is problematic for modeling, because it derives partly from transport from outside the modeling domain and partly from geogenic and biogenic emissions within the domain. Furthermore, PM models are not optimized to simulate clean days. Rather, in order to predict or evaluate urban PM episodes, PM models are designed to simulate those days with the worst conditions. Testing remains to be done to determine how well those models perform when concentrations are representative of natural background or the days with best visibility.

9.4 ATMOSPHERIC PARTICLES AFFECT THE GLOBAL RADIATION BALANCE

Particles in the atmosphere influence the balance between the solar radiation that reaches the earth's surface and the infrared radiation that is transmitted back into space. This process occurs both directly (direct forcing), as the particles scatter and absorb incoming solar and outgoing infrared radiation, and indirectly (indirect forcing), as the particles influence cloud formation and precipitation, which in turn affect the radiation balance. Changes in the radiation balance due to changes in the amount and type of PM, natural or anthropogenic, are believed to contribute to climate change through a net cooling of the atmosphere. This process is described in detail in a report of the Intergovernmental Panel on Climate Change (IPCC, 2002).

During the past decade, significant steps forward have been made by the climate-change scientific community in elucidating the role of aerosols in the global climate balance. The second assessment report of the IPCC (1996) documented a high level of uncertainty associated with attributing radiative forcing to natural and anthropogenic aerosols. The main causes were a lack of knowledge of the vertical and horizontal spatial distributions of the particles in the atmosphere, the particle-size distributions, their chemical composition, and the arrangement of components within individual particles.

The IPCC's Third Assessment Report (IPCC, 2002) indicates that progress has been made in addressing these issues. Although much uncertainty remains concerning the contributions of atmospheric aerosols, progress has occurred with emission inventories, scaling-up of measured data, and development of global scenario modeling capacity. Field studies, network monitoring, and satellite analyses have provided both process-level understanding and a descriptive understanding of the aerosols in different regions. Three-dimensional global-scale aerosol models have been developed for carbonaceous particles from biomass and fossil fuel burning, as well as for dust, sea salt and NO_3^- and NH_4^+ in secondary particles. Indirect forcing remains subject to much larger uncertainties than direct forcing, which reflects uncertainties in the understanding of the effects of particles on the size distribution and number of droplets within a cloud.

Perturbations to the energy balance of the planet are usually expressed as radiative forcing in units of Watts per square meter (W/m^2), with positive values representing warming. Figure 9.7 summarizes the current understanding and level of uncertainty attached to the influences of greenhouse gases and atmospheric particles over the last 250 years. Carbon dioxide and the other well-mixed greenhouse gases represent the greatest forcing at +2.4 W/m^2 with a high level of scientific understanding. Net increases in PM concentrations are estimated to counter this forcing (i.e., they contribute to cooling) by anywhere from a negligible amount to more than about –2.5 W/m^2. As this large range suggests, the uncertainties relating to aerosol radiative forcings remain very large, mainly because the values rely to a great extent on estimates from global modeling studies that are difficult to verify.

Figure 9.7 presents a first-order perspective on a global, annual-mean scale. In actuality, all the forcings have distinct spatial and seasonal features. For example, Figure 9.8 illustrates that the warming trend in the temperature of the atmosphere in North America tends to be less in the areas with the greatest PM concentration levels, especially sulfates. The figure shows that long-term average temperatures in the North Atlantic region have decreased while the average temperatures have increased over most of Canada and the United States, with the greatest increases over the central plains of Canada and the arctic regions. Temperatures have remained constant or have cooled slightly in most of the southern part of the United States and northern Mexico. Comparing Figure 9.8 with Figure 9.2 reveals that there is a tendency for temperature increases to have been greatest in areas with relatively good visibility (i.e., low particulate matter levels) and less in areas with poorer visibility (such as the southeastern United States).

Many of the issues that are important to, and being addressed by, the radiation-balance community are identical or similar to those that are important to the description of horizontal visibility. There are important differences, however. Light scattering and absorption both degrade visibility. Absorption tends to contribute to warming (as indicated by the black carbon portion of the fossil fuel bar in Figure 9.7). The direct effect of light scattering is cooling (as reflected in the sulfate, organic carbon, and biomass burning bars), albeit by a small amount compared to the warming effect of greenhouse gases. Therefore, strategies to mitigate visibility impairment may not offer benefits in mitigating climate change. Since the most severe visibility effects tend to occur on scales that are relatively localized compared to the overall geographic scale of the global radiation balance, it should be possible to develop pollution-management strategies that take into account both visibility and radiation balance, but attention will have to be paid to developing an integrated strategy.

Table 9.2, which summarizes the responses of climate and visibility to changes in emissions of various pollutants, illustrates the areas where the climate and visibility responses differ. (See Table 3.2 for similar assessment of responses in other air-quality metrics.)

Emission-inventory activities in support of continental or global outlook analysis and strategy development are underway at the international level. Examples of inventory analysis include those by the Canada-U.S. Air Quality Agreement, the North American Free Trade Agreement (NAFTA) environmental protocols, the United Nations Economic Commission for Europe, and the Great Lakes Water Quality Agreement, as well as the IPCC. These inventories are also of benefit to understanding the background level of haze.

The IPCC (2002) document makes five specific recommendations directed at the atmospheric aerosol issue, which are paraphrased below. They may benefit the air-pollution community in North America as well.

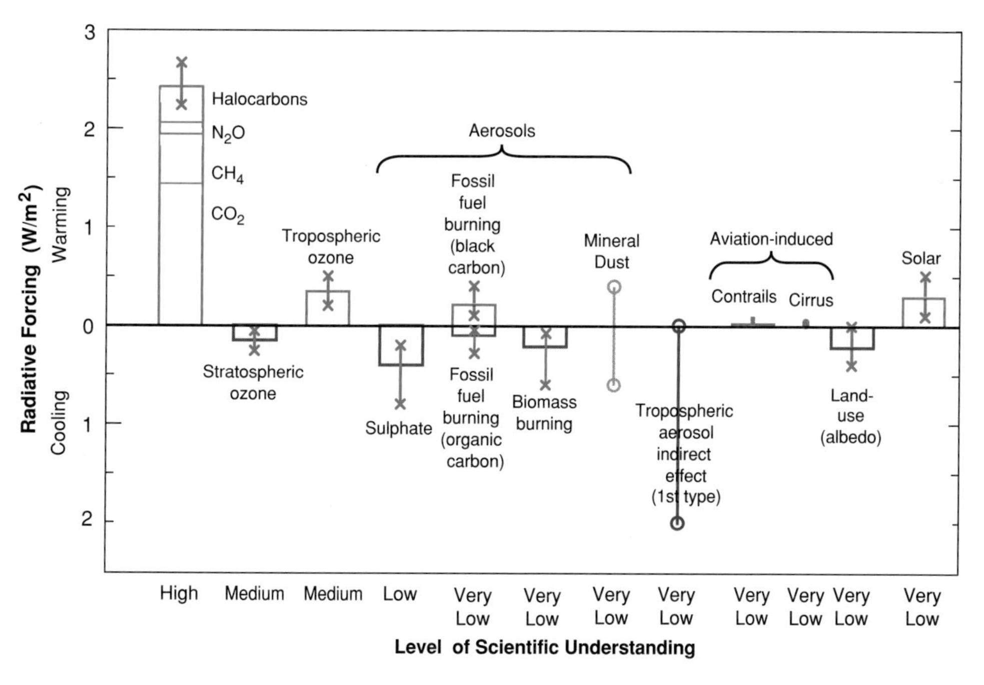

Figure 9.7. Global, annual mean radiative forcings (W/m^2) for the period from pre-industrial (1750) to the present. The height of each rectangular bar denotes a best estimate value while its absence denotes no best estimate is possible. The vertical line about the rectangular bar with "x" delimiters indicates an estimate of the uncertainty range. A vertical line without a rectangular bar and with "o" delimiters denotes a forcing for which no central estimate can be given owing to large uncertainties. A level of scientific understanding is accorded to each forcing, with H, M, L and VL denoting high, medium, low and very low levels, respectively. The figure shows that the dominant effect of atmospheric particles is a negative forcing (i.e., cooling). (IPCC, 2002; reproduced with permission).

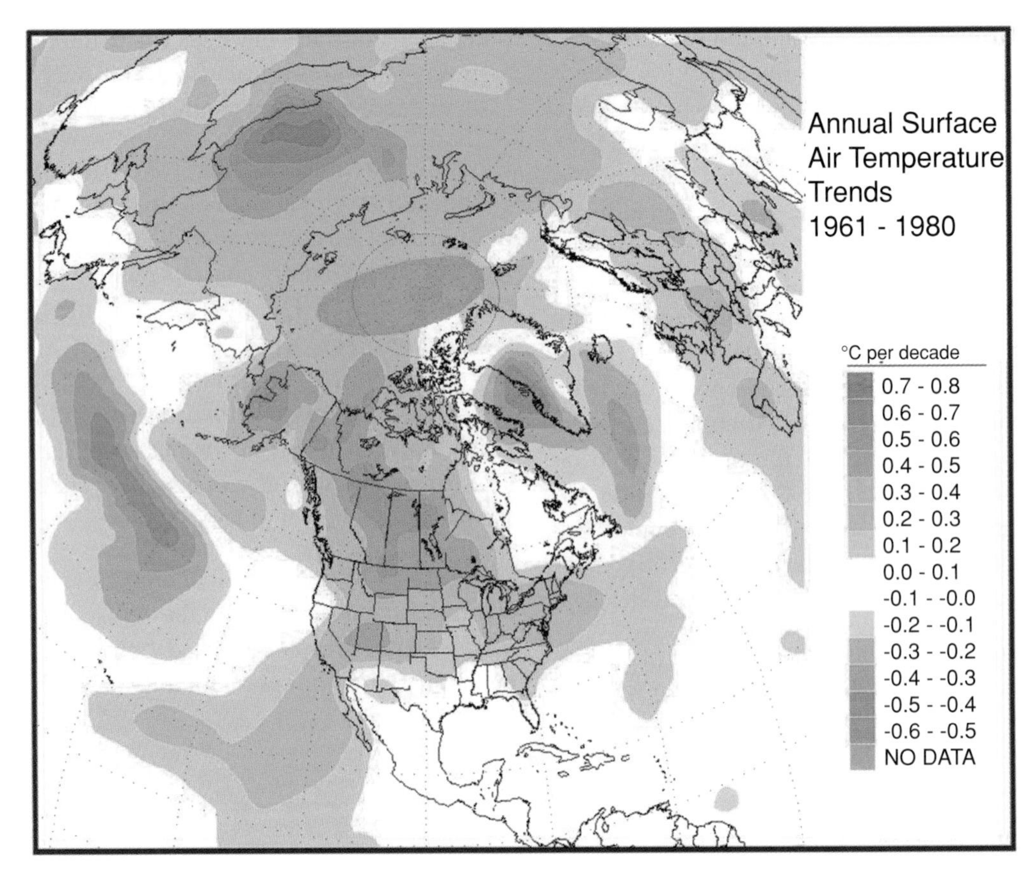

Figure 9.8. Measured change in annual surface air temperature 1961-1980. (Environment Canada, 1995).

Table 9.2. Responses of regional haze and climate to reductions in the emissions of secondary PM precursors and primary PM from present-day levels.

	Change In Associated Issue	
Pollutant Emitted	**Regional Haze**[a]	**Climate Impact**[c]
SO_2	↓	↑
NO_x	↑↓ [b]	↑↓
VOC	↑↓ [b]	↑↓
NH_3	↓	↑ [d]
Black Carbon	↓	↓
Primary Organic Compounds	↓	↑
Other primary PM (crustal, metals, etc.)	↓	↑

[a] Direction of arrow indicates increase (↑) or decrease (↓) and color signifies undesirable (red) or desirable (blue) impact; size of arrow signifies magnitude of change. Small arrows signify possible or small change.
[b] No change if little NH_3 available in atmosphere.
[c] Direct effects only; indirect effects through clouds and precipitation are highly uncertain. Note that the extent and possibly the scale of climate impacts for listed pollutants is quite different from those for CO_2 and CH_4, Direction of arrow indicates warming ↑ or cooling ↓.
[d] More accurately, decreased aerosol-induced cooling

1. Countries of the world need to develop and support a network of systematic ground-based observations of physical and chemical properties of atmospheric aerosols that are closely coordinated with satellite observations of aerosols.

2. Measurements of size-segregated PM physical, chemical, and optical properties are required through the atmospheric profile.

3. Characterization of aerosol processes through integrated experiments is needed in specified locations to quantitatively constrain models of aerosol dynamics and chemistry.

4. Studies to elucidate the indirect forcing of aerosols through solid understanding of cloud and precipitation processes are needed for development of a detailed microphysical representation of clouds for atmospheric models.

5. Measurements of aerosol characteristics from space could yield measurements of aerosol optical depth over land, reducing the uncertainties in current aerosol models.

Several examples illustrate the benefit of the interplay between research on visibility and radiation balance. For instance, although the radiation-balance work is more concerned with the distribution of aerosols through the depth of the atmosphere, the characterization of the transport of aerosols across the boundary layer means that information is also available on the layer or layers nearest the earth's surface, where visibility is an issue. Also, models that produce descriptions of regional particle burdens and atmospheric measurements of mass concentration, scattering and absorption coefficients, and water-uptake effects have direct implications for both radiation balance and visibility. Furthermore, global climate models are employing the calculation of optical properties from size distributions (Vignati et al., 2000), a process that is also of interest for visibility modeling.

Because of such linkages, enhanced scientific interactions between the radiation-balance and atmospheric-visibility communities would be of benefit to both communities. Such interactions could include coordinated measurement and modeling programs and joint scientific conferences (such as the specialty conferences held about once every three years under the joint sponsorship of the Air & Waste Management Association and the American Geophysical Union).

9.5 VISIBILITY MANAGEMENT ISSUES AND APPROACHES

9.5.1 What Is Being Done to Manage Visibility?

Management of air quality for the purpose of controlling visibility is being handled differently in the three North American countries. Of the three countries, the United States has the most advanced program for visibility management. Visibility has been formally recognized as an air-quality resource in several states, starting with an ambient air-quality standard for visibility in California in 1971. That standard was set at a level intended to protect visibility in the most polluted urban areas.

An organized nationwide strategy for managing visibility in Class-I areas (national parks and wilderness areas) was initiated with the U.S. Clean Air Act Amendments of 1977, which set a national goal of eliminating anthropogenic visibility impairment in those areas and established procedures for assessment of visibility as an "air-quality related value." Regulations adopted by the U.S. EPA in 1980 established procedures for determining and mitigating plume blight from single sources or small groups of sources. The scope of the visibility program was expanded substantially in 1999, when the U.S. EPA promulgated regulations intended to mitigate regional haze at Class-I areas over a period of 65 years.

In addition, protection of visibility in other areas, including cities, is an objective of a secondary (i.e., non-health) National Ambient Air Quality Standard for $PM_{2.5}$ (which is currently at the same level as the primary, health-protective standard).

Canada relies on its environmental-impact assessment process, outlined in the Canadian

Environmental Assessment Act, to address visibility impacts. This is generally applied in a source-by-source manner, but recently efforts have been made to accommodate regional approaches in areas such as the Oil Sands Region of northern Alberta and the diamond mining operations in the Northwest Territories. A first nationwide assessment of visibility impacts due to particulate matter in Canada has just been completed (McDonald, 2002).

Formal programs for mitigating either rural or urban visibility impairment have not been developed in Mexico. Visibility is being used as an indicator of progress in ongoing programs to improve air quality in Mexico City, however.

9.5.2 Alignment of Visibility and PM Control Programs

For the public, the ability to see through the atmosphere is perceived as an indicator of good air quality. Consequently, control programs for PM mass may benefit by considering the improvement of visibility as a public measure of success. Tracking improvements in visibility is one way that jurisdictions have of demonstrating improvement in particulate pollution. This immediate measure of progress is considerably more accessible and uncomplicated than human-health indicators, but the integrity of visibility as an indicator of progress relies on the PM-control program addressing the size fraction of particles that is assumed to affect health.

There is one situation where the linkage between the management of PM for health and for visibility may not be as straightforward. Strategies required to protect visibility on the clearest 20 percent of days in national parks and wilderness areas, as required by the U.S. regional-haze program, may be different than those required for protection of human health. One hurdle will be the impression that money is being spent to improve air quality at times when it is already pristine and is not likely to be adverse to health.

Because of the close linkage between $PM_{2.5}$ and visibility, air-pollution programs to address both are likely to be more efficient and successful when they operate in a coordinated fashion. As a specific example, the U.S. Congress has required that the programs for implementing the 1997 National Ambient Air Quality Standards for PM and the 1999 Regional Haze Rule be synchronized in time, principally to reduce the administrative and financial burdens on the states and on the owners of pollution sources.

In Canada, the management of visibility is not as closely coordinated with the management of PM. Environment Canada has the responsibility to ensure that transboundary issues, including both PM and visibility, are addressed under the Canadian Environmental Protection Act. Heritage Canada's Parks Service shares responsibility for managing visibility issues within the National Parks system, while the provincial governments are generally responsible for other air issues under the Canadian Environmental Assessment Act. Coordination with regard to visibility issues, similar to that in the United States, is not available nationally, but has taken place in localized settings, such as Kejimkujik National Park and the Lower Fraser Valley around Vancouver, British Columbia.

It is important to note that the levels of fine-particle mass concentrations established as air-quality management standards for the protection of health may not be effective for protecting or improving scenic visibility. Specifically, meeting the U.S. 24-hour $PM_{2.5}$ ambient standard of 65 mg/m^3 will enhance visibility in only a few areas. Visual range in some eastern major urban centers and much of California will benefit by achieving the 24-hour standard. In other areas the standard is higher than currently observed mass concentrations, hence visibility will not be improved unless the particulate-matter load is reduced even though it already achieves the level of the standard. The number of places where visibility will be improved by meeting the annual $PM_{2.5}$ ambient standard of 15 mg/m^3 is slightly larger, but again these sites are found predominantly in the eastern states and California. In general, the national standards for $PM_{2.5}$ are not going to contribute significantly to the improvement or protection of visibility in much of the United States.

In Canada, most locales experience 24-hour mass concentrations at or above the 30 $\mu g/m^3$ Canada-Wide Standard level only occasionally. The 98th percentile of mass measurements has potential to exceed the standard only in urban centers of the Golden Triangle and Upper Canada. Therefore, meeting the Canada-Wide Standard at these sites will improve local visibility to a small degree. However, for the rest of Canada, including all of the west and the Maritimes, maintaining the Canada-Wide Standard allows increasing the current PM mass concentrations and could produce a degradation in visibility (McDonald, 2002). Therefore, the Canada-Wide Standard for $PM_{2.5}$ does not promote the protection or improvement of visibility over much of the country. This is especially so for the most pristine regions, including national parks and wilderness areas.

9.5.3 Regional Planning Organizations

Because regional haze is a result of long-range transport and can cover several states or provinces, the United States has established five regional planning organizations for the purpose of performing the analyses and modeling needed to implement the regional haze rule. The concept of regional planning organizations for visibility originated with the Grand Canyon Visibility Transport Commission (GCVTC), which was prescribed by Congress in the Clean Air Act Amendments of 1990. The GCVTC, which encompassed 9 states in the western end of the country, took on a broader charter of assessing strategies for protecting visibility throughout the 16 Class-I areas of the Colorado Plateau, of which Grand Canyon National Park is one. A comprehensive regional air-quality model was developed and was applied to evaluate the visibility consequences of a variety of air-quality management options. The final product of the GCVTC was a report to the U.S. EPA with recommendations for an approach to air-quality management.

The GCVTC has been superseded by the Western Regional Air Partnership (WRAP), which is actively engaged in formulating an air-quality management program for 13 states; it is framed around the GCVTC recommendations and conforms to the subsequent Regional Haze Rule. Four other regional planning organizations have been formed to perform similar functions over the rest of the country.

In addition to the regional planning organizations, other regional consortia have performed visibility research and attribution analyses over multi-state regions. A notable example is the Southern Appalachian Mountains Initiative (SAMI), which addressed visibility as one of its concerns.

9.5.4 Point-Source Control Programs

Section 9.2.1 described several short-term studies that were concerned with the contributions of electric-power generating stations to visibility impairment at Class-I areas or, in one case, in an urban area. The emission controls that have been installed, or will be installed, are the first examples of the management of visibility impairment attributable to point sources. (Although the emissions of a single source are involved, these were not the typical cases of plume blight, because long transport distances were involved and secondary pollutants were the dominant concern.) Each of these instances involved a field-measurement and air-quality modeling study that cost many millions of dollars.

The Regional Haze Rule prescribes that the visibility effects of many major sources be assessed and the feasibility of applying Best Available Retrofit Technology be evaluated. It does not seem feasible to carry out a major field program and modeling study for each of these facilities, so less data- and resource-intensive means of making these assessments are needed.

9.5.5 International Programs

Some cooperative efforts at studying PM and visibility have occurred along the U.S.-Canadian border and the U.S.-Mexican border. The joint U.S.-Canadian efforts have addressed 1) a pair of contiguous national parks across the border from each other, the Waterton Lakes (Alberta) and Glacier (Montana) National Parks, and 2) the Acadia National Park in Maine (Hoff et al., 1997).

A joint U.S.-Mexican effort studied visibility-related PM on both sides of the border in the vicinity of Big Bend National Park in Texas (Big Bend Air Quality Work Group, 1999). A subsequent study to apportion the observed aerosol and visibility impairment to sources in the two countries (the BRAVO Study, described in Table 9.1) has been carried out without Mexican participation, however.

Although trans-boundary transport of air pollution is known to occur, no actions have yet been taken to regulate emissions in one country for the purpose of mitigating visibility impairment in another country. However, Canada and the United States have signed an Air Quality Accord that includes provisions for each nation to consider the transboundary impact of air pollution on pristine sites in the other country and to implement any needed controls by 2010. To the south, the GCVTC analyses, the BRAVO Study, and other studies have investigated transport of visibility-related pollutants between the United States and Mexico through the modeling of transport fluxes.

9.6 SUMMARY AND CONCLUSIONS

Fine particles in the atmosphere are not individually visible to the human eye, but their presence is visible through their contribution to the hazy appearance of polluted air and the impairment of visibility. The effects of particles on visibility can be described theoretically or using empirical relationships, given the chemical composition and physical characteristics of the aerosol derived from measurements or application of a mathematical model. Particles also affect the radiation balance of the earth through mechanisms of light scattering and absorption similar to those that affect visibility, as well as through indirect mechanisms that involve the influence of PM on clouds.

Most of the particles of importance to visibility and the earth's radiation balance are secondary particles. These are formed in the atmosphere through physical and chemical processes involving gaseous precursors – both natural and anthropogenic. The optical effects of particles depend primarily on size and chemical composition, and secondary particles tend to be of the size range (particle diameters of 0.3 to 1.0 μm) that is most effective at scattering visible light. Relative humidity is a particularly important atmospheric factor, since particle light scattering at high relative humidity is many times that of the dry aerosol for hygroscopic species, such as sulfates, nitrates, and some organics. Much of the current understanding of relationships between PM and visibility comes from a broad range of short-term field studies at both urban and pristine locations and from the long-term IMPROVE monitoring program. Most studies have taken place in the United States, some have occurred along the southern tier of Canada, a few have taken place in Mexico City, and one has included measurements in northeastern Mexico. Consequently, visibility and its relationship to PM are currently best characterized in the United States.

These studies demonstrate that the spatial distribution of light extinction (the sum of light scattering and absorption) generally matches that of fine PM concentration, except that a given PM concentration causes greater extinction in a very humid environment than in a dry environment. Consequently, outside of urban areas, extinction tends to be greatest (and visibility poorest) in the eastern portions of Canada and the United States. Visibility is best in the arid intermountain portion of the western United States. Also, because of declining fine-particle concentrations throughout much of the United States, visibility has been improving in most areas during the past two decades. Because of the lack of data, trends in Canada and Mexico are not known.

Formulas that represent light extinction as the sum of weighted chemical concentrations are useful for estimating the optical effects of the aerosol. Each concentration, derived from measurements or modeling, is multiplied by an empirical scaling factor that represents the light scattering or absorption efficiency of that chemical species. More rigorous calculations of light extinction, based on theoretical descriptions of the interaction of light with particles, require further information about the size distribution of the particles (either from specialized measurements or from advanced aerosol models) and are computationally more demanding.

Based on such analyses, non-urban visibility impairment in eastern North America is predominantly due to $SO_4^=$ particles, with organic particles generally second in importance. In the West, the contributions of $SO_4^=$ and OC are comparable, and NO_3^- plays a significant role in the populated areas of California. Carbonaceous material, either OC or BC, is an important contributor in some urban areas of North America. Soil particles can be important contributors to visibility impairment in areas susceptible to wind-blown dust.

These relationships suggest that efforts to mitigate visibility impairment will require control of emissions of different chemical species in different geographic areas. Also, because visibility in the clearest areas is sensitive to even minute increases in PM concentrations, strategies to preserve visibility on the clearest days may require stringent emission limitations,and dilute impacts from very distant sources can still be important.

Modeling of the visibility effects of PM is based on prediction of the aerosol composition by a PM chemical-transport model, which then is followed by the additional step of calculating light extinction by an empirical or theoretical formula. Thus the quality of the visibility prediction depends greatly on the skill of the PM model. As described in Chapter 8, models that predict secondary PM on a regional basis have significant limitations for some species, and the performance of the latest PM models has not yet been evaluated extensively. Consequently the quality of modeled representations of light extinction based on these PM calculations is currently somewhat uncertain.

The situation is similar with the modeling of the earth's radiation balance. Much remains to be learned about the important aerosol processes in the upper atmosphere and modeling appears to have significant uncertainties. Extensive work is underway to learn more about the particles and their characteristics and to improve models.

A major gap in knowledge is the role of organic compounds in particles important to visibility and the radiation balance. Currently all organic species are lumped into single category labeled "OC," although different organic compounds have varying degrees of reactivity, volatility, and solubility in water. There is a need to better characterize both the various constituents of the volatile organic compounds that are emitted from sources and the resulting organic PM, and to develop mechanistic theories that explain the transition. Furthermore, because of the importance of water uptake to light scattering, the hygroscopic behavior of the organic matter needs to be better explained.

Other enhancements are required for the PM modeling that underlies visibility calculations. As discussed in Section 8.10, the most important are an improved characterization of the meteorological field and refinements in emission inventories. For visibility, meteorological information that is especially important includes the wind and humidity fields and the descriptions of cloud locations and characteristics. Shortcomings in these areas have accounted for much of the error in recent efforts at modeling visibility impacts, both from isolated sources and for regional emission distributions.

Despite some gaps in understanding, uncertainties in characterizing visibility and the effects of PM management on it are significantly smaller than for other effects, such as human health and global climate change. Direct measurements of light extinction in the field are an independent check on methods to construct extinction from particle information or by other model approaches. In fact, visibility measurements have been used as a surrogate for PM measurements and provide a basis for evaluating regional PM models. In particular, the IMPROVE network has produced a consistent and long-term data set of PM composition that is not yet matched by others available for human-health and radiation-balance purposes. One should note, though, that deciphering a change in a small value (such as the visibility impairment in a pristine region) may cause more difficulties analytically and statistically than looking at trends in the elevated values of PM that are of relevance for health purposes

Data-quality control is important. Mundane issues such as missing observations impose challenges in calculating truly representative hourly, daily or annual values through averaging of instantaneous measurements, and appropriate uncertainties need to be defined as part of the data-analysis process. Questions about the spatial representativeness of measurements (typically at a point) or model endpoints (typically an average for a grid cell) provide perhaps the greatest uncertainty. Current air-quality management approaches that consider regional management zones are designed to reduce this uncertainty.

As with PM itself, research will be required over the long-term to decrease uncertainties in visibility-relevant fundamental science issues, such as the roles of size distributions, volatile organics, hygroscopic growth, and extinction efficiencies. The uncertainties resulting from lack of PM process understanding are potentially quite large, but again the ability to directly measure visibility provides the capability to test updated hypotheses as the science progresses.

Recognizing limitations in understanding and, specifically, in the ability to model visibility, the U.S. National Academy of Sciences concluded that existing knowledge and tools appeared to be adequate to develop strategies for managing visibility (NRC, 1993). On this basis, the United States has initiated a 65-year program for mitigating regional haze in national parks and wilderness areas. This regional-haze program is separately addressing the needed PM emission reductions in five regions of the country, with appropriate attention to transport between planning regions. The initial emission-management actions that may be implemented under that program are not likely to be very sensitive to uncertainties in current models. Better tools will be required to make the more sensitive and costly strategy decisions further downstream, in a few decades, but one can assume that the models and data will be improved by then.

Comparable programs to address visibility have not been proposed by Canada and Mexico, although efforts at limiting and reducing PM concentrations for health reasons can be expected to provide benefits to visibility. Furthermore, the measurement networks needed to characterize visibility and its trends nationwide do not exist in either country, so the visibility situation there is not well documented. The United States and Canada have agreed, though, to limit the transboundary impacts of air pollution (including effects on visibility) on pristine sites in the other country, which has beneficial visibility implications for near-border scenic areas in both countries.

9.7 POLICY IMPLICATIONS

The key points of this chapter and their policy implications are summarized below.

Visibility impairment is sensitive to the chemical composition of $PM_{2.5}$, and also depends strongly on ambient relative humidity. Secondary particles tend to be of the size range (diameters of 0.3 to 1.0 μm) that is most effective at scattering visible light. Relative humidity is a particularly important atmospheric factor, with light scattering of fine PM at high relative humidity being many times that of the dry aerosol. *Implication:* Controlling the precursors of secondary PM, especially the precursors of $SO_4^=$ and NO_3^-, will be effective in improving regional haze as these particles grow in relation to relative humidity.

Non-urban visibility impairment in eastern North America is predominantly due to $SO_4^=$ particles, with organic particles generally second in importance. In the West, the contributions of $SO_4^=$ and OC are comparable, and NO_3^- plays a significant role in the populated areas of California. In some urban areas across North America, BC is an important contributor. Soil particles can be important contributors to visibility impairment in areas susceptible to windblown dust. Wildfires, prescribed burning, and agricultural fires can be dominant episodic contributors to regional haze in all areas. *Implication:* Efforts to mitigate visibility impairment will require control of emissions of different chemical species in different geographic areas.

Visibility in the clearest areas is sensitive to even minute increases in PM concentrations. *Implication:* Strategies to preserve visibility on the clearest days may require stringent limitations on upwind emissions growth, and dilute impacts from very distant sources can be important.

Theoretical or empirical algorithms can be used to estimate visibility impairment attributable to measured PM concentrations of $SO_4^=$, NO_3^-, OC, BC, and soil at a specific relative humidity, although the estimates are more uncertain at extreme levels of relative humidity (> 90 percent). Modeling future visibility changes due to PM-concentration changes relies on the prediction of aerosol composition by a chemical-transport model (CTM), followed by the additional step of calculating light extinction using an empirical or theoretical formula. The quality of the visibility predictions is greatly dependent on the performance of the CTM. Currently this modeling has significant limitations for NO_3^- and carbonaceous PM species. *Implication:* Chemical-transport models can be used now for semi-quantitative projections of regional-haze improvements resulting from reductions in PM and PM precursors, with the caveat that the current quality of modeled representations of species concentrations is somewhat uncertain.

Particles affect the radiation balance of the earth through mechanisms of light scattering and absorption similar to those that affect visibility, as well as through indirect effects on clouds. While scattering and absorption both degrade visibility, they can have counterbalancing effects on the aerosol contribution to the radiative balance. Specifically, light absorption by particles tends to increase surface heating while light scattering by particles tends to decrease surface heating, although the direct effect is estimated to be small compared to the effects of greenhouse gases. The indirect effect may be larger, but is highly uncertain. *Implication:* Strategies for managing aerosol chemical-component concentrations to mitigate visibility impairment and to reduce aerosol forcing of climate change may not be wholly consistent, but the different spatial scales of haze and of climate-forcing aerosol mean that it should be possible to address both concerns with an integrated strategy.

9.8 REFERENCES

Andrews, E., Saxena, P., Musarra, S., Hildemann, L.M., Koutrakis, P., McMurry, P.H., Olmez, I., White, W.H., 2000. Concentration and composition of atmospheric aerosols from the 1995 SEAVS experiment and a review of the closure between chemical and gravimetric measurements. Journal of the Air & Waste Management Association 50, 648-664.

Big Bend Air Quality Work Group, 1999. Big Bend National Park regional visibility preliminary study. Report by Big Bend Air Quality Work Group, Dallas and Mexico City.

Binkowski, F.S., Roselle, S.J., Eder, B.K., Mebust, M.R., 2001. A preliminary evaluation of Models-3 CMAQ using visibility parameters. American Meteorological Society Annual Meeting, Albuquerque, New Mexico, January.

Chow, J.C., Watson, J.G., Edgerton, S.A., Vega, E., 2002. Chemical composition of $PM_{2.5}$ and PM_{10} in Mexico City during winter 1997. The Science of the Total Environment 287, 177-201.

Ciudad de México, 1999. Towards an air quality programme 2000-2010, for the Mexico City Metropolitan Area. http://www.sma.df.gob.mx/publicaciones/aire/prog_cal_aire00_10/2000/ing/capitulo02.pdf.

Eidels-Dubovoi, S., 2002. Aerosol impacts on visible light extinction in the atmosphere of Mexico City. The Science of the Total Environment 287, 213-220.

Environment Canada, 1995. The state of Canada's climate: Monitoring variability and change. SOE Report 95-1.

FLAG, 2000. Federal Land Managers' Air Quality Related Values Workgroup (FLAG) Phase I Report (December 2000). http://www.aqd.nps.gov/ard/flagfree.

Flocchini, R.G., Cahill, T.A., Pitchford, M.L., Eldred, R.A., Feeney, P.J., Ashbaugh, L.L., 1981. Characterization of particles in the arid West. Atmospheric Environment, 15, 2017-2030.

Gebhart K.A., Malm W.C., and Flores M. 2000. A preliminary look at source-receptor relationships in the Texas-Mexico border area. Journal of the Air & Waste Management Association 50, 858-868.

Green, M.C., Kuhns, H., Pitchford, M.L., 2000. An overview of the Big Bend Regional Aerosol and Visibility Observational (BRAVO) Study. Paper 206, Proceedings (CD-ROM), 93rd Annual Meeting of the Air & Waste Management Association, Salt Lake City, 18-22 June.

Hansen, D.A., Edgerton, E., Hartsell, B., 2000. The Southeastern Aerosol Research and Characterization Study (SEARCH): An overview. Paper 10AS.2 in Proceedings (CD-ROM) of A&WMA Specialty Conference on PM2000: Particulate Matter and Health – The Scientific Basis for Regulatory Decision-making, Charleston, South Carolina, 24-28 January.

Henry, R.C., 1999. Perception of color images of simulated haze. Report to the Electric Power Research Institute, Palo Alto, California.

Hoff, R., Guise-Bagley, L., Moran, M., McDonald, K., Nejedly, Z., Campbell, I., Pryor, S., Golestani, Y., Malm, W., 1997. Recent visibility measurements in Canada. Proceedings (CD-ROM), 90th Annual Meeting of the Air & Waste Management Association, Toronto, 8-13 June.

Husar, R.B., Husar, J.D., Martin, L., 2000. Distribution of continental surface aerosol extinction based on surface visual range data. Atmospheric Environment, 34, 5067-5078.

IPCC (Intergovernmental Panel on Climate Change), 1996. Climate Change 1996. Four volumes. Cambridge University Press, Cambridge, UK.

IPCC (Intergovernmental Panel on Climate Change), 2002. Climate Change 2001: The Scientific Basis. Cambridge University Press, Cambridge, UK.

Jansen, J.J., Edgerton, E., Hartsell, B., Hansen, A., 2001. Particulate matter composition: Preliminary results from SEARCH. Proceedings (CD-ROM), Electric Utilities Environmental Conference, Tucson, Arizona, 8-12 January.

Karamchandani, P., Santos, L., Sykes, I., 1999. SCICHEM: A new generation plume-in-grid model. TR-113097, EPRI, Palo Alto, California.

Magliano, K.L., McDade, C., 2001. The California Regional $PM_{10}/PM_{2.5}$ Air Quality Study (CRPAQS): Field study description and initial results. Proceedings (CD-ROM), A&WMA/AGU Specialty Conference on Regional Haze and Global Radiation Balance — Aerosol Measurements and Models: Closure, Reconciliation, and Evaluation. Bend, Oregon, 2-5 October 2001.

Malm, W.C., Johnson, C.E., 1984. Optical characteristics of fine and coarse particulates at Grand Canyon, Arizona. Atmospheric Environment 18, 1231-1237.

Malm, W.C., Molenar, J.V., 1984. Visibility measurements in national parks in the western United States. Journal of the Air Pollution Control Association 34, 899-904.

Malm, W. , Gebhart, K., Latimer, D., Cahill, T., Eldred R., Pielke R., Stocker R., Watson J., 1989. National Park Service report on the Winter Haze Intensive Tracer Experiment. Final Report by National Park Service, Denver.

Malm, W.C., Sisler, J.F., Huffman, D., Eldred, R.A., Cahill, T.A., 1994. Spatial and seasonal trends in particle concentration and optical extinction in the United States. Journal of Geophysical Research 99, 1347-1370.

Malm W.C., 2000a. Introduction to Visibility. Cooperative Institute for Research in the Atmosphere (CIRA), Ft. Collins, Colorado.

Malm, W. C., 2000b. Spatial and seasonal patterns and temporal variability of haze and its constituents in the United States: Report III: Cooperative Institute for Research in the Atmosphere (CIRA), Ft. Collins, Colorado

McDade, C., Tombach, I., Seigneur, C., Mueller, P. K., Saxena, P., 2000. Study of the relationship of distant SO_2 emissions to Dallas-Fort Worth Winter Haze. Journal of the Air & Waste Management Association 50,826-834.

McDonald, K., 2002. Review and characterization of visibility impacts related to particulate matter in Canada. Report En40-773/2002E-IN, Environment Canada.

Molenar, J. V., Malm, W.C., Johnson, C.E., 1994. Visual air quality simulation techniques. Atmospheric Environment 28, 1055-1063.

Mora, V.R., Melgar, E.M., Pascual, C., Ruiz, M.E., 2001. Visibility trends in two Mexico City sites. Proceedings (CD-ROM), 94th Annual Meeting of the Air & Waste Management Association, Orlando, 24-28 June.

Mueller, P.K., Hansen, D.A., Watson, J.G., 1986. The Subregional Cooperative Electric Utility, Department of Defense, National Park Service, and EPA Study (SCENES) on visibility: An overview. Report EA-4664-SR, Electric Power Research Institute, Palo Alto, California.

National Research Council (NRC) 1990. Haze in the Grand Canyon: an evaluation of the Winter Haze Intensive Tracer Experiment. National Academy Press, Washington, D.C.

National Research Council (NRC) 1993. Protecting visibility in national parks and wilderness areas. National Academy Press, Washington, D.C.

Pitchford, M., Malm, W.C., 1994. Development and application of a standard visual index. Atmospheric Environment 28, 1049-1054.

Pitchford, M., Green, M., Kuhns, H., Tombach, I., Malm, W., Scruggs, M., Farber, R., Mirabella, V., 1999. Project MOHAVE Final Report. Prepared for U. S. Environmental Protection Agency and Southern California Edison Company. http://www.epa.gov/region09/air/mohave/report.html.

Pitchford, M., Frank, N., Damberg, R., Scruggs, M., Malm, W., Silva, S., Fisher, R., Archer, S., Poirot, R., Ely, D., Lebens, R., Moore,T., Eldred, R., Ashbaugh, L., 2000. Expanded IMPROVE network for regional haze monitoring. Proceedings (CD-ROM), 93rd Annual Meeting of the Air & Waste Management Association, Salt Lake City, 18-22 June.

Richards, L.W., Blanchard, C.L., Blumenthal, D.L., 1991. Navajo Generating Station visibility study final report (Draft Number 2). STI-90200-1124-FRD2, Sonoma Technology Inc, Santa Rosa, California.

Seigneur, C., Wu, X.A., Constantinou, E., Gillespie, P., Bergstrom, R.W., Sykes, I., Venkatram, A., Karamchandani P., 1997. Formulation of a second-generation reactive plume visibility model. Journal of the Air & Waste Management Association 47, 176-184.

Shah, J., 1981. Measurements of carbonaceous aerosol across the U.S.: Sources and role in visibility degradation. Ph.D. thesis, Oregon Graduate Center, Beaverton, Oregon.

Sisler, J. F., Malm, W. C., 2000. Trends of $PM_{2.5}$ and reconstructed visibility from the IMPROVE network for the years 1988-1998. Paper 442, Proceedings (CD-ROM), 93rd Annual Meeting of the Air & Waste Management Association, Salt Lake City, 18-22 June.

Tang I., N., 1996. Chemical and size effects of hygroscopic aerosols on light scattering coefficients. Journal of Geophysical Research 101, 19,245-19,250.

Tombach, I.H., Allard D.W., Drake R.L., Lewis R.C., 1987. Western regional air quality studies: visibility and air quality measurements: 1981-1982. Report EA-4903, Electric Power Research Institute, Palo Alto, California.

Tombach, I., Seigneur C., McDade C., Heisler S., 1996. Dallas-Fort Worth winter haze project. Report EPRI TR-106775 (3 volumes), Electric Power Research Institute, Palo Alto, California.

Trijonis, J.C., Malm W.C., Pitchford M., and White W.H., 1990. Visibility: existing and historical conditions– causes and effects. Acidic Deposition: State of Science and Technology, Report 24, National Acid Precipitation Assessment Program, Washington, DC.

Vignati, E., Wilson, J., Feichter, J. 2000. Modelling size resolved mixed aerosol fields in the ECHAM GCM, Journal of Aerosol Science 31, S148.

Watson, J.G., Blumenthal, D., Chow, J., Cahill, C., Richards, L.W., Dietrich, D., Morris, R., Houck, J., Dickson, R.J., Anderson, S., 1996. Mt. Zirkel Wilderness Area reasonable attribution study of visibility impairment. Final report by Desert Research Institute, Reno, Nevada.

Watson, J.G., 2002. Visibility: Science and regulation. Journal of the Air & Waste Management Association. 52, 628-713.

White, W.H., 1985. Optical characteristics of fine and coarse particulates at Grand Canyon, Arizona. Atmospheric Environment 19, 1021-1022.

White, W.H., 1986. On the theoretical and empirical basis for apportioning extinction by aerosols: A critical review. Atmospheric Environment 20, 1659-1672.

White, W.H., Macias, E.H., 1989. Carbonaceous particles and regional haze in the western United States. Aerosol Science and Technology 10, 111-117.

WinHaze, 2002. Software available from Air Resource Specialists, Ft. Collins, Colorado, USA.

CHAPTER 10

Conceptual Models of PM for North American Regions

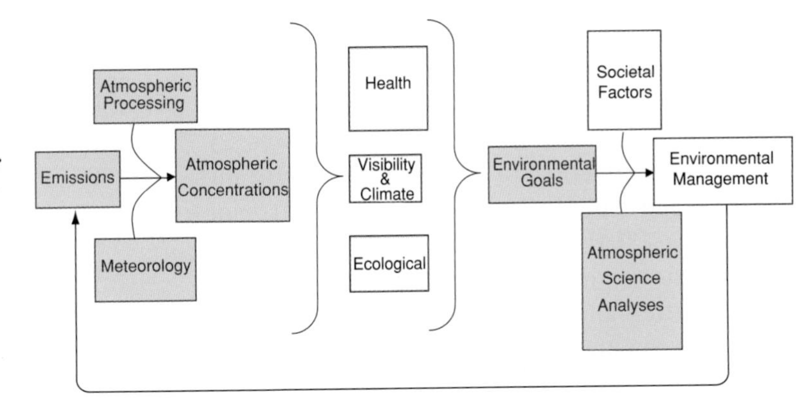

Technical Editor: James Vickery

10.1 OVERVIEW

As indicated in the discussions in Chapters 6 and 7, there are substantial differences in ambient PM concentrations and chemical composition across North America. These differences reflect the varying sources of PM and the atmospheric processes involved. Accordingly, it is unlikely that a single, unified approach for PM reduction will be universally successful. Initial consideration of approaches that are compatible with current PM standards and level of scientific understanding suggests that PM-reduction plans should be based on specific local and regional PM characteristics. Scientists recently have begun to develop conceptual models of the nature and origins of PM for several North American locations. These models generally form a foundation for developing PM-management strategies.

A conceptual model can be used to identify the limiting processes: those aspects of the PM problem, which if addressed will most effectively reduce the ambient mass or chemical concentration (Pun and Seigneur, 1999; CARB CRPAQS, 1999). Conceptual models are representations of the best understanding of the influence of emissions, meteorology, and atmospheric processes on ambient concentrations for any given region or airshed at any given time. Figure 10.1 illustrates how the application of science tools using the best current understanding of the atmospheric environment can provide a state-of-science understanding in the form of a conceptual model for use by policy makers. The use of corroborating evidence from multiple science tools is integral to the development of conceptual models.

Conceptual models can guide the development and application of CTMs that can be used to develop PM-management strategies based on the evaluation of

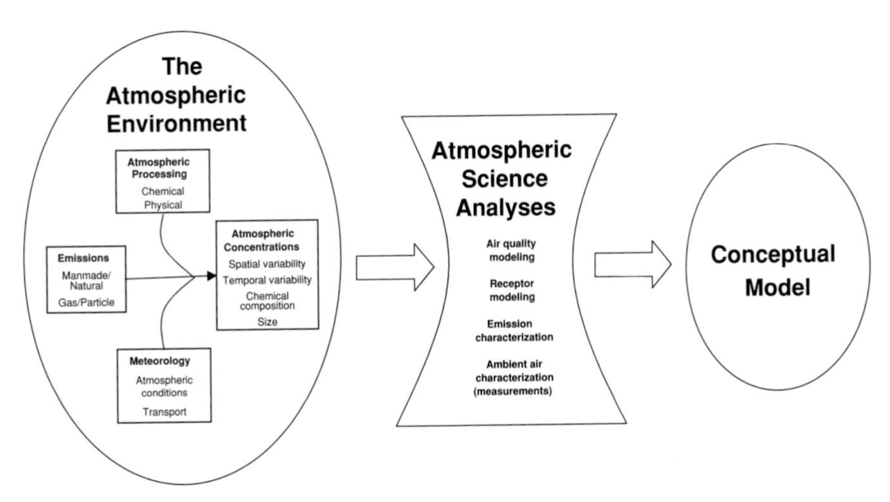

Figure 10.1 Development of a conceptual model by applying atmospheric science analyses to the best current understanding of an area's atmospheric environment.

future alternative scenarios. They can be used to: 1) define the important components of an air-quality management strategy for effective control of PM concentrations by identifying limiting processes, 2) guide data collection to characterize important processes and to fill key knowledge gaps, and 3) point out where CTMs should be used to examine opportunities to maximize multi-pollutant control opportunities and minimize potential counter-productive copollutant interactions (see Figure 10.2).

Conceptual models have been prepared for nine widely different areas of North America (see Figure 10.3) and are presented here. The nine are:

1) The San Joaquin Valley of California, an enormous, variably populated valley, surrounded by mountains, exposed to the inflow of air from the eastern Pacific coastal-marine environment, with a variety of urban and rural sources, and having meteorological conditions that promote photochemical processing in summer, intermittent transport, and intermittently severe stagnation conditions especially in winter;
2) The Los Angeles area of Southern California, a huge metropolitan area in a basin along the Pacific Coast with a classical photochemical pollution environment where evening temperature inversions, especially in the late summer and early fall, produce low mixing depths that trap pollutants close to the surface;
3) Mexico City and its suburbs, one of the largest metropolitan areas in the world, lying in a tropical region at an altitude of more than 2 km above sea level, in an arid valley surrounded by mountains that are well ventilated overnight, with PM pollution heavily influenced by a combination of non-vegetated areas and unpaved roads, motor vehicle traffic, a variety of local and regional combustion processes, and animal sources of NH_3;
4) The southeastern United States, where major post-World War II industrialization and population growth have taken place, and warm humid conditions persist during the summer when periodic highs can lead to relatively slow movement of air through the region;
5) The northeastern United States, exemplifying the historical industrial heartland of the United States, and a region known to be affected by both local and regional sources of pollution;
6) The Windsor-Quebec City Corridor, the most populated and industrialized area of Canada, subject to westerly airflow that travels across remarkably different upwind regions, from the remote north to the heavily populated trans-border area just to the southwest, where periodic stationary high pressure and light southwesterly winds mix the Corridor's local PM and precursor emissions with aged polluted air to produce high PM episodes;
7) The U.S. Upper Midwest-Great Lakes area, representing the northern tier of industrialization in the United States and the combined agricultural and industrial activity of southern Canada;
8) The Canadian Southern Prairies and U.S. Northern Plains, exemplifying the relatively sparsely populated, mid-latitude, mid-continent

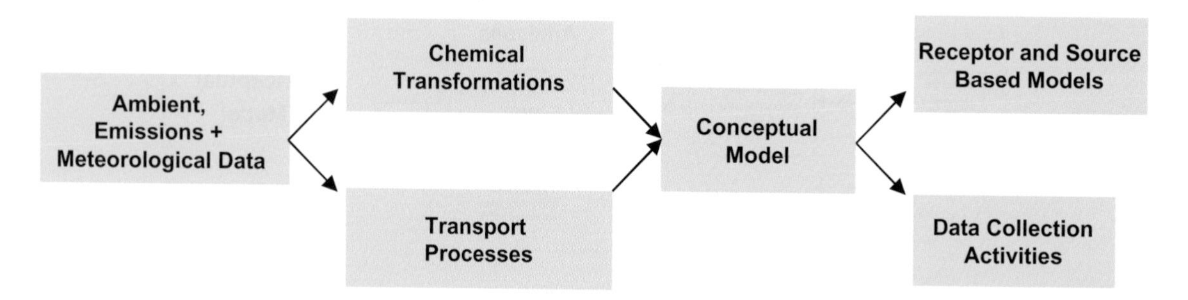

Figure 10.2. The use of conceptual model to identify needed ambient concentration information and CTM applications.

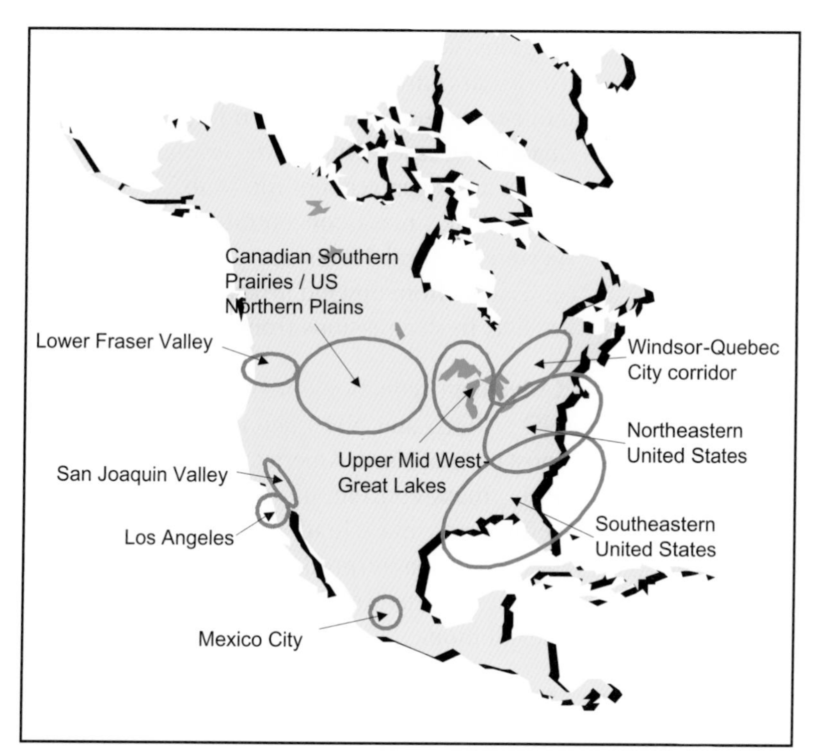

Figure 10.3. North American areas covered by conceptual models.

conditions under the westerly air flow across the United States and Canada;

9) The Lower Fraser Valley of British Columbia, a north-central Pacific coastal area, home to fewer than 2 million people and generally isolated from outside pollution, which experiences summertime inversions.

Differences in the information provided for each area reflect individual areal differences in monitoring history and field studies. Some areas, such as Los Angeles and the San Joaquin Valley of California, have been studied extensively. Others, such as Mexico City and the Lower Fraser Valley of British Columbia, have not. Newly instituted monitoring and research programs are underway that will add substantially to these conceptual models over the next few years. As they now stand, each conceptual model offers useful insights into the PM issue for its area and those with similar topography, meteorology, and contributing source categories.

The conceptual descriptions that follow are highly summarized; the PM issue in any one of these areas is more complicated than presented here. Readers wishing more detail for a particular area should refer to the data and studies cited in the references, Section 10.12. What follows is the best information and collective judgment of the authors of this Assessment describing conditions in the selected areas, based on contemporary theoretical precepts for tropospheric aerosols.

The descriptions for each conceptual model are organized according to the framework used to inform PM management (refer to Chapter 1 and its discussion of Figure 1.1) and have been produced following the process described in Figure 10.1. That is, current atmospheric science analyses and tools have been applied to produce the best understanding of the atmospheric environment over each area. Each model presents insights on five topics: three in relation to the atmospheric environment (atmospheric concentrations, atmospheric processing, and meteorology), and two giving the outcome of atmospheric-science analysis (sources and policy implications).

10.2 SUMMARY

Figures 10.4 through 10.12 graphically summarize the conceptual models for each of the nine areas. More complete discussions are contained in Sections 10.3 through 10.11. Some of the key, potentially cross-cutting insights for the nine areas that emerge from viewing these figures and reading the conceptual model discussions are:

- Observed PM concentrations exceed applicable $PM_{2.5}$ standards more frequently than they exceed PM_{10} standards. Elevated concentrations are much more frequently observed in relation to annual $PM_{2.5}$ standards than the daily $PM_{2.5}$ standards, where both standard forms exist. Most major urban areas of the eastern United States and southeastern Canada are likely to experience levels above $PM_{2.5}$ standards. Most urban areas of the San Joaquin Valley and the Los Angeles basin of California also experience concentrations above the $PM_{2.5}$ standards. The U.S. northern upper Midwest, Canadian Southern Prairie and U.S. Northern Plains portions of the continent have locations with PM levels of concern over some large urban areas. Mexico City's PM_{10} levels are frequently above both daily and annual standards. Mexico presently has no $PM_{2.5}$ standards.

- The eastern and western coastal regions of the United States and Canada show marked seasonality in high $PM_{2.5}$ concentrations, while the central interior regions generally do not. Mexico City's PM_{10} levels show marked seasonality.

- The largest fraction of eastern U.S. $PM_{2.5}$ on average is $SO_4^=$, particularly during periods of high PM concentration in the summer. Organic carbon is the second largest fraction and in some eastern U.S. urban areas is equal to or greater than $SO_4^=$. Sulfate is a major fraction of southeastern Canada's $PM_{2.5}$, though NO_3^- and OC also make up large fractions. Nitrate in $PM_{2.5}$ tends to be more important in West than in the East on the average. $PM_{2.5}$ sampled in the interior continental areas tends to be primarily OC and $SO_4^=$. This also is the case for a sampling of Mexico City's $PM_{2.5}$ composition.

  > There are important ties between $PM_{2.5}$ and oxidant-forming ozone processes. The eastern United States and southeastern Canada experience strong co-seasonality of their high PM season and summer oxidant season. There is also some co-seasonality during the late summer and early fall in Los Angeles and the Lower Fraser Valley. Even in the San Joaquin Valley, where periods of maximum $PM_{2.5}$ are nearly seasonally opposite to periods of high ozone, processes involving the same gaseous precursors are important factors in $PM_{2.5}$ production.

  > More than two-thirds of $PM_{2.5}$ mass concentration can attributed to anthropogenic sources in most locations of North America.

- Regional contributions are an important addition to local emissions when ambient $PM_{2.5}$ concentrations are being interpreted, in the majority of cases. Rural $PM_{2.5}$ levels surrounding urban areas can account for 50 to 75 percent of urban $PM_{2.5}$ mass concentrations during peak periods. Rural levels are composed of aged emissions from upwind urban and rural areas as well as fresh emissions from local sources.

- Coincident reductions of precursors (i.e., SO_2, NO_x, VOC, and NH_3) should be beneficial in most parts of North America in achieving desired $PM_{2.5}$ mass concentrations, but some of these reductions may lead to temporary and/or localized counterproductive impacts in some areas, for instance, due to NO_3^- for $SO_4^=$ substitution.

- At present there is insufficient information on the OC fraction of PM to warrant any guidance other than generally recognizing the importance of controlling emissions from both primary organic and VOC sources. Primary sources include transportation and industrial activity, as well as burning of wood and other organic matter.

Area-specific insights based on the conceptual descriptions presented in this Assessment include the following points:

- In Los Angeles, $PM_{2.5}$ concentrations exceed the annual standard by a factor of two. Much of the excess derives from intense and frequent episodes during the late summer and early fall. The San Joaquin Valley's mass concentrations are noticeably greater than annual standards and most severe in intensity and frequency during the fall and winter. The largest fraction of annual $PM_{2.5}$ in both Los Angeles and the San Joaquin Valley is NH_4NO_3; the next largest fraction is OC. Crustal/geological material can be a significant fraction of the $PM_{2.5}$ in Los Angeles.

- During the winter periods of peak $PM_{2.5}$ concentrations in Los Angeles and the San Joaquin Valley, NH_4NO_3 is the dominant component and is HNO_3 limited. HNO_3 can be reduced via VOC and NO_x emission reductions. The possibility of seasonal strategies that emphasize VOC controls for $PM_{2.5}$ mass in winter and NO_x for ozone in summer requires optimization with the assistance of CTMs and receptor models.

- The Lower Fraser Valley of the Pacific Northwest experiences its maximum $PM_{2.5}$ in the late summer or fall. Mobile-source emissions, agricultural emissions (NH_3) and road dusts account for about 70 percent of $PM_{2.5}$ mass.

- Mexico City's $PM_{2.5}$ levels appear to be largely dominated by primary BC and OC, with secondary components being potentially an important factor at times. More complete characterization with year-round monitoring is needed to add information about seasonal and annual variability of $PM_{2.5}$ concentrations and composition. Planned ecological restoration and preservation along with road paving are expected to be effective in reducing PM_{10}. Control of diesel emissions is important to reducing both PM_{10} and $PM_{2.5}$.

- The Canadian Southern Prairie – U.S. Northern Plains, and U.S. Upper Midwest – Great Lakes rural regions generally have PM mass concentrations below or near Canada-wide and U.S. annual standards, respectively. There is little seasonal variation in PM concentrations. Sulfate and OC are principal components, with contributions of both local and regional transport being important.

- Summer regional $PM_{2.5}$ concentrations in the northeastern United States and the Windsor-Quebec City Corridor (WQC) on average are twice winter concentrations. However, in large urban areas such as Philadelphia, New York City, and those of the eastern WQC, peak $PM_{2.5}$ occurs in winter when mass concentrations on average are slightly higher than summer. Summer regional $SO_4^=$ in the northeastern United States is more than twice the next nearest component, OC, and more than four times the NO_3^- and BC combined. Winter urban OC and $SO_4^=$ in Philadelphia and New York City each account for about a third of $PM_{2.5}$ mass. Nitrate is a significant component of northeastern U.S. and WQC urban $PM_{2.5}$ during the winter. Winter urban concentrations of OC, $SO_4^=$, and NO_3^- can be twice regional concentrations, indicating the importance of local source contributions. In the WQC, summer $SO_4^=$ is also an important fraction though OC and NO_3^- are also significant fractions throughout the year. Both local and regional $SO_4^=$ and OC are important in both regions.

- The southeastern United States experiences $PM_{2.5}$ concentrations above the annual and sometimes the 24-hr standards in urban areas. Summer $PM_{2.5}$ concentrations are one to three times higher than winter concentrations. As with the northeastern United States, summer $SO_4^=$ and OC concentrations dominate, but in contrast to the Northeast these two are nearly balanced in the Southeast. Both local and regional contributions are important.

10.3 CONCEPTUAL MODEL OF PM OVER THE SAN JOAQUIN VALLEY OF CALIFORNIA

The San Joaquin Valley (SJV) is an enormous, variably populated valley, surrounded by mountains and exposed to the inflow of air from the eastern Pacific coastal-marine environment. The SJV contains a variety of urban and rural sources of PM, NO_x, VOC, and NH_3 from agricultural and petroleum-production operations and a variety of transportation, industrial, commercial, and domestic sources. Meteorological conditions promote photochemical processing in summer, intermittent transport from the Los Angeles, San Francisco, and Sacramento areas, and intermittently severe stagnation conditions, especially in winter.

10.3.1 Annual and Seasonal Levels of $PM_{2.5}$ and PM_{10} in Relation to Mass-Based Standards

Annual $PM_{2.5}$ levels in excess of 15 µg/m^3 are expected in the coming years for urban areas of the SJV. Available data for 1999 show annual means exceeding 20 µg/m^3 for most urban areas monitored, and occasional 24-hr levels considerably greater than 65 µg/m^3. Annual levels are dominated by elevated

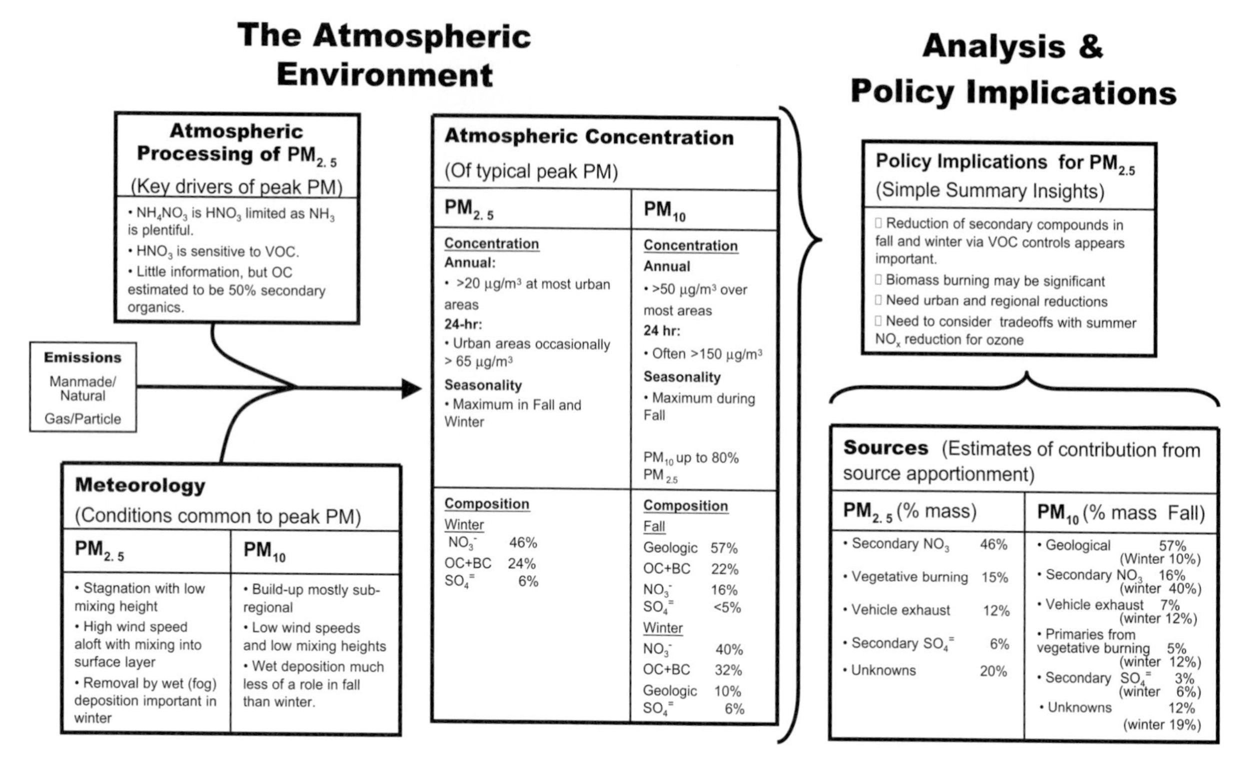

Figure 10.4. Simplified conceptual model for the San Joaquin Valley.

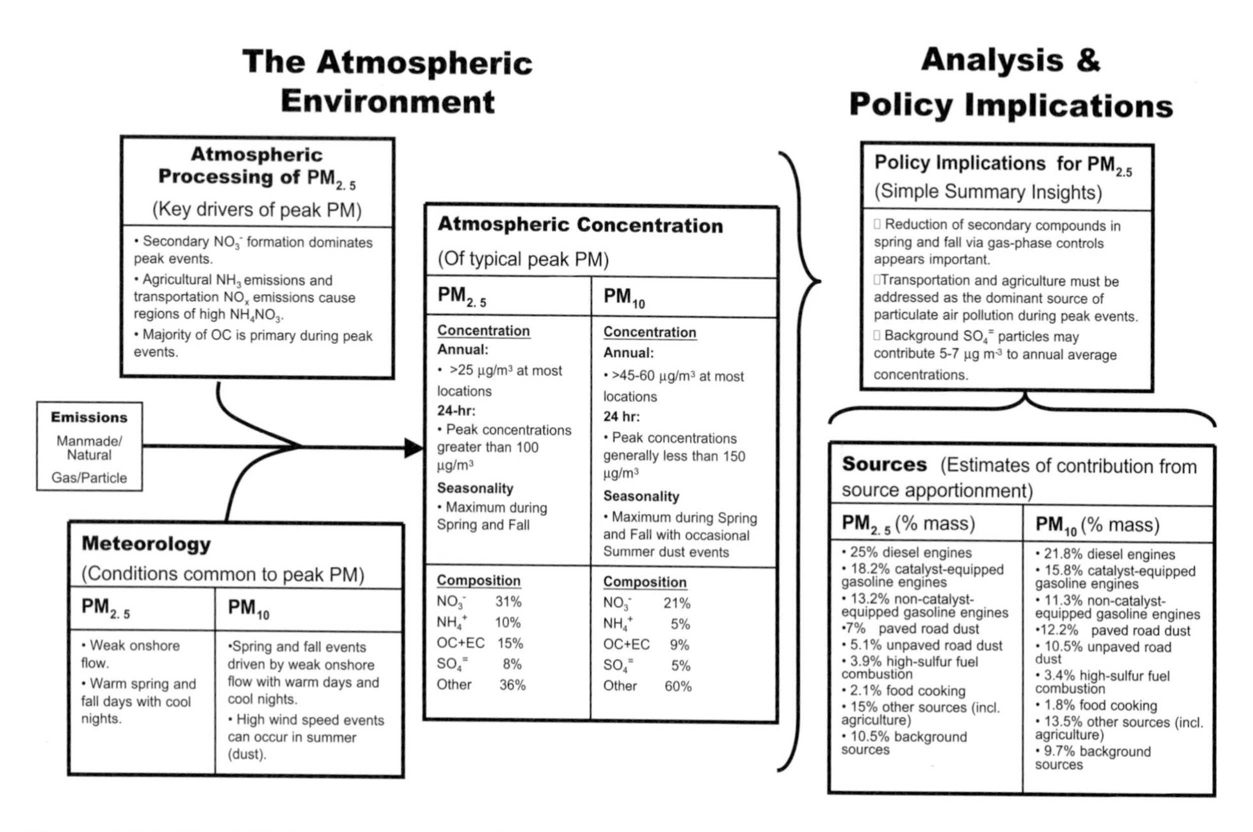

Figure 10.5. Simplified conceptual model for the Los Angeles Air Basin.

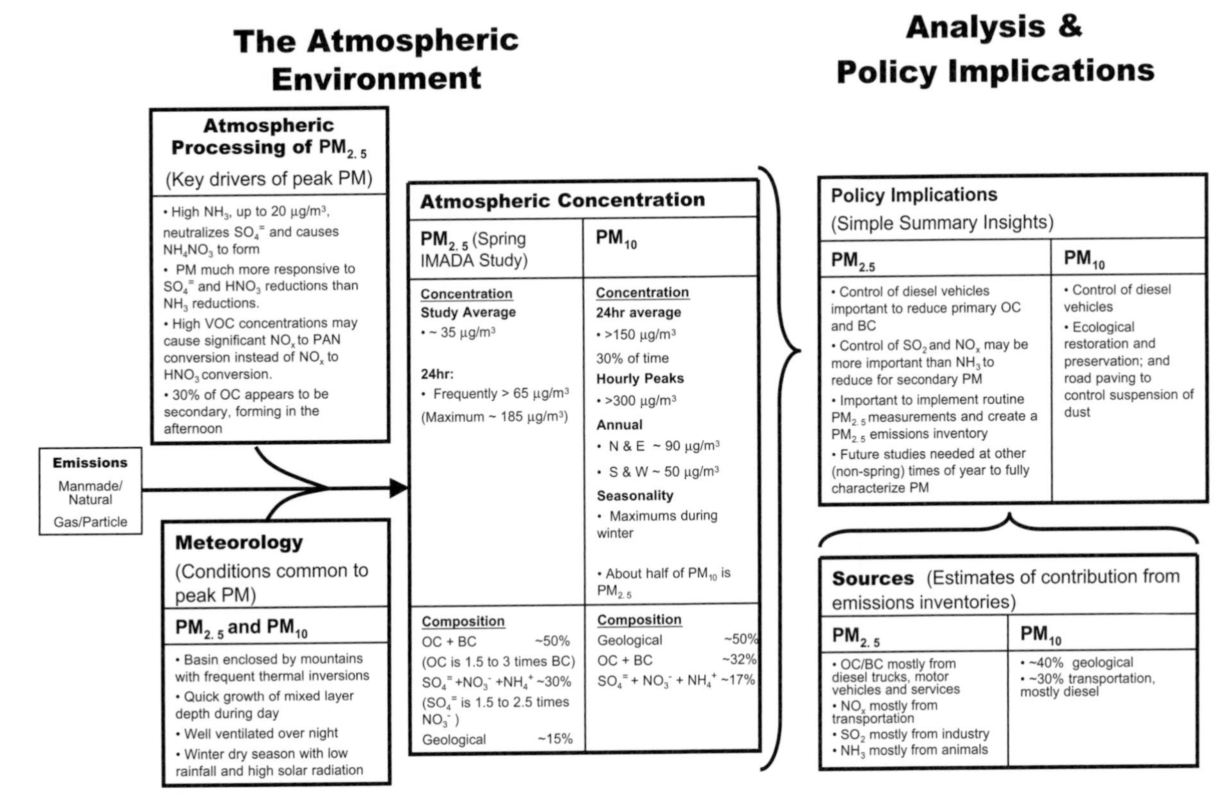

Figure 10.6. Simplified conceptual model for Mexico City.

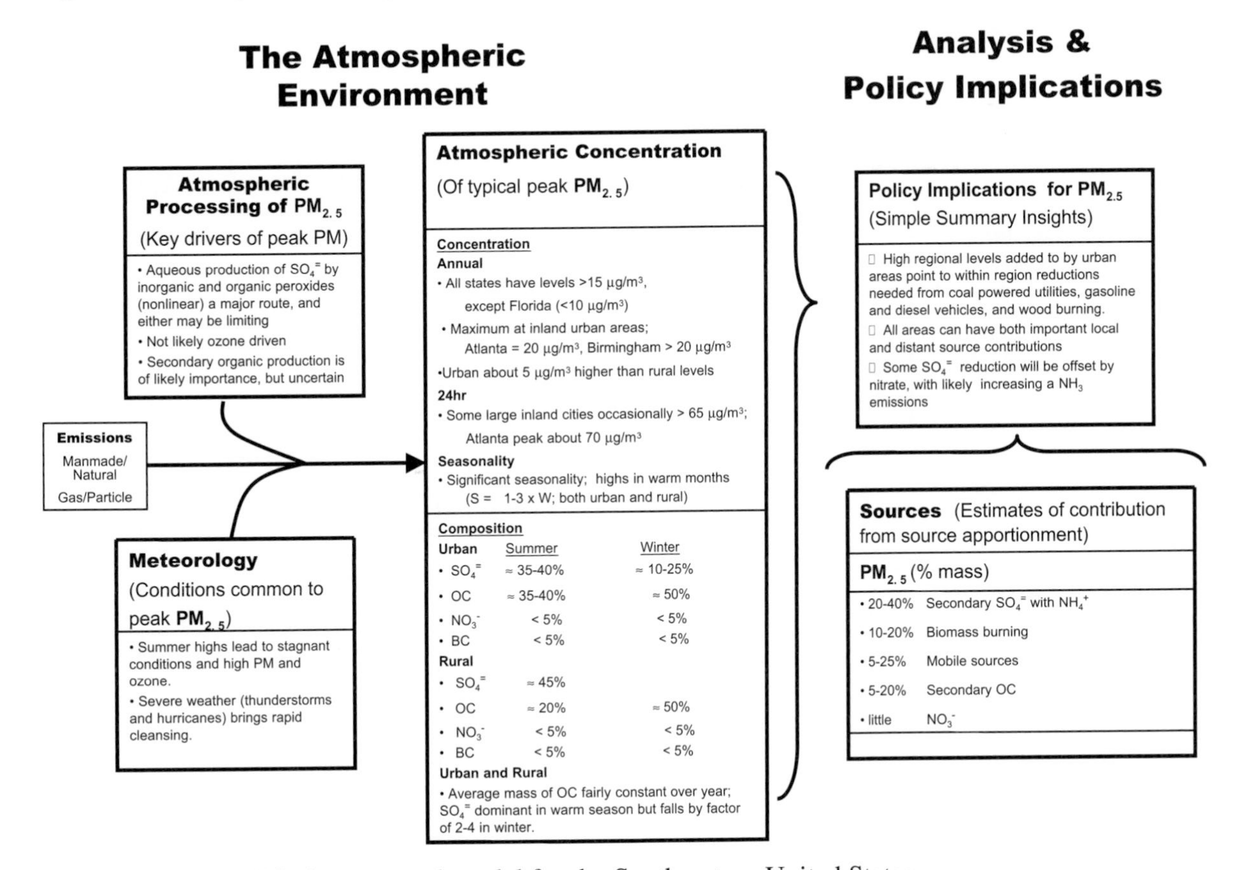

Figure 10.7. Simplified conceptual model for the Southeastern United States.

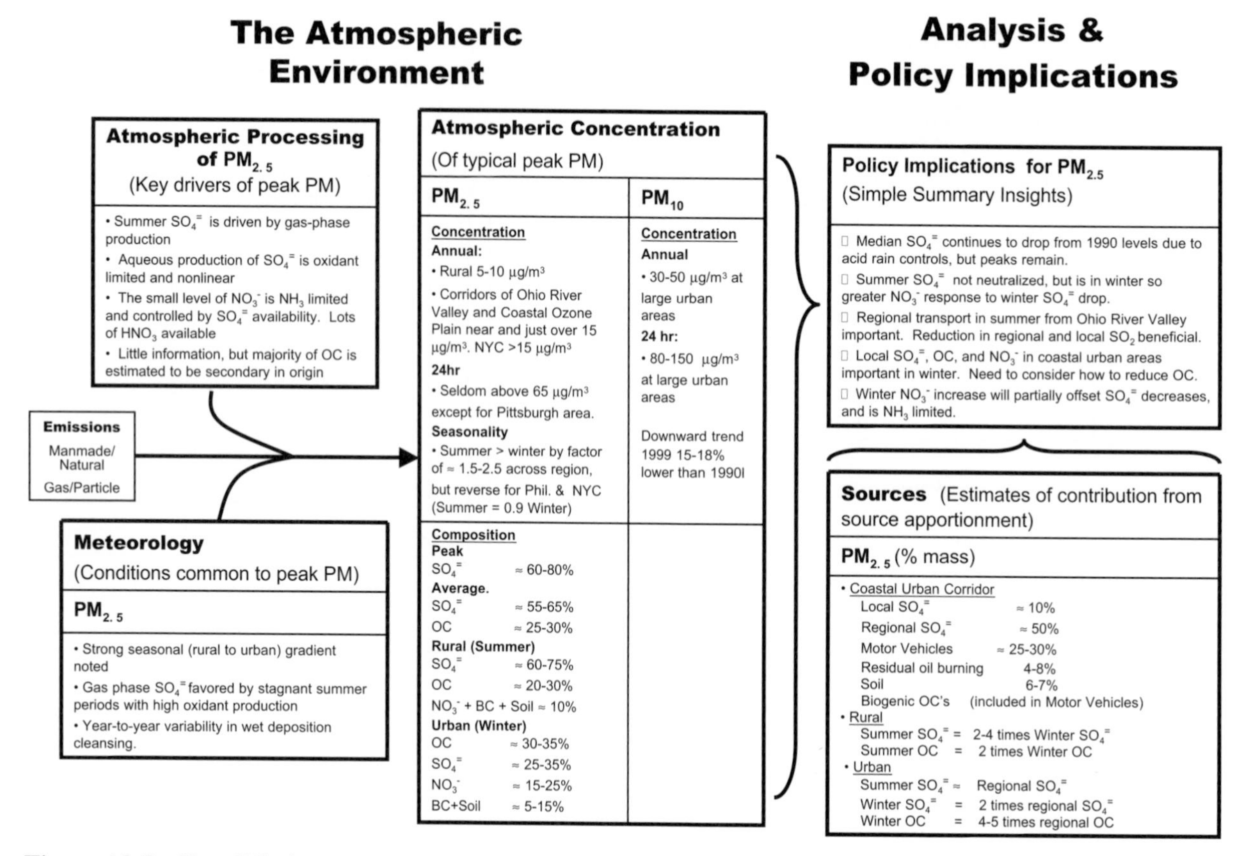

Figure 10.8. Simplified conceptual model for the Northeastern United States

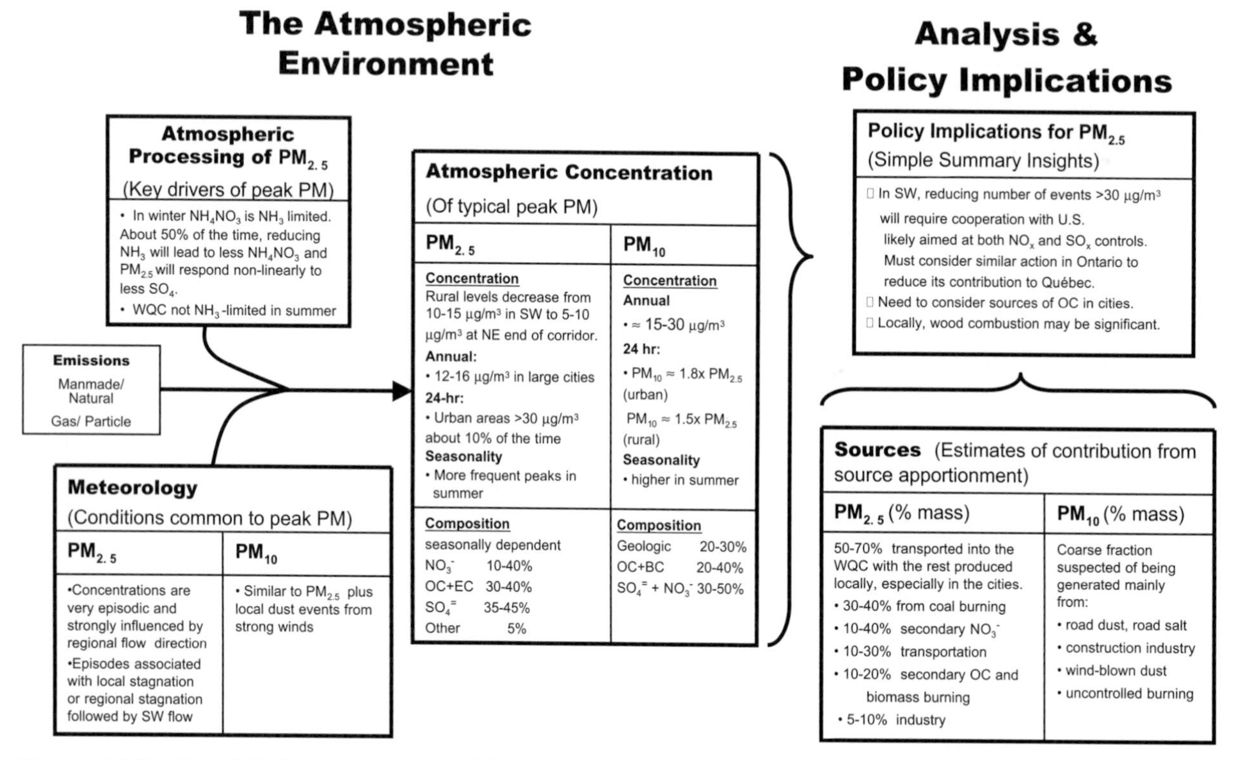

Figure 10.9. Simplified conceptual model for the Windsor - Quebec City Corridor

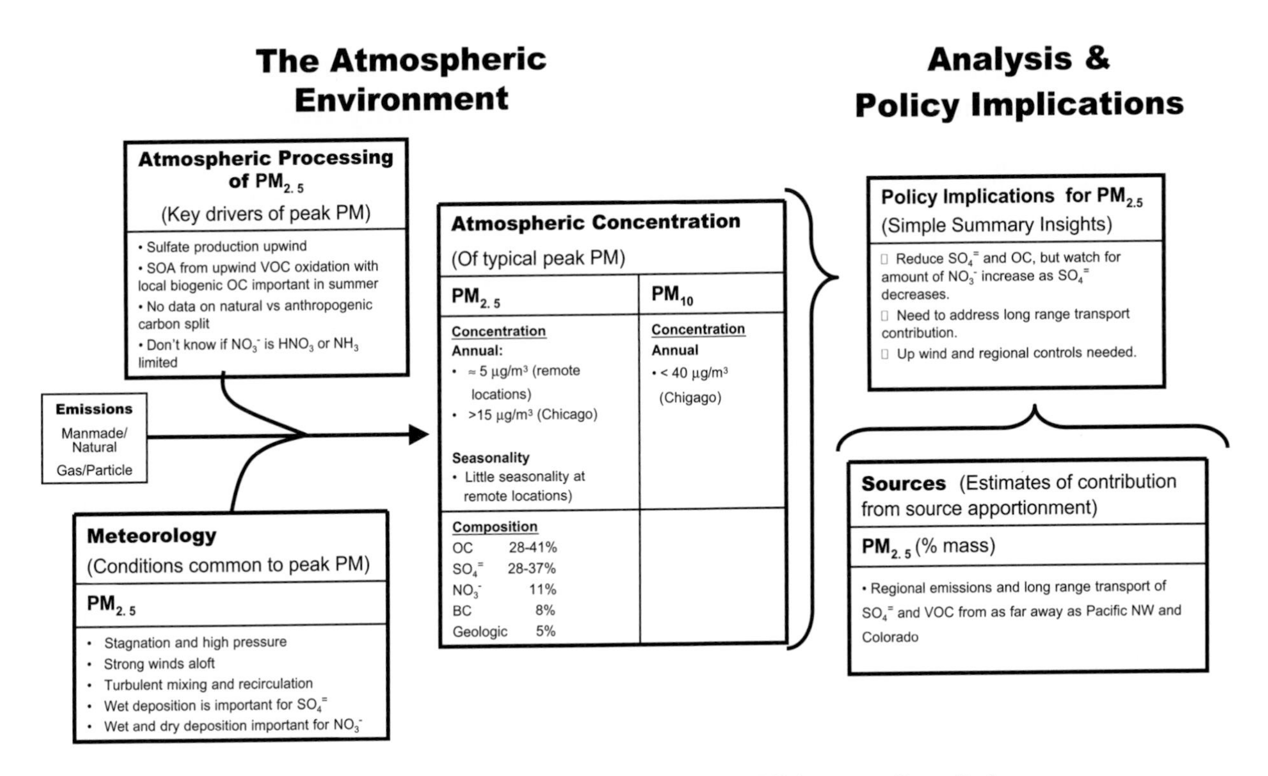

Figure 10.10. Simplified conceptual model for the U.S. Upper Midwest— Great Lakes.

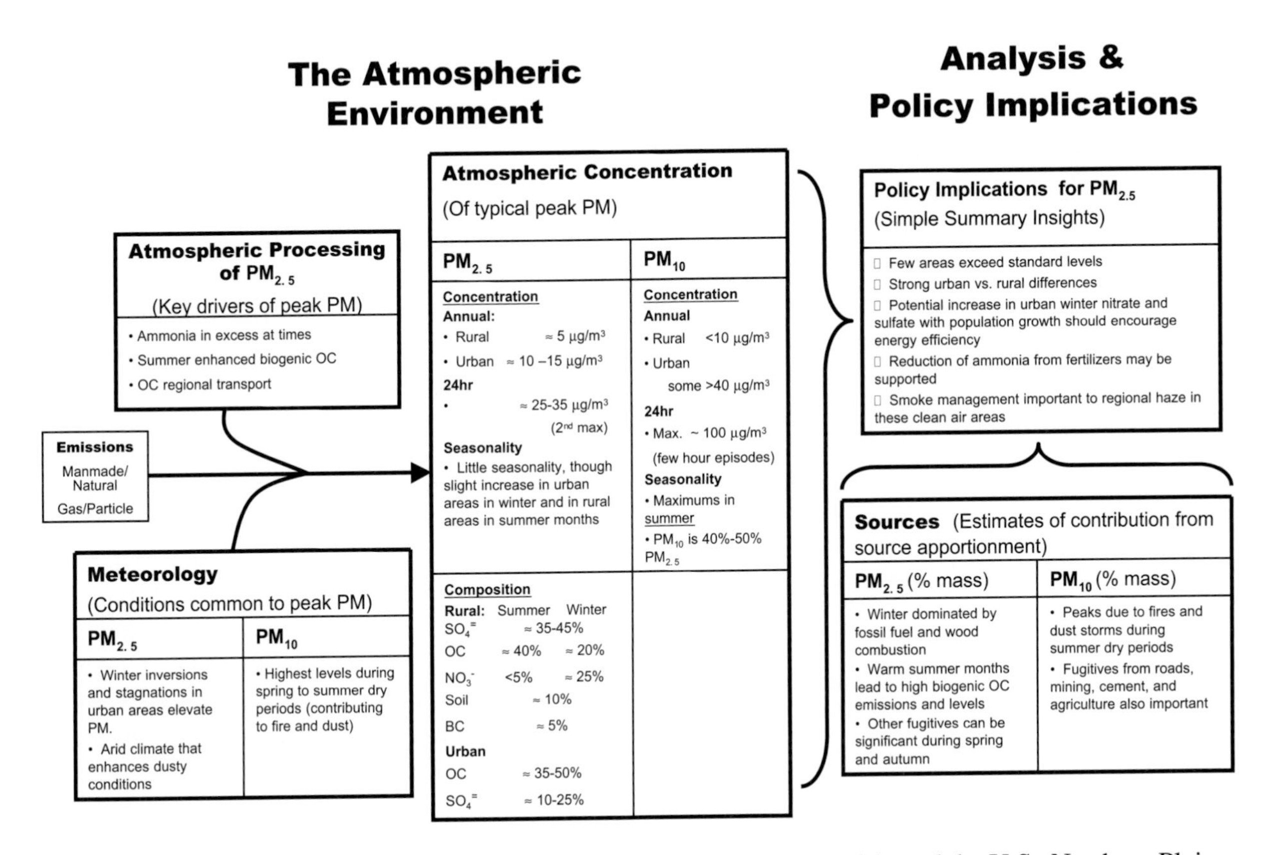

Figure 10.11 Simplified conceptual model for the Canadian Southern Prairie and the U.S. Northern Plains.

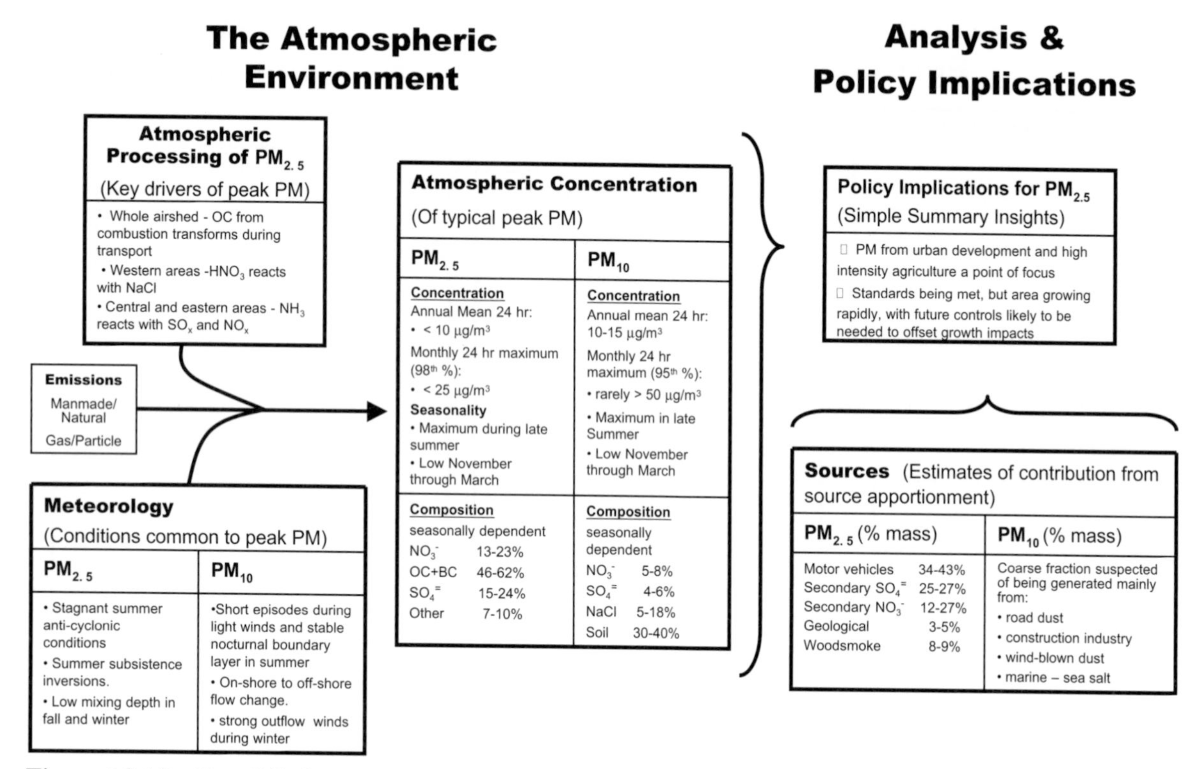

The Atmospheric Environment
Analysis & Policy Implications
Atmospheric Processing of PM2.5
(Key drivers of peak PM)
• Whole airshed - OC from combustion transforms during transport
• Western areas -HNO3 reacts with NaCl
• Central and eastern areas - NH3 reacts with SOx and NOx
Emissions
Manmade/ Natural
Gas/Particle
Meteorology
(Conditions common to peak PM)
PM2.5
• Stagnant summer anti-cyclonic conditions
• Summer subsistence inversions.
• Low mixing depth in fall and winter
PM10
•Short episodes during light winds and stable nocturnal boundary layer in summer
• On-shore to off-shore flow change.
• strong outflow winds during winter
Atmospheric Concentration
(Of typical peak PM)
PM2.5
Concentration
Annual Mean 24 hr:
• < 10 µg/m3
Monthly 24 hr maximum (98th %):
• < 25 µg/m3
Seasonality
• Maximum during late summer
• Low November through March
Composition
seasonally dependent
NO3- 13-23%
OC+BC 46-62%
SO4= 15-24%
Other 7-10%
PM10
Concentration
Annual mean 24 hr: 10-15 µg/m3
Monthly 24 hr maximum (95th %):
• rarely > 50 µg/m3
• Maximum in late Summer
• Low November through March
Composition
seasonally dependent
NO3- 5-8%
SO4= 4-6%
NaCl 5-18%
Soil 30-40%
Policy Implications for PM2.5
(Simple Summary Insights)
PM from urban development and high intensity agriculture a point of focus
Standards being met, but area growing rapidly, with future controls likely to be needed to offset growth impacts
Sources (Estimates of contribution from source apportionment)
PM2.5 (% mass)
Motor vehicles 34-43%
Secondary SO4= 25-27%
Secondary NO3- 12-27%
Geological 3-5%
Woodsmoke 8-9%
PM10 (% mass)
Coarse fraction suspected of being generated mainly from:
• road dust
• construction industry
• wind-blown dust
• marine – sea salt

Figure 10.12. Simplified conceptual model for the Lower Fraser Valley.

daily concentrations during the fall and winter months. Levels of 70 to 100 µg/m^3 over 24 hours are common at Bakersfield (25 percent of 1st Qtr.1999 daily levels) and Fresno (5 days out of a 25-day period in 1995).

Annual PM_{10} levels in excess of 50 µg/m^3 are also expected for much of the SJV based on past data. Annual mean levels in 1999 were above 50 µg/m^3 for most areas monitored. Daily levels in the fall months drive these annual averages and range from 110 to 135 µg/m^3 for residential, agricultural, and industrial sites. For example, daily levels exceeded 150 µg/m^3 many times (3 of 4 sampled) at Corcoran during November 1995 (see Figure 10.13).

10.3.2 Compositional Analysis of PM

For winter $PM_{2.5}$ episodes, NO_3^- is dominant (46 percent on average, less in urban areas than in rural areas), with OC and BC making up 34 percent and $(NH_4)_2SO_4$ being 6 percent according to winter 1995 data. A small fraction (<1 percent) is geological material, and 13 percent is unidentified other material, including water. PM_{10} is composed of geological material (57 percent), BC and OC (22 percent), NO_3^- (16 percent), $SO_4^=$ (3 percent), and unidentified other material (2 percent), according to November 1995 data.

10.3.3 Meteorological Influences

Both urban-scale distribution of primary source emissions and regional-scale distribution of secondary compounds influence $PM_{2.5}$ in the SJV during the winter. Air stagnation is a key reason for the buildup of PM, with atmospheric diffusion often replacing advection as the dominant transport mode due to low wind speeds. Low mixing heights during episodes also contribute to $PM_{2.5}$ buildup. Transport of pollutants aloft occurs with wind speeds typically higher than those near the surface; subsequent mixing of these pollutants from aloft into the surface layer during the day may be an important means of distributing pollutants within the valley during multi-day episodes. The primary removal processes for trapped pollutants during these episodes are wet and dry deposition. Wet deposition, especially during fog

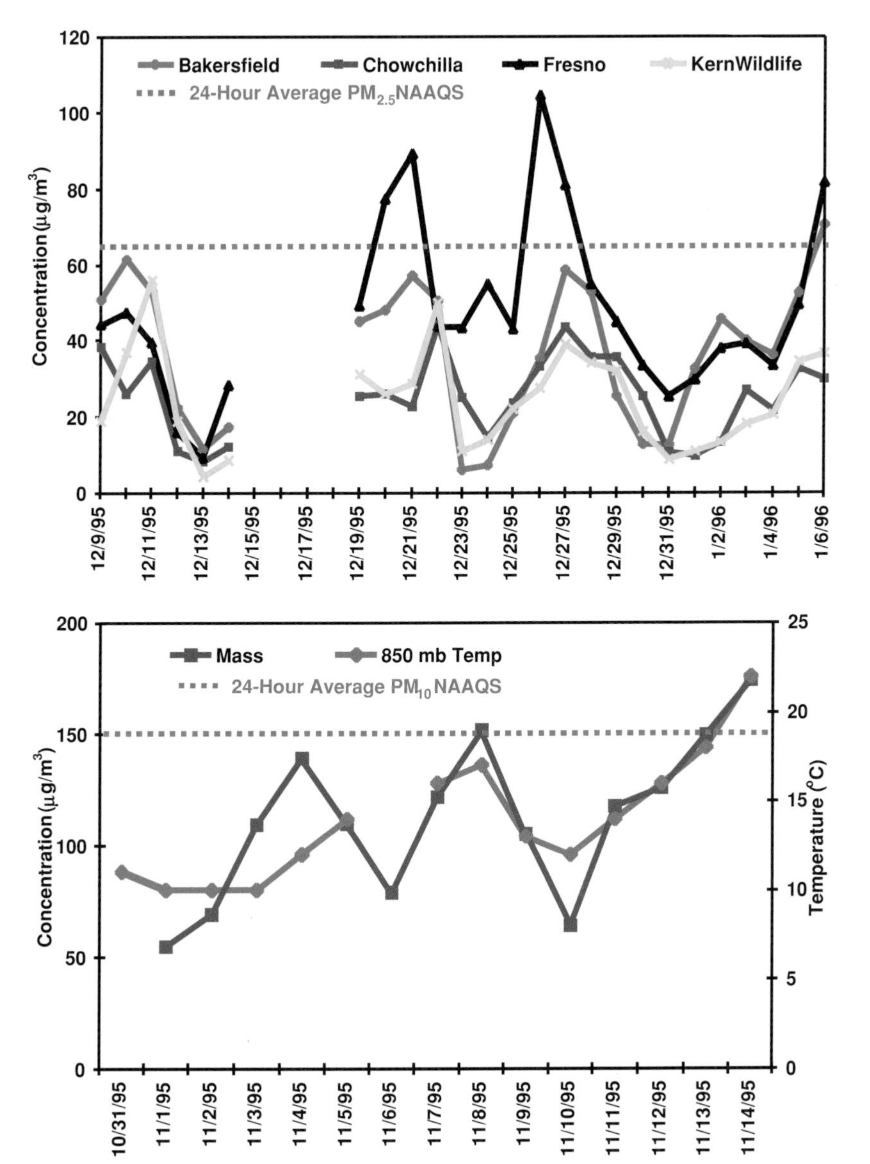

Figure 10.13. Temporal variations of PM during IMS95. Top: $PM_{2.5}$ mass concentrations during the winter study. Bottom: PM_{10} mass during the fall study, Corcoran, CA. Dotted lines indicate the 24-hr $PM_{2.5}$ NAAQS value of 65 $\mu g/m^3$ and the 24-hr PM_{10} NAAQS of 150 $\mu g/m^3$ (source: Pun and Seigneur, 1999).

events, is likely to be the most important removal process during winter for soluble species such as NO_3^-, NH_4^+, and $SO_4^=$.

High PM_{10} concentrations are observed primarily during fall episodes. Secondary components are more uniformly distributed than primary components. The buildup of PM during fall episodes is mostly at sub-regional scales (5 to 15 km). Again low wind speeds and low mixing heights are major factors in elevated PM levels. Wet deposition plays much less of a removal role during fall than winter due to lower precipitation and patchy and mild fogs in the fall.

10.3.4 Atmospheric Processes Contributing to PM

High spatial representativeness of $PM_{2.5}$ concentrations at rural sites relative to urban sites points to the dominance of secondary compounds. The relatively high contribution of primary PM in urban samples is indicative of limited dispersion of emissions from urban centers. The regional distribution of secondary compounds (e.g., NH_4NO_3) is the result of a combination of factors including widespread sources (e.g., NH_3) and regional transport.

Ammonium nitrate is the most dominant compound in the $PM_{2.5}$ fraction. It is formed by the combination of NH_3 and HNO_3 ($NH_3 + HNO_3 = NH_4NO_3$). In the SJV, the limiting factor for this reaction is the availability of HNO_3. There is typically an abundance of NH_3 present. Nitric acid is not emitted directly but depends on several other reactions needing the availability of NO_2 and oxidants. Oxidants are formed by the reactions of VOC and NO_x. Wintertime NO_3^- formation may be influenced by both daytime chemistry (reaction between NO_2 and OH) and nighttime chemistry (reaction sequence initiated by the reaction between NO_2 and ozone). Temperature, water vapor and sunlight are also important. The formation of oxidants can be limited by the availability of one of two components, VOC or NO_x. In the case of secondary winter NO_3^- in the SJV, oxidant formation (and NO_3^- formation) is thought to be sensitive to VOC concentrations in many urban areas, e.g., Fresno. During the summer, when ozone is the pollutant of concern, oxidant formation in the non-urban areas of the SJV is more sensitive to NO_x.

Organic compounds account for 35 percent of $PM_{2.5}$ mass at urban sites. A notable feature of organic compounds is that some arise from primary emissions and some occur from secondary formation. Unfortunately, measurements are not available to differentiate between primary and secondary OC. Using levoglucosan, syringol, resin acids, PAHs, alkanoic and alkenore acids as PM tracers, Schauer and Cass (2000) were able to perform a detailed OC source-apportionment study. They estimated the maximum secondary fraction at 16 percent on a 24-hr average basis. Strader et al. (1999) estimated that 15 to 20 μg/m^3 of secondary organic PM (or about half of the organic fraction) can be formed during certain late-afternoon periods.

10.3.5 Sources and Source Regions Contributing Principal Chemicals of Concern

Sources: The dominant winter source contributions of $PM_{2.5}$ in the SJV are secondary NO_3^- (46 percent) (see discussion of governing precursors above), vegetative burning (15 percent), diesel- and gasoline-vehicle exhaust (12 percent), secondary $SO_4^=$ (6 percent), soil and road dust (1 percent), and unknowns of roughly 20 percent. In urban areas, such as Fresno and Bakersfield, the contribution of primary sources tended to be higher during IMS95. Vegetative burning and mobile sources accounted for 21 percent and 14 percent, respectively, of $PM_{2.5}$ in urban areas but only 9 percent and 10 percent, respectively, in rural areas. On the other hand, NH_4NO_3 accounted for 32 percent of $PM_{2.5}$ in urban areas and 60 percent in rural areas.

The dominant source contributions to PM_{10} in the SJV in the fall are primary geological material (57 percent), secondary NH_4NO_3 (16 percent), primary motor-vehicle exhaust (7 percent), primary vegetative burning (5 percent), secondary $(NH_4)_2SO_4$ (3 percent), with unknowns of roughly 12 percent. In winter, the fraction of geological material was lower (10 percent), replaced by higher contribution from secondary NO_3^- (40 percent). The other contributions are from primary mobile sources and vegetative burning (12 percent each), $SO_4^=$ (6 percent), and unknown (19 percent). Mechanical generation, rather than windblown dust, is the more likely contribution of geological material.

Local emissions vs. long-range transport: There is a strong urban contribution to $PM_{2.5}$ levels on top of high regional levels. For example, on days that exceeded the 24-hr NAAQS for $PM_{2.5}$, urban areas in the SJV on average had 83 percent higher $PM_{2.5}$ mass than did rural areas, with 4.6 times the vegetative-burning material (consistent with domestic fireplace usage) and 2 times the mobile source contribution. Absolute NO_3^- contributions were similar. Average $PM_{2.5}$ mass on episode days was 57 μg/m^3 at urban sites and 31 μg/m^3 at rural sites. Another example shows that peak daily levels over Fresno and Bakersfield were 80 to100 μg/m^3 while at the nearby rural locations of Chowchilla and Kern Wildlife levels were around 45 μg/m^3. The regional distribution of secondary compounds likely results from widespread area sources with a combination of diffusive transport and long-range transport aloft.

Man-made vs. natural: As discussed, during winter $PM_{2.5}$ episodes, $NH_4NO_3$and organic compounds dominate. Ammonium nitrate is influenced primarily by HNO_3 which is likely limited by VOC emissions; the relative contribution of anthropogenic and biogenic VOC emissions to oxidant formation during winter is unknown at this point. Primary organic compound levels likely arise principally from vegetative burning and mobile sources, and secondary organic compounds result from both man-made and naturally generated precursors. Vegetative-burning contributions have been shown to be consistent with domestic fireplace use, and as evidenced by high levels during evenings in urban areas, particularly during the holiday season. Rural levels, consisting mainly of secondary components, remain small and essentially constant, indicating regional distributions of $PM_{2.5}$.

During fall PM_{10} episodes, geological material (57 percent average) is primarily mechanically generated through processes such as entrainment of road dust and agricultural tilling. Ammonium nitrate (16 percent average) may be more sensitive to daytime chemistry that produces high levels of HNO_3 than during $PM_{2.5}$ wintertime episodes.

10.3.6 Implications for Policy Makers

Reducing secondary compounds during fall and winter months appears centrally important to reducing $PM_{2.5}$ levels, although the source contributions of biomass burning (fireplace usage) to urban $PM_{2.5}$ pollution may be significant. As the components that make up better than three-fourths of secondary compounds are NO_3^- and organics, and both of these are influenced most by volatile and semivolatile organic-compound levels, sources of these would be worth pursuing. Both urban and regional reductions in these sources will be needed.

Strategies for reducing ozone during summer may focus on NO_x emission reductions since ozone formation within the SJV is then NO_x limited. Reductions in NO_x may not be the best course of action for reducing NO_3^- in the possibly VOC-sensitive wintertime condition. Box-model simulations indicate that NO_x reductions may have the counter-intuitive effect of increasing NO_3^- formation during winter (Pun and Seigneur, 2001). Therefore, coordinated efforts will be required to formulate control strategies beneficial to both ozone and PM air quality.

10.4 CONCEPTUAL MODEL OF PM OVER LOS ANGELES, CALIFORNIA

The city of Los Angeles is located on the Pacific coast of southern California between San Diego and San Francisco. The South Coast Air Basin that encompasses the greater Los Angeles area has an area of 10,743 square miles with a population of more than 15 million people (approximately half the population of California). It is the second most populous urban area in the United States. More than 9 million motor vehicles, thousands of businesses, and industries emit significant amounts of primary PM and precursor gases. Mountains on three sides of the air basin combine with strong thermal inversions to trap pollutants close to the Earth's surface where they can undergo photochemical transformation leading to extremely polluted conditions.

10.4.1 Annual and Seasonal Levels of $PM_{2.5}$ and PM_{10} in Relation to Mass-Based Standards

Monthly $PM_{2.5}$ concentrations throughout the South Coast Air Basin surrounding Los Angeles during 1993 are shown in Figure 10.14 (Christoforou et al., 1999). The regional $PM_{2.5}$ concentration in 1993 averaged across all sites and months was 28 $\mu g/m^3$, while the peak 24-hr $PM_{2.5}$ concentration recorded was 139 $\mu g/m^3$ at Rubidoux, CA. Monthly $PM_{2.5}$ concentrations measured at sites located in the South Coast Air Basin between August 1998 and July 1999 are shown in Figure 10.14 (Kim et al., 2000a, 2000b). The regional $PM_{2.5}$ concentration during this 12-month period averaged across all sites and months was 25 $\mu g/m^3$, while the peak 24-hr $PM_{2.5}$ concentration recorded was 98 $\mu g/m^3$ at Fontana, CA. See Figure 6.8 for the spatial distribution of annual-average $PM_{2.5}$ concentrations in the South Coast Air Basin.

The $PM_{2.5}$ trends illustrated in Figures 10.14 and 10.15 clearly indicate that Los Angeles experiences annual $PM_{2.5}$ levels that are approximately double the annual-average $PM_{2.5}$ NAAQS of 15 $\mu g/m^3$. Likewise, the peak 24-hr average $PM_{2.5}$ concentrations detected during 1993 and 1998-99 indicate that the 24-hr average $PM_{2.5}$ NAAQS of 65 $\mu g/m^3$ is also exceeded in Los Angeles.

Monthly PM_{10} concentrations throughout the South Coast Air Basin between August 1998 and July 1999 are shown in Figure 10.16 (Kim et al., 2000a,b). Annual-average PM_{10} concentrations in the eastern end of the air basin (Fontana, Ontario, Rubidoux) during this time period were 62 $\mu g/m^3$, while annual-average concentrations in the central and western portions of the air basin (Central LA, Diamond Bar, Anaheim, Long Beach) were 45 $\mu g/m^3$. The peak

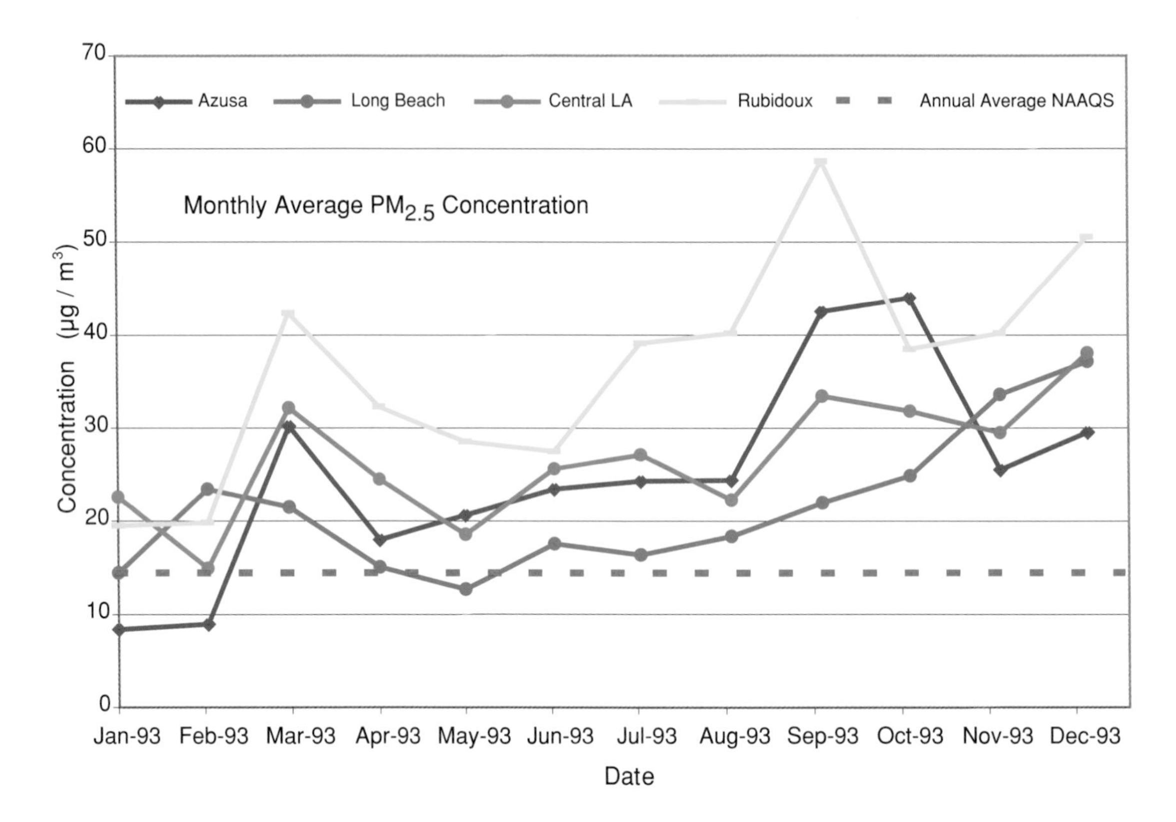

Figure 10.14. Temporal variations of monthly-average $PM_{2.5}$ at sites in the South Coast Air Basin surrounding Los Angeles during 1993. Dotted line indicates the annual-average $PM_{2.5}$ NAAQS value of 15 µg/m^3 (source: Christoforou et al., 2000).

recorded 24-hr average PM_{10} concentration between August 1998 and July 1999 was 187 µg/m^3 (Ontario). Annual PM_{10} concentrations in the South Coast Air Basin are also available for the year 2000 (AIRS database 2001). The average PM_{10} concentration at Rubidoux, CA, during this year was 59 µg/m^3, with lower concentrations once again measured in the central and western portions of the air basin (Azusa: 21 µg/m^3, central Los Angeles: 40 µg/m^3, Long Beach: 31 µg/m^3). See Figure 6.5 for the spatial distribution of annual-average PM_{10} concentrations in the South Coast Air Basin.

The PM_{10} trends described above and illustrated in Figure 10.16 indicate that the eastern portion of the South Coast Air Basin experiences PM_{10} concentrations that exceed the annual average PM_{10} NAAQS of 50 µg/m^3. Peak 24-hr average PM_{10} concentrations do not routinely exceed the 24-hr NAAQS of 150 µg/m^3.

Annual $PM_{2.5}$ concentrations averaged at all monitoring sites between August 1998 and July 1999 accounted for approximately 50 percent of the annual-average PM_{10} concentrations. In contrast, $PM_{2.5}$ concentrations accounted for 84 percent of PM_{10} during the peak $PM_{2.5}$ concentration event that occurred on January 18, 1999.

10.4.2 Compositional Analysis of PM

Contributing chemicals: The annual-average composition of $PM_{2.5}$ in the South Coast Air Basin during 1993 and between August 1998-July 1999 is shown in Figure 10.17. Slightly more than 50 percent of the annual-average $PM_{2.5}$ concentrations in the South Coast Air Basin during both years was accounted for by primary pollutants including organic compounds, BC, and "other" material (composed of crustal and metal elements). Soluble ions including NO_3^-, $SO_4^=$, NH_4^+, Na^+ and Cl^- accounted for slightly less than 50 percent of the annual-average $PM_{2.5}$ concentrations. Note that previous source-apportionment studies have shown that the majority

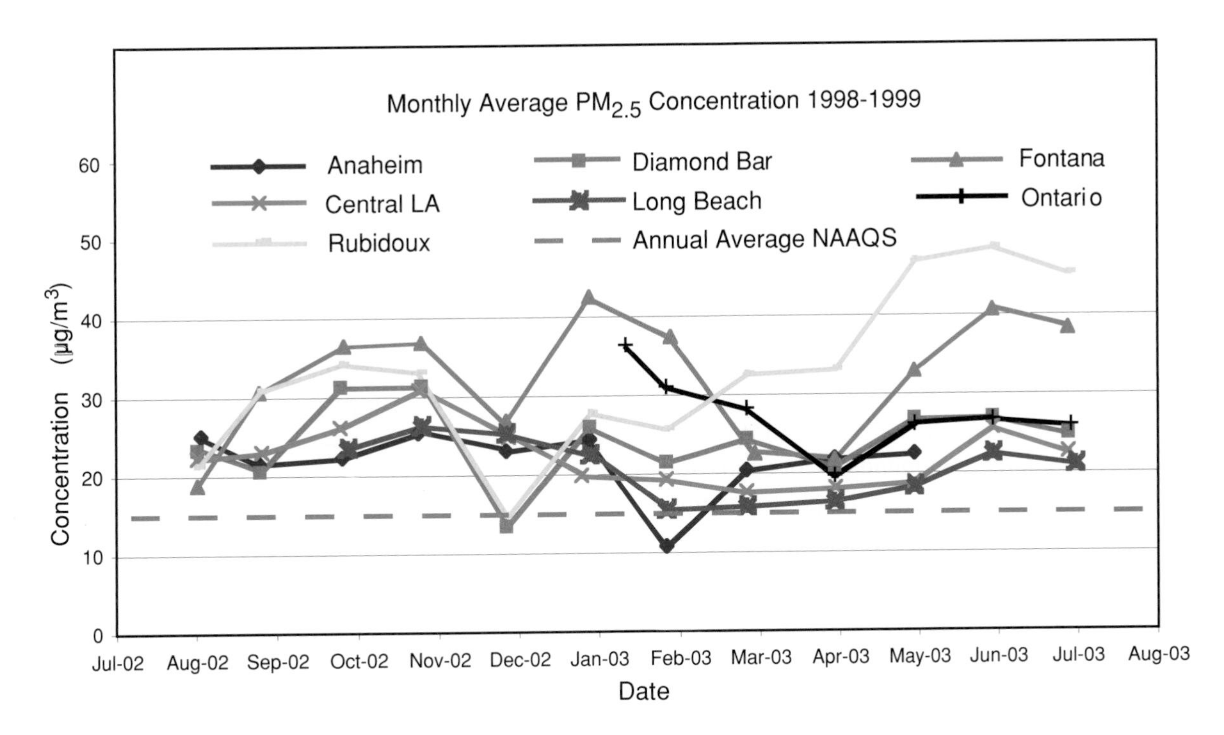

Figure 10.15. Temporal variations of monthly-average $PM_{2.5}$ at sites in the South Coast Air Basin surrounding Los Angeles between August 1998 and July 1999. Dotted line indicates the annual-average $PM_{2.5}$ NAAQS value of 15 μg/m^3 (source: Kim et al. 2000a,b).

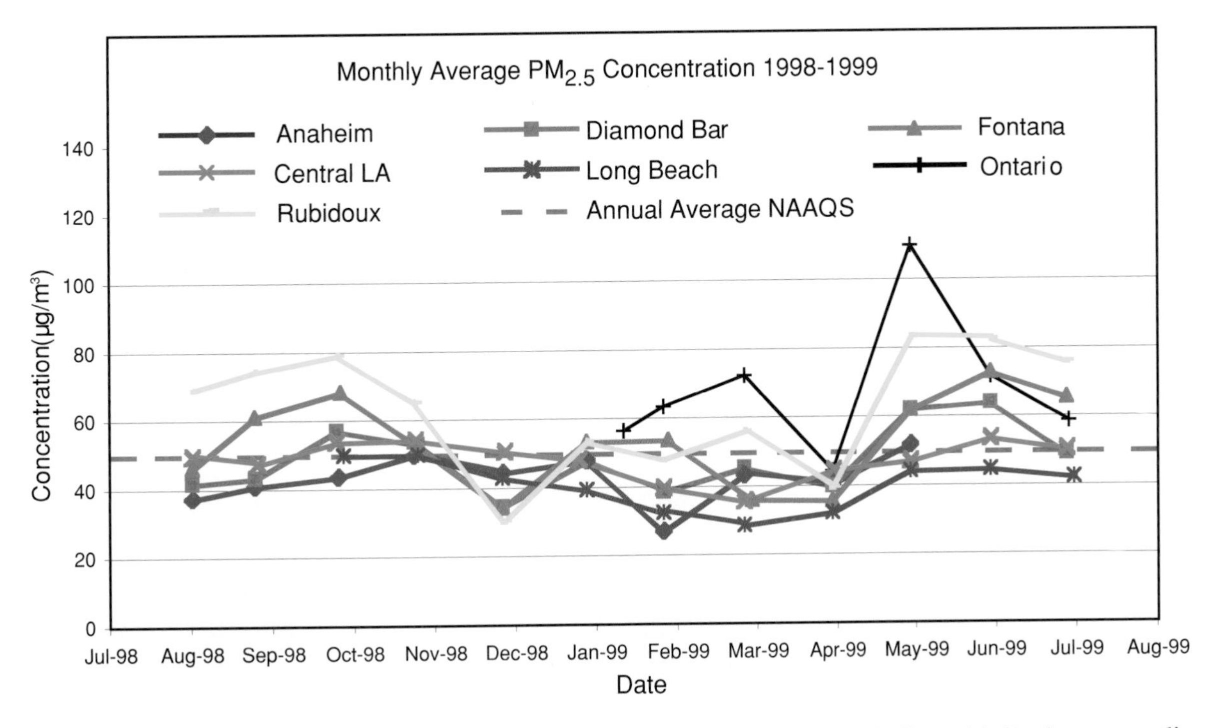

Figure 10.16. Temporal variations of monthly-average PM_{10} at sites in the South Coast Air Basin surrounding Los Angeles between August 1998 and July 1999. Dotted line indicates the annual-average PM_{10} NAAQS value of 50 μg/m^3 (source: Kim et al., 2000a,b).

of $SO_4^=$ in the South Coast Air Basin is associated with background particles, not local SO_x emissions (Kleeman et al., 1999).

Approximately 5 percent of the organic compounds contained in airborne particles in the Los Angeles area during 1993 have been identified using GC/MS techniques. Commonly occurring compounds include n-alkanes, branched alkanes, alkanols, alkanals, alkanoic acids, alkanoic diacids, steranes, hopanes, diterpenoid acids, polycyclic aromatic hydrocarbons (PAH), oxyPAH, nitroPAH, and heterocyclic aromatic compounds. Concentrations of secondary OC associated with photochemical reactions are higher in the eastern (downwind) portion of the South Coast Air Basin.

Although the majority of the annual-average $PM_{2.5}$ mass is associated with primary particle emissions, peak-concentration events that occur in the eastern portion of the South Coast Air Basin result largely from the formation of secondary PM such as NH_4NO_3. Secondary NO_3^-, NH_4^+, and $SO_4^=$ accounted for over 50 percent of the $PM_{2.5}$ concentrations in the greater Los Angeles area during the 16 days, with the highest recorded $PM_{2.5}$ concentrations between August 1998 and July 1999 (Kim et al., 2000a,b).

10.4.3 Meteorological Influences

The South Coast Air Basin surrounding Los Angeles provides a textbook example of how meteorology and terrain can contribute to air-quality problems. The air basin is surrounded on three sides by mountains, with the western edge open to the Pacific Ocean. The prevailing wind direction is west to east, resulting in generally higher concentrations in the downwind eastern portion of the air basin (see Figures 6.8, 10.14 and 10.15). During the late-summer and early-fall months, temperature inversions form during the evening hours as the atmosphere cools from below. This condition produces low mixing depths and effectively puts a "lid" on the box created by the surrounding terrain. Pollutants are trapped close to the earth's surface leading to high concentrations of primary $PM_{2.5}$ and the formation of large amounts of secondary $PM_{2.5}$.

The primary removal pathways for $PM_{2.5}$ are dry deposition to the Earth's surface and transport out of the South Coast Air Basin. Heavy fogs are not as common in Los Angeles as they are in California's central valley, and so wet deposition is not a major removal process. Particulate air pollution from Los Angeles has been identified as a potential cause for air-quality problems and visibility reduction in neighboring states.

10.4.4 Atmospheric Processes Contributing to PM

Airborne PM in the South Coast Air Basin surrounding Los Angeles can be divided into three general categories: background particles advected into the air basin, primary PM emitted in the air basin, and secondary PM that forms in the air basin. Each of these particle categories interacts with atmospheric processes to produce the observed $PM_{2.5}$ trends.

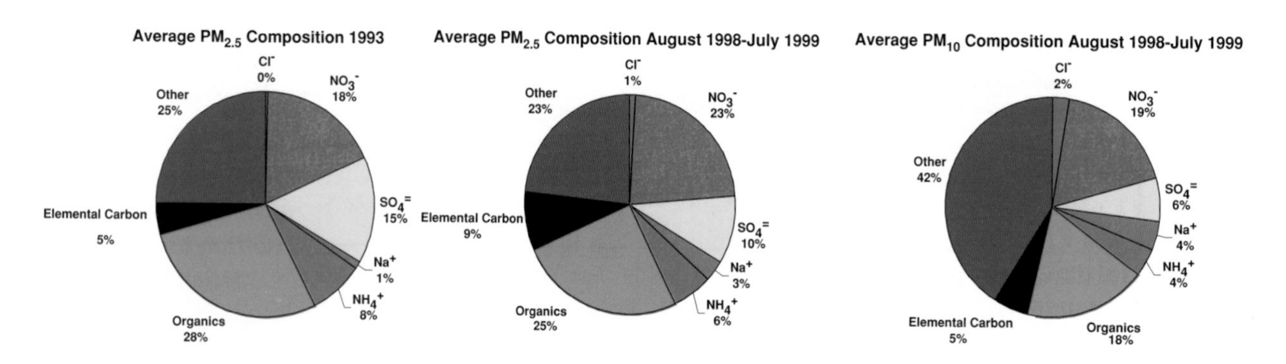

Figure 10.17. Annual $PM_{2.5}$ and PM_{10} composition averaged across all monitoring sites in the South Coast Air Basin.

Background particles including sea spray and fine-particle $SO_4^=$ are advected into the South Coast Air Basin from the Pacific Ocean. Sea-spray particles are produced by the mechanical action of breaking waves. Particles of this type are generally larger than 1 μm diameter, and so the majority of sea-spray particles contribute to PM_{10} concentrations with smaller contributions to $PM_{2.5}$. Background $SO_4^=$-containing particles with a mass distribution that peaks below 1 μm particle diameter also are advected into the South Coast Air Basin. The origin of these particles is currently unknown, but their size distribution suggests biological processes that convert DMS to $SO_4^=$, or combustion sources that burn high sulfur-content fuel. It is unlikely that background $SO_4^=$ particles originate from combustion sources in California since high-sulfur fuel is not used in that state. One possibility is that large-scale weather systems transport airborne particles out over the ocean and then into the Los Angeles area. Another possibility is that ships burning high-sulfur fuel off California's west coast may produce $SO_4^=$ particles which are transported into the Los Angeles area.

The annual-average concentration of fine background particles measured at San Nicolas Island (off the west shore of the South Coast Air Basin) was 5 to 6 μg/m^3 during 1993 and 7 μg/m^3 during 1995 (Kim et al. 2000a,b). Since particles in the $PM_{2.5}$ size-range can be transported for long distances, it is likely that background particles accounted for approximately one-third to one-half of the annual $PM_{2.5}$ NAAQS (15 μg/m^3) in Los Angeles. Background particles also have been shown to play a significant role in the formation of secondary PM during peak $PM_{2.5}$-concentration events. Source-oriented mechanistic air-quality calculations have demonstrated that hygroscopic background particles act as ideal sites for the formation of particulate NH_4NO_3 (Kleeman et al., 1999). Note that monthly-average $PM_{2.5}$ concentrations in parts of the South Coast Air Basin exceed 40 μg/m^3, clearly indicating that anthropogenic sources account for the majority of the particulate air-pollution problem.

Primary particles are emitted by a variety of sources in the South Coast Air Basin including combustion processes (mobile and stationary), food preparation, paved and unpaved road travel, and agricultural activities. Primary particle emissions are broadly distributed in the air basin, reflecting a general pattern of urban sprawl. The concentration of primary PM in the atmosphere is a direct balance between emission rate, deposition to the surface, and mixing aloft. The data record for 1993 and 1998-99 indicates that that the combination of background and primary PM is large enough to violate the annual-average $PM_{2.5}$ NAAQS at most locations in the South Coast Air Basin. Note that statistical source-receptor calculations have shown that the majority of OC in the South Coast Air Basin is primary rather than secondary in origin (Schauer et al., 1996; Hannigan, 1997).

Secondary PM forms chiefly in the downwind eastern portion of the South Coast Air Basin after precursor gases have time to react in the atmosphere. Ammonium nitrate is the most dominant secondary compound in the $PM_{2.5}$ fraction. It is formed by the reaction $NH_3 + HNO_3 = NH_4NO_3$. Agricultural operations in the eastern portion of the air basin emit large amounts of NH_3, leading to high NH_4NO_3 concentrations around Rubidoux and Riverside, CA. The formation of secondary NH_4NO_3 drives the highest 24-hr $PM_{2.5}$ values recorded during 1993 and 1998-99.

Nitric acid is a product of the photochemical reaction system that forms ozone in the atmosphere. The concentration of HNO_3 is influenced by the same factors that affect ozone concentrations, including precursor emissions (NO_x, VOC), sunlight, and temperature. It is also important to note that NH_4NO_3 is semivolatile; NH_4NO_3 partitions back to gas-phase HNO_3 and NH_3 as temperature increases. For this reason, particulate NH_4NO_3 concentrations are typically highest in fall and winter months, not during summer months when temperatures are much hotter.

10.4.5 Sources and Source Regions Contributing Principal Chemicals of Concern

Sources: A number of studies have calculated how different sources contribute to PM concentrations in the South Coast Air Basin. A detailed source apportionment of PM was carried out at Riverside, CA, on September 25, 1996 (U.S.EPA, AIRS database 2001a,b,c). Secondary PM made a large contribution to $PM_{2.5}$ concentrations during this episode (NO_3^- 47 percent, NH_4^+ 15 percent, $SO_4^=$ 3 percent). The majority of this secondary PM accumulated onto hygroscopic background particles advected into the air basin from the Pacific Ocean. By tracking primary and secondary PM formation within a mechanistic air-quality model, it is possible to calculate source contributions to total (primary + secondary) PM concentrations. Figure 10.18 shows how different sources contributed to the size distributions of total airborne PM at Long Beach, Fullerton, and Riverside, California on September 24-25, 1996. These locations transect the greater Los Angeles region along the approximate path of an air parcel during the periods of weak onshore flow that typically occur during the fall months when $PM_{2.5}$ concentrations are highest. Crustal sources and paved road dust dominate the PM size distribution above 2.5 μm at all sites. Background $SO_4^=$ particles contribute significantly to $PM_{2.5}$ mass at Long Beach where concentrations are relatively low. Primary and secondary PM associated with transportation sources such as diesel engines and catalyst-equipped gasoline engines make the largest contribution to $PM_{2.5}$ mass at the heavily polluted Riverside site.

The sources that contributed to total (primary + secondary) $PM_{2.5}$ concentrations during this episode were diesel engines (25 percent), catalyst-equipped gasoline-powered engines (18.2 percent), non-catalyst-equipped gasoline-powered engines (13.2 percent), paved-road dust (7 percent), unpaved-road dust (5.1 percent), high-sulfur fuel combustion (3.9 percent), food cooking (2.1 percent), other anthropogenic sources (15 percent) and background sources (10.5 percent).

The source-apportionment results for PM_{10} at Riverside, CA, on September 25, 1996 are similar to those described above, since 85 percent of the airborne PM was calculated to be in the $PM_{2.5}$ size range. The sources that contributed to total (primary + secondary) PM_{10} concentrations were diesel engines (21.8 percent), catalyst-equipped gasoline-powered engines (15.8 percent), non-catalyst-equipped gasoline-powered engines (11.3 percent), paved road dust (12.2 percent), unpaved road dust (10.5 percent), high-sulfur fuel combustion (3.4 percent), food cooking (1.8 percent), other anthropogenic sources (13.5 percent) and background sources (9.7 percent). Primary and secondary PM

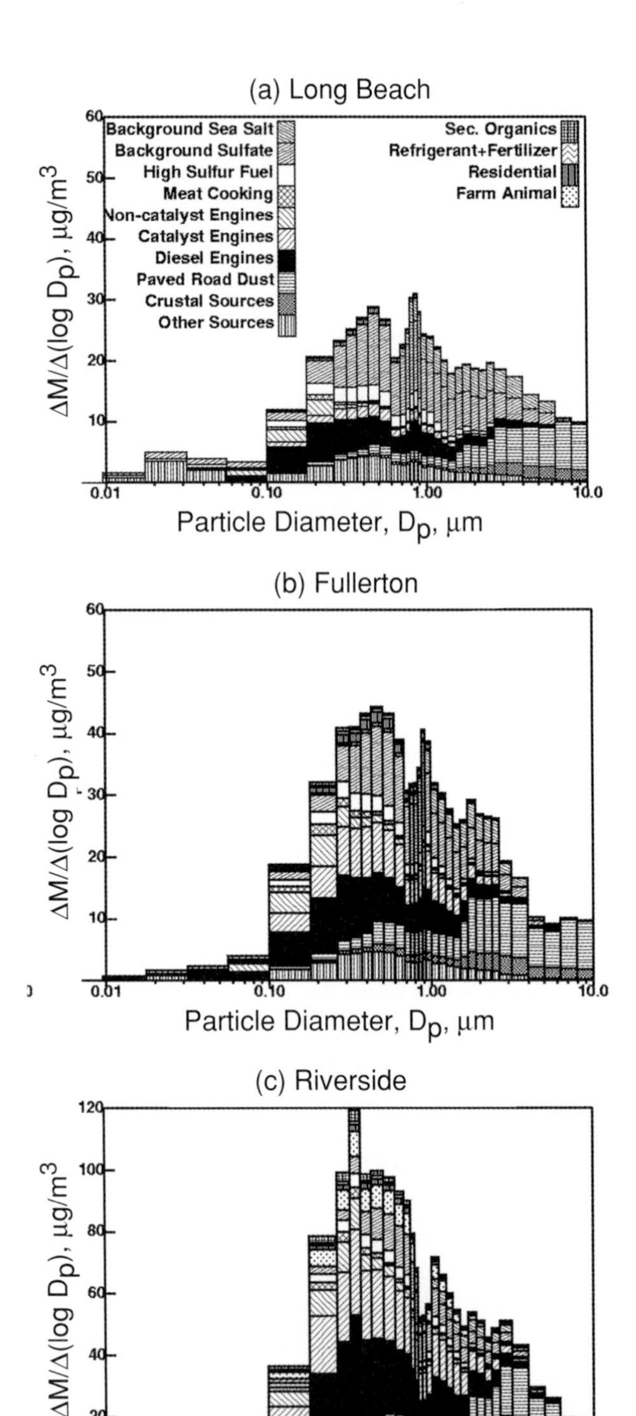

Figure 10.18. Individual source contributions to the 24-hr averaged size distribution of airborne PM at receptor locations in the South Coast Air Basin between September 24-25, 1996. The primary and secondary PM associated with each source is added to show its overall contribution to the airborne particle distribution (Mysliwiec and Kleeman et al. 2002).

released from transportation sources appear to account for approximately 70 percent of the airborne PM during a typical peak pollution event in the Los Angeles area.

Local emissions vs. long-range transport: As discussed previously, background particles advected into the South Coast Air Basin accounted for approximately 5 to 7 μg/m^3 of $PM_{2.5}$ mass averaged over the entire year. These background particles contribute to the annual-average concentration and they also act as sites for the formation of secondary PM. Local emissions of PM and precursor gases account for the majority of current $PM_{2.5}$ in Los Angeles, but background particles will play an increasing important role in meeting annual-average air-quality standards.

Man-made vs. natural: Organic compounds and NH_4NO_3 currently dominate $PM_{2.5}$ mass concentrations. The majority of OC found in the Los Angeles atmosphere is released directly from man-made sources. A lesser fraction of the OC (15 to 35 percent) may be produced by chemical reactions involving precursor gases (Schauer et al., 1996; Hannigan, 1997). Biogenic emissions may account for some part of this secondary organic PM; however, the exact fraction is not clear at this time.

The precursor gases that react to eventually form NH_4NO_3 are almost exclusively man-made. The NO_x that forms HNO_3 is released primarily from stationary and mobile fuel combustion. The largest single source of NH_3 in the area arises from agricultural operations directly upwind of Rubidoux where peak NH_4NO_3 concentrations are observed. The measurements that have been taken in Los Angeles to date suggest that the majority of the current particulate air-quality problem is caused by man-made emissions.

10.4.6 Implications for Policy Makers

Transportation sources appear to account for the majority of the secondary NO_3^- that contributes to peak airborne PM concentrations. A reduction of airborne-particle concentrations to meet the current 24-hr NAAQS for $PM_{2.5}$ will require a reduction of the precursor-gas emissions that lead to the production of NH_4NO_3. Strategies adopted to achieve this goal should also reduce ozone concentrations. Note that control measures adopted to reduce NO_3^- concentrations will require reductions in emissions of VOCs and NO_x. NO_x reductions alone can actually enhance gas-phase reactions for NO_3^- production.

A reduction in airborne particle concentrations to meet the current annual NAAQS for $PM_{2.5}$ will require a reduction of primary particle emissions in the South Coast Air Basin and possibly a reduction in "background" $SO_4^=$ particles advected into the air basin. Control of the background $SO_4^=$ particles may be possible since the size and composition distribution of these particles suggests that they may be released by an upwind combustion source.

10.5 CONCEPTUAL MODEL OF PM OVER MEXICO CITY

The Mexico City Metropolitan Area (MCMA) has 18 million inhabitants, who are exposed to air pollutants generated by more than 3 million vehicles and 35,000 industries. The MCMA is located in the tropics at a high elevation (2240 m above sea level), lying in an enclosed basin of about 5000 km^2, which is surrounded by mountains. Although air-pollution controls enacted in the 1990s have successfully reduced ambient concentrations of Pb, SO_2, and CO, concentrations of ozone and PM remain among the highest in the world. PM concentrations are generally highest during the dry winter months, creating a highly visible haze over the urban area (Molina and Molina, 2002). The rapid growth in population, vehicle ownership, and energy consumption, as well as the rapid outward urban expansion, will continue to pose important challenges for urban infrastructure, mobility, and poverty, as well as air quality, putting air-quality management in the context of sustainable development.

10.5.1 Annual and Seasonal Levels of $PM_{2.5}$ and PM_{10} in Relation to Mass-Based Standards

Since 1995, PM_{10} has been measured hourly at 10 stations around the MCMA, using TEOMs.

Measurements from 1995 to 1999 show that the Mexican 24-hr standard for PM_{10} (150 μg/m^3) was violated on about 30 percent of days at five stations representing five areas of the city (INE, 2000; SMA-GDF, 2000). Peak hourly concentrations of PM_{10} greater than 300 μg/m^3 have been observed. Figure 10.19 shows the annual-average concentrations of PM_{10} at five stations.

Overall, the annual-average standard for PM_{10} (50 μg/m^3) was violated at 8 of the 10 measuring stations in 2000. Concentrations are consistently high at stations north and east of the city, with averages of 95 μg/m^3 at Xalostoc and 91 μg/m^3 at Nezahualcoyotl between 1995 and 1999; concentrations are lower southwest of the city, with an average concentration of 50 μg/m^3 at Pedregal. Figure 10.20 shows that PM_{10} is highest in the winter months, while the hourly-average PM_{10} distribution (Figure 10.21) shows dual peaks in the morning and afternoon. In addition to these TEOM measurements, gravimetric measurements of PM_{10} and TSP have been taken at several sites every six days since 1986.

Currently, no measurements of $PM_{2.5}$ are taken routinely in the monitoring network, although plans are currently underway to add continuous $PM_{2.5}$ monitors at several existing monitoring stations. There is likewise no standard for $PM_{2.5}$ currently in place in Mexico.

Measurements of $PM_{2.5}$ and PM_{10} were taken at several sites during the IMADA field campaign in February and March of 1997 (Edgerton et al., 1999). The average 24-hr concentration of $PM_{2.5}$ at six core sites during this one-month period was 39 μg/m^3, ranging from 25 μg/m^3 at Pedregal (southwest) to 55 μg/m^3 at Nezahualcoyotl (east). The maximum 24 hour $PM_{2.5}$ concentration measured during the sampling period was 184 μg/m^3 (Chow et al., 2002), and the level of 65 μg/m^3 was exceeded four times during this study. Figure 10.22 shows the 24-hr averaged $PM_{2.5}$ concentrations at the 6 core sites over the sample period. The average 24-hr concentration of PM_{10} at six core sites during this one-month period was 72 μg/m^3, indicating that $PM_{2.5}$ is about half of the PM_{10}. $PM_{2.5}$ was observed to be a higher fraction of the PM_{10} during the morning hours.

10.5.2 Compositional Analysis of PM

The best understanding of the composition of aerosols in Mexico City comes from measurements during the IMADA campaign; other aerosol measurements were reviewed by Molina et al. (2002) and Raga et

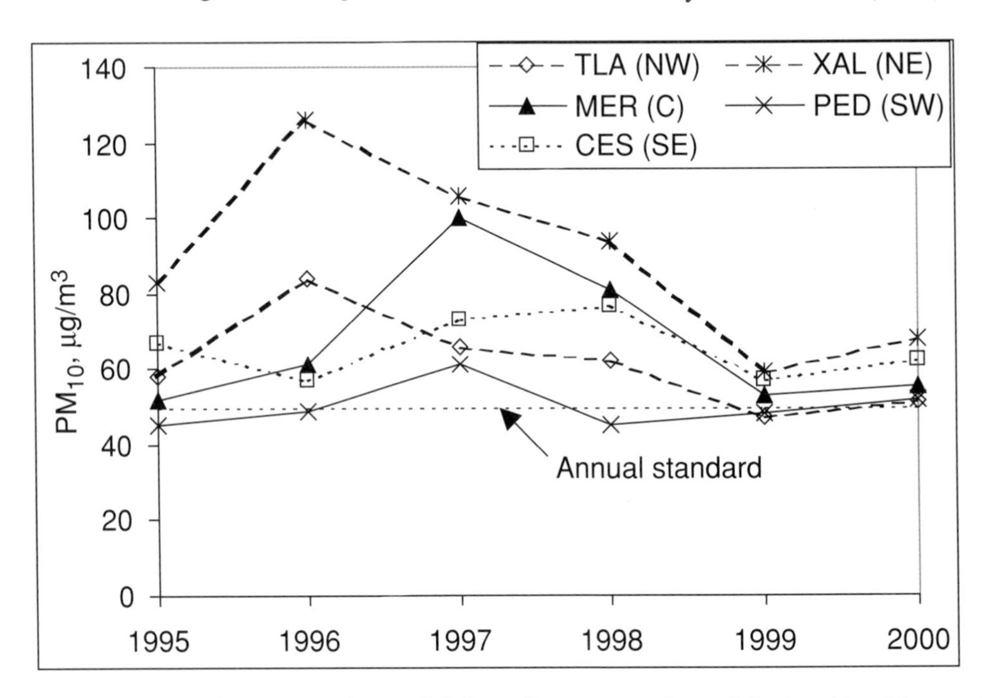

Figure 10.19. Average annual concentrations of PM_{10} at five stations in the Mexico City Metropolitan Area (data from Instituto Nacional de Ecología, Mexico, 2000).

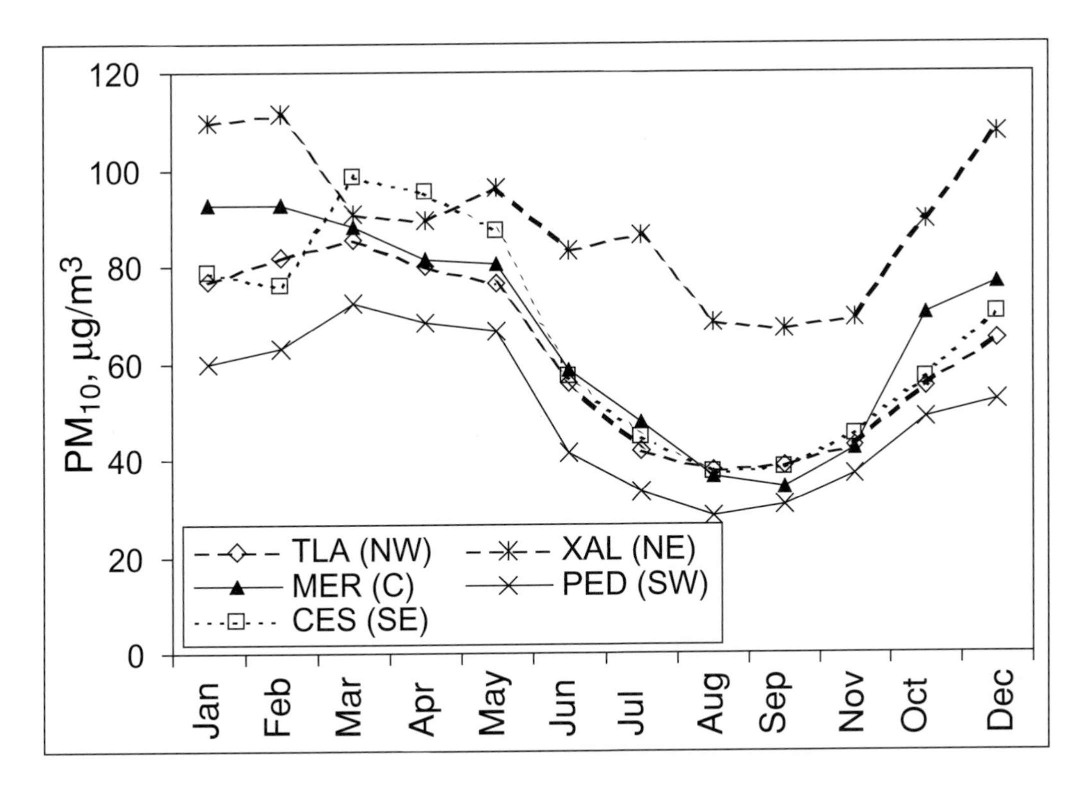

Figure 10.20. Monthly average concentrations of PM_{10} at five stations in the Mexico City Metropolitan Area, for 1995-2000 (data from Instituto Nacional de Ecología, Mexico, 2000).

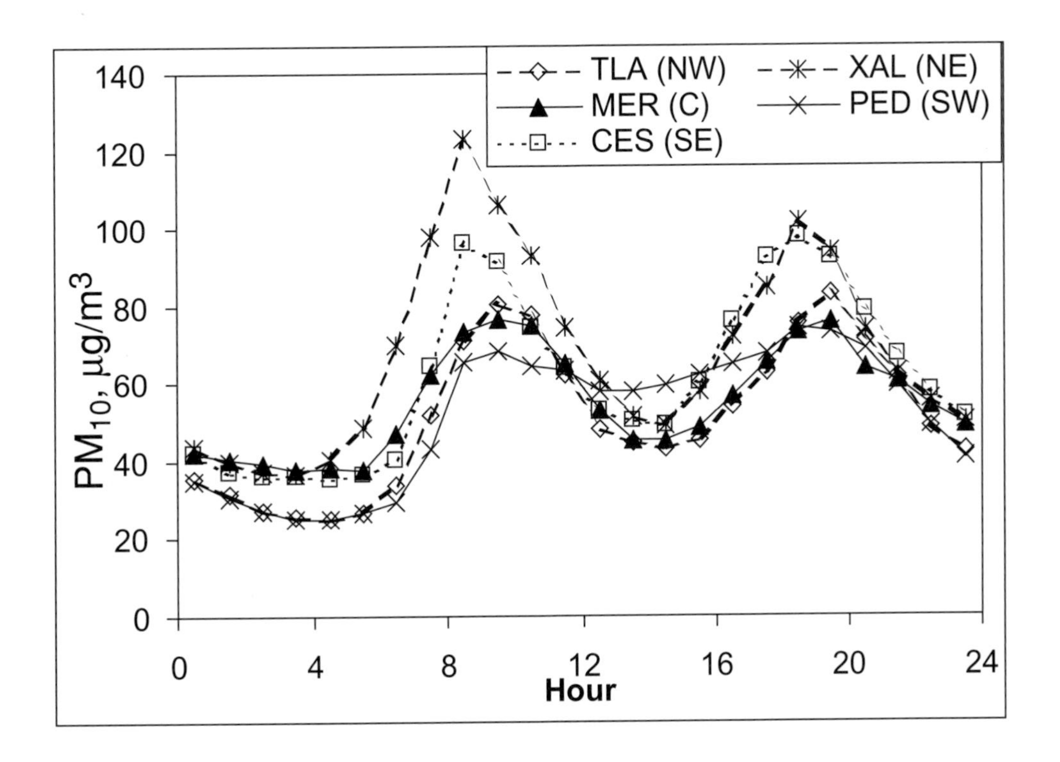

Figure 10.21. Hourly average concentrations of PM_{10} at five stations in the Mexico City Metropolitan Area in 2000 (data from Instituto Nacional de Ecología, Mexico, 2000).

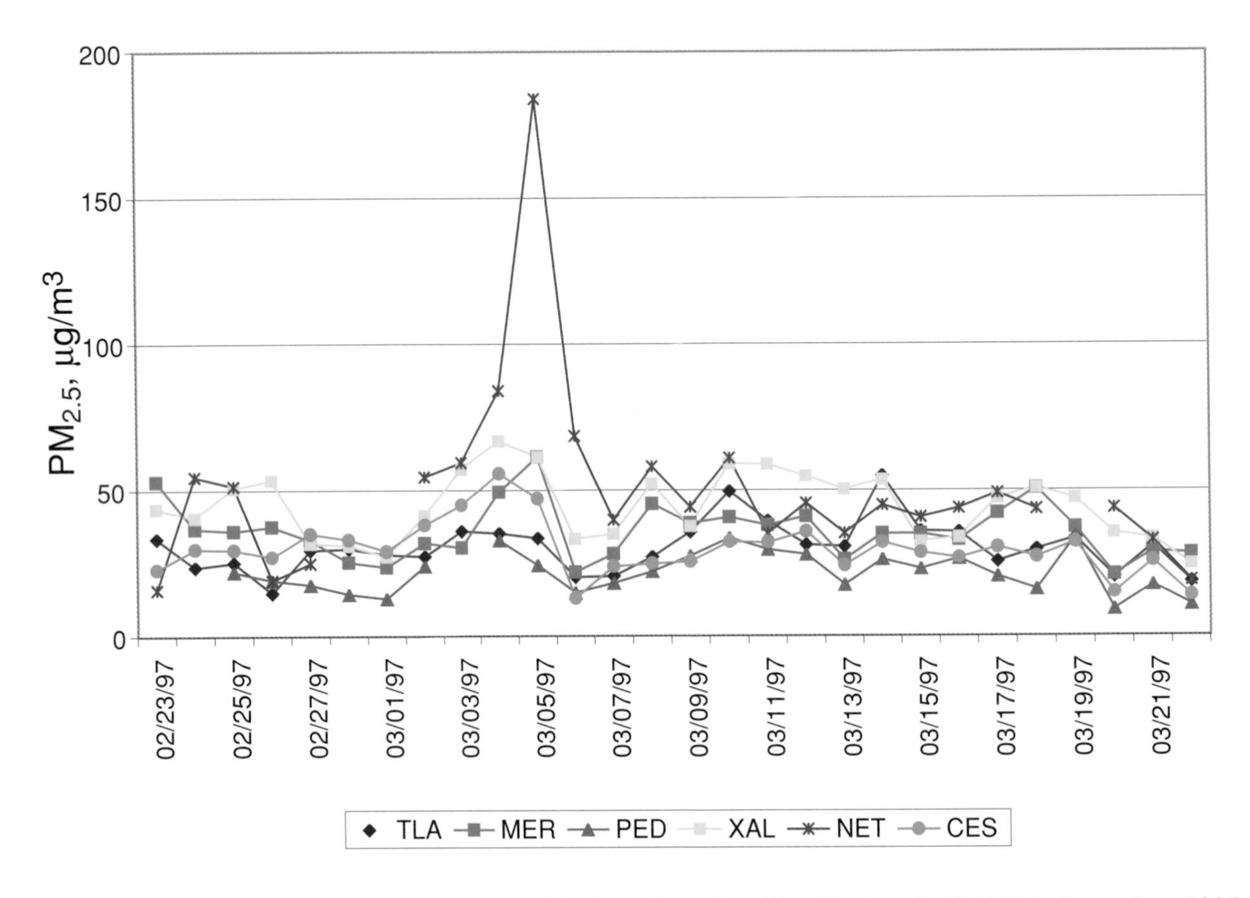

Figure 10.22. Concentrations of $PM_{2.5}$ at the Six Core Sampling Sites During the IMADA Campaign, 1997 (Edgerton et al., 1999).

al. (2001). IMADA measurements indicate that the largest component of $PM_{2.5}$ was from OC and BC, contributing ~50 percent. Secondary $SO_4^=$, NO_3^-, and NH_4^+ contributed ~30 percent of the $PM_{2.5}$, and geologic material (dust) contributed ~15 percent. PM_{10} was observed to be ~50 percent geologic material, ~32 percent OC and BC, and ~17 percent $SO_4^=$, NO_3^- , and NH_4^+. More than three-fourths of the OC, BC, $SO_4^=$, NO_3^- , and NH_4^+ were observed to be in the $PM_{2.5}$ fraction (Chow et al., 2002).

Table 10.1 shows that the contribution of geologic material to the $PM_{2.5}$ is greatest northeast and east of Mexico City (also for PM_{10}), indicating that the sources of dust are from this region. In contrast, secondary inorganic PM ($SO_4^=$, NO_3^- , and NH_4^+) is observed to be a higher fraction of the total in the southwest. Concentrations of $SO_4^=$ at regional outlying sites were two-thirds of the those measured in the urban area (~6 µg/m^3). There is little variability in $SO_4^=$ within the urban area, indicating that much of it may result from regional-scale transport. The concentrations of OC and BC were estimated to be about twice as high in the Valley as measured at regional outlying sites (Chow et al., 2002; Sosa et al., 2000). No measurements of the chemical speciation of the organic PM have yet been published.

10.5.3 Meteorological Influences on PM

The geography of the Mexico City basin is important for the meteorology that controls pollutant transport. Pollutant concentrations are highest in the dry season (winter), when thermal inversions are most common and there is less removal of pollutants by rainfall. High solar radiation promotes photochemical reactions, although PM may significantly reduce photochemical reaction rates (Raga and Raga, 2000).

Thermal inversions are commonly observed during the morning hours. These inversions trap pollutants near the surface, causing the observed morning peak in PM_{10} (Figure 10.21). Solar heating of the surface

Table 10.1. Average chemical composition of $PM_{2.5}$ at four sites, as percent of total, from March 2 to March 19, 1997 (Chow et al., 2002).

	Xalostoc (NE)	Nezahualcoyotl (E)	Merced (C)	Pedregal (SW)
Geological	15	24	14	8
OC	31	22	34	37
BC	20	15	15	12
NH_4NO_3	7	5	12	16
$(NH_4)_2SO_4$	17	12	19	25
Salt	3	1	0	0
Unidentified	4	19	3	0
Elements	3	2	3	2
Total $PM_{2.5}$ ($\mu g/m^3$)	48.4	58.2	38.1	24.6

and mountain slopes causes the inversion to lift during daylight hours to a maximum mixing layer height of up to 4000 m (Doran et al., 1998). This causes pollutants to disperse and mix vertically, as pollutants are created by photochemical production. The afternoon peak in PM_{10} is likely due to increased dust suspension by stronger winds, as well as the formation of secondary particles.

During daylight hours, predominant winds transport pollutants from the industrial area in the northeast across the city center to the residential areas in the southwest, but exceptions to this general pattern are common. Meteorological measurements and modeling (Doran et al., 1998; Fast and Zhong, 1998) show that the Valley of Mexico is well-ventilated overnight, and so peak daytime concentrations of pollutants are generally the result of same-day emissions, rather than the buildup of pollutants over several days. However, the few airplane measurements available show elevated pollutant concentrations aloft, which may be a residual of the previous day's pollution and may affect air quality at the surface (Nickerson et al., 1992; Pérez Vidal and Raga, 1998).

10.5.4 Atmospheric Processes Contributing to PM

In the IMADA campaign, secondary inorganic PM ($SO_4^=$, NO_3^-, and NH_4^+) composed ~30 percent of the $PM_{2.5}$, and there was generally sufficient NH_3 present to fully neutralize the $SO_4^=$ and cause NH_4NO_3 aerosol to form. Gas-phase measurements of NH_3 and HNO_3 were taken at only one site during the IMADA campaign, but these measurements show that NH_3 concentrations are very high, up to 20 $\mu g/m^3$. Preliminary modeling using an equilibrium box model shows that because of this excess NH_3, PM shows little response to changes in NH_3 concentration, and a large response to changes in HNO_3 and $SO_4^=$ concentrations (Sosa et al., 2000). This suggests that if the high NH_3 concentrations are representative of the metropolitan area, reducing NH_3 emissions might be less effective at reducing PM than elsewhere in North America. High measurements of PANs (Gaffney et al., 1999) suggest that the high concentrations of VOCs may cause a substantial fraction of NO_X to be converted to PANs, rather than to HNO_3 and NH_4NO_3 aerosols.

For organic aerosols, a preliminary analysis assuming a linear relationship between emissions of BC and OC suggests that about 30 percent of the organic PM is secondary, and that this secondary material is most apparent in the afternoon (Sosa et al., 2000).

10.5.5 Sources and Source Regions Contributing Principal Chemicals of Concern

The higher concentrations of PM north and east of Mexico City are believed to be due to the suspension of dust from dry, unvegetated areas and unpaved roads in this region. They may also result from the heavy industry and associated transportation emissions (e.g. from trucks) in the north, as OC and BC concentrations are highest in the northeast. This spatial pattern of PM is not likely due to the formation of secondary PM, as predominant winds are from the northeast, and other secondary pollutants such as ozone show peak concentrations in the southwest (INE, 2000). Since poorer populations live east and northeast of the city, the poor are exposed to the highest PM_{10} concentrations, but this difference is mainly due to dust; secondary components show less spatial variability.

Table 10.2 shows the official 1998 emission inventory for the MCMA. Primary PM emissions are expressed in the emission inventory as PM_{10}; no inventory of primary $PM_{2.5}$ has yet been constructed. Recent emission inventories for Mexico City were reviewed by Molina et al. (2000), who found that differences between successive inventories reflected changes in methodology more than real changes in emissions; the inventories are therefore not of sufficient quality to track changes in emissions.

According to the emission inventory, 40 percent of the primary PM_{10} results from the suspension of dust by the wind, while 36 percent is emitted by transportation sources. Because of uncertainties in the activity data and in determining emission factors appropriate for Mexico City, the uncertainty in PM_{10} emissions is considerable. Clearly, the estimate of PM_{10} emissions from soil is highly uncertain and is highly variable in time and space, depending on soil moisture and wind speed. These soil emissions are also likely to be predominantly in the form of coarse particles. In contrast, transportation emissions are likely to fall mostly in the $PM_{2.5}$ range. According to the emission inventory, 29 percent of the total PM_{10} emissions come from diesel vehicles (mainly trucks), which can be expected to be mainly OC and BC. Although there is no inventory for primary $PM_{2.5}$, it can be inferred that diesel vehicles are the largest contributor to primary $PM_{2.5}$. However, experience in the United States indicates that in addition to large combustion sources and motor vehicles, OC and BC can derive from a number of unique sources, including operations involving tar, wood combustion, road dust, and meat cooking (Gray and Cass, 1998); these may not be well characterized in the emission inventory, and may not be considered in pollution control programs.

For the precursors of secondary PM, NO_X emissions are dominated by transportation sources and SO_2 emissions by industry. VOC emissions are split between a variety of services (the largest being solvent consumption) and transportation emissions. A separate emission inventory for NH_3 in the MCMA (Osnaya and Gasca, 1998) reports that domestic animals (dogs and cats) account for 39 percent, cattle account for 19 percent, sewage accounts for 18 percent, and other animals (mainly rats and mice) account for 15 percent.

The emission inventory is estimated for the metropolitan area, which is fairly well defined by mountains. The effect of emissions from a ring of smaller cities around the MCMA on air quality within the Mexico City basin is not well understood.

10.5.6 Implications for Policy Makers

Past measurements have succeeded in determining PM concentrations, their daily and seasonal patterns, and the chemical composition of PM in the dry season when pollution is typically most severe. Organic and

Table 10.2. 1998 Emission Inventory for the MCMA (metric tons/year) (CAM, 2001)

Sources	NO_x	CO	SO_x	HC	PM_{10}
Industry	26,988	9,213	12,422	23,980	3,093
Services	9,866	25,960	5,354	247,599	1,678
Transportation	165,838	1,733,663	4,760	187,773	7,133
Vegetation and Soil	3,193	0	0	15,669	7,985
Total	205,885	1,768,836	22,466	475,021	19,899

black carbon make up about half of the $PM_{2.5}$ in Mexico City. The sources of this carbonaceous material are not well understood, but preliminary analysis suggests that it is mostly primary. It can be inferred from the emission inventory of PM_{10} that the largest sources of OC and BC are diesel vehicles. Reducing emissions from diesel vehicles will be effective at reducing $PM_{2.5}$, and reducing emissions of highly reactive VOCs, particularly aromatics, is expected to be effective at reducing secondary organic PM. Reductions in VOC emissions, as well as reductions in NO_x emissions, are also expected to reduce ozone formation.

Sulfate, NO_3^-, and NH_4^+ together account for ~30 percent of the total $PM_{2.5}$. Because NH_3 appears to be in abundance, emission reductions of SO_2 and NO_x can be expected to be effective at reducing $PM_{2.5}$, while emission reductions of NH_3 may be less effective. Emission reductions of all three of these species should be evaluated further, using model analyses. Windblown dust makes up about half of the PM_{10} in Mexico City, with the largest contribution in the winter. Although emissions of dust are not quantified with certainty, it is clear that much of this dust derives from dry and degraded ecosystems north and east of the city. Proposed actions to address this dust, including ecological preservation and restoration and paving unpaved roads, will be effective at reducing PM_{10}.

Measurements during the IMADA campaign are sufficient to support modeling studies now underway; however, these measurements are limited spatially, temporally, and in the chemical species measured. Future PM research in Mexico City should aim to: characterize particulates at other times of year; estimate the chemical composition of the large OC fraction; measure gas-phase precursors concurrently with PM composition; perform source-apportionment studies (currently underway); measure meteorology and air quality near the boundary of the airshed and above the surface; and use models extensively to test the emission inventory and the response of ambient concentrations to changes in emissions. Finally, creating an emission inventory for $PM_{2.5}$ and implementing routine $PM_{2.5}$ measurements are clearly important recommendations for Mexico City.

10.6 CONCEPTUAL MODEL OF PM OVER THE SOUTHEASTERN UNITED STATES

The region of North America covered by this conceptual model is the southeastern United States, approximately south of Maryland and east of the Mississippi River. This area is known for long, warm, humid summers, relatively large expanses of forests, and, recently, rapidly growing populations in many of the urban areas. Likewise, this region is known for summertime humidity and associated hazy conditions.

10.6.1 Annual and Seasonal Levels of $PM_{2.5}$ and PM_{10} in Relation to Mass-Based Standards

The IMPROVE monitoring network, and recently a network of $PM_{2.5}$ mass monitors and specialized study results, found that the Southeast stands out as having regionally high levels. (See, for example, EPA, 2001: Initial Summary of Preliminary 1999 Fine PM Monitoring Data, Figs. 1 and 2 as well as the map of PM levels and composition from the IMPROVE network: IMPROVE, 2000.) Although the highest one-day maximum and annual-average $PM_{2.5}$ levels are found in and around Los Angeles, the southeastern United States region as a whole has the next highest levels, and not just downwind of a major city or two, but over the central core of the region. Recent data from the new $PM_{2.5}$ network (e.g., U.S. EPA 2000), preliminary analysis of $PM_{2.5}$ show that, except for Florida, every state in the region is experiencing annual levels above 15 µg/m^3. On occasion, 24-hr average levels also go above 65 µg/m^3. The highest levels appear to be in the largest cities away from the coasts.

Atlanta, the largest city central to the Southeast, may have some of the highest $PM_{2.5}$ levels. In-town sites had annual averages of about 20 µg/m^3 for 2000 and about 19 µg/m^3 in 1999. The monitor with the highest annual average was just above 21 µg/m^3. Peak daily levels were above 70 µg/m^3 and there is evidence that some of those high levels were due, in part, to localized emissions and/or measurement artifacts (e.g., Butler, 2000). As shown in Figure 10.23, the seasonal levels appear to peak in the summer and autumn (again, with only two years of data, such a finding is preliminary, though the longer monitoring by IMPROVE supports this.). Other cities in the Southeast also have similarly high levels, e.g., one site in Birmingham recorded an annual average above 23 μg/m^3. (Results for 1999 from the SEARCH network also indicate Birmingham has slightly higher $PM_{2.5}$ levels than Atlanta, though the rural Alabama site is somewhat lower than the rural Georgia site (Edgerton, 2001).) Regions in Florida have levels more on the order of 10 μg/m^3.

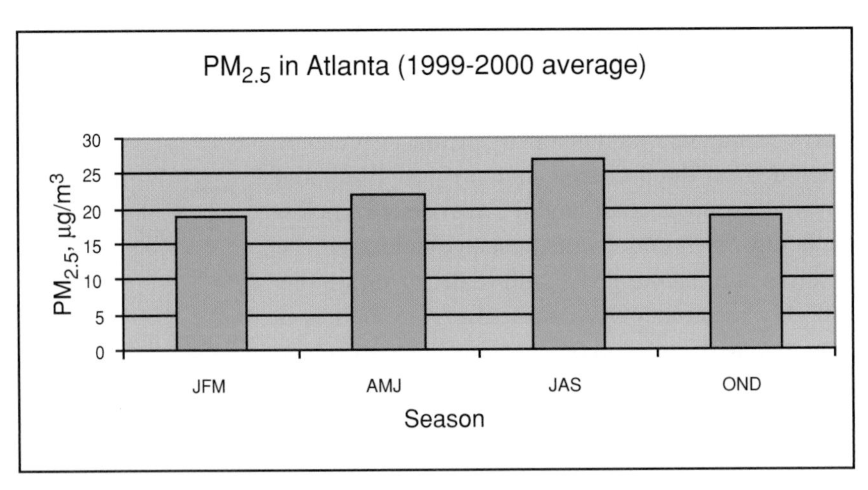

Figure 10.23. Seasonal $PM_{2.5}$ at the Fire Station monitor in Atlanta. (JFM: January, February March, etc.) While the average $PM_{2.5}$ is shown to peak in the summer quarter, the highest average month at that station was in October 2000 (34.9 μg/m^3), suggesting that other periods can have sustained high levels of fine $PM_{2.5}$. The highest daily value was 78.8 μg/m^3 in November 2000.

10.6.2 Compositional Analysis of PM

The largest component of the summertime fine $PM_{2.5}$ in Atlanta is $SO_4^=$ (with the associated NH_4^+), with OC following. In the winter, when mass concentrations decrease, the amount of $SO_4^=$ decreases and the fractions of OC and BC increase (Figure 10.24), such that, on an annual basis, OC (with the related hydrogen and oxygen) is a larger fraction of the total. Nitrate is very small during the summer, and increases in the winter, though it is not a major constituent.

Urban impacts are clearly demonstrated by recent results from the SEARCH network (Edgerton, 2001). As part of the SEARCH design, urban-rural pairs of PM monitors were deployed in three states, Georgia, Alabama and Mississippi, in the Southeast. Results for 1999 showed that the urban locations (Atlanta and Birmingham) in the two states with complete data had similar $SO_4^=$ levels as the rural monitor, but significantly higher OC and BC (Table 10.3). Recent, detailed organic analysis of $PM_{2.5}$ data from SEARCH suggests that regionally, primary emissions from biomass burning and motor vehicles dominate carbonaceous $PM_{2.5}$ levels, which constitute between 15 and 40 percent of the total $PM_{2.5}$ (Zheng et al., 2002). The increase in the urban areas noted above was due, in part, to increased levels from motor vehicles. Secondary $SO_4^=$ constituted between about

Table 10.3. Difference in annual average (1999) $PM_{2.5}$ levels in the urban-rural pairs of the SEARCH network. Delta values shown are the annual monitored level for the urban sites minus the corresponding values for their rural counterparts. This indicates the relative impact of urban sources, though it should be noted that the rural pair is also impacted by urban emissions.

Site Pair	Delta $SO_4^=$ (μg/m^3)	Delta OC (μg/m^3)	Delta EC (μg/m^3)	Delta $PM_{2.5}$ mass (μg/m^3)
Georgia (Atlanta-Yorkville)	0.6	2.1	1.2	4.2
Alabama (Birmingham-Centreville)	0.9	2.9	2.0	6.2

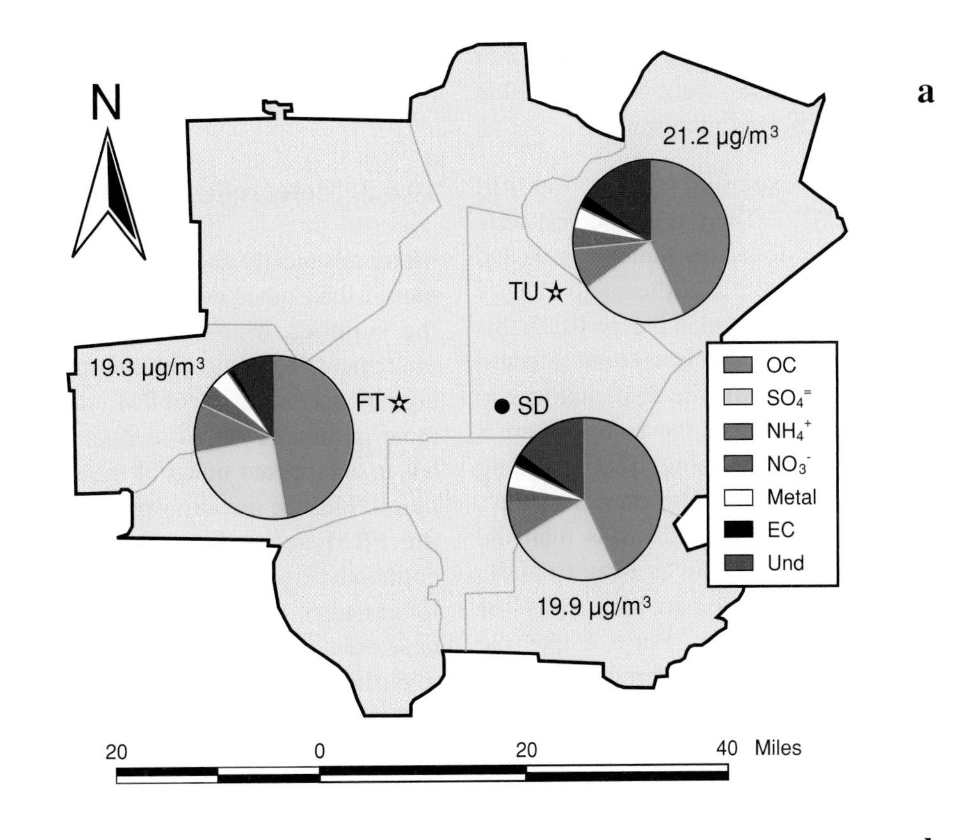

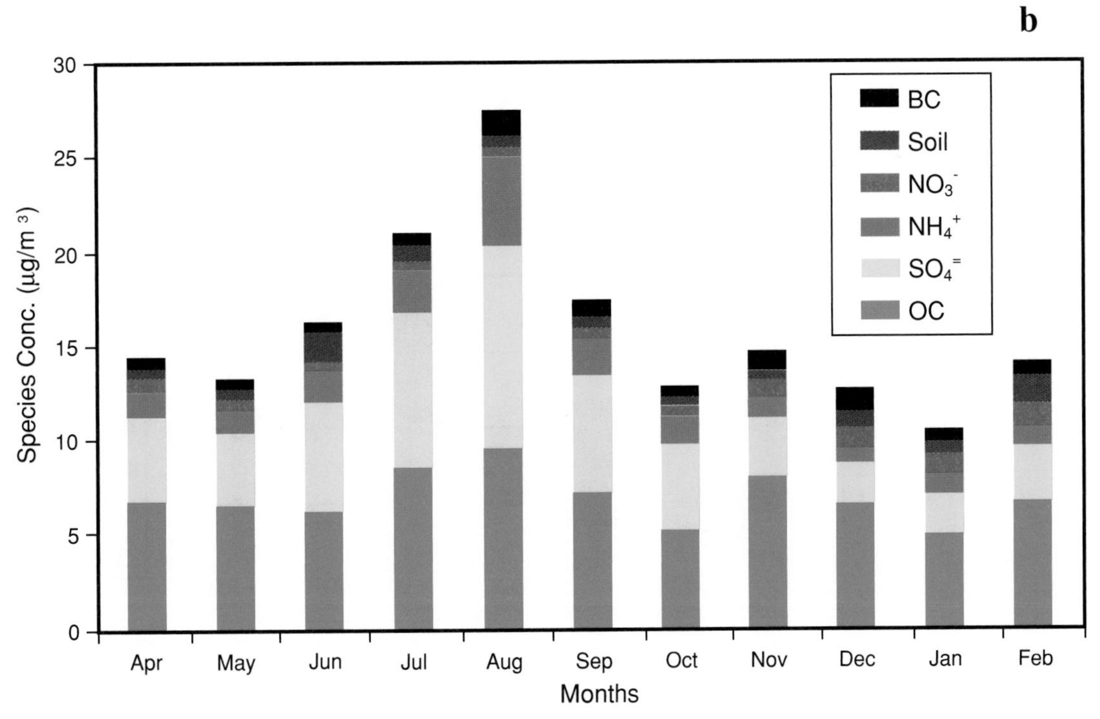

Figure 10.24. a) Annual-average PM fine composition at three sites in Atlanta and b) monthly variation in composition. Data are for April 1999 to February 2000 (Butler, 2000). While $SO_4^=$ is the single largest constituent in the summer, annually OC is greater. There is relatively little variation in the composition spatially, though the site is expected to be most directly impacted by freeway emissions does find somewhat greater BC levels.

10 and 30 percent of the $PM_{2.5}$ mass. While not dominating on a yearly basis, there is evidence that secondary OC is significant in the summer.

Non-urbanized areas experience lower, but still elevated, levels of $PM_{2.5}$. There is a relatively long record at Great Smoky Mountains National Park, and other Class I areas in the Southeast that have IMPROVE monitors. As shown in Figure 10.25, this "pristine" area also records levels near the standard (approximately 14 $\mu g/m^3$ for the four quarters in 1999). Again, in the summer the composition is largely $SO_4^=$, with an increasing fraction being carbonaceous in other seasons. This rural site shows more seasonal variability in total mass than the monitors in Atlanta. As one moves away from the central core of the Southeast to the coasts (or westward), PM levels decrease. For example, the 1999 annual PM fine levels at Okefenokee Swamp, in southeast Georgia, were 11.7 $\mu g/m^3$.

Given the apparent extent of the problem, the next question is "what factors lead to these elevated levels, both the seasonally variation as well as the composition?"

10.6.3 Meteorological Influences on PM

Meteorologically, the Southeast is warmer and more humid than other portions of the country. During the summer, highs can lead to relatively slow movement of air through the region. This leads to higher concentrations of $PM_{2.5}$, as shown above, and other pollutants such as ozone. Recently a drought has encompassed much of the region. The lack of rain to cleanse the atmosphere over the years since the FRM network was put in place may have contributed to high measured levels. However, longer-term measurements from the IMPROVE network suggest that the levels would be high regardless. Severe weather systems (e.g., strong storm fronts) that impact the region can lead to rapid and thorough cleansing.

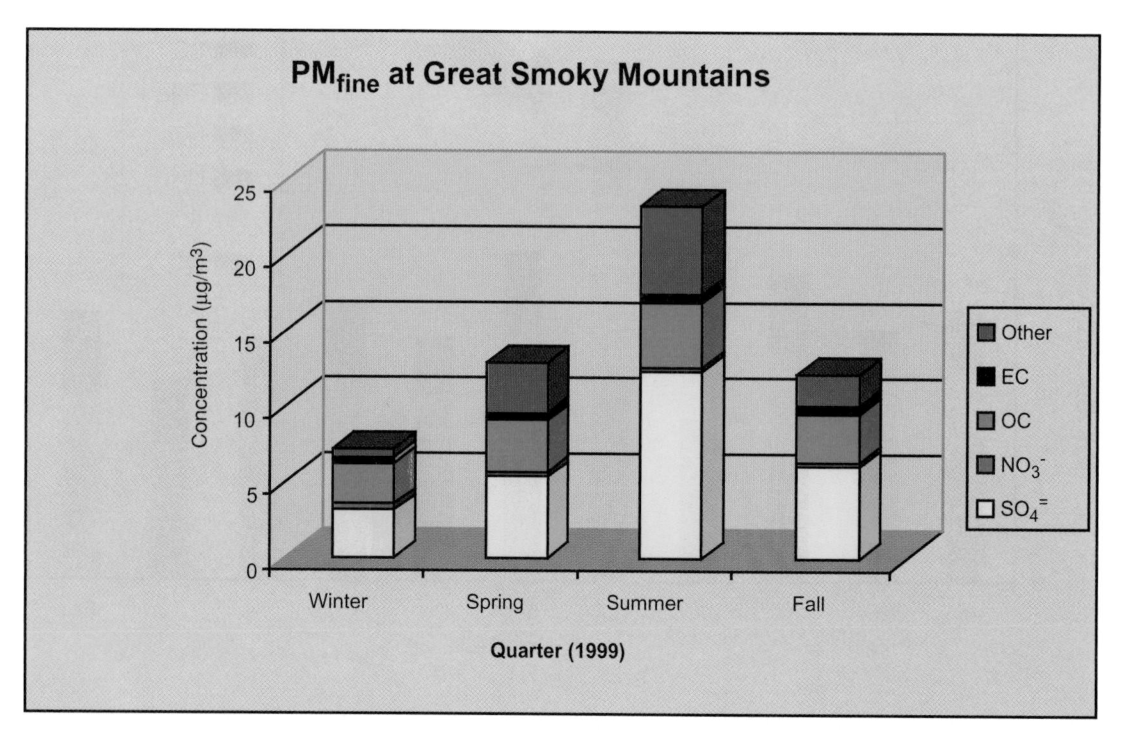

Figure 10.25. $PM_{2.5}$ composition at Great Smoky Mountains during the four quarters of 1999. (Note: the quarters are spring (March, April and May 1999) through winter (December 1999, January and February 2000.) Sulfate dominates in the summer and decreases dramatically in the other quarters, while OC remains fairly constant. (Data obtained from http://improve.cnl.ucdavis.edu.)

10.6.4 Atmospheric Processes Contributing to PM

Given that $SO_4^=$ is mainly a secondary particulate species, it is of interest to characterize the chemical pathways leading to its formation from SO_2 emissions. Homogeneous gas-phase oxidation by the ubiquitous hydroxyl radical is important, though not always dominating. Aqueous-phase oxidation by inorganic and organic peroxides in water-containing droplets (e.g., clouds, fog, haze) appears to be a major route. At present, the aqueous-phase oxidation by ozone appears to be less significant because this route is very pH sensitive, and most of the droplets are acidic, effectively removing this pathway. However, as sulfur emissions decline and NH_3 emissions increase, $SO_4^=$ formation in the aqueous phase by ozone oxidation may become more significant. Southern Appalachian Mountains Imitative (SAMI) modeling finds that in some instances aqueous-phase $SO_4^=$ formation is often limited by the availability of peroxides, at others by the amount of SO_2. When peroxides are limiting species, reductions in sulfur emissions will have a smaller than linear impact in the region. Secondary organic PM formation may also be of importance (air-quality modeling suggests this is the case), particularly of biogenic organics. However, estimates of biogenic emissions, especially the higher molecular-weight molecules, are very uncertain, and thus are the estimates of secondary OC formation.

10.6.5 Sources and Source Regions

A large number of coal-fired power plants are in the region, leading to significant SO_2 emissions. Approximately 5.4 million tons of SO_2 are emitted per year (using the 1996 NET inventory; Pacific Environmental Services, 2001) in the nine-state Southeast region including West Virginia/Kentucky and south and Mississippi/Tennessee and east. Primary $PM_{2.5}$ emissions in those states are 1.4 million tons per year. Midwestern states, e.g., Ohio, Illinois, can contribute still more. Cities in the region experience higher than average motor traffic (i.e., the per capita VMT is higher than average for the rest of the United States), and evidence suggests that vehicle speeds are higher as well. These two factors lead to increased vehicular emissions. The Southeast has also seen rapid growth, with the associated increased emissions (e.g., construction related, including dust, part of which is in the fine fraction). A final piece of the puzzle is that much of the Southeast is covered by forest. This leads to emissions from growing plants, as well as from wood burning (wildfires, prescribed burns and fireplace/wood heating). Regional emission inventories, CTM modeling (e.g., Boylan et al., 2000), the composition of the measured PM, and receptor modeling (Butler, 2000) suggest that all of these sources can play significant roles.

The picture that comes out of the discussion above, and reinforced by recent modeling results, is that the Southeast has an elevated regional level of fine PM, with higher levels in and downwind of the major cities. While a not insignificant fraction is transported into the region, a large part of the regional cloud is self-generated from biogenic emissions (including wood burning) and coal-fired electricity generation. Increased levels in the urban areas arise from a variety of activities, particularly vehicle traffic. For example, while $SO_4^=$ levels at the Yorkville site outside of Atlanta and an in-town site are similar; in-town levels of OC and BC are increased, suggesting a significant contribution from traffic, including diesels.

The question of source region of impact, and the difficulty in answering such a question, is highlighted by an analysis performed for SAMI. As part of that effort, the sensitivity of $SO_4^=$ PM to SO_2 emission reductions in different regions of the modeling domain were quantified. Preliminary results show that the contribution of each of the source regions on $SO_4^=$ levels varied significantly from day to day (Figure 10.26). While much of the time the dominant source regions appear local, on other days longer-range transport dominates.

10.6.6 Implications for Policy Makers

Looking into the future, Clean Air Act controls will lead to decreased $SO_4^=$ levels. While this should reduce the overall levels, the likely response is not linear. Air-quality modeling, as part of SAMI, suggests that the decreased $SO_4^=$, as well as the increasing NH_3 emissions in the region, will lead to

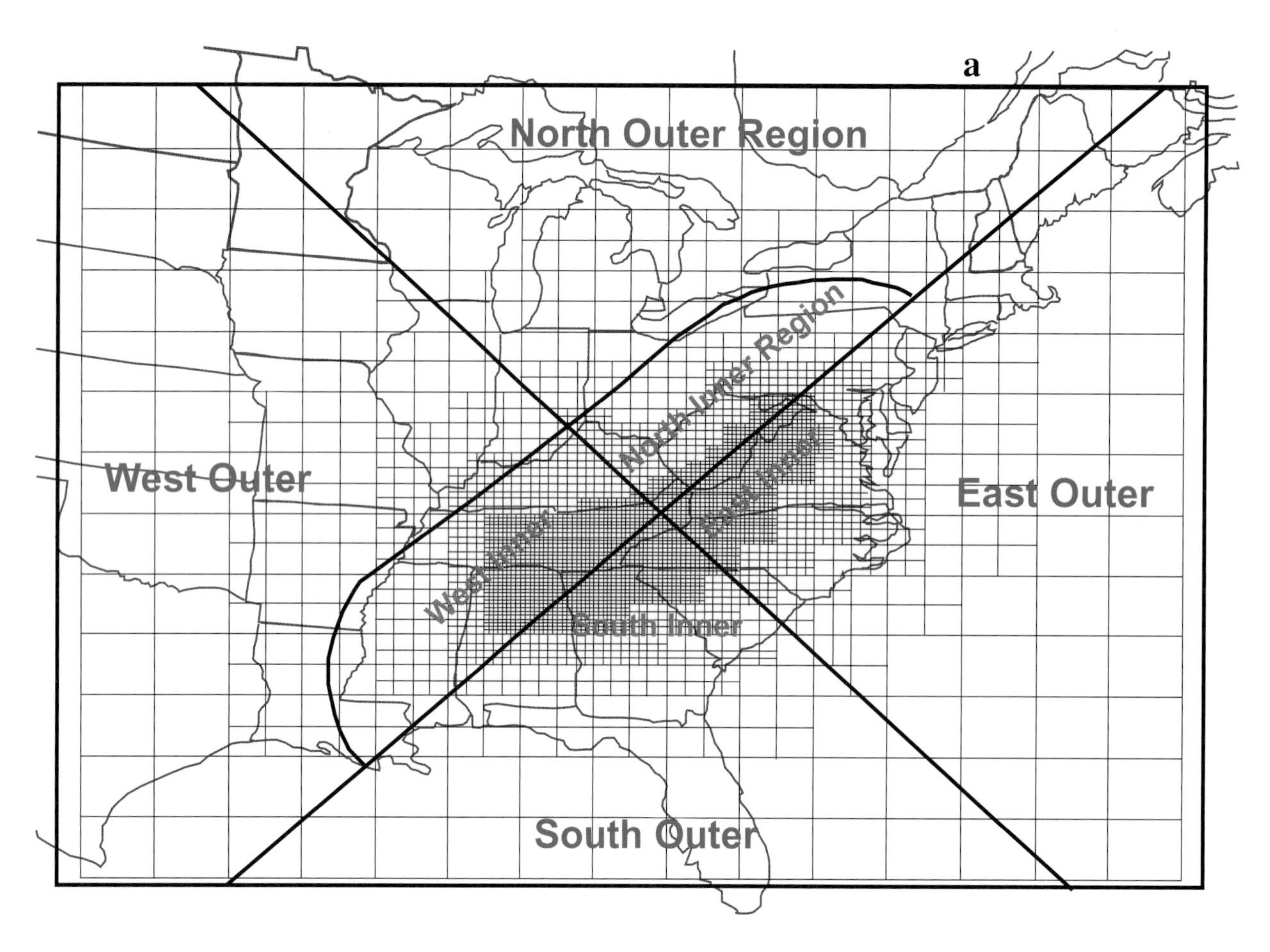

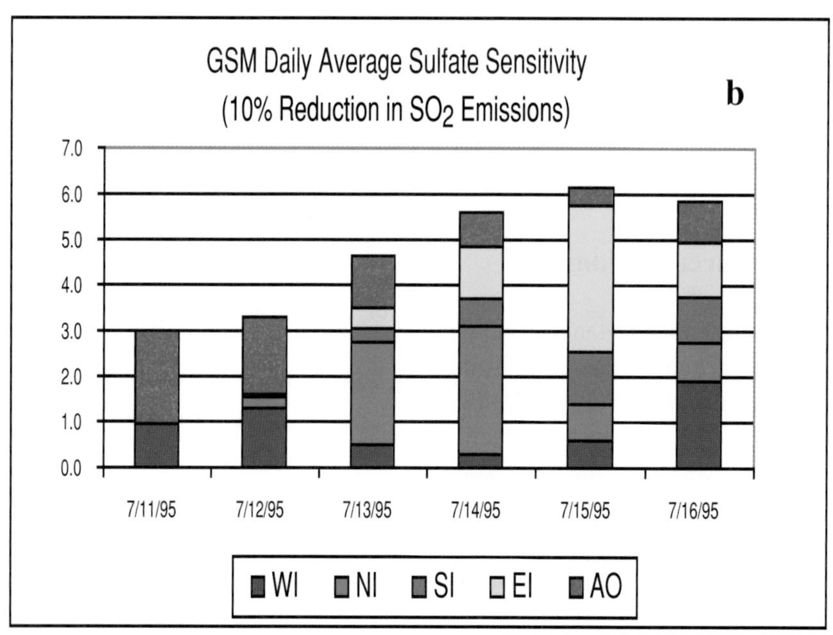

Figure 10.26. Preliminary source-area impact analysis of $SO_4^=$ PM at Great Smoky Mountain National Park. The modeling domain was divided into five regions as shown in (a): West Inner (WI), North Inner (NI), South Inner (SI), East Inner (EI) and All Outer (AO). Over the six days assessed, (b): the impact from each region on the receptor location varies significantly depending on the meteorology.

greater NO_3^- levels, offsetting some of the reductions in the $SO_4^=$ levels. Also, SO_2 oxidation can be limited by the availability of oxidants such as H_2O_2. Further analysis is required to assess how OC levels will evolve in the region.

10.7 CONCEPTUAL MODEL OF PM OVER THE NORTHEASTERN UNITED STATES

The northeastern United States (NE) conceptual model domain covers all of the Northeast and Atlantic seaboard states plus New York, Pennsylvania, Maryland, and Virginia (as a transition state). The states to the west-southwest (Ohio, Michigan, West Virginia and Kentucky) are important upwind sources of transported pollutants. The Windsor-Quebec corridor forms the northern border of the NE domain. The NE includes New York City and its environs, and the highly populated "northeast corridor" from Washington, D.C. to Boston, MA. Roughly two-thirds of the population of the NE lives in this urban corridor, which fits in less than ten percent of its land area. This region is known as an "ozone corridor," producing very high levels of ozone when the winds line up, transporting ozone and its precursor emissions from one urban area to the next.

Major eastern-U.S. utility sources of SO_2 emissions lie within and to the west-southwest of the NE domain, notably along the Ohio River Valley. These are also major sources of NO_x emissions, but they are more than balanced by NO_x emissions along the northeast corridor, particularly from mobile sources. The anthropogenic primary OC emissions of the NE are highly concentrated along the northeast corridor. Modeling analyses suggest a large portion of the primary plus secondary OC mass along the populated northeast corridor is anthropogenic in origin. Outside the urban corridor biogenic OC plays an important role. These characteristics suggest both local sources and pollution transport contribute to the PM levels across the NE domain, particularly for the urbanized areas.

10.7.1 Annual and Seasonal Levels of $PM_{2.5}$ and PM_{10} in Relation to Mass-based Standards

10.7.1.1 Annual Mean Concentrations of $PM_{2.5}$

Rural monitoring sites: None of the rural Northeast sites exceed the U.S. annual $PM_{2.5}$ standard for the 1996-98 period (Malm, 2000). Annual values are around 10 µg/m^3 in the high SO_2-emission region and decrease to around 5 µg/m^3 in upper New England. A downward trend in $SO_4^=$ has been quantified at the IMPROVE (Malm, 2000) and CASTNet (Holland et al.,1999) rural sites in response to the SO_2 reductions associated with Phase 1 of Title IV of the 1990 Clean Air Act Amendments. At the CASTNet sites the relative reduction in $SO_4^=$ concentrations is smaller than that for SO_2. Interestingly, at Shenandoah there is a downward trend in median $SO_4^=$ concentration, but little to no trend in the peak (90th percentile) $SO_4^=$ concentrations (Malm, 2000).

Urban monitoring sites: Many urban locations across the Northeast have the potential to exceed the U.S. annual $PM_{2.5}$ standard (U.S. EPA AIRS database, 2001). This conclusion is based on only two years of data with many sites in 1999 having incomplete coverage. Contiguous geographic regions with levels potentially above the annual standard surround the Ohio River Valley, reaching out to Indianapolis, IN and Columbus, OH. Pittsburgh, PA, Cleveland, OH and Detroit, MI are connected urban areas with levels potentially above the annual standard. Cities along the Washington D.C./Baltimore-to-New York ozone corridor, including Newark, NJ, constitute a second region with concentrations showing a potential for violations, although their annual $PM_{2.5}$ values appear to be lower than in the region to the west. These northeast-corridor urban areas are slightly below or above the annual standard of 15 µg/m^3. Washington D.C. and Philadelphia are just below the U.S. annual $PM_{2.5}$ standard (14.5 µg/m^3 and 14.7 µg/m^3, respectively) and New York is above (17.3 µg/m^3) (Malm, 2000; U.S. EPA AIRS database 2001).

10.7.1.2 24-hr-Mean Concentration of $PM_{2.5}$

Based only on two years of EPA FRM data, few sites will exceed the U.S. 24-hr $PM_{2.5}$ standard, mostly around Pittsburgh. Thus, the U.S. annual $PM_{2.5}$ standard is expected to be the standard most often exceeded in the northeastern United States (U.S. EPA AIRS database, 2001).

10.7.1.3 Annual and Daily PM_{10}

PM_{10} standards are not now exceeded in the northeastern United States and PM_{10} concentrations in the Northeast have downward trends. Concentrations in 1999 are 15 to 18 percent lower than those in 1990. At most of the monitoring sites, the PM_{10} concentrations are less than half of either the annual or the 24-hr standard. A number of large urban areas have annual means in the range of 30 to 50 µg/m^3. Locations of these areas are similar to those with the potential to exceed the annual $PM_{2.5}$ standard. These areas, like New York City, Philadelphia, Pittsburgh, Cleveland, Columbus and Detroit, also had 24-hr mean PM_{10} concentrations in the range of 80 to 150 µg/m^3 in 1998 and 1999. With the downward trend in concentrations, PM_{10} is not expected to become an issue in the northeastern United States (EPA National Air Quality and Emission Trends Report, 1998 and 1999; EPA AIRS Data as of 7/10/2000).

10.7.1.4 Seasonal-Mean Concentrations of $PM_{2.5}$

Close to the Ohio River Valley, the regional concentrations of $PM_{2.5}$ at the IMPROVE sites increase, on average, by a factor of 2.6 from winter to summer (Dolly Sods and Shenandoah), showing strong seasonality (see Figure 10.27). For sites far from the Ohio River Valley, the seasonal increase of summer over winter is a little less than a factor of 2, on average 1.9 (Lye Brook and Acadia). For Washington, D.C. and Brigantine, the seasonal increase in summer over winter is only a factor of 1.35 and 1.47, respectively (weak seasonality). For the major urban areas of Philadelphia and New York, winter is actually slightly higher than summer; the seasonal "increase" in summer over winter is a factor of 0.85 on average (inverse seasonality). The difference in seasonality across sites is mostly caused by the winter $PM_{2.5}$ concentrations: summer concentrations at Dolly Sods, Shenandoah, Washington D.C., Brigantine, Philadelphia, and New York are all rather similar; whereas winter concentrations in Washington D.C. are at least a factor of 2 higher than winter concentrations at Dolly Sods and Shenandoah, and winter concentrations in Philadelphia and New York are at least a factor of 3 higher. This indicates local sources are important in winter.

Rural areas are high in summer, low in winter (Figure 10.28a). Sulfate levels by and large control this behavior. In the Ohio River Valley region of high SO_2 emissions, summer $SO_4^=$ is about 3.5 (3.4 to 3.6) times higher than winter $SO_4^=$. In upper New England, summer $SO_4^=$ is about 2.3 (2.0 to 2.5) times higher than winter $SO_4^=$. Except for Dolly Sods, regional OC also goes down in winter to about half the summer values. In the source region summer averages of $PM_{2.5}$ are 15 to 16 µg/m^3; the winter averages of $PM_{2.5}$ for the rural source region sites are between 4 and 6 µg/m^3, which keeps the annual average down to about 10 µg/m^3.

10.7.2 Compositional Analysis of PM

Urban Sites: The few urban areas with seasonal data have levels of $PM_{2.5}$ that are more often higher in winter than in summer (Figure 10.28b). Winter $SO_4^=$ is lower than in summer, with a strong north-south or urban-size gradient (factors of 0.4, 0.7, and 0.95 for Washington D.C., Philadelphia, and New York, respectively). Nitrate is significantly higher in winter than in summer (factors of 3, 3.2, and 3.7 for the 3 cities in same order). OC is slightly higher in winter than in summer (7 percent, 14 percent, and 16 percent higher for the 3 cities in same order). In general, winter increases in NO_3^- and OC offset winter decreases in $SO_4^=$. The overall urban mass increase of winter $PM_{2.5}$ levels over summer is modest, being 2 to 4 µg/m^3.

Regional Sites: The majority of the $PM_{2.5}$, averaged over 1996 to 1998, at regional (IMPROVE) sites is $SO_4^=$: from 64 percent close to the Ohio River Valley to 57 percent away from the Ohio River Valley. Organic carbon is the second most important, at around 25 percent except for Acadia with 29 percent.

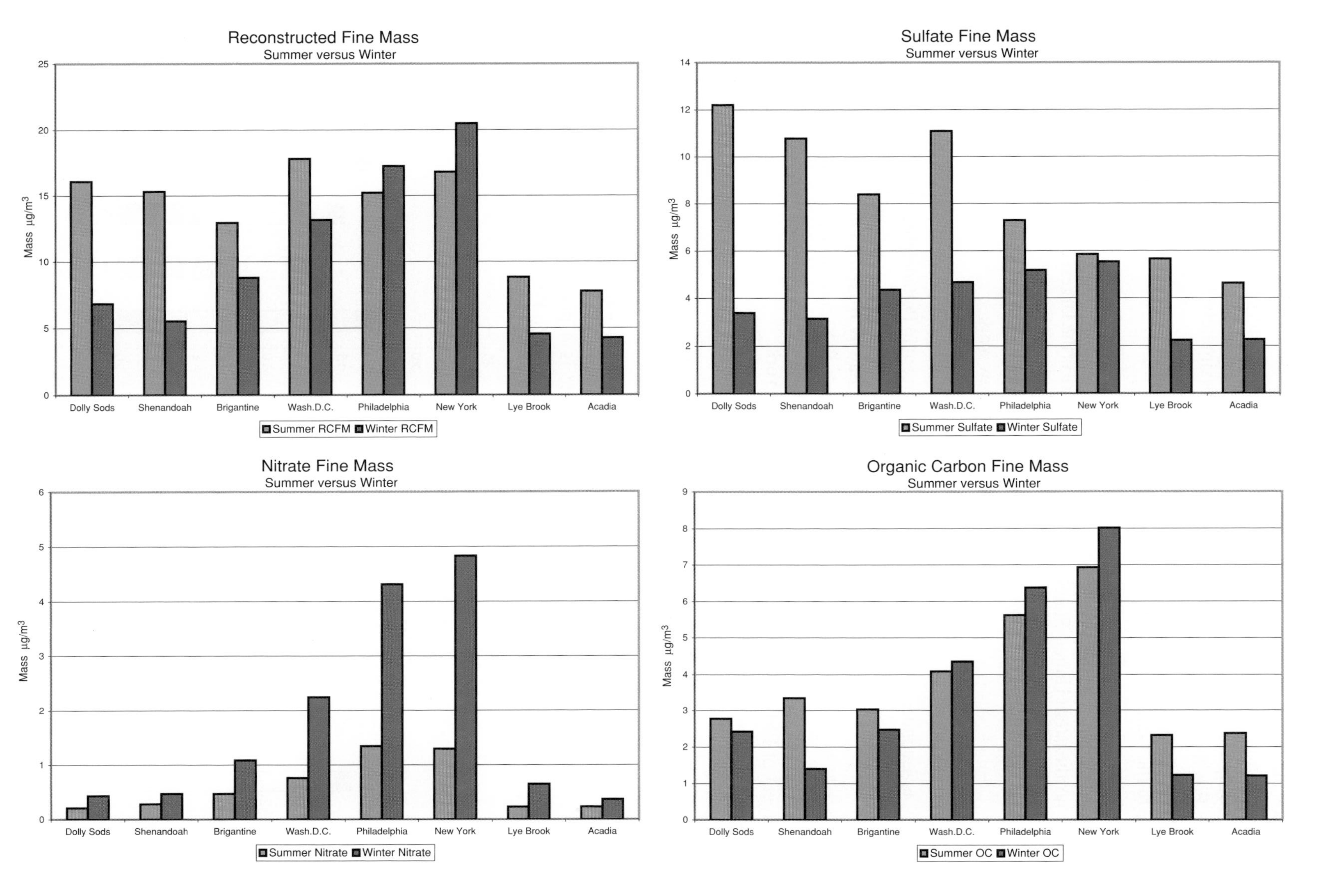

Figure 10.27. Seasonal behavior (summer/winter) of the Reconstructed Fine Mass and three of its most important components, $SO_4^=$, NO_3^- and OC, at the 8 sites with speciation data across the northeastern United States.

Top 20 percent of Regional Annual Concentration Distribution: For the highest $PM_{2.5}$ concentrations of the annual distribution, the $SO_4^=$ fraction increases substantially compared to that for the mean concentrations. For the highest 20 percent of the 3-year data, the $SO_4^=$ contribution increases to 80 to 84 percent close to the Ohio River Valley and to 60 to 70 percent away from the Ohio River Valley (using the 90th percentile fraction). Sulfate is clearly very important in the northeastern United States.

10.7.2.1 Seasonal Mean $PM_{2.5}$ Composition

In the northeastern United States, the influence of primary emissions on $PM_{2.5}$ is small. The great majority of $PM_{2.5}$ is secondary in origin.

Rural Sites

Summer:

- 60 to 75 percent of $PM_{2.5}$ is $SO_4^=$. Sulfate is an important summer issue.
- 20 to 30 percent of $PM_{2.5}$ is OC.
- Approximately 10 percent is NO_3^- + BC + Soil.

Winter:

- 50 to 55 percent of $PM_{2.5}$ is $SO_4^=$. Sulfate still dominates in winter.
- Approximately 30 percent of $PM_{2.5}$ is OC.
- Approximately 10 percent is NO_3^-.
- Approximately 10 percent is BC + Soil.

Urban Sites

Summer:

- Sulfate is the largest contributor in summer, with OC second for Washington D.C. and Philadelphia, but OC is the largest contributor for New York.
- Sulfate is 62 percent in Washington D.C., 48 percent in Philadelphia, and 35 percent in New York. The $SO_4^=$ fraction declines as one gets farther away from the Ohio River Valley region of high SO_2 emissions, although the absolute magnitude is the same between Philadelphia and New York.
- OC is 23 percent in Washington D.C., 37 percent in Philadelphia, and 41 percent in New York. However, there is about an equal increase in absolute OC mass from Washington D.C. to Philadelphia and then again to New York.
- Nitrate is 4 percent in Washington D.C., and 8-9 percent in Philadelphia and New York.

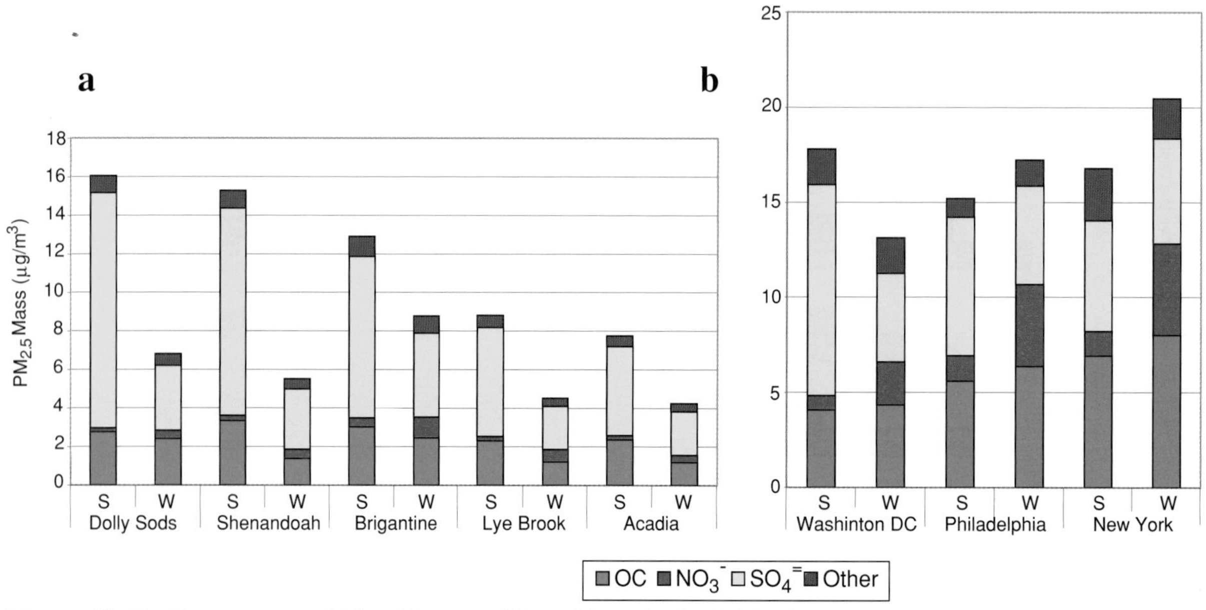

Figure 10.28. Reconstructed Fine Mass partitioned into the individual components: (a) rural sites (left) and (b) urban sites (right), where S=summer and W=winter.

- BC + Soil is approximately 10 percent in all three cities.

Winter:

- OC is the largest contributor in winter, with $SO_4^=$ a close second and NO_3^- an important third for Philadelphia and New York. Sulfate is the largest contributor in winter with OC a close second for Washington D.C.
- Sulfate is 36 percent in Washington D.C., 30 percent in Philadelphia and 27 percent in New York.
- OC is 33 percent in Washington D.C. (higher fraction than summer) and ~38 percent in Philadelphia and New York (same fraction as summer).
- Nitrate is 17 percent in Washington D.C. and ~25 percent in Philadelphia and New York. Nitrate increases significantly in importance in the winter.
- BC + Soil is 14 percent in Washington D.C., 6 percent in Philadelphia and 11 percent in New York with EC contributing two-thirds or more.

10.7.3 Meteorological Influences on PM

Meteorology is a factor in the annual-average long-range transport of fine PM. The climatology of transport and fine-particle lifetimes favor pollutant transport to the northeast and east directions. Emission sources will have their largest reach of influence in these directions; thus, much of the northeast coast of the United States is downwind of the sources in the Ohio River Valley.

Meteorology contributes importantly to inter-annual variability of fine PM and involves meteorological aspects that are not directly chemical in nature, in particular precipitation cleansing. The year-to-year variation in monthly-mean precipitation is an important contributor to inter-annual variability and an important factor in the resultant PM budget.

Annual averages encompass broad meteorological patterns. Periods that favor gas-phase production of $SO_4^=$ and subsequent transport under conditions of little rainfall will be important to annual averages. This will include high-pressure systems, such as the Bermuda high, and stagnant days conducive to photochemical production.

10.7.4 Atmospheric Processes Contributing to PM

Sulfate is produced from gas-phase and aqueous-phase pathways. RADM model analyses estimated that, on an annual average basis, more $SO_4^=$ is produced by the aqueous-phase pathway than the gas-phase one (McHenry and Dennis, 1994). More recent analyses with the CMAQ model suggest gas-phase produced $SO_4^=$ accounts for at least half or more of the overall $SO_4^=$ budget. Based on CMAQ model analyses, aqueous-phase production in the northeast appears to be oxidant limited and hence very nonlinear. Thus, $SO_4^=$ summer peaks are driven by conditions conducive to a dominance of the gas-phase production pathway. Nonetheless, there is an overall less-than-proportional reduction in $SO_4^=$ to reductions in SO_2 emissions, which is controlled by the aqueous pathway.

Analysis of CASTNet data indicates $SO_4^=$ is not neutralized across the northeastern United States in summer, but is neutralized in winter. Thus, the NO_3^- response to a reduction in $SO_4^=$ is expected to be proportionately greater in winter than in summer.

Total NO_3^- (NO_3^- + HNO_3) is a termination product of photochemistry involving VOC and NO_X precursor gases. Nitrate particle production in the northeastern United States is NH_3-limited and controlled by the availability of $SO_4^=$, especially along the east coast. Modeling studies show NO_3^- is not HNO_3-limited in the Northeast. There is an abundance of HNO_3 available. There are large seasonal swings in NO_3^- and in NO_3^--to-HNO_3 partitioning because of the seasonality of $SO_4^=$ production and NH_3 emission rates; nonetheless, in most areas, the modeling analyses indicate NO_3^- represents only a fraction of the total $PM_{2.5}$ in the Northeast, even in winter, especially along the eastern seaboard. The regional picture is supported by CASTNet data. More study with urban data and finer-scale modeling is needed.

OC has both biogenic and anthropogenic sources and can be primary or secondary in origin. The source of the OC is an area of large uncertainty. Data are needed to be able to assess the relative contribution of primary versus secondary and anthropogenic versus biogenic to overall production of OC. Extrapolation of data from the Southeast (Lewis, 2002) using models suggests that biogenic sources of OC are important. Although there is little information, recent analyses suggest a majority of OC is secondary in origin.

10.7.5 Sources and Source Regions Contributing Principal Chemicals of Concern

Long-range transport of $SO_4^=$ is very important, particularly in summer (Figure 10.27). There is a clear gradient of $SO_4^=$ from the Ohio River Valley to upper New England. However, local or urban production of fine PM is also important, particularly in winter and along the northeast corridor. At the urban speciation sites summer $SO_4^=$ appears to follow regional levels, implying transport is important in summer. However, even though winter $SO_4^=$ is lower than in summer, winter $SO_4^=$ in the urban areas is noticeably higher than the regional background $SO_4^=$, being almost double background levels. This indicates local urban sources of winter $SO_4^=$ of comparable magnitude to the regional $SO_4^=$. The other major seasonality factor is a significant increase in NO_3^- in the winter, especially at the urban sites. The large increase at the urban sites relative to the rural sites suggests the urban increase is mostly local. This is consistent with large local NO_x emissions. Most NO_x emissions are associated with the population and mobile sources of the NE region. While OC is important, it is roughly the same in summer and winter (slightly higher in winter). Summer OC is about double regional OC and winter OC is four to five times regional levels, indicating local urban sources of OC are important, especially in winter.

A source-apportionment analysis of Philadelphia with UNMIX and Positive Matrix Factorization (Norris, 2000) provides additional insight on sources for the urban areas of the East-Coast urban corridor. The following apportionment of the $PM_{2.5}$ mass was found:

Local SO_2 and $SO_4^=$	~ 10 percent
Regional $SO_4^=$	~ 50 percent
Residual oil	8 to 4 percent
Soil	6 to 7 percent
Motor vehicles	25 to 30 percent

Biogenic sources are not accounted for in this analysis and most likely are embedded in the motor-vehicle fraction.

10.7.6 Implications for Policy Makers

Conclusions drawn from only one or two years of data cannot be considered firm.

Sulfate is a key particulate to reduce in the northeastern United States. The percent reductions in $SO_4^=$ air concentrations are expected to be smaller than the percent reductions in SO_2 emissions. Nonetheless, significant reductions in $SO_4^=$ should be possible.

Long-range transport and local production of pollutants are both important, especially along the eastern seaboard. Thus, one must look beyond regional $SO_4^=$ in the northeastern United States. During winter, consideration also needs to be given to reducing urban sources of SO_2.

Although OC is very important in urban areas, more information is needed. The anthropogenic fraction and the sources need to be quantified.

Nitrate is an issue for the urban areas along the East Coast, and possibly for urban areas near the Ohio River Valley. As $SO_4^=$ decreases, NO_3^- will likely increase, particularly in winter when $SO_4^=$ is fully neutralized by NH_3. Nitrate is more a winter issue, but even in urban areas in the summer, NO_3^- should show some increase. The question is how much. There are NO_x emission reductions on the horizon (NO_x SIP call and Tier II tailpipe standards); however, NO_3^- is not expected to be HNO_3-limited until NO_x emissions have been significantly reduced. Thus, it will be important to explore ways to decrease NH_3 emissions.

10.8 CONCEPTUAL MODEL OF PM OVER THE WINDSOR-QUEBEC CITY CORRIDOR

The Windsor-Quebec City Corridor (WQC) covers a long, narrow area (roughly 1200 km x 200 km) of southeastern Canada. It extends from Windsor/Detroit in the southwest to Montreal and Quebec City in the northeast, following the north shores of Lakes Erie and Ontario and then straddling the St. Lawrence River through southern Quebec. The entire corridor borders the United States from Michigan to Maine, lying to the east of the U.S. Upper Midwest/Great Lakes region (Section 10.9) and to the north of the northeastern U.S. region (Section 10.7).

The WQC is the most densely populated region in Canada: over 85 percent of the population of Ontario and 65 percent of the population of Quebec, Canada's two most populous provinces, reside in the WQC (e.g., CCME, 1990). As a consequence, the WQC contains significant sources of PM and its precursors. It is also a transitional region in terms of both emissions and air quality, because major U.S. emission-source regions are located to its west and south but few sources are located to its north and east (see Chapter 4). Thus, both local sources and long-range transport contribute to WQC ambient PM levels (e.g., Olson et al., 1983; Yap et al., 1988; Brook et al., 2002). These characteristics lead to the WQC experiencing Canada's most significant population exposures and potential public-health problems related to air quality. Furthermore, there are many environmentally sensitive areas that extend from within the WQC northwards that are impacted by ambient air pollution.

10.8.1 Annual and Seasonal Levels of $PM_{2.5}$ and PM_{10} in Relation to Mass-Based Standards

Annual-average 24-hr $PM_{2.5}$ concentrations in the WQC are highest in the southwest portion of the region (~10-15 µg/m^3) and generally decrease northeastward (~5-10 µg/m^3), though with local maxima (~12-16 µg/m^3) in the two largest WQC urban areas: the Hamilton-Toronto conurbation and Montréal (see Figure 10.29 and MSC (2001)). This pattern reflects both the local PM and precursor emission-density distribution within the WQC and the increasing distance from high-emission areas in the United States to the west and south (i.e., Great Lakes states and Ohio Valley).

As shown in Figure 10.29, much of the WQC may be in exceedance of the Canadian 24-hr standard for $PM_{2.5}$ of 30 µg/m^3 based on a three-year average of 98th percentile 24-hr values. High levels of $PM_{2.5}$ are most often measured in the summer, but they also occur in spring and autumn and even in the winter (MSC, 2001). For example, the highest 24-hr concentration of $PM_{2.5}$ observed in Toronto in the 2000-2002 time period, 57 µg/m^3, occurred in November 2001. This event was dominated by NO_3^- and OC with relatively low $SO_4^=$.

Going from left to right, sites are ordered in this figure roughly from west to east. The solid line shows the 24-hr Canada Wide Standard for $PM_{2.5}$ of 30 µg/m^3, which is evaluated based upon annual 98th percentile values averaged over a consecutive three-year period (from MSC, 2001).

Average seasonal $PM_{2.5}$ concentrations in the WQC tend to be highest in the summer and lowest in the winter or spring, although the difference is relatively small and is negligible in cities from Ottawa eastward. At a GAViM-network urban site in Toronto and a rural site about 75 km north of Toronto at Egbert, the ratios of summer-to-winter average $PM_{2.5}$ values were 1.3 and 1.4, respectively. At other sites in the WQC, this ratio ranged from slightly less than one (i.e., slightly higher winter average) in a few urban areas in the eastern half of the region to close to two in rural southwestern areas (Brook et al., 1997a; MSC, 2001).

Annual-average 24-hr ambient concentrations of PM_{10} in the WQC ranged from 11 to 42 µg/m^3 in the period 1984-1995, with most values falling in the 20 to 32 µg/m^3 range (Brook et al., 1997a; Health Canada/Environment Canada, 1999). Corresponding values for the 1995-1998 period ranged from 15.3 µg/m^3 at Egbert, Ontario, to 26.2 µg/m^3 at Hamilton, Ontario (MSC, 2001). And like $PM_{2.5}$, coarse-fraction ($PM_{10-2.5}$) concentration values tended

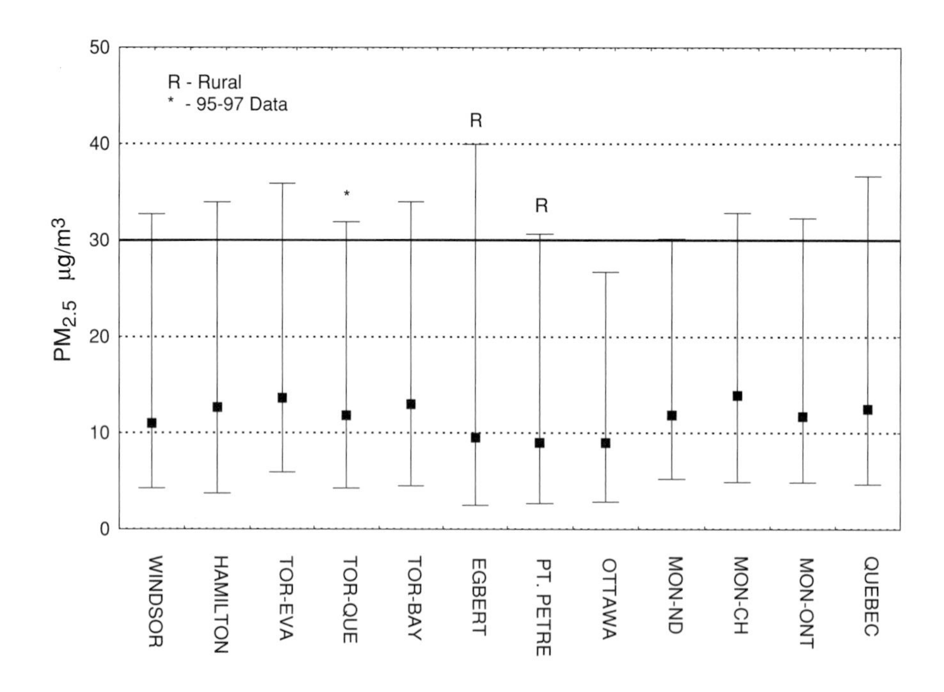

Figure 10.29. Three-year mean, 10th and 98th percentile 24-hr $PM_{2.5}$ concentrations (µg/m^3) for NAPS dichotomous sampler sites in the WQC. Data are from the period 1997 to 1999, except at the Toronto-QUE site, where the period is 1995-1997. The data are from urban sites except for the two rural sites marked with an 'R'.

to be slightly higher in the summer than the winter (e.g., Brook et al., 1997a).

Median 24-hr $PM_{2.5}$-to-PM_{10} mass ratios in the WQC for the period 1994-1998 ranged from 50 to 60 percent in urban areas and from 60 to 70 percent in rural areas, suggesting an important contribution from long-range transport to PM_{10} mass through its contribution from the (dominant) $PM_{2.5}$ fraction (MSC, 2001).

10.8.2 Compositional Analysis of PM

The three largest chemical components of $PM_{2.5}$ in the WQC on average are $SO_4^=$, NO_3^-, and OC (see Figure 10.30 and MSC (2001)). Due to relatively high SO_2 emissions within and to the west and south of the region, $SO_4^=$ plus associated NH_4^+ contributes a relatively large portion (~25 to 35 percent) of the $PM_{2.5}$ in the WQC, particularly during periods of high concentrations. During clean periods $SO_4^=$ may only explain 10 to 20 percent of the $PM_{2.5}$ mass, but during PM episodes this fraction may rise to over 50 percent (e.g., Figure 10.31b). The highest $SO_4^=$ mass concentrations occur during the summer due to enhanced SO_2 oxidation and the prevailing weather patterns.

Nitrate can vary from less than 5 percent to over 50 percent of the $PM_{2.5}$ mass depending upon season and origin of the air mass. Its spatial pattern seems to reflect more strongly the presence or absence of urban centers (e.g., NO_3^- contributes more on an annual basis than $SO_4^=$ in Toronto) and the availability of NH_3. Both its highest relative contributions and mass concentrations (5 to 15 µg/m^3) occur in the late fall, winter, and early spring (Brook and Dann, 1999; MSC, 2001), and, as shown in Figure 10.31a, the contribution of NO_3^- dominates wintertime $PM_{2.5}$ episodes. In absolute terms, both $SO_4^=$ and NO_3^- mass decrease in the WQC from west to east.

As in most other regions of North America, the carbonaceous component (OC and BC, excluding carbonates) is a large (~25 to 50 percent) and poorly understood fraction of the $PM_{2.5}$. On average, it contributes proportionately in the winter, due mostly to the decrease in the importance of $SO_4^=$ in the

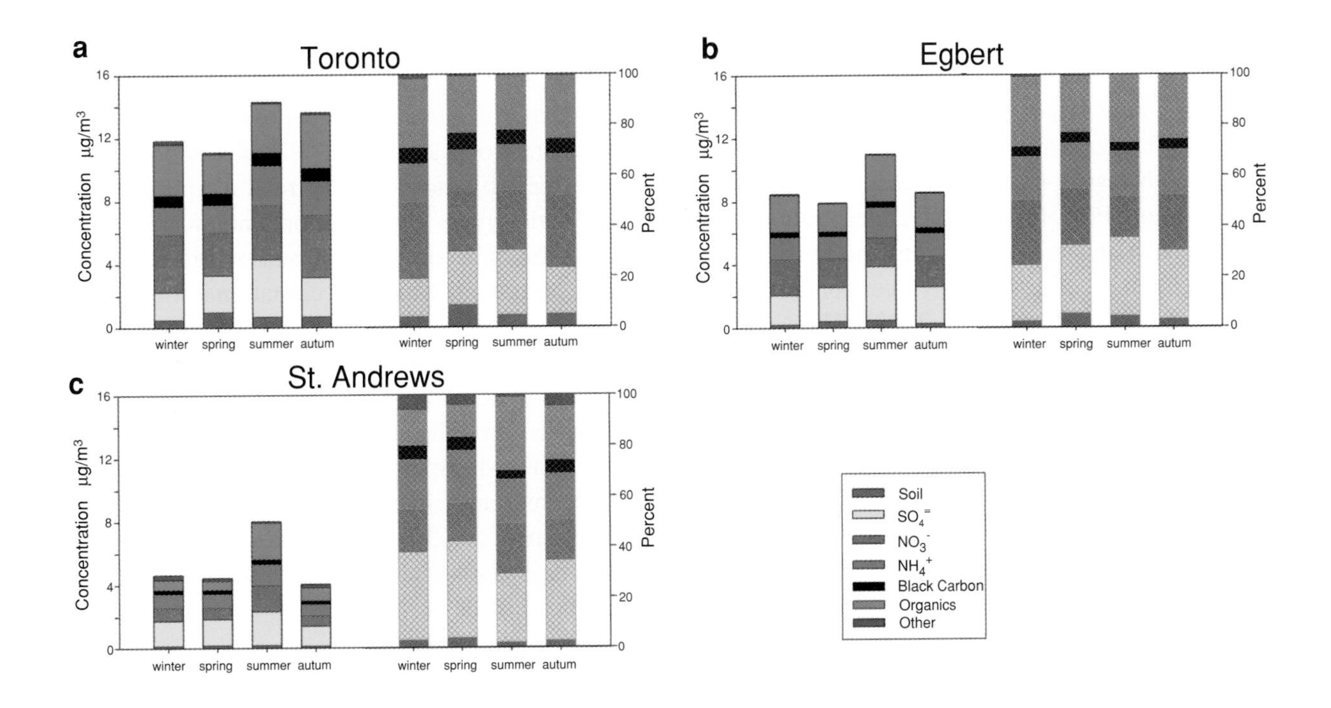

Figure 10.30. Bar charts of seasonal variations in $PM_{2.5}$ chemical composition in terms of component mass (µg/m^3: left side) and percentage contribution (right side) at three GAViM sites in eastern Canada: (a) Toronto, Ont. (1997-99, urban); (b) Egbert, Ont. (1994-99, rural); and (c) St. Andrews, N.B. (1994-97, rural).

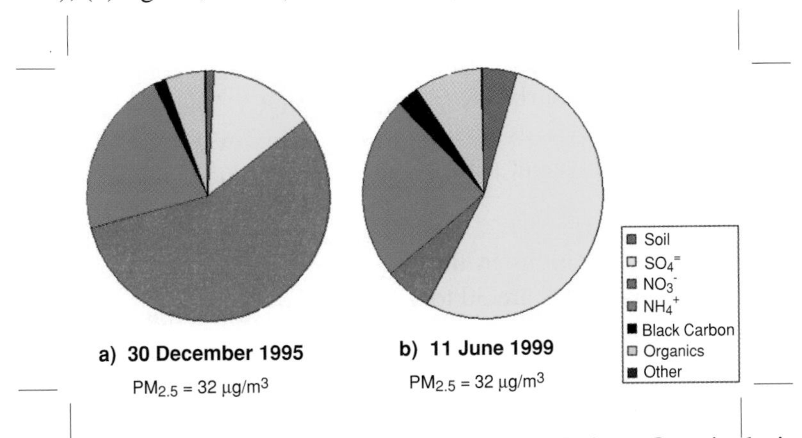

Figure 10.31. The relative composition (percent) of $PM_{2.5}$ mass at Egbert, Ontario during (a) the wintertime $PM_{2.5}$ episode of December 30, 1995 and (b) the summertime episode of June 11, 1999. The data are from the GAViM monitoring program (MSC, 2001).

winter, but the highest carbonaceous mass concentrations tend to occur in the summer. In addition, OC and BC tend to be higher in the WQC cities (Brook et al., 2002). Recent measurements in the WQC suggest that there can be considerable secondary OC mass in the summer, the peak photochemical period, due to both transport of aged air masses and local formation during stagnation conditions.

Soil or crustal material makes up the largest component of the coarse fraction in the WQC, contributing 50 to 65 percent of the coarse-fraction mass (and 20 to 30 percent of the PM_{10} mass). This is followed by the carbonaceous component, which contributes about 15 to 30 percent (20 to 40 percent), major inorganic ions, contributing 5 to 20 percent (30 to 50 percent), and sea salt (dominated by road salt in winter), which contributes 2 to 10 percent (1 to 4 percent) (MSC, 2001).

10.8.3 Meteorological Influences on PM

There is a rather sharp south-to-north gradient in the regional background PM concentrations as well as in the southwest-to-northeast direction. For example, the incoming $PM_{2.5}$ background concentration for air masses entering the WQC from the north is 4 to 6 μg/m^3 compared to higher values for flows from other directions (Brook et al., 2002). These patterns have been well documented for $SO_4^=$ and NO_3^- by the CAPMoN and NAPS networks (see Chapter 6). Measurements also suggest that $SO_4^=$ is the most uniform $PM_{2.5}$ chemical component regionally (e.g., Brook et al., 1997b).

One common feature of "being on the gradient" is that $PM_{2.5}$ concentrations in the WQC are very episodic. The most common meteorological condition leading to PM episodes in the WQC is the presence of a quasi-stationary high-pressure system located over the eastern United States and southeastern Canada and accompanied by light southerly or southwesterly winds or by stagnant conditions (e.g., Yap et al., 1988; MSC, 2001). The extent of northward penetration of the regional haze created by these transport conditions varies from episode to episode, with more southerly areas experiencing more frequent excursions. This in turn leads to the existence of the sharp gradient across the area.

Comparisons of rural $PM_{2.5}$ concentrations in the WQC with urban values show that levels are 30 to 50 percent higher in the two major cities of Toronto and Montreal than regional levels, depending upon the meteorological conditions and the location of measurement in the city (e.g., Brook et al., 1997a). Brook et al. (2002) have estimated that during periods of high $PM_{2.5}$ concentration in the greater Toronto area, regional transport accounts for 65 to 70 percent and local emissions for 30 to 35 percent of local $PM_{2.5}$.

In addition to controlling atmospheric-transport patterns, meteorology can influence local PM concentrations through its effects on emissions, diffusion, chemistry, and removal. PM_{10} episodes can be driven by strong winds. Particulate NO_3^- is more likely to be present at lower temperatures (Section 3.5.2). Near the middle of the WQC in Ottawa, daily mean temperatures typically range from about –10° C in January to 20° C in July. Studies in the WQC have suggested that more $SO_4^=$ is removed by wet deposition than by dry deposition on an annual basis (e.g., Ro et al., 1988). PM wet deposition is influenced by both precipitation amount and precipitation type. Precipitation amounts in eastern Canada are quite similar from season to season, but from December to March precipitation is more likely to occur as snow than as rain. Below-freezing temperatures and the accumulation of snow produce a snow-covered landscape. The presence of snow and frozen ground reduces both wind-erosion and traffic-induced suspension of primary PM (inorganic and organic) and affects the removal of PM via dry deposition (e.g., Zhang et al., 2001). Residential wood combustion is also correlated with temperature, i.e., much higher in winter.

10.8.4 Atmospheric Processes Contributing to PM

Given current data, one apparent difference between PM in the WQC and the upper Midwestern and northeastern United States (cf. Sections 10.9 and 10.7) is the greater contribution of NO_3^- to $PM_{2.5}$ mass in the WQC. The reason for this is not yet clear, though lower temperatures and greater NH_3 availability (due to higher local NH_3 emissions (see Figure 4.4) and lower $SO_4^=$ levels) are hypothesized to play a role. As Figure 6.14 shows, the WQC is within a belt of higher NO_3^- extending through parts of Missouri, Iowa, Illinois, Wisconsin, Indiana, Ohio, Michigan and into southern Ontario. This "NO_3^- belt" lies on the northern edge of the high-$SO_4^=$ region and has both considerable agriculture (i.e., sources of NH_3) and the potential for lower temperatures (i.e., continental interior). Although the processes involved are complex, the patterns in Figure 6.14 suggest that there may be an interaction between the amounts of $SO_4^=$ and NO_3^- that can form; that is, where SO_2 emissions are high, acidic $SO_4^=$ consumes more of the available NH_3, leaving less for NO_3^- formation.

For the WQC Brook and Dann (1999) found that in the May-September period, about one-third of available NO_3^- in the WQC partitioned to the particle phase on average vs. two-thirds to HNO_3, whereas in the October-April period more than half of available NO_3^- was present in the particle phase. Calculations of the gas ratio of total "free NH_3" to total NO_3^- (see Section 3.6) for seven WQC sites indicate a nonlinear PM response to $SO_4^=$ reductions about 50 percent of the time in the winter, suggesting an NH_3-controlled system under some winter conditions (MSC, 2001).

10.8.5 Sources and Source Regions Contributing Principal Chemicals of Concern

The WQC contains significant sources of NO_x, VOC, and PM associated with industry and with its large population. Coal-fired electric power generation and non-ferrous smelting are major sources of SO_2. The WQC also contains significant sources of NH_3 associated with agriculture, in contrast to the northeastern United States (see Chapter 4 or MSC (2001)). Motor vehicles contribute significantly to the higher urban levels and to within-city spatial variability, as suggested by the higher percent contribution of BC near major roads (Brook et al., 2001). Other neighborhood-scale sources are also important, such as residential wood combustion in east Montreal, heavy industry in Hamilton, and refineries, cement plants, and power plants in the Greater Toronto Area (GTA) as well as other areas (e.g., Sarnia, Montreal).

The sources of $PM_{2.5}$ during episodes vary from west to east across the WQC. Local sources clearly have an impact, especially in the colder months when more NO_3^- forms and there is more residential wood combustion, but different upwind regions also play a role in different portions of the WQC. The urban plume of the Detroit Metropolitan Area can impact Windsor and the east shore of Lake St. Clair, leading to locally high levels. For example, the highest annual-average concentration for the period from 1988-93 across all NAPS $PM_{2.5}$ sites in Canada, 17.6 $\mu g/m^3$, was observed east of Detroit at a rural site on Walpole Island in the mouth of the St. Clair River (Brook et al., 1997a). The Upper Midwest region plus the Ohio River Valley contribute the most to southern Ontario, but the relative influence of the northeastern United States increases in southwestern Quebec. In areas east of the WQC, such as southern New Brunswick and Nova Scotia, the impact from the northeastern United States is much more substantial.

In terms of source apportionment, receptor methods applied to the WQC have suggested that coal combustion contributes 30 to 40 percent of $PM_{2.5}$ mass on average, secondary NO_3^- contributes 10 to 40 percent, transportation sources contribute 10 to 30 percent (plus a portion of the NO_3^- due to NO_x emissions from mobile sources), secondary OC and biomass combustion (including residential wood smoke) combined contribute 10 to 20 percent, and industrial emissions contribute 5 to 10 percent (Brook, 2002; Lee, 2002; Lillyman 2001).

10.8.6 Implications for Policy Makers

As discussed above, $PM_{2.5}$ in the WQC has many sources, both local and long-range, whose relative importance varies by season. On a rural basis, speciation measurements suggest that secondary sources contribute more to ambient $PM_{2.5}$ mass than do primary sources, although primary sources may dominate in urban areas, where the highest concentrations are found. This suggests that any management program will of necessity have to be multi-faceted and will require both local measures and cooperation with other jurisdictions to the west and south of the WQC. Control measures to reduce SO_2, NO_x, and major local primary $PM_{2.5}$ sources will likely be fruitful, but more research will be required before control measures related to VOC and NH_3 emissions are recommended for reducing PM levels. The relatively high concentrations of NO_3^- during the colder months, the apparent interplay between NO_3^- and $SO_4^=$, and the potential for NH_3-limited conditions suggest that reducing SO_2 emissions to decrease ambient $SO_4^=$ may result in a slight increase in wintertime NO_3^- levels. Investigation of sources of OC in the Toronto and Montreal metropolitan areas may be worthwhile given elevated urban $PM_{2.5}$ levels, and residential wood burning may be a source category of interest. Control measures also may vary, depending on whether the goal is to reduce peak episodic levels or annual-average levels, since PM composition during episodes differs from average PM composition.

10.9 CONCEPTUAL MODEL OF PM OVER THE U.S. UPPER MIDWEST – GREAT LAKES AREA

The Upper Midwest-Great Lakes conceptual model was formulated for several Class I areas in the U.S. Upper Midwest; e.g., Boundary Waters Canoe Area, Minnesota; Voyageurs National Park, Minnesota; and Isle Royale National Park and Seney National Wildlife Refuge, Michigan; but should also be applicable to nearby Canadian Wilderness and Natural Environment Areas, e.g., Quetico Provincial Park and Algonquin Provincial Park. The region is bounded to the south by the Chicago urban area.

10.9.1 Annual and Seasonal Levels of $PM_{2.5}$ and PM_{10} in Relation to Mass-Based Standards

Annual $PM_{2.5}$ concentrations are low (e.g., about 5 to 6 µg/m^3) in Class I areas. There is little seasonal variation in $PM_{2.5}$ concentrations at these rural locations although highest concentrations tend to occur during winter and summer months. Figure 10.32a depicts $PM_{2.5}$ measurements at Boundary Waters Canoe Area, Minnesota from 1991 to 1999. This lack of rural seasonal variation distinguishes the $PM_{2.5}$ problem in the U.S. upper Midwest from that in the rural East where summer $PM_{2.5}$ concentrations dominate the annual-average values (see Figure 10.32b as one example and see the conceptual models for the northeastern United States, Windsor-Quebec City Corridor, and southeastern United States).

Annual $PM_{2.5}$ concentrations in Chicago, Illinois are above 15 µg/m^3 at most monitoring sites and some seasonality is observed. Annual PM_{10} concentrations, on the other hand, are below 40 µg/m^3 at all monitoring sites.

10.9.2 Compositional Analysis of PM

In remote areas, $PM_{2.5}$ is composed mostly of $SO_4^=$ and OC. For example, the average $PM_{2.5}$ composition at Boundary Waters Canoe Area, Minnesota between 1991 and 1999 showed that $SO_4^=$ contributed between 28 and 37 percent and OC contributed between 28 and 41 percent to $PM_{2.5}$ mass. Black carbon contributed 8 percent, NO_3^- 11 percent and crustal material 6 percent. The contribution of NO_3^- is greater during the winter months and lower during the summer months (Kenski and Koerber, 2001) because the HNO_3/NO_3^- partitioning is temperature-dependent. Chemical speciation was not determined for the urban areas, as limited data are available from routine $PM_{2.5}$ monitoring, which started in 1999.

10.9.3 Meteorological Influences

The highest $PM_{2.5}$ concentrations are observed when surface winds in the region are stagnant or nearly stagnant. This is typically the case when a high-pressure system is centered over the upper Midwest. Thus, horizontal transport in the boundary layer is limited. However, stronger winds tend to be observed aloft in the days leading to an episode of high $PM_{2.5}$ concentrations (see discussion of sources of PM below). Wet deposition dominates the removal of $SO_4^=$; for HNO_3/NO_3^-, both dry and wet deposition contribute to atmospheric removal.

10.9.4 Atmospheric Processes Contributing to PM

When high pressure resides over the Midwest, pollutants advected aloft from the west can be transported to the surface by the combined action of subsidence and turbulent mixing. On the other hand, pollutants emitted in the Upper Midwest (urban areas) may be recirculated within the boundary layer (about 1000 to 1500 m deep) thereby contributing to the $PM_{2.5}$ levels.

A review of SO_2 and $SO_4^=$ concentrations in the Upper Midwest suggests that some $SO_4^=$ production takes place in upwind locations; those $SO_4^=$ particles may therefore be transported over long distances into the Upper Midwest.

Organic PM may be of primary or secondary origin. Secondary OC formation may result from anthropogenic or biogenic VOC emissions. In the remote areas of the Upper Midwest, there should be no significant anthropogenic VOC emissions and it is likely that reactive anthropogenic VOC from urban areas already has been oxidized. Therefore, any ambient concentrations of anthropogenic SOA likely are transported to those locations. However, local biogenic emissions are likely to be important, especially in the summer. Currently there are no data available to assess the relative contribution of primary vs. secondary and anthropogenic vs. biogenic OC.

Available data at two sites in Wisconsin and Michigan show HNO_3 concentrations in the range of 1 to 2 µg/m^3, and NO_3^- concentrations in the same range. However, no NH_3 data are available and it is not possible at this point to state whether NO_3^- formation is NH_3- or HNO_3-limited. If NO_3^- formation is NH_3-limited, then future trends of NO_3^- may be affected by strategies to reduce $SO_4^=$, which may free NH_3 for the formation of NO_3^-.

10.9.5 Sources and Source Regions Contributing Principal Chemicals of Concern

Since $SO_4^=$ and OC dominate the rural annual $PM_{2.5}$ chemical composition, SO_2 and VOC are the two main primary precursors of concern. As mentioned above, high rural $PM_{2.5}$ levels in the upper Midwest are likely to result from a combination of regional emissions and long-range transport.

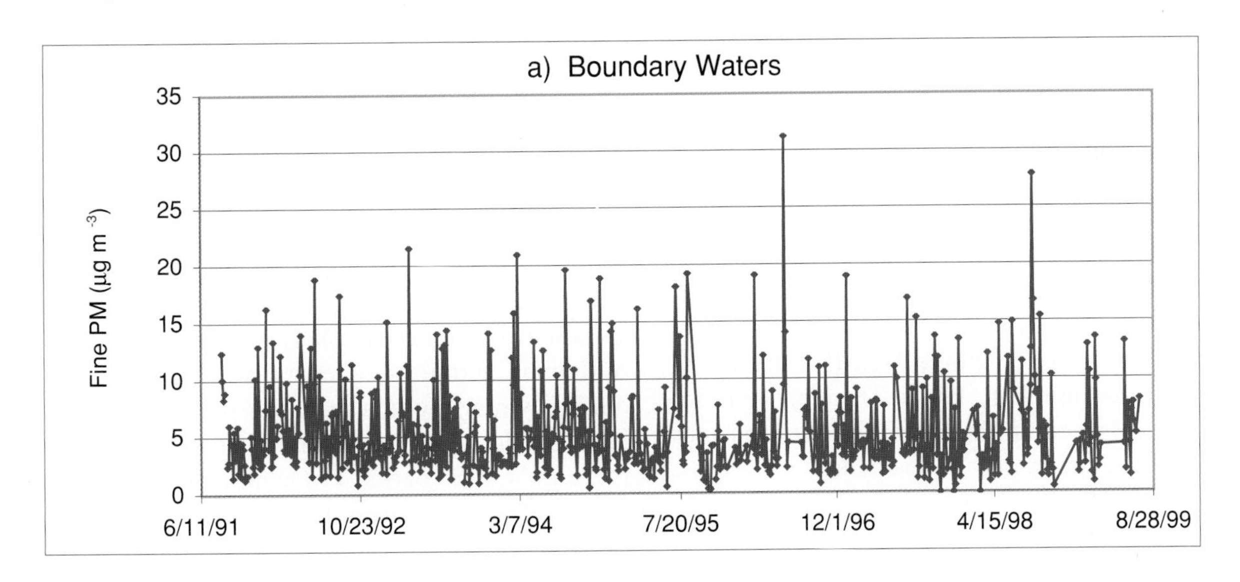

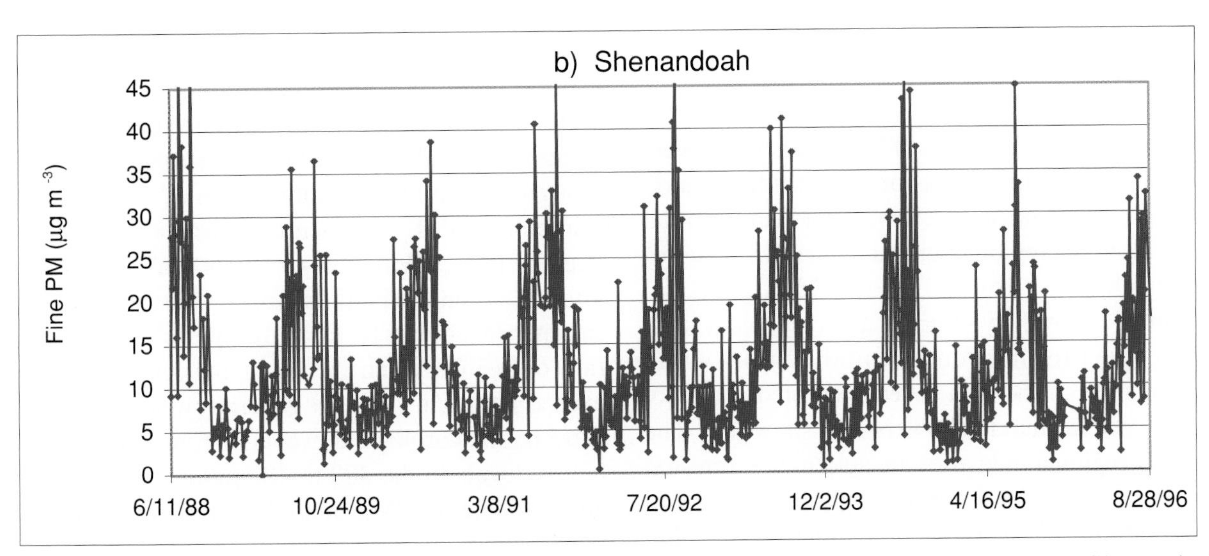

Figure 10.32. $PM_{2.5}$ measurements (a) at Boundary Waters Canoe Area from 1991 to 1999 (b) at Shenandoah National Park from 1988 to 1996 (source: Pun et al., 2001).

An analysis of wind patterns was used to infer likely source regions associated with high $PM_{2.5}$ concentrations (Pun et al., 2001). High $PM_{2.5}$ concentrations are most frequently associated with 1) air masses from the Pacific Northwest during summer and 2) air masses from Colorado during the other seasons. Clean-air conditions are typically associated with air masses coming from the north (i.e., Canada). It is anticipated that the $PM_{2.5}$ in the major cities of this region will be primarily a result of local emissions with some contribution from transported emissions.

10.9.6 Implications for Policy Makers

Reducing $SO_4^=$ and OC seems to be the most effective approach to reducing rural $PM_{2.5}$ levels and improving/maintaining visibility in the upper Midwest. However, one should keep in mind that as $SO_4^=$ levels decrease, more NH_3 could become available to react with HNO_3, and NO_3^- levels may increase. It is likely that long-range transport of pollutants from areas west and southwest of this upper Midwest region contribute significantly to the relatively low rural $PM_{2.5}$ levels. Therefore, reducing emissions of precursors in upwind regions, in addition to regional sources, may be desirable.

10.10 CONCEPTUAL MODEL OF PM OVER THE CANADIAN PRAIRIE AND U.S. CENTRAL PLAINS

The Great Plains region of North America covers the southernmost parts of the Canadian prairie provinces (Alberta, Saskatchewan and Manitoba) and the heart of the region bounded by the U.S. states of North Dakota, South Dakota, Montana, and Wyoming. Warm, dry summers and very cold winters influence this mid-continental region while precipitation to the area is received predominantly as convective thundershowers and snowfall. The primary regional industries are agriculture, forestry, mining, coal-fired power generation, and transportation. Urban centers are relatively small in population with the largest cities at nearly one million people. This area experiences some of the cleanest air quality in North America and is highly sensitive to regional haze.

10.10.1 Annual and Seasonal Levels of $PM_{2.5}$ and PM_{10} in Relation to Mass-Based Standards

Annual $PM_{2.5}$ mass concentrations in remote regions of the Great Plains range between 3 and 5 μg/m^3 and are spatially consistent, indicating a typical widespread, regional condition. Urban regions show more variability with annual levels of 7 to 15 μg/m^3, indicating some potential of exceeding the U.S. annual standard of 15 μg/m^3. The second maximum 24-hr observations are typically in the 25 to 35 μg/m^3 range with hourly levels in the 40 to 70 μg/m^3 range. This indicates some potential of exceeding the Canada-wide Standard of 30 μg/m^3 for the 98th percentile over three years. AIRS, NAPS and regional-network reporting information are summarized in Table 10.4. Seasonal fine-mass variation is minimal especially in areas not affected by human activities. Generally, $PM_{2.5}$ increases during winter months in urban regions while it increases during summer months in remote areas. In the area's U.S. parks, PM concentrations have generally decreased over the last decade though not as to the extent seen in the eastern United States (IMPROVE, 2000).

Annual PM_{10} mass concentrations range from less than 10 μg/m^3 in rural unaffected regions to more than 40 μg/m^3 in some urban settings. At present there is little risk of U.S. areas exceeding the 50 μg/m^3 annual standard broadly across the plains, though individual communities may have some challenges. PM_{10} shows more seasonal variability than the $PM_{2.5}$. Winter 24-hr levels can be less than 4 μg/m^3 when the earth is frozen. The peak 24-hr summer levels can be up to 100 μg/m^3. For those rare occasions when the 24-hr concentrations do exceed 100 μg/m^3, they are generally associated with extreme meteorological events or forest fires, with hourly levels in the 300 to 400 μg/m^3 range. Data are available for PM_{10} in all the jurisdictions on the plains, although urban centers are more studied than rural or remote regions.

Table 10.4. Long-term average values reported by national and state or provincial agencies for $PM_{2.5}$ and PM_{10} annual average concentrations. Compiled from AIRS (1996-2001), NAPS (1997-1999) and available data from regional networks. Note: North Dakota is presently operating two $PM_{2.5}$ stations in Saskatchewan as part of a transboundary air-quality study.

Region	Number of Stations Reporting in 99/00		$PM_{2.5}$ range of annual averages ($\mu g/m^3$)	PM_{10} range of annual averages ($\mu g/m^3$)
	$PM_{2.5}$	PM_{10}		
Alberta	2	4	3.2 - 18.0	8.8 - 34.6
Saskatchewan	2*	2	6.2 - 8.7	16 - 23
Manitoba	1	6	6.4 - 9.5	9.0 - 47.2
Montana	0	20	na	3.5 - 36.2
North Dakota	9	4	3.1 - 8.7	5.6 - 30.4
South Dakota	0	5	na	19.0 - 46.0
Wyoming	0	12	na	5.2 - 54.9

10.10.2 Compositional Analysis of PM

Compositional analysis of $PM_{2.5}$ in remote regions indicates that $SO_4^=$ continues to be a significant proportion of the fine fraction, ranging from 33 percent at the Canadian location, Esther (shown in Table 10.5), to 43 percent in the South Dakota location, Badlands National Park (shown in Figure 10.33). Figure 10.33 demonstrates that the proportion of $SO_4^=$ is relatively consistent throughout the seasons with a small increase in the summer. The contribution of NO_3^- and OC, however, can change dramatically with the seasons depending on the location. For Badlands National Park, in the summer months, NO_3^- is very low while OC approaches levels similar to those of $SO_4^=$. In the winter months, NO_3^- levels increase to levels similar to those of OC, with both being less than the $SO_4^=$ fraction. In the more western remote regions (Glacier, Yellowstone, and Bridger Parks), OC exceeds $SO_4^=$, NO_3^- remains low and somewhat constant throughout the year, while the soil component is increased in summer.

Sulfate proportions of 10 percent to 25 percent (Cheng et al., 1998) and OC of 35 percent to 50 percent (EC-MSC, 2001) appear to be typical for urban areas of the plains regions of both Canada and the United States (as shown in Table 10.5 for Bismarck, ND). In one rural-urban comparison (Table 10.5), OC appears to offset $SO_4^=$ as a proportion of $PM_{2.5}$ mass when moving from a rural setting to an urban center. More speciated data are required over a wider area to address this question properly.

10.10.3 Meteorological Influences on PM

On the Great Plains, warm, dry summers and very cold winters influence air quality. Meteorology plays an important role in PM concentration in the region where many contributing sources are affected by ambient conditions. For example, soil in the air can increase with increased wind speeds, NO_3^- can increase with decreasing temperatures, and biogenic organic emissions and NH_4^+ increase with increasing temperatures. The extremely high one-hour mass concentrations of both PM_{10} and $PM_{2.5}$ observed throughout the region demonstrate this variability.

In the winter months, these northern regions can experience extremely low mixing depths with concurrent stagnant conditions resulting in a trapped layer of air near the surface (Myrick et al., 1997). When urban emissions are increased in cold temperatures, the combination results in elevated $PM_{2.5}$ concentrations in urban areas. With the frozen surface, there is little geological contribution in the winter months and so the rural regions show low levels with little variation at that time of year unless affected by the urban emissions (McDonald et al., 1999). In the summer months, enhanced biogenic emissions and regional transport appear to contribute to slightly increased $PM_{2.5}$ levels evident in the pristine regions of Figure 10.33.

Table 10.5. Comparison of urban and rural $PM_{2.5}$ speciation measurements. Data from monitoring stations operated in Bismarck (North Dakota Department of Health, Dan Harmon, private communications) and Esther (McDonald et al., 1998 and 2000).

Bismarck ND Summer 2000	**Urban Annual**	**Esther, Alberta 1995-99**	**Remote Annual**
$PM_{2.5}$ mass, $\mu g/m^3$	6.29	$PM_{2.5}$ mass, $\mu g/m^3$	4.4
$SO_4^=$ percent	20.3	$SO_4^=$ percent	33.5
NO_3^- percent	9.1	NO_3^- percent	na
OC percent	39.7	Organics percent	29.4
BC percent	3.7	BC percent	6.9
Soil percent	8.1	Soil percent	10.8
Unaccounted percent	19.1	Unaccounted percent	19.5

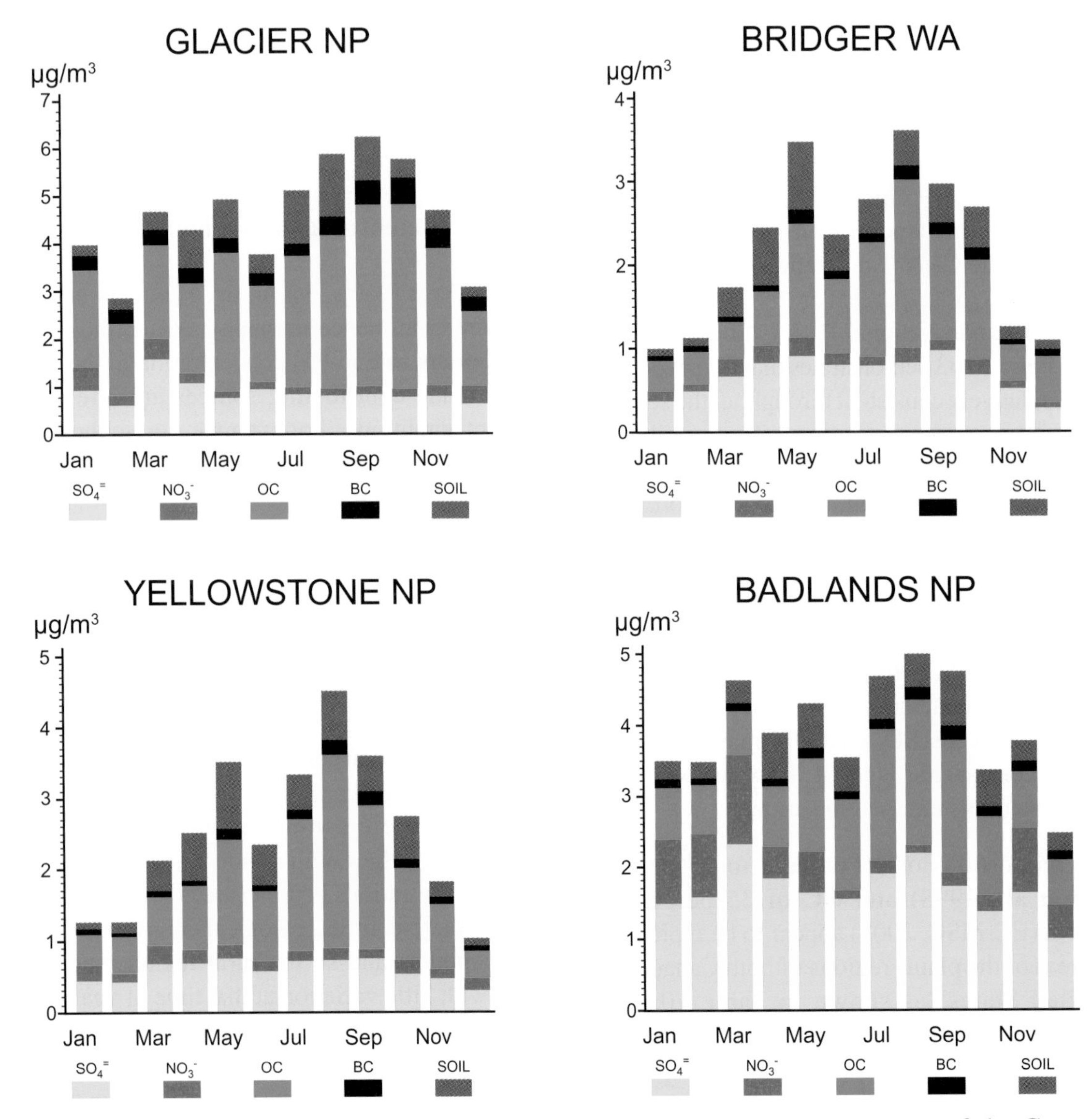

Figure 10.33. Chemical speciation of $PM_{2.5}$ in national parks and wilderness areas of the Great Plains (IMPROVE, 2000).

10.10.4 Atmospheric Processes Contributing to PM

In both remote and urban areas of the plains, NO_3^- reaches its maximum levels in the winter months – some locations with 5 to 8 times more than in the warm summer months. It may be that the very low winter temperatures are affecting the relative contributions of NO_3^- through gas-to-particle partitioning (Peake et al,, 1988). Organic carbon contributions to PM levels are significant in this region. Biogenic emissions are low in the winter season but increase substantially in summer months. More complete understanding of the atmospheric chemical processes associated with organic precursors will assist in the characterization of PM in this region. In addition, the role of NH_3 as a precursor in gas-to-particle conversion and the large magnitude of NH_3 emissions from the area (EC-MSC, 2001) suggest that PM mass may be enhanced by these processes.

10.10.5 Sources and Source Regions Contributing Principal Chemicals of Concern

Sources of PM and precursor gases are similar throughout the plains region. Concentrations of $PM_{2.5}$ in urban areas in winter are predominately affected by fossil-fuel and wood-fuel combustion, including that by individual vehicles, residences, and power-generation facilities. In spring and autumn, levels are further affected by activities such as vehicle traffic (road dust), agricultural processes, vegetative burning, including prescribed forest burns, and fugitive dust from cement and mining operations. These latter activities enhance PM_{10} concentrations to a greater extent than $PM_{2.5}$. Other naturally occurring conditions contribute to the highest observed concentrations, especially of PM_{10}. For example, forest fires and dust storms can contribute to cases where PM_{10} levels reach 300-400 μg/m³.

Investigation of the urban-rural differences in PM in Alberta, Canada has been performed (Cheng et al., 2000) to determine a regional background level and to identify the potential for transport of PM from urban and industrial areas into the rural communities. The results are presented in Table 10.6, demonstrating that more than 50 percent of the mass in rural areas near urban or industrial sources may be contributed by those sources. Significantly more PM of both size fractions was observed at rural industrial sites compared to urban centers.

Organic carbon emissions are also significant in this region (EC-MSC, 2001). Where biogenics decrease in the winter, anthropogenic sources for heating and transportation may increase, keeping the annual contribution of organics to $PM_{2.5}$ high relative to other chemical components.

10.10.6 Implications for Policy Makers

Policy implications are similar across the plains region where few areas will have concentrations exceeding any national standards. There are strong urban-rural differences in the region. Some urban centers may experience long-term $PM_{2.5}$ mass concentrations approaching the U.S. annual standard or the Canada Wide Standard. In addition, many parts of the region are experiencing industrial development and population growth with the inherent risk of increasing the PM levels.

The observed seasonal variability in the NO_3^- component offers some insight for future development in the region. If precursors of NO_3^- or $SO_4^=$ are increased, high $PM_{2.5}$ concentrations during winter-inversion conditions will increase. For this

Table 10.6. Period-of-record mean levels measured in Alberta, Canada for the types of sites as indicated (Cheng et al., 2000).

Alberta Mean Levels	**$PM_{2.5}$ μg/m³**	**PM_{10} μg/m³**	**$PM_{2.5}$ / PM_{10} ratio**
Rural Clean Sites	3.2	8.8	0.38
Rural Influenced Sites	7.9	16.8	0.49
Rural Industrial Sites	18.0	34.6	0.52
Urban Sites	11.2	27.7	0.42

reason, energy-efficient processes are encouraged even in regions with low SO_2 or NO_2 concentrations.

Regional haze is an important driver of PM reduction for the cleanest areas in the region (Malm, 1999). Smoke management is a consideration in many areas affected by burning due to agricultural and forest-management practices. Wood burning produces visible emissions even though the contribution to ambient PM mass concentration is small. Many jurisdictions have already recognized the importance of decreasing emissions from wood burning, home heating and cooking.

A reduction in NH_3 emissions from the production and use of fertilizers (EC-MSC, 2001) also may be supported. The importance of NH_3 as a contributor to regional haze and local visibility issues has been recognized in other jurisdictions (Pryor et al., 1997) and similar issues may apply in the Great Plains.

10.11. CONCEPTUAL DESCRIPTION OF PM OVER THE LOWER FRASER VALLEY AIRSHED

The Lower Fraser Valley airshed of southwestern British Columbia and northwestern Washington State is bounded by the Coast Mountains to the north, rising to approximately 2000 meters, and by the North Cascades that lie in Washington State to the south. The valley is confined to the west by the Strait of Georgia and then Vancouver Island.

The airshed is home to roughly 2.5 million people and comprises three jurisdictions; the Greater Vancouver and Fraser Valley Regional Districts and Whatcom County in Washington State. The main urban center of greater Vancouver is situated to the northwest. The valley becomes increasingly rural as one moves eastward. An industrial area extends along the western portion of Whatcom County, the southwestern portion of the airshed that ends at the city of Bellingham.

The topography of the area often controls the flow of air pollutants. The proximity of the ocean creates a land/sea breeze, marine inversions, and sources of natural emissions for particle formation. A large portion of the central and eastern Valley is used for agriculture with considerable areas of natural vegetation. Both contribute anthropogenic and biogenic emissions. The "funnel" shape of the Valley promotes the occasional development of strong outflow winds that cause crustal material from exposed riverbeds and agricultural fields to be suspended and carried westward into the main areas of the airshed.

10.11.1 Annual and Seasonal Levels of $PM_{2.5}$ and PM_{10} in Relation to Mass-Based Standards

Annual and Seasonal $PM_{2.5}$: Chilliwack, located in the eastern part of the Lower Fraser Valley, has the longest record of continuous TEOM $PM_{2.5}$ measurements in the valley. Data from Chilliwack (1996-99) presented in Figure 10.34 indicate annual mean 24-hr $PM_{2.5}$ concentrations in the range of 7.4 to 8.2 μg/m^3. Figure 10.34 also shows that annual $PM_{2.5}$ levels at Chilliwack are well below the Canada Wide Standard. Mean 24-hr concentrations reach a maximum during July, August and September as indicated in Figure 10.35.

The REVEAL II study that used IMPROVE samplers and protocols, conducted between April 1994 and June 1995 (Pryor and Barthelmie, 1999), found a mean concentration of $PM_{2.5}$ of 7.9 μg/m^3 at Clearbrook and 8.8 μg/m^3 at Chilliwack. Clearbrook and Chilliwack represent the middle and eastern parts, respectively, of the Lower Fraser Valley. Maximum concentrations of $PM_{2.5}$ at these two sites were 19.0 and 20.4 μg/m^3 respectively. Data collected under the NAPS program between 1984 and 1995 at two dichotomous sampler sites in Vancouver show a 24-hr average $PM_{2.5}$ concentration between 12.9 and 17.3 μg/m^3 (WGAQOG, 1999).

Annual and seasonal PM_{10}: Six years of TEOM data (1994 to 1999) aggregated from 13 stations in the Greater Vancouver Regional District (GVRD) monitoring network have an annual-mean 24-hr concentration of PM_{10} of roughly 13 μg/m^3, with a range in annual means of 11.8 to 15.2 μg/m^3. The annual-average 95th percentile of 24-hr concentrations

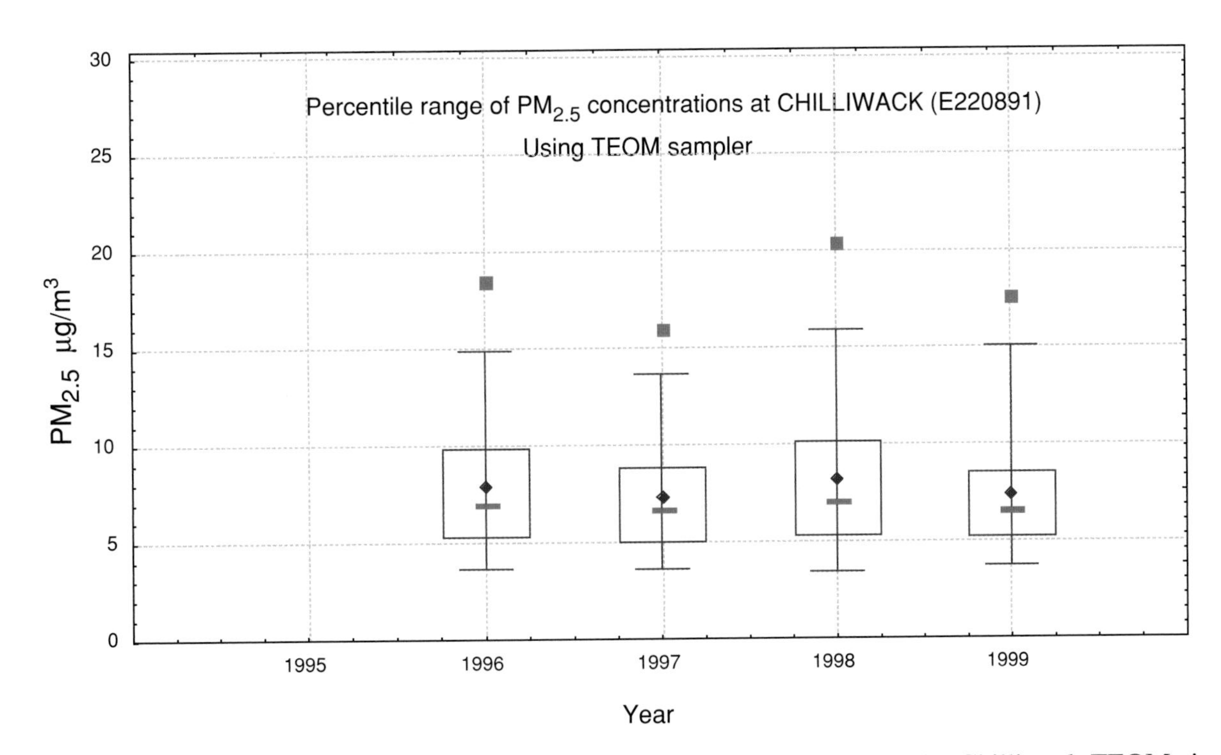

Figure 10.34. Annual variations in 24-hr average $PM_{2.5}$ measurements from the Chilliwack TEOM site. Box boundaries indicate the 1st and 3rd quartiles (25th and 75th percentiles) and whiskers indicate the 5th and 95th percentiles. Symbols represent the median (▬), mean (◆), and 98th percentile (■).

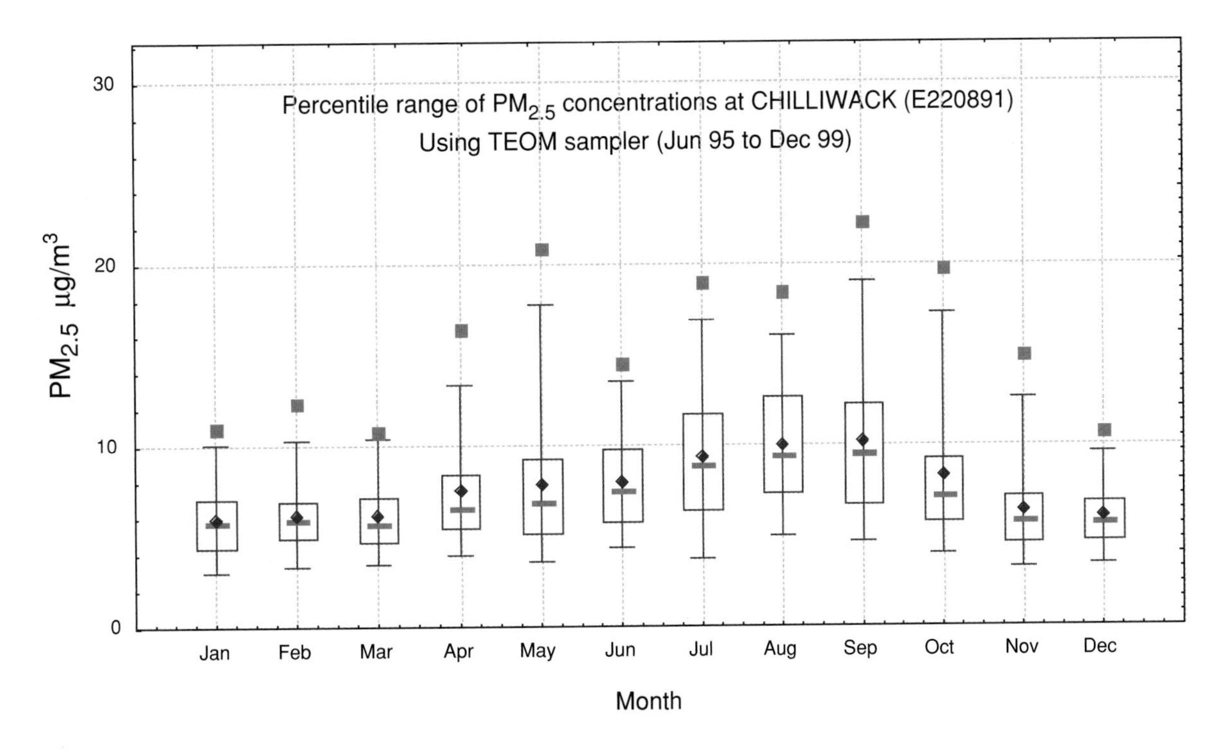

Figure 10.35. Seasonal variations in 24-hr average $PM_{2.5}$ measurements from the Chilliwack TEOM site Box boundaries indicate the 1st and 3rd quartiles (25th and 75th percentiles) and whiskers indicate the 5th and 95th percentiles. Symbols represent the median (▬), mean (◆), and 98th percentile (■).

aggregated for the valley is roughly 26 μg/m^3 while the mean annual maximum is approximately 46 μg/m^3.

Figure 10.36 shows mean annual 24-hr PM_{10} concentrations at Chilliwack, located in the eastern part of the valley, in the range of 11.3 to 15.7 μg/m^3. The provincial standard of 50 μg/m^3 is rarely exceeded at this site. Figure 10.37 shows mean PM_{10} concentrations at Chilliwack to be highest in late summer and early fall.

Concentrations of PM_{10} in the Lower Fraser Valley are substantially lower than those found at sites in the interior of the province. A study by the provincial ministry (ARB, 1997) based on six years of data (1990 –95) showed province-wide 24-hr average concentrations ranged from less than 15 μg/m^3 to greater than 50 μg/m^3. Urban PM_{10} concentrations in the Lower Fraser Valley are relatively low compared to other Canadian cities and large urban centers such as the Los Angeles basin. Elevated PM_{10} concentrations were observed in the Lower Fraser Valley during April and May 1998 at the time of the Asian dust storm (McKendry et al., 2001).

10.11.2 Compositional Analysis of PM

The composition of $PM_{2.5}$ changes from urban areas, to suburban areas, to more rural or agricultural areas. Particles in western portions of the airshed exhibit characteristics related to natural emissions from the marine environment, such as sea salt.

Figure 10.38 shows the composition of particles typical of the urban areas of Vancouver. Particles are dominated by carbonaceous material with OC representing 46 percent of the total. Sulfate and NO_3^- concentrations are important portions of the $PM_{2.5}$ fraction.

Chemical analysis during REVEAL II (Pryor and Barthelmie, 1999) showed the fine fraction to consist of inorganics, mainly $SO_4^=$ and NO_3^-, and a large contribution from organic compounds. Other constituents include BC, soil, and marine particles (Figures 10.39 and 10.40).

The Clearbrook site is located in the central region of the airshed whereas Chilliwack is located in the extreme eastern reaches of the valley. The percentage contribution from OC and BC decreases through central areas of the airshed. However, the dominance of $SO_4^=$ and NO_3^- continues throughout the entire airshed. Peak levels of fine-particle NO_3^- (NH_4NO_3) tend to be observed in these central and eastern areas due to greater emissions of NH_3. However, overall, a greater amount of NO_3^- throughout the Lower Fraser Valley is found in the coarse fraction of PM_{10}. This is predominantly related to $NaNO_3$ forming from the reaction of HNO_3 and sea salt (NaCl).

The fractional contribution from organic aerosols is highest in fall and winter, possibly due to reduced mixing depths. Highest BC concentrations were observed in winter due to combustion of wood and other fuels for heating. Highest soil concentrations were observed in winter perhaps due to the influence of strong outflow events and the large source areas of uncultivated fields exposed during the winter season.

10.11.3 Meteorological Influences on PM

During summer, anti-cyclonic conditions produce subsidence inversions that effectively place a cap on the valley, thus inhibiting the dispersion of pollutants. Surface winds are predominantly easterly throughout the year, however, a daytime sea breeze pattern is more characteristic of summer resulting in a westerly, up-valley flow of pollutants. Over the southern end of the Straight of Georgia near the Gulf Islands and San Juan Islands there is a greater frequency of light winds. In the summer this favors the buildup of pollutants from a variety of sources, particularly those near the coast in BC and Washington, and marine emissions (i.e., ships). When sea breezes and mountain upslope flows develop, this relatively polluted air mass, which at times may have 2 or 3 days to 'age' over the Strait, is drawn inland. This can lead to higher particle levels as far east as Chilliwack, especially when accompanied by a stronger westerly inflow through the Strait of Juan de Fuca (Snyder et al., 2002). Subsequent outflows (land breeze), which tend to be weaker than the inflow, can bring this polluted air mass back out over the Strait of Georgia, having picked up additional pollutants emitted over the mainland. Such a cycle

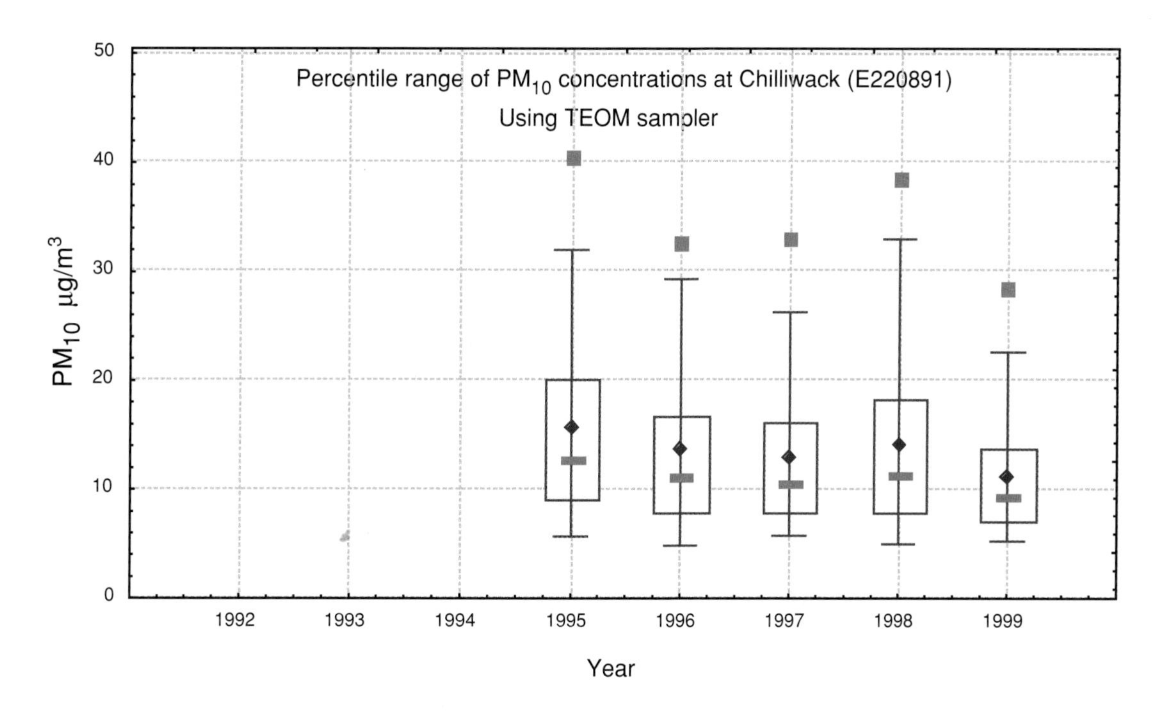

Figure 10.36. Annual variations in 24-hr average PM_{10} measurements from the Chilliwack TEOM site. Box boundaries indicate the 1st and 3rd quartiles (25th and 75th percentiles) and whiskers indicate the 5th and 95th percentiles. Symbols represent the median (▬), mean (◆), and 98th percentile (■).

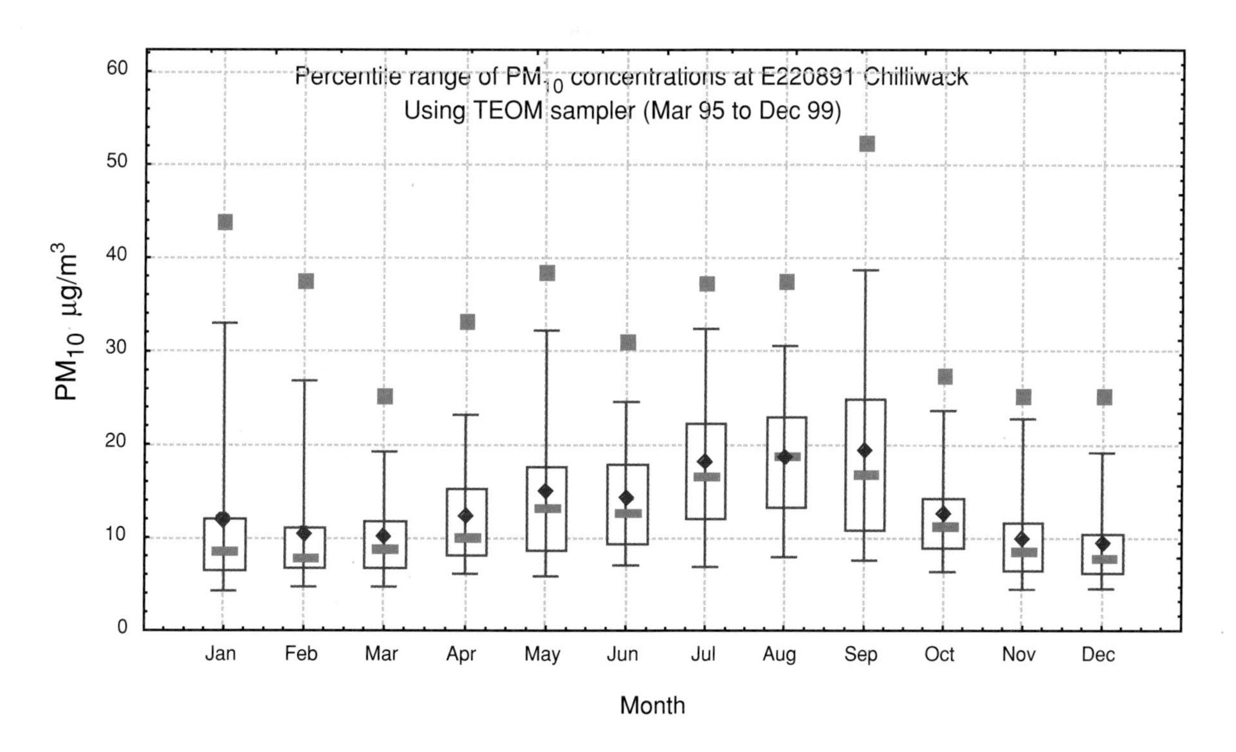

Figure 10.37. Seasonal variations in 24-hr average PM_{10} measurements from the Chilliwack TEOM site. Box boundaries indicate the 1st and 3rd quartiles (25th and 75th percentiles) and whiskers indicate the 5th and 95th percentiles. Symbols represent the median (▬), mean (◆), and 98th percentile (■).

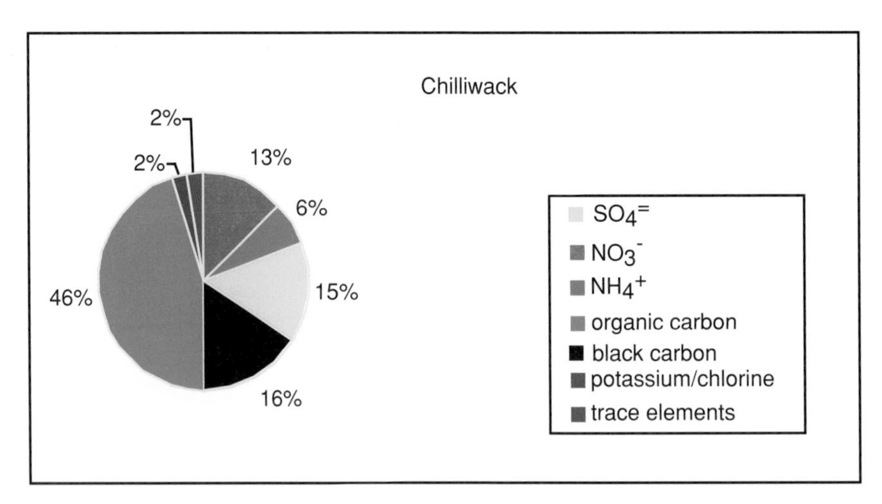

Figure 10.38. The percentage distribution of the main $PM_{2.5}$ chemical constituents (urban Vancouver).

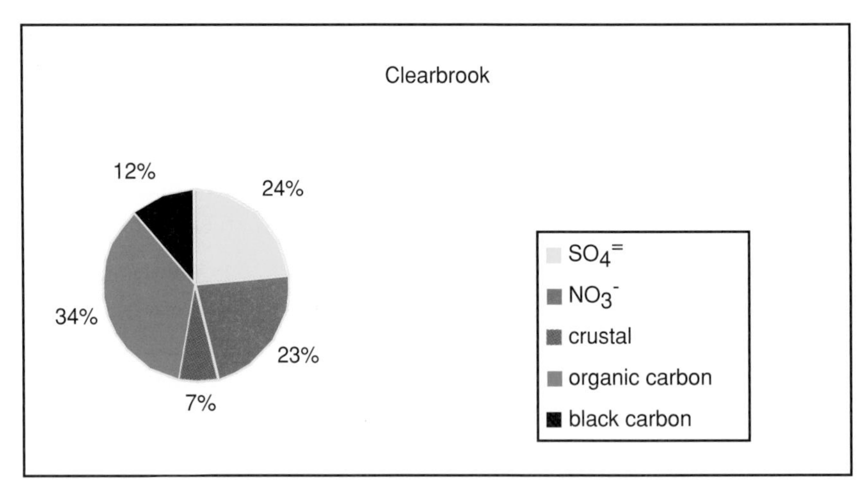

Figure 10.39. Percent contribution to reconstructed fine mass from the five dominant modes of fine particle composition for Clearbrook .

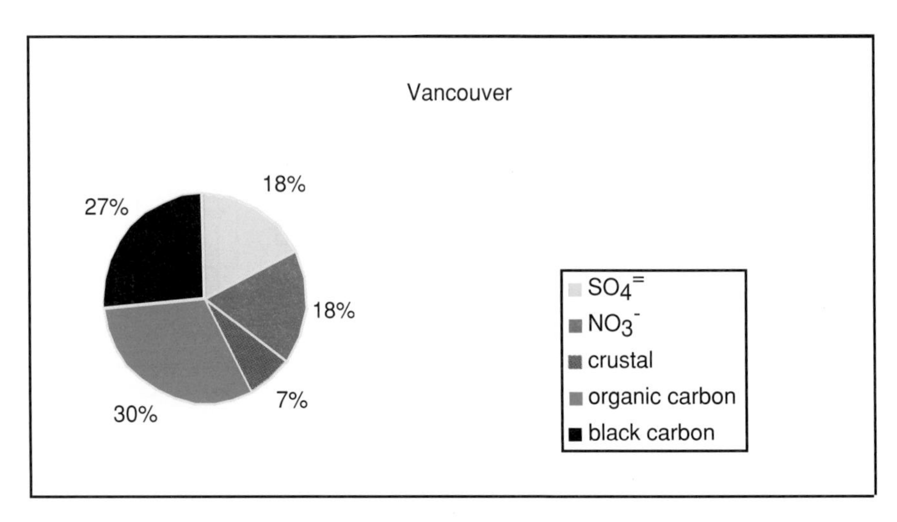

Figure 10.40. Percent contribution to reconstructed fine mass from the five dominant modes of fine particle composition for Chilliwack.

can repeat itself for multiple days. If this persists for more than three days in the summer, then significant inland visibility degredation is expected. The length of time that this pattern persists and the strength of the overlying subsidence inversions dictate the overall extent of particle buildup.

These same stagnant summertime anti-cyclonic conditions, with a land-sea interaction, are also known to produce ground-level ozone episodes. During these periods daytime concentrations of PM_{10} may reach 50 to 75 $\mu g/m^3$ and peak $PM_{2.5}$ concentrations of 20 to 30 $\mu g/m^3$ are not uncommon. However, there is only a weak correlation between ground-level ozone and PM_{10} concentrations (McKendry, 2000). Secondary particle formation, which has been linked to visibility degradation (Pryor and Barthelmie, 1999), becomes very active during summertime subsidence inversions and progresses even further during prolonged periods of stagnation and inflow/outflow conditions.

During the winter, Arctic-outflow winds raise dust from farmland soil and from exposed riverbeds. Elevated PM levels (~100 $\mu g/m^3$) may persist for several days and are mostly limited to the eastern end of the valley.

10.11.4 Atmospheric Processes Contributing to PM

Measurements of the chemical components of fine PM indicate that the formation of carbonaceous particles dominate in the airshed. A large portion of the carbonaceous material is OC, primarily resulting from the combustion of fossil fuels. It has been observed that the composition of the organic particles alters during transport, producing different species depending on the distance from the formation source.

The close proximity of the marine environment influences particle formation in western areas of the airshed. This is also the location in the airshed where emissions from urban and industrial sources produce HNO_3. Sodium chloride combines with HNO_3 to form $NaNO_3$ particles. The abundance of $NaNO_3$ decreases further inland where NH_3 emissions, primarily from agricultural sources, begin to dominate the formation chemistry resulting in the production of both $(NH_4)_2SO_4$ and NH_4NO_3 particles. This change in chemical composition and therefore relative size of the particles may partially explain the apparent difference in appearance of haze layers from western to eastern areas of the airshed.

Local emissions of NH_3 from agricultural practices contribute to poor visibility in the eastern part of the valley. Ammonia reacts with H_2SO_4 and HNO_3 in the atmosphere to form $(NH_4)_2SO_4$, NH_4HSO_4, and NH_4NO_3. When humidity is high, these salts grow to form particles whose diameters are efficient at scattering light. Ammonium sulfate and NH_4NO_3 are highest during summer when there is a sufficient source of NH_3 from agricultural operations.

10.11.5 Sources and Source Regions Contributing Principal Chemicals of Concern

Measurements from an IMPROVE sampler were analyzed using CMB to estimate the relative contributions of sources to the $PM_{2.5}$ mass (Lowenthal et al., 1994). The results indicated that motor vehicles (34 to 43 percent) were a dominant source with secondary $SO_4^=$ and NO_3^- contributing 25 to 27 percent and 12 to 27 percent respectively. Wood smoke was present at 8 percent and geological sources contribute 3 to 5 percent.

An emission inventory recently completed for the year 2000 (Greater Vancouver Regional District, 2002) provides source estimates for the entire Lower Fraser Valley airshed. The airshed includes the Greater Vancouver and Fraser Valley Regional Districts in British Columbia along with Whatcom County in Washington State.

Figure 10.41 indicates that area sources make the largest contribution to emissions of both PM_{10} and $PM_{2.5}$. Agricultural emissions make up 40 percent of the PM_{10} but only 4 percent of the $PM_{2.5}$ whereas space heating is responsible for 9 percent of the PM_{10} and 41 percent of the $PM_{2.5}$ emissions. The PM_{10} and $PM_{2.5}$ from point sources is dominated by bulk shipping terminals in the Greater Vancouver Regional District with primary metal industries and petroleum products being dominant in Whatcom County. For the mobile sector, non-road sources contribute 40 percent to the PM_{10} and 41 percent to the $PM_{2.5}$

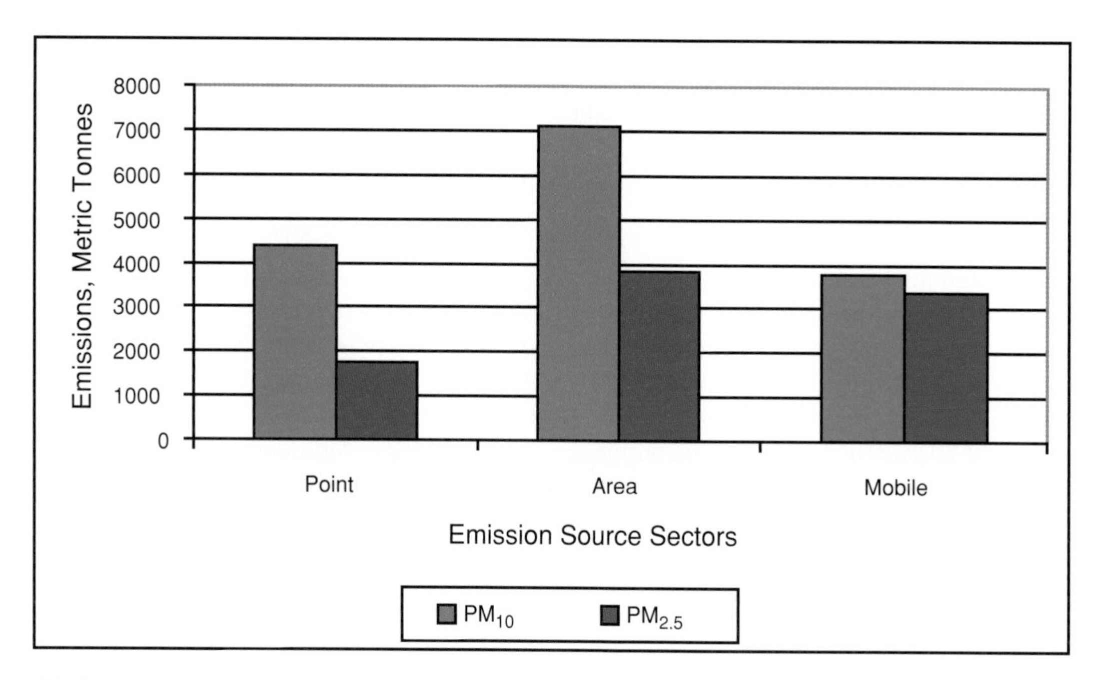

Figure 10.41. Year 2000 Emission Inventory for PM_{10} and $PM_{2.5}$ for the Lower Fraser Valley (GVRD, 2002).

emissions. Road dust contributes 45 percent of the PM_{10} and 17 percent of the $PM_{2.5}$ emissions to the Lower Fraser Valley airshed.

The airshed is subjected to a variety of sources depending on location. Urban centers such as Vancouver are dominated by emissions from motor vehicles, with emissions of OC and NO_x being the most important. Wood combustion and secondary OC emissions were found to be important, as well as a non-source-specific NH_4NO_3 factor (Brook, 2002).

There have been documented cases of trans-Pacific transport of Asian dust into the Pacific Northwest. This was referenced earlier as an episode in late April and early May of 1998 and other episodes have also been observed, such as in April-May 2001 (Gong et al., 2002).

10.11.6 Implications for Policy Makers

PM in the Lower Fraser Valley airshed is associated with urban development and high-intensity agriculture. The Vancouver metropolitan area is one of the top 10 fastest growing metropolitan areas in North America and its population is expected to reach 3 million by 2020, almost double the population in 1991. The number of vehicles is expected to increase 60 percent to about 2 million over this period, while the average trip distance and trip time will increase at even faster rates.

The Lower Fraser Valley airshed is presently in attainment with current and proposed air-quality standards and objectives. The complex nature of emission patterns and sources indicate that a strategy to maintain and improve air quality will need to be carefully crafted. Emissions from light-duty vehicles will be significantly reduced over the next decade; however, these emission reductions will likely be minimized by other mobile transportation-sector sources such as marine vessels and railways.

The airshed is composed of industrial and agricultural emission sources. Control and reduction strategies must account for the differences in these sectors. Actions appropriate to control and reduce NH_3 emissions from agricultural enterprises will be different than those needed to deal with emissions of NO_x. The marine sector is largely controlled through international jurisdiction making local initiatives difficult to deliver, but is currently an issue of great interest.

10.12 REFERENCES

ARB (Air Resources Branch), 1997. Air Quality Report for British Columbia: Fine Particulate (PM_{10}) levels (1990-1995). British Columbia Ministry of Environment, Lands, and Parks.

Boylan, J.W., Wilkinson J.G., Yang Y.J., Odman M.T., Russell A.G., 2000. Particulate matter and acid deposition modeling for the Southern Appalachian Mountains Initiative. In Proceedings of the 93rd Annual A&WMA Conference.

Brockton, J.H., Jalaludin, B., McClure, E., Cobb, N., Johnson, C.A., 1994. Surveillance for dust storms and respiratory diseases in Washington state, 1991. Archives of Environmental Health 49, 170-175.

Brook J.R., 2002. Health effects study with enhanced characterization of the urban air pollutant mix (SHEMP). Final Report to the Toxic Substances Research Initiative, Health Canada.

Brook, J.R., Lillyman, C.D., Shepherd, M.F., Mamedov, A., 2002. Regional transport and urban contributions to fine particle concentrations in southeastern Canada, Journal of the Air & Waste Management Association 52, 855-866.

Brook, J.R., Li. S.-M., Lu, G., Lillyman, C., 2001. Spatial variation in fine particulate organic and elemental carbon in a downtown urban core. In: Proc. American Association for Aerosol Research (AAAR) Annual Meeting, Portland, OR, October.

Brook, J.R., Dann, T.F., 1999. Contribution of nitrate and carbonaceous species to $PM_{2.5}$ observed in Canadian cities. Journal of the Air & Waste Management Association 49, 193-199.

Brook, J.R., Dann, T.F., Burnett, R.T., 1997a. The relationship among TSP, PM_{10}, $PM_{2.5}$, and inorganic constituents of atmospheric particulate matter at multiple Canadian locations. Journal of the Air & Waste Management Association 47, 2-19.

Brook, J.R., Wiebe, A.H., Woodhouse, S.A., Audette, C.V., Dann, T.F., Callaghan, S., Piechowski, M., Dabek-Zlotorzynska, E., Dloughy, J.F., 1997b. Temporal and spatial relationships in fine particle strong acidity, sulphate, PM_{10} and $PM_{2.5}$ across multiple Canadian locations. Atmospheric Environment 31, 4223-4236.

Butler, A., 2000. Ph.D. Thesis, School of Civil and Environmental Engineering, Georgia Institute of Technology, Atlanta

California Air Resources Board (1999)—CRPAX Study.

CCME, 2000. Canada-Wide Standards for Particulate Matter (PM) and Ozone. Canadian Council of Ministers of the Environment, June 5-6, 2000.

Cheng, L., Sandhu, H.S., Angle, R.P., McDonald, K.M., Myrick, R.H., 2000. Rural particulate matter in Alberta, Canada. Atmospheric Environment 34, 3365-3372.

Cheng, L., Sandhu, H.S., Angle, R.P., Myrick, R.H., 1998. Characteristics of inhalable particulate matter in Alberta cities. Atmospheric Environment 32, 3835-3844.

Chow, J.C., Watson, J.G., Edgerton, S.A., Vega E., 2002. Chemical composition of PM_{10} and $PM_{2.5}$ in Mexico City during winter 1997, Science of the Total Environment 287, 177-201.

Christoforou, C.S., Salmon, L.G. Hannigan, M.P., Solomon, P.A., Cass, G.R., 1999. Trends in fine particle concentration and chemical composition in southern California. Journal of the Air and Waste Management Association 50, 43-53.

Comisión Ambiental Metropolitana (CAM), 2001. Inventario de emisiones de la Zona Metropolitana del Valle de México, 1998.

Doran, J. C., and 35 others, 1998. The IMADA-AVER boundary layer experiment in the Mexico City area. Bulletin of the American Meteorological Society 79 (11), 2497-2508.

Edgerton, E., 2001. Personal Communication with A. Russell May 7, 2001. Also Attachments to a letter from A. Hansen to S. Mathias, August 3, 2000, as provided to A. Russell, May 7, 2001.

Edgerton, S.A., Arriaga, J.L., Archuleta, J., Bian, X., Bossert, J.E., Chow, J.C., Coulter, R.L., Doran, J.C., Doskey, P.V., Elliot, S., Fast, J.D., Gaffney, J.S., Guzman, F., Hubbe, J.M., Lee, J.T., Marley, N.A., McNair, L.A., Neff, W., Ortiz, E., Petty, R., Ruiz, M., Shaw, W.J., Sosa, G., Vega, E., Watson, J.G., Whiteman, C.D., Zhong, S., 1999. Particulate Air Pollution in Mexico City: A Collaborative Research Project. Journal of the Air and Waste Management Association 49,1221-1229.

Fast, J. D., Zhong, S., 1998. Meteorological factors associated with inhomogeneous ozone concentrations within the Mexico City basin. Journal of Geophysical Research, 103(D15), 18,927-18,946.

Gaffney, J.S., Marley, N.A., Cunningham, M.N., Doskey, P. V., 1999. Measurements of peroxyacyl nitrates (PANs) in Mexico City: implications for megacity air quality impacts on regional scales. Atmospheric Environment 33, 5003-5012.

Gong S.L., Zhang X.Y., Zhao T.L., McKendry I.G., Jaffe D.A., Lu N.M., 2002. Characterization of soil dust distribution in China and its transport during ACE-Asia 2: model simulation and validation. Journal of Geophysical Research (submitted).

Gray, H.A., Cass, G.R., 1998. Source contributions to atmospheric fine carbon particle concentrations. Atmospheric Environment, 32(22), 3805-3825.

GVRD, 2002. 2000 Emission inventory for the Lower Fraser Valley airshed. http://www.gvrd.bc.ca/services/air/emissions/reports/LFV2000EI.pdf.

Hannigan, M.P., 1997. Mutagenic Particulate Matter in Air Pollutant Source Emissions and in Ambient Air. Doctoral Thesis, California Institute of Technology, Pasadena, CA.

Health Canada/Environment Canada, 1999. National Ambient Air Quality Objectives for Particulate Matter: Part I. Science Assessment Document, Report by CEPA/FPAC Working Group on Air Quality Objectives and Guidelines, ISBN 0-662-267/15-X, Environmental Health Directorate, Health Canada, Ottawa, and Climate and Atmospheric Science Research Directorate, Environment Canada [Also available from http://www.msc-smc.ec.gc.ca/saib/smog_e.cfm].

Holland, D.M., Principe, P.P., Sickles II, J.E., 1999. Trends in atmospheric sulfur and nitrogen species in the eastern United States for 1989-1995. Atmospheric Environment 33, 37-49.

IMPROVE, 2000. Fig. 3, page 6 of The IMPROVE Newsletter, Vol.9, #2. Published by Air Resources Specialists, Inc. Fort Collins, CO.

Instituto Nacional de Ecologia (INE), 2000. Almanaque de Datos y Tendencias de la Calidad del Aire en Ciudades Mexicanas.

Kenski, D.M., Koerber, M., 2001. Regional haze in the Midwest: Composition, seasonality, and source region identification. Lake Michigan Air Directors Consortium, Des Plaines, IL.

Kim, B., Teffera, S., Zeldin, M. 2000b. Characterization of PM2.5 and PM10 in the South Coast Air Basin of southern California: Part 2 - Temporal variations. Journal of the Air and Waste Management Association 50, 245-259.

Kim, B., Teffera, S., Zeldin, M., 2000a. Characterization of PM2.5 and PM10 in the South Coast Air Basin of southern California: Part 1 - Spatial variations. Journal of the Air and Waste Management Association 50, 234-244.

Kleeman, M.J., Hughes, L.S., Allen, J.O., Cass, G.R. 1999. Source contributions to the size and composition distribution of atmospheric particles: Southern California in September 1996. Environmental Science and Technology 33, 4331-4341.

Lee, P.K.H., 2002. Receptor Modeling on Canadian Atmospheric Fine Particulate Matter ($PM_{2.5}$) by Positive Matrix Factorization. M.Sc. Thesis, Dept. of Chemistry, University of Toronto, Toronto.

Lewis, Charles, 2002. Private communication, U.S. EPA.

Lillyman, C.D., 2001. The Quantification of Mobile Source Contributions to Fine Particulate Matter in the Greater Toronto Area. M. Sc. Thesis, Dept. of Geography, University of Toronto.

Lowenthal, D.H., Wittorff, D., Gertler, A.W., 1994. CMB Source Apportionment During REVEAL (Regional Visibility Experiment in the Lower Fraser Valley). Report to the Air Resources Branch, Ministry of Environment, Lands and Parks, ENV. 484410/10/94.

Magliano, K.L., Hughes, V.M., Chinkin, L.R., Coe, D.L., Haste, T.L., Kumar, N., Lurmann, F.W., 1999. Spatial and temporal variations of PM_{10} and $PM_{2.5}$ source contributions and comparison to emissions during the 1999 Integrated Monitoring Study. Atmospheric Environment 33, 4757-5773.

Malm, W.C., 2000. Spatial and seasonal patterns and temporal variability of haze and its constituents in the United States. IMPROVE Report III, ISSN: 0737-5352-47, Cooperative Institute for Research in the Atmosphere, Colorado State University, Fort Collins, CO.

McDonald, K.M., Hudson, A., McArthur, B., Campbell, I., Nejedly, Z., 2000. Radiative effects of mid-continental aerosol at northern latitude. European Aerosol Conference, Dublin, Ireland, September 3-8.

McDonald, K.M., de Groot, E., Chapman, R., 1999. Source-receptor analysis of fine aerosol concentrations at Elk Island National Park. American Association of Aerosol Research, Tacoma, Washington, October 11-15.

McDonald, K.M., Lieu, W., Wu, S., Prince, D., Nejedly, Z., Campbell, I., 1998. Fine aerosol chemistry at dissimilar non-urban sites. International Global Atmospheric Chemistry Conference, Seattle, Washington, August 19-25.

McHenry, J.N., Dennis, R.L., 1994. The relative importance of oxidation pathways and clouds to atmospheric ambient sulfate production as predicted by the Regional Acid Deposition Model. Journal of Applied Meteorology 33, 890-905.

McKendry, I.G., 2000. PM_{10} levels in the Lower Fraser Valley, British Columbia, Canada: an overview of spatiotemporal variations and meteorological controls. Journal of the Air & Waste Management Association 50, 443-45.

McKendry, I.G., Hacker, J.P., Stull, R., Sakiyama, S., Mignacca, D., Reid, K., 2001. Long-range transport of Asian dust to the Lower Fraser Valley, British Columbia, Canada. Journal of Geophysical Research 106, 18,361-18,370.

McNaughton, D.J., Vet, R.J., 1996. Eulerian Model Evaluation Field Study (EMEFS): A summary of surface network measurements and data quality. Atmospheric Environment 30, 227-238.

Molina, L.T., Molina, M.J., eds., 2002. Air quality in the Mexico Megacity: an integrated assessment. Kluwer, Dordrecht.

Molina, M.J., Molina, L.T., Sosa, G., Gasca, J., West, J. J., 2000. Analysis and diagnostics of the inventory of emissions of the Metropolitan Area of the Valley of Mexico. MIT-Integrated Program on Urban, Regional, and Global Air Pollution, Report No. 5, 59 pages.

Molina, M.J., Molina, L.T., West, J.J., Sosa, G., Sheinbaum Pardo, C., San Martini, F., Zavala, M.A., McRae, G., 2002. Air pollution science in the MCMA: Understanding source-receptor relationships through emissions inventories, measurements and modeling, Chapter 5 in Air quality in the Mexico Megacity: an integrated assessment. Molina L.T., Molina, M.J., eds., Kluwer, Dordrecht.

MSC, 2001. Precursor Contributions to Ambient Fine Particulate Matter in Canada. Internal report, August, Air Quality Research Branch, Meteorological Service of Canada, Downsview, Ontario [Also available from http://www.msc-smc.ec.gc.ca/saib/].

Myrick, R.H., Sakiyama, S.K., Angle, R.P., Sandhu, H.S., 1997. Seasonal mixing heights and inversions at Edmonton, Alberta. Atmospheric Environment 28, 723-729.

Mysliwiec M.J., Kleeman M.J., 2002. Source apportionment of secondary particulate matter in a polluted atmosphere. Environ. Science and Technology, in press.

Nickerson, E.C., Sosa, G., Hochstein, H., McCaslin, P., Luke, W., Schanot, A., 1992. Project Águila: in situ measurements of Mexico City air pollution by a research aircraft. Atmospheric Environment 26B(4), 445-451.

Olson, M.P., Voldner, E.C., Oikawa, K.K., 1983, Transfer matrices from the AES-LRT model. Atmosphere-Ocean 31, 344-361.

Osnaya R.P., Gasca, J.R., 1998. Inventario de amoniaco para la ZMCM, Revision de Diciembre de 1998. Instituto Mexicano del Petróleo, Project DOB-7238, 23 p.

Pacific Environmental Services, 2001. Assessment of emissions inventory needs for regional haze plans. Report prepared for the Ozone Transport Commission and the Southeast States Air Resource Managers.

Peake, E., MacLean, M.A., and Sandhu, H.S., 1988. Total inorganic nitrate (particulate nitrate and nitric acid) in the atmosphere of Edmonton, Alberta, Canada. Atmospheric Environment 22, 2891-2893..

Pérez Vidal, H., Raga, G.B., 1998. On the vertical distribution of pollutants in Mexico City. Atmósfera 11, 95-108.

Pryor, S.C., Barthelmie, R.J., 1999. REVEAL II Characterizing fine aerosols in the Fraser Valley. Report to Fraser Valley Regional District.

Pryor, S.C., Barthelmie, R.J., Hoff, R.M., Sakiyama, S., Simpson, R., Steyn, D., 1997. REVEAL: Characterizing fine aerosols in the Fraser Valley, BC. Atmosphere-Ocean 35, 209-227.

Pun, B., Seigneur, C., 2001. Sensitivity of particulate matter nitrate formation to precursor emissions in the California San Joaquin Valley. Environmental Science and Technology 35, 2979-2987.

Pun, B.K., Leidner, M., Seigneur, C., 2001. Conceptual model for regional haze in the Upper Midwest. Proceedings of the Regional Haze and Global Radiation Balance – Aerosol Measurements and Models: Closure, Reconciliation and Evaluation, October 2-5, 2001, Bend, OR.

Pun, B.K., Seigneur, C., 1999. Understanding particulate matter formation in the California San Joaquin Valley: conceptual model and data needs. Atmospheric Environment 33, 4865-4875.

Raga, G.B., Baumgardner, D., Castro, T., Martínez-Arroyo, A., Navarro-González, R., 2001. Mexico City air quality: a qualitative review of measurements (1960-2000). Atmospheric Environment 35, 4041-4058.

Raga, G.B., Raga, A.C., 2000. On the formation of an elevated ozone peak in Mexico City. Atmospheric Environment 34, 4097-4102.

Ro, C.U., Tang, A.J.S., Chan, W.H., Kirk, R.W., Reid, N.W., Lusis, M.A., 1988. Wet and dry deposition of sulfur and nitrogen compounds in Ontario. Atmospheric Environment 22, 2763-2772.

Schauer, J.J. Cass, G.R., 1999. Source apportionment of winter-time gas-phase and particle-phase air pollutants using organic compounds as tracers. Environmental Science and Technology, submitted for publication.

Schauer, J.J., Rogge, W.F., Hildemann, L.M., Mazurek, M.A., Cass, G.R., 1996. Source apportionment of airborne particulate matter using organic compounds as tracers. Atmospheric Environment 30, 3837-3855.

Secretaría del Medio Ambiente, Gobierno del Distrito Federal (SMA-GDF), 2000. Compendio Estadístico de la Calidad del Aire, 1986-1999.

Snyder, B., Brook, J.R., Strawbridge K., 2002. Pollutant build-up over the Strait of Georgia and the role of mesoscale flow patterns (A71A – 0089). American Geophysical Union Annual Meeting, December 2002, San Francisco, USA.

Sosa, G., West, J.J., San Martini, F., Molina, L. T., Molina, M.J., 2000. Air quality modeling and data analysis for ozone and particulates in Mexico City. MIT-Integrated Program on Urban, Regional, and Global Air Pollution, Report No. 15.

Strader, R., Lurmann, F., Pandis, S.N., 1999. Evaluation of secondary organic aerosol formation in winter. Atmospheric Environment 33, 4849-4863.

U.S. EPA AIRS database, as of April 12, 2001 (FRM Summaries).

U.S. EPA AIRS database, as of April 2001 (Speciation data).

U.S. EPA AIRS database, May 9, 2001.

Vega, E., García, I., Apam, D., Ruíz, M.E., Barbiaux, M., 1997. Application of a chemical mass balance receptor model to respirable particulate matter in Mexico City. Journal of the Air & Waste Management Association 47, 524-529.

West, J.J., Sosa, G., San Martini, F., Molina, M.J., Molina, L.T., Sheinbaum, C., 2000a. Air pollution science in Mexico City: understanding source-receptor relationships for informing decisions. MIT-IPURGAP Report No. 9.

West, J.J., Sosa, G., San Martini, F., Molina, M.J., Molina, L.T., 2000b. Air quality modeling and data analysis for ozone and particulates in Mexico City. MIT-IPURGAP Report No. 15.

WGAQOG, 1999. National Ambient Air Quality Objectives for Particulate Matter. Part 1: Science Assessment Document. A report by the CEPA/FPAC Working Group on Air Quality Objectives and Guidelines. Minister, Public Works and Government Services.

Yap, D., Ning, D.T., Dong, W., 1988. An assessment of source contributions to the ozone concentrations in southern Ontario, 1979-1985. Atmospheric Environment 22, 1161-1168.

Zhang, L., Gong, S., Padro, J., Barrie, L., 2001. A size-segregated particle dry deposition scheme for an atmospheric aerosol module. Atmospheric Environment 35, 549-560.

Zheng, M., Cass, G.R., Schauer, J.J., Edgerton, E.S., 2002. Source apportionment of fine particle air pollutants in the Southeastern United States using solvent-extractable organic compounds as tracers. Environmental Science and Technology 36, 2361-2371.

CHAPTER 11

Recommended Research to Inform Public Policy

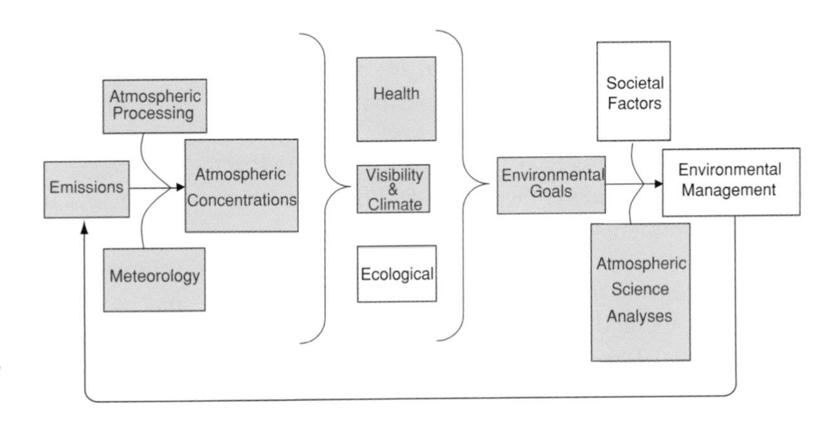

Principal Author: Peter McMurry

Advocates on all sides of the environmental-science debate agree that good decisions should be informed by a sound understanding of science and its uncertainties. This chapter summarizes recommendations for future research that will lead to an improved scientific basis for policy decisions. These recommendations emerged through discussions among the lead authors of this Assessment and are consistent with the thrust of NARSTO's Fine Particle Research Plan (NARSTO, 1998).

Detailed discussions of scientific work needed to establish a more thorough understanding of phenomena that determine the chemical and physical properties, behavior, and spatial/temporal distributions of atmospheric PM have been published previously [AQRS (1998, 1999), DOE (2001), NARSTO (1998), NRC (1998, 1999, and 2001), Meteorological Service of Canada (2001)]. The recommendations in this Assessment focus on relevance and importance to NARSTO stakeholders, with a particular emphasis on recommendations related to filling science gaps that will have the greatest impact for policy makers as they implement current mass-based PM standards. For example, PM effects may depend on some property other than mass concentrations. For example, light scattering, which affects visibility and the earth's radiative energy balance, depends on particle shape, refractive index and size distribution. Also, the health-science community has identified hypotheses regarding particle properties that might be responsible for reported health effects, and work aimed at determining the sensitivity of human health to those properties is underway. Several of the recommendations in this chapter are aimed toward obtaining data that could be used to test those hypotheses.

Development of a new scientific tool typically precedes its availability for use by policy makers by about a decade. During this time, proof of concept, acceptance, adoption, and application on a sufficiently broad scale to have an impact take place. Some of the recommendations made in this chapter apply to tools, such as instruments for semi-continuous measurements of aerosol composition, that are or will soon be ready to deploy. Others, such as work on origins and properties of primary and secondary organic PM, apply to problems that will require a much longer time to solve. Progress over the past several decades has led to a much sounder basis for current PM management strategies. Similarly, the work recommended here will lead to better PM management strategies in the future.

The recommendations presented in this chapter fall into six broad themes:

1 Developing a better understanding of carbonaceous PM.

2 Performing long-term monitoring of mass, aerosol composition, gas-phase precursors of secondary PM, copollutants, and size distributions in parallel with measurements of appropriate health indices. The importance of using compatible measurement strategies across North America is emphasized.

3 Continuing the development and evaluation of chemical-transport models for PM.

4 Improving the characterization of source emissions, refining emission inventories, and developing more detailed emission models.

5 Investing more resources in archiving, evaluating, and synthesizing data from ambient measurements.

6 Developing more systematic approaches for integrating diverse types of knowledge on origins, properties, and effects of atmospheric PM to assist with the development of management strategies and the measurement of progress toward protecting health. These strategies must account for uncertainties.

In this chapter, a synopsis of the policy relevance for each of these broad recommendations is followed by a list of specific science needs and corresponding brief rationale. The recommendations are deliberately not prioritized, because many of them are interdependent and work must progress simultaneously in all areas in order to move forward. The chapter concludes with recommendations for future NARSTO PM assessments

Table 11.1 lists the recommendations and associated science needs. The table includes a crosslink to the scientific discussions in this assessment that support the recommendations and to similar science needs identified in NARSTO's PM Science Plan (NARSTO, 2001).

11.1 RECOMMENDATIONS

RECOMMENDATION 1. **Improve the understanding of carbonaceous aerosols.** Work should include the development of improved measurement methods, improved source characterization, and improved scientific understanding of chemical and physical properties

Table 11.1. Summary of Recommendations

Science Need	Statement	Further Discussion	NARSTO PM Science Plan (NARSTO, 2001)
Rec. 1	**Improve the understanding of carbonaceous aerosols.**		
1.1	Improve methods for measuring organic carbon (OC) and black carbon (BC) mass concentrations.	Ch. 3 5.1.2	
1.2	Obtain more information on the ratio of carbon mass to total mass for OC.	5.1.2 5.1.3	
1.3	Characterize organic PM properties and its gas-phase precursors, including factors that govern the gas-particle partitioning of semivolatile organic compounds, the hygroscopicity of particulate organic compounds, and the proclivity of gas phase organic precursors to form secondary organic aerosols.	3.4 3.5.3 5.2 8.4.1 9.1.3 Ch. 3	3.2.1.3 3.2.1.4 4.4 5.4 5.5
1.4	Develop cost-effective techniques for obtaining more detailed spatial and temporal information on carbonaceous PM species.	Ch. 2 3.5.3 4.1 5.1.2 6.2.4 Ch. 3	
1.5	Determine composition of primary and secondary organic species associated with anthropogenic and biogenic emissions.	3.5.3 4.1 5.1.2 5.1.3 7.2 7.5.1 8.4.1 Ch. 3	

Table 11.1. Summary of Recommendations (cont.)

Science Need	Statement	Further Discussion	NARSTO PM Science Plan (NARSTO, 2001)
Rec. 2	**Perform long-term monitoring of concentrations of PM components and mass as well as gas-phase precursors and copollutants of particles in parallel with studies of health impacts.**		
2.1	Monitor PM mass concentrations, PM composition, gas-phase precursors of secondary particulate species, and gas-phase copollutants.	Ch. 2 6.5	3.2.3.1 3.2.1.5 3.2.4.1
2.2	Assess trends; requires that monitoring be carried out using consistent, comparable and accurate techniques over a sufficiently long period of time and over sufficiently varied types of airsheds.	6.5 9.1.1	
2.3	Measure collocated gas and particle phases for semivolatile species in sampling networks.	3.6.3 3.6.5 8.4.3	
2.4	Measure particle size distributions and size-resolved chemical composition at selected sites.	Ch. 2 8.4.3	3.2.1.1 5.3
2.5	Replace filter samplers with instruments for real-time measurements of PM mass and composition, as these methodologies mature.	5.1.3	
2.6	Measure novel PM properties to test hypotheses regarding causal agents for human-health effects.	Ch. 2	
2.7	Design monitoring networks that provide compatible information for Canada, the U.S. and Mexico, and meet diverse needs with maximum efficiency.	Ch. 2 5.3.1 6.5 Ch. 2	3.2.4.1 3.2.4.2
2.8	Make readily available optical extinction data obtained with the Automated Surface Observing System (ASOS) and the Automated Weather Observation System (AWOS) networks that replaced human observers at airports in 1994.	9.2.1	
2.9	Carry out long-term studies of human-health effects in parallel with air-quality monitoring.	Ch. 2 Ch. 3	3.2.4.3 2.8
Rec. 3	**Continue to invest resources in evaluating and further developing the performance of chemical-transport models for PM.**		
3.1	Design intensive atmospheric field studies and long-term monitoring programs that can be used to evaluate and refine chemical-transport models for particulate matter.	8.6	7.3.1 7.3.2
3.2	Conduct model-evaluation studies to compare model predictions with observations for particulate species concentrations and size distributions.	8.4.3	7.3.2
3.3	Establish centers to facilitate communication regarding best PM modeling practices among regulatory agencies in North America.	8.1.1	
3.4	Improve our ability to model certain atmospheric processes. Of particular importance are cloud processing, processes that involve carbonaceous aerosols, and the formation of new particles by homogeneous nucleation.	3.2 3.5.3 3.6.4 8.6 9.3.3	4.4 4.5 5.1 5.4
3.5	Obtain more detailed information on the size-resolved composition of primary particulate emissions from various sources.	7.2.1 8.2.5	5.2 5.3
3.6	Develop approaches for using data from satellites or real-time sensors to obtain boundary conditions for models.	8.4.2	

Table 11.1. Summary of Recommendations (cont.)

Science Need	Statement	Further Discussion	NARSTO PM Science Plan (NARSTO, 2001)
Rec. 4	**Improve emission inventories and emission models.**		
4.1	Develop improved and standardized approaches for developing national emission inventories for Canada, Mexico and the United States, which provide compatible data.	4.2.1	
4.2	Update and expand emission inventories to include new types of information including size distributions and size-resolved composition, organic speciation, and updated source profiles.	4.4.1 7.2.7 8.2.5	3.2.5.1 4.1 4.3 4.4 4.5 5.2
4.3	Develop approaches that can be used to reconcile emission inventories with measurements of species concentrations observed in the atmosphere.	4.3 7.3.2.4 7.4.2 8.2.5	4.7 7.2.5
4.4	Extend emission inventories to include all available information from particle sources typically viewed as unmanageable, such as forest fires, and dust.	4.3 8.2.5	
4.5	Improve emission models for dust.	4.4.2 7.4.2 8.2.5	4.2
4.6	Develop emission models for NH_3.	4.4.2 8.2.5	4.6
4.7	Develop emission models that account for condensation or nucleation of semivolatile compounds, as hot emissions mix with cool ambient air.	3.2 4.4.2	3.2.4.4 5.2
4.8	Continue to improve the timeliness of emission inventories.	4.5	
4.9	Develop approaches for obtaining emission information with adequate resolution in space, time, and composition for atmospheric modeling and exposure assessments. These approaches will likely involve combinations of emission models and emission data with higher temporal resolution than is currently included in inventories.	Ch. 2 4.5	
4.10	Prepare national emission inventories over a period of years in parallel with long-term monitoring of health effects and ambient concentrations of PM and its precursors to facilitate analyses of trends and benefits.	Ch. 2 4.5	4.8
Rec. 5	**Commit adequate resources to the evaluation, synthesis and archiving of data obtained in current and future atmospheric aerosol studies, and to fostering interactions between the communities of scientists working on atmospheric science, health effects, and global climate change.**		
5.1	Establish procedures at funding agencies which ensure that atmospheric aerosol data are systematically archived in a manner that provides convenient access to all interested parties.	Ch. 5	3.2.6.1
5.2	Develop a comprehensive plan for data analysis, integral to measurement programs, and commit adequate resources to carry out such analyses.	Ch. 2	

Table 11.1. Summary of Recommendations (end table 11.1)

Science Need	Statement	Further Discussion	NARSTO PM Science Plan (NARSTO, 2001)
5.3	Foster collaborations among the communities of scientists working on PM pollution, health effects, and global climate change.	Ch. 2 9.4	3.2.5.2 4.3 7.2.6
Rec. 6	**Implement more systematic approaches for integrating diverse types of knowledge on origins, properties, and effects of atmospheric PM to assist with the development of management strategies and the measurement of progress towards protecting health.**		
6.1	Develop a more systematic approach for integrating knowledge gained from measurements, receptor models, and chemical-transport models to make optimal PM-management decisions.	1.0 7.1	
6.2	Develop an accountability framework that will enable measurement of progress towards the goal of protecting human health.	Ch. 2 1.8	

and behavior in the atmosphere. This work should also include research on gas/particle partitioning, water uptake, and the chemical mechanisms of secondary OC formation from important precursor gases.

POLICY RELEVANCE

Carbonaceous aerosols, which include organic and black carbon, make up roughly one fifth to one half of the average annual $PM_{2.5}$ mass concentration. Organic aerosols include a complex mixture of many compounds, and an understanding of organic aerosols is primitive relative to the understanding of the other major fine-particle species ($SO_4^=$, NO_3^-, and dust). Limitations in the ability to measure the composition and properties of carbonaceous particles limit the ability to identify their origins (which include both biogenic and anthropogenic sources), to develop PM models, to assess trends, and consequently to develop optimal PM control strategies. In contrast, secondary $SO_4^=$,and NO_3^- are relatively easy to measure, are produced from a few gas-phase precursors that are emitted by a few types of major sources, and they are transformed to secondary PM in the atmosphere by a small number of primary reaction pathways. An improved understanding of carbonaceous aerosols would lead to a better understanding of their sources, relative contributions of primary vs. secondary carbon to PM mass, their effects on visibility, and the linkages between PM and ozone management. Understanding of the health effects of carbonaceous aerosols will also improve with improved measurements. Furthermore, because primary organic PM often contributes significantly to PM mass, reducing primary OC emissions may, at times, be a viable approach toward controlling PM mass concentrations. Finally, it has been hypothesized that organic PM includes toxic compounds that contribute to PM health effects. If so, reductions in concentrations of those compounds could lead to improved human health.

SCIENTIFIC NEEDS

Recommendation 1.1 *Improve methods for measuring OC and BC mass concentrations.*

Rationale. Carbonaceous particles include those species that are identified as OC and BC. Because some tarry organic compounds are black, and because distinctions between OC and BC depend somewhat on analytical method, the distinction between OC and BC is not always clear. An issue that remains unresolved is the accuracy with which OC and BC can be measured. Organic PM measurements are especially prone to error due to sampling artifacts that include both the volatilization of semivolatile compounds from collected deposits of particles and the adsorption or absorption of organic gases on sampling substrates or collected particle deposits during sampling. It is likely that both positive and negative artifacts occur during sampling, and there is no consensus as to the net magnitude of these artifacts. It is likely that some compounds contribute to net negative artifacts while others contribute to net positive artifacts, and the relative magnitudes of these effects may vary with sampling location and time. An improved understanding of carbonaceous aerosols will require more detailed information on chemical characteristics. Ideally, this would include information on compounds and their thermodynamic properties. It is unlikely, however, that techniques will be available in the near future to provide complete chemical characterizations of carbonaceous aerosols with the frequency that would be required for air-quality studies, so improved techniques for OC/BC measurements are needed for the near term.

Recommendation 1.2 *Obtain more information on the ratio of carbon mass to total mass for OC.*

Rationale. The most commonly used approaches for determining organic PM mass concentrations involve measuring the amount of CO_2 released when a collected PM sample undergoes oxidation. Particulate organics contain carbon plus other elements such as oxygen and hydrogen. Because these other elements are typically not measured, it is commonly assumed that organic PM mass concentrations can be inferred by multiplying the measured carbon mass by a factor that accounts for other associated elements. This factor is typically assumed to be between 1.2 and 1.4. Recent work (Turpin, 2001) has shown that 1.4 may be a minimum value for freshly emitted urban PM, and that the mass-conversion factor may exceed 2.0 for aged aerosols that are more highly oxidized. Because this factor can significantly affect estimates of OC contributions to total PM mass concentrations, an effort should be made in the near term to assess appropriate values in different source environments and seasons.

Recommendation 1.3 *Characterize organic PM properties and its gas-phase precursors, including factors that govern the gas-particle partitioning of semivolatile organic compounds, the hygroscopicity of particulate organic compounds, and the proclivity of gas-phase organic precursors to form secondary organic aerosols.*

Rationale. Many of the compounds that make up OC are semivolatile (i.e., they are present to an appreciable extent in both the gas and condensed phases.) The relative amounts of a given compound that are found in the gas and condensed phases vary with thermodynamic properties of the compound (vapor pressure, adsorption or absorption equilibrium in particulate species, etc.) and on whether or not sufficient time has elapsed to achieve equilibrium. Because the gas-particle distribution is likely to depend on atmospheric properties including temperature and relative humidity, gas-particle distributions are likely to vary diurnally. Furthermore, because pressure, temperature and relative humidity in sampling devices typically are not identical to values in the atmosphere, changes in gas-particle distributions can be (and typically are) induced by measurement. An improved understanding of gas/particle partitioning will lead to an improved understanding of how aerosols evolve as they are transported from the source to a receptor, how humans or other organisms will be exposed to a particular compound, and how to design samplers to collect representative samples. Also, the development of improved chemical-transport models will require more information on reactivity of gases and their tendency for form secondary particulate products. Finally, recent work has shown that the hygroscopicity of organic PM is variable, and tends to increase with the length of time particles have been in the air. Hygroscopicity influences a particle's tendency to be removed by wet deposition or in the lungs as well as its optical properties. An understanding of the behavior, lifetime, and effects of organic PM will require a better understanding of its tendency to absorb water. Because these questions are complex, work over a period of one to two decades will be needed.

Recommendation 1.4 *Develop cost-effective techniques for obtaining more detailed spatial and temporal information on carbonaceous PM species.*

Rationale. Carbonaceous PM encompasses thousands of compounds displaying a vast range of properties. The most sophisticated analytical methods available to date can identify only 10 to 20 percent of these compounds, and such measurements are expensive and require highly trained personnel. Furthermore, only a few laboratories in North

America are equipped to carry out such analyses. Identifying the origins of organic PM, understanding its transformations in the atmosphere, and understanding its effects will require measurements of speciation, not simply OC/BC. Good ideas for work in this area should be pursued.

Recommendation 1.5 *Determine composition of primary and secondary organic species associated with anthropogenic and biogenic emissions.*

Rationale. A limited amount of work to date has shown that organic PM formed by chemical transformations in the atmosphere and emitted by different primary sources has distinct chemical signatures. Therefore, measurements of organic speciation at the source and in the atmosphere can provide valuable insights into the origins of ambient OC. Such measurements are currently expensive, can be applied to only a limited number of samples, require sampling by techniques that are inherently prone to artifacts due to gas adsorption on filters or the evaporation of semivolatile components from particles during sampling, and (to date) have provided only very limited information on composition as a function of particle size. Nevertheless, the information that has been obtained from these measurements is tremendously rich relative to what can be learned by merely measuring the OC and BC concentrations. More work of this type in the near term is recommended. Efforts should be made to use similar methodologies throughout North America, and to use the same methodologies to characterize emissions at the source and concentrations in the atmosphere. Such work would provide new information on the sources of carbonaceous PM, would support receptor modeling, and would improve process understanding needed to develop PM chemical-transport models. Quantitative measurements of targeted species concentrations are needed to quantify relationships between health responses and species exposures.

RECOMMENDATION 2. **Perform long-term monitoring of concentrations of PM components and mass as well as gas-phase precursors and copollutants of particles in parallel with studies of health.** Where possible, instruments that provide continuous, real-time information should be used. Similar instruments and measurement strategies should be used throughout North American networks to the extent possible to facilitate the sharing of data among Canada, the United States, and Mexico. Provision should be made to optimize monitoring networks to meet diverse information and needs. And most importantly, long-term air-quality monitoring data must be linked with human-health data for defined populations to characterize human-health impacts. A similar recommendation was made in the NARSTO Ozone Assessment (NARSTO, 2000).

POLICY RELEVANCE

Assessing the success of emission-control strategies on PM composition and concentrations and health or environmental impacts requires accurate data from monitoring networks. Measurements over a period of 5 to 10 years are typically required to assess trends. Monitoring data are also needed to evaluate transboundary transport to assess contributions of different jurisdictions to local PM and haze levels. Chemically speciated and highly time-resolved data from monitoring networks for different meteorological conditions/seasons are also needed to evaluate PM models. The availability of empirical information on trends and improved models with known accuracy would reduce uncertainties in predicted future trends. Thus, the efficacies of policy decisions regarding emission scenarios could be estimated with greater confidence.

SCIENTIFIC NEEDS

Recommendation 2.1 *Monitor PM mass concentrations, PM composition, gas-phase precursors of secondary particulate species, and gas-phase copollutants.*

Rationale. Connections exist between the chemical pathways that lead to the production of secondary gaseous pollutants (including ozone) and secondary particulate products (including NO_3^-, $SO_4^=$ and OC). These associations raise two issues of concern to the policy maker 1) To what extent will efforts to control one pollutant lead to increases or decreases in the concentrations of a copollutant? and 2) To what extent are the effects attributed to one pollutant actually due to the presence of another covarying pollutant, or to the mixture of copollutants? Long-term monitoring of the multipollutant mixture would help to ensure the availability of data required to validate chemical-transport models for PM and to assess health and ecological effects of PM and its copollutants. These issues illustrate the need to devise strategies for monitoring and emission control that take these interrelationships into account. Care also should be taken to ensure that monitoring networks provide adequate spatial coverage. There is currently a deficiency of monitoring data (either temporally, spatially, or both) in Alaska, Northern Canada, and Mexico. The conduct of epidemiological studies of air pollution requires knowledge of all major constituents in the air, thereby allowing a simultaneous assessment of the contribution of individual constituents to the overall health impacts that may be observed. Significant advances in measurement capabilities have been made over the past decade, and plans should be made now to implement these new capabilities in future monitoring networks.

Recommendation 2.2 *Assess trends; requires that monitoring be carried out using consistent, comparable and accurate techniques over a sufficiently long period of time and over sufficiently varied types of airsheds.*

Rationale. The effects of emission-control strategies on ambient concentrations of pollutants typically require about a decade to be assessed. For example, the first phase of the 1990 U.S. Clean Air Act Amendments led to changes in spatial emissions of SO_2 and NO_x that occurred over the past decade. Also, the time required to substantially change the composition of the vehicle fleet when new emission standards are enforced is about a decade. Year-to-year meteorological fluctuations cause variations in the species concentrations that add to the difficulty of assessing temporal trends. In order to obtain data that can provide information on trends, it is essential that monitoring programs include continuous, uninterrupted long-term measurements with consistent methodologies to enable the unambiguous determination of trends on the time scale of a decade. Measurement strategies should take advantage of recently-developed measurement techniques, and should be implemented now, so as to assess the impacts of emission-controls strategies that are currently being implemented. Measurements across North American networks should be designed to enable the determination of spatial and temporal trends of the multipollutant mixtures. Issues including reference standards, measurement methodologies and sampling strategies need to be considered in assessing the compatibility of sampling networks. Measurement uncertainties should be as small as possible and should be reported routinely. A similar recommendation for ozone and its precursors was made in the NARSTO Ozone Assessment.

Recommendation 2.3 *Measure collocated gas and particle phases for semivolatile species in sampling networks.*

Rationale. Ammonium nitrate and a portion of OC are semivolatile (i.e., significant fractions of these species are found in both the gas and particle phases.) The gas/particle distributions of these compounds vary with temperature, relative humidity, PM composition and the total concentration of the semivolatile compounds. In order to determine whether the formation of NH_4NO_3 is limited by the availability of NH_3 or HNO_3, monitoring networks should include measurements of NH_3 and HNO_3 gases as well as NH_4^+, NO_3^-, and $SO_4^=$. Filter-based techniques for measuring the gas- and particle-phase fractions of these inorganic compounds are available, and semi-continuous methods are being developed and may soon be available for routine measurements. It would be desirable to obtain such data to assess the impact of SO_2 control strategies on ambient NO_3^- concentrations. Although it would also be desirable to have data on gas- and particle-phase concentrations of semivolatile organic compounds, methodologies for routine measurements of such compounds are not yet available. Efforts should be made to develop such measurement methods.

Recommendation 2.4 *Measure particle-size distributions and size-resolved chemical composition at selected sites.*

Rationale. Continuous measurements of particle-size distributions could provide information on particle concentrations in all size ranges from 0.003 μm to 10 μm with a time resolution of about five minutes. Such measurements would be valuable for model-evaluation studies, for studies of PM health effects, and for understanding the relationships between emissions and emission-control strategies on aerosol properties. For example, recent evidence suggests that PM emission controls that are being adapted for new-generation diesel engines are leading to enhanced emissions of ultrafine particles. It is important that data be acquired that will enable us to determine whether these changes in emission patterns are leading to significant trends in ambient size distributions. Recent work has shown that size distributions can be measured accurately and routinely. Such measurements should be done in several locations where they can provide the greatest benefit. In addition, information on size-resolved composition is needed to evaluate the performance of chemical-transport models for PM, and to assess the contributions of species to optical extinction and climate forcing. Selected measurements of size-resolved composition should be included in network design.

Recommendation 2.5 *Replace filter samplers with instruments for real-time measurements of PM mass and composition, as these methodologies mature.*

Rationale. Real-time instruments can be less expensive to operate than filter samplers and can provide valuable information on diurnal trends. Information on short-term variations may be useful in assessing health effects and for model-evaluation studies that have been previously constrained by the availability of only daily filter measurements. Significant progress has been made in the past several years on developing instrumentation for real-time measurements of PM mass and composition. Some of these instruments have been commercialized, and it is likely that others will be commercialized in the future.

Recommendation 2.6 *Measure novel aerosol properties to test hypotheses regarding causal agents for human-health effects.*

Rationale. The health-effects community has proposed several hypotheses regarding PM properties that might be responsible for the observed association between health and particulate matter. These properties should be monitored over the long term in conjunction with health-effects studies to test these hypotheses. In some cases it will be necessary to develop new measurement methods. For example, cells are exposed only to the surfaces of insoluble particles. Therefore, any biological effects due to insoluble particles would likely depend on surface rather than on bulk composition. Because the surface and bulk composition may be very different, it is necessary to have information on surface composition. Similarly, if reactive transition metals or organic compounds were responsible for health effects, then it would be important to have measurement techniques for them. Decisions about which measurement methods to develop and how they should be deployed should be made through collaborative interactions of health and atmospheric scientists.

Recommendation 2.7 *Design monitoring networks that provide compatible information for Canada, the United States, and Mexico, and meet diverse needs with maximum efficiency.*

Rationale. Monitoring networks fill a variety of needs including the measurement of long-term trends, the measurement of baseline pollutant levels, the assessment of impacts, and the acquisition of data for model validation. The data set required typically varies with its application, and an effort should be made to ensure that existing monitoring networks meet diverse needs with maximum efficiency. Furthermore, similar instruments and measurement strategies should be used throughout North American networks to the extent possible to facilitate the sharing of data among Canada, the United States, and Mexico. The availability of monitoring data from harmonized networks is crucial to the conduct of large-scale epidemiological studies as illustrated by the recent National Morbidity, Mortality, and Air Pollution Study (NMMAPS) (Samet et al., 2000 a and b).

Recommendation 2.8 *Make readily available optical extinction data obtained with the Automated Surface Observing System (ASOS) and the Automated Weather Observation System (AWOS) networks that began to replace human observers at airports beginning in the 1990s.*

Rationale. For decades, observations of visual range taken hourly by observers at airports have been used to document the spatial and temporal pattern of atmospheric haziness and as a surrogate for trends in fine-particle mass. The automated ASOS and AWOS networks began to replace human observers in the 1990s, and include sensors in both the United States and Canada. These systems include sensors that can automatically monitor the haziness with high dynamic range, accuracy, and precision. The ASOS and AWOS systems also monitor temperature, dew point, precipitation, etc., which allows the separation of fog, rain, and snow from dust, smoke, and haze. Hence, it is also possible to relate the Sensor Equivalent Visibility (SEV) reported by ASOS to the in-situ concentrations of fine-particle dust, smoke, and haze. There are currently more than 900 ASOS stations operated by the FAA, the NWS and the DOD throughout the United States. Unfortunately, the full-resolution ASOS visibility data are not routinely reported, which significantly diminishes the value of these data for analysis of air-quality trends. The raw data from only a subset (~200 stations) of these stations are being archived at NOAA's National Climatic Data Center. Furthermore, the instruments are not being maintained or calibrated with sufficient diligence to enable collection of data with the accuracy that could be provided by these instruments. Clearly, the ASOS visibility measurements and the ancillary meteorological data represent a valuable resource for air-quality investigations. Resources should be made available to the agencies that operate and evaluate the ASOS network to make the data more widely available and in a form that represents the full dynamic range of the instruments. Optimal calibration procedures might require a coordinated effort between the U.S. NWS and other interested agencies.

Recommendation 2.9 *Carry out long-term studies of human-health effects in parallel with air-quality monitoring.*

Rationale. Long-term measurement of appropriate health indices must be carried out in parallel with measurements of air quality in order to conduct epidemiological studies on the effects of specific PM constituents. Identification of specific PM constituents as having a role in producing adverse effects will provide the opportunity for development of control strategies that target those constituents. There is a need to build on past successful collaborations between aerosol scientists and health specialists to design and conduct epidemiological studies to test the hypothesized role of certain constituents. There is a particular need for increased work directed to understanding relationships between PM and other pollutants in ambient air, in homes, and work places and ultimately, in the breathing zone (personal exposure) of people during their daily activities. This may require additional studies on the relationship between measurements made using personal-exposure monitors, and measurements of air quality made using conventional instrumentation.

RECOMMENDATION 3. **Continue to invest resources in evaluating and further developing the performance of chemical-transport models for PM.**

POLICY RELEVANCE

PM models are powerful and necessary tools for the policy maker to determine the impact of various specific and generic sources to PM in a given region or airshed, and to assess the impact of various control strategies on PM concentrations at that location. Significant progress has been made in recent years on chemical-transport models for PM that account for emissions, atmospheric transformations, transport, and removal. Significantly more work is required, however, to improve scientific understanding of certain atmospheric processes so that they can be credibly described by models and to establish the credibility of models for use by policy makers. Existing chemical-transport models do a reasonably good job of predicting the directions in which the concentrations of gases and inorganic particle species would change if the emission rates of a given precursor species were changed; however, the extents of these changes are still uncertain. Chemical-transport models are also less successful at this time at predicting absolute concentrations of species as a function of time and location, although values averaged over time can be calculated with more confidence. More work is needed to develop such models into reliable tools. In the foreseeable future, chemical-transport models need to be used by policy makers in conjunction with other supporting information in making decisions regarding PM management. As PM models improve, policy makers will rely increasingly on the predictions from such models.

SCIENTIFIC NEEDS

Recommendation 3.1 *Design intensive atmospheric field studies and long-term monitoring programs that can be used to evaluate and refine chemical-transport models for PM.*

Rationale. Model-validation studies with varied temporal and spatial scales will be needed to develop models into credible tools for policy makers. The optimal design of such studies will require close collaboration between experimentalists and modelers. A variety of types of studies will be needed. Intensive, short-term campaigns are often well suited to evaluate the performance of chemical-transport models at describing aerosol processes or microscale particle properties. There is also a need for "operational evaluations" designed to determine whether model predictions agree with observations, and for "diagnostic studies" to determine whether models get the right answers for the right reasons. Operational and diagnostic studies will require measurements over extended periods that cover differing seasons and a range of meteorological conditions, and typically cannot be carried out during short (~4-week), intensive measurement campaigns. It should also be kept in mind that models are often designed for conditions where PM concentrations are in the range of standards. Additional model-evaluation work will be needed to evaluate model performance under conditions pertinent to regional haze, where the focus is often on protecting the most pristine conditions.

Recommendation 3.2 *Conduct model-evaluation studies to compare model predictions with observations for PM species concentrations and size distributions.*

Rationale. Models that predict size-resolved composition of aerosols are being developed, and studies that are specifically designed to evaluate this aspect of their performance are needed. Such studies have been carried out in Los Angeles and in the San Joaquin Valley and as part of the EPA Supersite program in other areas. Studies of this type can help both in evaluating the model predictions, and by iteration, the quality of emission inventories and emission models used by chemical-transport models. Size-resolved measurements of temporal and spatial variabilities of the major PM chemical species and their gas-phase precursors are needed for such model evaluations, and measurements made with high temporal resolution will be particularly valuable. Because vertical distributions can vary significantly, vertical mixing in the atmosphere can significantly affect time-dependent ground-level concentrations, and measurements of vertical distributions are required for model-validation studies. Furthermore, more meteorological data, especially for complex terrain, are necessary for running and evaluating models. Priority should be given to carrying out a relatively small number of comprehensive, multi-investigator studies that include collaborations among experimentalists and modelers.

Recommendation 3.3 *Establish centers to facilitate communication regarding best PM modeling practices among regulatory agencies in North America.*

Rationale. The same model, when used by different modelers, will produce different results because the modeler must frequently apply judgments based on previous experience in configuring the model and its inputs for a particular application. As experience in using PM models increases, modelers will develop a greater awareness of how they should be used to obtain credible results. Comparisons with observations will also provide insights into the accuracy with which PM properties can be predicted with models. At present, chemical-transport models are evolving rapidly as scientific understanding improves. It is likely that such models soon will be applied by regulatory agencies, and when this occurs the users may typically be less familiar with the intricacies of the scientific underpinnings of the models. It will be necessary to ensure that individuals using these models are properly trained. Establishing a network of modelers will help to facilitate communication regarding best modeling practices. In this regard, the U.S. Western Regional Air

Partnership (WRAP) has established a regional modeling center at a university. The purpose of this center is to perform regional PM modeling in connection with managing regional haze and to train personnel to apply these models for their local needs. It would be beneficial if this approach were applied more broadly to support the use of models by policy makers.

Recommendation 3.4 *Improve the ability to model certain atmospheric processes. Of particular importance are cloud processing, processes that involve carbonaceous aerosols, and the formation of new particles by homogeneous nucleation.*

Rationale. The ability to model atmospheric PM requires an adequate scientific understanding of relevant atmospheric processes. Current understanding of some phenomena is reasonably good. For example, the chemical pathways responsible for transforming SO_2 and NO_x into secondary particle products are reasonably well understood both for the homogeneous atmosphere and for clouds and fogs. The ability to predict cloud processing is compromised somewhat, however, by the difficulty of predicting the extent to which clouds will form and process polluted air and possibly remove PM via wet deposition. The scientific basis for modeling carbonaceous aerosols is probably the greatest limitation of current chemical-transport models; detailed knowledge of anthropogenic and biogenic sources of primary carbonaceous emissions and their gas-phase precursors is not yet adequate. Furthermore, modeling carbonaceous compounds requires a good understanding of the chemical pathways by which secondary organic PM is formed as well as the physical properties of those products that influence their gas/particle distributions. Also, while recent measurements have shown that the formation of new particles by nucleation from the gas phase is an important source of ultrafine particles, the chemistry and physics of such nucleation phenomena are not sufficiently well understood to enable nucleation to be credibly included in models. Although research on these issues is underway, substantial additional work is needed.

Recommendation 3.5 *Obtain more detailed information on the size-resolved composition of primary particulate emissions from various sources.*

Rationale. The effects of PM on visibility, as well as the deposition location and efficiency of particles in the respiratory system, are also dependent on size. Much work is being done to develop chemical-transport models that predict the size-resolved composition of atmospheric PM as well as fine- and coarse-particle mass contributions. In order to bring such models to their full potential, more detailed information on the size-resolved composition of primary particulate emissions from various sources is needed. This will provide valid initial conditions for calculations, and will facilitate evaluating the differing behaviors of different particle types during transport through the atmosphere.

Recommendation 3.6 *Develop approaches for using data from satellites or real-time sensors to obtain boundary conditions for models.*

Rationale. Obtaining accurate results from models requires accurate boundary and initial conditions. This can be problematic when, for example, sources that are typically not included in emission inventories affect PM concentrations. Examples of such sources are forest fires and the intercontinental transport of dust. When this occurs, data obtained in real time from satellites or real-time sensors might be used to provide initial or boundary conditions for chemical-transport models. Approaches for making optimal use of such information should be developed and implemented when possible.

RECOMMENDATION 4. **Improve emission inventories and emission models.** North American emission inventories should be expanded to include more information on organic speciation and size-resolved composition. Information on emissions from sources typically viewed as "unmanageable" is also needed. Efforts should be continued to refine models that describe the dependence of emissions of, for example, NH_3 and soil dust on time-dependent environmental conditions such as temperature, wind speed, and soil type/moisture content. Emission models that predict size distributions formed when hot exhaust containing vapors that nucleate or condense upon mixing with cooler ambient air are also needed. This recommendation would require that significant additional resources be directed toward the characterization of emissions.

POLICY RELEVANCE

Emissions pertinent to ambient PM include primary particles, precursor gases of secondary PM, and gases that participate in chemical processes that ultimately produce secondary PM. Emissions originate from biogenic and anthropogenic sources, and include emissions from point, mobile and open sources. Appropriate information on emissions from all significant sources is required for receptor models and for chemical-transport models. Ultimately, these models are extended to estimate the personal exposure of populations of individuals. Such comprehensive information on emissions would enable policy makers to examine the efficacy of various emission-control strategies on PM management.

SCIENTIFIC NEEDS

Recommendation 4.1 *Develop improved and standardized approaches for developing national emission inventories for Canada, the United States, and Mexico, which provide compatible data.*

Rationale. Canada and the United States currently have national emission inventories. Until very recently, Mexico's emission inventory was restricted to several major cities and sources. Consistency in approaches used for emission inventories in these countries would facilitate the use of emission data for initiating and evaluating PM models. At present, it is common practice for modelers to reprocess inventory data for compatibility with their model. This process could be streamlined if a unified approach were taken to design emission inventories for North America.

Recommendation 4.2 *Update and expand emission inventories to include new types of information including size distributions and size-resolved composition, organic speciation, and updated source profiles.*

Rationale. Emission inventories should be continually updated and expanded to include information on nontraditional sources known to be important, such as cooking, vehicular emissions during cold start, and high-emitting vehicles. Planning should be initiated to investigate the feasibility of expanding emission inventories to include detailed information on size-resolved composition. Particular attention should be given to providing more detailed information on organic emissions from both anthropogenic and biogenic sources, including the speciation of both particulate and gaseous organics. Such information would provide information on organic gases that can react to form secondary organic PM in the atmosphere, or that could condense as hot emissions mix with cool air near the source. The emission of organic gases from sewers, which is known to be significant in Mexico, should also be included. The lack of information on organic species in emission inventories compromises the performance of aerosol chemical-transport models and source-apportionment models, since organics comprise a significant fraction of the fine-particle mass. Finally, many source profiles used for source-apportionment studies date from the 1970s. Changes in control technologies since then have led to significant changes in source profiles. There is a need to provide updated source profiles for current work of this type.

Recommendation 4.3 *Develop approaches that can be used to reconcile emission inventories with measurements of species concentrations observed in the atmosphere.*

Rationale. While understanding of emissions has improved markedly in the past decade, uncertainties about emissions still contribute significantly to uncertainties in ambient PM source apportionment. Much can be learned through attempts to reconcile emission inventories with species concentrations measured in the atmosphere. For example, emission data can be used with receptor models and/or chemical-transport models to determine whether or not ambient PM concentrations are consistent with expectations. Studies to reconcile emissions with ambient measurements should be carried out for a variety of representative regions in North America.

Recommendation 4.4 *Extend emission inventories to include all available information from particle sources typically viewed as unmanageable, such as forest fires and dust.*

Rationale. Ways to incorporate emission-inventory information on dust or smoke from forest fires should be found. When such species are transported over long ranges, useful information can be obtained from satellite images. This may require the sharing of information and information-gathering techniques across agencies that might not normally communicate in this way. In the United States, for example, input from NASA (satellite photographs), the EPA (controlled sources), and the National Park Service and the National Forest Service (controlled and uncontrolled forest fires) would be needed. Incorporating such diverse information would enhance the usefulness of emission inventories. Such inventories would be useful for short-term forecasting, for providing input that could be used to determine whether or not controlled burns or vehicular traffic would lead to unhealthy conditions during a specified period.

Recommendation 4.5 *Improve emission models for dust.*

Rationale. Work should be done to improve models used to predict dust emissions in emission inventories. Dust emissions depend on particle size, soil type and moisture content, terrain, and wind speed. Because they are emitted close to the surface, coarse particles (i.e., PM_{10}) tend to deposit near the source. Therefore, they are less likely to be transported between model grid cells than small particles. These dependencies need to be built into emission models to enhance the likelihood that dust concentrations, which can contribute a significant portion of $PM_{2.5}$ and an even larger portion of PM_{10}, are calculated correctly.

Recommendation 4.6 *Develop emission models for NH_3.*

Rationale. Ammonia plays an important role in regulating the concentration and composition of $SO_4^=$ and NO_3^-. Incomplete understanding of the anthropogenic and biogenic sources and sinks of NH_3 hinders the ability to perform refined modeling of $SO_4^=$ and NO_3^- formation. Plants and soils can be a sink for NH_3, but can release NH_3 depending on climate, air-chemistry, plant, or soil conditions including moisture content. A better understanding of these biological and physiological processes will contribute to refined estimates and understanding of $SO_4^=$ and NO_3^- formation. However, animal husbandry operations, especially animal waste handling and disposal, are predominant sources of NH_3. Better information is needed on the chemical and biological processes and agricultural practices that influence the timing and amounts of NH_3 released to the atmosphere.

Recommendation 4.7 *Develop emission models that account for condensation or nucleation of semivolatile compounds as hot emissions mix with cool ambient air.*

Rationale. Hot emissions often contain vapors that nucleate or condense on preexisting particles as they mix with cooler ambient air. These vapors include species such as organics in automotive exhaust and SO_3 in power-plant stacks. Such processes are not accounted for in emission inventories, nor are they included in chemical-transport models. It is necessary, therefore, to develop emission models that describe condensation/nucleation near the source. Such models will help to ensure that primary PM emissions are not underestimated, and will establish the connection between vehicular emissions and concentrations of ultrafine PM.

Recommendation 4.8 *Continue to improve the timeliness of emission inventories.*

Rationale. In the United States, recent efforts by the EPA have reduced the time required to incorporate emission data into inventories to about 30 months following the calendar year of interest, while in Canada the goal is to provide annual national inventories that combine the common air-contaminant information and toxics information on a two-year lag. Timely inventories are important in reconciling emissions with ambient concentrations and for forecasting PM concentrations. More up-to-date information would be useful to establish and validate protocols for determining the effects of emission controls on ambient air quality. Although including real-time emission data is probably not a realistic goal for the near future, it might be possible to update emission inventories more quickly than is currently done.

Recommendation 4.9 *Develop approaches for obtaining emission information with adequate resolution in space, time, and composition for atmospheric modeling and exposure assessments. These approaches will likely involve combinations of emission models and emission data with higher resolution in time and composition than are currently included in inventories.*

Rationale. Emission inventories are used for a variety of purposes including permitting and regulatory-compliance programs, tracking trends in emissions or control technologies, and providing source data for chemical-transport models. Chemical-transport models often require information on emissions that is highly resolved in space, time, and composition. Information in emission inventories is commonly aggregated across space and time. For example, area-source emissions in the United States are typically reported as annual emissions averaged over an entire county. It is common for modelers to spend significant effort to reprocess emission data to provide hourly speciated emission estimates with spatial scales consistent with those of their models. Reprocessing also often involves providing estimates of size-resolved composition. Furthermore, it is likely that future work will involve the development of models that can provide estimates of human exposure. Such models will require detailed information on local emissions. Efficiencies would be significantly improved if, in the future, emission inventories were modified to include information that is required for emission models and exposure models. Efforts to integrate this type of information into emission models will require close collaboration among experts on emission inventories, atmospheric modeling, and exposure assessment. Increased attention is being given to linking ambient atmospheric models to assessments of personal exposure. This includes individuals working in the ambient outdoor environment. In addition, these models are being joined with emission information on the "built" environment (e.g., homes and offices, where people spend substantial time) to assess overall personal-exposure. A major challenge exists in developing personal exposure profiles applicable to the diverse lifestyles of large populations.

Recommendation 4.10 *Prepare national emission inventories over a period of years in parallel with long-term monitoring of health effects and ambient concentrations of PM and its precursors to facilitate analyses of trends and benefits.*

Rational. A major deficiency in North America is the lack of a record of emission trends over an extended period of time at the population-center level. The emission-inventory record of PM and its gaseous precursors (except SO_2) is so limited that it is virtually impossible to trace the response of PM to changes in emissions over the years in urban locations, let alone the patterns of change in urban PM levels over the lifetimes of most people living in and around North American cities. This is an important aspect to interpreting patterns of change in ambient pollution concentrations and associated benefits to human health.

RECOMMENDATION 5. **Commit adequate resources to the evaluation, synthesis and archiving of data obtained in current and future atmospheric aerosol studies, and to fostering interactions between the communities of scientists working on atmospheric science, health effects, and global climate change.**

POLICY RELEVANCE

Ambient measurements are expensive. Optimal benefit from measurement programs would accrue if adequate resources were routinely committed to the analysis of data and the application of uniform standards for archiving data. The effort required to evaluate and synthesize data from major field studies is typically one-fourth to one-half of the effort expended on the total project, and adequate support for such work often is not provided. In the long term, policy makers could make better use of limited resources if they were to establish a system for archiving data, and require all investigators supported by funding from their agencies make use of that system.

SCIENTIFIC NEEDS

Recommendation 5.1 *Establish procedures at funding agencies to ensure that atmospheric aerosol data are systematically archived in a manner that provides convenient access to all interested parties.*

Rationale. All too often data obtained in atmospheric field studies are not exploited to full potential. This occurs, in part, because adequate attention is not given to archiving the data for unambiguous interpretation by groups that were not directly involved with data collection. Archiving aerosol data is a particular challenge, since such measurements involve complex data from a wide variety of instruments that require informed judgment to be properly interpreted. Uniform standards should be established for characterizing, documenting and archiving data and metadata from studies of atmospheric aerosols. Such archiving should be completed in a timely manner. In some cases, it may be possible to link this atmospheric data with human-health statistics, as was done with monitoring data in the NMMAPS study (Samet et al., 2000 a, b) to improve an understanding of the role of specific atmospheric constituents in causing adverse health effects.

Recommendation 5.2 *Develop a comprehensive plan for data analysis, integral to measurement programs, and commit adequate resources to carry out such analyses.*

Rationale. It is a commonly held view that it is easier to obtain funding for new measurement programs than to complete the analyses of data from previous studies. Given the expense and effort required for field measurement programs, this would be unfortunate if true. Care should be taken to ensure that when measurement programs are designed, adequate resources are provided for data interpretation and analysis. Opportunities to gain additional insights from the re-analysis of existing data sets need to be continually re-appraised as new analytical tools are developed. Examples of data-analysis needs include developing tools to track long-term trends in PM air quality, quantifying uncertainties in the various observation-based source-receptor models described in Chapter 7, and elucidating the impacts of the multipollutant mix on human health.

Recommendation 5.3 *Foster collaborations among the communities of scientists working on PM pollution, health effects, and global climate change.*

Rationale. Close collaboration between health-effects researchers and atmospheric scientists will be needed to establish cause-effect relationships for pollutant health effects. Such collaboration will help to ensure that atmospheric-research programs provide the data required to test health-effects hypotheses, and will also help to ensure that hypotheses tested reflect the complex realities of the multipollutant mix, as they are understood by atmospheric scientists. Furthermore, while the disciplinary skills required to investigate pollution problems overlap substantially with those required for studies of global climate change, the communities of scientists working on these problems do not interact as much as might be desirable. Research programs that are designed and carried out collaboratively are an ideal vehicle for fostering such interactions.

RECOMMENDATION 6. **Develop more systematic approaches for integrating diverse types of knowledge on origins, properties, and effects of atmospheric PM to assist with the development of management strategies and the measurement of progress towards protecting health.** The challenge here is akin to that encountered in risk assessment (NRC, 1996). Policy makers are faced with the need to make decisions based on imperfect understanding of the atmosphere and of human-health responses. These uncertainties are compounded by inherent variabilities in, for example, emissions, climate, and individual responses. There is a need to systematically account for these uncertainties and variabilities when making policy decisions, and to assess and communicate the benefits of those decisions.

POLICY RELEVANCE

The primary goal of PM management is to protect human health. Systematic approaches that would enable policy makers to integrate information from diverse sources to assist them with selecting management options do not presently exist. This problem is exacerbated by the increasing sophistication and types of science tools available for use by policy makers. Furthermore, each type of information contains inherent uncertainty, and quantifying that uncertainty and taking it into account when choosing management strategies is also problematic. An even bigger challenge is to assess relationships between management strategies, PM properties and human health. PM is only one of many variables that can affect human health, and quantifying the benefits of PM management strategies is a daunting task. Nevertheless, efforts to establish those benefits are essential.

SCIENTIFIC NEEDS

Recommendation 6.1 *Develop a more systematic approach for integrating knowledge gained from measurements, receptor models, and chemical-transport models to make optimal PM-management decisions.*

Rationale. The policy maker is faced with an increasingly complex array of information to use in managing PM. Informed judgment is required to know how best to use this information, which may arise from methods with which the policy maker has no direct personal experience and which may even appear to be contradictory. Systematic approaches for integrating diverse information from measurement networks, receptor models, and chemical-transport models would help to ensure that optimal use is made of existing knowledge when establishing PM management strategies. These approaches must account for inherent uncertainties in each type of information.

Recommendation 6.2 *Develop an accountability framework that will enable measurement of progress towards the goal of protecting human health.*

Rationale. Epidemiological studies have identified a statistical association between increased levels of PM as well as other pollutants and increased morbidity and mortality, especially cardiac and respiratory diseases. Cardiorespiratory morbidity and mortality results from a multitude of factors, several of which have greater impact than air pollution. This makes it difficult to quantify the effects of PM or specific PM constituents and other air pollutants and especially to identify positive trends in improved cardiorespiratory health and to relate these to improved emission controls and air quality. At this time it is easier to quantify improvements in air quality than it is to show by measurement the improvements in health linked to improved air quality. It is important to devise methods to measure the benefits of PM management strategies to human health so that the benefits of those strategies can be assessed. This will require an understanding of the relationship between ambient PM properties, personal exposure, and health. Such methods will need to be applied to large populations so that the wide variety of individuals exposures and responses to PM can be taken into account. This is a difficult task for which no paradigm currently exists.

11.2 FUTURE NARSTO PM ASSESSMENTS

Understanding of atmospheric PM and its effects is evolving rapidly, and significant new information is anticipated to be available in the near future. Furthermore, PM regulations intended to protect health and welfare are evolving as new information becomes available. Recognizing that this will likely be the first of a series of PM assessments, the co-chairs and authors of this report have discussed ways in which future PM assessments might evolve. NARSTO should determine how this and other assessments have been used, and should ensure that future assessments continue to meet needs that have been met by these earlier documents. Also, as understanding of particle properties that affect health becomes better understood, NARSTO should broaden the scope of future PM assessments to explicitly focus on those properties. Future assessments should go further in exploring relationships among PM and other atmospheric constituents, such as ozone, NH_3, NO_x, and anthropogenic and biogenic VOCs (especially secondary organic PM precursors). Furthermore, NARSTO should examine the value of broadening the scope to include trends in energy-utilization scenarios and developments in emission-control technologies. The possibility of using the assessment process as a vehicle to facilitate communications among the measurement, modeling, and environmental-effects communities (especially health effects and global climate change) should be considered.

Currently available sources of information enable semi-continuous assessments of spatial and temporal trends. With the development and deployment of new instruments for continuous measurements of ambient PM mass and species concentrations it soon will be possible to develop more nearly continuous assessments of these phenomena. NARSTO should address the value of examining information from such networks on a continuous or periodic basis.

Given the rapid rate at which new information on PM and its effects is becoming available, the next NARSTO PM assessment is recommended to be initiated in about 2008.

11.3 REFERENCES

AQRS, 1998. Air Quality Research Subcommittee Strategic Plan. Committee on Environment and Natural Resources, National Science and Technology Council, Washington, DC.

AQRS, 1999. Federal Air Quality Research 1998-2000. Committee on Environment and Natural Resources, National Science and Technology Council, Washington, DC.

DOE, 2001. TAP program plan. DOE/SC-0034, U.S. Department of Energy, Office of Science, Office of Biological and Environmental Research, Washington, DC.

Meteorological Service of Canada, 2001. Precursor contributions to ambient fine particulate matter in Canada. ISBN: 0-662-30650-3, Catalogue No. En56-167/2001E, http://www.msc-smc.ec.gc.ca/saib/.

NARSTO, 1998. NARSTO Executive Steering Committee atmospheric particulate matter: research needs.

NARSTO, 2000. An Assessment of Tropospheric Ozone Pollution - A North American Perspective, July 2000.

NARSTO, 2001. Part 4: Science plan for suspended particulate matter. Draft for review, February 2001, http://www.cgenv.com/Narsto/.

NRC, 1996. Science and Judgment in Risk Assessment. National Academy Press, Washington, DC.

NRC, 1998. Research Priorities for Airborne Particulate Matter I. Immediate Priorities and a Long Range Research Portfolio. National Academy Press, Washington, DC.

NRC, 1999. Research Priorities for Airborne Particulate Matter II. Evaluating Research Progress and Updating the Portfolio (1999). National Academy Press, Washington, DC.

NRC, 2001. Research Priorities for Airborne Particulate Matter III. Early Research Progress (2001). National Academy Press, Washington, DC.

Samet, J.M., Dominici, F., Zeger, S. L., Schwartz, J., Dockery, D. W., 2000a. The National Morbidity, Mortality, and Air Pollution Study. Part I: Methods and methodologic issues. Research Report, Health Effects Institute 5-14; discussion 75-84.

Samet, J. M., Zeger, S. L., Dominici, F., Curriero, F., Coursac, I., Dockery, D. W., Schwartz, J., Zanobetti, A., 2000b. The National Morbidity, Mortality, and Air Pollution Study. Part II: Morbidity and mortality from air pollution in the United States. Research Report, Health Effects Institute 94, 5-70; discussion 71-79.

Turpin, B., 2001. Species contributions to PM2.5 mass concentrations: Revisiting common assumptions for estimating organic mass, Aerosol Science and Technology 35, 602-610.

GLOSSARY

ACRONYMS AND ABBREVIATIONS

AAS Atomic absorption spectroscopy

AC Automated colorimeter

AIM Aerosol Inorganic Model

AIRS U.S. Aerometric Information Retrieval System for ambient air quality observations

ASOS Automated surface observing system

AURAMS A Unified Regional Air-quality Modeling System developed by Meteorological Service of Canada, Environment Canada

AUSPEX Atmospheric Utility Signatures, Predictions, and Experiments

AWOS Automated weather observing system

BC Black carbon

BEIS Biogenic Emission Inventory System

CAC Canadian Common Air Contaminants emission inventory

CAAA (U.S.) Clean Air Act and its Amendments

CAPMoN Canadian Acid Precipitation Monitoring Network

CARB California Air Resources Board

CASTNet (U.S.) Clean Air Status and Trends Network

CEM Continuous emission monitor

CEPA Canadian Environmental Protection Act

CEPS Canadian emission processing systems

CMAQ Comprehensive Multi-scale Air Quality model (CTM) with detailed aerosol chemistry

CMB Chemical mass balance

CPI Carbon Preference Index

CRPAQS California Regional PM Air Quality Study

CTM Chemical-transport model

CWS Canada Wide Standard

EC Elemental carbon

EIIP (U.S.) Emission inventory improvement program

EKMA Empirical kinetic modeling approach

EPA (U.S.) Environmental Protection Agency

EPS Emission processing system

ER Emission reduction factor

FRM (U.S.) Federal Reference Method

GATOR Gas, Aerosol, Transport and Radiation Model

GAViM Guelph Aerosol and Visibility Monitoring program sponsored by Environment Canada

GC-MS Gas chromatograph - mass spectroscopy

GR Gas ratio

HAP Hazardous air pollutant

IC Ion chromatography

ICP-MS Inductively coupled plasma combined with mass spectroscopy

IMPROVE	Interagency Monitoring of PROtected Environments
INAA	Instrumental neutron activation analysis
INE	Instituto Nacional de Ecologia - Mexico
IMP	Instituto Mexicano del Petroleo - Mexico
LAC	Light-absorbing carbon
LC-MS	Liquid chromatography and mass spectroscopy
LPI	Low pressure impactor
MSC	Meteorological Service of Canada, Environment Canada
NAAQS	(U.S.) National Ambient Air Quality Standards
NAPS	Canadian National Air Pollution Surveillance network
NARSTO	A tri-national cooperative research entity for policy-relevant study of tropospheric ozone and PM phenomena
NATA	(U.S.) National Air Toxics Assessment
NEI	(U.S.) National Emission Inventory
NIOSH	(U.S.) National Institute of Occupational Safety and Health
NOM	Normas Oficiales Mexicanes – Environmental quality guidelines for Mexico.
NPRI	Canadian National Pollutant Release Inventory
OC	Organic carbon
PAH	Polycyclic aromatic hydrocarbon
PAN	Peroxyacetyl nitrate
PEMB	Piezoelectric microbalance
PIXE	Proton Induced X-ray Emission spectroscopy
PM	Particulate matter
PMF	Positive matrix factorization
QA/QC	Quality assurance / quality control
RADM	Regional Acid Deposition Model
RH	Relative humidity
SCAQMD	South Coast Air Quality Management District
SCAQS	Southern California Air Quality Study
SEM	Scanning electron microscope
SEM/XRF	Scanning electron microscope combined with x-ray fluorescence
SIP	(U.S.) State implementation plans
SJVAQS	San Joaquin Valley Air Quality Study
SMOKE	An emission processing model used to provide inputs to chemical transport models.
SOA	Secondary organic aerosol
SRM	Source-receptor matrix
TEOM®	Tapered Element Oscillating Microbalance
TOMS	Total ozone mapping spectrometer
TOR	Thermal/optical reflectance
TOT	Thermal/optical transmission
TSEM	Tagged Species Engineering Model
TSP	Total suspended particles
UAM-AERO	Urban Airshed Model version IV with aerosol chemistry
UV	Ultraviolet light or energy
VMT	Vehicle miles traveled

VKT	Vehicle kilometers traveled
VOC	Volatile organic compounds
WQC	Windsor-Quebec City Corridor
XAD	An adsorption material used to coat denuder tubes and remove gas-phase organic compounds from an air sample.
XRF	X-ray fluorescence spectroscopy

DEFINITIONS

Activity factors Emission characterization factors derived for consumption or use rates and source operating conditions that are used to estimate emission rates for specific source categories.

Aerosol A mixture of suspended particulate matter and its gaseous suspending medium.

Airshed A geographic region defined by topographical features that result in near-uniform atmospheric transport influences throughout. The airshed may be as small as an isolated mountain valley or as large as an atmospheric transport corridor, for example northeastern North America.

Anthropogenic emissions Emissions resulting from human activities, which are either directly emitted as particles, or are particles formed in the atmosphere from pollutant precursor gases.

Black carbon (BC) and Elemental Carbon (EC) Light-absorbing carbonaceous material in atmospheric particles. These terms,as well as "soot" and "graphitic carbon" sometimes are used interchangeably. BC and EC often are used to indicate optical and thermal measurement methods, respectively. In this Assessment, we use BC generically and use other terms only when required by context

Chemical mass balance A receptor-based model used to estimate PM source apportionment based on a material balance of chemical components in ambient air compared with the chemical components in sources.

Conceptual description The best available qualitative compilation of the physical and chemical processes that govern the formation of PM for a given airshed.

Conceptual model The qualitative compilation which, to the extent possible, is supported by quantitative information of atmospheric processes that affect spatial and temporal distribution of particles and the degree to which different precursors and processes limit or enhance particle formation varies for a given airshed, season, or meteorological condition.

Deciview An index of atmospheric haziness based on the logarithm of the light extinction coefficient. A given change in deciviews is assumed to be perceived approximately the same by a human observer, independent of the absolute level of the haziness.

Dosage (Dose) The relationship between the concentration of pollutants encountering an exposed individual and the amount of time the individual experiences the concentration.

Emission factor The average rate per unit process input of a specific source category, which to the extent possible takes into consideration many variables such as process parameters, effluent temperature, ambient temperature, wind speed, and soil moisture.

Emission processing systems Computer models used to refine a national annual aggregate emission information into spatially gridded, time-resolved data required as inputs by the chemical transport models.

Emission reduction factor A number that accounts the impact of emission controls employed on a source including 1) various effluent exit devices such as bag-house filters, or electrostatic precipitators for removal of particles, 2) scrubbers for SO_2 removal, 3) low NO_x combustors in boilers or selective NO_x reduction technologies, and 4) VOC absorbers or effluent gas combustors.

Epidemiology The empirical, statistics-based study of the relationships between exposure to a human stress and the human physiological or disease-based response.

Enrichment factor A term used in a type of receptor model that accounts for increases in ratios of chemical concentration in the atmosphere relative to the same ratios in a reference material associated with a source.

Exposure The concentration of gas or particles breathed by a human over a period of time, determined from ambient air, indoor air combination, in combination with individual activity patterns.

Extinction coefficient A measure of the fraction of light (b_{ext}) that is lost from a beam of light as it traverses a unit distance through a uniform atmosphere.

Gas ratio For example, $[NH_3]^F$ / $[HNO_3]^T$ where $[NH_3]^F$ is the "free" ammonia, that is ammonia not associated with sulfate nor nitrate, and $[HNO_3]^T$ is the total nitrate (gas and aerosol).

Gravimetric filter-based system The accepted method for determining an estimate of particle mass concentration in the troposphere.

Hazardous air pollutants Toxic air pollutants including heavy metals and persistent organic pollutants defined for regulatory purposes by the U.S. Clean Air Act Amendments.

Integrated denuder system A sampling system for aerosol particles that provides for removal of reactive gases prior to filtration which may interfere with the gravimetric estimates of PM mass concentration.

Linearity In the context of source-receptor relationships, linearity implies that given reductions in emissions will result directly in proportionate changes in ambient concentration over a general area, taking into account the presence of a baseline or background level which is assumed to be irreducible.

Manageable emissions Emissions that can readily be identified with industrial sources, commercial operations, power plants, residential dwellings, and transportation as well as certain fugitive sources such as dust from roadways, slash burning of vegetation, etc.

Organic carbon (OC) Organic material in particles. OC includes both primary emissions and secondary organic particles produced in the atmosphere by chemical transformation of volatile organic compounds (VOC). Primary OC emissions include directly emitted particles as well as those formed by nucleation or condensation of high-molecular-weight organic vapors in the immediate vicinity of the source.

Particulate matter Any non-gaseous material (liquid or solid) which, owing to its small gravitational settling rate, remains suspended in the atmosphere for appreciable time periods.

Plume blight Impairment of visibility by a discernible plume. When a plume is viewed against the sky, plume blight is reflected by discoloration of part of the sky. When a terrestrial object is viewed through the plume, plume blight impairs the ability to see that object. Plume blight is distinguished from "regional haze."

Positive matrix factorization A receptor modeling technique relying on a statistical analysis of the chemical composition of particles compared with expectations from sources.

Primary particles Particles that enter the atmosphere via direct injection from a source or in-situ formation from the gas phase (nucleation).

Proportionality In the context of source-receptor relationships, a proportional relationship between emission change and ambient concentration is rarely observed at a single site because of the influence of random variations associated with meteorological fluctuations.

Radiative balance The net energy flux into or out of a system, in this context, Earth's atmosphere. Specifically, the equilibrium between the visible and infrared radiation that reaches the earth from the sun and the similar radiation that is emitted from the earth into space. The balance between incoming and outgoing radiation governs the temperature of the earth.

Receptor models Receptor models or receptor-based techniques use the differences in chemical composition, particle size, and concentration patterns in space and time to identify source types and to quantify source contributions that affect particle mass concentrations, light extinction, or deposition. They provide a theoretical and mathematical framework for quantifying source contributions.

Reconstructed mass Estimation of bulk PM mass from the sum of individual chemical component species mass, often after the components are adjusted for missing elements, e.g., oxygen in iron oxide, or oxygen and hydrogen in organic material.

Regional haze A regulatory term, defined in terms of impairment of visibility that is distributed over such a large area (tens or hundreds of kilometers) that it appears to be relatively uniform to an observer within that area. Regional haze is distinguished from "plume blight."

Respiratory system The human anatomical system that facilitates breathing and absorption of oxygen.

Secondary particles The solid or liquid particles that form as a result of chemical transformations of precursor gases (SO_2, NH_3, NO_x, VOCs) in the atmosphere.

Semivolatile species Species found in the troposphere that partition at equilibrium between the vapor and condensed phase, including NO_3^- and certain VOCs.

Single-particle property A physical or chemical property of single aerosol particles that depends, for example, on size, crystalline nature, or the nature of chemical mixing of the condensed phase.

Source apportionment identifies semi-quantitatively general source types or categories (e.g., vehicle exhaust, industrial processes, power plants, or fugitive dust) responsible for the observed PM.

Source attribution Estimates the PM sources contributing to a given PM sample(s); essentially the same as the source apportionment.

Source-based air quality models Source-oriented or chemical transport models combine source emission rates with meteorological transport and chemical changes to estimate PM concentrations and composition at a receptor site.

Source profiles The chemical composition and particle size distribution information for particles emitted from specific sources.

Speciation/speciate The identification of component chemical species making up the particle mass.

Standard reference material A chemical that provides a reference composition which can be used to calibrate PM instrumentation.

Total suspended particles Typically refers to the mass concentration to all tropospheric aerosol particles less than 30 to 40 μm aerodynamic diameter, as measured by a gravimetric method.

Toxicological response The response of humans creating a pre-disease or disease condition that is attributable to a n exposure or dose of a chemical.

Trajectory models Use grid-point fields of meteorological variables to compute air mass transport associated with horizontal winds either ending at a "receptor," or originating from a source.

Ultraviolet Radiation beyond the visible spectral range having a wavelength shorter than visible light and longer than those of X-rays.

Unmanageable emissions Emissions that result from volcanic eruptions, windblown sea spray, dust storms from remote arid areas, and forest or brush fires initiated by lightning strikes or spontaneous combustion. Other unmanageable emissions are gases and vapors that react in the atmosphere to produce particles, including sulfur gases from terrestrial and marine sources, nitrogen oxides from soil respiration and lightning strikes, and organic vapors from vegetation.

Valley of Mexico The region immediately surrounding Mexico City.

Visibility The ability to see through the atmosphere. Visibility also represents the quality of that seeing, as reflected in the color and contrast of objects in the view, and in the visual appearance of the atmosphere itself. The term "visibility" is sometimes used to indicate the distance one can see. (See "visual range.")

Visual range When looking horizontally through the atmosphere, the greatest distance at which one can discern a dark object against the horizon sky.

Volatile organic compounds Organic compounds found primarily in the vapor phase at ambient temperature and pressure, which participate in atmospheric photochemical reactions forming oxidants or PM.

Windsor-Quebec City Corridor The geographic region extending along the southern edges of Ontario and Quebec between the cities of Windsor and Quebec City.

APPENDIX A. EMISSION CALCULATIONS AND INVENTORY LISTINGS

This appendix contains two components that supplement the discussion in Chapter 4. The first component is a summary of the methods for calculating emissions, including uncertainties in the methods. The second component is a listing of the emission inventories for the United States and Canada by detailed source category.

A.1 HOW ARE EMISSIONS CALCULATED?

The methods for emission-inventory preparation are generally similar in the three nations. Historically they derive from work begun in the 1960s in the United States, represented in three early documents that set the stage for continuing development and refinement of present-day inventories (U.S. HEW, 1965; U.S. HEW, 1966; U.S. HEW, 1968). The basic method for estimating emissions follows a simple conceptual formula described by:

$$E = A \times EF\ (1\text{-}ER/100),$$

where E = emission rate,

A = the activity rate,

EF = the emission factor, and

ER = overall emission reduction factor (%).

A.1.1 Emission and Emission Reduction Factors

The emission factors (EF) can be as simple as an average rate per unit process input. In most cases, EFs depend on many variables such as process parameters, effluent temperature, ambient temperature, wind speed, and soil moisture. In these cases the formula is applied to estimate emissions for a particular set of conditions. Under some circumstances, such as the current inventorying of $PM_{2.5}$, EFs and estimation techniques are applied for analyses other those for which there were developed. This arises most frequently when emerging air-quality issues overtake the rate of test-data acquisition. Emission factors represent an average of a range of emission rates corresponding to a specific technology under various operating conditions, or to engineering estimation of facility operations or processes within facilities. The EFs are derived for units of different capacity, and often processes are reviewed over a period of years to capture changes in operating conditions and modifications in the technology either from maintenance, or from wear and tear during use. The accuracy and representativeness of the EF are determined by the reliability of the testing methodology, how uniformly it is applied across sources, or the engineering process information used to derive the EFs. Differences arise between nations, and for different geographical or climatic conditions. Examples of such differences include factors for motor vehicles, fugitive-dust generation from unpaved roads, and dust or NH_3 emissions from agricultural operations. Data on representative EFs are available from national regulatory sources; e.g., U.S. EPA (2001); EPS (1999); EPS (2000).

Spatial and temporal resolutions of point and areal sources are limited in the current inventories, which are compiled typically on an annual-census division basis. Improved spatial resolution is accomplished by assigning a location of major sources according to conventional survey mapping, the use of population census, GIS positioning, and characteristic meteorological conditions, using stack geometry, effluent temperature and velocity, combined with ambient conditions, including temperature, plume buoyancy, and wind speed.

Emission factors for some emission categories are more reliable than others. In some cases an EF may not be available for a source category because of insufficient or unacceptable data for generalization across source type. Often it is difficult to determine precisely what the certainty in the EF is. Thus, the application of EFs requires subjectivity and judgment from cognizant technical staff for the application of concern. While documented EFs are used to complete areally coarse inventories for various purposes, the generic values may be quite unreliable and are not necessarily recommended by federal or state agencies for finer-scale applications needed to study the impacts of specific sources. Users of EFs in national, regional- and urban-scale studies should be cognizant of their potential limitations, and other techniques should be considered to improve the confidence in the emission inventories. Several such approaches have been developed: continuous emission-monitoring sensors (CEMs, e.g., for SO_2 and NO_x from large utility boilers), material balances, specialized source profiling for composition and compositional material balances, source sampling to obtain improved particle-size distributions and location-specific emission rates, near-source ambient characterization, and (especially) source-apportionment techniques as discussed in Chapter 7.

The reliability of EF estimation decreases when only a few source tests are used as the basis of the factor, or when judgmental decisions are made from analogy between technologies. Differences in EF estimates also can develop if the current operations or processes are significantly different from those upon which the original EFs were derived. A good example of this is the recent re-evaluation of hazardous air pollution (HAP) emissions from power plants, which showed major differences between EF literature values and measurements from current plants (EPRI, 1994). As a result of the application of the newly acquired data, the risk of exposure to HAPs from current power plants was shown generally to be less than indicated by the original EFs. However, the cost of obtaining the emission information for representative U.S. plants was very high for EPRI and the U.S. Department of Energy, amounting to more than $40 million over about four years.

The emission-reduction factor (ER) accounts for emission controls employed on a source. These include 1) various effluent exit devices such has bag house filters and electrostatic precipitators for removal of particles, 2) scrubbers for SO_2 removal, 3) low-NO_x combustors in boilers or selective NO_x-reduction technologies, and 4) VOC absorbers, or effluent-gas combustors. Emission-control devices have pollutant ERs for PM that generally exceed 98 percent for precipitators and filters, but can be much lower for scrubbers and cyclones. Post-combustion devices for SO_2, VOC and NO_x have ERs in the range of 50 to over 95 percent depending on their design and performance.

Like other process equipment, emission controls have variable operating performance depending on their design, maintenance, and the nature of the process controlled. Thus like EFs, values of ERs are overall averages for specific processes and emission-control designs based on limited testing. Actual values of ERs vary in time and by process in an undocumented manner, adding uncertainty to emission estimates.

A.1.2 Activity Patterns

Activity patterns (A) describe the average temporal operating characteristics of a process, including estimates of the down time for maintenance or process failure. Values of A for point, area, and transportation sources are each obtained in different ways owing to the differing nature of the sources.

Most point sources or industrial sources operate with local permits, and these require information about process emissions, including temporal characteristics. For sources with CEMs installed for monitoring SO_2, NO_x, and opacity (roughly proportional to fine PM loading), such as large electric utility boilers, real-time data are available to derive activity patterns, and deduce emission variability over extended time periods. Furthermore, point sources keep (and report) records of output during operating periods, and maintenance or other down times. One of the difficulties in obtaining point-source information in Canada and Mexico, is these countries' restriction on confidentiality of operating data. In the United States, these data are generally more available for regulatory analyses.

There is a great deal of complexity in acquiring activity data for area sources, which are diverse in character, individually small, and often intermittent, but collectively significant. Though such sources are difficult to characterize, they are generally important to PM emission estimation because their aggregated mass emissions can be large and their chemical composition may be important for estimating source attribution. One example of such sources is commercial frying or char-broiling, which has been found to be a surprising large contributor to $PM_{2.5}$ and VOC emissions in such cities as Los Angeles (e.g., Hildemann et al., 1994). Another areal source is fugitive-dust generation from road traffic, or agricultural operations, which again is a large contribution to primary $PM_{2.5}$. Forest and brush fires, as well as agricultural burning, are other potentially large sources of ambient primary PM. Another areal source is fugitive-dust generation from road traffic, construction, and agricultural operations, which again are large contributions to primary $PM_{2.5}$. Forest and brush fires, burning of land-clearing debris, and agricultural burning are also potentially large sources of ambient primary PM.

Temporal resolution depends on allocation of emissions aggregated seasonally, weekly, daily or by diurnal variation, depending on use and industry activity patterns. The temporal allocations allow for improved approximation of the actual temporal patterns that can be important not only for precise annual averaging using seasonal or daily allocations, but also for short-term health exposures or visibility impairment taken over periods of 24 hours or less. "Typical" temporal variations for different sources have been developed from surveys, activity analyses and expert consensus. These temporal models are approximations that may deviate substantially from actual emissions in a given location. Depending on the requirements for precision in estimations, local testing through observations and activity data may be required, not only for large point or area sources, but for smaller ones that may be of special interest.

For areal sources, emissions can be estimated coarsely from "top-down" measures of activity at the provincial- state- or national-level demographics, land use, and economic activity. The construction industry, for example, is based on the total annual expenditures at the province or (U.S. EPA) region level. These estimates are then allocated by census unit or county, using a procedure linked with construction costs and estimated area under construction. The construction-industry emission estimates in Canada, for example, are based on the total annual expenditures at the province level. These estimates are then allocated by census unit or county using a procedure linked with construction costs and estimated area under construction. This approach clearly results in approximations of varying degrees of accuracy at the county level. In contrast, in the United States, these data are obtained at the county level and are thus more spatially reliable.

Because of their potential importance as PM sources, considerable effort has been devoted recently to the characterization of emissions and activity patterns for areal sources, both in Canada and the United States. The Canadian effort as applied to forest fires and fugitive dust from agricultural sources contains more detail than the U.S. counterpart. In Mexico, one of the recognized important limitations in emission estimation to date is fugitive dust.

Estimation of emissions from fires depends upon knowledge of the time, location and areal extent of the burn, fuel loading, types of combustible material and moisture content. Several federal, state, and provincial agencies, as well as private landowners, conduct managed burns, and accidental wild land fires occur frequently. Thus far, the knowledge base is incomplete and a concentrated effort is needed to compile a timely database of these data.

Residential wood burning also is an important local source. Quantification of emissions from this source category has been approached through acquisition of data on how much fuel is burned in fireplaces and stoves using national consumption estimates. Where this source is especially large, local surveys of firewood use may be conducted. In some areas, residential or industrial open burning may occur; this is handled in a similar way with local surveys to augment larger scale estimates (e.g., The Valley of Mexico).

Mobile source-emission estimation is a complex issue in urban areas. The EF for vehicles, for example, is given in terms of mass per vehicle-mile traveled, or mass per mass of fuel consumed. It is obviously impractical to trace vehicle-use patterns, operational differences, and fleet emissions in every city. Therefore, characteristic information has been compiled using standardized vehicle laboratory simulations that represent average vehicle-use behavior. A similar complexity is apparent in the EFs and usage of non-road equipment, e.g., tractors, construction equipment, lawn mowers. These EF and activity data have been synthesized using national mobile activity models such as the MOBILE and Non Road (NR) emission models.

Despite continued refinements to MOBILE and NR, mobile-source emission estimation remains one of the largest areas of controversy in emission characterization. Part of this issue stems from vehicle EFs, which are known to change with the model, age and maintenance of the vehicle, and part stems from the assumed activity profile. The variation over a period of time is poorly documented for both of the activity patterns and EFs. Driving patterns in urban areas vary somewhat by season, by workday vs. weekend, and by specific city. The emission model adopts a "universal" activity profile based on a combination of average driving patterns and the results of laboratory engine testing, both of which cannot capture local or national practices. The mobile-source models are continuously being modified and upgraded using new data. However, their ability to quantitatively capture PM and reactive-gas emissions from a changing fleet of vehicles used in many different ways remains problematic. With the exception of California, which uses the EMFAC model, the MOBILE model is used universally in the United States, Canada, and Mexico for making emission estimates of light-duty, on-road vehicles. However, both the EFs and the activity component are unlikely to apply without modification for cross-national driving conditions. Canada has made certain modifications (MOBILE VC) for its use. Mexico has used a modified version of MOBILE V for application in border cities like Juarez. However, mobile-source emissions remain a major uncertainty for all urban areas in Mexico, which currently is working on additional modifications of MOBILE that will better reflect its driving and fleet conditions.

A.1.3 Spatial Allocation

The national, provincial, and state emission inventories have spatial disaggregation to census level or to county. For many spatially resolved analyses, however, emissions are required in a uniform geographic grid that necessitates assignment of emissions into the grid from the county or census unit data. This can be done readily for point sources, which are identified by latitude and longitude or some other specific locator. However, the majority of area, mobile, and natural-emission categories are not as geographically discrete as point sources. Thus, spatial allocation is required to estimate their location within a domain of interest. This is accomplished through emission-processing systems (EPS). To facilitate this, each category is assigned a spatial surrogate based on a probability of occurrence for each source category. For example, water surface area is typically thought to be a spatial surrogate for allocating shipping and marine pleasure craft. Many categories use population as their allocation surrogate. However, a number of categories don't have highly correlated spatial surrogates, e.g., residential wood combustion and land clearing.

Natural emissions currently are based on spatially resolved data, or estimated from emission models. These sources are typically based on emission factors applied to land use and vegetative cover defined at the census or county level, or within a CTM grid. Currently the biogenic emission-inventory system (BEIS) has been adopted by all three nations to estimate VOC and soil-NO_x emissions (e.g., Geron et al., 2000). In the United States, for example, this model can be used on a one-kilometer resolved vegetation database. The natural NO_x and VOC emissions are generally calculated with spatial resolution within the CTM framework, and are not generally reported in the national inventories. Of most concern in recent modeling for secondary $PM_{2.5}$ production are terpenes and other high molecular-weight volatile compounds of biogenic origin.

Many mobile-source categories have geographically based surrogates, or linear transportation networks (e.g., road maps), for which emissions are assigned. For these sources, emissions can be adjusted to represent activity occurring along each particular segment of the link. Highway vehicles are assigned to road networks, commercial marine vessels are assigned to shipping lanes, and locomotives are assigned to railway routes. Each of these networks is reviewed for updating when new roads, channels, or other linkages are deployed.

A.1.4 Processing for Model Applications.

As noted above, the emission input for air-quality modeling (CTMs) generally requires manipulation of the national inventory data to prepare appropriate temporal and spatial resolution for model computations. The CTMs basically yield four-dimensional numerical solutions for the coupled mass-balance equations for the major contributing species making up PM; these include hourly, and spatial grid-resolved estimates to 50 km or less. To convert the conventional inventory into the needed input for the CTMs, the EPS software is required, including primary PM sources and gaseous precursors, as well as estimations of reactive chemical species, such as VOC categories by source type (e.g., Dickson and Oliver, 1991). Chemical speciation is particularly important for VOCs, where the inventories report only total emissions while the CTMs require more detailed emission information for individual VOCs or groups of VOC species. The ability of the EPS to bridge the differences between the emission-inventory characteristics and the CTM requirements makes them a useful means for merging and analyzing emission information.

A number of EPSs are available, including CEPS, MEPS and SMOKE. These have certain common features including:

- Estimation of the variation of emissions by season defined in terms of a sequence of three calendar months with winter beginning in December.
- Estimation of species and source categories outside the inventory listing to accommodate missing information from inventory files, for example, NH_3 emissions and on-road mobile-source emissions for SO_2, $PM_{2.5}$ and PM_{10}.
- Estimation of emissions as a function of space and time, including the effective height of injection into the atmosphere.
- Estimations of biogenic VOC and NO_x emissions from vegetation and soils, including the influence of meteorological conditions and land use.
- Estimation of potential VOC contributions to PM.

Typically biogenic VOC and NO_x emissions are estimated hourly for mid-season months, January, April, July and October. These are based on three-hour gridded, objectively analyzed meteorological fields for North America for each day of the four months, and then are summed for each month and multiplied by three to scale to annual emissions. While a simplification, the emission estimates from BEIS are believed to be sufficiently representative to provide a relative comparison with emissions of anthropogenic source types. While biogenic sulfur emissions are known to exist and can be large locally, particularly in coastal areas, or inter-tidal zones, they are not currently estimated in the EPS.

To facilitate estimation of potential VOC contributions to PM mass concentration, a detailed CTM species set is used in the EPS. Five gas-phase groups, SO_2, NH_3, NO_x, VOC, and CO are currently tabulated by the CEPS into 80 to 90 gas-phase species for the AURAMS model, including individual or aggregated VOC species. The higher molecular-weight fraction or the aromatic grouping then can be separated from the oxidant, producing components to estimate PM formation.

A.1.5 Limitations and Uncertainties

Certain fundamental issues limit the application of estimation methods and should be kept in mind. With regard to the EFs, average emissions differ significantly from source-to-source even within the same category and, therefore, EFs may not provide adequate estimates of average emissions for a specific source. The extent of between-source variability that exists, even among similar individual sources, can be large depending on source type. Although the causes of this variability are considered in EF development, this type of information is seldom included in emission-test reports used to develop factors for the U.S. EPA AP-42 emission-inventory manual (U.S. EPA, 2001). As a result, some EFs are derived from tests that may vary by an order of magnitude or more. Even when the major process variables are accounted for, the EFs developed may be the result of averaging tests that differ by factor of five or more.

Air-pollution control techniques also may cause differing emission characteristics, both in rate and in variation in emitted size distributions and chemical composition. These variations are not well documented even for large point sources in the AP-42 manual. Often the nature of the variables are not documented in test reports, at least in a form conducive to detailed analysis of parameter variation, and thus cannot be accounted for the EF estimates. In the case of fugitive emissions, the type and volume of traffic, road-cleaning techniques, the use of different roadbed material for unpaved roads, soil type and moisture all influence the material suspended. In the case of windblown dust, a "reservoir effect" where loose materials are removed initially can limit the duration of a wind event. This effect is also a factor in paved-road emissions.

Using EFs to estimate short-term emissions adds further uncertainty to emission estimates: short-term emissions from a specific source, within-source variability, fluctuations in process operating conditions, control-device operating conditions, raw material characteristics, ambient conditions and other factors. EFs generally are developed to represent long-term average emissions, so testing is usually conducted under normal operating conditions. Parameters that can cause short-term fluctuations are generally avoided in testing and are taken into account in test evaluation. Thus, using EFs to estimate short-term emissions will create increased uncertainty in the inventory. To assess within-source variability and the range of short-term emissions from a source, a number of tests are needed over an extended period of time, or continuous monitoring is required. This kind of testing is rarely done without continuous monitoring because of the costs.

Uncertainty in activity patterns for a given source category varies as well, with type of source, averaging time, and normal vs. upset conditions, including shutdowns for maintenance. The idealized diurnal, weekly, or seasonal patterns can deviate substantially from real conditions so that estimates are usually considered long-term representations of normal operating conditions. The extent of random or non-random deviations is rarely documented in emission inventories, even in cases where short-term modeling is done.

The information resulting in EF and activity profiles, as well as profiles of chemical composition, are obviously a concern, representing large uncertainty in the inventories. In addition to the possibility that technology improvements or emission controls may have changed, it is often not known whether or not comparable testing methods were used for equivalent emission estimates. In most cases the testing methods or equipment add uncertainties in themselves, and these usually are not well documented in time. This is particularly the case of carbonaceous material. In one example, VOC (C_2-C_{12}) profiles available in the inventories are responsive to photochemical modeling for oxidant formation. Above carbon number 6, the range of compounds increases dramatically, and many found in the atmosphere have not been identified. The VOC profiles focus on low molecular-weight species affecting the total VOC volume concentration rather than the smaller volume fraction representing high molecular-weight material relevant to particles. These organic compounds of carbon number 6 and higher are particularly difficult to identify and quantify by routine analytical methods. Nevertheless, observations of at least some of the higher molecular-weight organic vapors are now underway in research programs of Canada and the United States.

Canada and the United States have made sustained progress in assigning accurate spatial and temporal profiles for most point sources and many area (transportation) sources, as has Mexico for the case of Mexico City. However, EFs for many area and non-road sources remain to be characterized, especially in Mexico. To improve spatial and temporal resolution, continued investigation of such categories as domestic and small commercial enterprises to open sources, such as agricultural activity for dust and NH_3, open burning or vegetation fires, dust emissions from different road configurations, and windblown dust releases are important to all three countries. Improvements are warranted for these sources at the local, provincial and state levels as well as in the national inventories.

A.2 EMISSION INVENTORIES BY DETAILED SOURCE CATEGORY

This appendix section provides a detailed listing of 1999 national emissions for the United States (Table A.1) and Canada (Table A.2). The listing is intended to supplement the summary emissions found in Chapter 4 and to provide the reader with knowledge about the emission categories used by the two countries, so that they may be compared.

Table A.1. 1999 U.S national emissions for PM and related pollutants (thousand short tons).

Source Category	NO_x	VOC	PM_{10}	$PM_{2.5}$	SO_2	NH_3
FUEL COMB. ELEC. UTIL	**5,715**	**56**	**225**	**128**	**12,698**	**7**
Coal	4,935	29	194	102	11,856	0
bituminous	3,229	0	131	61	8,806	0
subbituminous	1,504	0	46	30	2,427	0
anthracite & lignite	202	0	17	11	623	0
other	0	29	0	0	0	0
Oil	202	5	5	4	657	3
residual	199	0	5	4	651	v
distillate	3	0	0	0	6	0
other	0	5	0	0	0	3
Gas	385	9	1	1	12	4
natural	367	0	0	0	0	0
process	18	0	0	0	0	0
other	0	9	1	1	12	4
Other	26	1	7	3	115	0
Internal Combustion	167	11	19	19	58	0
FUEL COMB. INDUSTRIAL	**3,136**	**178**	**236**	**151**	**2,805**	**34**
Coal	542	7	74	24	1,317	0
bituminous	370	0	44	18	890	0
subbituminous	46	0	5	3	64	0
anthracite & lignite	18	0	1	0	57	0
other	108	7	23	2	306	0
Oil	214	8	43	24	757	4
residual	129	0	35	20	574	0
distillate	73	0	7	4	159	0
other	11	8	1	0	24	4
Gas	1,202	60	43	39	576	25
natural	985	0	28	25	0	0
process	214	0	14	14	0	0
other	3	60	0	0	576	25
Other	118	35	60	49	135	0
wood/bark waste	82	0	53	43	0	0
liquid waste	8	0	1	0	0	0
other	28	35	6	5	135	0
Internal Combustion	1,059	69	17	15	20	5
FUEL COMB. OTHER	**1,175**	**670**	**568**	**487**	**588**	**7**
Commercial/Institutional Coal	37	1	17	7	196	0
Commercial/Institutional Oil	80	3	9	4	246	2
Commercial/Institutional Gas	266	15	8	7	11	1
Misc. Fuel Comb. (Except Residential)	28	10	81	81	6	0
Residential Wood	40	608	431	374	6	0
fireplaces	0	512	359	359	0	0
woodstoves	0	39	36	14	0	0
other	40	57	35	0	6	0

Table A.1. 1999 U.S national emissions for PM and related pollutants (thousand short tons) (continued).

Source Category	NO_x	VOC	PM_{10}	$PM_{2.5}$	SO_2	NH_3
FUEL COMB. OTHER (continued)						
Residential Other	723	34	22	14	123	4
distillate oil	175	0	0	0	98	0
natural gas	433	0	0	0	18	0
other	116	34	22	14	6	4
CHEMICAL & ALLIED PRODUCT MFG	**131**	**395**	**66**	**40**	**262**	**133**
Organic Chemical Mfg	21	138	30	12	4	0
ethylene oxide mfg	0	0	0	0	0	0
terephthalic acid mfg	0	11	0	0	0	0
phenol mfg	0	2	0	0	0	0
ethylene mfg	0	5	0	0	0	0
charcoal mfg	0	32	0	0	0	0
socmi reactor	0	29	0	0	0	0
socmi distillation	0	4	0	0	0	0
socmi batch	0	0	0	0	0	0
socmi air oxidation processes	0	1	0	0	0	0
socmi fugitives	0	42	0	0	0	0
other	21	12	30	12	4	0
Inorganic Chemical Mfg	6	3	4	3	179	0
pigments; TiO2 chloride process: reactor	0	0	0	0	0	0
sulfur compounds	0	0	0	0	177	0
other	6	3	4	3	2	0
Polymer & Resin Mfg	3	124	3	2	1	0
polypropylene mfg	0	2	0	0	0	0
polyethylene mfg	0	17	0	0	0	0
polystyrene resins	0	3	0	0	0	0
synthetic fiber	0	83	0	0	0	0
styrene/butadiene rubber	0	7	0	0	0	0
leaks from polymer mfg.	0	0	0	0	0	0
other	3	13	3	2	1	0
Agricultural Chemical Mfg	53	8	9	6	1	118
Ammonium Nitrate/Urea Mfg.	0	0	0	0	0	44
other	53	8	9	6	1	73
Paint, Varnish, Lacquer, Enamel Mfg	0	8	1	0	0	0
paint & varnish mfg	0	6	0	0	0	0
other	0	2	1	0	0	0
Pharmaceutical Mfg	0	8	0	0	0	0
Other Chemical Mfg	47	107	19	17	76	15
carbon black mfg	0	28	0	0	0	0
printing ink mfg	0	1	0	0	0	0
fugitives unclassified	0	13	0	0	0	0
carbon black furnace: fugitives	0	0	0	0	0	0
other	47	65	19	17	76	15

Table A.1. 1999 U.S national emissions for PM and related pollutants (thousand short tons) (continued).

Source Category	NO_x	VOC	PM_{10}	$PM_{2.5}$	SO_2	NH_3
METALS PROCESSING	**88**	**77**	**147**	**103**	**401**	**5**
Non-Ferrous Metals Processing	12	20	35	23	272	0
aluminum anode baking	0	0	0	0	0	0
prebake aluminum cell	0	0	0	0	0	0
aluminum	0	0	0	0	56	0
lead production	0	0	2	2	114	0
copper production	0	0	6	4	97	0
zinc production	0	0	2	1	0	0
lead battery manufacture	0	0	0	0	0	0
lead cable coating	0	0	0	0	0	0
other	12	20	25	15	5	0
Ferrous Metals Processing	70	46	93	67	113	5
basic oxygen furnace	0	0	0	0	0	0
carbon steel electric arc furnace	0	0	0	0	0	0
coke oven door & topside leaks	0	6	0	0	0	0
coke oven by-product plants	0	5	0	0	0	0
coke oven charging	0	0	0	0	0	0
gray iron cupola	0	0	0	0	0	0
iron ore sinter plant windbox	0	0	0	0	0	0
coke manufacturing	0	0	0	0	0	0
ferroalloy production	0	0	0	0	0	0
iron production	0	0	0	0	0	0
steel production	0	0	0	0	0	0
gray iron production	0	0	0	0	0	0
primary	0	0	67	50	0	0
secondary	0	0	26	17	0	0
other	70	36	0	0	113	5
Metals Processing NEC	6	10	19	14	17	0
PETROLEUM & RELATED INDUSTRIES	**143**	**424**	**29**	**17**	**341**	**17**
Oil & Gas Production	88	271	1	1	90	0
natural gas	0	0	0	0	89	0
other	88	271	1	1	1	0
Petroleum Refineries & Related Industries	48	149	17	12	244	17
vacuum distillation	0	3	0	0	0	0
cracking units	0	16	12	8	162	17
petroleum refinery fugitives	0	27	0	0	0	0
other	48	101	5	4	82	0
process unit turnarounds	0	2	0	0	0	0
Asphalt Manufacturing	7	4	11	4	7	0
OTHER INDUSTRIAL PROCESSES	**470**	**449**	**343**	**191**	**418**	**45**
Agriculture, Food, & Kindred Products	5	111	61	22	5	4
vegetable oil mfg	0	10	0	0	0	0
whiskey fermentation: aging	0	16	0	0	0	0
bakeries	0	43	0	0	0	0
soybean mills	0	0	7	3	0	0
wheat mills	0	0	2	1	0	0
other grain mills	0	0	5	3	0	0
country elevators	0	0	5	1	0	0
terminal elevators	0	0	2	0	0	0
feed mills	0	0	3	1	0	0
other	5	51	36	14	5	4

Table A.1. 1999 U.S national emissions for PM and related pollutants (thousand short tons) (continued).

Source Category	NO_x	VOC	PM_{10}	$PM_{2.5}$	SO_2	NH_3
OTHER INDUSTRIAL PROCESSES (continued						
Textiles, Leather, & Apparel Products	1	10	1	0	0	0
Wood, Pulp & Paper, & Publishing Products	93	167	80	56	109	1
sulfate pulping: rec. furnace/evaporator	0	0	41	33	0	0
sulafte (kraft) pulping: lime kiln	0	0	0	0	0	0
other	93	167	39	23	109	1
Rubber & Miscellaneous Plastic Products	0	52	4	2	1	0
rubber tire mfg	0	6	0	0	0	0
green tire spray	0	2	0	0	0	0
other	0	44	4	2	1	0
Mineral Products	356	32	168	93	288	0
cement mfg	213	0	24	11	183	0
surface mining	0	0	17	7	0	0
glass mfg	78	0	0	0	0	0
stone quarrying/processing	0	0	24	9	0	0
other	65	32	103	66	106	0
Machinery Products	3	12	5	2	0	0
Electronic Equipment	0	1	1	1	0	0
Transportation Equipment	0	4	0	0	0	0
Construction	0	0	0	0	0	0
Miscellaneous Industrial Processes	12	61	22	15	14	40
SOLVENT UTILIZATION	**3**	**4,825**	**6**	**6**	**1**	**0**
Degreasing	0	371	0	0	0	0
open top	0	4	0	0	0	0
conveyorized	0	2	0	0	0	0
cold cleaning	0	11	0	0	0	0
other	0	354	0	0	0	0
Graphic Arts	1	293	1	1	0	0
letterpress	0	6	0	0	0	0
flexographic	0	15	0	0	0	0
lithographic	0	13	0	0	0	0
gravure	0	44	0	0	0	0
other	1	214	1	1	0	0
Dry Cleaning	0	168	0	0	0	0
perchloroethylene	0	63	0	0	0	0
petroleum solvent	0	97	0	0	0	0
other	0	8	0	0	0	0
Surface Coating	2	2,136	5	4	0	0
industrial adhesives	0	148	0	0	0	0
fabrics	0	10	0	0	0	0
paper	0	51	0	0	0	0
large appliances	0	22	0	0	0	0
magnet wire	0	2	0	0	0	0
autos & light trucks	0	106	0	0	0	0
metal cans	0	113	0	0	0	0
metal coil	0	49	0	0	0	0

Table A.1. 1999 U.S national emissions for PM and related pollutants (thousand short tons) (continued).

Source Category	NO_x	VOC	PM_{10}	$PM_{2.5}$	SO_2	NH_3
SOLVENT UTILIZATION (continued)						
Surface Coating (continued)						
wood furniture	0	130	0	0	0	0
metal furniture	0	58	0	0	0	0
flatwood products	0	18	0	0	0	0
plastic parts	0	16	0	0	0	0
large ships	0	19	0	0	0	0
aircraft	0	5	0	0	0	0
misc. metal parts	0	40	0	0	0	0
steel drums	0	4	0	0	0	0
architectural	0	483	0	0	0	0
traffic markings	0	93	0	0	0	0
maintenance coatings	0	85	0	0	0	0
railroad	0	4	0	0	0	0
auto refinishing	0	104	0	0	0	0
machinery	0	20	0	0	0	0
electronic & other electrical	0	82	0	0	0	0
general	0	107	0	0	0	0
miscellaneous	0	32	0	0	0	0
thinning solvents	0	54	0	0	0	0
Other	2	282	5	4	0	0
Other Industrial	0	113	0	0	1	0
rubber & plastics mfg	0	40	0	0	0	0
other	0	72	0	0	1	0
Nonindustrial	0	1,743	0	0	0	0
cutback asphalt	0	147	0	0	0	0
other asphalt	0	46	0	0	0	0
pesticide application	0	412	0	0	0	0
adhesives	0	250	0	0	0	0
consumer solvents	0	883	0	0	0	0
other	0	5	0	0	0	0
Solvent Utilization NEC	0	2	0	0	0	0
STORAGE & TRANSPORT	**16**	**1,240**	**85**	**31**	**5**	**1**
Bulk Terminals & Plants	2	203	0	0	1	0
fixed roof	0	6	0	0	0	0
floating roof	0	11	0	0	0	0
variable vapor space	0	0	0	0	0	0
efr with seals	0	2	0	0	0	0
ifr with seals	0	3	0	0	0	0
underground tanks	0	2	0	0	0	0
area source: gasoline	0	157	0	0	0	0
other	2	22	0	0	1	0

Table A.1. 1999 U.S national emissions for PM and related pollutants (thousand short tons) (continued).

Source Category	NO_x	VOC	PM_{10}	$PM_{2.5}$	SO_2	NH_3
STORAGE & TRANSPORT (continued)						
Petroleum & Petroleum Product Storage	8	108	1	0	0	1
fixed roof gasoline	0	1	0	0	0	0
fixed roof crude	0	10	0	0	0	0
floating roof gasoline	0	11	0	0	0	0
floating roof crude	0	2	0	0	0	0
efr / seal gasoline	0	9	0	0	0	0
efr / seal crude	0	3	0	0	0	0
ifr / seal gasoline	0	3	0	0	0	0
ifr / seal crude	0	1	0	0	0	0
variable vapor space gasoline	0	0	0	0	0	0
area source: gasoline	0	0	0	0	0	0
area source: crude	0	0	0	0	0	0
other	8	68	1	0	0	1
Petroleum & Petroleum Product Transport	0	120	0	0	2	0
gasoline loading: normal / splash	0	3	0	0	0	0
gasoline loading: balanced / submerged	0	7	0	0	0	0
gasoline loading: normal / submerged	0	14	0	0	0	0
gasoline loading: clean / submerged	0	0	0	0	0	0
marine vessel loading: gasoline & crude	0	34	0	0	0	0
other	0	62	0	0	2	0
Service Stations: Stage I	0	320	0	0	0	0
Service Stations: Stage II	0	412	0	0	0	0
Service Stations: Breathing & Emptying	0	45	0	0	0	0
Organic Chemical Storage	4	25	1	1	0	0
Organic Chemical Transport	0	5	0	0	0	0
Inorganic Chemical Storage	0	1	1	0	0	0
Inorganic Chemical Transport	0	0	0	0	0	0
Bulk Materials Storage	2	1	82	30	2	0
storage	0	0	27	11	0	0
transfer	0	0	54	18	0	0
combined	0	0	0	0	0	0
other	2	1	0	0	2	0
Bulk Materials Transport	0	0	0	0	0	0
WASTE DISPOSAL & RECYCLING	**91**	**586**	**587**	**525**	**37**	**88**
Incineration	58	51	92	47	30	0
conical wood burner	0	0	0	0	0	0
municipal incinerator	0	0	0	0	0	0
industrial	0	0	0	0	7	0
commmercial/institutional	0	0	0	0	0	0
residential	0	0	63	31	0	0
other	58	51	28	16	24	0
Open Burning	30	356	492	476	5	0
industrial	0	0	0	0	0	0
commmercial/institutional	0	0	0	0	0	0
residential	0	149	188	173	0	0
other	30	207	303	303	5	0

Table A.1. 1999 U.S national emissions for PM and related pollutants (thousand short tons) (continued).

Source Category	NO_x	VOC	PM_{10}	$PM_{2.5}$	SO_2	NH_3
WASTE DISPOSAL & RECYCLING (continued)						
Wastewater treatment	0	0	0	0	0	87
other	0	50	0	0	0	0
Industrial Waste Water	0	21	0	0	0	0
POTW	0	50	0	0	0	87
TSDF	0	42	0	0	0	0
industrial	0	0	0	0	0	0
other	0	42	0	0	0	0
Landfills	2	36	3	2	1	0
industrial	0	0	0	0	0	0
other	2	36	3	2	0	0
Other	1	30	1	0	0	0
ON-ROAD VEHICLES	**8,590**	**5,297**	**295**	**229**	**363**	**260**
Light-Duty Gas Vehicles & Motorcycles	2,859	2,911	59	34	137	174
light-duty gas vehicles	2,846	2,870	58	34	136	0
motorcycles	13	42	0	0	0	0
Light-Duty Gas Trucks	1,638	1,722	36	22	91	76
ldgt1	1,110	1,132	25	15	68	0
ldgt2	529	589	11	7	24	0
Heavy-Duty Gas Vehicles	459	375	12	8	17	4
Diesels	3,635	289	189	166	118	6
hddv	3,620	284	186	164	0	0
lddt	6	2	1	1	0	0
lddv	8	3	1	1	0	0
NON-ROAD ENGINES AND VEHICLES	**5,515**	**3,232**	**458**	**411**	**936**	**10**
Non-Road Gasoline	187	2,593	89	82	28	1
recreational	8	185	3	3	0	0
construction	6	51	2	2	0	0
industrial	13	30	0	0	0	0
lawn & garden	78	845	20	19	0	0
farm	4	15	0	0	0	0
light commercial	34	172	2	2	0	0
logging	5	369	23	21	0	0
airport service	0	0	0	0	0	0
railway maintenance	0	0	0	0	0	0
recreational marine vessels	37	924	39	36	0	0
other	0	0	0	0	28	1
Non-Road Diesel	2,707	372	253	233	507	3
recreational	5	1	1	1	0	0
construction	1,247	185	128	118	0	0
industrial	237	39	33	30	0	0
lawn & garden	83	17	12	11	0	0
farm	906	94	54	50	0	0
light commercial	123	20	15	14	0	0
logging	610	8	6	5	0	0

Table A.1. 1999 U.S national emissions for PM and related pollutants (thousand short tons) (continued).

Source Category	NO_x	VOC	PM_{10}	$PM_{2.5}$	SO_2	NH_3
NON-ROAD ENGINES AND VEHICLES (continued)						
Non-Road Diesel (continued)						
airport service	10	2	1	1	0	0
railway maintenance	4	1	1	0	0	0
recreational marine vessels	31	5	2	2	0	0
other	0	0	0	0	507	3
Aircraft	175	183	38	27	12	4
Marine Vessels	1,007	34	46	40	273	1
coal	0	1	3	1	0	0
diesel	995	33	42	38	0	0
residual oil	0	0	0	0	0	0
gasoline	12	1	1	0	0	0
other	0	0	0	0	273	1
Railroads	1,204	49	30	27	113	1
Other	235	0	2	2	3	0
liquified petroleum gas	206	0	1	1	0	0
compressed natural gas	29	0	0	0	0	0
other	0	0	0	0	3	0
NATURAL SOURCES	**0**	**0**	**0**	**0**	**0**	**35**
Biogenic	0	0	0	0	0	35
MISCELLANEOUS	**320**	**716**	**20,634**	**4,454**	**12**	**4,322**
Agriculture & Forestry	0	8	4,888	948	0	4,322
agricultural crops	0	0	4,298	860	0	3,552
agricultural livestock	0	0	590	89	0	769
other	0	8	0	0	0	0
Other Combustion	320	702	1,007	872	12	0
structural fires	0	15	21	19	0	0
agricultural fires	0	53	104	94	0	0
slash/prescribed burning	0	311	621	526	0	0
forest wildfires	0	319	261	233	0	0
other	320	3	0	0	12	0
Catastrophic/Accidental Releases	0	5	0	0	0	0
Repair Shops	0	0	0	0	0	0
Health Services	0	1	0	0	0	0
Cooling Towers	0	1	3	3	0	0
Fugitive Dust	0	0	14,736	2,631	0	0
unpaved roads	0	0	9,360	1,411	0	0
paved roads	0	0	2,728	682	0	0
construction	0	0	1,956	391	0	0
other	0	0	692	146	0	0
TOTAL ALL SOURCES	**25,393**	**18,145**	**23,679**	**6,773**	**18,867**	**4,964**

Table A.2. 1999 Canadian emission inventory for PM and related pollutants (Ktonnes/yr).

CATEGORY / SECTOR	PART	PM_{10}	$PM_{2.5}$	SO_x	NO_x	VOC	NH_3
Industrial Sources							
Abrasives Manufacture	0.8	0.4	0.3	2.8	0.2	1.5	0.0
Aluminum Industry	11.8	7.8	5.3	46.2	1.1	1.0	0.2
Asbestos Industry	0.1	0.0	0.0	0.8	0.2	0.0	0.0
Asphalt Paving Industry	32.9	5.5	2.0	2.4	2.0	3.3	0.0
Bakeries					0.0	6.0	0.0
Cement and Concrete Industry	21.1	8.5	3.8	34.0	32.2	0.4	0.2
Chemicals Industry	4.5	2.6	1.4	6.4	24.1	9.4	20.8
Clay Products Industry	2.6	0.6	0.2	0.0	0.1	0.0	0.0
Coal Mining Industry	11.7	8.8	6.3	5.3	3.2	1.8	0.0
Ferrous Foundries	0.7	0.4	0.4	1.7	0.0	1.8	0.0
Grain Industries	58.3	11.7	1.7	0.0	0.0	0.0	
Iron and Steel Industries	20.7	10.8	7.1	62.8	25.5	28.3	0.4
Iron Ore Mining Industry	39.4	21.3	7.6	54.6	7.8	0.8	0.0
Mining and Rock Quarrying	86.0	11.5	3.2	20.8	14.6	0.7	0.2
Non-Ferrous Mining and Smelting Industry	15.6	13.2	9.8	891.7	3.5	0.1	0.3
Oil Sands	3.9	1.8	1.4	160.9	16.5	0.1	0.8
Other Petroleum and Coal Products Industry	0.3	0.1	0.1	0.6	0.4	0.1	0.0
Paint & Varnish Manufacturing	0.1	0.1	0.0		0.0	2.0	0.0
Petrochemical Industry	1.3	0.7	0.3	1.3	11.6	16.5	0.2
Petroleum Refining	6.5	5.0	3.3 1	41.1	26.9	47.7	1.7
Plastics & Synthetic Resins Fabrication	0.2	0.1	0.1	0.3	0.4	6.7	0.0
Pulp and Paper Industry	74.4	50.8	39.3	77.0	58.1	23.3	6.1
Upstream Oil and Gas Industry	2.1	2.0	1.9	387.3	314.9	689.4	0.0
Wood Industry	153.7	86.0	52.6	2.6	16.0	47.1	1.7
Other Industries	72.6	37.5	23.8	49.0	60.9	53.0	2.1
Total Industrial Sources	**621.2**	**287.3**	**171.8**	**1,949.6**	**620.4**	**940.8**	**34.7**

Table A.2. 1999 Canadian emission inventory for PM and related pollutants (Ktonnes/yr) (continued).

CATEGORY / SECTOR	PART	PM_{10}	$PM_{2.5}$	SO_x	NO_x	VOC	NH_3
Non Industrial Fuel Combustion							
Commercial Fuel Combustion	3.4	3.0	2.7	13.0	29.3	1.7	0.5
Electric Power Generation (Utilities)	78.8	34.9	18.6	534.3	255.0	3.0	1.0
Residential Fuel Combustion	4.8	4.0	3.7	17.3	36.7	2.3	0.6
Residential Fuel Wood Combustion	137.8	137.3	131.8	1.8	12.2	400.1	0.8
Total Non Industrial Fuel Combustion	**224.9**	**179.1**	**156.9**	**566.4**	**333.2**	**407.1**	**2.9**
Transportation							
Air Transportation	2.0	1.1	0.8	2.3	34.0	11.6	0.0
Heavy-duty diesel vehicles	32.1	32.1	29.5	32.8	378.3	48.5	0.1
Heavy-duty gasoline trucks	0.5	0.5	0.4	0.6	15.1	11.8	0.1
Light-duty diesel trucks	1.3	1.3	1.2	1.5	5.6	2.6	0.0
Light-duty diesel vehicles	0.4	0.4	0.3	0.6	2.0	0.7	0.0
Light-duty gasoline trucks	2.6	2.5	2.0	4.4	112.4	142.4	5.4
Light-duty gasoline vehicles	4.9	4.7	3.3	11.0	273.4	355.9	13.3
Marine Transportation	8.4	8.1	7.4	58.0	118.6	37.4	0.1
Motor cycles	0.0	0.0	0.0	0.0	0.6	2.0	0.0
Off-road use of diesel	17.1	17.1	15.7	16.1	209.2	22.6	0.5
Off-road use of gasoline	4.4	3.9	3.4	1.0	25.4	93.1	0.2
Rail Transportation	19.5	19.5	17.9	7.2	115.6	5.6	0.0
Tire wear & Brake lining	4.4	4.3	1.4				
Total Transportation	**97.6**	**95.5**	**83.3**	**135.7**	**1,290.2**	**734.4**	**19.7**

Table A.2. 1999 Canadian emission inventory for PM and related pollutants (Ktonnes/yr) (continued).

CATEGORY / SECTOR	PART	PM_{10}	$PM_{2.5}$	SO_x	NO_x	VOC	NH_3
Incineration							
Crematorium	0.0	0.0	0.0	0.0	0.0		
Industrial & Commercial Incineration	0.1	0.1	0.0	0.6	0.8	0.7	0.1
Municipal Incineration	0.4	0.4	0.4	0.5	1.3	0.7	0.1
Wood Waste Incineration	1.8	1.0	0.7	0.0	0.3	4.6	0.2
Other Incineration & Utilities	0.2	0.0	0.0	0.1	0.2	0.3	0.1
Total Incineration	**2.5**	**1.5**	**1.1**	**1.3**	**2.5**	**6.3**	**0.5**
Miscellaneous							
Cigarette Smoking	1.0	1.0	1.0		0.0	0.0	0.0
Dry Cleaning					0.0	7.8	0.0
Fuel Marketing	0.0	0.0	0.0	0.0	0.3	98.5	
General Solvent Use						274.9	0.0
Human							7.6
Marine Cargo Handling Industry	3.1	1.4	0.4			0.0	
Meat Cooking	1.6	1.6	1.6				
Pesticides and Fertilizer Application	10.5	5.2	1.5		0.8	0.1	180.1
Printing						29.1	0.0
Structural Fires	5.3	5.2	4.8		0.0	5.1	
Surface Coatings						134.2	
Total Miscellaneous	**21.5**	**14.4**	**9.2**	**0.0**	**1.1**	**549.7**	**187.6**

Table A.2. 1999 Canadian emission inventory for PM and related pollutants (Ktonnes/yr) (continued).

CATEGORY / SECTOR	PART	PM_{10}	$PM_{2.5}$	SO_x	NO_x	VOC	NH_3
Open Sources							
Agriculture (Animals)	248.7	141.0	22.3		13.0	294.1	
Agriculture Tilling and Wind Erosion	1,754.4	848.4	20.7				
Construction Operations	2,402.1	528.4	10.7				
Dust from Paved Roads	2,549.5	511.2	129.5				
Dust from Unpaved Roads	6,833.7	2,020.7	300.6				
Forest Fires	835.4	706.1	585.0	0.5	211.0	902.4	14.0
Landfills Sites	4.7	0.4	0.1		5.1		
Mine Tailings	46.9	3.7	0.9				
Prescribed Burning	41.4	33.0	26.9	0.1	5.6	16.3	0.3
Total Open Sources	14,716.9	4,792.9	1,096.8	0.6	216.6	936.9	308.4
NATIONAL TOTAL	15,685	5,371	1,519	2,654	2,464	3,575	554

Note: Numbers may not add to totals, due to rounding.

A.3 REFERENCES

Dickson, R., Oliver, W., 1991. Emissions models for regional air quality studies. Environmental Science and Technology 25, 1533-1535.

Environmental Protection Service (Canada) (EPS), 1999. 1995 Criteria Air Contaminants Emissions Inventory Guidebook. Environment Canada, Preliminary Draft. Hull, QC.

EPRI, 1994. Electric Utility Trace Substances Synthesis Report. Report TR-104614, EPRI, Palo Alto, CA.

EPS, 2000. Canadian Emissions Inventory of Air Contaminants (1995). (Draft). Pollution Data Branch, Environment Canada, Hull, QC.

ERG/Radian, 2000. Proposed Protocol for Development of Mexico Emissions Inventory.

Geron, C., Guenther, A., Pierce, T., 2000. An imporved model for estimating emissions of VOC from forests in the eastern United States. Journal of Geophysical Research 99, 12,773-12,792.

Hildemann, L., Kleindinst, D., Klouda, G., Currie, L., Cass, G., 1994. Sources of urban contemporary aerosol. Environmental Science and Technology 28, 1565-1576.

U.S. EPA, 2001. Compilation of Air Pollution Emission Factors. Report AP-42. Internet Edition.

U.S. Dept. of Health Education and Welfare (U.S. HEW), 1965. Compilation of Air Pollutant Emission Factors for Combustion Processes, Gasoline Evaporation, and Selected Industrial Processes. U.S. Dept. of Health, Education and Welfare, Cincinnati, OH.

U.S. HEW, 1966. Atmospheric Emissions from Coal Combustion. U.S. Dept. of Health, Education and Welfare, Cincinnati, OH.

U.S. HEW, 1968. Compilation of Air Pollution Emission Factors. Report AP-42. U.S. Dept. of Health, Education and Welfare, Durham, NC.

APPENDIX B. MEASUREMENTS

B.1. APPLICATIONS OF DATA FROM AIR-QUALITY MEASUREMENTS

As described in Chapter 5, measuring PM, PM precursors, and copollutants serves a variety of user-community information needs. These communities seek to understand the impact of PM on health and visibility; to determine compliance with existing standards; to establish the sources of compounds that contribute to elevated levels of PM; to determine trends in air-quality; to evaluate the effectiveness of air-quality management strategies (program evaluation referred to in Table B.1); and to provide the scientific understanding required to reliably predict the effectiveness of new management strategies and to forecast air quality using advanced air-quality modeling systems. These information needs are listed in Table B.1.

Three levels of user requirements are indicated in Table B.1: *required* indicates that the measurement is mandated by current air-quality standards (for PM or other copollutants and pollutant precursors); *needed* identifies measurements that supply information to answer current PM-related air-quality questions; and *useful* indicates that the measurement will supply ancillary information required to best address PM-related issues. Table B.1 also indicates the three levels of time resolution (duration of a single measurement) required to meet measurement needs: 24 hours, 1 hour to 6 hours, and less than 1 hour. The appropriate time resolution for the measurement reflects the user's data requirements. For example, a time resolution of 24 hours may be stipulated for compliance monitoring. However, shorter-duration measurements may be required to adequately resolve daily activity for human-exposure monitoring and health-effect research. Likewise, time resolution that is commensurate with meteorological variability would be most useful for source apportionment ($\leq$ 1 hour), while information for model development and testing would be best satisfied by measurement times that capture the characteristic time-scales for chemical and physical evolution of particles and meteorological variability ($\leq$ 1 hour to hours) (Seigneur et al., 1999). With the exception of airborne measurements for assessment, model development, and testing, and for understanding pollutant transport on hourly and daily time scales, most measurement needs are satisfied by ground-based measurements. For airborne measurements, where the sampling platform moves through the atmosphere at 100 m/s or more, measurement time-response of one second or less is desirable. In addition, advances in remote sensing may allow some aloft data to be collected from surface instruments or from those on satellites.

B.2. CURRENTLY AVAILABLE TECHNOLOGY AND INSTRUMENT CAPABILITIES

B.2.1. Inlets

Because inlets define the range of particle sizes that are sampled, they play a critical role in all phases of aerosol measurements. Chow (1995), Hering (1995), Mark (1998), and Marple and Olson (2001) have summarized a variety of sampling inlets that are used for ambient, workplace, and personal monitoring. Watson and Chow (2001) identify the PM_{10} and $PM_{2.5}$ inlets in use for ambient aerosol sampling. All of these devices rely on some form of inertial separation of large (heavy) particles from small (light) particles and the quoted sizes are the diameter at which the collection efficiency falls to 50 percent (usually called the 50 percent cut-point). The effectiveness of the inlets to accurately determine sizes of particles can deteriorate if the inlet becomes dirty.

B.2.2 Integrated Denuder and Filter Systems (substrate- and absorbent-based measurements) for Mass and Composition Sampling

Time-integrated filter-based ambient aerosol sampling systems used to obtain data on PM mass and the chemical components of mass require special sampling methods to minimize sampling artifacts and

APPENDIX B

Table B.1. Quantifiable properties for particle and particle-related measurements along with the appropriate time resolution for those measurements that are required, needed, or useful to provide information to user communities to address PM air-quality issues. The table also indicates if measurements are needed at other elevations in the atmosphere. Size units are diameter, typically μm = 1 • 10^{-6} meters.

Size

Property	Health Effects	Exposure	Compliance	Visibility Impairment	Source Attribution	Program Evaluation	Scientific Understanding
Size cut; dia. ≤ 10 μm	Needed hours	Needed hours	Required 24 hr	Useful ≤1 hr	Needed ≤ 1 hr	Needed 24 hr	Needed ≤1 hr surface/airborne [a]
Size cut; dia. ≤ 2.5 μm	Needed hours	Needed hours	Required 24 hr	Useful ≤1 hr	Needed ≤ 1 hr	Needed 24 hr	Needed ≤ 1 hr surface/airborne [a]
Ultrafine; dia. ≤ 0.1 μm	Needed hours	Useful ≤1 hr					Useful ≤ 1 hr surface/airborne [a]

[a] 1 hr sufficient for surface, much less than 1 hr is needed for measurements aboard aircraft.

Distribution

Property	Health Effects	Exposure	Compliance	Visibility Impairment	Source Attribution	Program Evaluation	Scientific Understanding
Distribution 0.1 to 1 μm	Needed ≤ 1 hr	Needed hours		Useful ≤1 hr	Useful ≤1 hr		Needed ≤ 1 hr surface/airborne
Distribution 0.001 to 0.1 μm	Needed ≤ 1 hr	Needed hours			Needed ≤ 1 hr		Needed ≤ 1 hr surface/airborne
Distribution 2.5 to 10 μm	Needed ≤ 1 hr				Useful ≤1 hr		Needed ≤ 1 hr surface/airborne

Mass

Property	Health Effects	Exposure	Compliance	Visibility Impairment	Source Attribution	Program Evaluation	Scientific Understanding
Mass ≤ 10 μm	Required hours	Required hours	Required 24 hr	Required ≤ 1 hr	Required ≤ 1 hr	Required 24 hr	Required ≤ 1 hr surface/airborne
Mass ≤ 2.5 μm	Required hours	Required hours	Required 24 hr	Required ≤ 1 hr	Required ≤ 1 hr	Required 24 hr	Required ≤ 1 hr surface/airborne
Mass PM_{coarse} [b]							

[b] $PM_{coarse} = PM_{10} - PM_{2.5}$, and a new standard will be proposed in 2004 defining a PM_{coarse} NAAQS, in which case compliance may also apply.

Table B.1. Quantifiable properties for particle and particle-related measurements, continued.

PM Composition

Property	Health Effects	Exposure	Compliance	Visibility Impairment	Source Attribution	Program Evaluation	Scientific Understanding
Acidity	Needed hours	Needed hours					Useful ≤ 1 hr surface/airborne
Sulfate ($SO_4^{=}$)	Needed hours	Needed hours		Needed ≤1 hr	Needed ≤1 hr	Needed 24 hr	Needed ≤ 1 hr surface/airborne
Nitrate (NO_3^-)	Needed hours	Needed hours		Needed ≤1 hr	Needed ≤1 hr	Needed 24 hr	Needed ≤ 1 hr surface/airborne
Chloride (Cl^-)					Needed ≤1 hr		Needed ≤ 1 hr surface/airborne
Ammonium (NH_4^+)				Useful ≤1 hr	Needed ≤1 hr	Needed 24 hr	Needed ≤ 1 hr surface/airborne
Organic acids	Useful hours	Useful hours			Needed ≤1 hr		Useful ≤ 1 hr surface/airborne
Organic carbon	Needed hours	Needed hours		Needed ≤1 hr	Needed ≤1 hr	Useful ≤ 1 hr surface/airborne	Needed ≤ 1 hr surface/airborne
Elemental carbon	Needed hours	Needed hours		Needed ≤1 hr	Needed ≤1 hr	Useful ≤ 1 hr surface/airborne	Needed ≤ 1 hr surface/airborne
Organic (speciated)	Needed hours	Needed hours			Needed ≤1 hr		Needed ≤ 1 hr surface/airborne
Major soil elements				Useful ≤1 hr	Needed ≤1 hr	Useful 1 hr	Needed 1 hr surface/airborne
Trace elements	Required 24 hr	Required 24 hr			Needed ≤1 hr		Useful ≤ 1 hr surface/airborne
Biological aerosols	Needed hours	Needed hours					Needed ≤ 1 hr surface/airborne

Optical properties

Property	Health Effects	Exposure	Compliance	Visibility Impairment	Source Attribution	Program Evaluation	Scientific Understanding
Light scattering				Needed ≤1 hr	Needed ≤1 hr		Needed ≤ 1 hr surface/airborne
Light absorption				Needed ≤1 hr	Needed ≤1 hr		Useful ≤ 1 hr surface/airborne
Light extinction				Needed, ≤1 hr	Needed ≤1 hr		Needed ≤ 1 hr surface/airborne
Solar Radiation (Sun-tracking photometers)							
Aerosol Backscatter (LIDAR)					Useful ≤1 hr long-path		Needed ≤ 1 hr long path

APPENDIX B

Table B.1. Quantifiable properties for particle and particle-related measurements, continued.

Fog (units are mass per unit volume, $\mu g/m^3 = 1 \cdot 10^{-6}$ grams per cubic meter)

Property	Health Effects	Exposure	Compliance	Visibility Impairment	Source Attribution	Program Evaluation	Scientific Understanding
Water content						Useful, ≤1 hr	Needed, ≤ 1 hr, event
Composition		Needed, ≤1 hr				Useful, ≤1 hr	Needed, ≤ 1 hr, event

Gas phase compounds

Property	Health Effects	Exposure	Compliance	Visibility Impairment	Source Attribution	Program Evaluation	Scientific Understanding
Ozone (O_3)	Needed, ≤1 hr	Required, hours	Required, ≤ 1 hr		Needed ≤ 1 hr	Needed, ≤ 1 hr, surface/airborne	Needed, ≤ 1 hr, surface/airborne
NO, NO_2 (NO_x)	Needed, ≤1 hr	Required, hours	Needed, ≤1 hr	Useful, ≤1 hr	Needed, ≤1 hr	Needed, 24 hr	Needed, ≤ 1 hr, surface/airborne
PAN						Needed, ≤1 hr	Needed, ≤ 1 hr, surface/airborne
Nitric acid (HNO_3)	Useful, ≤1 hr	Useful, ≤1 hr				Needed, ≤1 hr, long path	Needed, ≤ 1 hr, surface/airborne
Nitrogen oxides (NO_y)						Needed, 24 hr	Needed, ≤ 1 hr, surface/airborne
Sulfur dioxide (SO_2)	Required, hours	Required, hours	Required, ≤ 1 hr		Needed, ≤1 hr	Needed, 24 hr, surface	Needed ≤ 1 hr, surface/airborne
Carbon monoxide (CO)	Needed, hours	Needed, hours	Required, ≤ 1 hr		Needed, ≤1 hr	Needed, 24 hr	Needed ≤ 1 hr. surface/airborne
H_2O_2 and other peroxides		Useful, hours					Useful, ≤ 1 hr, surface/airborne
VOCs	Useful, hours	Useful, hours			Useful, ≤1 hr	Needed, 24 hr	Needed, ≤ 1 hr, surface/airborne
Odd-hydrogen radicals (OH, HO_2, RO_2)							Useful, ≤ 1 hr, surface/airborne
Nitrate radical (NO_3)							Useful, ≤ 1 hr, surface/airborne

Table B.1. Quantifiable properties for particle and particle-related measurements, continued.

Meteorological variables

Property	Health Effects	Exposure	Compliance	Visibility Impairment	Source Attribution	Program Evaluation	Scientific Understanding
Wind speed (meter/sec.)	Needed $\leq$1 hr	Needed $\leq$1 hr			Needed $\leq$1 hr	Needed $\leq$1 hr	Needed $\leq$ 1 hr surface/airborne
Wind direction (° from north)	Needed $\leq$1 hr	Needed $\leq$1 hr			Needed $\leq$1 hr	Needed $\leq$1 hr	Needed $\leq$ 1 hr surface/airborne
Temperature (degree)	Needed $\leq$1 hr	Needed $\leq$1 hr		Needed $\leq$1 hr	Needed $\leq$1 hr	Needed $\leq$1 hr	Needed $\leq$ 1 hr surface/airborne
Relative humidity (%)	Needed $\leq$1 hr	Needed $\leq$1 hr		Needed $\leq$1 hr	Needed $\leq$1 hr	Needed $\leq$1 hr	Needed $\leq$ 1 hr surface/airborne
Boundary layer height (meters)					Needed $\leq$1 hr	Needed $\leq$1 hr	Needed $\leq$ 1 hr surface/airborne

other interferences. Chow (1995) specified the requirements of these types of sampling systems, tabulated available sampler components that can meet the specified requirements, and described compliance and research-grade sampling systems. These sampling systems typically include: 1) size-selective inlets (see section B.2.1), 2) sampling surfaces, 3) denuders, 4) filter holders, 5) filters, and 6) flow measurements and controllers. A simplified schematic diagram of the integrated denuder and filter sampling system is given in Figure 5.3.

Some of the elements of these samplers, with their long history of use, have become standard. Commercially available multi-channel speciation samplers include many of these components: 1) RAAS (Reference Ambient Air Sampler, Thermo-Anderson Samples, Inc.), 2) MASS (Mass Aerosol Speciation Sampler, URG), 3) SASS (Spiral Ambient Speciation Sampler, Met One), 4) Partisol Speciation Sampler (R&P), and 5) IMPROVE (Interagency Monitoring of Protected Visual Environments). These instruments employ a denuder and filter-pack approach, to acquire particle and/or precursor-gas concentrations concurrently on a manual or sequential basis (U.S. EPA, 1999). In the United States, the National $PM_{2.5}$ Speciation Trends Network (STN) uses the denuder and filter-pack method only for collection of PM NO_3^-, without concern for gas-phase precursors, NH_4^+, or particle phase semivolatile organic material. However, once routine monitoring capabilities become available for these species , those samplers may be incorporated into the network. The principal sampler components for several systems currently in use are given in Table B.2.

Denuders

Denuders are placed in the sampling system to remove more than 95 percent of selected gases while transmitting more than 95 percent of particles (Kitto and Colbeck, 1999). Denuders can be used as part of, or immediately behind, size-selective inlets to remove gases that might interfere with the collection of particles on the filter (positive interference). Denuders also can be used to quantify directly the concentrations of gases that are precursors to secondary PM formation (e.g., HNO_3 and NH_3, which are precursors of NH_4NO_3). In addition, denuders may be placed after the filter to capture semivolatile gases that may evaporate from particles collected on the filters (negative interference). In the latter two cases, the denuders are extracted and the extract analyzed to determine the concentration of the species of interest in the gas phase.

Denuders are fabricated from stainless steel, aluminum, Teflon, etched glass, and cloth, and are coated with substances that retain the gas-phase compounds of interest (e.g., MgO or Na_2CO_3 to collect acidic gas-phase species, such as HNO_3). To maximize the surface-to-volume ratio, denuder

Table B.2. Major components of selected integrated particulate samplers as operated in either the EPA National $PM_{2.5}$ Chemical Speciation Trends Network or in the IMPROVE Network..

SAMPLER ELEMENT AND SPECIES	**EPA National $PM_{2.5}$ Chemical Speciation Trends Network Protocols**				**IMPROVE Protocol**
	RAAS (Reference Ambient Air Sampler, *Anderson Samples, Inc.*)	**MASS (Mass Aerosol Speciation Sampler, 400 & 450, *URG*)**	**SASS (Spiral Ambient Speciation Sampler, *Met One*),**	**Partisol Speciation Sampler (*R&P*)**	**IMPROVE Sampler**
Inlet/$PM_{2.5}$ Fractionator	PM_{10} Size Selective/ $PM_{2.5}$ Cyclone	PM_{10} Size/ Selective/$PM_{2.5}$ WINS	$PM_{2.5}$ Sharp-cut cyclone	$PM_{2.5}$ Harvard Impactor	$PM_{2.5}$ or PM_{10} Cyclone
Denuders [a]	Glass annular denuder inside an Al tube	Glass annular denuder inside an Al tube	Aluminum honeycomb.	Glass honeycomb	Anodized aluminum
HNO_3 [d] **(Denuders used to remove HNO_3 and other acidic gases)**	MgO Changed after 30 sampling periods.	MgO Changed after 30 sampling periods.	MgO Changed after 30 sampling periods.	Na_2CO_3 Changed daily.	Na_2CO_3 Changed once or twice per year.
Filters [b]	Four filter channels; Teflon, nylon, quartz, blank	Two filter channels; Teflon/nylon filter pack, quartz	Three filter channels; Teflon, nylon, quartz	Three filter channels; Teflon, quartz, nylon	Four filter channels; Teflon ($PM_{2.5}$ and PM_{10}). Nylon, quartz
PM_{10} Mass	Not Measured	Not Measured	Not Measured	Not Measured	Teflon: gravimetric
$PM_{2.5}$ Mass	Teflon: gravimetric	Teflon: gravimetric	Teflon: gravimetric	Teflon: gravimetric	Teflon: gravimetric
Elements [c]	Teflon: XRF (same filter as used for mass)	Teflon: XRF (same filter as used for mass)	Teflon: XRF (same filter as used for mass)	Teflon: XRF (same filter as used for mass)	Teflon: PIXE, XRF, PESA (same filter as used for mass)
Particulate NO_3^- [b], $SO_4^=$, NH_4^+ [b], other ionic species. (denuder/filter)	Nylon: IC, water extraction	Teflon/nylons: IC, water extraction	Nylon: IC, water extraction	Nylon: IC, water extraction	Nylon: IC, water extraction
Organic [b] and elemental carbon	Quartz: TOT/STN Protocol	Quartz: TOT/STN Protocol	Quartz: TOT/ STN Protocol	Quartz: TOT/STN Protocol	Quartz: TOR/IMPROVE Protocol

Note: Ammonia is not collected, removed, or measured in any of these samples.

[a] Used to remove gas phase interferences, acid gases only (e.g., HNO_3). The denuders are not analyzed routinely in these networks.
[b] Combination of denuder and filer measurements are needed to quantify semivolatile species, but in these samplers it is only for particulate NO_3^-.
[c] As described in the text, other methods are available, PIXE, INAA, and ICP-MS, but XRF is the most commonly used method and recommended in EPA's National $PM_{2.5}$ Chemical Speciation Trends Network
[d] HNO_3, while not measured currently in either of the networks, could be obtained by the denuder difference method or extraction of Na_2CO_3-coated denuders followed by IC analysis of the extract.

designs include tubular, parallel-plate, annular, and honeycomb configurations. The capacity and efficiency of denuders for inorganic acidic gases have been reasonably well documented, but much more effort is needed to determine and document the efficiency and capacity of denuders for collecting gas-phase semivolatile organic material.

Filters

Filtration is the principal method used in integrated sampling systems for collecting atmospheric PM. Selection of filters for collecting PM for gravimetric and chemical analysis is determined by the characteristics that are imposed by the types of chemical analyses to be undertaken (Chow, 1995; Chow and Watson, 1998; Solomon et al., 2001; and U.S. EPA, 2001). Chow (1995) and Chow and Watson (1998) summarize filter substrates and their chemical/physical properties. Chemical analysis (discussed at the end of this section) includes bulk analysis (c.f., Solomon et al., 2001), and single-particle analysis methods (e.g., scanning electron microscopy) (Fletcher et al. 2001).

Stable species, such as $SO_4^=$ and trace elements (e.g., Al, Si, Ca, Fe, Mn, Ti, etc), are collected with minimal sampling bias on inert filters (e.g., Teflon) if proper inlets, transport tubes, filter substrates, and flow controls are used in the sampler. Labile species, such as NH_4NO_3 and semivolatile organic compounds that exist in both the gas and particle phases, require specialized sampling protocols, typically using denuders and reactive collection substrates (Chow, 1995). For these species, the denuder is used to remove gas-phase material while a reactive backup filter is used to collect the semivolatile particulate-phase species of interest. Most denuder/filter-based systems use either a single absorbent or a chemically treated reactive filter (i.e., either a filter is impregnated with a reactive material or the filter itself is absorbent), or a filter pack. The latter consists of a top inert filter (e.g., Teflon or quartz-fiber) followed by a reactive filter. For example, NO_3^- is collected on a nylon filter or a filter impregnated with Na_2CO_3, behind either an MgO coated denuder or a denuder coated with Na_2CO_3, respectively. Organic material is the most problematic to collect using reactive filters, because denuders and reactive filters are not yet well characterized for these compounds. Currently, the organic fraction of particles is collected using either quartz-fiber filters or filters that are impregnated with charcoal, or coated with finely ground XAD resin or KOH. The denuders used prior to these filters are coated with the same reactive material as the filter, to minimize biases in sampling due to the efficiency of removing the reactive material. While, quartz-filters are widely used for OC and BC analysis (e.g., in STN and IMPROVE networks), they also can have high blank values for OC due to sorption of organic gases. This sorption occurs during sampling as well as during storage and handling, so denuders do not fully solve the problem.

Impactors

Cascade impactors represent a special class of substrate- and absorbent-based techniques for composition sampling. These devices use a series of impactor stages (as described in the inlet section above) to collect a series of size-resolved particle samples (c.f., Marple and Olsen, 2001). The most recent advances have dealt with extending impactor capability to smaller particle sizes (e.g., below 0.2 μm. Examples of available cascade impactors are the Micro-Orifice Uniform Deposit Impactor (MOUDI), the Berner impactor, the Davis rotating unit for monitoring (DRUM), and the low-pressure impactor. These instruments have been used to obtain size-resolved measurements in ambient chemistry and visibility studies. For example, it is feasible, using the DRUM impactor, to obtain 2-hour concentration data for PM mass and trace elements. In the DRUM sampler, Mylar substrates are mounted on cylinders that rotate below the impaction jet so that particle deposits can be monitored as a function of time and analyzed for mass by beta attenuation, for elements by proton-induced x-ray emission (PIXE), and for particle morphology by scanning electron microscopy/x-ray fluorescence (SEM/XRF). The Low Pressure Impactor (LPI) system allows for the separation and collection of particles down to 0.05 μm.

Chemical Analysis Methods for PM Collected on Filters

During the last half-century, a wide variety of chemical compounds has been identified and quantified using filter-based methods. The chemical composition of PM provides clues to its origin and transformations that may have occurred during atmospheric transport. Health studies also are indicating certain chemical components as potential causal factors in observed adverse health effects. In general, the majority of mass is accounted for by just six components ($SO_4^=$, NO_3^-, NH_4^+, OC, BC, and crustal material [e.g., oxides of Al, Si, Ca, Ti, Mn, Fe, and Zn]). However, in detail, PM is composed of hundreds and even thousands of compounds (e.g., OC is composed of many individual organic compounds). The six major components also are the primary species measured in the STN. Chemical-analysis methods for the STN have been summarized by Solomon et al. (2001). After collection, filters are either analyzed directly or extracted in an appropriate solvent and the extract is analyzed to determine concentrations of the species present. As discussed in the previous sections, special sampling protocols are required to minimize bias and interferences for the collection of semivolatile compounds.

Typically, PM mass is obtained from an inert Teflon filter, which is weighed before and after collection under controlled conditions. The difference in the mass collected on the filter divided by the sampled volume gives the atmospheric concentration. Anions ($SO_4^=$, NO_3^-, and Cl^-) and cations (NH_4^+ and water-soluble Na, Mg, K, and Ca) typically are extracted with deionized distilled water (DDW) or another appropriate solvent. Their concentrations in the extract are determined using, for example, ion chromatography (IC; Chow and Watson (1999)). NH_4^+ also can be determined using an automated colorimeter (AC), while flame and graphite-furnace atomic-absorption spectrometry (AAS) are used to quantify amounts of many trace elements in the collected PM. After back calculation to the amount of species on the filter, the atmospheric concentration is determined by dividing by the sampled volume.

Trace elements are commonly measured directly on the Teflon or other filter substrates using various methods of direct excitation to induce a characteristic atomic or nuclear fluorescence. The main approaches are XRF, PIXE, or instrumental neutron activation analysis (INAA) Watson et al. (1999). Alternatively, the elements can be extracted from the filter by acid digestion, volatilized, and ionized in an inductively coupled radio-frequency plasma, and measured by inductively coupled plasma mass-spectrometry (ICP-MS) or the extract directly analyzed by flame AAS or flameless graphite-furnace AAS. In-vitro and in-vivo studies have focused on the effects of air-toxic elements such as the soluble fractions of transition metals and/or hazardous air pollutants. Thus, soluble metal fractions can be obtained by extracting the filters in DDW followed by graphite-furnace AAS, instead of flame AAS, to accommodate the low sample loadings. Other methods for analysis of metals are available.

The OC, EC, carbonate, and total carbon (TC) components in PM collected on the filters are analyzed by thermal combustion, thermal manganese oxidation, and organic extraction/thermal methods. The IMPROVE thermal/optical reflectance protocol (TOR, Chow et al., 1999, 2001) and the NIOSH (Method 5040) protocol that is based on thermal/optical transmission (TOT, after Birch and Cary, 1996) are used in the IMPROVE and EPA STN networks, respectively. Intercomparisons (Countess, 1990; Schmid et al., 2001; Fung et al., 2002; Sharma et al., 2002) of different carbon-analysis methods reported good agreement (±10 percent) on TC, but resulted in factor of two or higher differences in OC and EC concentrations. With the two thermal-optical methods, Chow et al. (2001) and Norris et al. (2002) reported differences of a factor-of-two in OC when comparing TOR and TOT methods using the IMPROVE and NIOSH protocols, respectively, on the same filters owing to the different temperature programs. EPA's speciation program uses the TOT method with a slightly modified version of the NIOSH temperature protocol, similar to that described by Norris et al. (2002). Primary standards for OC and BC are needed to minimize these discrepancies. However, work is underway at the National Institute of Standards and Technology (NIST) to resuspend bulk soil (Klouda and Parish, 2000) and bulk PM to establish filter-based carbon standards. As well, work is underway at NIST and EPA to understand the OC and BC STN and IMPROVE protocols in more detail and to optimize the methods to minimize potential artifacts (Conny et al., 2002).

Individual organic compounds also can be measured by extracting quartz-fiber and Teflon filters using a variety of organic solvents followed by GC-MS. This type of analysis can provide a great deal of information regarding the sources and formation processes of carbonaceous particles (Schauer et al., 1996). Although speciation is desirable, it is not easy to perform because there is no single analytical method that can be used to analyze all classes of organics. Currently, analysis is conducted to obtain non-polar compounds, and to date typically only 10-20 percent of the total OC mass has been identified as individual compounds. However, more advanced methods are beginning to look at polar organic materials using GC-MS and LC-MS methods. As with OC and BC, considerable effort is needed to develop calibration and reference standards and a standard set of organic markers (e.g., alkenes, alkenes, alcohols, acids, PAHs) that can explain the major portion of total organic material in air. Work is underway at EPA in conjunction with NIST to develop a series of reference and calibration standards and to evaluate differences among laboratories that perform organic species analyses for a wide variety of important marker compounds.

Overall, measurements of OC, BC, and OC species are challenging and prone to errors due to PM volatility, interferences from gaseous organic species, limitations with analytical methods, and lack of accepted primary and transfer standards. However, significant advances are expected in the near future permitting more widespread measurements of OC, BC, and organic species.

While not included in the analyses for the STN, filters collected from these samplers or others can be analyzed by single-particle techniques to obtain the chemical compositions of single particles (Fletcher et al., 2001). These analyses are extremely useful for source-apportionment studies. While this is a time-consuming analysis, recent advances using computer-controlled SEM/XRF are allowing these methods to be more practical (Mamane et al., 2001). In SEM/XRF, atoms in the collected PM are excited and the x-rays from the florescence are detected. The method is semi-quantitative for composition, but it provides morphological information of the collected particles, which gives clues to their source and fate in air.

B.2.3 Continuous and Semi-continuous Real-time Measurements

Mass and Mass Equivalent

Continuous or semi-continuous methods obtain PM mass concentrations by measuring changes associated with 1) the inertia imparted to a dynamic element by collected particles; 2) electron attenuation through a surface by particles collected on the surface, and 3) pressure drop across small pores in a filter due to collected particles. Continuous and semi-continuous methods for mass are reviewed by Watson et al. (1998). Table B.3a summarizes the methods discussed here.

Inertial Methods

The inertial methods in use currently include the TEOM (tapered element oscillating microbalance) RAMS (real-time ambient mass sampler), and piezoelectric crystals. The TEOM measures directly the inertial mass of particles collected on a filter, 0.5 cm in diameter, mounted on top of a tapered hollow glass element that oscillates when an electric field is applied. The inlet temperature is maintained at 30 to 50° C to minimize thermal expansion of the tapered element and to control relative humidity. However, this heating can result in the loss of semivolatile species, such as water, NH_4NO_3, and SVOCs. These losses can be in the range of 10 percent to 30 percent or more in the reported PM measurements (i.e., dry mass or volatile reduced-mass) as compared to filter-based gravimetric measurements. Among recent developments is a "differential TEOM system," which is aimed at acquiring accurate PM mass at ambient temperature. The technique places a diffusion drier behind the size-selective inlet to remove particle-bound water. This is followed by a parallel pair of electrostatic precipitators (ESP) that alternately switch on and off. Frequency data are collected for both TEOM sensors on a continuous basis. The technique allows the correction for volatilization and absorbtion (Patashnick et al., 2001).

The RAMS combines TEOM principles with diffusion denuder technology (Eatough et al., 2001; Obeidi and Eatough, 2000) to measure $PM_{2.5}$, including NH_4NO_3 and semivolatile organic material.

The RAMS uses five denuders in tandem: 1) a triethanolamine-coated (TEA) annular denuder to remove NO_2 and O_3; 2) a Nafion diffusion dryer to remove particle-bound water, followed by a particle concentrator to reduce the amount of gas-phase organics that must be removed by diffusion denuders; 3) a BOSS (Brigham Young University Organic Sampling System) diffusion denuder to remove gas-phase organic compounds; 4) an additional TEA denuder; and 5) an additional Nafion dryer. The Teflon-coated filter is backed up with a charcoal-impregnated filter (CIF) (to retain semivolatile species lost from those collected on the filter) and mounted at the tip of the tapered oscillating element of a TEOM. Field comparison with PC-BOSS (Particle Concentrator-BOSS) showed that $PM_{2.5}$ mass, including semivolatile fine particulate NO_3^- and organic species, can be continuously and accurately monitored with RAMS (Eatough et al., 2001).

Piezoelectric microbalances determine PM mass concentrations by measuring the changes in natural resonant frequency of the crystal caused by the deposition of particles (either by electrostatic precipitation or inertial impaction) on a piezoelectric quartz crystal disk. The changing frequency of the sampling crystal is electronically compared to a clean reference crystal, generating a signal that is proportional to the collected mass. Work needs to be done to establish the equivalence between piezoelectric microbalances and filter-based mass measurements.

Pressure-Drop Method

The CAMM (continuous ambient mass monitoring) system measures the pressure drop across a porous membrane (Fluoropore) filter (Koutrakis et al., 1996; Sioutas et al., 1999). The pressure drop is linearly correlated to the particle mass deposited on the filter under properly chosen conditions (Koutrakis et al., 1996; Wang, 1997; Sioutas et al., 1999). The Andersen CAMM is equipped with a Nafion dryer to remove particle-bound water. This is followed by a concentrator prior to sampling on the membrane filter (Babich et al., 2000).

Electron-Attenuation Method

The BAM (beta attenuation monitor) measures the attenuation of beta rays from a radioactive source after they pass through a filter where particles have been deposited. BAM technology also has been used to measure aerosol liquid water content (Spear et al., 1997). Watson et al. (1998) showed that measured differences with other methods appear mostly during wintertime and when semivolatile species are the major component. Differences are minimized during summertime, however. Chow and Watson (1997; Chow et al., 1998a) and Dutcher et al. (1999) compared some collocated PM_{10} filter-based and continuous BAM measurements. Equivalent measurements can be achieved if the instrument is adequately calibrated and maintained.

Size Distribution and Mobility

A wide range of techniques and protocols has been used for particle-size measurements. Since the range of particle sizes of interest can vary over more than four orders of magnitude, from <0.003 μm to >10 μm, different techniques are used to measure different size distributions. The techniques to measure size distribution fall into three broad categories: 1) Aerodynamic techniques, measured by inertial methods such as impactors and cyclones, depend on particle shape, density and size; 2) Electrical mobility techniques, obtained by electrostatic mobility analyzers, depend on particle shape and size, and are obtained by the rate of migration of charged particles in an electrostatic field; and 3) Optical techniques, obtained by light-scattering detectors, depend on particle refractive index, shape, and size. Table B.3b summarizes the methods discussed here.

Recent developments in instruments for obtaining continuous or semi-continuous measurements of particle-size sampling show promise. Chief among these is the scanning mobility particle sizer (SMPS), which measures the size distribution of particles that range from 0.003 to 0.05 μm or 0.01 to 0.5 μm, depending on its configuration. The SMPS provides adequate time response (<5minutes) for ambient sampling and resolves particle sizes into a user-selected number of fractions, or bins. Collocating

Table B.3. Summary of real-time particle monitoring techniques.
a. Mass [a] (c.f., McMurry, 2000).

Observables	Methodology	Specifications	Availability	Comments
$PM_{2.5}$ or PM_{10} mass by Tapered element Oscillating Microbalance (TEOM)	Particles are continuously collected on a filter mounted on the tip of a hollow glass element that oscillates in an applied electric field. The resonant frequency of the element decreases as mass accumulates on the filter, directly measuring inertial mass. Temperatures are maintained at a constant value, typically 30 C or 50 C, to minimize thermal expansion of the tapered element.	Particle mass. Detection limit ~ 5 μg/m^3 for a 5-minute average.	Commercial	The typical signal averaging period is 10 minutes. Volatile species can evaporate due to the heated airflow. This results in low mass concentrations compared to gravimetric methods. A differential TEOM system that acquires PM mass at ambient temperatures is commercially available. A new method, the Filter Dynamics Measurement System, also is commercially available and appears to provide an even better estimate of PM mass by accounting for both volatilization and absorption.
$PM_{2.5}$ or PM_{10} mass by Real-Time Total Ambient Mass Sampler (RAMS)	Measures mass concentration (including volatilized species) using a combination of TEOM principles with diffusion denuder technology. The sampling system consists of two TEA-coated annular denuders, a Nafion dryer, a BOSS carbon denuder, a third TEA-coated denuder, and a final Nafion dryer.	Particle mass. Detection limit ~ 5 μg/m^3 for a 1-hour average.	Research	
$PM_{2.5}$ or PM_{10} mass by Continuous Ambient Mass Monitor System (CAMMS)	Measures the pressure drop across a porous (Fluoropore) membrane filter. For properly chosen conditions, the pressure drop is linearly correlated to the particle mass deposited on the filter.	Particle mass. Detection limit ~ 2 μg/m^3 for a 1-hour average.	Commercial	
$PM_{2.5}$ or PM_{10} mass by Beta Attenuation Monitor (BAM)	Beta rays (electrons with energies in the 0.01 to 0.1 MeV range) are attenuated according to an approximate exponential (Beer's Law) function of particulate mass, when they pass through deposits on a filter tape. Filter tape samplers measure the attenuation through unexposed and exposed segments of tape. The blank-corrected attenuation readings are converted to mass concentrations.	Particle mass. Detection limit ~ 2 μg/m^3 for a 1-hour average.	Commercial	Averaging times can be as short as 30 minutes. Encounters interference under high relative humidity.
$PM_{2.5}$ or PM_{10} mass by Piezoelectric Microbalance	Particles are deposited by inertial impaction or electrostatic precipitation onto the surface of a piezoelectric quartz crystal disk. The natural resonant frequency of the crystal decreases as particle mass accumulates. The clean reference crystal, used for comparison, allows for temperature compensation.	Particle mass. Detection limit ~ 10 μg/m^3 for a 1-minute average.	Commercial	

[a] Continuous mass methods have not been approved as Federal Equivalent Methods (FEM), and therefore, have not been used in compliance monitoring.

Table B.3. Summary of real-time particle monitoring techniques, continued.
b. Number density and size distribution (c.f., McMurry, 2000).

Observables	Methodology	Specifications	Availability	Comments
Condensation Particle Counter (CPC)	Total particle number determined by exposure to high super-saturation of a working fluid such as alcohol. Droplets are subsequently nucleated and detected by light scattering.	Total particle number	Commercial.	Particle size cut off at 3-20 nm depending on model. Multiple models can be operated in parallel to give course size resolution in the Aitken mode.
Particle number concentration (0.003 to 0.5 μm) by Scanning Mobility Particle Sizer (SMPS) that consists of Differential Mobility Analyzer and (usually) a CPC.	Particles are charged and selected based on their electrical mobility. Particle number measured with a CPC, but possibly with an aerosol electrometer.	Particle number by size in user selected number of size fractions.	Commercial.	Extremely accurate technique in constant aerosol conditions. More expensive commercial unit reports equivalent electric mobility size.
Particle number concentration (0.3 to 10 μm) by Aerodynamic Particle Sizer.	Parallel laser beams measure the velocity lag of particles suspended in accelerating airflow.	Number of particles in different size ranges.	Commercial.	Reports aerodynamic diameter. Suffers inaccuracies at high concentrations.
Fine (0.1 to 2.5 μm) and coarse (2.5 to 10 μm) particle number concentration by Optical Particle Counter/Size Spectrometer.	Light scattered by individual particles traversing a light beam is detected at various angles; these signals are interpreted in terms of particle size via calibrations.	Number of particles in the 0.1 to 50 μm size range.	Commercial.	Reports optical diameter. Manufacturer specifies number of size ranges and size fractions for each instrument. Sensitive to refractive index (i.e., composition and shape)
Electrical Low Pressure Impactor (ELPI).	Particles are charged and separated by their inertia onto impactor stages. The electrical charge that accumulates on each impactor stage is measured. Particle sizes are obtained from 0.03 to 10 μm in 13 size classes.	0.03 to 10 μm in 13 size classes.	Commercial.	Measures aerodynamic size with relatively coarse size resolution. Suffers inaccuracies at low particle concentrations.

Table B.3. Summary of real-time particle monitoring techniques, continued.
c. Chemical composition (Watson et al., 1998).

Observables	Methodology	Specifications	Availability	Comments
OC, EC, and TC by Ambient Particulate Carbon Monitor (APCM)	Measurement of particulate carbon by automatic thermal evolution of CO_2 at 340 °C (adjustable) for OC and 750 °C for TC. The carbon collected on a ceramic impactor plate oxidizes at elevated temperatures after sample collection is complete. A CO_2 monitor measures the amount of carbon released as result of sample oxidation.	Detection limit, ~ 0.3 $\mu g/m^3$ for a 30-minute average.	Commercial	
OC and EC by In-Situ Thermal/Optical Carbon Analyzer	This sampler provides on-line thermal/optical analysis of exposed quartz-fiber filters; OC and EC are acquired by thermal evolution in He and He/O_2 atmospheres, respectively. Light transmission through the filter is used to correct for charring (pyrolysis) of OC during analysis.	Detection limit, ~ 0.2 $\mu g/m^3$ for a 1-hour average.	Commercial	The analysis principles follow thermal/optical analysis. Prototype was tested at the St. Louis Supersites project.
Total carbon (TC) by Total Carbon Analyzer	TC is measured using flash volatilization of particles collected by impaction onto a filament followed by non-dispersive infrared (NDIR) detection of CO_2.	Detection limit, 1 $\mu g/m^3$ for a 10-minute average.	Commercial	Prototype was tested during summer 2000 at the Houston Supersites project.
Total carbon (TC) by Particles in Liquid	TC is measured by collection of particles into purified water and analyzed in aqueous systems.	Detection limit, ~ 0.5 $\mu g/m^3$ for 10-second to 5-minute averages.	Research	This system shows potential for short-time-resolution carbon measurements, but is restricted to the analysis of soluble organic components.
Particulate NO_3^- by Automated Particle NO_3^- Monitor	Particles are humidified prior to collection by impaction followed by flash vaporization and detection of the evolved gases in a chemiluminescence NO_x analyzer.	Detection limit, ~ 0.5 $\mu g/m^3$ for a 12-minute average.	Commercial	
Particulate NO_3^-, $SO_4^=$, NH_4^+, and gases (HCl, HNO_2, HNO_3, SO_2, NH_3) by Continuous Gas and Particle Speciation Monitor	Samples acquired in collection vessels or continuously, followed by on-line ion chromatographic analysis with conductivity detector. This system uses wet denuders to collect gases into a liquid buffer solution for gas measurement and a chamber to grow particles for cyclone collection in a liquid effluent for ion measurements.	Detection limit, ~ 0.5 $\mu g/m^3$.	Commercial	Several IC-based units, that analyze liquid obtained from denuders, provide concentrations of a variety of acidic and basic gases and aerosol components.
Particulate $SO_4^=$ by Automated Particulate $SO_4^=$ Monitor	Particles are humidified prior to collection by impaction followed by flash vaporization and detection of the evolved gases in an ultraviolet absorption of SO_2.	Detection limit, 0.5 $\mu g/m^3$ for a 12-minute average.	Commercial	

Table B.3. Summary of real-time particle monitoring techniques, continued.
c. Chemical composition (Watson et al., 1998), continued.

Observables	Methodology	Specifications	Availability	Comments
Particulate $SO_4^=$ by Continuous SO_2 Analyzer	Thermally convert $SO_4^=$ to SO_2 at high temperature (1,000 C) using a high-efficiency (>95%) stainless steel converter followed by high-sensitivity pulsed fluorescence analysis of SO_2.	Detection limit, ~ 0.5 μg/m^3 for a 1-hour average.	Research	Uses a diffusion denuder upstream of the converter to remove gaseous sulfur compounds. Options to add NH_3-coated denuder tube to remove H_2SO_4. Prototype was tested at the St. Louis Supersites project.
Particulate sulfur by Flame Photometric Detection (FPD)	Sulfur species are combusted in a hydrogen flame, creating excited sulfur dimers (S_2*). Fluorescence emission near 400 nm is detected by a photomultiplier. The photomultiplier current is related to the concentration of sulfur in all species.	Detection limit, 0.1 μg/m^3 for a 1-hour average.	Research	Four out of five FPD systems agreed to within ± 5% in a 1-week ambient sampling intercomparison.
$PM_{2.5}$ and PM_{10} elements by Streaker	Particles are continuously collected on two impaction stages and a Nuclepore polycarbonate-membrane after-filter followed by particle-induced x-ray emission (PIXE) analysis for multiple elements.		Commercial	
$PM_{2.5}$ elements by Semi-continuous Elements in Aerosol System (SEAS)/ High Frequency Aerosol Slurry Sampler (HFASS)	Particles containing Al, Ca, Fe, Cu, Cr, Mn, Zn, Cd, As, Sb, Pb, Ni, V, and Se are collected in an aqueous slurry after condensing with water vapor.	Detection limit, potentially 0.1 ng/m^3 (see text) depending on element, for 30- to 60-minute averages.	Research	System was used at the Baltimore Supersite, St. Louis Supersite, and Pittsburgh Supersites project.

an SMPS with optical particle counters and/or aerodynamic particle sizers allows the acquisition of particle-size distributions over a complete range from 0.003 to >10 μm (c.f. Woo et al., 2001). Among other recently developed techniques, the electrical low-pressure impactor (ELPI) produces particle-size distributions with very fast time response (<5 seconds) over a wide size range (0.007 to 10 μm), but with relatively crude particle-size resolution. Techniques for determining the size distributions of ultrafine particles using condensation of a vapor and light scattering have been used (Saros et al., 1996), but may be more suitable for specific process studies than for routine monitoring.

Bulk Chemical Composition Methods

Many advances in PM species-specific continuous and semi-continuous measurement methods have occurred recently (c.f., Weber et al., 2003). The following section reviews some of the current methods and further summarizes them in Table B.3c.

Black Carbon (BC) and Organic Carbon (OC)

The carbonaceous component of atmospheric particles consists of BC, which is often described as elemental carbon (EC) or soot[1], and OC, which is composed of hundreds and even thousands of individual organic compounds. Both of these terms are operationally defined (Chow et al., 2001). No real-time technique is available that can adequately speciate OC. Hence, measurements usually are restricted to the major carbon fractions and are labeled accordingly, OC and BC. Typically, OC and BC (or EC) measurements involve collecting particles on a substrate (e.g., a filter or impaction plate) for a period of time sufficient to achieve the desired sensitivity. This is typically several minutes to an hour depending on the concentration of carbon-containing compounds in the collected particles. The substrate or plate is then heated in a controlled atmosphere to first volatilize the OC and then further heated to burn off the EC. The OC and EC evolve to CO_2 gas, which can be measured directly by a non-disperse infrared detector (NDIR), or converted to CH_4 and measured by flame ionization detection (FID). Several methods use absorption or transmittance to correct for pyrolysis of OC that would otherwise be reported as EC.

The R&P 5400 Ambient Carbon Particulate Monitor (Rupprecht & Patashnick, Albany, NY) collects particles by impaction on a ceramic plate and measures OC and EC by determination of the CO_2 that is evolved upon heating to 350 °C (adjustable) and 750 °C (Rupprecht et al., 1995). Temperature programmable research-grade instruments for real-time OC and BC/EC measurements have been developed that use transmittance (rather than reflectance) of a 638 nm laser to adjust for OC pyrolysis (c.f., Birch and Cary, 1996). Stolzenburg and Hering (2000) acquire TC through flash vaporization of particles collected by impaction onto a filament followed by NDIR detection of CO_2. Total water-soluble OC also is measured by collecting particles into purified water (Khlystov et al., 1995; Weber et al., 2001) and analyzed in aqueous systems (e.g., Sievers Instruments Inc, Boulder CO, OI Corp. Houston, TX). Although these techniques are under development, they have the potential for time resolution on the order of minutes.

The PAS2000 real-time PAH monitor (EcoChem Analytics, League City, TX) measures particle-bound PAH based on the ionization of particles and their detection with an electrometer. The photoionization method for particle-bound PAH is reproducible, but at present it can only be related to absolute concentrations of particle-bound PAH via collocated filter samples(Watson and Chow, 2002a; 2002b). A method for real-time determination of BC content of single particles using laser-induced incandescence is also under development and shows promise.

Ionic Component of Aerosol Particles

Ion chromatography (IC) permits speciation and quantitative measurement for a range of ions, including cations (e.g. NH_4^+, Na^+, K^+), inorganic anions (e.g., $SO_4^=$, NO_3^-,Cl^-), and organic acids (e.g., formate, acetate, and oxalate) (Chow and Watson, 1999). Efforts have been made to collect ambient particles directly into purified water for on-line IC analysis. In all cases, diffusion denuders need to be placed upstream of the instrument to remove potentially interfering gases such as SO_2, NH_3, and HNO_3.

Recent research has involved the measurement of NO_3^- by using a diffusion denuder to remove the gaseous nitrogen-containing compounds, heating the sample, and converting the evolved gaseous NO_3^- to NO, which is measured by chemiluminescence. Similar techniques are being developed for the measurement of NH_4^+. These techniques have high sampling rates (1 Hz) but generally result in lower sensitivities. Continued developmental work is required for reliable real-time inorganic NO_3^- measurements.

Variations on real-time $SO_4^=$ detectors, based on the use of a flame-photometric detector (FPD), were developed originally for measuring gaseous sulfur compounds. It was found that $SO_4^=$ could be measured directly with this detector if diffusion denuders were used to remove SO_2 prior to detection. Work is under way (George Allen, personal communication) to refine continuous $SO_4^=$ measurements. This approach is based on thermally converting $SO_4^=$ to SO_2 at high temperature (1,000° C) using a stainless steel converter oven with high efficiency (>95 percent) followed by high-sensitivity pulsed-fluorescence SO_2 analysis. A diffusion denuder is used upstream of the converter to remove gas-phase sulfur compounds. If needed, an NH_3-coated permeation tube can be added to convert H_2SO_4 to $(NH_3)_2SO_4$ to prevent any reduced response to acidic $SO_4^=$.

[1] In concordance with the practice in earlier chapters the term BC shall be used here as a general term describing the sooty carbon component of PM, unless distinction by measurement technique (e.g., optical vs. thermal evolution) necessitates specific reference to BC or EC, respectively.

A commercially available continuous NO_3^- and $SO_4^=$ analyzer (Rupprecht & Patashnick, Albany, NY) involves the flash vaporization of particles collected by impaction onto a filament, followed by chemiluminescence detection of nitrogen oxides for NO_3^- by ultraviolet absorption of SO_2 for $SO_4^=$, (Stolzenburg and Hering, 2000) or by NDIR for OC.

Particulate Metals

Many of the techniques that have been developed to identify the metals contained in PM collected on filters are being adapted for continuous time-resolved measurement. Detection methods include XRF and PIXE. In each case, the metals are identified by the X-ray emissions of the elements that are excited in the sample (Cooper, J., Cooper Environmental Services, Portland, OR, personal communication, 2002).

In addition, collection and flash vaporization - followed by detection using atomic-emission spectrometry, emission spectrometry, and mass spectrometry - are being developed to measure metals in aerosols. Also, the same detectors are used to distinguish the soluble metals contained in aerosols sampled into a solvent stream. The application of single-particle analysis using mass spectrometry for detection has great potential to give semi-quantitative, size-resolved real-time measurement of PM metals. On-line elemental analysis, developed by the University of Maryland and Harvard University (semi-continuous elemental analysis system SEAS), provides for semi-continuous analysis of as many as 6 trace metals simultaneously. The PM is collected by water condensation into an aqueous slurry and then analyzed by graphite furnace AAS (Kidwell et al., 2001; Ondov and Kidwell, 2000a, 2000b). The averaging time is 30 to 60 minutes for Al, Ca, Fe, Cu, Cr, Mn, Zn, Cd, As, Sb, Pb, Ni, V, and Se measured with detection limits of ~1 to 5 ng/m^3. An improved version uses the high frequency aerosol slurry sampler (HFASS) with a dynamic aerosol concentrator. The HFASS grows particles by steam condensation to facilitate separation of the particles from the air stream. $PM_{2.5}$ samples are acquired at a high flow rate (200 L/min) and delivered to either a graphite AAS for on-line analysis or to an auto-sequencing sample collector for off-line analysis. This new system provides higher sensitivity, with 5-minute average sample duration and allows for the determination of elements at lower (0.1 ng/m^3) detection limits.

B.2.4. Single-Particle Measurements

Single-particle mass spectrometers (SPMS) allow real-time detection of the chemical composition of individual particles. These recently developed systems detect the basic chemical and physical properties of individual particles and, hence, promise to provide the most information concerning the sources of the particles and the history of atmospheric processes that determine their composition. Of particular value is their ability to provide information on the mixing of different particle components. The salient features these instruments are listed in Table B.4 (Middlebrook et al., 2003).

These instruments decompose individual particles, ionizing their components and providing mass spectra that give elemental identification and some molecular structure information. Most importantly, these in-situ instruments can largely avoid the artifact problems associated with the collection of semivolatile species by filter/denuder methods, such as NH_4NO_3. Some specific analytical achievements made with these techniques include the detection of internal mixing of NO_3^- and $SO_4^=$ in remote continental accumulation mode particles (Murphy and Thomson, 1997), the measurement of the conversion of NaCl to $NaNO_3$ as sea-salt particles traversed the Los Angeles region (Gard et al., 1997), and the measurement of substantial amounts of organic material in upper-tropospheric particles (Murphy et al., 1998).

To date, SPMS methods are qualitative; however, some level of quantification has been obtained using chemical composition obtained simultaneously using a MOUDI and SPMS. This is the most significant limitation to the current generation of laser ionization instruments. Since most instruments use optical and aerodynamic sizing, there also is a deficiency in measurements at size ranges <0.4 μm, although this limit is being extended. However, instruments have been designed to measure the composition of single particles down to 10-20 nm (RSMS; Rhoads et al., 2003).

Finally, another type of mass-spectrometer instrument is becoming available, which evaporates particles and uses a more conventional electron-impact mass spectrometer for analysis (AMS, Jimenez et al., 2003; Tobias and Ziemann, 1999; Jayne et al., 2000). Its capabilities have not been fully evaluated, but it is likely that it will enable quantification of more molecular species than the laser-ionization instruments. These instruments can respond to a single chemical species in single particles but usually must average over many particles if more than one species is being analyzed. They may fill a middle ground between complete analyses of single particles and fully quantitative analyses of bulk samples.

B.2.5. Optical Properties of Aerosols and Long-Path Optical Measurements

Current methods allow for the real-time determination of the optical properties of aerosols at both point locations and along a path. Together these methods allow for total extinction to be measured directly and as individual components (PM and gas absorption and scattering). These measurements are extremely useful for understanding regional haze, and along with chemical measurements obtained using either integrated or continuous methods, allow visibility reduction to be apportioned back to its sources. Table B.5 summarizes the measurements of the optical properties of aerosols.

In-situ Measurements of Light Scattering and Light Absorption

In-situ properties of primary interest are the aerosol light-scattering and absorption coefficients. The particle light-scattering coefficient (b_{sp}) is measured directly with an integrating nephelometer. A number of vendors produce field-ready versions of this instrument suitable for monitoring applications. The most commonly used technique for monitoring the particle light-absorption coefficient (b_{ap}) measures the attenuation of light transmitted through a deposit of particles on a filter, such as the traditional paper-tape measurements of coefficient of haze. The commonly used aethalometer and the particle soot absorption photometer (PSAP) applies optical attenuation to estimate BC.

The methods described above are mostly adequate for routine measurements of the integral optical properties of aerosols related to visibility reduction. In most cases, air must be drawn from a sample inlet to the instrument, which means that care must be

Table B.4. Summary of real-time single particle measurement techniques. (Middlebrook et al., 2003).

Name	Methodology	Aerodynamic size (μm)	Comments
PALMS (particle analysis by laser mass spectrometry)	Laser detection and ionization at λ=193 nm. Detection by single polarity time-of-flight mass spectrometer.	0.35 - 2.5	Semi-quantitative. Real time detection. Laser-based instruments are broadly consistent and show similar trends as a function of size for organic/sulfate and mineral particles.
ATOFMS (Aerosol time-of-flight mass spectrometry)	Laser detection and ionization at λ=266 nm. Detection by dual polarity time-of-flight mass spectrometer.	> 0.2	Semi-quantitative. Real time detection. Laser-based instruments are broadly consistent and show similar trends as a function of size for organic/sulfate and mineral particles.
RSMS (Rapid single-particle mass spectrometer)	Laser detection and ionization at λ=193 nm. Detection by single polarity time-of-flight mass spectrometer.	0.015 - 1.3	Semi-quantitative. Real time detection. Laser-based instruments are broadly consistent and show similar trends as a function of size for organic/sulfate and mineral particles.
AMS (Aerosol mass spectrometer)	Collection on surface followed by vaporization at T ~ 550° C. Gases ionized by electron impact, with Detection by quadrupole mass spectrometer.	0.05 - 2.5	AMS provides size-resolved measurements of volatile aerosol components. Real time detection and analysis. Quantitative measurements are possible. Does not yield single-particle spectra.

taken to avoid particle losses. Changes in relative humidity can affect the measurements, requiring measurement and, in general, control of sample relative humidity. Particles larger than a few μm in diameter are difficult to characterize quantitatively, due to inlet effects and optical truncation. More work is needed to characterize measurement artifacts of the filter-based methods for light absorption, as well as development of methods to determine the RH-dependence of aerosol light absorption.

In that regard, new methods are emerging that can determine scattering, absorption and/or total particulate extinction. Methods such as cavity ringdown spectrometers and direct absorption cells, by virtue of their principle of operation, do not have the negative biases noted for the integrating nephelometer in Table B.5a. In addition, the photoacoustic absorption cell, which is a real-time method for determining the absorption coefficient, has shown promise in laboratory and surface field measurements and is being adapted for airborne studies.

Long-Path Measurement Techniques: Remote Sensing and Visibility

Sun-tracking photometers measure vertical optical depth by observing the transmission loss between the sun and the instrument. Measurements can be made only in daytime, and a cloud-free line-of-sight is essential if particles are to be measured. The effects of all species in the column are inseparably combined, and the molecular contribution must be subtracted to isolate the attenuation by particles.

Spectral sun-photometer measurements can be inverted to obtain information about the average size-distribution of the particles in the column. Interpretation of the size distributions should be limited to general rather than detailed features, because the results are typically quite sensitive to measurement errors and data-retrieval algorithms. Shadow-band radiometers measure both direct and diffuse radiation and provide similar information as a sun photometer, but with less accuracy. However, estimates of the single-scattering albedo of the particles in the column can be made from the diffuse-direct measurements. Automated spectral sun photometers and shadow-band radiometers are available commercially, and are standard instruments in some radiation monitoring networks. Radiative-transfer and climate issues usually motivate use of these instruments, but they also can contribute to pollution research, especially in combination with other instruments.

Telephotometers or teleradiometers have been used to measure the contrast reduction of a distant reference target due to the intervening haze. The results are closely applicable to human perception, but depend on lighting conditions. Scene cameras are even more attuned to visual perception and are less quantitative than telephotometers in terms of quantifying aerosol optical properties. Transmissometers measure one-way transmission loss, and hence the average extinction coefficient, along a path. The contribution of scattering and absorption of light by gas-phase molecules in the light path is also embedded in the signal, and must be subtracted to isolate the particulate portion. Transmissometers designed for long-term operational use are commercially available, e.g., as a standard instrument in the IMPROVE network. Transmissometers often have been used at airports, where the design is optimized for very low visibility (fog and dense haze). Transmissometers can operate continually, but measurement periods of minutes or longer are necessary to average the signal variations caused by atmospheric turbulence.

Long-path measurements are becoming available that can provide spatial resolution. Lidar (light detection and ranging) uses a laser to transmit a pulse of light into the atmosphere. The small portion that is scattered back from the air is detected as a function of time, or equivalently, of distance from the device. The information provided by current instruments is semi-qualitative, indicating the approximate distribution of aerosols in the atmosphere. This information is extremely valuable for locating aerosols and determining the thickness of the layers. Lidar can operate from the surface or from aircraft. Automated, eye-safe lidars are commercially available along with new emerging research prototypes.

Table B.5.a. In-situ measurements. Measurements of aerosol optical properties, c.f., Seigneur et al., 1998.

Method	Specifications	Comments
Integrating Nephelometer	Continuous. Measures optical scattering of sampled aerosols.	Commercially available. Size and refractive index of many aerosol particles change with relative humidity (RH), leading to a pronounced RH-dependence. Open-air nephelometers operate at ambient RH, heated nephelometers operate at a low RH, and humidity-controlled nephelometers allow direct measurement of dependence on RH. Nephelometers have significant negative biases for particles with diameters larger than 1 μm.
Aethalometer	Continuous Converts optical attenuation to BC. Collects on portion of moving tape. Unexposed part of the tape is used for blank	Commercially available. Difference in light attenuation between two parallel segments of tape is related to BC by comparison with thermal evolution methods. BC estimated by this method is operationally defined by the carbon method used for comparison. Method has considerable uncertainty.
Absorption Photometer (PSAP)	Collects particles on filter and convert optical attenuation through filter to BC.	Commercially available. Bond et al. (1999) demonstrated that filter interferences and light scattering by the deposited particles introduce considerable uncertainty. Nevertheless, the simplicity and sensitivity of filter-based methods make them the method of choice for present monitoring applications.

Table B.5.b. Long path measurements. Measurements of aerosol optical properties.

Method	Specifications	Comments
Sun-tracking Photometers	Continuous. Measures the vertical optical depth by observing the transmission loss between the sun and the instrument.	Commercially available. Standard instruments in radiation monitoring networks (e.g., AERONET). Measurements made only in daytime. Cloud-free line-of-sight is required to determine particle scattering. The molecular contribution must be subtracted to isolate the portion from particles. Spectral photometer measurements obtain information about average size distribution of particles in the column.
Shadow-band Radiometer	Continuous. Measures direct and diffuse radiation.	Commercially available. Provides similar information as a sun photometer, but with less accuracy.
Transmissometers	Continuous. Measures the one-way transmission loss, and hence the average extinction coefficient, along a path.	Accuracy is approximately 1 percent in transmission loss. The molecular scatter and absorption must be subtracted to isolate the particulate portion.
Teleradiometers	Continuous. Measures radiances of distant, preferably dark topographical features and the horizon sky, from which the apparent target-sky contrast is calculated.	Commercially available. Measure the contrast reduction of a distant reference target by the intervening haze. Results are closely applicable to human perception, but linkage to light extinction varies with lighting conditions
Lidar	Continuous. Detects as a function of time (distance) backscatter from pulsed laser. Multi-spectral lidar can provide profiles of the size distribution.	Commercially available. Can operate from the surface or from aircraft. Vertical profile of aerosol backscatter coefficients can be derived from calibration. High spectral resolution lidar (HSRL) separates the return from aerosol and molecules to obtain calibrated profiles.
DOAS (Differential Optical Absorption Spectrometer)	Continuous. Detects attenuation of a multi-spectral light source. Can infer information about scattering, absorption and size distribution. (see transmissometers, above)	Can detect light scattering and absorption by particles over a long path between light source and detector of between light source/detector and retro-reflector. Corrections must be made for molecular scattering and absorption.

With assumptions, or with the help of independent information and ancillary measurements, the optical extinction-coefficient profile can be estimated from the calibrated lidar backscatter profile. In addition, recent research has shown the value of data from various lidar systems for inferring profiles of the physical properties of particles, such as $PM_{2.5}$ and even the size distribution. The influence of relative humidity on the distribution, optical properties of aerosols and their behavior with height in the atmosphere can be studied using simultaneous lidar measurements of water vapor and aerosol backscatter or extinction. In addition, lidar is capable of determining the depolarization ratio of backscattered light, a source of information relatively untapped in aerosol studies. The depolarization ratio indicates the morphology of aerosols by specifying the degree to which particles are non-spherical. Finally, coupling lidar measurements with other measurements (e.g., a spectral sun photometer) can greatly enhance the information available. Additional development and evaluation of such methods should yield valuable dividends for future aerosol studies.

Satellite Measurements

Several techniques exist for detecting aerosols from space using passive remote-sensing methods (c.f., Kaufman et al., 1997 and King et al., 1999). Each method has its own advantages and disadvantages. Satellite techniques are at present most suited to define aerosol-distribution patterns, and may be useful in evaluating total aerosol optical depth for comparison with model computations. They can usually provide the only information available in remote areas and can offer global coverage.

Visible and near-infrared techniques also have been tested for detecting dust aerosols (e.g., Ackerman, 1997). The TOMS uses the ultraviolet spectrum and has the ability to distinguish between absorbing and non-absorbing aerosols (Hsu et al., 1996; Herman et al., 1997; Torres et al., 1998). A 20-year record is available globally, over land and ocean. A major advantage of this technique is the availability of aerosol coverage since the surface UV reflectivity is low and nearly constant over land and water. However, the TOMS aerosol product is not sensitive to aerosols below 1 km and the spatial resolution of the instrument is on the order of 40 by 40 km. Multi-angle measurements also can be used to detect aerosols (e.g., Kahn et al., 1997) while polarization techniques can be used to retrieve PM properties.

B.2.6. Gas-Phase Aerosol Precursors, Ozone, Ozone Precursors and Oxidants

The techniques available to measure the ambient atmospheric concentrations of gas-phase PM precursors, ozone, ozone-precursors (including the odd-hydrogen free radicals) and oxidation products of these compounds and their salient features are presented in Tables B.6a through B.6i. The tables are divided as follows: a) the sulfur compounds, SO_2 and H_2SO_4; b) NH_3; c) O_3; d) CO; e) speciated VOC; f) NO, NO_2 and total reactive nitrogen (NO_y); g) other nitrogen oxides (PAN, HNO_3, HONO, NO_3, and total gas-phase reactive nitrogen); h) peroxides (H_2O_2 and ROOH); and i) odd hydrogen species. Parrish and Fehsenfeld (2000) recently reviewed most of these techniques and discussed their capabilities and limitations.

The VOCs are of particular interest in connection with measurement uncertainties. VOC measurements are difficult due to the extreme complexity of the organic-compound mixtures that can be present in the atmosphere. Over 850 different VOCs have been detected in the vapor over gasoline, and over 300 different VOCs from vehicle exhaust have been identified. Natural hydrocarbons emitted by vegetation, which are estimated to account for approximately 50 percent of the VOCs emitted into the atmosphere in North America, are highly reactive olefins. Air samples obviously can contain a very great number of different VOCs of natural and anthropogenic origin, and the oxidation of each of these species creates additional VOCs that are reactive.

The difficulties associated with VOC measurement increase with the complexity of the compound, depending on whether its carbon bonds are saturated and if it is oxygenated. Apel et al., (1999) identified large measurement uncertainties for many of these VOCs. Serious problems have been observed for higher molecular-weight ($\geq C_8$) compounds which have a high probability of participating in particle formation. Clearly, one of the most critical needs of the ambient VOC measurement community is a rigorous field intercomparison of measurement techniques. Such an intercomparison will define measurement capabilities more clearly and identify prevalent problems that must be addressed.

B.2.7. Meteorological Measurements

Surface meteorological measurements are well established and allow for the determination of wind speed (WS), wind direction (WD), temperature (T), relative humidity (RH), barometric pressure (BP), and solar radiation. Many of these parameters are measured nationwide at numerous sites with needed accuracy, under most conditions. Surface methods are not reviewed here but summaries can be found in Seigneur et al. (1998).

Measurements aloft of some of these parameters can be obtained in near-real time up to 10 km, with vertical resolutions of less than 50 m. NOAA has a national network of radar profilers that measure WS and WD aloft. A series of tropospheric profilers in the central United States, make measurements from about 100 m or so up to 10 km, while along the east and west coasts a series of boundary-layer profilers make measurements from about 75 m or so up to about 5 km. Many of these profiler stations are augmented with radio acoustic sounding systems (RASS), which allow simultaneous temperature profiles to be measured.

RASS allows for determination of boundary-layer mixing height if measurements are obtained properly. Sodar also allows measurements of WS and WD aloft. Its measurements are limited to about 500 m; but it has better resolution within that range. Lidar uses backscatter of gases or aerosols and allows for determination of boundary-layer mixing height. It also allows observation of PM and ozone in the vertical direction, either from the surface looking up or from aircraft looking downward. While these determinations are qualitative, they provide considerable information about layering in the atmosphere (Solomon et al, 2000b). These and others relevant methods are described in Table B.7 and additional information can be found in Seigneur et al. (1998).

Table B.6. Summary of real-time gas-phase precursor measurements. **a. Sulfur compounds: sulfur dioxide (SO_2) and sulfuric acid (H_2SO_4)** [a].

Observable	Methodology	Specifications	Availability	Comments
SO_2	Pulsed-fluorescence: A gas sample is excited with radiation at 190- 230 nm and detects fluorescence at 220-400 nm	Continuous DL = 0.1 ppbv MT = 10 sec.	Commercially available	Currently employed in monitoring networks. The inlet usually has method to remove hydrocarbon interferences. Commercial instruments are adequate for most ground-based measurements.
	Differential optical absorption spectroscopy (DOAS)	Continuous DL = 0.5 ppbv MT = 1 min. PL = 1 km	Commercially available	Uses light and wavelength-dependent absorption features to determine the path-integrated concentration of many molecules including SO_2. Increasing the path length can reduce the detection limit.
H_2SO_4	Chemical ionization mass spectrometric (CIMS)	Continuous DL = 10^5 cm^{-3} MT = 10 sec.	Research grade instrument	Basis of the OH CIMS technique.

Note: DL is detection limit; MT is measurement (or integration) time required to achieve the stated detection limit.

[a] Sulfur dioxide (SO_2) is a major precursor to secondary particle formation. Sulfuric acid (H_2SO_4) produced from the oxidation of SO_2 is a very effective nucleating center for fine particles. Because of this role, the measurements of SO_2 are useful additions to most particle measurement sites in order to help determine fine particle sources.

Table B.6. Summary of real-time gas-phase precursor measurements, continued. **b. Ammonia (NH_3)** [a].

Methodology	Specifications	Availability	Comments
Coated filter and denuders, extracted, and analyzed.	Individual samples DL = 10 pptv MT = 2 hr.	Prescription for coating described in literature.	Methods are currently used in networks. Collecting surface coated with citric or oxalic acid. Analysis for NH_4^+ by ion chromatography after extraction. For comparison of techniques see Wiebe et al. (1990), Appel et al (1988). Incorporated in many integrated sampling systems. Methods require laborious post-exposure processing.
Photofragmentation/ two-photon laser-induced fluorescence. (LIF)	DL = 5 pptv MT = 5	Research grade instrument	Ammonia is photolysed by a UV laser and the resulting NH radical detected by LIF Schendel et al. (1990)
Chemical ionization mass spectrometry (CIMS)	Continuous DL = 0.1 ppbv MT = 10 sec.	Research grade instrument	Tested in ground-based comparison (Leibrock and Huey, 2000, Fehsenfeld et al., 2003)

Note: DL is detection limit; MT is measurement (or integration) time required to achieve the stated detection limit.

[a] Ammonia plays a critical role in the formation of particles. Because of this role, the measurements of ammonia are often incorporated in particle measurement systems. The development of new sampling techniques also must be accompanied by improved inlet design to reduce ammonia uptake on sampling lines.

Table B.6. Summary of real-time gas-phase precursor measurements, continued. **c. Ozone (O_3)** [a].

Methodology	Specifications	Availability	Comments
UV absorption at 254 nm	Continuous DL = 1 ppbv MT = 10 sec.	Available commercially	Widely used standard method. Absolute; but VOC interferences can be important at very low ozone concentrations. Calibration procedures should be used only to identify field instruments in need of cleaning and/or repair, never to alter the instrument measurement results. Currently employed in monitoring networks.
Chemiluminescence using NO	Continuous DL= 0.01 ppbv MT = 1 sec.		Not absolute but faster response than UV absorption (c.f., text) making technique especially suitable for aircraft measurements. Reaction vessel must be maintained at sub-ambient (20 torr) pressure. NO is toxic.
Chemiluminescence using ethylene	Continuous DL $\leq$ 1 ppbv MT = 10 sec.	Available commercially	Not absolute. Much less sensitive than NO chemiluminescence method above.
Differential optical absorption spectrometer (DOAS)	Continuous DL = 5 ppbv MT = 1 min. PL = 1 km	Available commercially	Longer path length, longer integration time can be used to improve detection limit.
Lidar	Continuous DL = 5 ppbv MT = 1 sec. Resolution = 0.1 km	Available commercially	Eye-safe technology currently available. Allows determination of ozone over a range of altitudes. Can be scanned from vertical to horizontal. Techniques have been improved sufficiently so that ozone profiling is becoming an increasingly attractive measurement tool. Typical accuracy of ozone lidar is 10 ppbv, but can be considerably better or worse depending on the particular instrument, spatial and temporal averaging, and distance from the lidar.

Note: DL is detection limit; MT is measurement (or integration) time required to achieve the stated detection limit.

[a] In-situ measurement of ozone can be made accurately and routinely with chemiluminescence and commercial UV absorption instruments. Wide-range ozone measurement technology exists to supply special measurement needs (e.g., airborne and long path). Recent developments of note are multi-wavelength lidars using the DIAL method (differential absorption of light), which can measure profiles of ozone (also, H_2O, and occasionally SO_2) from the ground or from aircraft (c.f., Alvarez et al., 1998).

Table B.6. Summary of real-time gas-phase precursor measurements, continued.

d. Carbon monoxide (CO) [a].

Methodology	Specifications	Availability	Comments
Non-dispersive infrared absorption (NDIA)	Continuous DL = 2 ppbv MT = 1 hr.	Available commercially	Widely used in monitoring networks. CO absorbs radiation from a wideband IR source. Selectivity obtained by use of a sample cell. The commercial instrument has proven to be satisfactory for most ground-based measurements but the basic instrument must be modified to improve zero and calibration methods. If this is not done the drift in the zero baseline of the instrument greatly increases the uncertainty in the measurements.
Vacuum UV resonance fluorescence (VUV)	Continuous DL = 1 ppbv MT = 1 sec.	Research instrument	Fast response, suitable for aircraft use.
Differential absorption CO measurement (DACOM)		Research instrument	Instrument relies on the differential absorption of two laser-generated infrared wavelengths by CO within the multi-pass cell of a laser spectrometer. Fast response, suitable for aircraft use.

Note: DL is detection limit; MT is measurement (or integration) time required to achieve the stated detection limit.

[a] There are presently a variety of instruments that are capable of measuring CO. The choice of instrument depends on the accuracy, precision and frequency demanded of the measurements.

Table B.6. Summary of real-time gas-phase precursor measurements, continued. **e. Speciated volatile organic compounds (VOCs)**[1].

Methodology	Specifications	Availability	Comments
Gas chromatograph with flame ionization detection (GC – FID)	Sample/analyze DL = 0.1 ppbvC for one cm^3 sample MT, see comment $C_2 - C_{10}$	Commercially available.	Gas Chromatography (GC) with flame ionization detection (FID). FID detects carbon in VOC and is therefore not compound specific. Identification set by compound passage through GC column. Samples either real time measurements in the field or collected in canisters with subsequent lab analysis. Detection limit depends on sample size. Sampling time set by rate of filling of canister or sample loop. In practice, the number of samples that can be taken is limited by the analysis time in the gas chromatograph. In complex mixture of VOCs interferences become increasing likely.
Gas chromatograph with mass spectrometric detection (GC – MS)	Sample/analyze DL = 100 pptv MT, see comment.	Commercially available.	Gas Chromatography (GC) with mass spectrometer detection (MS). MS is therefore compound specific. Samples either collected in canisters or with subsequent lab analysis or real time measurements at the site. Sampling time set by rate of filling of canister or sample loop. In practice, the number of samples that can be taken is limited by the analysis time in the gas chromatograph.
Proton Transfer mass spectrometry (PTR-MS)	Continuous, DL = 100 pptv MT = 10 sec.	Research grade	Detects compounds with proton affinity greater than H_2O. Sensitivity and specificity are limited for many compounds by the multitude of organic species and fragment ions with similar masses. (Hansel and Wisthaler, 2000)
Differential optical absorption spectrometer (DOAS)	Continuous, DL, see comments, MT = 1 min. PL = 1 km	Commercially available.	Detection limit depends on compound. Ranges from 1 ppbv for benzene to 0.1 ppbv for phenol. Has utility for measurements of aromatic hydrocarbons such as benzene and toluene and polycyclic aromatic compounds.
Chemical ionization mass Spectrometry (CIMS)	Continuous, DL = 100 pptv MT = 1 sec.	Research instrument	Specific ion chemistry allows fast-response sensitive detection of specific VOCs suitable for use on aircraft. Isoprene has been measured and undergone intercomparison studies (Leibrock and Huey 2000)

Note: DL is detection limit; MT is measurement (or integration) time required to achieve the stated detection limit.

[a] The VOCs constitute a very large class of compounds generally defined as volatile at STP. VOC measurements are difficult due to the extreme complexity of the hydrocarbon mixtures that can be present in the atmosphere.

Table B.6. Summary of real-time gas-phase precursor measurements, continued. **f. Nitric oxide (NO), nitrogen dioxide (NO_2), and total reactive nitrogen (NO_y).**

Observable	Methodology	Specification	Availability	Comments
Nitric Oxide (NO)	Chemiluminescence with ozone:	Continuous DL = 1 pptv MT = 1 sec.	Several commercial suppliers	Widely used standard method. Currently employed in monitoring networks. Commercial instruments adequate for most surface measurements. Research grade instruments have pptv detection limits and very fast response.
	Two photon laser-induced fluorescence (LIF)	Continuous DL = 1 pptv MT = 1 sec.	Research instrument	
Nitrogen Dioxide (NO_2)	Photolysis of NO_2 to NO by UV followed by chemi-luminescence detection of NO	Continuous DL = 10 pptv MT = 1 sec.	Commercial version available	Retrofitting existing NO measurements to detect NO_2 may allow NO_2 to be monitored in networks.
	Tunable diode laser absorption (TDLAS)	Continuous DL = 0.1 ppbv MT = 10 sec.	Research instrument	Absorption of infrared laser radiation in a low pressure, multi-pass cell
	Differential optical absorption spectrometer (DOAS)	Continuous DL ≤ 0.1 ppbv RT = 1 min. PL = 1 km	Commercial version available	Longer path length, longer integration time can be used to improve detection limit.
Total reactive Nitrogen (NO_y)	Conversion of NO_y to NO followed by Chemi-luminescence	Continuous DL = 1 pptv MT = 1 sec	Commercial versions of the MoO reduction instrument available.	Measures all reducible species, although the conversion of fine particulate NO_3^- is problematic. Instruments based on molybdenum oxide (MoO) or gold catalyzed reduction.

Note: DL is detection limit; MT is measurement (or integration) time required to achieve the stated detection limit.

[a] The oxides of nitrogen (NO and NO_2) are precursors of particles and ozone, and the total oxidized nitrogen family (NO_y) in an air parcel represents the total emissions of these precursors that remain in the atmosphere.

Table B.6. Summary of real-time gas-phase precursor measurements, continued. **g. Other nitrogen oxides (PAN, Nitric acid, nitrous acid, nitrogen trioxide and total gas-phase reactive nitrogen)** [a].

Observable	Methodology	Specifications	Availability	Comments
Peroxyacetyl nitrate (PAN)	Gas chromatograph with electron capture detection (GC-ECD)	Continuous DL = 10 pptv MT = 10 sec sample each 3 minutes.	Instrumentation readily available	Gas chromatography (GC) with electron capture detection (ECD). Calibration is difficult. Not measured in networks.
Nitric Acid (HNO_3)	Tunable diode laser absorption spectrometer (TDLAS)	Continuous DL = 0.1 ppb MT = 15 min.	Research grade instrument	Absorption of infrared laser radiation in a low pressure, multi-pass cell. (Fehsenfeld, et al., 1998).
	Filter collection, extraction, analysis	Individual filters DL = 10 pptv MT = 2 hr.	Nylon filters available commercially	Widely used standard method. Either nylon or base-impregnated filters used for collection. Laboratory analysis for NO_3^- by ion chromatography. Problems with sampling artifacts described in literature.
	Chemical ionization mass spectrometry: (CIMS)	DL = 10 pptv MT = 1 Hz.	Research grade instrument	Recently developed technique. Sensitive, fast response interference-free. Used in aircraft studies Huey et al. (1998).
Nitrous acid (HONO)	Differential optical absorption spectrometer (DOAS)	Continuous DL = 0.5 ppbv MT = 1 min. PL = 1 km	Commercial version available	Longer path length, longer integration time can be used to improve detection limit.
Nitrogen trioxide (NO_3)	DOAS	Continuous DL = 25 pptv MT = 1 min. PL = 1 km	Commercial version available	Uses light and wavelength-dependent absorption features to determine the path-integrated concentration of NO_3. Longer path length, longer integration time can be used to improve detection limit.
	Cavity ring-down spectrometer (CaRDS)	Continuous DL = 0.5 pptv MT = 5 sec.	Research grade instrument	Sensitive light-absorption method based on the time constant for the exponential decay of light intensity in an optical cavity.
Dinitrogen pentoxide (N_2O_5)	(CaRDS)	Continuous DL = 0.5 pptv MT = 5 sec.	Research grade instrument	Thermal dissociation of N_2O_5 to release NO_3 followed by measurement of the decay of light intensity in an optical cavity that is produced by the light absorption of NO_3.

Note: DL is detection limit, MT is measurement time, and PL is path length.

[a] The atmospheric chemistry involving these species is currently of research interest. PAN and nitric acid sequester NO and NO_2. Nitrous acid and NO_3 are important factors in nighttime chemistry. Reliable methods to calibrate these systems are problematic, owing to the low levels of calibration standards required, the ability to generate these standards in the field, and their stability.

Table B.6. Summary of real-time gas-phase precursor measurements, continued. **h. Peroxides (H_2O_2 and ROOH).** [a]

Observable	Methodology	Specifications	Comments
Hydrogen peroxide (H_2O_2)	Tunable diode laser absorption (TDLAS).	Continuous DL = 0.2 ppbv MT = 3 min.	
	Enzymatic derivatization with fluorometric detection:	Continuous DL = 0.05 ppbv MT = 1 min.	Collects gas phase H_2O_2 into solution, followed by derivatization for sensitive detection.
	Fenton derivatization with fluorometric detection is similar to method above	Continuous DL = 1 ppbv MT = 1 min.	
Multiple peroxides	High-pressure liquid chromatography with fluorometric detection.	Continuous DL = 1 ppbv MT = 1 min	Similar to methods above, except that the peroxides are separated by HPLC.

Note: DL is detection limit; MT is measurement (or integration) time required to achieve the stated detection limit.

[a] The peroxides play an important role the heterogeneous conversion of sulfur oxides to $SO_4^=$. Variation of these gas-phase techniques may be incorporated into particle composition measurements since the peroxides have been suggested as contributing to the adverse health effects of fine-particles. Peroxide measurements are not made in current networks and no commercially available instruments exist. Significant developments have been made in the measurement of peroxides. However, intercomparisons indicate that important problems remain especially with regard to differentiating between organic and inorganic peroxides. Through-the-inlet calibrations with standard additions of H_2O_2 and, other organic peroxides are essential to determine differentiation afforded by various techniques.

Table B.6. Summary of real-time gas-phase precursor measurements, continued. **i. Odd hydrogen species.** [a]

Observable	Methodology	Specifications	Comments
Hydroxyl radical (OH)	Chemical ionization mass spectrometry (CIMS)	Continuous DL = $2 \bullet 10^4$ cm^{-3} MT = 0.5 min	Intercomparisons have indicated that long path absorption (LPA), laser induced fluorescence (LIF-FAGE), and chemical ionization mass spectroscopy (CIMS) appear to be capable of measuring OH at levels observed in the troposphere.
	Differential optical absorption spectrometer (DOAS)	Continuous DL = 10^6 cm^{-3} MT = 1 min	See comment for CIMS.
	Laser induced fluorescence (LIF)	Continuous DL = 10^6 cm^{-3} MT = 1 min	See comment for CIMS.
Hydro Peroxy Radical (HO_2)	Chemical conversion to OH by reaction with NO; OH detected by LIF.	Continuous DL = 10^7 cm^{-3} MT = 1 min	
Peroxy radicals (RO_2)	Radical conversion to NO_2 with amplification	Continuous DL = 10 ppbv MT = 1 min	

Note: DL is detection limit; MT is measurement (or integration) time required to achieve the stated detection limit.

[a] Odd hydrogen measurements are not made in current networks and no commercially available instruments exist. These measurements are important for development and testing of diagnostic and predictive chemical models. Significant headway has been made in the development of methods to measure OH and HO_2. A measurement technique for OH and HO_2 using laser induced fluorescence detection has been developed and provided the first measurements of OH and HO_2 that are produced during the night by NO_3 and O_3 chemistry. In addition, an instrument has been developed to measure the OH loss rate due to reactions with other atmospheric chemicals. This new instrument, the Total OH Loss-rate Measurement (TOHLM), has been successfully deployed (Kovacs and Brune, 2001). Mauldin et al., (1999) developed a fast response instrument that measures OH and H_2SO_4 that has been used in ground-based and aircraft studies. Calibration of measurements remains a significant problem.

Table B.7 Instruments used to measure meteorological parameters over a long-path and/or above the surface.

Instrument	What is measured and how it is measured
Sodar (Acoustic Remote Sensing)	Transmitter broadcasts power at audio frequencies (a few thousand Hertz). Detects reflection from atmospheric turbulence. Continuous, height-resolved (typically, 1-30 meter resolution) measurement from surface to about 500 meters above the surface of the temperature distribution in the atmosphere. Can determine boundary layer height, turbulent mixing properties of the boundary layer, sensible heat flux and surface friction. With Doppler sodar wind speed and direction also can be determined.
Profilers (Radar with radio acoustic sounding system, RASS)	Radar, typically operating 915 Megahertz, senses refractive index variations, primarily humidity, in clear air with vertical resolution of approximately 60 meters. System measures wind speed and direction continuously from near the surface (≥ 100 meters) to several kilometers. RASS generates audio power that allows radar reflectance measurement of temperature profiles. Profilers are commercially available.
Lidar	Reflection of pulsed light beam allows space resolved (typically, 1 meter resolution) detection of particle distribution in the atmosphere in the atmosphere (see Section 4.4.5). Information can be used to infer boundary-layer height and turbulence. This provides valuable information concerning atmospheric layering.
Doppler lidar	Doppler shifted reflection of pulsed light beam, allows detection of particle distribution in the atmosphere. Measures the radial components of the wind, usually in a scanning mode. Can measure the details of boundary-layer mixing processes including boundary-layer height, turbulence, and winds.
Optical crosswind sensors	Observes a distant light source through two optical receivers. Correlates the time delay in the scintillation patterns and determines the wind component normal to the beam. Optical crosswind sensors measure the wind component normal to the path (path-lengths between 100 m and several kilometers between a transmitter and receiver. The instrument uses naturally occurring inhomogeneities in optical refractive index as a tracer for the wind. Provides a path-averaged wind with no mechanically limited threshold for low wind speeds. Major use has been to measure drainage flows channeled in narrow valleys.
In-situ (tethered balloon and kite borne) measurements	Tethered balloons and kites provide platforms for in-situ sensors with the advantage that the measurement devices are reusable. Commercial instruments are used to measure wind, temperature, relative humidity, pressure, and radiation. Sonic anemometers and fast-response hygrometers can also be used to measure the turbulent transfer of heat, moisture, and momentum through the atmospheric surface layer. The primary concern in the application of surface meteorological instruments in networks is adequacy of their location. The interference of tethers with aircraft operation is a clear concern and limitation.
Inert tracers	Inert tracer releases are used to evaluate conceptual and numerical models of atmospheric dispersion. Sulfur hexafluoride (SF_6) is an example of a frequently used short-range tracer. A number of perfluorocarbon tracers are also available for this application.

B.3. MEASUREMENT UNCERTAINTY AND VALIDATION

Numerous studies have been conducted to test the equivalency of methods for PM mass and its physical and chemical properties. The results from these studies are contained in several recent reports and reviews (c.f., Chow, 1995; Hering, et al., 1988; Lawson and Hering, 1990; Ito, et al., 1998; Seigneur et al., 1998; McMurry, 2000; Turpin et al., 2000; Solomon et al.; 2000a; Solomon et al. 2000b, Chameides and McMurry, 2001, Solomon et al., 2001; Watson and Chow, 2001; Solomon, et al., 2003; Middlebrook et al., 2003). Descriptions of other studies that compare and evaluate instruments can be found on the web (See http://www.epa.gov/ttn/amtic/supersites.html. This web site also provides links to the web pages of several other PM projects). Given the proliferation of methods being developed that measure the chemical components and the physical properties of PM, and the important role these measurements are playing in understanding effects of PM on health and within the regulatory arena, it is imperative that benchmark techniques, reference methods, and calibration standards be further developed to better evaluate the accuracy or bias of the methods used to measure fine particles.

B.4. REFERENCES

Ackerman, S.A., 1997. Remote sensing of aerosols using satellite infrared observations. Journal of Geophysical Research 102, 17069-17079.

Alvarez, R.J. II, Senff, C.J., Hardesty, R.M., Parrish, D.D., Luke, W.T., Watson, T.B., Daum, P.H., Gillani, N., 1998. Comparisons of airborne lidar measurements of ozone with airborne in situ measurements during the 1995 Southern Oxidants Study. Journal of Geophysical Research 103, 31,155-31,171

Apel, E.C., Calvert, J.C., et al., 1999. The Non-Methane Hydrocarbon Intercomparison Experiment (NOMHICE): Task 3. Journal of Geophysical Research 104, 26,069-26,086.

Appel, B.R., Tokiwa, Y., Kothny, E.L., Wu, R., Povard, V., 1988. Evaluation of procedures for measuring atmospheric nitric acid and ammonia. Atmospheric Environment 22, 1565-1573.

Babich, P., Davey, M., Allen, G., et al., 2000. Method comparisons for particulate nitrate, elemental carbon, and PM2.5 mass in seven US cities. Journal of the Air and Waste Management Association 50 (7), 1095-1105.

Birch, M.E., Cary, R.A., 1996. Elemental carbon-based method for monitoring occupational exposures to particulate diesel exhaust. Aerosol Science and Technology 25, 221-241.

Bond, T.C., Anderson, T.L., Campbell, D.E., 1999. Calibration and intercomparison of filter-based measurements of visible light absorption by aerosols. Aerosol Science and Technology 30, 582-600.

Chameides, W., McMurry, P., 2001. Report of the Atlanta '99 Supersite Science Team: Proposed Papers to Be Prepared Using the Integrated Data Set Generated from The Summer '99 Field Intensive in Atlanta. Prepared by Georgia Institute of Technology, Earth and Atmospheric Sciences, Atlanta, GA. (http://www-wlc.eas.gatech.edu/supersite/)

Chow, J.C., Watson, J.G., Pritchett, L.C., Pierson, W.R., Frazier, C.A., and Purcell, R.G. (1993). The DRI Thermal/Optical Reflectance carbon analysis system: Description, evaluation and applications in U.S. air quality studies. Atmospheric Environment 27A, 1185-1201.

Chow, J.C., 1995. Critical review: Measurement methods to determine compliance with ambient air quality standards for suspended particles. Journal of the Air and Waste Management Association 45, 320-382.

Chow, J.C., Watson, J.G., 1997. Imperial Valley/Mexicali Cross Border PM_{10} Transport Study. Report No. 4692.1D1. Prepared for U.S. Environmental Protection Agency, Region IX, San Francisco, CA, by Desert Research Institute, Reno, NV.

Chow, J.C., Watson, J.G., DuBois, D.W., Green, M.C., Lowenthal, D.H., Kohl, S.D., Egami, R.T., Gillies, J.A., Rogers, C.F., Frazier, C.A., Cates, W., 1998a. Middle- and neighborhood-scale variations of PM_{10} source contributions in Las Vegas, Nevada. In Proceedings, Health Effects of Particulate Matter in Ambient Air, Vostal, J.J., ed. Air and Waste Management Association, Pittsburgh, PA, pp. 443-460.

Chow, J.C., Watson, J.G., 1998. Guideline on speciated particulate monitoring. Prepared for U.S. EPA, Research Triangle Park, NC, by Desert Research Institute, Reno, NV.

Chow, J.C., Watson, J.G., 1999. Ion chromatography. In Elemental Analysis of Airborne Particles, Landsberger, S., Creatchman, M., eds. Gordon and Breach, Newark, NJ, pp. 97-137.

Chow, J.C., Watson, J.G., Crow, D., Lowenthal, D.H., Merrifield, T., 2001. Comparison of IMPROVE and NIOSH carbon measurements. Aerosol Science and Technology 34, 23-34.

Conny, J.M., Klineinst, D.B., Wight, S.A., Paulsen, J.L., 2003. Optimizing thermal-optical methods for measuring elemental(black) carbon: A response-surface study. Aerosol Science and Technology 37, 702-723.

Countess, R.J., 1990. Interlaboratory analyses of carbonaceous aerosol samples. Aerosol Science and Technology 12, 114-121.

Dutcher, D., Chung, A., Kleeman, M.J., Miller, A.E., Perry, K.D., Cahill, T.A., Chang, D.P.Y., 1999. Instrument intercomparison study, Bakersfield, CA 1998-1999. Report No. 97-536. Prepared for Calif. Air Resources Board, Sacramento, CA, by University of California, Davis, CA.

Eatough, D.J., Eatough, N.L., Obeidi, F., Pang, Y., Modey,W., Long, R., 2001. Continuous determination of $PM_{2.5}$ mass, including semi-volatile species. Aerosol Science and Technology 34, 1-8.

Fehsenfeld, F.C., Huey, L.G., Sueper, D.T., Norton, R.B., Williams, E.J., Eisele, F.L., Mauldin III, R.L., Tanner, D.J., 1998. Ground-based intercomparison of nitric acid measurement techniques. Journal of Geophysical Research 103, 3343-3354.

Fehsenfeld, F.C., Huey, L.G., Leibrock, E., Dissly, R., Williams, E., Ryerson, T.B., Norton, R., Sueper, D.T., and Hartell, B., 2003. Results from an informal intercomparison of ammonia measurement techniques. Journal of Geophysical Research, in press.

Fletcher, R.A., Small, J.A., Scott, J.H.J., 2001. Analysis of Individual Collected Particles. In Aerosol Measurement: Principles, Techniques, and Applications, Baron, P.A., Willeke, K., eds. John Wiley and Sons, New York, NY.

Gard, E.E., Mayer, J.E., Prather, K.A., 1997. Real-time analysis of individual atmospheric particles: Design and performance of a portable ATOFMS. Analytical Chemistry 69, 4083-4091.

Hansel, A., Wisthaler, A., 2000. A method for real-time detection of PAN, PPN, and MPAN in ambient air. Geophysical Research Letters 27, 895-898.

Hering, S.V., 1995. Impactors, cyclones, and other inertial and gravitational collectors. In Air Sampling Instruments - For Evaluation of Atmospheric Contaminants, 8th ed., Cohen, B., Hering, S.V., eds. Am. Conf. of Governmental Industrial Hygienists, Cincinnati, OH, pp. 279-321.

Hering, S.V., et al,1998. The nitric acid shootout: Field comparison of measurement methods. Atmospheric Environment, 22, 1519-1539.

Herman, J.R., Bhartia, P.K., et al. 1997. Global distribution of UV-absorbing aerosols from NIMBUS-7/TOMS data. Journal of Geophysical Research 102, 16911-16922.

Hsu, N.C., Herman, J.R., et al., 1996. Detection of biomass burning smoke from TOMS measurements. Geophysical Research Letters 23, 745-748.

Huey, L.G., Dunlea, E.J., Lovejoy, E.R., Hanson, D.R., Norton, R.B., Fehsenfeld, F.C., Howard, C.J., 1998. Fast time response measurements of HNO_3 in air with a chemical ionization mass spectrometer. Journal of Geophysical Research 103, 3355-3360.

Ito, K., Chasteen, C.C., Chung, H.K., Poruthoor, S.K., Genfa, Z., Dasgupta, P.K., 1998. A continuous monitoring system for strong acidity in aerosols. Analytical Chemistry 70, 2839-2847.

Jayne, J.T., Leard, D.C., et al., 2000. Development of an aerosol mass spectrometer for size and composition analysis of submicron particles. Aerosol Science and Technology 33, 49-70.

Jimenez, J.L., Jayne, J.T., Shi, Q., Kolb, C.E., Worsnop, D.R., Yourshaw, I., Seinfeld, J.H., Flagan, R.C., Zhang, X., Smith, K.A., Morris, J., Davidovits, P., 2003. Ambient aerosol sampling using the Aerodyne Aerosol Mass Spectrometer. JGR – Atmospheres, Special Issue for the Atlanta Supersites Project. Journal of Geophysical Research 108, in press, doi:10.1029/2001JD001213.

Kahn, R., West, R., et al. 1997. Sensitivity of multi angle remote sensing observations to aerosol sphericity. Journal of Geophysical Research 102, 16861-16870.

Kaufman, Y.J., et al., 1997. Passive remote sensing of tropospheric aerosol and atmospheric correction for the aerosol effect. Journal of Geophysical Research 102, 16815-16830.

Khlystov, A., Wyers, G. P., et al., 1995. The steam-jet aerosol collector. Atmospheric Environment 29, 2229-2234.

Kidwell, C.B., Ondov, J.M., 2001. Development and evaluation of a prototype system for collection of sub-hourly ambient aerosol for chemical analysis. Aerosol Science and Technology 35, 596-601.

King, M.D., Kaufman, Y.J., et al., 1999. Remote sensing of tropospheric aerosols from space: Past, present, and future. Bulletin of the American Meteorological Society 80(11), 2229-2260.

Kitto, A.M., Colbeck, I., 1999. Filtration and denuder sampling techniques. In Analytical Chemistry of Aerosols, Spurny, K.R., ed. Lewis Publishers, Boca Raton, FL, pp. 103-132.

Klouda, G.A., Parish, H.J., 2000. Resuspension of urban dust for production of a $PM_{2.5}$ filter standard reference material. In Tropospheric Aerosols: Science and Decisions in an International Community conference, Queretaro, Mexico. NARSTO, Pasco, WA.

Koutrakis, P., Wang, P.Y., Wolfson, J.M., Sioutas, C., 1996. Method and apparatus to measure particulate matter in gas. U.S. Patent No. 5,571,945 filed Nov. 5, 1996. Available: www.patents.ibm.com.

Kovacs, T.A., Brune, W.H., 2001. Total OH Loss Rate Measurement. Journal of Atmospheric Chemistry, 39, 105-122.

Lawson, D.R., Hering, S.V., 1990. The Carbonaceous Species Methods Comparison Study – An Overview. Special Issue, Aerosol Science and Technology 12(1), 1-2.

Leibrock, E., Huey, L.G., 2000. Ion chemistry for the detection of isoprene and other volatile organic compounds in ambient air. Geophysical Research Letters 19, 1763-1766.

Mamane, Y., Willis, R.D., and Conner, T.L., 2001. Evaluation of computer-controlled scanning electron microscopy applied to an ambient urban aerosol sample. Aerosol Science and Technology 34 (1), 97-107. EPA/600/J-01/081.

Mark, D. 1998. Atmospheric Aerosol Sampling. In Atmospheric Particles, Harrison, R.M., van Grieken, R.E., IUPAC Series on Analytical and Physical Chemistry of Environmental Systems, Volume 5. John Wiley and Sons, New York, NY.

Marple, V.A., Olson, B.A., 2001. Interial, gravitational, centrifugal, and thermal collection techniques. In Aerosol Measurement: Principles, Techniques and Applications, Second Edition, Baron, P.A., Willeke, K., eds. Van Nostrand Reinhold, New York, NY.

Mauldin III, R.L., Tanner, D.J., et al., 1999. Measurements of OH during PEM-Tropics A. Journal of Geophysical Research 104: 5817-5827.

McMurry, P.H. 2000. A review of atmospheric aerosol measurements. Atmospheric Environment 34, 1959-1999.

Middlebrook, A.M., et al., 2003. A comparison of particles mass spectrometers during the 1999 Atlanta Supersite experiment, Journal of Geophysical Research, 108, in press.

Murphy, D.M., Thomson, D.S., 1997. Chemical composition of single aerosol particles at Idaho Hill: Negative ion measurements. Journal of Geophysical Research 102: 6353-6368.

Murphy, D.M., Thomson, D.S., et al., 1998. In situ measurements of organics, meteoritic material, mercury, and other elements in aerosols at 5 to 19 kilometers. Science 282, 1664-1669.

Norris, G.A., Birch, E.M., Lewis, C.W., Tolocka, M.P., Solomon, P.A., 2002. Comparison of Particulate Organic And Elemental Carbon Measurements Made With the IMPROVE and NIOSH Method 5040 Protocols. Aerosol Science and Technology, submitted for publication.

Obeidi, F.D., Eatough, D.J., 2000. Continuous measurement of semi-volatile fine particulate mass in Provo, UT. Journal of Aerosol Science, submitted.

Ondov, J.M., Kidwell, C.B., 2000a. Improved system for sub-hourly measurement of elemental constituents of ambient aerosol particles. Presented at Measurement of Toxic and Related Air Pollutants, Research Triangle Park, NC, 12-14 September 2000. Air & Waste Management Assoc., Pittsburgh, PA.

Ondov, J.M., Kidwell, C.B., 2000b. Sub-hourly analysis of elements in ambient aerosol by atomic spectroscopy after dynamic preconcentration. Presented at PM2000: Particulate Matter and Health - The Scientific Basis for Regulatory Decision Making, Charleston, SC, 24-28 January 2000. Air & Waste Management Assoc., Pittsburgh, PA.

Parrish, D.D., Fehsenfeld, F.C., 2000. Methods for gas-phase measurements of ozone, ozone precursors and aerosol precursors. Journal of Geophysical Research 34, 1921-1957.

Patashnick, H., Rupprecht, G., Ambs, J.L., Meyer, M.B., 2001. Development of a reference standard for particulate matter mass in ambient air. Aerosol Science and Technology 34, 42-45.

Rhoads, K.P., Phares, D.J., Wexler, A.S., Johnston, M.V., 2003. Size-resolved ultrafine particle composition analysis, 1, Atlanta, Journal of Geophysical Research 108, in press, doi:10.1029/2001JD001211.

Rupprecht, E.G., Patashnick, H., Beeson, D.E., Green, R.E., Meyer, M.B., 1995. A new automated monitor for the measurement of particulate carbon in the atmosphere. In Proceedings, Particulate Matter: Health and Regulatory Issues, Cooper, J.A., Grant, L.D., eds. Air and Waste Management Association, Pittsburgh, PA, pp. 262-267.

Saros, M.T., Weber, R.J., Marti, J.J., McMurry, P.H., 1996. Ultrafine aerosol measurement using a condensation nucleus counter with pulse height analysis. Aerosol Science and Technology 25, 200-213.

Schauer, J.J., Rogge, W.F., Hildemann, L.M., Mazurek, M.A., Cass, G.R., Simoneit, B.R.T., 1996. Source apportionment of airborne particulate matter using organic compounds as tracers. Atmospheric Environment 30, 3,837-3,855.

Schendel, J.S., Stickel, R.E., van Dijk, C.A., Sandholm, S.T., Davis, D.D., Bradshaw, J.D., 1990. Atmospheric ammonia measurement using a VUV/Photofragmentation laser-induced fluorescence technique. Applied Optics 29, 4924-4937.

Schmid, H.P., Laskus, L., Abraham, H.J., Baltensperger, U., Lavanchy, V.M.H., Bizjak, M., Burba, P., Cachier, H., Crow, D.J., Chow, J.C., Gnauk, T., Even, A., ten Brink, H.M., Giesen, K.P., Hitzenberger, R., Hueglin, C., Maenhaut, W., Pio, C.A., Puttock, J., Putaud, J.P., Toom-Sauntry, D., , Puxbaum, H., 2001. Results of the "Carbon Conference" international aerosol carbon round robin test: Stage 1. Atmospheric Environment, 12, 2111-2121.

Seigneur, C., Pun, B., Prasad, P., Louis, J-F., Solomon, P.A., Koutrakis, P., Emery, C., Morris, R., White, W., Zahniser, M., Worshop, D., Tombach, I., 1998. Guidance for the Performance Evaluation of Three-Dimensional Air Quality Modeling Systems for Particulate Matter and Visibility, Appendix A: Measurement Methods. Prepared for the American Petroleum Institute, Washington, DC by Atmospheric Environmental Research, Inc., San Ramon, CA.

Seigneur, C., Pun, B., Pai, P., Louis, J-F., Solomon, P., Emery, C.E., Morris, R., Zahniser, M., Worsnop, D., Koutrakis, P., White, W., Tombach, I., 1999. Guidance for the Performance Evaluation of Three-Dimensional Air Quality Modeling Systems for Particulate Matter and Visibility. Journal of the Air and Waste Management Association 50, 588-599.

Sioutas, C., Koutrakis, P., Wang, P.Y., Babich, P., Wolfson, J.M., 1999. Experimental investigation of pressure drop with particle loading in Nuclepore filters. Aerosol Science and Technology 30, 71-83.

Solomon, P.A., Mitchell, W., Gemmill, D., Tolocka, M.P., Norris, G., Wiener, R., Eberly, S., Rice, J., Homolya, J., Scheffe, R., Vanderpool, R., Murdoch, R., Natarajan, S., Hardison, E., 2000a. Evaluation Of PM2.5 chemical speciation samplers for use in the U.S. EPA National PM2.5 chemical speciation network. Prepared for Office of Air Quality Planning and Standards, RTP, NC, by the Office of Research and Development, RTP, NC.

Solomon, P.A., Cowling, E., Hidy, G., Furiness, C., 2000b. Comparison of scientific findings from major ozone field studies in North America and Europe. Atmospheric Environment 34, 1885-1920.

Solomon, P.A., Baumann, K., Edgerton, E., Tanner, R., Eatough, D., Modey, W., Marin, H., Savoie, D., Natarajan, S., Meyer, M.B., Norris, G., 2003. Comparison of integrated samplers for mass and composition during the 1999 Atlanta Supersites Project. Journal of Geophysical Research 108, in press.

Solomon, P.A., Landis, M., Norris, G., Tolocka, M.P., 2001. Atmospheric sample analysis and sampling artifacts. In Aerosol Measurement: Principles, Techniques and Applications, Second Edition, Baron, P.A., Willeke, K., eds. Van Nostrand Reinhold, New York, NY.

Stolzenburg, M.R., Hering, S.V., 2000. Method for the automated measurement of fine particle nitrate in the atmosphere. Environmental Science and Technology 34, 907-914.

Tobias H.J., Ziemann P.J. 1999. Compound identification in organic aerosols using temperature-programmed thermal desorption particle beam mass spectrometry. Analytical Chemistry 71 (16), 3428-3435.

Torres, O., Bhartia, P.K., et al., 1998. Derivation of aerosol properties from satellite measurements of backscattered ultraviolet radiation: Theoretical basis. Journal of Geophysical Research 103, 17099-17110.

Turpin, B.J., Saxena, P., Andrews, E., 2000. Measuring and simulating particulate organics in the atmosphere: Problems and prospects. Atmospheric Environment, 34, 2983-3013.

U.S. EPA 1999. Particulate matter ($PM_{2.5}$) speciation guidance document. Third draft. Prepared by U.S. EPA, Monitoring & Quality Assurance Group, Emissions, Monitoring, & Analysis Division, Ofc. of Air Quality Planning & Standards, Research Triangle Park, NC.

U.S. EPA. 2001. Air quality criteria for particulate matter [second external review draft]. Research Triangle Park, NC: Office of Research and Development, National Center for Environmental Assessment-RTP Office; report no. EPA 600/P-99/002a,b. 2v.

Wang, P.Y., 1997. Continuous Aerosol Mass Measurement by Flow Obstruction. Prepared by Harvard University, School of Public Health, Boston, MA.

Watson, J.G., Chow, J.C., Moosmüller, H., Green, M.C., Frank, N.H., Pitchford, M.L., 1998. Guidance for using continuous monitors in $PM_{2.5}$ monitoring networks . Report No. EPA-454/R-98-012. Prepared by U.S. Environmental Protection Agency, Research Triangle Park, NC.

Watson, J.G., Chow, J.C., Frazier, C.A. 1998. X-ray fluorescence analysis of ambient air samples, in Elemental Analysis of Airborne Particles, Vol. 1. Landsberger, S., Creatchman, M., editors; Gordon and Breach Science: Amsterdam, 1999; Chapter 2, pp. 67-96.

Watson, J.G., Chow, J.C., 2001. Ambient air sampling. In Aerosol Measurement: Principles, Techniques and Applications, Second Edition, Baron, P.A., Willeke, K., eds. Van Nostrand, Reinhold, New York, NY, pp. 821-844.

Watson, J.G., Chow, J.C. 2002a. A Wintertime $PM_{2.5}$ Episode at the Fresno, CA, Supersite. Atmospheric Environment 36(3):465-475.

Watson, J.G., Chow, J.C. 2002b. Comaprison & evaluation of in situ & filter carbon measurements at the Fresno supersite. Journal of Geophysical Research, 107(D21): ICC 3-1 to ICC 3-15.

Weber, R.J., Orsini, D., Daun, Y., Lee, Y.N., Klotz, P., Brechtel, F., 2001. A new particle-in-liquid collector for rapid measurements of aerosol chemical composition. Journal of Aerosol Science and Technology 35, 718-727.

Weber, R., Orsini, D., Duan, Y., Baumann, K., Kiang, C. S., Chameides, W., Lee, Y. N., Brechtel, F., Klotz, P., Jongejan, P., ten Brink, H., Slanina, J., Boring, C. B., Genfa, Z., Dasgupta, P., Hering, S., Stolzenburg, M., Dutcher, D. D., Edgerton, E., Hartsell, B., Solomon, P., Tanner, R. 2003. Intercomparison of near real time monitors of $PM_{2.5}$ nitrate and sulfate at the U.S. Environmental Protection Agency Atlanta Supersite. Journal of Geophysical Research, 108, (D7) 10.1029.

Wiebe, H.A., Anlauf, K.G., Tuazon, E.C., Winer, A.M., Biermann, H.W., Appel, B.R., Solomon, P.A., Cass, G.R., Ellestad, T.G., Knapp, K.T., Peake, E., Spicer, C.W., Lawson, D.R., 1990. A comparison of measurements of atmospheric ammonia by filter packs, transition-flow reactors, simple and annular denuders and Fourier transform infrared spectroscopy. Atmospheric Environment, 24A:1019-1028.

Woo, K.S., Chen, D.R., Pui, D.Y.H., McMurry, P.H., 2001. Measurement of Atlanta size distributions: Observation of Ultrafine Particle Events 34, 75-87.

APPENDIX C. MONITORING DATA: AVAILABILITY, LIMITATIONS, AND NETWORK ISSUES

C.1. MONITORING PROGRAMS AND OBJECTIVES

PM data are currently available from a variety of monitoring programs and databases, which include

1. Global Atmosphere Watch (GAW) [global]
2. Interagency Monitoring of Protected Visual Environments (IMPROVE) [United States]
3. National Air Pollution Surveillance (NAPS) network [Canada]
4. Canadian Aerosol and Precipitation Monitoring Network (CAPMoN) [Canada]
5. Guelph Aerosol and Visibility Monitoring Network (GAViM) [Canada]
6. Federal Reference Method (FRM) fine particle monitoring network and U.S. National $PM_{2.5}$ speciation network [United States]
7. National PM Research Monitoring Network [United States]
8. Red Automatica de Monitoreo Atmosferico (RAMA) [Mexico City]
9. Clean Air Status and Trends Network (CASTNet) [United States]
10. U.S. National Park Service data [United States]
11. California Air Resources Board databases [United States]
12. U.S. EPA Aerometric Information Retrieval System (AIRS) databases [United States]
13. Special studies.

Different PM monitoring networks have been designed for different purposes. Any one or more of the following objectives may be of principal importance:

1. Determine PM levels
2. Monitor exposure to some size fraction of PM or to one or more components
3. Monitor visibility impairment
4. Ascertain compliance with regulatory standards
5. Develop estimates of source contributions
6. Evaluate how well control strategies are achieving goals
7. Provide information needed to advance scientific understanding.

In practice, monitoring networks are resource-limited and cost considerations typically limit their ability to address multiple objectives.

GAW sites are operated by individual investigators for scientific research purposes. Efforts are underway to establish an integrated archive of data.

The IMPROVE network was designed to monitor visibility impairment, with additional goals related to improving scientific understanding of the physical and chemical processes affecting visibility. Much of the nonurban data for the United States are from the IMPROVE network. IMPROVE fine-particle measurements are made in U.S. national parks and other locations that are typically representative of regional background PM concentrations. The IMPROVE network collected aerosol samples at ~30 sites since the mid-1980s, and was recently expanded to over 100 locations. Because of its focus on providing data related to visibility in Class I and other areas where visibility is protected, the IMPROVE network is concentrated in the western United States, but future measurements will provide broader geographical coverage.

The Canadian NAPS was designed to provide a large and geographically diverse database for characterizing PM. It presently covers all main urban areas of Canada and includes 15 urban and four rural locations. Sampling has been conducted using dichotomous samplers, and data collection commenced as early as the mid-1980s at some sites. Samples are normally collected over a 24-hour period once every six days. Both the coarse ($PM_{10-2.5}$) and fine ($PM_{2.5}$) fractions are analyzed for mass and over 50 elements.

The CAPMoN is a regional-scale air and precipitation monitoring network with multiple objectives. These objectives are to measure regional-scale spatial and temporal variations and long-term trends in the chemical composition of air and precipitation, to provide data for use in model development and testing, to provide data for process studies, and to provide a set of standard monitors across Canada. Sites are located outside urban areas and away from point or transportation emission sources and agricultural activities. The types of measurements made vary among sites, with, for example, 10 sites currently measuring SO_2 and 18 sites sampling precipitation chemistry. Data records extend back to the late 1970s for some sites. PM samples are collected daily. The size cutoff for particulate samples is not well defined, but is estimated to be about 8 μm (Nejedly et al., 1998).

The GAViM network is intended to improve scientific understanding of the physical and chemical processes affecting visibility in Canada. It has made speciated aerosol and optical measurements at five urban and rural locations across Canada twice per week, beginning in 1994. Measurements are made using IMPROVE-protocol samplers. Particle-size cutoffs are 2.5 and 10 μm.

The new U.S. fine-particulate FRM network is primarily a compliance-monitoring network, and numbers roughly 1000 sites nationwide. These sites are concentrated in urban areas. The U.S. National $PM_{2.5}$ Speciation Trends Network was designed to develop estimates of source contributions, monitor exposure to PM components, and track trends.

The RAMA network consists of 32 stations in Mexico City and has operated since 1986. Types of measurements vary among sites, but typically include surface meteorological parameters, gas-phase species (ozone, CO, SO_2, and NO_x), and PM_{10} mass. A variety of special studies provides additional data for Mexico City over a 40-year period (Raga et al., 2000), but present measurements of PM composition are limited to a one-month period in 1997 (Edgerton et al., 1999).

The U.S. EPA CASTNet network was designed to provide estimates of dry-deposition rates of sulfur and nitrogen compounds, which are combined with precipitation-chemistry measurements to estimate total sulfur and nitrogen loading rates. The standard CASTNet sites do not select a specific PM size, and they collect samples over a one-week time period. CASTNet sites are concentrated in the eastern United States, where IMPROVE coverage is limited. Eight eastern CASTNet sites have been instrumented with samplers operated according to the IMPROVE protocols (U.S. EPA, 2000).

An important ongoing special study is the Southeastern Aerosol Research and Characterization (SEARCH) project (Hansen et al., 2003). It began operating four urban-nonurban site pairs in mid-1998 in the southeastern United States (with urban sites in Atlanta GA, Birmingham AL, Gulfport MS, and Pensacola FL). Primary goals of the SEARCH network include aerosol research, measurement evaluation, characterization of urban-rural contrasts, and improved understanding of atmospheric processes. Both PM and gas-phase species are measured, and PM sampling includes continuous as well as filter-based methods.

C.2 NETWORK DESIGN

Network design begins with specification of network objectives. Contemporary networks emphasize different objectives, leading to a variety of general design specifications (Table C.1). Some of the specific key dimensions of PM monitoring programs include:

1. Size fractions
2. Chemical species
3. Sampling duration
4. Sampling frequency
5. Number and location of monitors
6. Length of monitoring program

7. Instrumentation required to accomplish each of the above accurately and affordably.

Size fractions and species measured, sampling duration and frequency, site location, sampler design, and other variables are typically optimized for the purpose of each network, and these differences add to uncertainty in the use of the data when applied for different purposes. An ongoing need is to provide

Table C.1 General specifications for PM observation and monitoring networks.

Objective	Measurements	Spatial Resolution	Temporal Duration and Resolution	Design Procedure	Comments
What's there	FRM [a] or other prescribed methods; specialized instruments	Variable targets-ground and aloft	Variable; usually short term campaign	Subjective-targets of opportunity	Historical guidance and direction from air chemistry; later, regulatory motivation
Exposure/ Dose	FRM and speciation measures	Mainly urban population oriented	Long term-multiyear	Relies on existing compliance networks or special process studies	Only considered explicitly in a few recent short term studies; e.g., PTEAM (Riverside, CA) and ARIES (Atlanta)
Visibility Variation	Optical properties or surrogate for human optical response	Designated air quality related value for pristine areas [b]	Prescriptive: linked with daytime visual perception.	Subjective with constraints on visual range	Recently driven by need to document light extinction combined with PM mass concentration and composition.
Compliance	FRM or other prescribed methods	Mainly urban/some regional non-urban	Prescriptive: long-term and linked with PM standards	Subjective or prescriptive (e.g. population based) [c]	Driven by regulatory needs for reporting community conditions-mostly urban focus. [d]
Source-Receptor Model	Specialized ground PM/precursor instrumentation supplemented with compliance networks; intermittent measures aloft; coupled with meteorological data.	Urban/regional scales covered by ground network, included nested grid for multiple spatial scale study.	Variable depending on model character; ranges from daily events to annual average considerations.	Subjective but recent objective concepts developed	Design varies with model methods; two regional studies followed semi-objective design; formal objective design approach attempted in concept.
Trends	FRM or other prescribed methods	Mainly urban/some regional non-urban	Long term-multi-year	Subjective; guided in U.S. as part of NAMS	Tied to emissions change; generally relies on compliance observations.
Processes	Specialized PM-precursor	Variable targets/designs range in scales from <1km-100km; aloft	Variable; usually short term campaign	Subjective-process specified	Highly focused; hypothesis generated; attention to secondary PM

[a] U.S. Federal Reference Method for PM.
[b] See, for example, the IMPROVE network.
[c] e.g., U.S. guidelines.
[d] For example, the NAMS, SLAMs networks in the U.S., NAPS in Canada, the RAMA network in Mexico City, or the California-Mexico border study.

complete documentation on network methods and comparability of data across networks.

Monitoring networks fill a variety of needs including the measurement of long-term trends, the measurement of baseline pollutant levels, the assessment of impacts, and the acquisition of data for model validation. The data set required typically varies with its application, and an effort should be made to ensure that existing monitoring networks meet diverse needs with maximum efficiency. Furthermore, North American networks should be harmonized to the extent possible to facilitate the sharing of data among Canada, the United States, and Mexico.

Objectives, measurements, and design criteria for existing national air-quality networks in Canada and the United States, including those that monitor PM, are summarized by Demerjian (2000). The Demerjian (2000) review also identifies the types of data reports produced by each network as well as the additional types of data analyses of interest from the standpoints of science or of air-quality management, delineates data-analysis techniques of use for the analyses of interest, and characterizes data limitations. A key conclusion of the Demerjian (2000) review is that the differences between the types of monitoring information needed for scientific purposes compared with air-quality management become critical for secondary pollutants. In the case of secondary pollutants, understanding the relative effectiveness of emission-control strategies requires analyses of data that are capable of revealing the relations between the production of secondary species and their precursor concentrations.

The U.S. EPA has been an active participant in various attempts to apply network-design methods to deposition-monitoring networks, including the National Atmospheric Deposition Program/National Trends Network (e.g., Seilkop and Finkelstein, 1987; Seilkop, 1987, Haas, 1990; Oehlert, 1993), the Environmental Monitoring and Assessment Program (Bromberg et al., 1989; U.S. EPA, 1989), the CASTNet, and, most recently, the networks included within the National Ambient Air Monitoring Strategy (National Monitoring Strategy Committee, 2001; U.S. EPA, 2002). However, past efforts to establish scientific bases for network configuration have not always yielded results that were implemented or that affected network designs in significant ways.

The new National Air Monitoring Strategy offers the prospect of integrating current, single-pollutant monitoring approaches to better address the management of linked, multi-pollutant air-quality issues (U.S. EPA, 2002). This monitoring strategy is specifically focused on the National Air Monitoring Stations (NAMS), the State and Local Air Monitoring Stations (SLAMS), Photochemical Air Monitoring Stations (PAMS), and IMPROVE. The National Air Monitoring Strategy is an initial step, and continuing efforts are needed to expand the focus of scientific network design and coordination to include air toxics and deposition networks. To date, analyses of ozone data from monitors nationwide have shown how to identify monitors that are effectively measuring the same ambient concentrations within local areas. This information will be used to redistribute monitors to use resources more efficiently. Completion of these types of analyses requires the availability of an existing database. In addition, the National Air Monitoring Strategy outlines a plan for developing a national core network of up to just under one hundred locations where a number of air pollutants would be monitored in a coordinated manner.

In California, the results of ozone and PM field studies and modeling have been used to design enhancements to the routine monitoring network, which could reduce future reliance on major field programs (Sweet et al, 2002). A critical component of this design effort was the use of twenty years of air-quality studies to identify the numbers of monitoring locations and types of measurements needed for providing ongoing data that will characterize mesoscale meteorological features of importance to air quality. Examples of such features include onshore and offshore coastal flows, mixing depths, upslope and downslope flows, and eddies and jets. A network capable of resolving such features could routinely provide the data needed for operating and evaluating three-dimensional Eulerian air-quality models for any set of days. This capability would then permit assessment of pollutant transport, emission-control measures, and other questions of importance for air-quality management for all days of interest, thus addressing continuing concerns about the representativeness of special field-study periods.

The problem of characterizing the spatial field, or the spatial-temporal field, of one or more measurements has in fact attracted a great deal of attention over the years in several areas of study. Much of this attention has focused on network design, specifically on the placement of monitoring sites. Examples from the study of precipitation amount (rainfall) include Rodriguez-Iturbe and Mejia (1974) and Bras and Rodriguez-Iturbe (1976). Such network-design techniques yield the number and configuration of monitoring stations that minimize an objective function of estimation error and cost. A significant body of literature on the design of air-quality networks exists and dates back many years (e.g., Seinfeld, 1972; Noll et al, 1977; Nakamori and Sawaragi, 1984; Liu et al, 1986; Langstaff et al., 1987).

C.3 NETWORK NEEDS

Specific instrumentation requirements are discussed in Chapter 5. Specification of instrumentation should ensure that the key measurements are made with necessary accuracy and time resolution. Recent developments in semi-continuous monitoring for mass and chemical PM components should be pursued so these methods become usable routinely in air monitoring and research programs.

The variety of types of measurement techniques currently in use raises questions about the accuracy and comparability of data obtained from different networks. These questions can be addressed only if adequate funding is devoted to comparisons (for comparisons of U.S. and Canadian data, see, e.g., Brook et al., 1997, Brook and Dann, 1999, Nejedly et al., 1998). Comparison of data from IMPROVE and nearby CASTNet sites indicate that the IMPROVE and CASTNet $SO_4^=$ measurements are comparable, but NO_3^- measurements are not (Malm, 2000). The differences in NO_3^- concentrations stem from the presence of coarse-particle NO_3^- and NO_3^- volatilization in the CASTNet samples. It has also been shown that the dichotomous samplers used in the Canadian NAPS experience losses of particulate NO_3^- (Brook and Dann, 1999). The U.S. EPA Federal Reference Method suffers from loss of NH_4NO_3 and semivolatile organic compounds. Losses can be significant and at times may represent over half the collected mass (Hering and Cass, 2000).

Measurements that are representative of both the mass and composition of particles in the ambient air are difficult to obtain with either filter-based or semi-continuous methods. Much of the fine PM mass is composed of secondary species (those formed in the air or condensed from the gas phase onto existing particles). Such species often have significant vapor pressures, and thus exist in both the gas and particle phases in quasi-equilibrium. Depending on the filter material used, collection of PM by filters is plagued by both positive and negative sampling artifacts that are exacerbated by the need to transport, store, and analyze the filter some time after collection. Since PM composition varies with location, season, and time of day, it is difficult to quantify the sampling artifacts. Denuders and reactive filters can be successfully used to obtain NO_3^- and NH_4^+ with minimal bias. These data can be used to correct the collected mass, but significant uncertainty still exists in the mass value. Organic species are even more difficult to collect and no method to date exists for collecting a relatively bias-free organic sample (see Chapter 5). Recent results from the 1999 Atlanta Supersite Project suggest lower interferences from sampling artifacts are observed with semi-continuous methods than with filter-based integrated methods.

Many networks measure a suite of PM components, including $SO_4^=$, NO_3^-, NH_4^+, OC, BC, and a variety of elements that represent crustal contributions. However, apart from limited special studies, virtually no speciated carbon measurements exist. Measurement of a variety of PM carbon compounds at emission sources and receptor locations holds the potential for substantially improving current knowledge of source contributions to ambient OC and BC concentrations. Similarly, gas-phase measurements are usually not included within most PM networks, yet they are necessary for fully understanding the behavior of semivolatile compounds (e.g., NH_4NO_3 and semivolatile organics). Because many networks make the same types of measurements, it is likely that duplication of effort exists in some specific locales. With appropriate harmonization of operating schedules and methods, more data could be shared across networks, freeing resources for measurements that are not now made by any network. *It is unlikely that techniques will be available in the near future to provide complete chemical characterizations of carbonaceous aerosols with the frequency that would be required for air-quality studies, so improved techniques for OC/BC measurements are needed for the near term. There is a need to learn more about the properties of organic compounds found in the gas and particulate phases. Moreover, an understanding of the behavior, lifetime, and effects of organic PM will require a better understanding of its tendency to absorb water.*

Continuing interaction with the health-effects and visibility research communities is needed to ensure appropriate specification of PM size fractions and species of interest. In the case of health-effects research, this interaction needs to be ongoing and iterative. Health-effects studies require monitoring information to conduct future epidemiological analyses, the conclusions from which may more fully identify the measurements of most importance. In contrast, the measurements needed for monitoring visibility impairment are generally well known as a result of past visibility studies.

C.4 REFERENCES

Bras, R.L., Rodriguez-Iturbe, I., 1976. Network design for the estimation of areal mean of rainfall events. Water Resources Research 12, 1185-1196.

Bromberg, S., Edgerton, E., Gibson, J., .Holland, D., 1989. Air/Deposition Monitoring for the Environmental Monitoring and Assessment Program (EMAP). Research Triangle Park: U.S. EPA, Atmospheric Research and Exposure Assessment Laboratory.

Brook, J.R., Dann, T.F., 1999. Contribution of nitrate and carbonaceous species to PM2.5 observed in Canadian cities. Journal of the Air and Waste Management Association 49, 193-199.

Demerjian, K.L., 2000. A review of national monitoring networks in North America. Atmospheric Environment 34, 1861-1884.

Edgerton, S.A., Bian, X., Doran, J.C., Fast, J.D., Hubbe, J.M., Malone, E.L., Shaw, W.J., Whiteman, C.D., Zhong, S., Arriaga, J.L., Ortiz, E., Ruiz, M., Sosa, G., Vega, E., Limon, T., Guzman, F., Archuleta, J., Bossert, J.E., Elliot, S.M., Lee, J.T., McNair, L.A., Chow, J.C., Watson, J.G., Coulter, R.L., Doskey, P.V., Gaffney, J.S., Marley, N.A., Neff, W., Petty, E., 1999. Particulate air pollution in Mexico City: a collaborative research project. Journal of the Air and Waste Management Association 49, 1221-1229.

Haas, T.C., 1990. Kriging and automated variogram modeling within a moving window. Atmospheric Environment 24A, 1759-1769.

Hansen, D.A., Edgerton, E.S., Hartsell, B.E., Jansen, J.J., Hidy, G.M., Kandasamy, K., 2003. The Southeastern Aerosol Research and Characterization Study (SEARCH): 1. Overview. Journal of the Air and Waste Management Association, In press. http://www.atmospheric-research.com/searchhome.htm

Hering, S., Cass, G., 2000. The magnitude of bias in the measurement of $PM_{2.5}$ arising from volatilization of particulate nitrate from Teflon filters. Journal of the Air and Waste Management Association 49, 725-733.

Hidy, G.M., 1994. Atmospheric Sulfur and Nitrogen Oxides: Eastern North America Source-Receptor Relationships. Academic Press, San Diego, CA.

Langstaff, J.E., Seigneur, C., Liu, M.K., Behar, J.V., McElroy, J.L., 1987. Design of an optimum air monitoring network for exposure assessments. Atmospheric Environment 21, 1393-1410.

Liu, M.K., Arvin, J., Pollack, R.I., Behar, J.V., McElroy, J.L., 1986. Methodology for designing air quality monitoring networks. I. Theoretical aspects. Environ. Monitoring and Assessment 6: 1-11.

Malm, W.C., 2000. Spatial and seasonal patterns and temporal variability of haze and its constituents in the United States. National Park Service, Washington, DC.

Nakamori, Y., Sawaragi, Y., 1984. Interactive design

of urban level air quality monitoring network. Atmospheric Environment 18, 793-799.

National Monitoring Strategy Committee, 2001. National Ambient Air Monitoring Strategy. U.S. Environmental Protection Agency, Research Triangle Park, NC.

Nejedly, Z., Campbell, J.L., Teesdale, W.J., Dlouhy, J.F., Dann, T.F., Hoff, R.M., Brook, J.R., Wiebe, H.A., 1998. Inter-laboratory comparison of air particulate monitoring data. Journal of the Air and Waste Management Association 48, 386-397.

Noll, K.E., Miler, T.L., Norco, J.E., Raufer, R.L., 1977. An objective air monitoring site selection methodology for large point sources. Atmospheric Environment 11, 1051-1059.

Oehlert, G.W., 1993. Regional trends in wet sulfate deposition. Journal of the American Statistical Association 88, 390-399.

Raga, G.B., Baumgardner, D., Castro, T., Martinez-Arroyo, A., Navarro-Gonzalez, R., 2000. Mexico City air quality: a qualitative review of gas and aerosol measurements (1960-2000).

Rodriguez-Iturbe, I., Mejia, J.M., 1974. The design of rainfall networks. Water Resources Research 10, 713.

Seilkop, S., 1987. Evaluation of the National Trends Network's Site Placement Design. Atmospheric Research and Exposure Assessment Laboratory, U.S. EPA, Research Triangle Park, NC.

Seilkop, S.K., Finkelstein, P.L., 1987. Acid precipitation patterns and trends in eastern North America, 1980-84. Journal of Climate and Applied Meteorology 26, 980-994.

Seinfeld, J.H., 1972. Optimal location of pollutant monitoring stations in an airshed. Atmospheric Environment 6, 847-858.

Sweet, J.W., Shipp, E., Tanrikulu, S., DeMandel, R., Ziman, S., 2002. Conceptual design of an enhanced multipurpose aerometric monitoring network in central California. Air Waste Management Assoc. Symposium on Air Quality Measurement Methods and Technology – 2002. San Francisco CA, November 13-15, 2002.

U.S. EPA. 1989. Design Report for the Environmental Monitoring and Assessment Program. U.S. EPA, Atmospheric Research and Exposure Assessment Laboratory, Research Triangle Park, NC. .

U.S. EPA. 2002. National Ambient Air Monitoring Strategy: Summary Document.: U.S. EPA., Office of Air Quality Planning and Standards. Research Triangle Park, NC. www.epa.gov/ttn/amtic/stratdoc.html and www.epa.gov/ttn/amtic/monitor.html .

APPENDIX D. GLOBAL AEROSOL TRANSPORT

This appendix* describes some of the key features of the global aerosol pattern. The most prominent aerosol plume is seen over the Atlantic, originating from West Africa and crossing the tropical Atlantic (Figure D.1). It clearly reaches the Caribbean Sea and Amazon delta, 6,000 km away. Measurements indicate that the plume is dominated by crustal material (soil or dust). The dust plume is most intense during the warm season (March-August) as shown in Figure D.1. In the winter, the West African dust plume is shifted south toward the Gulf of Guinea by the prevailing winds. Airborne sampling of cross-sections of the Atlantic aerosol plumes has provided clear evidence that a significant amount of aerosol is transported above and well separated from the boundary layer.

Particulate levels over Asia show extreme variations between the pristine clean air over the Tibet Plateau and the hazy low-lying valleys of the Indian subcontinent, Indochina and China, and the dusty regions of the Arabian Peninsula. The specific sources of the aerosol on the fringes of India, Indochina, and China are not known, but the hazy region is adjacent to the highest regional population density in the world. Continental Eastern Asia is known for its springtime dust sources, sulfur and other emissions from industrial sources, and for significant biomass burning.

Deforestation and agricultural burning produce thick layers of haze in the interior of South America. The PM levels are most pronounced over western Brazil and Bolivia, during the dry August, September, and October season when the biomass burning is most significant. Other aerosols of continental origin are evident off the coasts of Central America, from Mexico to Venezuela. They are most significant during the spring and summer and virtually disappear during fall and winter. These areas are also known for strong seasonal biomass burning.

A diffuse aerosol plume extends across the North Atlantic from eastern North America during the spring and summer seasons (Figure D.1). Measurements indicate the presence of small particles associated with the known $SO_4^{=}$-OC haze from urban industrial sources (e.g., Banic et al., 1996; Daum et al., 1996). Compared to other continents, North America and Europe show relatively low levels of aerosol optical thickness near their shores. Satellite imagery of the North African plume indicates transport reaching Central America and the U.S. Gulf Coast states, with the time of maximum impact occurring in July (Figure D.2). Measurements at surface sites along the transport path (e.g., Canary Islands, Bermuda) reveal that the dominant component of the North African PM plume is crustal material, also known as soil or dust. Surface-monitoring data from the IMPROVE network also show seasonal maximum dust concentrations in July (Figure D.3) and the plume is evident in daily PM_{10} concentrations, whose 90th percentile values in the southeastern United States occur during July.

The transported component may be more significant to the daily-average concentrations during summer episodes, possibly representing about 5 to 10 $\mu g/m^3$ on some days and up to 20 $\mu g/m^3$ at some locations (Figure D.3). For example, analyses by the Texas Natural Resource Conservation Commission indicate that the North African plume contributed as much as 15 to 20 $\mu g/m^3$ at sites in the Houston area on two days in June and August, 1997 (Price et al., 1998).

The transport of dust from Asia to North America also has been documented (e.g., Jaffe et al., 1999). For example, during April 1998, satellite data provide evidence of an aerosol plume reaching into North America (Figure D.4). Over the Pacific Ocean, the dust cloud followed the path of the springtime East-Asian aerosol plume shown by the optical thickness data. The $PM_{2.5}$ dust concentration data from the IMPROVE speciated aerosol network showed virtually no dust on April 25th, suggesting that the dust cloud had not reached the surface (Figure D.5). Higher $PM_{2.5}$ values then occurred over the West Coast on April 29 and further inland on May 2 (Figure D.5). Presently available data suggest that this event had an unusually strong effect on surface PM

* Material in this Appendix was contributed by R. Husar.

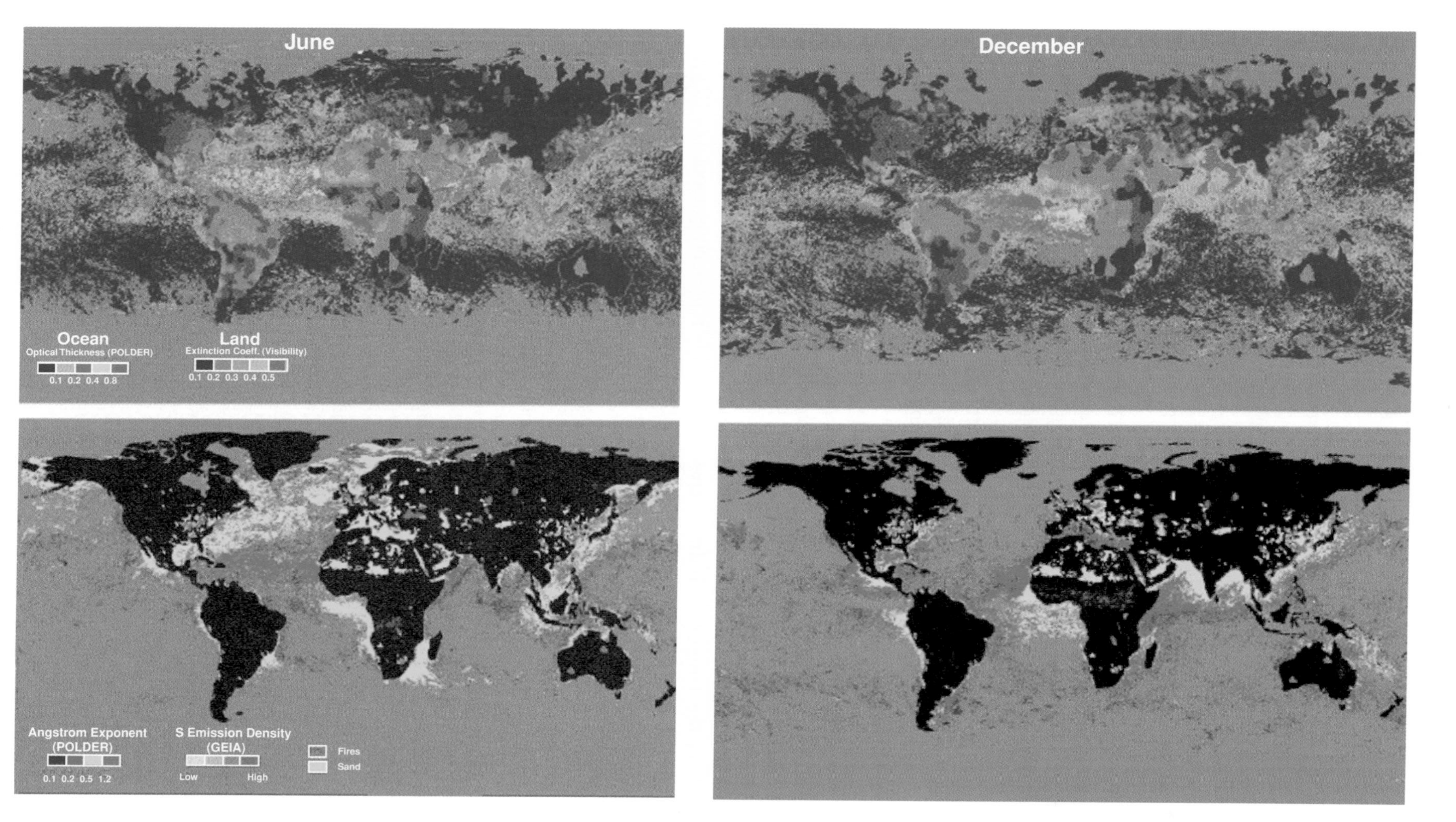

Figure D.1. Global distribution of aerosols and their likely source regions during June and December. (Source: R. Husar, pers. comm.).

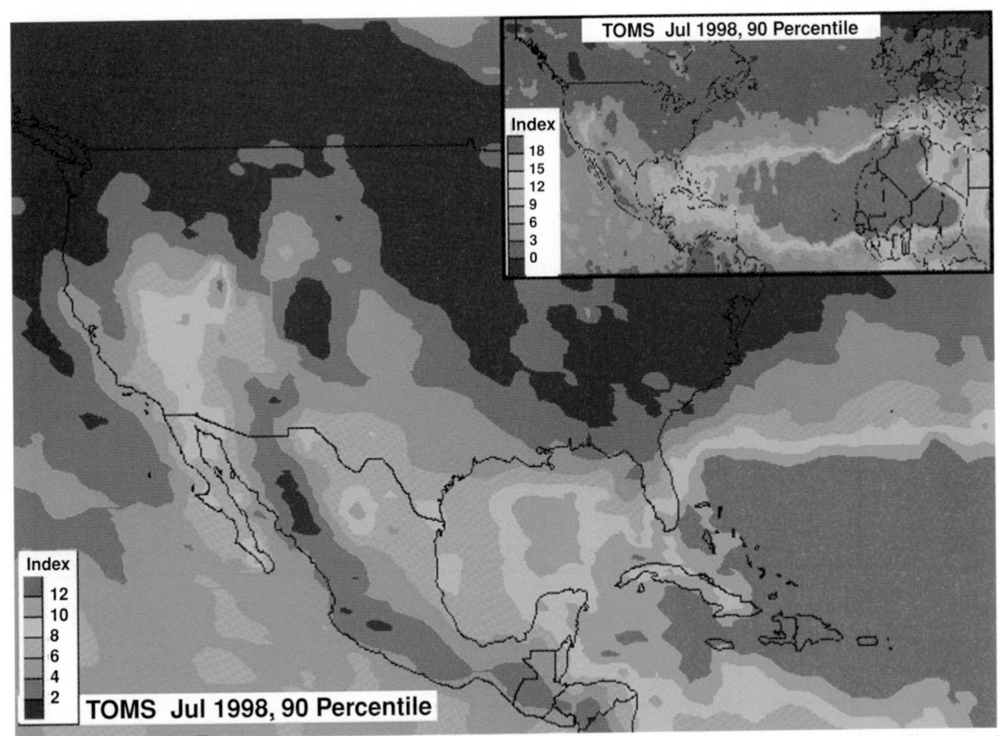

Figure D.2. Satellite data from July 1998 illustrating the extent of the Sahara dust plume. (Source: R. Husar, pers. comm.).

concentrations. For example, over the period 1988-98, the average $PM_{2.5}$ dust concentrations at three northern IMPROVE monitoring sites (Mt. Rainier in WA, Crater Lake in OR, and Boundary Waters in MN) were well below 1 $\mu g/m^3$, with occasional peaks of 1 to 3 $\mu g/m^3$. On April 29, 1998, these sites showed a simultaneous sharp rise to 3 to 11 $\mu g/m^3$. Evidently, the April 1998 Asian dust event caused dust concentrations 2 to 3 times higher than any other event during 1988-1998. Transport episodes of the magnitude of the April 1998 event therefore appear to occur less frequently than once per year and perhaps no more often than once per ten years.

In April and May of 1998, satellite data showed the presence of large numbers of fires in Central America and prominent aerosol clouds (Figure D.6). Such fires represent a significant source of internationally transported fine PM affecting considerable portions of Mexico. Typically, these aerosol clouds move northward along the coastal areas, with the central mountain regions less affected (Figure D.6). During the PM episodes of April and May 1998, PM_{10} concentrations at locations in Texas and the southeastern United States all increased (Figure D.7), indicating the presence of a regional aerosol affecting sites across a broad area. Speciated PM measurements from the IMPROVE site at Big Bend, Texas, indicated that about half the PM mass consisted of OC. On average days, OC may be one-fourth to one-third of PM mass.

As each of these examples of long-range transport indicates, the combined use of satellite and surface measurements provides evidence linking changes in surface PM concentrations to transported aerosol plumes. Further characterization of the frequency and magnitude of long-range aerosol transport requires systematic analysis of long-term databases of both satellite and surface measurements.

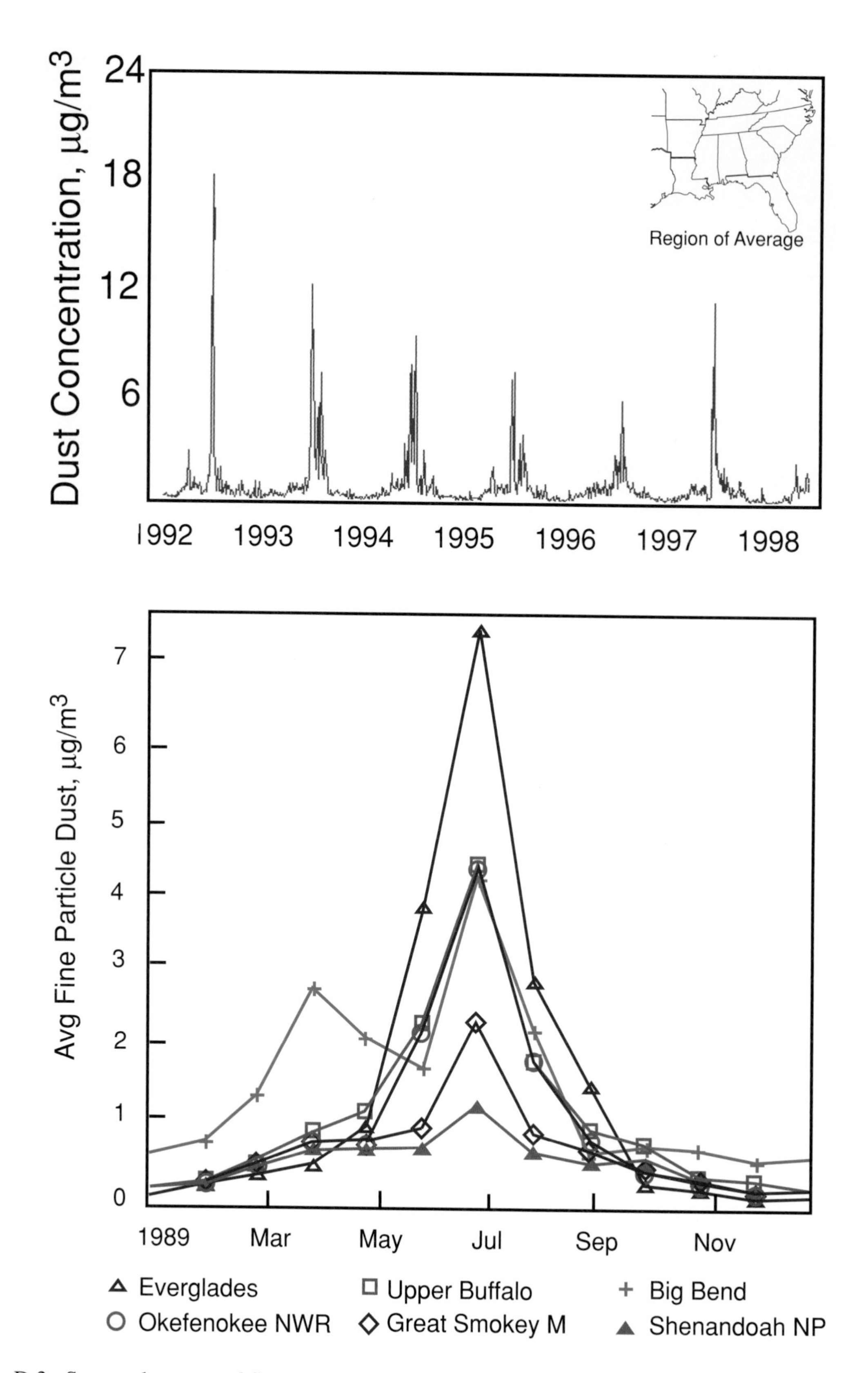

Figure D.3. Seasonal pattern of fine particulate soil (dust) concentration at IMPROVE monitoring sites in the southeastern United States. Top: Average of all sites versus date. Bottom: Monthly averages for six sites. (Source: R. Husar, pers. comm.).

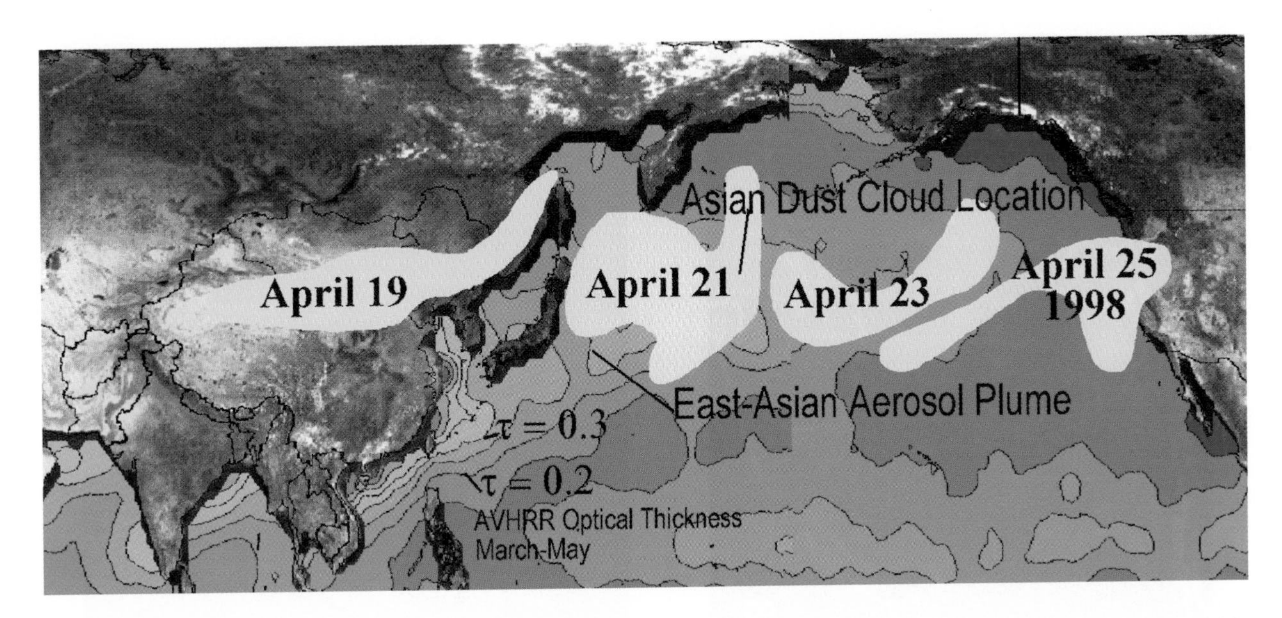

Figure D.4. Aerosol optical thickness and illustration of the positions of a dust plume during April 1998. (Source: R. Husar, pers. comm.).

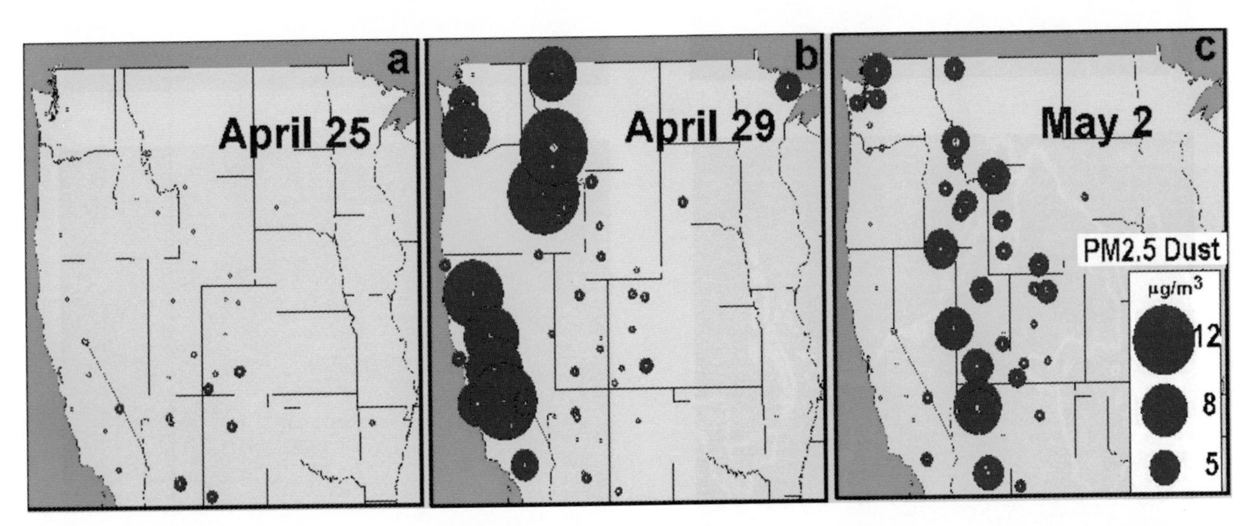

Figure D.5. $PM_{2.5}$ dust concentrations at IMPROVE monitoring sites on three dates during April and May of 1998. (Source: R. Husar, pers. comm.).

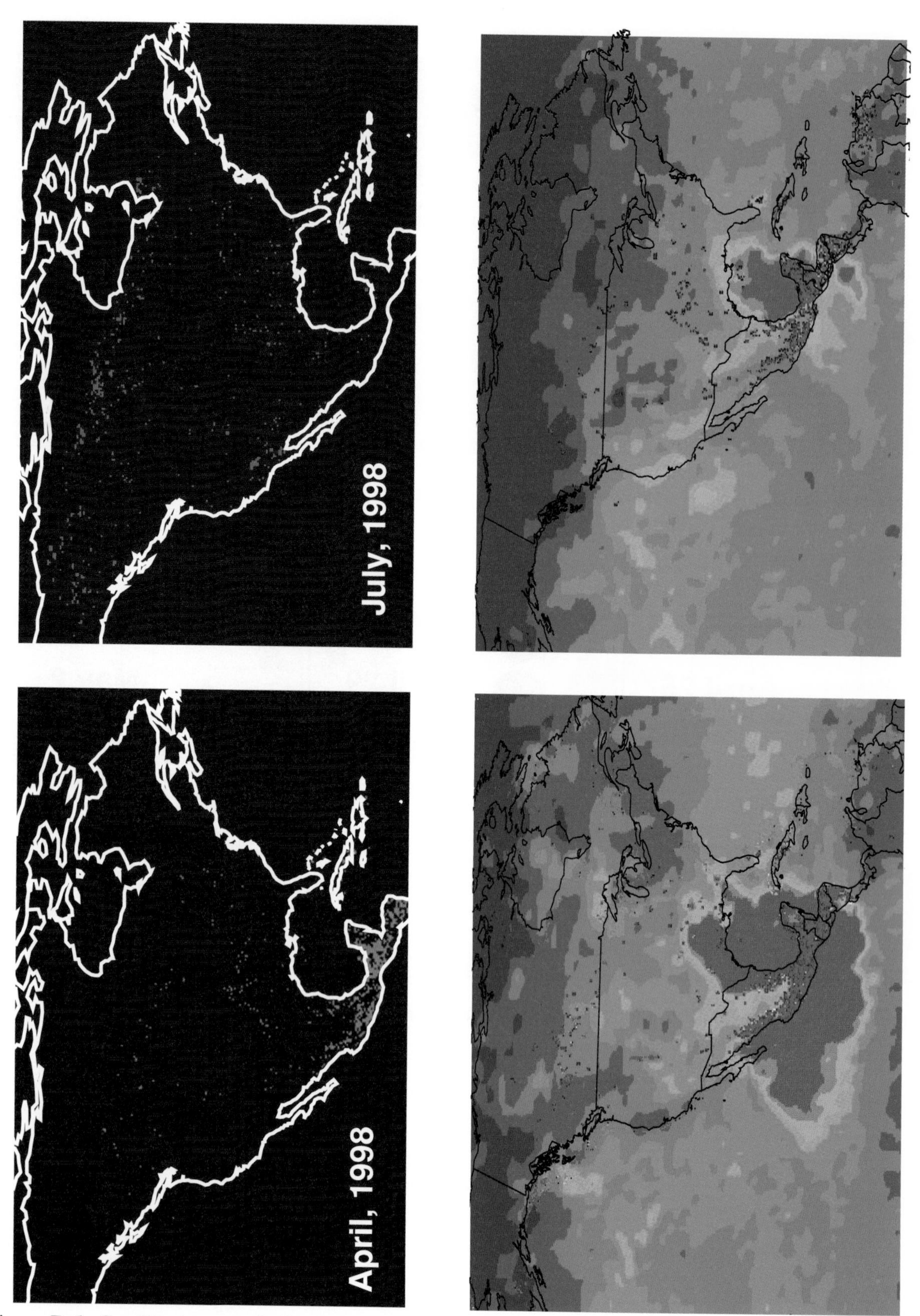

Figure D.6. Satellite imagery indicating the locations of major fires during April and July 1998 (left) and aerosol optical thickness during April and May 1998 (right). (Source: R. Husar, pers. comm.).

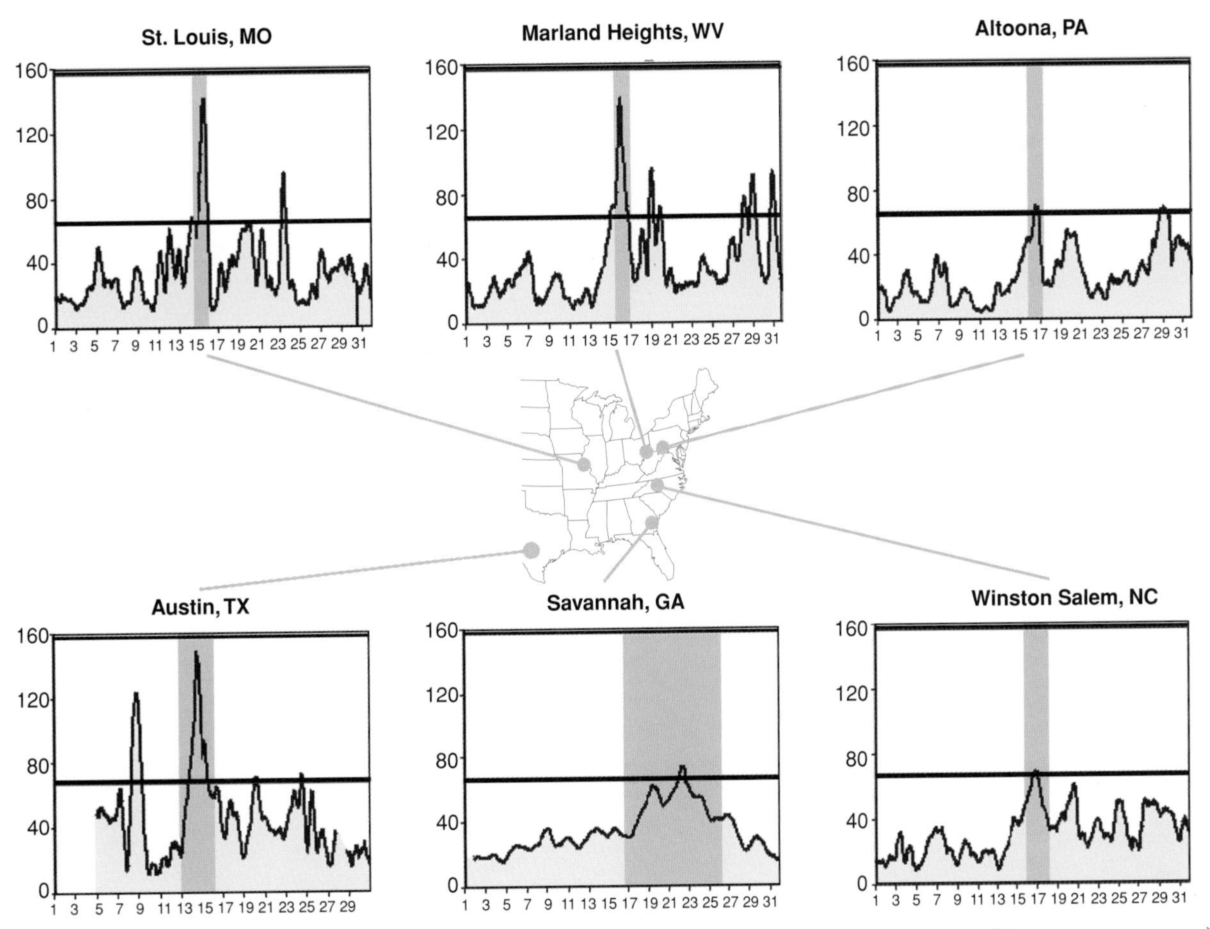

Figure D.7. Hourly PM_{10} concentrations at six locations during May 1998. (Source: R. Husar, pers. comm.).

REFERENCES

Banic, C.M., Leaitch, W.R., Isaac, G.A., Couture, M.D., Kleinman, L.I., Springston, S.R., MacPherson, J.I., 1996. Transport of ozone and sulfur to the North Atlantic atmosphere during the North Atlantic Regional Experiment. Journal of Geophysical Research 101, 29091-29104.

Daum, P.H., Kleinman, L.I., Newman, L., Luke, W.T., Weinstein-Lloyd, J., Berkowitz, C.M., Busness, K.M., 1996. Chemical and physical properties of plumes of anthropogenic pollutants transported over the North Atlantic during the North Atlantic Regional Experiment. Journal of Geophysical Research 101, 29029-29042.

Husar, R.B., Tratt, D.M., Schichtel, B.A., et al., 2001. Asian dust events of April 1998. Journal of Geophysical Research 106, 18317-18330.

Jaffe, D. et al., 1999. Transport of Asian air pollution to North America. Geophysical Research Letters 26, 711-714.

Price, J., Dattner, S., Lambeth, B., Kamrath, J., Aguirre, M., McMullen, G., Loos, K., Crow, W., Tropp, R., Chow, J., 1998. Preliminary results of early $PM_{2.5}$ monitoring in Texas: separating the impacts of transport and local contributions. In: J Chow and P Koutrakis, eds. $PM_{2.5}$: A Fine Particle Standard. Proceedings of a Specialty Conference, January 28-30, 1998. Pittsburgh PA: Air & Waste Management Association.

APPENDIX E. PREPARATION OF THIS ASSESSMENT

The development of *Particulate Matter Science for Policy Makers: A NARSTO Assessment* began in January 1999. The development process incorporated guidance from scientists, policy makers, industry, academia, and the public, with the goal of providing air-quality regulators with as useful and scientifically advanced tools as possible.

Interactions with the policy community began in January 1999 as well, at the *NARSTO Fine Particle Characterization and Atmospheric Process Research Planning Workshop* in Crystal City, Virginia. This workshop, which included both science and policy-program representatives, was held to assist NARSTO in the preparation of its PM Research Strategy and included a session on "Assessment and Linkages." That session culminated in a definition of the context, purpose, audience, and timing for a PM assessment.

From autumn 1999 through March 2000, the NARSTO Assessment and Analysis Team co-chairs prepared a draft set of policy questions to frame this Assessment. The questions were based upon a consolidation of questions from materials used at in the Crystal City Workshop, found in the NARSTO Strategy and Charter, and used in the NARSTO Ozone Assessment. This initial set of questions was part of the March 2000 Assessment Plan approved by the Executive Assembly and Executive Steering Committee.

The NARSTO Executive Assembly first commissioned the preparation of a PM science assessment in March 2000, and charged the Analysis and Assessment Team Co-Chairs with selecting the authors, outlining a scope, and proposing a process for completion of the assessment by the end of 2003. The Charge was subsequently approved by the NARSTO Executive Steering Committee. The Analysis and Assessment Co-Chairs identified as Lead Authors internationally recognized experts in the areas to be covered by this Assessment, and had the Lead Authors approved by the Executive Steering Committee. Subsequently, a three-person Assessment Co-Chair arrangement was established. These Assessment Co-Chairs and Lead Authors met during the spring of 2000 and outlined the technical guidelines for the assessment process. Among these guidelines was an underlying theme that the Assessment should focus on usefulness to policy makers. As such, it was to take advantage of recent discipline-specific reviews, present summary technical judgments and insights that are known to and scientifically defendable by the authors, present current data and information, and present information in a reader-friendly manner with the few references cited being centrally important to information or judgments.

Further comments on the policy questions from NARSTO members directed that they be more basic, and similar to those used in the Ozone Assessment. The Assessment Co-Chairs re-drafted the policy questions into a final draft form, and presented them to assessment authors at their first meeting in May 2000. At that time the co-chairs drafted a set of science questions for each policy question, to guide the authors in their writing.

During August 2000 through October 2000, hour-long interviews were held with a representative set (67) of senior policy makers at 49 federal, state, and provincial environment departments and private industries in Canada, the United States, and Mexico. Interview questions, stemming back to guidance from the U.S. National Research Council's peer review of the Ozone Assessment, sought to identify senior policy-maker goals in managing air quality, to determine the information required by them to meet their goals, and to gauge their impressions of uncertainties in the atmospheric science as of that time in relation to managing PM.

Responses to interview questions led to the confirmation of eight policy questions, which are fundamental to solving the PM-management problem in North America and useful in organizing the scope and content of the NARSTO PM Assessment:

PQ1. Is there a significant PM problem and how confident are we?

PQ2. Where there is a PM problem, what is its composition and what factors contribute to elevated concentrations?

PQ3. What broad, pollutant-based, approaches might be taken to fix the problem?

PQ4. What source-specific options are there for fixing the problem, given the broad control approaches above?

PQ5. What is the relationship between PM, its components, and other air-pollution problems on which the atmospheric science community is working?

PQ6. How can we measure our progress? How can we determine the effectiveness of our actions in bringing about emission reductions and air-quality improvements, with their corresponding exposure reductions and health improvements?

PQ7. When and how should we reassess and update our implementation programs to adjust for any weaknesses in our plan, and take advantage of advances in science and technology?

PQ8. What further atmospheric-sciences information will be needed in the periodic reviews of our national standards?

Generalizations from the interview results in relationship to the policy questions framing this Assessment are that 80 percent of government respondents explicitly agreed that identified policy questions would give them the information they needed to make decisions, 77 percent of industry respondents explicitly recognized that these policy questions describe the breadth of science needed to implement the standards, and no one said that these questions were inappropriate or incomplete in the context of implementation information needs.

The 67 senior policy makers interviewed in the first phase of this assessment represented a cross-section of officials and their principal advisors. They were selected because they were seen as being the highest-ranking members of their organization or the ones who would be expected to advise the ultimate decision maker on the organization's position regarding PM policies and programs. From the federal government (12 United States, 6 Canadian, and 3 Mexican), interviewees included an Assistant Administrator, Office Directors, Director Generals, Directors, and U.S. Legislative Chiefs of Staff. From state and provincial government (19 United States, 7 Canadian, and 4 Mexican), interviewees included Commissioners, a Board Chair, a Secretary, Air Directors, Assistant Directors, Executive Officers, Executive Directors, and Executive Coordinators. Industry representatives came from organizations most involved in past rule-making actions, and included auto makers, petroleum industries, chemical manufacturers, and forest- and paper-product manufacturers. Private-sector interviewees (12 United States and 3 Canadian) included Vice Presidents, Senior Legal Advisors, Directors of Regulatory Affairs, and Directors of Public Policy.

Drafts of this Assessment were thoroughly reviewed by the full NARSTO membership during July and August 2001. A revised draft, responding to NARSTO membership comments, was released in January 2001 for external review by the public and a tri-national panel of experts convened by the National Research Council of the United States, the Royal Society of Canada, and the United States-Mexico Foundation for Science. This final document addresses these comments and has been reviewed and approved by the NARSTO Executive Steering Committee.